Handbook of Exe

Sam Goldstein • Jack A. Naglieri
Editors

Handbook of Executive Functioning

Editors
Sam Goldstein
Neurology, Learning
  and Behavior Center
University of Utah School of Medicine
Salt Lake City, UT, USA

Jack A. Naglieri
Curry School of Education
University of Virginia
Charlottesville, VA, USA

ISBN 978-1-4614-8105-8 (Hardcover)   ISBN 978-1-4614-8106-5 (eBook)
ISBN 978-1-4939-0337-5 (Softcover)
DOI 10.1007/978-1-4614-8106-5
Springer New York Heidelberg Dordrecht London

Library of Congress Control Number: 2013949438

© Springer Science+Business Media New York 2014. First softcover printing 2014.
This work is subject to copyright. All rights are reserved by the Publisher, whether the whole or part of the material is concerned, specifically the rights of translation, reprinting, reuse of illustrations, recitation, broadcasting, reproduction on microfilms or in any other physical way, and transmission or information storage and retrieval, electronic adaptation, computer software, or by similar or dissimilar methodology now known or hereafter developed. Exempted from this legal reservation are brief excerpts in connection with reviews or scholarly analysis or material supplied specifically for the purpose of being entered and executed on a computer system, for exclusive use by the purchaser of the work. Duplication of this publication or parts thereof is permitted only under the provisions of the Copyright Law of the Publisher's location, in its current version, and permission for use must always be obtained from Springer. Permissions for use may be obtained through RightsLink at the Copyright Clearance Center. Violations are liable to prosecution under the respective Copyright Law.
The use of general descriptive names, registered names, trademarks, service marks, etc. in this publication does not imply, even in the absence of a specific statement, that such names are exempt from the relevant protective laws and regulations and therefore free for general use.
While the advice and information in this book are believed to be true and accurate at the date of publication, neither the authors nor the editors nor the publisher can accept any legal responsibility for any errors or omissions that may be made. The publisher makes no warranty, express or implied, with respect to the material contained herein.

Printed on acid-free paper

Springer is part of Springer Science+Business Media (www.springer.com)

I would like to recognize my parents Martha and Sam Naglieri for teaching how executive function works in the real world. Their example of considering how to do what you do was exemplary.

Jack A. Naglieri

I am grateful to all of the exceptional colleagues I have had the good fortune to work with and learn from over my 40-year career. This text is dedicated to them and to my dear wife Sherrie from whom I gather strength each day. This work is also dedicated to the memory of my son-in-law, Brandon Custer.

Sam Goldstein

We want to thank our authors for their willingness to contribute to this volume. As always we could not complete a volume such as this one without the organizational expertise of Ms. Kathleen Gardner.

Sam Goldstein and Jack A. Naglieri

Management is efficiency in climbing the ladder of success; leadership determines whether the ladder is leaning against the right wall.
Stephen Covey

Efficiency is doing things right; effectiveness is doing the right things.
Peter Drucker

Tell me and I forget. Teach me and I remember. Involve me and I learn.
Benjamin Franklin

Preface

In 1848, while working with Phineas Gage, a young man who miraculously survived a severe injury to his brain, physician John Martyn Harlow observed that Gage had lost the balance between his "intellectual faculties and animal propensities." He had difficulty making plans and his loss of control led him to be disrespectful and profane. Gage cared little as to how his behavior and actions affected others. He went from being a model railroad foreman to an out-of-work stable hand and eventually 12 years after his injury passing away at the age of 36 following a series of seizures.

It is now well accepted that the injury Gage suffered adversely impacted the frontal lobes governing the efficient operation of his brain. In the last 50 years, an interest in this part of the brain and its operation has come to the forefront for many researchers and clinicians. The frontal lobes have become increasingly conceptualized as a governor or executive. In the 1890s, Oppenheim associated personality changes with the orbital and mesial frontal lobes (Oppenheim 1890, 1891). The term "executive" was used some 40 years ago by Luria as he described the functions of the frontal lobes or his third functional unit as serving an executive role (Luria, 1980). Executive functioning has come to represent a number of mental processes which allows individuals to use thought to govern behavior and to perform complex activities involving planning, organizing, strategizing, controlling, and sustaining attention and self-management. Executive dysfunction has been documented in a diversity of conditions, including dementia, traumatic brain injury, white matter lesions, borderline personality disorder, substance abuse, multiple system atrophy, multiple sclerosis, schizophrenia, autism, attention deficit hyperactivity disorder, progressive supranuclear palsy, CADASIL, and Korsakoff syndrome. Ironically, individuals experiencing executive function problems, the result of either atypical development or trauma, often retain their memory and capacity to master academic skills but they struggle how to efficiently use what they know. They are inconsistent, unpredictable, and often poorly self-governed. They are inefficient in their ability to make plans, keep track of time, evaluate their behavior, and socialize appropriately. Typically they struggle in many critical aspects of life.

In this textbook, we have sought to bring together the leading theoreticians, researchers, and clinical practitioners involved with the scientific examination, assessment, and clinical and educational application of executive function. We have sought to provide a wide breadth and scope of theory and ideas but, most importantly, to provide ample resources to begin the process of creating efficient and effective strategies to help individuals across the life span struggling with executive function impairments.

Our book begins with a short history of executive function as a theoretical and clinical construct. Jin Chung and colleagues provide an overview in the next chapter of the physiology of executive function and the brain. Chapter 3 by respected scientist and researcher, Nick Goldberg, discusses executive function and the operations of the frontal lobe. The first part providing conceptualization of executive function ends with a chapter by Marilyn Welch and Bruce Pennington describing the normative developmental changes in executive function as children mature.

Part II provides an overview of issues related to what we have placed under an umbrella titled Practical Implications. Lisa Weyandt and her colleagues review the use of executive function tasks and externalizing and internalizing disorders. Cecil Reynolds and Arthur Horton provide an overview of the neuropsychology of executive function as it relates to the Diagnostic and Statistical Manual of the American Psychiatric Association. Kevin Antshel and Russell Barkley discuss executive function theory and ADHD. Hilde Geurts discusses executive function and autism. Finally, Melissa DeVries and Dana Princiotta describe executive function as a mediator of age-related cognitive decline in adults.

Part III, by far the largest part of this text, contains 12 chapters providing overviews of the most widely used neuropsychological tests and questionnaires to evaluate executive function. This part begins with a chapter by Andrew Livanis discussing evaluation and treatment integrity, an often overlooked but critical issue in clinical practice. Well-respected researchers and clinicians were invited to write chapters about the instruments they have developed. Peter Isquith and colleagues have provided contributions concerning their Behavior Rating Inventory of Executive Function. Russell Barkley has written about his Deficits in Executive Function scales, and Dawn Flannagan and Sam Ortiz have provided a summary chapter describing their cross battery approach and the utilization of diverse tools to measure EF. We provide a chapter on the Comprehensive Executive Function Inventory.

The text concludes with a part of six chapters, the result of our efforts to gather strategies and ideas to facilitate the development and functioning of executive function. Such programs are still in their infancy, with many frequently recommended strategies untested. This part begins with a chapter by Jack Naglieri covering psychometric issues and the evaluation of treatment effectiveness. Peg Dawson, Lynn Meltzer, Milt Dehn, Bonnie Aberson, and Kathleen Kryza have all provided a framework for the work they are doing to facilitate and develop executive function in children.

Richard Dawkins has written, "by all means let's be open minded but not so open minded that our brains drop out." The science of executive function is truly in its infancy. Theories and tests are many; however, scientific findings are only slowing emerging. It is our hope this volume adds to the breadth and scope of knowledge about executive function and provides a sourcebook for future researchers and clinicians.

Salt Lake City, UT, USA Sam Goldstein
Charlottesville, VA, USA Jack A. Naglieri

References

Luria, A. R. (1980). *Higher cortical functions in man*. New York: Consultants Bureau.
Oppenheim, H. (1890). Zur Pathologie der Grosshirngeschwülste. *Arch Psychiatrie Nervenkrankh*, 21, 560–587, 705–745.
Oppenheim, H. (1891). Zur Pathologie der Grosshirngeschwülste. *Arch Psychiatrie Nervenkrankh*, 22, 27–72.

Contents

Part I Conceptualizations of Executive Functioning

1. **Introduction: A History of Executive Functioning as a Theoretical and Clinical Construct**.................... 3
Sam Goldstein, Jack A. Naglieri, Dana Princiotta, and Tulio M. Otero

2. **The Physiology of Executive Functioning**.................... 13
Hyun Jin Chung, Lisa L. Weyandt, and Anthony Swentosky

3. **The Frontal Lobes and Executive Functioning**.................... 29
Tulio M. Otero and Lauren A. Barker

4. **The Development of Hot and Cool Executive Functions in Childhood and Adolescence: Are We Getting Warmer?**.................... 45
Eric Peterson and Marilyn C. Welsh

Part II Practical Implications

5. **A Review of the Use of Executive Function Tasks in Externalizing and Internalizing Disorders**.................... 69
Lisa L. Weyandt, W. Grant Willis, Anthony Swentosky, Kimberly Wilson, Grace M. Janusis, Hyun Jin Chung, Kyle Turcotte, and Stephanie Marshall

6. **The Neuropsychology of Executive Functioning and the DSM-5**.................... 89
Cecil R. Reynolds and Arthur MacNeill Horton Jr.

7. **Executive Functioning Theory and ADHD**.................... 107
Kevin M. Antshel, Bridget O. Hier, and Russell A. Barkley

8. **Executive Functioning Theory and Autism**.................... 121
Hilde M. Geurts, Marieke de Vries, and Sanne F.W.M. van den Bergh

9. **Executive Functioning as a Mediator of Age-Related Cognitive Decline in Adults**.................... 143
Dana Princiotta, Melissa DeVries, and Sam Goldstein

Part III Assessment of Executive Functioning

10 Assessment of Executive Function Using Rating Scales: Psychometric Considerations 159
Jack A. Naglieri and Sam Goldstein

11 The Cambridge Neuropsychological Test Automated Battery in the Assessment of Executive Functioning 171
Katherine V. Wild and Erica D. Musser

12 The Assessment of Executive Function Using the Cognitive Assessment System: Second Edition 191
Jack A. Naglieri and Tulio M. Otero

13 The Assessment of Executive Functioning Using the Delis-Kaplan Executive Functions System (D-KEFS) 209
Tammy L. Stephens

14 Using the Comprehensive Executive Function Inventory (CEFI) to Assess Executive Function: From Theory to Application 223
Jack A. Naglieri and Sam Goldstein

15 The Assessment of Executive Functioning Using the Barkley Deficits in Executive Functioning Scales 245
Russell A. Barkley

16 Assessment with the Test of Verbal Conceptualization and Fluency (TVCF) 265
Cecil R. Reynolds and Arthur MacNeill Horton Jr.

17 Examining Executive Functioning Using the Behavior Assessment System for Children (BASC) 283
Mauricio A. Garcia-Barrera, Emily C. Duggan, Justin E. Karr, and Cecil R. Reynolds

18 Assessment of Executive Functioning Using the Behavior Rating Inventory of Executive Function (BRIEF) 301
Robert M. Roth, Peter K. Isquith, and Gerard A. Gioia

19 Assessment of Executive Functioning Using Tasks of Executive Control 333
Peter K. Isquith, Robert M. Roth, and Gerard A. Gioia

20 The Assessment of Executive Functioning Using the Childhood Executive Functioning Inventory (CHEXI) 359
Lisa B. Thorell and Corinne Catale

21 The Assessment of Executive Functioning Using the Delis Rating of Executive Functions (D-REF) 367
Jessica A. Rueter

22 Cross-Battery Approach to the Assessment
of Executive Functions 379
Dawn P. Flanagan, Vincent C. Alfonso, and Shauna G. Dixon

Part IV   Interventions Related to Executive Functioning

23 Treatment Integrity in Interventions That Target
the Executive Function 413
Andrew Livanis, Ayla Mertturk, Samantha Benvenuto,
and Christy Ann Mulligan

24 Interventions to Promote Executive Development
in Children and Adolescents 427
Peg Dawson and Richard Guare

25 Teaching Executive Functioning Processes:
Promoting Metacognition, Strategy Use, and Effort 445
Lynn Meltzer

26 Working Memory Training and Cogmed 475
Peter C. Entwistle and Charles Shinaver

27 Supporting and Strengthening Working Memory
in the Classroom to Enhance Executive Functioning 495
Milton J. Dehn

28 Building Executive Functioning in Children
Through Problem Solving 509
Bonnie Aberson

29 Practical Strategies for Developing Executive Functioning
Skills for ALL Learners in the Differentiated Classroom 523
Kathleen Kryza

About the Editors 555

Index 557

Contributors

Bonnie Aberson Joe Dimaggio Children's Hospital, Hollywood, FL, USA

Vincent C. Alfonso Fordham University, New York City, NY, USA

Kevin M. Antshel SUNY—Upstate Medical University, Syracuse, NY, USA

Lauren A. Barker The Chicago School of Professional Psychology, Loyola University, Chicago, IL, USA

Russell A. Barkley Medical University of South Carolina, Charleston, SC, USA

Samantha Benvenuto Long Island University—Brooklyn, Brooklyn, NY, USA

Corinne Catale University of Liège, Liège, Belgium

Hyun Jin Chung University of Rhode Island, Kingston, RI, USA

Peg Dawson Center for Learning and Attention Disorders, Portsmouth, NH, USA

Milton J. Dehn Schoolhouse Educational Services, Onalaska, WI, USA

Marieke de Vries Department of Psychology, Dutch Autism & ADHD Research Center (d'Arc), University of Amsterdam, Amsterdam, The Netherlands

Melissa DeVries Neurology, Learning and Behavior Center, University of Utah School of Medicine, Salt Lake City, UT, USA

Emily C. Duggan University of Victoria, Victoria, BC, Canada

Shauna G. Dixon CEO, St. Johns University, Queens, NY, USA

Peter C. Entwistle Pearson Clinical Assessments, San Antonio, TX, USA

Dawn P. Flanagan CEO, St. Johns University, Queens, NY, USA

Mauricio A. Garcia-Barrera University of Victoria, Victoria, BC, Canada

Hilda M. Geurts Department of Psychology, Dutch Autism & ADHD Research Center (d'Arc), University of Amsterdam, Amsterdam, The Netherlands

Dr Leo Kannerhuis, Amsterdam/Doorwerth, The Netherlands

Gerard A. Gioia Children's National Medical Center, George Washington University School of Medicine, Rockville, MD, USA

Sam Goldstein Neurology, Learning and Behavior Center, University of Utah School of Medicine, Salt Lake City, UT, USA

Richard Guare Center for Learning and Attention Disorders, Portsmouth, NH, USA

Bridget O. Hier Syracuse University, Syracuse, NY, USA

Arthur MacNeill Horton Jr. Psychological Associates of Maryland, Towson, MD, USA

Peter K. Isquith Geisel School of Medicine at Dartmouth School, Lebanon, NH, USA

Grace M. Janusis University of Rhode Island, Kingston, RI, USA

Justin E. Karr University of Victoria, Victoria, BC, Canada

Kathleen Kryza CEO, Infinite Horizons and Inspiring Learners, Ann Arbor, MI, USA

Andrew Livanis Long Island University—Brooklyn, Brooklyn, NY, USA

Stephanie Marshall University of Rhode Island, Kingston, RI, USA

Lynne Meltzer Research Institute for Learning and Development (ResearchILD), and Harvard Graduate School of Education, Lexington, MA, USA

Ayla Mertturk Long Island University—Brooklyn, Brooklyn, NY, USA

Christy Ann Mulligan Long Island University—Brooklyn, Brooklyn, NY, USA

Erica D. Musser Oregon Health and Science University, Portland, OR, USA

Jack A. Naglieri University of Virginia, Charlottesville, VA, USA

Samuel O. Ortiz St. Johns University, Queens, NY, USA

Tulio M. Otero The Chicago School of Professional Psychology, Loyola University, Chicago, IL, USA

Eric Peterson University of Northern Colorado, Greeley, CO, USA

Dana Princiotta Neurology, Learning and Behavior Center, School of Medicine, University of Utah, Salt Lake City, UT, USA

Jessica A. Rueter University of Texas at Tyler, Tyler, TX, USA

Cecil R. Reynolds Texas A&M University, College Station, TX, USA

Robert M. Roth Geisel School of Medicine at Dartmouth School, Lebanon, NH, USA

Charles Shinaver Pearson Clinical Assessments, San Antonio, TX, USA

Tammy L. Stephens Pearson Clinical Assessments, San Antonio, TX, USA

Anthony Swentosky University of Rhode Island, Kingston, RI, USA

Lisa B. Thorell Karolinska Institutet, Stockholm, Sweden

Kyle Turcotte University of Rhode Island, Kingston, RI, USA

Sanne F.W.M. van den Bergh Department of Psychology, Dutch Autism & ADHD Research Center (d'Arc), University of Amsterdam, Amsterdam, The Netherlands

Dr Leo Kannerhuis, Amsterdam/Doorwerth, The Netherlands

Marilyn C. Welsh University of Northern Colorado, Greeley, CO, USA

Lisa L. Weyandt University of Rhode Island, Kingston, RI, USA

Katherine V. Wild Oregon Health and Science University, Portland, OR, USA

Kimberly Wilson University of Rhode Island, Kingston, RI, USA

W. Grant Willis University of Rhode Island, Kingston, RI, USA

Part I
Conceptualizations of Executive Functioning

Introduction: A History of Executive Functioning as a Theoretical and Clinical Construct

Sam Goldstein, Jack A. Naglieri, Dana Princiotta, and Tulio M. Otero

Introduction

Executive function (EF) has come to be an umbrella term used for a diversity of hypothesized cognitive processes, including planning, working memory, attention, inhibition, self-monitoring, self-regulation, and initiation carried out by prefrontal areas of the frontal lobes.

Although the concept of EF was first defined in the 1970s, the concept of a control mechanism was discussed as far back as the 1840s. Phineas Gage offers perhaps one of the most fascinating case studies associated with EF. In 1840, as a railroad construction foreman, Phineas was pierced with a large iron rod through his frontal lobe (see Ratiu & Talos, 2004). This accident destroyed a majority of his left frontal lobe. Phineas survived and after a period of recovery changes in Phineas' behavior and personality became apparent. Phineas was described as "disinhibited" or "hyperactive," which suggested a lack of inhibition often found in those with damage to the prefrontal cortex (Pribram, 1973). This case and others prompted early brain researchers to further investigate the role of the frontal lobes and the concept of executive function.

By the 1950s, psychologists and neuroscientists became more interested in understanding the role of the prefrontal cortex in intelligent behavior. British psychologist Donald Broadbent (1953) described differences between automatic and controlled processes. This distinction was further elaborated by Shifrin and Schneider (1977). These authors introduced the notion of selective attention to which EF is closely related. In 1975, psychologist Michael Posner coined the term "cognitive control" in a book chapter titled "Attention and Cognitive Control." Posner proposed that there is a separate executive branch of the attentional system responsible for focusing attention on selected aspects of the environment. Alan Baddeley proposed a similar system as part of his model of working memory, arguing there must be a component which he referred to as the "central executive" allowing information to be manipulated in short-term memory. Shallice (1988) also suggested that attention is regulated by a "supervisory system which can over-ride automatic responses in favor of scheduling behavior on the basis of plans or intentions." Consensus slowly emerged that this control system is housed

S. Goldstein, Ph.D. (✉)
Neurology, Learning and Behavior Center, University of Utah School of Medicine, 230 South 500 East, Suite 100, Salt Lake City, UT 84102, USA
e-mail: info@samgoldstein.com

J.A. Naglieri, Ph.D.
University of Virginia, Charlottesville, VA 22904, USA
e-mail: jnaglieri@gmail.com

D. Princiotta, Ph.D.
Neurology, Learning and Behavior Center, School of Medicine, University of Utah, Salt Lake City, UT 84102, USA

T.M. Otero, Ph.D.
The Chicago School of Professional Psychology, Loyola University, Chicago, IL, USA

in the most anterior portion of the brain, the prefrontal cortex.

Pribram (1973) was one of the first to use the term "executive" when discussing matters of prefrontal cortex functioning. Since then at least 30 or more constructs have been included under the umbrella term, EF, making the concept hard to operationally define. Many authors have made attempts to define the concept of executive function using models that range from one to multiple components. Lezak (1995) suggested that EFs consisted of components related to volition, planning, purposeful action, and effective performance. It has been hypothesized that each component involves a distinct set of related behaviors. Reynolds and Horton (2006) suggested that EFs are distinct from general knowledge. They suggest that executive functions represent the capacity to plan, to do things, and to perform adaptive actions, while general knowledge related to the retention of an organized set of objective facts. They further hypothesized that EF involves decision making, planning actions, and generating novel motor outputs adapted to external demands rather than the passive retention of information. Naglieri and Goldstein (2013) based their view of the behavioral aspects of executive function on a large national study of children. They suggest that executive function is best represented as a single phenomena, conceptualized as the efficiency with which individuals go about acquiring knowledge as well as how well problems can be solved across nine areas (attention, emotion regulation, flexibility, inhibitory control, initiation, organization, planning, self-monitoring, and working memory).

A Review of EF Definitions

Anderson (2002): "Processes associated with EF are numerous, but the principal elements include anticipation, goal selection, planning, initiation of activity, self-regulation, mental flexibility, deployment of attention, and utilization of feedback." (p. 71)

Banich (2009): … "providing resistance to information that is distracting or task irrelevant, switching behavior task goals, utilizing relevant information in support of decision making, categorizing or otherwise abstracting common elements across items, and handling novel information or situations." (p. 89)

Barkley (2011a): "EF is thus a self-directed set of actions intended to alter a delayed (future) outcome (attain a goal for instance)." (p. 11)

Baron (2004): "Executive functioning skills "allow an individual to perceive stimuli from his or her environment, respond adaptively, flexibly change direction, anticipate future goals, consider consequences, and respond in an integrated or commonsense way." (p. 135)

Best, Miller, and Jones (2009): "Executive function (EF) serves as an umbrella term to encompass the goal-oriented control functions of the PFC [prefrontal cortex]." (p. 180)

Borkowski and Burke (1996): "EF coordinates two levels of cognition by monitoring and controlling the use of the knowledge and strategies in concordance with the metacognitive level." (p. 241)

Burgess (1997): "a range of poorly defined processes which are putatively involved in activities such as "problem-solving," … "planning" … 'initiation' of activity, 'cognitive estimation,' and 'prospective memory.'" (p. 81)

Corbett et al. (2009) "Executive function (EF) is an overarching term that refers to mental control processes that enable physical, cognitive, and emotional self-control." (p. 210)

Crone (2009): "For example, during childhood and adolescence, children gain increasing capacity for inhibition and mental flexibility, as is evident from, for example, improvements in the ability to switch back and forth between multiple tasks." (p. 826)

Dawson and Guare (2010): "Executive skills allow us to organize our behavior over time and override immediate demands in favor of longer-term goals." (p. 1)

Delis (2012): "Executive functions reflect the ability to manage and regulate one's behavior in order to achieve desired goals." (p. 14)

Delis (2012): "Neither a single ability nor a comprehensive definition fully captures the conceptual scope of executive functions; rather, executive functioning is the sum product of a collection of higher level skills that converge to enable an individual to adapt and thrive in complex psychosocial environments." (p. 14)

Denckla (1996): "EF has become a useful shorthand phrase for a set of domain-general control processes…." (p. 263)

Friedman, Haberstick, Willcutt, Miywake, Young, et al. (2007): "… a family of cognitive control processes that operate on lower-level processes to regulate and shape behavior." (p. 893)

Funahashi (2001): "Executive function is considered to be a product of the coordinated operation of various processes to accomplish a particular goal in a flexible manner." (p. 1)

Fuster (1997): EF "…is closely related, if not identical, to the function of temporal synthesis of action, which rests on the same subordinate functions. Temporal synthesis, however, does not need a central executive." (p. 165)

Gioia, Isquith, Guy, and Kenworthy (2000): "The executive functions are a collection of processes that are responsible for guiding, directing, and managing cognitive, emotional, and behavioral functions, particularly during active, novel problem solving." (p. 1)

Gioia and Isquith (2004): "The executive functions serve as an integrative directive system exerting regulatory control over the basic, domain-specific neuropsychological functions (e.g., language, visuospatial functions, memory, emotional experience, motor skills) in the service of reaching an intended goal." (p. 139)

Hughes (2009): "The term executive function' (EF), therefore, refers to a complex cognitive construct encompassing the whole set of processes underlying these controlled goal-directed responses to novel or difficult situations, processes which are generally associated with the prefrontal cortex (PFC)." (p. 313)

Lezak (1995): "Executive functioning asks how and whether a person goes about doing something." (p. 42)

Lezak (1995): "Executive functions refer to a collection of interrelated cognitive and behavioral skills that are responsible for purposeful, goal-directed activity, and include the highest level of human functioning, such as intellect, thought, self-control, and social interaction." (p. 42)

Luria (1966): "…Syntheses underlying own actions, without which goal-directed, selective behavior is impossible." (p. 224)

Luria (1966): "…besides the disturbance of initiative and the other aforementioned behavioral disturbances, almost all patients with a lesion of the frontal lobes have a marked loss of their 'critical faculty,' i.e., a disturbance of their ability to correctly evaluate their own behavior and the adequacy of their actions." (p. 227)

McCloskey (2011): "It is helpful to think of executive functions as a set of independent but coordinated processes rather than a single trait." (p. 2)

McCloskey (2006): "Executive Functions can be thought of as a diverse group of highly specific cognitive processes collected together to direct cognition, emotion, and motor activity, including mental functions associated with the ability to engage in purposeful, organized, strategic, self-regulated, goal directed behavior." (p. 1)

Miller and Cohen (2001): [our theory] "suggests that executive control involves the active maintenance of a particular type of information: The goals and rules of a task." (p. 185)

Oosterlaan, Scheres, and Sergeant (2005): "EF encompasses meta-cognitive processes that enable efficient planning, execution, verification, and regulation of goal directed behavior." (p. 69)

Pribram (1973): "… the frontal cortex is critically involved in implementing executive programmes where these are necessary to maintain brain organization in the face of insufficient redundancy in input processing and in the outcomes of behavior." (p. 301)

Robbins (1996): "Executive function is required when effective new plans of action must be formulated, and appropriate sequences of responses must be selected and scheduled." (p. 1463)

Roberts and Pennington (1996): EF "refers to a collection of related but somewhat distinct abilities such as planning, set maintenance, impulse control, working memory, and attentional control." (p. 105)

Stuss and Benson (1986): "*Executive functions* is a generic term that refers to a variety of different capacities that enable purposeful, goal-directed behavior, including behavioral regulation, working memory, planning and organizational skills, and self-monitoring." (p. 272)

Vriezen and Pigott (2002): "Executive function has been defined in a variety of ways but is generally viewed as a multidimensional construct encapsulating higher-order cognitive processes that control and regulate a variety of cognitive, emotional and behavioral functions." (p. 296)

Welsh and Pennington (1988): "Executive function is defined as the ability to maintain an appropriate problem-solving set for attainment of a future goal." (p. 201)

A Brief Review of EF Models

Conceptualizations of EF have been largely driven by observations of individuals having suffered frontal lobe damage. Groups of such individuals were first described by Luria and reported to exhibit disorganized actions and strategies for everyday tasks. Initially this came to be referred to as dysexecutive syndrome. Such individuals tended to perform normally when clinical- or laboratory-based tests were used to assess more fundamental cognitive processes such as memory, learning, language, and reasoning, It was therefore determined that there must be some overarching system responsible for coordinating these other cognitive resources that appeared to be working inefficiently in patients with frontal lobe injuries. Recent functional neuroimaging studies have supported the theory of the PFC as responsible for EF, demonstrating that two parts of the prefrontal cortex, the ACC and DLPFC, appear to be particularly important for completing tasks thought to be sensitive to EF. In this section we will provide a brief chronological overview of the theories that appear to have driven our appreciation, definition, and understanding of EF.

Automatic and Controlled Processes

Donald Broadbent's (1953) model of automatic and controlled processes, otherwise referred to as the filter model, proposed that a filter serves as a buffer that selects information for conscious awareness (Broadbent, 1958). When discussing competing stimuli, the filter determines which information must be distinguished as relevant or irrelevant (Barkley 2011a). In other words, select information will pass through the filter (as relevant), while the remaining information is ignored (irrelevant) (Broadbent, 1958). Under this model, terminologies such as "sensory store" and "sensory filter" are utilized to explain the instrument in which processing of stimuli occurs at the pre-attentive level, focusing on properties such as the sex of the speaker or type of sound (Driver, 2001). Through a visual diagram, the processing of information could be represented with parallel lines up to a point in which processing is then managed with the filter (Schiffrin & Schneider, 1977), resembling a bottleneck, an additional name for Broadbent's model (bottleneck theory)

(Driver, 2001). If not for this filter/buffer, Broadbent believed that the system would become inundated or overloaded with information (Broadbent, 1958; Driver, 2001).

Cognitive Control

Posner and Snyder (1975) expanded upon the work of Broadbent and previous researchers with his "cognitive control" model (Posner & Snyder, 1975). This conceptualization utilized the bottleneck theory postulated by Broadbent by furthering the examination of the role of attention during specific higher-level tasks, including visual searches, for example (Posner & Snyder, 1975). However, Posner also suggested that cognitive control is needed to manage thoughts and emotions (Rueda, Posner, & Rothbart, 2004). By cognitive control, Posner refers to processes that guide behaviors, analogous to working definitions of executive functioning today. According to Posner & Snyder (1975) cognitive control was regarded as responsible in overwriting automatic responses, illustrating the selective nature of the model as well as the inhibitory nature (Posner and Snyder 1975). In this model, cognitive control allows one to adapt from situation to situation depending upon the goals of the individual (Checa, Rodriguez-Bailon, & Rueda, 2008).

Controlled Processes

Schiffrin and Schneider (1977) proposed that because our ability to attend is limited, certain stimuli must be favored over opposing stimuli. They studied the strength of a controlled processes theory of detection, search, and attention by comparing automatic detection with controlled search and concluded that by learning categories, controlled search performance also improved (Schiffrin & Schneider, 1977). In this dual processing theory, automatic processing activates a learned sequence of elements and proceeds automatically, while controlled processing entails a temporary activation of a sequence of elements that can be established rapidly, but they do require attention, nonetheless (Schiffrin & Schneider, 1977). Automatic processes are "effortless, rapid, unavailable to consciousness, and unavoidable; permanent connections that are developed with practice or training" (Schiffrin & Schneider, 1977, p. 2). Without a need for active attention or active control, an individual is thus engaged in an automatic process. Controlled processes are "slow, effortful, and completely conscious; a temporary sequence of nodes activated under control of, and through attention by the subject" (Schiffrin & Schneider, 1977, p. 2). With repeated practice, skills that were controlled can become automated, meaning that skills will not require as much attention resources to be completed (Schneider & Chein, 2003).

Supervisory Attentional System

Shallice (2002) constructed a model of the executive system called the contention scheduling/supervisory attentional system model. Contention scheduling refers to the controlling mediator of inhibition of competing actions when selecting an action to be performed. The supervisory attentional system is a mediator for nonroutine situations in which inhibition may be necessary to make a decision during a novel encounter (Shallice, 1988, 2002). When deficits exist in this supervisory attentional system, Shallice argues that executive disorders are possible (e.g., disinhibition) (Shallice, 2002).

Central Executive

Baddeley, Sala, and Robbins's (1996) central executive hypothesis views the executive as a unified system with multiple functions, a homunculus of sorts. The central executive oversees the phonological loop, visuospatial sketchpad, and an episodic buffer. Below the central executive, Baddeley envisioned and described the following functions: time-sharing, selective attention, temporary activation of long-term memory, and switching of retrieval plans (Baddeley, 1986).

Cross Temporal Model

Fuster's 1997 model of cross-temporal synthesis is based on three concepts: interference control, planning, and working memory. The theory proposed that the main goal of executive functions lie within organizing behavior (Barkley, 2011a). Contrasting from previous models, especially Baddeley's central executive model, Fuster does not "place a ghost in the machine" (Barkley, 2011a, p. 12). There is no central executive or single component within Fuster's theory; rather, temporal mediation captures the interaction between short-term memory and the attention set (Fuster, 2000). In Fuster's terminology, "new or recently learned behavior, sensory impulses are processed along the sensory hierarchy and into the motor hierarchy. Sensory information is thus translated into action, processed down the motor hierarchy to produce changes in the environment."

Integrative Model

Miller and Cohen's (2001) model focused on cognitive control and particularly the activities that represent maintenance of goals. They also refer to executive functioning as an umbrella term of cognitive processes under goal-directed behavior. In their model executive functioning is a top-down system serving to encourage sensory and motor processing areas into interacting with each other (Miller & Cohen, 2001). Maps are created between the inputs and outputs in this model, wherein bias signals guide activities along the neural pathways (Miller & Cohen, 2001).

Cascade of Control

Banich (2009) proposed that sequential cascade of brain areas attributed to maintaining attentional sets. According to Banich (2009) the DLPFC is the first to act using top-down attention to activate brain regions involved, and other regions of the cortex determine what information is necessary for an appropriate response. Finally, the posterior dorsal cingulate may serve as a catch all for the problems associated with selection thus far in this model (Banich, 2009).

Extended Phenotype

Barkley (2011a) summarizes executive functioning with the term self-regulation composed of (1) working memory, (2) management of emotions, (3) problem solving, and (4) analysis and synthesis into new behavioral goals. Processes include working memory, planning, problem solving, self-monitoring, interference control, and self-motivation (Barkley, 2011b).

A Developmental Perspective of EF

An important foundation for understanding the development of EF can be found in the works of Luria (1963, 1966, 1973). Luria's neurodevelopmental model postulated specific developmental stages related to stages of higher cortical maturation. Luria suggested that various stages of mental development encountered as children mature provide a unique opportunity to study how EFs develop (Horton, 1987).

Luria (1966) postulated a number of stages by which neuropsychological functions critical for intelligence and EF are developed. These stages were thought to interact with environmental stimuli based on Vygotsky's cultural and historical theory (Van der Veer and Valsiner, 1994). Vygotsky developed a complex theory related to language and thought processes. He postulated that environmental and/or cultural influences were important in understanding the development of neurological structures responsible for higher-level mental abilities, such as abstraction, memory, and attention. Luria expanded Vygotsky's original theories (Vygotsky, 1997a, 1997b, 1997c, 1997d).

In 1966, Luria postulated that higher cortical functions involving EF required interaction of normal neurological development and specific environmental stimuli of a cultural, historical, and social nature of development. In this way, Luria's thoughts are very consistent with current theory suggesting that particular phenotypes are shaped by environmental experience, leading to multi-finality or multiple endophenotypes. Thus, the result of the optimal interaction of neurological development and environmental stimuli would result in more efficient cortical functioning related to abilities such as language, attention, memory, intelligence, and EF.

In 1980, Luria proposed five stages of human development:

- Stage One: This stage begins in the first year of life and involves development of the brain stem structures such as the reticular activating system.
- Stage Two: This stage involves the activation of the primary sensory areas for vision, hearing, and tactile perception and the primary motor areas of gross motor movement during the second year of life. This is consistent with Piaget's stage of sensorimotor operations.
- Stage Three: This stage involves development of single modalities in the secondary association areas of the brain as children enter their preschool years. The child's mind recognized and reproduces various symbolic materials and develops the ability to model physical movement. This stage is consistent with Piaget's concept of preoperational functioning.
- Stage Four: This stage begins as the child enters first or second grade (7–8 years of age) as the tertiary areas of the parietal lobes are activated. The tertiary parietal lobes, the temporal parietal and occipital lobes join anatomically and involve coordination of the three major sensory input channels. During this stage, the child's mind begins to make sense of sensory input and environmental stimulation. It is particularly important for the development of complex mental abilities. This stage fits Piaget's concept of concrete operations.
- Stage Five: During this stage, the brain becomes activated beginning at approximately 8 years of age, through adolescence and adulthood. This operation involves the frontal lobes; the area anterior to the central sulcus is crucial to the development of complex mental abilities involving abstract thinking, intentional memory, as well as the execution monitoring and evaluating for complex learning (Stuss & Benson, 1984). This stage fits Piaget's concept of formal operations.

Beyond Luria's stage theory of brain development, his theoretical account of dynamic brain function is perhaps one of the most complete of all theorists (Lewandowski, Lovett, Gordon, & Codding, 2008). Luria conceptualized four interconnected levels of brain-behavior relationships and neurocognitive functioning including (1) the structure of the brain, (2) the functional organization based on structure, (3) syndromes and impairments arising in brain disorders, and (4) clinical methods of assessment (Korkman, 1999). Luria's theoretical formulations, methods, and ideas are well articulated in his books, *Higher Cortical Functions in Man* (1966, 1980) and *The Working Brain* (1973). Luria viewed the brain as a functional mosaic, the parts of which interact in different combinations to subserve different cognitive processes (Luria, 1973). No single area of the brain functions without input from other areas; thus, integration is a key principle of brain function within a learning framework. Thought, problem solving, EF, and intelligent behavior result from interaction of complex brain activity across various areas. Luria's (1966, 1973, 1980) research on the functional aspects of brain structures forms the basis for the development of the planning, attention, simultaneous, and successive processes (PASS) theory, described by Das, Naglieri, and Kirby (1994) and operationalized by Naglieri, Das, and Goldstein (2013).

In the Lurian framework of intellectual function, attention, language, sensory, perception, motor, visuospatial facilities, learning, and memory are complex, interrelated capacities. They are composed of flexible and interactive subcomponents, mediated by an equally flexible interaction neural network (Luria, 1962, 1980). These cognitive functions as conceptualized by Luria are modulated by three separate but connected functional units that provide the four basic psychological processes. These three brain "systems" are referred to as functional units because their neuropsychological mechanisms work in separate but interrelated systems. Multiple brain systems mediate complex cognitive functions. For example, multiple brain regions interact to mediate attentional processes (Mirsky, 1996; Castellanos et al., 2003). The executive functions managed by the third functional unit, as described by Luria, regulate the attentional processes of the first functional unit in sustaining the appropriate level of arousal and vigilance necessary for the detection of selection of relevant details from the environment. Consider the example of response inhibition. Inhibitory behavior allows a child to resist or inhibit responding to saline by irrelevant details during a task. This improves task performance. Response inhibition allows the child to focus over time on task-relevant features.

Prefrontal areas of the frontal lobes of the brain are associated with the third functional unit (Luria, 1980). The prefrontal cortex is well connected with every distinct functional unit of the brain (Goldberg, 2009). This unit is most likely responsible for planning and is involved with most behaviors we typically consider associated with executive function and executive function capacity (McCloskey, Perkins, and Van Divner, 2009). The third functional unit is also further differentiated into three zones with the primary zone in the motor strip of frontal lobe being concerned with motor output. The secondary zone is responsible for the sequencing of motor activity and speech production, whereas the tertiary zone is primarily involved with behaviors typically described as executive function. Damage to any of several areas of the frontal regions has been related to difficulties with impulse control, learning from mistakes, delay of gratification, and efficient attention. Because the third functional unit has rich connections with other parts of the brain, cortical and subcortical, there are often forward and backward influences to and from other regions such as the thalamic and hypothalamic and limbic areas. This set of connections, consistent with evolutionary theory, is reflecting a building of the brain over billions of years from a brain stem forward to the frontal lobes. Additionally a growing body of evidence points to a network of connected regions in the adjacent frontal and parietal lobes which have been implicated in higher auto-processing such as attention, decision making, and intelligent behavior (Kolb and Whishaw, 2009).

Luria wrote that the frontal lobe synthesized the information about the outside world and is the means whereby the behavior of the organism is regulated in conformity with the effect produced by its actions (Luria, 1980, p. 263). The frontal lobes provide for the programming, regulation, and evaluation of behavior and enable the child to ask questions, develop strategies, and self-monitor (Luria, 1973). Other responsibilities of the third functional unit include the regulation of voluntary activity, conscious impulse control, and various linguistic skills such as spontaneous conversation. The third functional unit provides the most complex aspects of human behavior, including personality and consciousness (Das, 1980). A reciprocal relationship exists between the first and third functional units. The higher cortical systems both regulate and work in collaboration with the first functional unit while also receiving and processing information from the external world and determining an individual's dynamic activity (Luria, 1973). This unit is also influenced by the regulatory effects of the cortex. Ascending and descending systems of the reticular formation enable this relationship by transmitting impulses from lower parts of the brain to the cortex and vice versa. Thus, damage to the prefrontal area can alter this reciprocal relationship so that the brain may not be sufficiently aroused for complex behaviors requiring sustained attention. In 2009, Goldberg described a breakdown in any portion of this complex, loop-like interaction between the prefrontal ventral brain stem and posterior cortex as producing systems of attention deficit. Castellanos et al. (2001) further hypothesize that the right prefrontal cortex and organs at the basal ganglia such as the substantia nigra and the cerebellum form a critical set of connections he described as "brain's braking system." These interconnections innervate and come online when inhibition, attention, and self-regulation are required.

The connection between units also links the psychological processes that are routed in each of the functional units. For PASS theory this means that the psychological processes of attention and planning are necessarily strongly related because planning often has conscious control of attention. In other words, one's limited attentional resources are dictated by the plan for one's behavior. The combination of attention and planning offer a functional description of executive function. However, attention and other PASS processes are influenced by many variables beyond planning. One of the influences is the environment. Novel encounters within daily life demand that individuals act in one way or another. The interaction of knowledge and several PASS processes are involved as individuals make judgments about similarities and differences between past situations and present demands, while estimating possible outcomes of action, even as acting. Humans are uniquely the only species capable of simultaneously thinking, evaluating, and acting. As Bromhill (2004) notes humans are able to think one thing while saying and or doing something else.

Luria's organization of the brain into functional units was not an attempt to map out the precise locations with specific areas of higher cognition taking place. In fact, Luria believed no part of the brain works by itself; thus, no cognitive task solely requires simultaneous, successive planning or attention processing, or any other processes, but rather it is a matter of emphasis. Luria stated "perception of memorizing gnosis and praxis, speech and thinking, writing, reading and arithmetic cannot be regarded as isolated or even indivisible faculties" (Luria, 1973, p. 29). Thus, an attempt to identify a fixed cortical location for any complex behavior would be considered a mistaken endeavor. Instead the brain should be conceptualized as a functioning whole composed of units that provide purpose.

Conclusion

Over the last 150 years, significant and critical advancements have been made in our understanding of the manner in which the brain regulates, manages, organizes, and helps organisms interface with their environment. It has now been well documented that to function effectively the brain requires an executive system. This EF system controls and manages other systems, abilities, and processes. Prefrontal areas of the frontal lobes primarily carry out this operation. These are parts of the brain that from an evolutionary perspective are more recently evolved. Thus, it is not surprising that human beings possess a complex EF system. Future research will continue to define, understand, and develop strategic and clinical strategies and interventions to facilitate the development and operation of the EF system.

References

Baddeley, A. D. (1986). *Working memory*. Oxford: Oxford University Press.

Baddeley, A., Sala, S. D., & Robbins, T. W. (1996). Working memory and executive control. *Philosophical Transactions of the Royal Society of London. Series B, Biological Sciences, 351*(1346), 1378–1388.

Banich, M. T. (2009). Executive function: The search for an integrated account. *Current Directions in Psychological Science, 18*, 89–94.

Barkley, R. A. (2011a). *Executive functioning and self-regulation: Integration, extended phenotype, and clinical implications*. New York: Guilford Press.

Barkley, R. A. (2011b). *Barkley Deficits in Executive Functioning Scale (BDEFS) manual*. New York: Guilford.

Broadbent, D. E. (1958). *Perception and communication*. London: Pergamon.

Bromhill, C. (2004). *The eternal child*. Sydney, Australia: Edbury Press.

Castellanos, F. X., Giedd, J. N., Berquin, P. C., Walter, J. M., Sharp, W., Tran, T., et al. (2001). Quantitative brain magnetic resonance imaging in girls with Attention deficit/hyperactivity disorder. *Archives of General Psychiatry, 58*, 289–295.

Castellanos, F. X., Lee, P. P., Sharp, W., Jeffries, N. O., Greenstein, D. K., Clasen, L. S., et al. (2003).

Developmental trajectories of brain volume abnormalities in children and adolescents with attention deficit/hyperactivity disorder. *Journal of the American Medical Association, 288*, 1740–1748.

Checa, P., Rodriguez-Bailon, R., & Rueda, M. R. (2008). Neurocognitive and temperamental systems of early self-regulation and early adolescents' social and academic outcomes. *Mind Brain and Education, 2*, 177–187.

Das, J. P., Naglieri, J. A., & Kirby, J. R. (1994). *Assessment of cognitive processes*. Needham Heights, MA: Allyn & Bacon.

Driver, J. (2001). A selective review of selective attention research from the past century. *British Journal of Psychology, 92*, 53–78.

Fuster, J. M. (2000). Executive frontal functions. *Experimental Brain Research, 133*, 66–70.

Kolb, B., & Whishaw, I. Q. (2009). *Fundamentals of human neuropsychology* (6th ed.). New York: Freeman-Worth.

Lewandowski, L., Lovett, B., Gordon, M., & Codding, R. (2008). Symptoms of ADHD and academic concerns in college students with and without ADHD diagnoses. *Journal of Attention Disorders, 12*, 156–161.

Luria, A. R. (1963). *Restoration of function after brain injury*. New York: Pergamon Press.

Luria, A. R. (1966). *Human brain and psychological processes*. New York: Harper and Row Publishers.

Luria, A. R. (1973). *The working brain*. New York: Basic Books.

Luria, A. R. (1980). *Higher cortical functions in man* (2nd ed.). New York: Basic Books.

McCloskey, G., Perkins, L. A., & Van Divner, B. (2009). *Assessment and intervention for executive function difficulties*. School-based practice in action series. New York: Routledge.

Naglieri, J., Das, J. P., & Goldstein, S. (2013). *Cognitive assessment system—II rating scale examiner's manual*. Austin, TX: PRO-ED Publishers.

Naglieri, J., & Goldstein, S. (2013). *Comprehensive executive functioning inventory technical manual*. Toronto, Canada: Multi-Health Systems.

Posner, M. I., & Snyder, C. R. R. (1975). Attention and cognitive control. In R. Solso (Ed.), *Information processing and cognition: The Loyola symposium* (pp. 55–85). Hillsdale, NJ: Lawrence Erlbaum.

Ratiu, P., & Talos, I. F. (2004). Images in clinical medicine: The tale of Phineas Gage. *New England Journal of Medicine, 351*(23), e21.

Reynolds, C. R., & Horton, A. M. (2006). *Test of verbal conceptualization and fluency*. Austin, TX: Pro-Ed.

Rueda, M. R., Posner, M. I., & Rothbart, M. K. (2004). Attentional control and self-regulation. In R. F. Baumeister & K. D. Vohs (Eds.), *Handbook of self-regulation: Research, theory, and applications* (pp. 283–300). New York: Guilford Press.

Schiffrin, R. M., & Schneider, W. (1977). Controlled and automatic human information processing: Perceptual learning, automatic attending and a general theory. *Psychological Review, 84*(2), 127–190.

Shallice, T. (1988). *From neuropsychology to mental structure*. Cambridge: Cambridge University Press.

Shallice, T. (2002). Fractionation of the supervisory system. In D. T. Stuss & R. T. Knight (Eds.), *Principles of frontal lobe function* (pp. 261–277). New York: Oxford University Press.

Shifrin, R. M., & Schneider, W. (1977). Controlled and automatic human information processing: Perceptual learning, automatic attending and general theory. *Psychological Review, 84*, 127–190.

The Physiology of Executive Functioning

Hyun Jin Chung, Lisa L. Weyandt, and Anthony Swentosky

Executive function (EF) has been defined as a multifaceted construct that involves a variety of high-level cognitive abilities (De Frias, Dixon, & Strauss, 2006). For most of the last century, studies of executive functions originated from neuropsychological research that focused on adults with frontal lobe damage (Stuss & Benson, 1986). Results of these studies suggested that lesions in the prefrontal cortex are associated with difficulties in tasks that require the ability to control impulses, plan strategically, and inhibit behaviors (Luria, 1972). Over the years, major features of executive functions have been identified, and these include abilities such as inhibitory control, attention shifting, working memory, goal-directed behavior, and strategic planning (Barkley, 1997; Miyake et al., 2000; Zelazo & Müller, 2002). Although essential components, such as response inhibition and goal-directed behavior, have been identified as important facets of executive function (Weyandt, 2009), to date, there is no agreed upon definition for this construct (Jurado & Rosselli, 2007).

Despite the fact that there is no universal definition of executive function, many studies have attempted to examine the underlying physiological features of executive functions. The purpose of this chapter is to examine the physiological underpinning of executive functions, as well as the methodological limitations associated with these studies. Specifically, structural neuroimaging studies that have examined changes across development will be examined, followed by a discussion of functional neuroimaging studies that have focused on five constructs of executive function—planning, verbal fluency, working memory, response inhibition, and set shifting. In addition, common limitations associated with neuroimaging studies and suggestions for future research.

The articles presented in this chapter were obtained by searching two databases, namely, PsycArticles and ScienceDirect. The lists of reference were reviewed for the purpose of the study. Keywords such as executive function (or specific executive functions such as planning, verbal fluency, working memory, response inhibition, and set shifting) and structural imaging *or* functional imaging were used. In order for the article to be included in this review, the study had to be (a) published in a peer-reviewed journal between the years 1991 and 2012. In addition, the study had to (b) use neuroimaging techniques and (c) include a sample size larger than ten to examine the physiology of executive functions.

Physiological Underpinning of Executive Functions

Past research has created a false belief that the physiological underpinning of executive functions were allocated to the frontal lobes based on case studies with individuals who had sustained

H.J. Chung (✉) • L.L. Weyandt • A. Swentosky
University of Rhode Island, Kingston, RI, USA
e-mail: hyunjin_chung@my.uri.edu

damage to the frontal lobes. These individuals often displayed deficits on a range of tasks purported to measure executive functioning; hence, it was presumed that damage to the frontal lobes would result in low performance on executive function tasks (Alvarez & Emory, 2006; Collette, Hogge, Salmon, & Van der Linden, 2006). More recently, however, with the advancement of technology, various methods (e.g., MRI, fMRI, PET) have supported that executive functioning relies on various distributed networks, which include frontal and posterior regions of the cerebral cortex, as well as subcortical regions (Collette et al., 2006; Jurado & Rosselli, 2007; Marvel & Desmond, 2010).

Structural Neuroimaging Findings

A handful of structural neuroimaging studies have provided support that prefrontal and parietal regions are involved in executive functions (Badre & Wagner, 2007; Collette, Olivier et al., 2005; Gilbert, Bird, Brindley, Frith, & Burgess, 2008; Jacobs, Harvey, & Anderson, 2011; Keller, Baker, Downes, & Roberts, 2009; Raposo, Mendes, & Marques, 2012; Rypma, 2006; Tamm, Menon, & Reiss, 2003; Tamnes et al., 2010; Van Petten et al., 2004). For example, structural differences in the prefrontal cortex have been investigated. Keller et al. (2009) found volume atrophy in the dorsal prefrontal cortex with individuals with temporal lobe epilepsy, and performance on tasks of executive functioning (i.e., working memory index of the Wechsler Memory Scale and the Controlled Oral Word Association Test) was positively correlated with the volume of the dorsal prefrontal cortex. It is important, however, to note that results differ substantially among different age groups. For example, Jacobs et al. (2011) recently reported that along with the prefrontal cortex, the *entire brain* (p. 810) may play a crucial role in performing executive functioning tasks in childhood. On the other hand, studies conducted with older adults have also found that the prefrontal cortex appears to play a crucial part in executive functioning task performance. Specifically, some researchers have found positive correlations between prefrontal lobe volumes and executive functioning task performance (Gunning-Dixon & Raz, 2003; Salat, Kaye, & Janowsky, 2002).

In 2010, Tamnes and colleagues studied neuroanatomical correlates of executive functions in Norwegian children and adolescents (50 males/48 females), ages 16–19. In the study, the relationships between three executive functions— namely, updating, inhibition, and shifting—and cortical thickness were examined via magnetic resonance imaging (MRI). During childhood and adolescence, cortical maturation is believed to be associated with thinning of the gray matter (Shaw et al., 2006), so it was hypothesized that rapid thinning would be associated with greater cognitive gains. Therefore, the primary research question focused on whether cortical maturation of the prefrontal cortex was associated with higher levels of executive functioning. Specifically, the researchers hypothesized that there would be a negative relationship between cortical thickness and executive functions and higher levels of performance would reveal stronger negative associations with cortical thickness and age.

In the study, six different executive function tasks were used (keep track task, letter memory task, plus–minus task, Trail-making test, antisaccade task, and Stroop task). Updating was assessed by keep track task (adapted by Miyake et al., 2000) and the letter memory task (also adapted by Miyake et al., 2000). Both tasks required the participant to update their working memory by recalling the last few words or letters from a sequence of words/letters. Two tasks were used to measure shifting, namely, the plus-minus task (adapted by Miyake et al., 2000) and the D-KEFS Trail-making test (Delis, Kaplan, & Kramer, 2001). In these tasks, the participant had to shift their attention to follow directions. For the former task, the participants were asked to complete a number of mathematical problems by adding 3 and then another problem set that required them to subtract 3. After these two problem sets, participants were given the third problem set, which required alternating between adding 3 and subtracting 3. For the latter task, three conditions—number sequencing, letter sequencing, and number-letter sequencing—were administered. Specifically, participants were instructed to connect the numbers in numerical order in the number sequencing task. Similarly, participants were asked to link the letters in alphabetical order in the letter sequencing task. In the number-letter sequencing task, the participant had to connect both numbers and letters in ascending order (e.g., 1-A–2-B). Finally, inhibition was measured by the antisaccade task (adapted by Miyake et al., 2000) and the Stroop task (Delis et al., 2001). Both tasks required the participant to inhibit reflexive responses and focus on the target stimuli.

Before controlling for age, cortical thinning was observed across most parts of the cortical mantle, and negative associations were found between EF tasks (keep track, letter memory, antisaccade task) and cortical thickness. However, after controlling for age, results revealed that the keep track task (updating) was associated with cortical thinning in the parietal and frontal regions of the brain. In addition, thinning in the areas of the left inferior frontal gyrus (LIFG) and the right superior medial parietal areas was associated with better working memory updating performance. These results are consistent with functional magnetic resonance imaging (fMRI) findings showing that working memory is associated with the prefrontal cortex, anterior cingulate, and parietal and occipital regions of the brain (Honey, Bullmore, & Sharma, 2000). The antisaccade task (inhibition) was related to more thinning in the occipital (posterior) and parietal regions. The authors suggested that the antisaccade task might tap into visual detection and attention processes than inhibition ability in children and adolescence. Finally, there was no evidence supporting the hypothesis that individual differences in levels of executive functioning were related to structural maturation differences in the prefrontal cortex. The researcher speculated that the occipital and parietal regions of the brain were associated with basic cognitive processes that would not vary among individuals, whereas the prefrontal circuits, being highly associated with strategic thinking, would vary across participants (Collette, van der Linden et al., 2005).

There were several limitations associated with this study. First, cross-sectional data was used to examine the relationship between executive functioning tasks and structural brain maturation. Ideally, longitudinal studies would be used to investigate this relationship by including multiple time points and mapping developmental and maturational trajectories within participants. Next, individuals who participated in this study revealed relatively high cognitive functioning, which may not be representative of the general population. In addition, the executive functioning tasks used in the research was only limited to six tasks. Therefore, different results might emerge when different tasks are used. Finally, there was some possibility that other cognitive processes may have influenced task performance. For instance, the researchers did not control for non-executive abilities such as motor and processing speed that may have differed across age. Collectively, these studies suggest that improvement on executive functioning tasks is associated with structural maturation of the brain, with regional development of the cerebral cortex, subcortical structures, and white matter showing ongoing development from early childhood to adulthood (Giorgio et al., 2010).

Recently, Burzynska et al. (2011) examined the relationship between cortical thickness and executive function performance. Specifically, Burzynska et al. examined the relationship between cortical thickness and executive functioning as assessed by performance on the Wisconsin Card Sorting Test (WCST; Heaton, Chelune, Talley, Kay, & Curtiss, 1993). The WCST is a neuropsychological card sorting task that requires attention, inhibition, and set-shifting skills. In this study, researchers hypothesized that cortical thickness would be positively associated with WCST performance. This hypothesis was based on the theory that cortical thickness in adulthood may involve more neurons and synaptic connections, high degree of complex circuitry and myelination, and higher metabolic efficiency in the brain (Deary & Caryl, 1997). Seventy-three healthy young adults (32 women/41 men) between ages 20 and 32 and 56 healthy older adults (27 women/29 men) between ages 60 and

71 participated in the study. All participants achieved at least 8 years of education and had no history of neurological or psychiatric disease. Structural neuroimaging results (MRI) revealed that higher accuracy on the WCST was related to thicker cortex in the lateral prefrontal and parietal regions. Specifically, thicker cortices in bilateral middle frontal gyrus (MFG), right inferior frontal gyrus (RIFG), postcentral gyrus (PCG), precentral gyrus (preCG), and the superior parietal gyrus (SPG) were associated with higher percentage of correct responses on the WCST. The results of this study contradict the findings of Tamnes et al. (2010), which limited their research findings to young children. Studies that have investigated cortical changes in childhood agree that cortical thinning during this period is associated with better performance on executive functioning tasks, as well as academic outcomes (Shaw et al., 2006; Sowell et al., 2004). However, during adulthood, Miller, Alston, and Corsellis (1980) have suggested that the human brain undergoes a gradual reduction in volume. Perhaps the fact that cortical thinning is related to better performance on executive functioning tasks in childhood and adolescence no longer holds for older adults, since these individuals are experiencing reductions in brain volume. Therefore, with older adults, the maintenance of cortical thickness could be associated with *better* executive functioning. These ideas are speculative, of course, and warrant empirical investigation.

To further explore executive functions in the elderly population, Weinstein et al. (2011) investigated how aerobic fitness may impact executive functioning outcomes. In this study, participants completed two executive functioning tasks: the Stroop task and the spatial working memory assessment. Aerobic fitness was measured by maximal graded exercise test (VO_2 max), which is an indicator of cardiorespiratory fitness (CRF) (American College of Sports Medicine, 1991). To assess CRF, participants were asked to speed walk on a motor-operated treadmill within 2 weeks after the completion of the executive functioning tasks. Results of the study indicated that higher CRF levels were associated with better outcomes on the Stroop task and the spatial working memory task. In addition, individuals with higher CRF level had greater gray matter volume in the dorsolateral prefrontal cortex (DLPFC). Specifically, the volume of the right IFG and preCG mediated the relationship between fitness level and Stroop interference, whereas non-overlapping regions of the DLPFC mediated the association between fitness level and spatial working memory.

This study had several strengths in that it used a relatively large homogeneous sample, which allowed the researchers to test mediation models. In addition, this study used two validated cognitive tasks to examine the hypothesis. However, the cross-sectional design does not allow for causal inferences and longitudinal studies are needed. Moreover, other variables such as genetic factors that affect the production of neurotrophins may in turn influence executive functioning performance.

In summary, a number of neuroimaging studies suggest that broad areas of the anterior and posterior regions of the brain are likely associated with executive functions (Perry et al., 2009). Although the specific regions of activation differed across tasks (and studies), preliminary studies support that increased activation in the DLPFC, as well as the parietal regions (i.e., SPG), is associated with better executive functioning performance on tasks including the Stroop task, spatial working memory, and the WCST.

Functional Neuroimaging Findings

Numerous studies of executive functions have been conducted with functional neuroimaging techniques, i.e., those that assess regional cerebral blood flow (rCBF) or glucose metabolism (Weyandt, 2006). Most of these studies have used a cognitive subtraction method to deduce which particular regions of the brain are associated with the executive processes (Salmon & Collette, 2005). Specifically, this method compares regions of brain activity while participants engage in an executive functioning task compared to when the participant solves a nonexecutive control task.

By measuring regional brain activation patterns between executive and nonexecutive tasks, the activation patterns specific to the executive tasks are believed to represent the brain regions specifically recruited for executive processes (Collette et al., 2006). To improve on the cognitive subtraction methodology, several studies have extended these findings by applying "conjunction" analyses (Collette & Van der Linden, 2002; Collette, Oliver et al., 2005), which measures the common regional activation associated with performance on multiple tasks purported to measure the same executive function.

Jurado and Rosselli (2007) provided a review of the brain correlates of executive functions using single-photon emission computerized tomography (SPECT) and MRI. Results revealed that studies exploring strategic planning ability using the Tower of London task generally found that the DLPFC, anterior cingulate cortex (ACC), supramarginal gyrus (SMG), and angular and right and left prefrontal cortex were areas of increased activation (Goethals et al., 2004; Lazeron et al., 2000; Morris, Ahmed, Syed, & Toone, 1993). Additionally, various studies have reported that attentional control as measured by the Hayling task, Stroop task, and Wisconsin Card Sorting Test was related to increased activation in DLPFC (Collette et al., 2001; Gerton et al., 2004; Kaufmann et al., 2005; Lie, Specht, Marshall, & Fink, 2006) and the PFC (Collette et al., 2001; Fassbender et al., 2004). Verbal and nonverbal fluency performances were also associated with increased activation in various frontal regions (e.g., LIFG, ACC, and superior frontal sulcus) including the DLPFC (Frith, Friston, Liddle, & Frackowiak, 1991; Jahanshahi, Dirnberger, Fuller, & Frith, 2000). In the section that follows, neuroimaging findings exploring five executive functions—planning, verbal fluency, working memory, response inhibition, and set shifting—will be covered in more detail.

Planning

Planning is a complex construct, making it difficult to narrow down a specific set of brain regions or networks underlying this ability. For example, planning has been defined as a large category of responses and processes including, but not limited to, decision-making, judgments, and evaluation of one's own behaviors and the behaviors of others (Das & Heemsbergen, 1983). Various executive function tasks including variations of the Tower of London test and maze completion test have been used to assess planning (Purdy, 2002; Welsh & Huizinga, 2001). Research using fMRI and positron emission tomography (PET) has found consistent brain activation patterns during participant performance on planning tasks. For example, using fMRI, Unterrainer et al. (2004) assessed the performance of college students on a computerized version of the Tower of London test as a measure of planning ability. Individuals classified as "better problem-solvers" based on overall task performance demonstrated increased activation in the right DLPFC, right superior temporal region, and right inferior parietal region compared to those classified as "worse problem-solvers." Similarly, across the entire sample, better performance on the planning phase of the Tower of London test was associated with increased DLPFC activation. In addition, increased activation of the ACC was associated with erroneously solved trials. This increase in ACC activation during incorrectly solved trials is consistent with other neuroimaging studies that have found ACC activation to be associated with overriding responses, response-conflict, and errors of commission (Li et al., 2008).

Owen, Doyon, Petrides, and Evans (1996) used PET to examine regional activation during easier and more difficult versions of the Tower of London planning test with 12 healthy adults. Again, increased activation as measured by increased rCBF in the left DLPFC was associated with performance on the more difficult Tower of London task compared to a control task that consisted of identical visual stimuli and motor responses but was considered to be free of planning demands. In addition, statistically significant increased rCBF in the caudate and thalamus was also associated with performance on the more difficult version of the Tower of London test, implicating the involvement of a frontostriatal network during planning. Using fMRI with a sample of 22 healthy adults aged 21–49 years old, Van den Heuvel et al. (2003) also found increased blood oxygenated levels (BOLD) within the DLPFC, striatum, premotor cortex, supplementary association area, precuneus, and inferior parietal cortex to be associated with planning activity as measured by a variant of the Tower of London test. These studies, as well as others (Dagher, Owen, Boecker, & Brooks, 1999; Newman, Carpenter, Varma, & Just, 2003), consistently demonstrate increased activation in the DLPFC and frontostriatal networks during executive planning tasks.

Verbal Fluency

Verbal fluency refers to the ability to recall and produce words associated with a particular prespecified category or beginning with a particular letter. Phelps, Hyder, Blamire, and Shulman (1997) used fMRI and found that the LIFG, ACC, and superior frontal sulcus demonstrated statistically significant increased activation during a verbal fluency task. In a meta-analysis, Costafreda et al. (2006) also found statistically significant increased activation in the LIFG, with increased BOLD response in more dorsal regions associated with phonological verbal fluency as compared to semantic verbal fluency. Costafreda et al. (2006), however, did not find evidence of significant BOLD responses within the anteroposterior or medial-lateral areas of the LIFG during these verbal fluency tests. Using PET, Frith et al. (1991) found increased activity in the left DLPFC and decreased activation in the bilateral temporal cortices. In a more recent fMRI study, Birn et al. (2010) found that increased activation in the LIFG during the letter fluency as compared to the categorical fluency. Alternatively, categorical fluency was more strongly associated with left fusiform and left MFG activity as compared to the letter fluency.

Although multiple brain regions appear to be associated with performance on verbal fluency tasks, these neuroimaging studies are consistent with others that suggest the LIFG, as well as temporal and parietal regions, underlies performance on verbal fluency tasks (Gourovitch et al., 2000; Mummery, Patterson, Hodges, & Wise, 1996).

Working Memory

According to Baddeley (1992), working memory is the brain system that temporarily provides storage and manipulation of information. Working memory (WM) is usually involved in complex cognitive tasks such as language comprehension, learning, and reasoning. Some constructs of working memory that have been examined in the neuroimaging literature include selection of item representation, selection and updating, updating memory content, rehearsal, and coping with interference (Bledowski, Kaiser, & Rahm, 2010).

Neuroimaging studies examining the physiology of working memory have found both common and unique brain regions associated with working memory performance across different working memory tasks and task parameters (Lepsien, Griffin, Devlin, & Nobre, 2005; Marvel & Desmond, 2010; Rowe & Passingham, 2001; Rowe, Toni, Josephs, Frackowiak, & Passingham, 2000). Research has shown that increases in brain activation in the prefrontal cortex are associated with increased working memory demands (Braver et al., 1997; Bunge, Klinberg, Jacobson, & Gabrieli, 2000). For example, Barch et al. (1997) showed that the DLPFC, the left inferior frontal cortex (IFC), and an area within the left parietal cortex showed significantly increased activation during long-delay (8-s) task conditions compared to short-delay (1-s) task conditions on a modified version of a continuous performance task. This increased activation during long-delay conditions suggests that these regions showing increased activation are specifically associated with the maintenance of information in working memory. Furthermore, because activation of these regions did not show increased activation during task conditions not purported to contain working memory demands, these findings further support the unique role of the DLPFC, left IFC, and a left parietal region in working memory task performance. Along with the prefrontal cortex, Bunge et al. (2000) detected increased activation in the lateral prefrontal cortex (DLPFC) when participants were engaging in complex task (e.g., reading sentences and trying to retain target words). In 2004, Osaka and colleagues examined the neural substrates of executive functions with individuals who differed in working memory capacities. In this study, the authors hypothesized that the ACC and the LIFG would be the general neural basis for the central executive with reading span test (RST). Ten young adults aged 20–27 were divided into two subject groups: high-span subject (HSS) and low-span subject (LSS). Results of the fMRI indicated that increased activation was found in the ACC and LIFG when both groups were performing the complex RST. In addition, increased signal intensity in the ACC and IFG regions was detected for the HSS group. The cross correlation of signal change between IFG and ACC was higher for the HSS, which suggests that the network system between these two regions were more activated in the HSS than the LSS. The results of this study imply that the HSS have a more active working attention controlling system than the LSS group.

Recently, Fassbender et al. (2011) examined working memory in children with ADHD and found that these children lack specialization of brain function. In this study, the researchers hypothesized that there would be diminished activation in the prefrontal cortex, which is traditionally associated with WM. Researchers recruited 13 participants (ranging from 8 to 14 years) with ADHD combined type and typically developing controls matched on age, IQ, and SES. The Visual Serial Addition Task (VSAT) was used in conjunction with fMRI to examine working memory processes in these children. In this study, the authors hypothesized that participants with ADHD would reveal diminished activation in the prefrontal region of the brain and excess activation in areas that are associated with primary responding instead of higher cortical responding. The fMRI results indicated that the typically developing group had significantly greater activation in the bilateral MFG, right MFG extending into ACC, preCG, bilateral PCG, and the right cingulate when engaging in VSAT, whereas the ADHD group had significantly greater activation in regions of the brain that were not specific to working memory (i.e., medial prefrontal cortex and bilateral insula extending into basal ganglia). Both children in the typically developing group and ADHD group showed significant activation in the right MFG and the right precuneus. It is important to note that this study has several limitations and the results should be interpreted accordingly. For example, similar to many other neuroimaging studies, this study also had a relatively small sample size which compromises the statistical power of the study as small- and possibly medium-sized effects are unlikely to be detected. In addition, the average IQ level was relatively higher in both the ADHD group and control group, which may limit the generalizability of the study's results. Moreover, some of the participants with ADHD had a history of stimulant medication treatment, which potentially may have long-term effects on specific patterns of brain region activation. In conclusion, research suggests that the prefrontal cortex, specifically, the dorsolateral and parietal regions of the brain (Bledowski et al., 2010), show consistent activation while individuals perform working memory tasks.

Response Inhibition

Goldman-Rakic, Thierry, Glowinski, Goldman-Rakic, and Christen (1994) defines inhibition as the ability to reject an automatic tendency in a given situation. Inhibition is often considered an executive functioning ability or process (Barkley, 1997; Best & Miller, 2010; Miyake et al., 2000). Several authors have suggested that inhibition is a fractionated construct comprised of several similar yet distinct inhibitory processes (Friedman & Miyake, 2004). For example, some authors (Gray, 1982; Nigg, 2000, 2001) have distinguished between different types of inhibition such as response or motor inhibition, cognitive inhibition, interference control, motivational inhibition, and automatic inhibition of attention. Given the possibility of numerous related but distinct inhibitory processes, it is not surprising that weak correlations are typically found between measures that tap each of these abilities (Huizinga, Dolan, & van der Molen, 2006; Hull, Martin, Beier, Lane, & Hamilton, 2008). Furthermore, the existence of multiple types of inhibitory process would suggest the likelihood of overlapping yet possibly distinct brain regions underlying these multiple inhibitory processes.

Response inhibition has received considerable interest and research has shown that this is associated with increased activation of the dorsomedial prefrontal cortex, lateral prefrontal cortex, parietal cortex, insular cortex, bilateral precuneus, left angular gyrus, and right middle temporal gyrus (Blasi et al., 2006; Bunge, Dudukovic, Thomason, Vaidya, & Gabrieli, 2002; Mostofsky & Simmonds, 2008). Specifically, Blasi et al. (2006) examined response inhibition and interference monitoring and suppression in 57 healthy adults. In their study, neuroimaging results revealed that performance on a response inhibition task was associated with greater activation in DLPFC, ventrolateral prefrontal cortex (VLPFC), and parietal cortex (PC) as compared to performance on an interference monitoring and suppression task. Bunge et al. (2002) also have found that different inhibitory processes were in fact associated with differential brain region activation patterns. Better performance on an interference control task was associated with increased statistically significant activation of the VLPFC and insular cortex in both children and adults, although children showed increased activation in the left VLPC and insula, while adults showed increased activation in the right VLPFC and insula. Alternatively, regions associated with performance on a response inhibition task included the bilateral precuneus, left angular gyrus, and right middle temporal gyrus, and the right MFG for both children and adults. In adults, the bilateral VLPFC, bilateral DLPFC, and the anterior and posterior cingulate cortices were also significantly activated during task performance. In 2004, Aron, Robbins, and Poldrack found that the DLPFC, IFC, and the orbital frontal cortex are associated with inhibition tasks.

Recently, Carmona et al. (2011) examined response inhibition in medication-naïve adults with ADHD using a within-subject case–control design. Based on previous research, the authors hypothesized that the unmedicated adults with ADHD would reveal decreased activation in the IFG during response inhibition tasks. Twenty-three right-handed male adults with ADHD and 23 healthy controls participated in the study and the Go/NoGo task was used to measure response inhibition. Contrary to the study's hypothesis, results of the fMRI did not find differences in the bilateral IFG activation during Go/NoGo task performance. These results are consistent with other fMRI studies that have found no difference in brain activation during response inhibition task (Dillo et al., 2010) but are inconsistent with studies that have either found increased or decreased activation in the IFG when compared to controls (Epstein et al., 2007; Kooistra et al., 2010). Hence, the results of the neuroimaging findings warrant further investigation.

It should be noted, however, that several limitations characterized the study (Carmona et al., 2011). For example, the selectivity of the sample could bias the generalization of the results. Specifically, participants in this study were carefully screened for comorbidity and had to have an IQ that fell within one standard deviation of the mean. In addition, the sample size was relatively small in this study. Perhaps the study lacked the power to detect the group differences in the IFG activation due to the small sample size.

In addition, results from multiple studies have also implicated right lateralized fronto-striatal circuits in effective response inhibition, including the right inferior prefrontal cortex (Aron, Robbins, & Poldrack, 2004; Durston, Thomas, Worden, Yang, & Casey, 2002), presupplementary association area, and the striatum (Congdon et al., 2010). For example, in an event-related fMRI study by Rubia, Smith, Brammer, and Taylor (2003), effective response inhibition measured by performance on the stop task was primarily associated with statistically significant increased activation of the right inferior prefrontal cortex, while poorer task performance was associated with statistically significant increased activation of the ACC and bilateral inferior parietal lobes. In a different study that used the same stop task, Zandbelt and Vink (2010) also found that successful performance on stop trials was significantly correlated with increased activation of the right inferior prefrontal cortex, as well as the presupplementary motor area and the striatum. In addition to the previously mentioned brain regions, other neuroimaging studies have found statistically significant increased activation of parietal, cerebellar, and thalamic regions during inhibition tasks (Boehler, Appelbaum, Krebs, Hopf, & Woldorff, 2010; Rubia, Smith, Taylor, & Brammer, 2007).

In conclusion, these studies suggest that different inhibitory processes may be associated with different brain region activation; however, the VLPFC and the IFG may be involved across various inhibitory processes (Aron, Fletcher, Bullmore, Sahakian, & Robbins, 2004; Bunge et al., 2002). These findings are consistent with other neuroimaging studies showing increased VLPFC activation during the performance of both response inhibition and interference control tasks (Hazeltine, Poldrack, & Gabrieli, 2000; Rubia et al., 2001). Although a number have studies have examined the neural substrates of response inhibition, the delineation of the physiological substrates associated with different types of inhibitory processes remains a much needed area of future research.

Set-Shifting

Set-shifting is referred to as the ability to flexibly switch back and forth between tasks, operations, or mental sets (Miyake et al., 2000). Neuroimaging studies have shown that activation across prefrontal, parietal, and subcortical structures have been associated set-shifting ability (Salmon & Collette, 2005). For example, Wilkinson, Halligan, Marshall, Büchel, and Dolan (2001) used fMRI and found that performance on a set-shifting task (i.e., local-global task) was associated with statistically significant increased activation of the bilateral inferior parietal cortex, motor and premotor cortex, bilateral putamen, as well as a more general frontoparietal network. It should be noted that these regions showed differential degrees of activation across varying task parameters. In addition, Zakzanis, Mraz, and Graham (2005) obtained fMRI measures during participant performance on the Trail-making test. These authors found statistically significant increased activation in the left DLPFC, medial prefrontal cortex, and left middle and superior temporal gyrus during the shifting trial compared to the non-shifting trial.

In order to minimize the visuospatial demands inherent to most set-shifting tasks, Moll, de Oliveira-Souza, Moll, Bramati, and Andreiuolo (2002) obtained fMRI measures during participant performance on a variant of the Trail-making test that was intended to minimize visuospatial demands while increasing the verbal requirement. Verbal set-shifting ability was associated with significant increases in BOLD response in the left DLPFC, left supplementary motor area, and bilateral activation of the intraparietal sulci. Other neuroimaging studies using set-shifting tasks requiring minimal visual and spatial cognitive abilities have implicated the superior posterior parietal cortex (Gurd et al., 2002).

In 2004, Wager, Jonides, and Reading conducted a meta-analysis with 31 fMRI and PET studies to examine the neuroimaging studies of set-shifting. As a result, the researchers found that seven regions that showed significant activation across various set-shifting tasks. Specifically, the regions associated with set-shifting were the medial prefrontal cortex, right premotor cortex, bilateral posterior intraparietal sulcus, bilateral anterior intraparietal sulcus, and the left occipital region. Although both posterior (parietal and occipital) and frontal (DLPFC and anterior insula) regions were involved in set-shifting, the involvement of DLPFC was weaker than expected.

In conclusion, there were no specific brain regions that showed activation during set-shifting tasks. Instead, neuroimaging studies revealed that set-shifting is extended to multiple regions of the brain. Increased activation of the parietal cortex has been commonly reported, however, which may suggest that this region of the brain may play a core role in set-shifting (Gurd et al., 2002; Wager, Jonides, & Reading, 2004; Zakzanis et al., 2005).

Limitations and Future Directions

During the past decade, neuroimaging studies have provided additional information about brain structures and areas of functioning that may be involved with executive functions. This body of work is not without methodological problems, however, that ultimately limit the extent to which solid conclusions can be deduced.

First, as noted previously, there is no universally accepted definition of executive function (Jurado & Rosselli, 2007). Second, task impurity is a serious issue as many tasks that are commonly used as measures of executive function lack acceptable validity and reliability (Weyandt, 2009). Indeed, Rabbitt (1997) has expressed concerns regarding the low internal and test-retest reliability among executive function measures, and work by Tate, Perdices, and Maggiotto (1998) supports Rabbitt's concerns. For example, Tate et al. examined the temporal stability of the Wisconsin Card Sorting Test and found that the stability coefficient was in the .30–.40 range. Others have argued that various executive functioning tasks purported to measure a single construct have low intercorrelations and in many cases are statistically nonsignificant (Barkley, 2011; Collette et al., 2006; Greve et al., 2002; Humes, Welsh, Retzlaff, & Cookson, 1997; Salmon & Collette, 2005). Moreover, executive functioning tasks that are commonly used not only tap into a particular executive function but also other abilities such as general cognitive skills (Barkley, 2011) or nonexecutive skills (Collette et al., 2006). In addition, many executive functioning tasks have very low ecological validity (Ardila, 2008). Specifically, scholars have argued that these tasks are poorly correlated with daily life activities. Some scholars have suggested that the use of rating scales may be a better method of assessing executive functions (Barkley, 2011).

There are also a number of significant limitations associated with brain imaging techniques. First, neuroimaging studies typically involve small sample sizes (less than 20), which often compromises statistical power, and effect sizes are rarely reported. Confounding factors such as comorbidity are rarely acknowledged (Jacobs et al., 2011) as are potential medication effects (Anderson, Northam, Hendy, & Wrennall, 2001). Replication and reliability studies are virtually nonexistent. Another major limitation of neuroimaging studies is the use of cross-sectional data instead of longitudinal data. Specifically, most studies do not measure the brain activation of individuals across time nor do they measure this with short or long temporal delays (Collette et al., 2006). This one-time one-shot approach may compromise the reliability of the image. Furthermore, methods across studies vary greatly including the type of mathematical algorithms employed, colors representing activation levels, and statistical analysis procedures (Reeves, Mills, Billick, & Brodie, 2003; Weyandt, 2006; Weyandt & Swentosky, 2013), which may ultimately further complicate the interpretation of the results. In addition, researchers often fail to report the baseline activity in their studies and factors such as age, sex, emotional state, and health also could influence the results of neuroimaging studies; however, most studies have not yet considered these factors (Weyandt & Swentosky, in press).

Finally, it is important to note that in many of the previously cited studies (e.g., Li et al., 2008), it is unclear whether or not decisions were made a priori regarding the brain regions to be analyzed. In cases where the entire brain is analyzed, statistically significant activation patterns may simply be the result of the large number of regions analyzed (i.e., type I error) (Salmon & Collette, 2005). It should also be noted that in most of the studies previously cited, only significant findings were reported. Therefore, specific brain regions that did not show statistically significant levels of activation were not explicitly described. Furthermore, it is unlikely that specific brain regions are exclusively related to specific types of executive functions as performance on tasks purported to measure different executive functions often shows overlapping regions of activation. For example, significant activation of the LIFG has been found to be associated with performance on both verbal fluency and response inhibition tests

(Birn et al., 2010; Osaka et al., 2004). Significant activation of the DLPFC and the ACC has been found to be associated with performance on both planning and verbal fluency tests (Frith et al., 1991; Phelps et al., 1997; Unterrainer et al., 2004). Lastly, it is crucial to keep in mind that neuroimaging studies are correlational in nature and do not reveal causal relationships between executive functions and areas of increased brain activation.

Conclusion

During the past decade, the use of neuroimaging techniques to explore the physiological substrates of executive functions has increased substantially. In general, studies suggest that the physiology of executive function is not limited to the prefrontal cortex as hypothesized in previous studies (e.g., Birn et al., 2010; Fassbender et al., 2011; Newman et al., 2003; Osaka et al., 2004; Unterrainer et al., 2004). Instead, a wide range of brain structures and regions appear to be involved and these vary depending on the executive function measure employed. In general, these findings support that executive function is both a unitary and multifaceted construct. Future studies should show attempt to address the methodological limitations that exist in the current literature. For instance, mixed methodologies (e.g., longitudinal designs, neuroimaging subtraction, and conjunction analyses) and larger sample sizes would be beneficial as would attention to sample characteristics (e.g., IQ, sex, ethnicity). Lastly, further refinement of the conceptualization of the construct of executive functioning and the use of psychometrically sound executive functioning measures will contribute to a greater understanding of the neurophysiological substrates of executive functioning.

References

Alvarez, J. A., & Emory, E. (2006). Executive function and the frontal lobes: A meta-analytic review. *Neuropsychology Review, 16*, 17–42.

American College of Sports Medicine. (1991). *Guidelines for exercise testing and prescription*. Philadelphia: Lea & Febiger.

Anderson, V., Northam, E., Hendy, J., & Wrennall, J. (2001). *Developmental neuropsychology: A clinical approach*. East Sussex: Psychology Press.

Ardila, A. (2008). On the evolutionary origins of executive functions. *Brain and Cognition, 68*, 92–99.

Aron, A. R., Fletcher, P. C., Bullmore, E. T., Sahakian, B. J., & Robbins, T. W. (2004). Stop-signal inhibition disrupted by damage to right inferior frontal gyrus in humans. *Nature Neuroscience, 6*, 115–116.

Aron, A. R., Robbins, T. W., & Poldrack, R. A. (2004). Inhibition and the right inferior frontal cortex. *Trends in Cognitive Sciences, 8*(4), 170–177.

Baddeley, A. (1992). Working memory. *Science, 255*(5044), 556–559.

Badre, D., & Wagner, A. D. (2007). Left ventrolateral prefrontal cortex and the cognitive control of memory. *Neuropsychologia, 45*, 2883–2901.

Barch, D. M., Braver, T. S., Nystrom, L. E., Forman, S. D., Noll, D. C., & Cohen, J. D. (1997). Dissociating working memory from task difficulty in human prefrontal cortex. *Neuropsychologia, 35*(10), 1373–1380.

Barkley, R. A. (1997). Behavioral inhibition, sustained attention, and executive functions: Constructing a unifying theory of ADHD. *Psychological Bulletin, 121*(1), 65–94.

Barkley, R. A. (2011). Is executive functioning deficient in ADHD? It depends on your definitions and your measures. *The ADHD Report, 19*(4), 1–10.

Best, J. R., & Miller, P. H. (2010). A developmental perspective on executive function. *Child Development, 81*(6), 1641–1660.

Birn, R. M., Kenworthy, L., Case, L., Caravella, R., Jones, T. B., Bandettini, P. A., et al. (2010). Neural systems supporting lexical search guided by letter and semantic category cues: A self-paced overt response fMRI study of verbal fluency. *NeuroImage, 49*(1), 1099–1107.

Blasi, G., Goldberg, T. E., Weickert, T., Das, S., Kohn, P., Zoltick, B., et al. (2006). Brain regions underlying response inhibition and interference monitoring and suppression. *European Journal of Neuroscience, 23*, 1658–1664.

Bledowski, C., Kaiser, J., & Rahm, B. (2010). Basic operations in working memory: Contributions from functional imaging studies. *Behavioural Brain Research, 214*(2), 172–179.

Boehler, C. N., Appelbaum, L. G., Krebs, R. M., Hopf, J. M., & Woldorff, M. G. (2010). Pinning down response inhibition in the brain–conjunction analyses of the Stop-signal task. *NeuroImage, 52*(4), 1621–1632.

Braver, T. S., Cohen, J. D., Nystrom, L. E., Jonides, J., Smith, E. E., & Noll, D. C. (1997). A parametric study of PFC involvement in human working memory. *NeuroImage, 5*, 49–62.

Bunge, S. A., Dudukovic, N. M., Thomason, M. E., Vaidya, C. J., & Gabrieli, J. D. (2002). Immature frontal lobe contributions to cognitive control in children: Evidence from fMRI. *Neuron, 33*(2), 301–311.

Bunge, S. A., Klinberg, T., Jacobson, R. B., & Gabrieli, D. E. (2000). A resource model of the neural basis of executive working memory. *Proceedings of the National Academy of Sciences of the United States of America, 97*, 3573–3578.

Burzynska, A. Z., Nagel, I. E., Preuschhof, C., Gluth, S., Bäckman, L., Li, S., et al. (2011). Cortical thickness is linked to executive functioning in adulthood and aging. *Human Brain Mapping, 33*, 1607–1620. doi:10.1002/hbm.21311.

Carmona, S., Hoekzem, E., Ramos-Quiroga, J. A., Richarte, V., Canals, C., Bosch, R., et al. (2011). Response inhibition and reward anticipation in medication-naïve adults with attention-deficit/hyperactivity disorder: A within-subject case-control neuroimaging study. *Human Brain Mapping, 33*, 2350–2361. doi:10.1002/hbm.21368.

Collette, F., Hogge, M., Salmon, E., & Van der Linden, M. (2006). Exploration of the neural substrates of executive functioning by functional neuroimaging. *Neuroscience, 139*, 209–221.

Collette, F., Olivier, L., van der Linden, M., Laureys, S., Delfiore, G., Luxen, A., et al. (2005). Involvement of both prefrontal and inferior parietal cortex in dual-task performance. *Cognitive Brain Research, 24*, 237–251.

Collette, F., & Van der Linden, M. (2002). Brain imaging of the central executive component of working memory. *Neuroscience and Biobehavioral Reviews, 26*, 105–125.

Collette, F., Van der Linden, M., Delfiore, G., Degueldre, C., Luxen, A., & Salmon, E. (2001). The functional anatomy of inhibition processes investigated with the Hayling task. *Neuroimage, 14*, 258–267.

Collette, F., Van der Linden, M., Laureys, S., Delfiore, G., Degueldre, C., Luxen, A., et al. (2005). Exploring the unity and diversity of the neural substrates of executive functioning. *Human Brain Mapping, 25*, 409–423.

Congdon, E., Mumford, J. A., Cohen, J. R., Galvan, A., Aron, A. R., Xue, G., et al. (2010). Engagement of large-scale networks is related to individual differences in inhibitory control. *NeuroImage, 53*(2), 653–663.

Costafreda, S. G., Fu, C. H., Lee, L., Everitt, B., Brammer, M. J., & David, A. S. (2006). A systematic review and quantitative appraisal of fMRI studies of verbal fluency: Role of the left inferior frontal gyrus. *Human Brain Mapping, 27*(10), 799–810.

Dagher, A., Owen, A. M., Boecker, H., & Brooks, D. J. (1999). Mapping the network for planning: A correlational PET activation study with the Tower of London task. *Brain: A Journal of Neurology, 122*, 1973–1987.

Das, J. P., & Heemsbergen, D. B. (1983). Planning as a factor in the assessment of cognitive processes. *Journal of Psychoeducational Assessment, 1*, 1–15.

De Frias, C. M., Dixon, R. A., & Strauss, E. (2006). Structure of four executive functioning tests in healthy older adults. *Neuropsychology, 20*(2), 206–214.

Deary, I. J., & Caryl, P. G. (1997). Neuroscience and human intelligence differences. *Trends in Neurosciences, 20*, 365–371.

Delis, D. C., Kaplan, E., & Kramer, J. H. (2001). *The Delis-Kaplan executive function system*. San Antonio, TX: Psychological Corporation.

Dillo, W., Göke, A., Prox-Vagedes, V., Szycik, G. R., Roy, M., Donnerstag, F., et al. (2010). Neuronal correlates of ADHD in adults with evidence for compensation strategies—a functional MRI study with a Go/No-Go paradigm. *German Medical Science, 8*, Doc09, DOI: 10.3205/000098, URN: urn:nbn:de:0183-0000987

Durston, S., Thomas, K. M., Worden, M. S., Yang, Y., & Casey, B. J. (2002). The effect of preceding context on inhibition: An event-related fMRI study. *NeuroImage, 16*(2), 449–453.

Epstein, J. N., Casey, B. J., Tonev, S. T., Davidson, M. C., Reiss, A. L., Garrett, A., et al. (2007). ADHD- and medication-related brain activation effects in concordantly affected parent-child dyads with ADHD. *Journal of Child Psychology and Psychiatry, 48*, 899–913.

Fassbender, C., Murphy, K., Foxe, J., Wylie, G., Javitt, D., Robertson, I., et al. (2004). A topography of executive functions and their interactions revealed by functional magnetic resonance imaging. *Brain Research. Cognitive Brain Research, 20*, 132–143.

Fassbender, C., Schweiter, J. B., Cortes, C. R., Tagamets, M. A., Windsor, T. A., Reeves, G. M., et al. (2011). Working memory in attention deficit/hyperactivity disorder is characterized by a lack of specialization of brain function. *PLoS One, 6*(11), 1–11.

Friedman, N. P., & Miyake, A. (2004). The relations among inhibition and interference control functions: A latent-variable analysis. *Journal of Experimental Psychology, 133*(1), 101–135.

Frith, C., Friston, K., Liddle, P., & Frackowiak, R. (1991). A PET study of word finding. *Neuropsychologia, 29*, 1137–1148.

Gerton, B., Brown, T., Meyer-Lindenberg, A., Kohn, P., Holt, J., & Olsen, R. (2004). Shared and distinct neurophysiological components of the digits forward and backward tasks as revealed by functional neuroimaging. *Neurpsychologia, 42*, 1781–1787.

Gilbert, S. J., Bird, G., Brindley, R., Frith, C. D., & Burgess, P. W. (2008). Atypical recruitment of medial prefrontal cortex in autism spectrum disorders: An fMRI study of two executive function tasks. *Neuropsychologia, 46*, 2281–2291.

Giorgio, A., Watkins, K. E., Chadwick, M., James, S., Winmill, L., Douaud, G., et al. (2010). Longitudinal changes in grey and white matter during adolescence. *NeuroImage, 49*, 94–103.

Goethals, I., Audenaert, K., Jacobs, F., van der Wiele, C., Pyck, H., Ham, H., et al. (2004). Application of a neuropsychological activation probe with SPECT: The 'Tower of London' task in healthy volunteers. *Nuclear Medicine Communications, 25*, 177–182.

Goldman-Rakic, P. S., Thierry, A. M., Glowinski, J., Goldman-Rakic, P. S., & Christen, Y. (1994). *Motor and cognitive function of the prefrontal cortex*. Berlin: Springer.

Gourovitch, M. L., Kirkby, B. S., Goldberg, T. E., Weinberger, D. R., Gold, J. M., Esposito, G., et al. (2000). A comparison of rCBF patterns during letter and semantic fluency. *Neuropsychology, 14*(3), 353–360.

Gray, J. A. (1982). *The neuropsychology of anxiety: An enquiry into the functions of the septo-hippocampal system*. New York: Oxford University Press.

Greve, K. W., Love, J. M., Sherwin, E., Mathias, C. W., Houston, R. J., & Brennan, A. (2002). Temporal stability of the Wisconsin Card Sorting Test in a chronic traumatic brain injury sample. *Assessment, 9*, 271–277.

Gunning-Dixon, F. M., & Raz, N. (2003). Neuroanatomical correlates of selected executive functions in middle-aged and older adults: A prospective MRI study. *Neuropsychologia, 41*, 1929–1941.

Gurd, J. M., Amunts, K., Weiss, P. H., Zafiris, O., Zilles, K., Marshall, J. C., et al. (2002). Posterior parietal cortex is implicated in continuous switching between verbal fluency tasks: An fMRI study with clinical implications. *Brain, 125*, 1024–1038.

Hazeltine, E., Poldrack, R., & Gabrieli, J. D. (2000). Neural activation during response competition. *Journal of Cognitive Neuroscience, 12*, 118–129.

Heaton, R. K., Chelune, G. J., Talley, J. L., Kay, G. G., & Curtiss, G. (1993). *Wisconsin Card Sorting Test manual revised and expanded*. Odessa, FL: Psychological Assessment Resources.

Honey, G. D., Bullmore, E. T., & Sharma, T. (2000). Prolonged reaction time to a verbal working memory task predicts increased power of posterior parietal cortical activation. *Neuroimage, 12*, 495–503.

Huizinga, M., Dolan, C. V., & van der Molen, M. W. (2006). Age-related change in executive function: Developmental trends and a latent variable analysis. *Neuropsychologia, 44*, 2017–2036.

Hull, R., Martin, R. C., Beier, M. E., Lane, D., & Hamilton, A. C. (2008). Executive function in older adults: A structural equation modeling approach. *Neuropsychology, 22*(4), 508–522.

Humes, G., Welsh, M., Retzlaff, P., & Cookson, N. (1997). Tower of Hanoi and London: Reliability and validity of two executive function tasks. *Assessment, 4*, 249–257.

Jacobs, R., Harvey, A. S., & Anderson, V. (2011). Are executive skills primary mediated by the prefrontal cortex in childhood? Examination of focal brain lesions in childhood. *Cortex, 47*, 808–824.

Jahanshahi, M., Dirnberger, G., Fuller, R., & Frith, C. (2000). The role of the dorsolateral prefrontal cortex in random number generation: A study with positron emission tomography. *NeuroImage, 12*, 713–725.

Jurado, M. B., & Rosselli, M. (2007). The elusive nature of executive functions: A review of our current understanding. *Neuropsychology Review, 17*, 213–233.

Kaufmann, L., Koppelstaetter, F., Delazer, M., Siedentopf, C., Rhomberg, P., Golaszewski, S., et al. (2005). Neural correlates of distance and congruity effects in a numerical Stroop task: An event-related fMRI study. *NeuroImage, 15*, 888–898.

Keller, S. S., Baker, G., Downes, J. J., & Roberts, N. (2009). Quantitative MRI of the prefrontal cortex and executive function in patients with temporal lobe epilepsy. *Epilepsy & Behavior, 15*, 186–195.

Kooistra, L., van der Meere, J. J., Edwards, J. D., Kaplan, B. J., Crawford, S., & Goodyear, B. G. (2010). Preliminary fMRI findings on the effects of event rate in adults with ADHD. *Journal of Neural Transmission, 117*, 655–662.

Lazeron, R. H., Rombouts, S. A., Machielsen, W. C., Scheltens, P., Witter, M. P., Uylings, H. B., et al. (2000). Visualizing brain activation during planning: The tower of London test adapted for functional MR imaging. *American Journal of Neuroradiology, 21*, 1407–1414.

Lepsien, J., Griffin, I. C., Devlin, J. T., & Nobre, A. C. (2005). Directing spatial attention in mental representations: Interactions between attentional orienting and working-memory load. *NeuroImage, 26*(3), 733–743.

Li, C. S., Huang, C., Yan, P., Bhagwagar, Z., Milivojevic, V., & Sinha, R. (2008). Neural correlates of impulse control during stop signal inhibition in cocaine-dependent men. *Neuropsychopharmacology, 33*(8), 1798–1806.

Lie, C., Specht, K., Marshall, J., & Fink, G. (2006). Using fMRI to decompose the neural processes underlying the Wisconsin Card Sorting Test. *NeuroImage, 15*, 1038–1049.

Luria, A. R. (1972). *The man with a shatter world*. New York: Basic Books.

Marvel, C. L., & Desmond, J. E. (2010). Functional topography of the cerebellum in verbal working memory. *Neuropsychology Review, 20*(3), 271–279.

Miller, A. K., Alston, R. L., & Corsellis, J. A. (1980). Variation with age in the volumes of grey and white matter in the cerebral hemispheres of man: Measurements with an image analyser. *Neuropathology and Applied Neurobiology, 6*, 119–132.

Miyake, A., Friedman, N. P., Emerson, M. J., Witzki, A. H., Howerter, A., & Wager, T. D. (2000). The unity and diversity of executive functions and their contributions to complex frontal lobe tasks: A latent variable analysis. *Cognitive Psychology, 41*, 49–100.

Moll, J., de Oliveira-Souza, R., Moll, F. T., Bramati, I. E., & Andreiuolo, P. A. (2002). The cerebral correlates of set-shifting: An fMRI study of the trail making test. *Arquivos de Neuro-Psiquiatria, 60*(4), 900–905.

Morris, R., Ahmed, S., Syed, G., & Toone, B. (1993). Neural correlates of planning ability: Frontal lobe activation during the Tower of London test. *Neuropsychologia, 31*, 1367–1378.

Mostofsky, S. H., & Simmonds, D. J. (2008). Response inhibition and response selection: Two sides of the same coin. *Journal of Cognitive Neuroscience, 20*, 751–761.

Mummery, C. J., Patterson, K., Hodges, J. R., & Wise, R. J. (1996). Generating 'tiger' as an animal name or a word beginning with T: Differences in brain activation. *Proceedings of the Royal Society of Biological Psychiatry, 263*(1373), 989–995.

Newman, S. D., Carpenter, P. A., Varma, S., & Just, M. A. (2003). Frontal and parietal participation in problem solving in the Tower of London: fMRI and computational modeling of planning and high-level perception. *Neuropsychologia, 41*(12), 1668–1682.

Nigg, J. T. (2000). On inhibition/disinhibition in developmental psychopathology: Views from cognitive and personality psychology and a working inhibition taxonomy. *Psychological Bulletin, 126*(2), 220–246.

Nigg, J. T. (2001). Is ADHD an inhibitory disorder? *Psychological Bulletin, 127*(5), 571–598.

Osaka, N., Osaka, M., Kondo, H., Morishita, M., Fukuyama, H., & Shibasaki, H. (2004). The neural basis of executive function in working memory: An fMRI study based on individual difference. *NeuroImage, 21*, 623–631.

Owen, A. M., Doyon, J., Petrides, M., & Evans, A. C. (1996). Planning and spatial working memory: A positron emission tomography study in humans. *European Journal of Neuroscience, 8*, 353–364.

Perry, M. E., McDonald, C. R., Hagler, D. J., Gharapetian, L., Kuperman, J. M., & Koyamae, A. K. (2009). White matter tracts associated with set-shifting in healthy aging. *Neuropsychologia, 47*, 2835–2842.

Phelps, E. A., Hyder, F., Blamire, A. M., & Shulman, R. G. (1997). FMRI of the prefrontal cortex during overt verbal fluency. *Neuroreport, 8*, 561–565.

Purdy, M. (2002). Executive function ability in persons with aphasia. *Aphasiology, 16*, 549–557.

Rabbitt, P. (1997). Introduction: Methodologies and models in the study of executive function. In P. Rabbitt (Ed.), *Methodology of frontal and executive function* (pp. 1–38). Hove: Psychology Press.

Raposo, A., Mendes, M., & Marques, J. F. (2012). The hierarchical organization of semantic memory: Executive function in the processing of superordinate concepts. *NeuroImage, 59*, 1870–1878.

Reeves, D., Mills, M. J., Billick, S. B., & Brodie, J. D. (2003). Limitations of brain imaging in f forensic psychiatry. *The Journal of the American Academy of Psychiatry and the Law, 31*(1), 89–96.

Rowe, J. B., & Passingham, R. E. (2001). Working memory for location and time: Activity in prefrontal area 46 relates to selection rather than maintenance in memory. *NeuroImage, 14*(1), 77–86.

Rowe, J. B., Toni, I., Josephs, O., Frackowiak, R. S., & Passingham, R. E. (2000). The prefrontal cortex: Response selection or maintenance within working memory? *Science, 288*(5471), 1656–1660.

Rubia, K., Russell, T., Overmeyer, S., Brammer, M. J., Bullmore, E. T., Sharma, T., et al. (2001). Mapping motor inhibition: Conjunctive brain activations across different versions of go/no-go and stop tasks. *NeuroImage, 13*(2), 250–261.

Rubia, K., Smith, A. B., Brammer, M. J., & Taylor, E. (2003). Right inferior prefrontal cortex mediates response inhibition while mesial prefrontal cortex is responsible for error detection. *NeuroImage, 20*(1), 351–358.

Rubia, K., Smith, A. B., Taylor, E., & Brammer, M. (2007). Linear age-correlated functional development of right inferior fronto-striato-cerebellar networks during response inhibition and anterior cingulate during error-related processes. *Human Brain Mapping, 28*(11), 1163–1177.

Rypma, B. (2006). Factors controlling neural activity during delayed-response task performance: Testing a memory organization hypothesis of prefrontal function. *Neuroscience, 139*, 223–235.

Salat, D. H., Kaye, J. A., & Janowsky, J. S. (2002). Greater orbital prefrontal volume selectively predicts worse working memory performance in older adults. *Cerebral Cortex, 12*, 494–505.

Salmon, E., & Collette, F. (2005). Functional imaging of executive functions. *Acta Neurologica Belgica, 105*(4), 187–196.

Shaw, P., Greenstein, D., Lerch, J., Clasan, L., Lenroot, R., Gogtay, N., et al. (2006). Intellectual ability and cortical development in children and adolescents. *Nature, 440*, 677–679.

Sowell, E. R., Thompson, P. M., Leonard, C. M., Welcome, S. E., Kan, E., & Toga, A. W. (2004). Longitudinal mapping of cortical thickness and brain growth in normal children. *The Journal of Neuroscience, 22*(38), 8223–8231.

Stuss, D. T., & Benson, D. S. (1986). *The frontal lobes*. New York: Raven Press.

Tamm, L., Menon, V., & Reiss, A. L. (2003). Abnormal prefrontal cortex function during response inhibition in Turner syndrome: Functional magnetic resonance imaging evidence. *Biological Psychiatry, 53*, 107–111.

Tamnes, C. K., Østby, Y., Walhovd, K. B., Westlye, L. T., Due-Tønnessen, P., & Fjell, A. M. (2010). Neuroanatomical correlates of executive functions in children and adolescents: A magnetic resonance imaging (MRI) study of cortical thickness. *Neuropsychologia, 48*, 2496–2508.

Tate, R. L., Perdices, M., & Maggiotto, S. (1998). Stability of the Wisconsin Card Sorting Test and the determination of reliability of change in scores. *The Clinical Neuropsychologist, 12*, 348–357.

Unterrainer, J. M., Rahm, B., Kaller, C. P., Ruff, C. C., Spreer, J., Krause, B. J., et al. (2004). When planning fails: Individual differences and error-related brain activity in problem v solving. *Cerebral Cortex, 14*(12), 1390–1397.

Van den Heuvel, O. A., Groenewegen, H. J., Barkhof, F., Lazeron, R. H., van Dyck, R., & Veltman, D. J. (2003). Frontostriatal system in planning complexity: A parametric functional magnetic resonance version of Tower of London task. *NeuroImage, 18*(2), 367–374.

Van Petten, C., Plante, D., Davidson, P. S. R., Kuo, T. Y., Bajuscak, L., & Glisky, E. L. (2004). Memory and executive function in older adults: Relationships with temporal and prefrontal gray matter volumes and white matter hyperintensities. *Neuropsychologia, 42*, 1313–1335.

Wager, T. D., Jonides, J., & Reading, S. (2004). Neuroimaging studies of shifting attention: A meta-analysis. *NeuroImage, 22*, 1679–1693.

Weinstein, A. M., Voss, M. W., Prakash, R. S., Chaddock, L., Szabo, A., White, S. M., et al. (2011). The association between aerobic fitness and executive function is mediated by prefrontal cortex volume. *Brain, Behavior, and Immunity, 26*, 811–819. doi:10.1016/j.bbi.2011.11.008.

Welsh, M. C., & Huizinga, M. (2001). The development and preliminary validation of the Tower of Hanoi-revised. *Assessment, 8*, 167–176.

Weyandt, L. (2006). *The physiological bases of cognitive and behavioral disorders*. Mahwah, NJ: Lawrence Erlbaum.

Weyandt, L. L. (2009). Executive functions and attention deficit hyperactivity disorder. *The ADHD Report, 17*(6), 1–7.

Weyandt, L. & Swentosky, A. (2013). Neuroimaging and ADHD: fMRI, PET, DTI findings and methodological limitations. Mind & Brain: The Journal of Psychiatry.

Wilkinson, D. T., Halligan, P. W., Marshall, J. C., Büchel, C., & Dolan, R. J. (2001). Switching between the forest and the trees: Brain systems involved in local/global changed-level judgments. *Neuroimage, 13*(1), 56–67.

Zakzanis, K. K., Mraz, R., & Graham, S. J. (2005). An fMRI study of the Trail Making Test. *Neuropsychologia, 43*(13), 1878–1886.

Zandbelt, B. B., & Vink, M. (2010). On the role of the striatum in response inhibition. *PLoS One, 5*(11), 1–11.

Zelazo, P. D., & Müller, U. (2002). Executive function in typical and atypical development. In U. Goswami (Ed.), *Handbook of childhood cognitive development* (pp. 445–469). Oxford: Blackwell.

The Frontal Lobes and Executive Functioning

Tulio M. Otero and Lauren A. Barker

No sensible decision can be made any longer without taking into account not only the world as it is but also the world as it will be.

—Isaac Asimov

Introduction

The frontal lobes are often referred to as the seat of cognition and higher-order processing that play a role in virtually all domains of neuropsychological functioning; however, the examination of this mysterious cortical area is often plagued with dubiety. The frontal lobes have fascinated and perplexed scientists who study human behavior for decades, yet still remain largely understood (Filley, 2010). They play a role in virtually all neurological and psychiatric disorders (Levine & Craik, 2012) as well as in theories of development in children and adults. The frontal lobes regulate higher-order "executive" cognitive functions needed to successfully perform complex tasks in the environment. They include a number of psychological processes, including the selection and perception of pertinent information; maintenance, retrieval, and manipulation of information in working memory; self-directed behavior, planning, and organization; behavioral regulation and control in response to a changing environment; and appropriate decision-making on the basis of positive and negative outcomes. Dysfunction in the frontal lobes can result in a variety of deficits including distractibility and perseveration, social irresponsibility, lack of initiation, impulsivity, and disinhibition (Chudasama & Robbins, 2006).

Historically, researchers and theoreticians have believed that the expansion of the neocortex is what makes us "human" (Freeman & Watts, 1941; Stuss, 1991) and that executive functions such as problem-solving and goal-directed behavior are capacities that make us unique as humans (Baumeister, Schmeichel, & Vohs, 2007). Current research has proposed several alternatives to this hypothesis and will be discussed throughout this chapter. Despite technological advances in neuroimaging studies focusing on frontal lobe lesions, there are still many different theories about the functions of the frontal lobes and what is executive function and their relationship to neuropsychological deficits (Burgess, Simons, Dumontheil, & Gilbert, 2005; McCloskey and Perkins, 2013).

Although this chapter focuses on the relationship between the frontal lobes and executive function, it is important to begin by stating that no single part of the brain works in isolation; rather, we view its functioning as a complex integration of various neural circuits that run between many different areas within the brain. Understanding brain functioning requires a shift in mindset, that is, moving away from a cortico-cortical (horizontal) way of thinking to a subcortical–cortical (vertical) paradigm (Divac & Oberg, 1992; Koziol & Budding, 2009). Horizontal views often rely on localization and

T.M. Otero • L.A. Barker (✉)
The Chicago School of Professional Psychology,
Loyola University, Chicago, IL, USA
e-mail: TOtero@thechicagoschool.edu

discrete neuroanatomical areas to provide a model for cognition, thus failing to account for the complex interactions between cortical and subcortical areas, which are the primary orientation of vertical models.

The frontal lobes play a large role in executive function, but they do not facilitate higher-order thinking alone. Executive function is a result of complex interactions between many areas of the brain, and thus, the frontal lobes do not equal a central executive system and represent only one functional category within the frontal lobes. These frontal functions are domain general, possibly because of the extensive reciprocal connections with virtually all other brain regions, integrating information from these regions. Further integration of these processes with emotional and motivational processes allows the most complex behaviors to be executed (Stuss & Alexander, 2007). Research on social-emotional components of executive function is just beginning to come to fruition, and the majority of studies of executive function have focused on the cognitive aspects rather than the affective aspects (Damasio, Anderson, & Tranel, 2011). This is likely due to the fact that these constructs are more difficult to measure using the narrow range of assessment tools currently available to assess this domain of neuropsychological functioning.

A helpful metaphor for understanding the role of the frontal lobes in relation to executive function is that of driving a car. The driver of a car has all of the control and, simply put, uses the various parts of the car to engage in the action of driving. The complex action of driving a car cannot be done without many different components interacting together, such as using the mirrors to see the exterior of the vehicle, the pedal to propel forward, and the brakes to stop. It is through a complex interaction of all of the parts of the car being controlled by the driver that the action of driving the car can occur. There are also automatic, subcortical processes at work while driving. Think about when you are driving and someone cuts you off. Usually, we are able to make a quick decision to avoid an accident. This decision-making is *automatic*, it happens without conscious thought. This is similar to the workings of the brain. Slamming on the brakes is a stimulus-based response; we do this to ensure our survival and are able to do this with a high reaction speed based on the eminent danger posed by the car cutting us off. We will return to this driving analogy throughout the chapter to help reinforce the relationship between the frontal lobes and executive function as well as illustrate the complexity of the relationship between the two.

Evolutionary Theories of the Frontal Lobes

We are, by definition, *thinking* humans. In evolutionary terms, our thinking style has evolved as well. The human brain has evolved significantly since the beginning of human civilization. This is evidenced when comparing the physiological characteristics and changes therein over time. Historically, scientists have believed that the human frontal lobes were larger than in primates. Early work by Brodmann (1909) and Blinkov and Glezer (1968) indicated that the frontal lobes were almost 30 % smaller in chimpanzees than in humans, which was remarkable considering the similarities in all other areas of physical development between the two species over the course of evolution. This work was later refuted (see below), likely due to the development of more precise methods to assess brain volume in both humans and primates (Risberg, 2006).

Semendeferi and Damasio (2000) demonstrated that the frontal lobes in humans are more than 3 times larger than in great apes; however, when compared to overall brain volume, these differences were not considered disproportionate. The volume of white matter in human frontal lobes is also unremarkable compared to apes; however, humans appear to have greater white matter volume in the ridges of the cortex (Schenker, Desgouttes, & Semendeferi, 2005) and the most anterior parts of the prefrontal cortex (Schoenemann, Sheehan, & Glotzer, 2005).

Over time, the human prefrontal cortex has undergone significant rewiring and neural reorganization, as well as growth compared to other primates (Semendeferi, Armstrong, Schleicher, Zilles, & Van Hoesen, 1998, 2001). Semendeferi and colleagues identified several locations in the human frontal lobes that were larger (Brodmann Area 10) and smaller (Brodmann Area 13) in humans. Although functional imaging research cannot directly identify the boundaries of Brodmann Area 10, the terms *anterior prefrontal, rostral prefrontal cortex,* and *frontopolar prefrontal cortex* are used to refer to the area in the most anterior part of the frontal cortex that principally covers BA 10. The rostral prefrontal cortex (BA 10) is not just the largest part of the prefrontal cortex, but it is also larger in humans than in primates (Burgess, Gilbert, & Dumontheil, 2007). The functions of this area are the least understood of all cytoarchitectonic brain areas, although its role in higher-order cognition cannot be denied. The structure and organization of BA 10 is unique in humans in that it has a lower cell density, which results in space for more neural connections to occur between other association areas throughout the brain (Semendeferi et al., 2001) and is almost exclusively connected to areas within the prefrontal cortex and elsewhere, particularly areas responsible for integrating data from multiple sensory sources (Ramnani & Owen, 2004).

The rostral prefrontal cortex also demonstrates increased spine density compared to other cortical areas (Jacobs et al., 2001; Semendeferi et al., 2001). Brodmann Area 10 has been associated with biasing attention toward sensory input and internally generated thoughts, or one's ability to remain alert to the environment, deliberately concentrate on one's thoughts, and/or consciously shift between these states, which would require higher-order executive function abilities (i.e., shifting). It has also been hypothesized that Brodmann Area 10 acts as a "gateway" that determines which information is the priority to attend to at a given time (Burgess et al., 2005).

Evolutionarily, physiological differences in this area compared to primates may have been required to adjust to the increased physical size or as responses to a changing environment. Living in complex and changing environments has been recognized as a considerable factor in the evolution of cognition (Teffer & Semendeferi, 2012). Additionally, the timing of human brain development is prolonged and occurs more slowly than in other primates. Structural and neurophysiological features of the brain may require more time to form and fully develop. This is apparent when considering the density of synaptic connections, which also appear to follow a developmental trajectory and the traditional rise and fall pattern (Johnson & de Haan, 2011).

Divisions of the Frontal Lobes

The frontal lobes are the largest region of the brain and account for almost one-third of the cerebral cortex (Blumenfeld, 2010; Damasio et al., 2011). They are located at the most anterior region of the brain and are comprised of lateral, medial, and orbitofrontal surfaces. Substructures of the frontal lobes include the primary motor cortex, premotor cortex, supplementary motor cortex, motor speech area of Broca, frontal eye fields, and the prefrontal cortex. The prefrontal cortex is considered the primary influence on cognitive control and is subdivided into the dorsolateral, medial frontal (anterior cingulate), and orbitofrontal areas. The orbital and medial regions are involved in emotional behavior and have connections to the brainstem and limbic areas of the brain. The lateral region, which is maximally developed in humans, provides the cognitive support to the temporal organization of behavior, speech, and reasoning. These regions are part of various cortico-subcortical circuits that involves the basal ganglia and other areas. Connectivity problems in these circuits often result in behavioral manifestations that result in disinhibition (OFC), executive dysfunction (DLPFC), and apathy (MFC), referred to by Filley (2010) as the frontal lobe syndromes. Ardila (2008) proposed a model for classifying various functions of the frontal lobes based on cognitive and emotional behavioral manifestations. These two classifications can be referred to metacognitive executive functions and emotional/motivational executive functions.

Defining Executive Function: A Trick Question?

Before discussing the role of the frontal lobes in executive function, we must first define what we mean by the term *executive function*. Executive function (EF) is best understood as an umbrella term used for a diversity of hypothesized cognitive processes carried out by prefrontal areas of the frontal lobes; they include planning, working memory, attention, inhibition, self-monitoring, self-regulation, and initiation (Goldstein, Naglieri, Princiotta, & Otero, 2013).

Numerous cognitive processes are labeled as "executive," but some of these processes overlap and are highly interdependent. Thus, theoretical models of this complex multidimensional construct are required to provide a framework for the selection of assessment tools, interpreting test performance and everyday behavior, and understanding executive function development (Anderson, 2002; Garon, Bryson, & Smith, 2008). Various conceptual models of executive function have been proposed, although none has been universally adopted. A recent review identified over 30 different definitions (Goldstein et al., 2013) includes a plethora of higher-order cognitive constructs. These definitions have evolved over the course of several decades. McCloskey and Perkins (2013) provide a concise summary of the various models, definitions, and elements of executive function. The following examples are just a few examples of several of these definitions. Baddeley and Hitch (1974) referred to a central executive system that coordinates information processing through the phonological loop and the visual-spatial sketchpad. Welsh and Pennington (1988) described executive function as the ability to maintain appropriate problem-solving sets to attain future goals. Gioia, Isquith, Guy, and Kenworthy (1996) described executive functions as processes responsible for guiding, directing, and managing cognitive, emotional, and behavioral functions, often requiring novel problem-solving abilities. Delis, Kaplan, and Kramer (2001) defined executive function as involving various constructs including flexibility of thinking, inhibition, problem-solving, planning, impulse control, concept formation, abstract thinking, and creativity. Lezak, Howieson, Loring, Hannay, and Fischer (2004) described four components of executive function: (1) volition, (2) planning, (3) purposive action, and (4) effective performance. Each of these components involves a distinct set of behaviors that are necessary for socially appropriate behavior. Diamond (2006) describes three "core" executive functions that provide a base for more complex executive skills to develop. In her model, the prefrontal cortex plays a significant role in the neural circuitry required for mental health, academic achievement, and life success. These three "core" executive functions are inhibitory control, working memory, and cognitive flexibility.

McCloskey and Perkins (2013) has developed a theory which includes over 30 different constructs that are part of his definition of executive function. McCloskey, Perkins, and Van Divner (2009) provided the following operational definition of executive function, which is based on six interrelated concepts:

1. Executive functions are multiple in nature; they do not represent a single, unitary trait.
2. Executive functions are directive in nature, that is, they are mental constructs that are responsible for cueing and directing the use of other mental constructs.
3. Executive functions cue and direct mental functioning differentially within four broad construct domains: perception, emotion, cognition, and action.
4. Executive functions use can vary greatly across four arenas of involvement: intrapersonal, interpersonal, environment, and symbol system use.
5. Executive functions begin development very early in childhood and continue to develop at least into the third decade of life and most likely throughout the life span.
6. The use of executive functions is reflected in the activation of neural networks within various areas of the frontal lobes.

In an attempt to synthesize this broad array, Goldstein et al. (2013) defined executive function as an umbrella term that encompasses many different abilities that are mediated by prefrontal areas of the frontal lobes. These abilities include, but are not limited to, planning, working memory, attention, inhibition, self-monitoring, self-regulation, and initiation. Although definitions of executive function vary widely in scope, there are many common threads woven throughout them all. One shared theme is that executive function follows a developmental trajectory. We believe, as do many others, that the developmental trajectory of executive function also parallels the development of the brain, specifically, the frontal lobes.

Developmental Trajectories: Neuroanatomical Findings

Changes in the brain typically follow a cycle characterized by periods of active development followed by static periods. This cycling is sometimes referred to as "rises and falls" during development and occurs from infancy through young adulthood (Johnson & de Haan, 2011). An example of this is seen with Brodmann Area 10, which demonstrates one of the highest rates of brain growth between 5 and 11 years (Sowell et al., 2004). During adolescence, specific changes in neural architecture occur, most notably around the onset of puberty. Furthermore, reductions in gray matter density continue to occur during adolescence through early adulthood (Sowell, Thompson, Holmes, Jernigan, & Toga, 1999) in sensorimotor areas that spread during late adolescence into "higher-order" cortical regions, including the dorsolateral prefrontal cortex (Gogtay et al., 2004). In one study, frontal cortical thinning was related to improved ability to retain and retrieve verbal and spatial information (Sowell, Delis, Stiles, & Jernigan, 2001). Structural magnetic resonance imaging (MRI) studies have shown that decreased gray matter and increased myelination in the frontal lobes continues up to the age of 30, and white matter volume continues to increase up to the age of 60 years or beyond (Sowell et al., 2003).

Myelination plays a significant role on the development of the frontal lobes, specifically, the prefrontal cortex. As neurons develop, surrounding glial cells provide a layer of myelin, which forms around the axon. Myelin acts as an insulator and massively increases the speed of transmission of electrical impulses from neuron to neuron. Increased myelination results in a quicker and more efficient transfer of impulses between synapses (up to 100 times faster). Although myelination in sensory and motor areas of the brain is completed during the first few years of life, myelination occurs last in the human prefrontal cortex, and this process is not complete until the latter portion of the second decade of life (Yakovlev & Lecours, 1967). This finding suggests that the transmission speed of neural information in the frontal cortex should increase from childhood throughout young adulthood (Blakemore & Choudhury, 2006). Brodmann Area 10 is one of the last areas of the brain to myelinate, which has been associated with complex functions such as executive function (Bonin, 1950; Fuster, 1997).

This typical brain development may be altered in developmental disorders that cause impairments in working memory, attention, or inhibition (e.g., attention deficit hyperactivity disorder, obsessive-compulsive disorder, Tourette's syndrome, and schizophrenia) (Cohen, Barch, Carter, & Servan-Schreiber, 1999; Merriam, Thase, Haas, Keshavan, & Sweeney, 1999; Park & Holzman, 1992; Purcell, Maruff, Kyrios, & Pantelis, 1998; Rucklidge & Tannock, 2002; Stevens, Quittner, Zuckerman, & Moore, 2002; Weinberger, Berman, & Zec, 1986). Interestingly, prefrontal white matter has been implicated in these disorders (Kates et al., 2002; Mac Master, Keshavan, Dick, & Rosenberg, 1999).

Developmental Trajectories: Executive Functioning

Executive function development also follows a developmental trajectory, beginning in infancy. From this point on, several critical periods exist throughout early childhood, adolescence, and young adulthood. Development of executive function appears to also follow the "rises and falls" pattern described by Johnson and de Haan (2011) in relation to the development of the brain. Hunter, Edidin, and Hinkle (2012) provide a thorough outline of the development of executive function skills over time, making note to account for the potential impact that developmental disabilities and psychopathology have on the development of various executive functions. In infancy and preschool age children, interactions with caregivers and the environment are the primary influence on the development of executive function. Interactions become more complex due to the development of language abilities and social behavior. Attention, impulse control and self-regulation, and working memory abilities are the primary skills developed during this period. Problem-solving skills also begin to emerge during this time. During infancy, executive function skills involving inhibition, shifting, and cognitive flexibility are also present. Kovács and Mehler (2009) demonstrated this when they studied infants at the age of 7 months who were exposed to bilingual input from both parents speaking two different languages exclusively. Their work suggested that these children exhibited more advanced executive function than peers exposed to only one language; however, it was specific to these three executive constructs.

During early childhood, improvements in inhibition, working memory, verbal fluency, and planning abilities help prepare preschoolers for more active learning and more advanced academic tasks. To be successful when increased demands are placed on the child, they must engage in appropriate behavior within the school environment as well as be able to problem-solve and work well with other children and take directives from adults (Tarullo, Milner, & Gunmar, 2011). It is not surprising that a large focus in preschool settings is on establishing classroom routines and teaching expected school behavior, both in academic and social contexts. As children progress into middle childhood, educators expect children to be aware of the behavioral and learning expectations of the school environment, and the focus is shifted to mastering academic content. During this time period, children must also be able to integrate their executive function skills to meet increased academic demands. Brocki and Bohlin (2004) and Brocki, Fan, and Fossella (2008) have suggested that inhibition is fully developed in children between the ages of 10 and 12 years (Brocki & Bohlin, 2004; Brocki et al., 2008). Increased processing speed, verbal fluency, shifting, and planning abilities also become further refined during the middle childhood years to help meet the increasing demands placed on executive function and the child's ability to integrate these functions for academic success (Brocki & Bohlin, 2004).

By the time children reach adolescence, it is assumed that they have developed the executive function skills necessary to be successful within the school environment; however, the ability to demonstrate these skills is often inconsistent during this time period. This finding makes sense when we consider that the development of cognitive skills follows the "rise and fall" pattern described previously. Periods of "falls" may sometimes suggest regression; however, it is more plausible that these are not as much setbacks as they are indicative of the various developmental trajectories at play with the different constructs that comprise executive function. Research methods that employed direct measures of executive function have demonstrated that adolescents' performance on tasks that measure inhibitory control (Leon-Carrion, Garcia-Orza, & Perez-Santamaria, 2004; Luna, Garver, Urban, Lazar, & Sweeney, 2004), processing speed (Luna et al., 2004), and working memory and decision-making (Hooper, Luciana, Conklin, & Yarger, 2004; Luciana, Conklin, Cooper, & Yarger, 2005) continue to develop throughout adolescence. Anderson, Anderson, Northam, Jacobs, and Catroppa (2001) examined how adolescents 11–17 years old performed on a variety of executive function tasks. Their research demonstrated improvement in performance on some tasks, such as selective attention, working memory, and problem-solving; however, improvement was not noted across all constructs assessed. The studies referenced above suggest that performance on tasks requiring various aspects of executive function is linked to the neurobiological processes of pruning and myelination in the frontal cortex that occur during adolescence (Blakemore & Choudhury, 2006). Therefore, it is likely that different aspects of executive function may also follow different developmental trajectories.

It is hypothesized that executive function skills are not fully developed until young adulthood, which is supported by the continued white matter development due to myelination that occurs through the third decade of life (Sowell, Thompson, Tessner, & Toga, 2001). It is during young adulthood that we are able to problem-solve most effectively and efficiently manage tasks that are nonnovel in an automatic fashion, building upon our previously learned executive function skills. Working memory, cognitive flexibility, planning, and problem-solving all reach their peak during this time period (Huizinga, Dolan, & van der Molen, 2006). When we think back to our driving example, it is not surprising that automobile insurance providers often lower their rates when young adults turn 25. Society expects that by this time, young adults have built a repertoire of executive function skills to be able to function independently and effectively within their environment and engage in good decision-making and self-monitoring of behavior.

Cognitive Aspects of the Prefrontal Cortex

Metacognitive executive functions include problem-solving, planning, and working memory and are primarily mediated by the dorsolateral prefrontal cortex. The dorsolateral prefrontal cortex comprises the lateral portions of Brodmann's Areas 9, 10, 11, and 12; Areas 45 and 46; and the superior part of Area 47 (Damasio, 1996). The dorsolateral prefrontal cortex has numerous cortical and subcortical connections that aide in the integration and regulation of information from various neurological regions, including the thalamus, basal ganglia (the dorsal caudate nucleus), hippocampus, and primary and secondary associative areas of the neocortex, including posterior temporal, parietal, and occipital areas (Fuster, 1997, 2001). The prefrontal cortex has also been intimately linked to specific executive function deficits, such as planning, problem-solving, decision-making, and shifting (Siddiqui, Chatterjee, Kumar, Siddiqui, & Goyal, 2008). Recent evidence reviewed by Van Snellenberg and Wager (2009) suggests that distinct regions of prefrontal cortex subserve discrete functions in cognition that operate together in a modular manner to allow for the successful performance of a range of cognitive tasks. The overall picture of prefrontal cortex function presented by the authors leads to a conceptualization of a cognitive processing hierarchy that proceeds along an anterior to posterior gradient, from (a) representations of stimulus value in the orbital frontal cortex and rostral medial prefrontal cortex, to (b) processing of internal goal and task-hierarchy representations in the anterior insula prefrontal cortex, (c) top-down biasing of stimulus representation in posterior cortices by dorsal lateral prefrontal cortex, (d) representation and updating of specific stimulus–response mapping rules in inferior frontal junction and lateral premotor cortex, (e) the motivated planning of overt motor behavior in pre-SMA and cingulate motor areas, and (f) the actual production of behavior in primary motor cortex. This notion of hierarchy is present in related forms in several current models of prefrontal function (e.g., Christoff & Keramatian, 2007; Koechlin, Ody, & Kouneiher, 2003).

Affective Aspects of the Prefrontal Cortex

Emotional/motivational executive functions are mediated by the orbitofrontal and anterior cingulate/medial circuits and are responsible for linking cognition and emotion (Fuster, 1997, 2001). The orbitofrontal cortex consists of both orbital (ventral) and medial regions of PFC, including the medial portions of Brodmann's Areas 9, 10, 11, and 12; Areas 13 and 25; and the inferior portion of Area 47 (Damasio, 1996). The orbitofrontal cortex is part of a frontostriatal circuit that has strong connections to the amygdala and other parts of the limbic system (Chudasama & Robbins, 2006), which aides in the integration of affective and cognitive information, and for the regulation of motivated and goal-oriented behavior (Rolls, 2004).

Developmental studies of amygdala function have revealed gender differences throughout adolescence. Killgore, Oki, and Yurgelun-Todd (2001) identified such differences in the functional maturation of affect-related prefrontal amygdala circuits and found opposite patterns for a region of the prefrontal cortex. They generalized these findings to suggest that this may be attributed to a greater increase in emotional regulation by females, which is mediated by prefrontal systems of the frontal lobes. Using direct assessment measures of executive function, Gyurak et al. (2009) verified the conclusions made by imaging studies that showed an inhibitory relationship exists between the frontal lobes and areas of the brain responsible for emotional processing, such as the medial prefrontal cortex and amygdala.

Gold et al. (2011) demonstrated the role of the medial prefrontal cortex in case of PTSD. Compared to people without PTSD, members of the clinical sample in their study exhibited decreased regional cerebral blood flow in the medial prefrontal cortex during mental imagery of trauma-unrelated stressful personal experiences.

The implications of brain development for executive functions and social cognition during puberty and adolescence are paramount, which is likely attributed to neuronal reorganization within the circuitry of the frontal lobes. These human-specific cognitive and affective functions are likely the result of this neuronal reorganization and not increases in volume or size as originally believed. They also appear to follow a developmental trajectory, as discussed previously.

Contemporary Understanding of the Frontal Lobes

Stuss and Alexander (2007) identified discrete categories of function of the frontal lobes, one being executive functioning. They indicated, however, that the executive category does not refer solely to a central executive system and that the frontal lobes do not have a unitary organizing role. Rather, they described impairments in a multitude of anatomical and functional attentional control processes that are interrelated. This argument supports the view that the frontal lobes work in tandem with many other areas of the brain. Miller and Wallis (2009) describe executive control as the ability to take charge of one's actions and direct them toward unseen aims while making predictions about available goals and what resources can be utilized to achieve them. Engaging in goal-directed behavior also requires an individual to select and coordinate automatic sensory, memory, and motor functions to perform a particular action.

If all parts of the brain work in tandem, then the idea of hemispheric specialization is misleading, as it would be unlikely that the right or left hemisphere works independently of each other as there are multiple connections between the frontal lobes and other brain regions. The role of the corpus callosum by definition refutes this idea, as it serves as the connection between the two and facilitates the exchange of information between the two. Neural circuits cross into ipsilateral brain hemispheres within the brain. The functions of each hemisphere and their overall importance have been an area of debate over the decades. Goldberg's (2009) discussion of hemispheric specialization also suggests that neuroscientists may share this belief, as this is not a primary focus of the research within this discipline. The left hemisphere, which is typically associated with language and routine behavior, has usually been deemed the more important of the two. This is further evidenced in the medical discipline and surgical practices in neurosurgery, in which surgery on the left hemisphere is often the less preferred of the two, as there are so many structures associated with language, as well as working with things that are learned, or routine, in this hemisphere (Goldberg, 2009). In an academic sense, the left hemisphere is often associated with the verbal knowledge needed to be successful in academic settings, and typically the area is primarily taxed on various neuropsychological assessment tools. This is apparent with nonverbal assessment tools, even though they are supposed to be measuring skills typically associated with the right hemisphere, such as perceptual reasoning through abstract, novel tasks. Nonverbal psychological assessments require the comprehension of language and the ability to comply with the routine of typical test-taking behavior, which are often considered left hemisphere tasks. To successfully participate in an assessment, one must know the appropriate behaviors to participate in the test and how to respond to the test stimuli and items, to receive the directions throughout the various subtests of the assessment. Clearly, left and right functional differences are not a new concept, as the linkage to routine (left hemisphere) and novelty (right hemisphere) was first described in the scientific literature over 30 years ago (Goldberg & Costa, 1981), although not much research focused on these differences until recent years. Through the use of neuroimaging, these associations can now be substantiated with functional neuroanatomical evidence. This has direct implications for learning. When things are new, or unlearned, they are processed primarily in the right hemisphere. As things are learned over time, they become automatic, and the left takes on a more active role. An example of this was found when examining fluid intelligence following frontal lobe lesions. Roca et al. (2010) found that association with lesions in the right anterior frontal cortex was associated with executive dysfunction on a relatively novel task, the Wisconsin Card Sorting test. Understanding of varying prefrontal lobe deficits may assist in designing future assessment tools to gauge the impact of frontal lobe deficits on executive function (Roca et al., 2010).

In humans and primates, many structural, cellular, and molecular hemispheric asymmetries are also present. Goldberg illustrated this fact using the example of language, which is unique to humans. The planum temporale and sylvian fissure both exhibit structural asymmetries in humans, gorillas (Gannon, Holloway, Broadfield, & Braun, 1998), and chimpanzees (Barricka et al., 2005); however, only humans have developed language or what we as humans consider to be language. Hemispheric asymmetries are also present in both humans and primates in cortical thickness, neurochemistry, and white matter organization (Goldberg, 2009). When considering areas of the brain that are unique in humans, the frontal lobes also have a large number of differences from other primates, which is clearly not the case with other neurological structures or cells. Similar to other primates, however, the frontal lobes appear to follow a right-to-left processing of information. Functional imaging studies have provided evidence of this theory as well. For example, Gold and colleagues (1996) examined the changes in regional cerebral blood flow patterns using PET imaging using a combination of delayed response and delayed alternation tasks. Imaging data was collected early in the study, when the tasks were considered novel, as well as later, once the tasks were learned. Although activation patterns in the frontal lobes were reported under both conditions, much greater activation occurred when the task was still novel. During the novel condition, regional cerebral blood flow was greater in the right hemisphere in the prefrontal cortex. Once the task was learned and was no longer considered novel, greater activation was noted in the left hemisphere of the frontal lobes.

Neurotransmitters, Neuromodulators, and Frontal Lobes

Neurotransmitter and neuromodulator are terms often used interchangeably; however, they are not the same thing. Neuromodulators are substances that do not directly activate ion-channel receptors but act in conjunction with neurotransmitters to enhance the excitatory or inhibitory responses of the receptors (Neuromodulator, 2012). Unlike fast-acting neurotransmitters, such as glutamate and GABA, neuromodulators are slow moving and include dopamine, norepinephrine, serotonin, and acetylcholine. They are controlled by nuclei in the brainstem and make connections to distant brain regions via long axons. Some neurotransmitters are present throughout the whole brain, whereas others are restricted to specific areas (Goldberg, 2009). This is the case with dopamine and the frontal lobes. Of all of the neurotransmitters and neuromodulators, dopamine has the most extensive history in regard to research with frontal lobe functions and relationships to executive function.

Diamond (2011) described the properties of the dopamine system in the prefrontal cortex as "unusual" and that the observed sensitivity to environmental and genetic variations is unique to this brain region and not observed elsewhere. Furthermore, these variations may have different expressions based on gender and also have a direct impact on neurodevelopmental disorders, such as ADHD and PKU. Dopamine neurons that project into the prefrontal cortex also have a higher firing rate; dopamine turnover has also been observed (Diamond, 2011).

Different types of dopamine receptors, such as D1, D2, and D4, have various influences in the executive functions mediated by the frontal lobes (Wang, Vijayraghavan, & Goldman-Rakic, 2004). Arnsten and Bao-Ming (2005) demonstrated that heightened and reduced levels of dopamine have a significant impact on working memory and attention. Robbins and Roberts (2007) found evidence of the effects of dopamine on shifting and sustained attention. Psychopharmacological interventions also enhance or disrupt their functions, as demonstrated by Vijayraghavan, Wang, Birnbaum, Williams, and Arnsten (2007) in their examination of working memory and D1 receptors in the prefrontal cortex. Their work demonstrated that D1 receptor stimulation in the prefrontal cortex produced an "inverted-U" dose–response, which resulted in impairments in spatial working memory. Additionally, gender differences may have a significant impact on dosages of medications that affect dopamine levels in the prefrontal cortex. These findings have important implications for young women, since increased estrogen levels have been associated with higher levels of dopamine in the prefrontal cortex; the opposite occurs when estrogen levels are low (Diamond, 2011). Kayser, Allen, Navarro-Cebrian, Mitchell, and Fields (2012) demonstrated that raising cortical dopamine levels reduced impulsive actions by changing corticostriatal function.

Dopamine is not the only neurobiological component at play within the frontal lobes. Norepinephrine has been linked to deficits in working memory and inhibition (Arnsten & Bao-Ming, 2005). Clarke, Dalley, Crofts, Robbins, and Roberts (2004) demonstrated how prefrontal serotonin depletion results in cognitive inflexibility. Serotonin also plays a role in the affective components of executive function as described by Cools, Roberts, and Robbins (2008), and other research has pointed to serotonin playing a role in attention and impulsivity (Chudasama, 2011) and the possible outcomes this may have when considering psychopharmacological interventions, such is the case with SSRI's. Chudasama (2011) also demonstrated a relationship between levels of acetylcholine and accuracy and inhibition levels in studies conducted on animals. Like anything else with the brain, neurotransmitters and neuromodulators often work together for executive function.

Contemporary Theories of EF: A Cortico-Subcortical Example of Working Memory

Subcortical regions of the brain, such as the basal ganglia and cerebellum, also impact executive function. Koziol and Budding (2009) discussed various deficits associated in working memory and the impeding cognitive implications, such as planning for goal-directed behavior, or even being able to maintain a mindset of said goal-directed behavior. Deficits such as this have far-reaching implications to higher cognitive functioning that require an integration of various executive functions, such as organization of thoughts, perceptions, or behaviors to perform more complex, higher-order cognitive tasks. Simply speaking, the core features of working memory, the ability to pay attention to something, keep it in our mind, and then do something and/or manipulate it, are tasks that can best be conceptualized through an understanding of the complex interactions between the cortex and the basal ganglia. A common metaphor for these interactions has been described as the role of a doorman or bouncer in a nightclub, in which the doorman is the basal ganglia and nightclub is the cortex (McNab & Klingberg, 2008). This is made possible by prefrontal-cortical connections, which help keep information "online" and basal ganglia-cortical interactions, which "gate" the manipulation of information while reducing distractors. These processes are mediated by direct and indirect pathways, which are made possible by outputs from the internal globus pallidus, which project to the thalamus (Koziol & Budding, 2009). These pathways are both afferent and efferent, allowing necessary information in through direct pathways and keeping distracting information out through indirect pathways. Koziol and Budding (2009) conclude that working memory is characterized by both cortical and subcortical processes, which is observed in the maintenance (cortical) and updating (subcortical) of information. Specific working memory deficits are also possible in either maintenance or updating capabilities, as is the case in Parkinson's disease, in which patients have more deficits in the updating component of working memory compared to the maintenance components (Koziol & Budding, 2009).

Neuroimaging: Integrating Neurobiological and Neuropsychological Data to Better Understand EF

As discussed throughout the chapter, the development and refinement of higher-order cognitive functioning appears to have paralleled the development of the human frontal lobes, specifically, the prefrontal cortex. Technological advances have now allowed researchers to utilize neuroimaging techniques to study the relationships between executive function and the genetic processes underlying the neuronal circuitry responsible for these cognitive processes. In an fMRI study, Greene, Braet, Johnson, and Bellgrove (2008) examined the neural correlations between sustained attention, working memory, and response inhibition. Their work suggests, however, that although various executive functions may be simply operationally defined through behavioral observation and performance on neuropsychological tasks, there are still many uncertainties at the genetic level.

Resting brain metabolism or resting states are the involuntary and ingrained brain activities that occur in the absence of a discernible task. In the driving analogy, resting states are synonymous with a car idling (Johnson & de Haan, 2011). Using functional fMRI imaging techniques with infants during sleep cycles, Fransson et al. (2007) identified several resting state networks that are associated with various cortical regions, including the medial and dorsolateral prefrontal cortex. This has important implications for both the development of the frontal lobes as well as cognitive functions such as executive function. Although neither is fully developed until later in life, both appear to follow a similar developmental trajectory that begins in infancy and continues throughout childhood, adolescence, and culminates in early adulthood. When considering our driving metaphor, this has serious implications regarding the age of the driver. In recent years, debate has arisen as to what is the appropriate age for states to issue drivers licenses. Typically, this occurs during late adolescence. Research on development of the frontal lobes and executive function appears to support the argument that this is too young as many aspects of executive function are not fully developed at this time!

Another example of a relationship between the developmental trajectories of the frontal lobes and executive function is found in working memory. Perseveration errors on object permanence tasks have been linked with maturation of the dorsolateral prefrontal cortex in studies with human infants, infant monkeys, and adult monkeys with prefrontal cortex lesions. Children and adolescents' working memory capacity has been linked with the maturation of the lateral prefrontal cortex through structural and functional neuroimaging studies (Johnson & de Haan, 2011). Crone and Westenberg (2009) found that children do not learn to make advantageous choices consistently until they are 16–18 years old. Imaging studies suggest that this behavioral pattern relates to a disconnect between subcortical reward-processing systems and frontal executive control systems in adolescents, who are more driven by reward systems and can lead to poor decision-making in social situations. The use of neuroimaging techniques, such as MRI, has also demonstrated changes in the frontal cortex during adolescence. When considering the synchronicity between the development of the physiological and psychological aspects of the frontal lobes, it is plausible that executive function abilities might also be expected to improve during this time. For example, selective attention, decision-making, and response inhibition skills, along with the ability to carry out multiple tasks at once, potentially improve during adolescence (Blakemore & Choudhury, 2006).

Neuroimaging has also demonstrated that the localization of cognitive skills associated with executive function displays a large amount of overlap with each other in the brain. An example of this overlap is found with working memory, sustained attention, and response inhibition. As discussed above, Greene et al. (2008) found that these three constructs all activated functional networks involving prefrontal and parietal regions.

Conclusions and Future Directions

The frontal lobes are often referred to as the seat of cognition and higher-order processing that play a role in virtually all domains of neuropsychological functioning. The frontal lobes regulate higher-order "executive" cognitive functions needed to successfully perform complex tasks in the environment. Although theories of frontal lobe and executive function abound, what seems certain is that no single part of the brain works in isolation; rather, it is functioning as a complex integration of various neural circuits that run between many different areas within the brain.

Information about developmental trajectories is important for addressing several issues that occur from infancy through early adulthood. For example, they clarify why children of different ages differ in the particular components of EF they find difficult to recruit. Inhibition is both essential and particularly challenging during early childhood; older children and adolescents are less susceptible to distraction and less impulsive. Moreover, the sequence in which the EF components emerge (e.g., inhibition before planning) suggests possible causal relations among components during development. Cognitive processing proceeds along an anterior to posterior gradient, while the cognitive components of the frontal lobes and executive function center around several metacognitive skills that are subserved by several cortical and subcortical regions and networks aiding in the encoding, integration, retrieval, and regulation of information. Advances in neuroimaging and neurochemistry continue to advance out knowledge of brain-behavior correlates and the excitatory or inhibitory responses of the receptors within the frontal lobes and other structures.

By further clarifying the murky waters of FL function and their relation to EF, we may be better able to assess, intervene, and perhaps prevent disorders characteristic of impaired EF. Designing better interventions based on neuropsychology and neuroscience for attention deficits, cognitive and behavioral impulsivity, emotional dysregulation, memory issues, and a plethora of academic achievement difficulties is the current buzz within the field and will likely remain an important area of investigation and applied practice.

References

Anderson, P. (2002). Assessment and development of executive function (EF) during childhood. *Child Neuropsychology, 8*(2), 71–82.

Anderson, V. A., Anderson, P., Northam, E., Jacobs, R., & Catroppa, C. (2001). Development of executive functions through late childhood and adolescence in an Australian sample. *Developmental Neuropsychology, 20*(1), 385–406.

Ardila, A. (2008). On the evolutionary origins of executive functions. *Brain and Cognition, 68*(1), 92–99.

Arnsten, A. F. T., & Bao-Ming, L. (2005). Neurobiology of executive functions: Catecholamine influences on prefrontal cortical functions. *Biological Psychiatry, 57*(11), 1377–1384.

Baddeley, A., & Hitch, G. (1974). Working memory. In G. H. Bower (Ed.), *The psychology of learning and motivation: Advances in research and theory* (pp. 47–89). Salt Lake City, UT: Academic.

Barricka, T. R., Mackayb, C. E., Primac, S., Maesd, F., Vandermeulend, D., Crowb, T. J., et al. (2005).

Automatic analysis of cerebral asymmetry: An exploratory study of the relationship between brain torque and planum temporale asymmetry. *NeuroImage, 24*, 678–691.

Baumeister, R. F., Schmeichel, B. J., & Vohs, K. D. (2007). Self-regulation and the executive function: The self as controlling agent. In A. Kruglanski & E. T. Higgins (Eds.), *Social psychology: Handbook of basic principles* (2nd ed., pp. 516–539). New York: Guilford Press.

Blakemore, S. J., & Choudhury, S. (2006). Development of the adolescent brain: Implications for executive function and social cognition. *Journal of Child Psychology and Psychiatry, 47*(3), 296–312.

Blinkov, S. M., & Glezer, I. I. (1968). *Das Zentralnervensystem in Zahlen und Tabellen*. Jena: Fischer.

Blumenfeld, H. (2010). *Neuroanatomy through clinical cases* (2nd ed.). Sunderland, MA: Sinaeur Associates.

Bonin, G. V. (1950). *Essay on the cerebral cortex*. Springfield, IL: Charles C. Thomas.

Brocki, K. C., & Bohlin, G. (2004). Executive functions in children aged 6 to 13: A dimensional and developmental study. *Developmental Neuropsychology, 26*(2), 571–593.

Brocki, K. C., Fan, J., & Fossella, J. (2008). Placing neuroanatomical models of executive function in a developmental context: Imaging and imaging-genetic strategies. In D. W. Phaff & B. L. Kieffer (Eds.), *Molecular and biophysical mechanisms of arousal, alertness, and attention* (pp. 246–255). Boston, MA: Blackwell Publishing.

Brodmann, K. (1909). Vergleichende Lokalisationslehre der Grosshinrinde in ihren Prinzipien dargestellt auf Grund des Zellenbaues. In J. M. Fuster (Ed.), *The prefrontal cortex: Anatomy, physiology, and neuropsychology of the frontal lobe* (3rd ed.). Philadelphia: Lippincott–Raven.

Burgess, P. W., Gilbert, S. J., & Dumontheil, I. (2007). Function and localization within rostral prefrontal cortex (area 10). *Philosophical Transactions of the Royal Society of London. Series B, Biological Sciences, 362*(1481), 887–899.

Burgess, P. W., Simons, J. S., Dumontheil, I., & Gilbert, S. J. (2005). The gateway hypothesis of rostral prefrontal cortex (area 10) function. In J. Duncan, L. Phillips, & P. McLeod (Eds.), *Measuring the mind: Speed, control, and age* (pp. 217–248). Oxford: Oxford University Press.

Christoff, K., & Keramatian, K. (2007). Abstraction of mental representations: Theoretical considerations and neuroscientific evidence. In S. A. Bunge & J. D. Wallis (Eds.), *Perspectives on rule-guide behavior* (pp. 107–126). Oxford: Oxford University Press.

Chudasama, Y. (2011). Animal models of prefrontal prefrontal-executive function. *Behavioral Neuroscience, 125*(3), 327–343.

Chudasama, Y., & Robbins, T. W. (2006). Functions of frontostriatal systems in cognition: Comparative neuropsychopharmacological studies in rats, monkeys and humans. *Biological Psychology, 73*(1), 19–38.

Clarke, H. F., Dalley, J. W., Crofts, H. S., Robbins, T. W., & Roberts, A. C. (2004). Cognitive inflexibility after prefrontal serotonin depletion. *Science, 304*(5672), 878–880.

Cohen, J. D., Barch, D. M., Carter, C., & Servan-Schreiber, D. (1999). Context-processing deficits in schizophrenia: Converging evidence from three theoretically motivated cognitive tasks. *Journal of Abnormal Psychology, 108*, 120–133.

Cools, R., Roberts, A. C., & Robbins, T. W. (2008). Serotoninergic regulation of emotional and behavioural control processes. *Trends in Cognitive Sciences, 12*(1), 31–40.

Crone, E. A., & Westenberg, P. M. (2009). A brain-based account of developmental changes in social decision-making. In M. de Haan & M. R. Gunnar (Eds.), *Handbook of developmental social neuroscience* (pp. 378–398). New York: Guilford Press.

Damasio, A. R. (1996). The somatic marker hypothesis and the possible functions of the prefrontal cortex. *Philosophical Transactions of the Royal Society of London. Series B, Biological Sciences, 351*, 1413–1420.

Damasio, A., Anderson, S. W., & Tranel, D. (2011). The frontal lobes. In K. M. Heilman & E. Valenstein (Eds.), *Clinical neuropsychology* (5th ed., pp. 417–465). New York: Oxford University Press.

Delis, D. C., Kaplan, E., & Kramer, J. (2001). *Delis-Kaplan executive function system*. San Antonio, TX: Psychological Corporation.

Diamond, A. (2006). The early development of executive functions. In E. Bialystok & F. I. M. Craik (Eds.), *Lifespan cognition: Mechanisms of change*. New York: Oxford University Press.

Diamond, A. (2011). Biological and social influences on cognitive control processes dependent on prefrontal cortex. In O. Braddick, J. Atkinson, & G. Innocenti (Eds.), *Progress in brain research* (Vol. 189, pp. 319–340). Burlington: Academic.

Divac, I., & Oberg, R. (1992). Subcortical mechanisms in cognition. In G. Vallar, S. F. Cappa, & C. W. Wallesch (Eds.), *Neuropsychological disorders associated with subcortical lesions* (pp. 42–60). New York: Oxford University Press.

Filley, C. M. (2010). Chapter 35: The frontal lobes. In M. J. Aminoff, F. Boller, & D. F. Swaab (Eds.), Handbook of Clinical Neurology (pp. 557–70). New York: Elsevier.

Fransson, P., Skiöld, B., Horsch, S., Nordell, A., Blennow, M., Lagercrantz, H., et al. (2007). Resting-state networks in the infant brain. *Proceedings of the National Academy of Sciences, 104*(39), 15531–15536.

Freeman, W., & Watts, J. (1941). The frontal lobes and consciousness of the self. *Psychosomatic Medicine, 3*(2), 111–119.

Fuster, J. M. (1997). *The prefrontal cortex: Anatomy, physiology, and neuropsychology of the frontal lobe*. Philadelphia, PA: Lippincott-Raven.

Fuster, J. M. (2001). The prefrontal cortex—An update: Time is of the essence. *Neuron, 30*(2), 319–333.

Gannon, P. J., Holloway, R. L., Broadfield, D. C., & Braun, A. R. (1998). Asymmetry of chimpanzee planum temporale: Humanlike pattern of Wernicke's brain language area homolog. *Science, 279*(5348), 220–222.

Garon, N., Bryson, S. E., & Smith, I. M. (2008). Executive function in preschoolers: A review using an integrative framework. *Psychological Bulletin, 134*(1), 31–60.

Gioia, G., Isquith, P., Guy, S., & Kenworthy, L. (1996). *Behavior rating inventory of executive function.* Lutz, FL: Psychological Assessment Resources.

Gogtay, N., Giedd, J. N., Lusk, L., Hayashi, K. M., Greenstein, D., Vaituzis, A. C., et al. (2004). Dynamic mapping of human cortical development during childhood through early adulthood. *Proceedings of the National Academy of Sciences of the United States of America, 101*(21), 8174–8179.

Gold, A. L., Shin, L. M., Orr, S. P., Carson, M. A., Rauch, S. L., Macklin, M. L., et al. (2011). Decreased regional cerebral blood flow in medial prefrontal cortex during trauma-unrelated stressful imagery in Vietnam veterans with post-traumatic stress disorder. *Psychological Medicine, 41*(12), 2563–2572.

Gold, J. M., Berman, K. F., Randolph, C., Goldberg, E., & Weinberger, D. R. (1996). PET validation of a novel prefrontal task: Delayed response alteration. *Neuropsychology, 10*, 3–10.

Goldberg, E. (2009). *The new executive brain: Frontal lobes in a complex world.* New York, NY: Oxford University Press.

Goldberg, E., & Costa, L. D. (1981). Hemisphere differences in the acquisition and use of descriptive systems. *Brain and Language, 14*, 144–173.

Goldstein, S., Naglieri, J. A., Princiotta, D., & Otero, T. M. (2013). Introduction: A history of executive functioning. In S. Goldstein & J. A. Naglieri (Eds.), *Handbook of executive functioning.* New York, NY: Springer.

Greene, C. M., Braet, W., Johnson, K. A., & Bellgrove, M. (2008). Imaging the genetics of executive function. *Biological Psychology, 79*(1), 30–42.

Gyurak, A., Goodkind, M. S., Madan, A., Kramer, J. H., Miller, B. L., & Levenson, R. W. (2009). Do tests of executive functioning predict ability to down regulate emotions spontaneously and when instructed to suppress? *Cognitive, Affective, & Behavioral Neuroscience, 9*(2), 144–152.

Hooper, C. J., Luciana, M., Conklin, H. M., & Yarger, R. S. (2004). Adolescents' performance on the development of decision-making and ventromedial prefrontal cortex. *Developmental Psychology, 40*(6), 1148–1158.

Huizinga, M., Dolan, C. V., & van der Molen, M. W. (2006). Age-related change in executive function: Developmental trends and a latent variable analysis. *Neuropsychologia, 44*(11), 2017–2036.

Hunter, S. J., Edidin, J. P., & Hinkle, C. D. (2012). The developmental neuropsychology of executive functions. In S. J. Hunter & E. P. Sparrow (Eds.), *Executive function and dysfunction* (pp. 17–36). New York: Cambridge University Press.

Jacobs, B., Schall, M., Prather, M., Kapler, E., Driscoll, L., Baca, S., et al. (2001). Regional dendritic and spine variation in human cerebral cortex: A quantitative golgi study. *Cerebral Cortex, 11*(6), 558–571.

Johnson, M. H., & de Haan, M. (2011). *Developmental cognitive neuroscience: An introduction* (3rd ed.). Malden, MA: Wiley-Blackwell.

Kates, W. R., Frederikse, M., Mostofsky, S. H., Folley, B. S., Cooper, K., Mazur-Hopkins, P., et al. (2002). MRI parcellation of the frontal lobe in boys with attention deficit hyperactivity disorder or Tourette syndrome. *Psychiatry Research, 116*(1–2), 63–81.

Kayser, A. S., Allen, D. C., Navarro-Cebrian, A., Mitchell, J. M., & Fields, H. L. (2012). Dopamine, corticostriatal connectivity, and intertemporal choice. *The Journal of Neuroscience, 32*(27), 9402–9409.

Killgore, C. A., Oki, M., & Yurgelun-Todd, D. A. (2001). Sex-specific developmental changes in amygdala responses to affective faces. *Neuroreport, 12*(2), 427–433.

Koechlin, E., Ody, C., & Kouneiher, F. (2003). The architecture of cognitive control in the human prefrontal cortex. *Science, 302*(5648), 1181–1185.

Kovács, A. M., & Mehler, J. (2009). Cognitive gains in 7-month-old bilingual infants. *Proceedings of the National Academy of Sciences of the United States of America, 106*, 6556–6560.

Koziol, L. F., & Budding, D. E. (2009). *Subcortical structures and cognition: Implications for neuropsychological assessment.* New York, NY: Springer.

Leon-Carrion, J., Garcia-Orza, J., & Perez-Santamaria, F. J. (2004). The development of the inhibitory component of the executive functions in children and adolescents. *The International Journal of Neuroscience, 114*(10), 1291–1311.

Levine, B., & Craik, F. I. (2012). Unifying clinical, experimental, and neuroimaging studies of the human frontal lobes. In B. Levine & F. I. Craik (Eds.), *Mind and the frontal lobes: Cognition, behavior, and brain imaging* (pp. 3–15). New York, NY: Oxford University Press.

Lezak, M. D., Howieson, D. B., Loring, D. W., Hannay, H. J., & Fischer, J. S. (2004). *Neuropsychological assessment* (4th ed.). New York: Oxford University Press.

Luciana, M., Conklin, H. M., Cooper, C. J., & Yarger, R. S. (2005). The development of nonverbal working memory and executive control processes in adolescents. *Child Development, 76*(3), 697–712.

Luna, B., Garver, K. E., Urban, T. A., Lazar, N. A., & Sweeney, J. A. (2004). Maturation of cognitive processes from late childhood to adulthood. *Child Development, 75*(5), 1357–1372.

Mac Master, F. P., Keshavan, M. S., Dick, E. L., & Rosenberg, D. R. (1999). Corpus callosal signal intensity in treatment-naive pediatric obsessive compulsive disorders. *Progress in Neuro-Psychopharmacology & Biological Psychiatry, 23*(4), 601–612.

McCloskey, G., & Perkins, L. A. (2013). *Essentials of executive functions assessment.* Hoboken, NJ: Wiley.

McCloskey, G., Perkins, L. A., & Van Divner, B. R. (2009). *Assessment and intervention for executive function difficulties.* New York: Routledge.

McNab, F., & Klingberg, T. (2008). Prefrontal cortex and basal ganglia control access to working memory. *Nature Neuroscience, 11*(1), 103–107.

Merriam, E. P., Thase, M. E., Haas, G. L., Keshavan, M. S., & Sweeney, J. A. (1999). Prefrontal cortical dysfunction in depression determined by Wisconsin Card Sorting Test performance. *The American Journal of Psychiatry, 156*(5), 780–782.

Miller, E. K., & Wallis, J. D. (2009). Executive function and higher-order cognition: Definition and neural substrates. In L. R. Squire (Ed.), *Encyclopedia of neuroscience* (Vol. 4, pp. 99–104). Oxford: Academic.

Neuromodulator. (2012). *Encyclopædia Britannica*. Retrieved from http://www.britannica.com/EBchecked/topic/410660/neuromodulator

Park, S., & Holzman, P. S. (1992). Schizophrenics show spatial working memory deficits. *Archives of General Psychiatry, 49*(12), 975–982.

Purcell, R., Maruff, P., Kyrios, M., & Pantelis, C. (1998). Neuropsychological deficits in obsessive-compulsive disorder: A comparison with unipolar depression, panic disorder, and normal controls. *Archives of General Psychiatry, 55*(5), 415–423.

Ramnani, N., & Owen, A. M. (2004). Anterior prefrontal cortex: Insights into function from anatomy and neuroimaging. *Nature Reviews Neuroscience, 5*(3), 184–194.

Risberg, J. (2006). Evolutionary aspects on the frontal lobes. In J. Risberg & J. Grafman (Eds.), *The frontal lobes: Development, function, and pathology*. New York, NY: Cambridge University Press.

Robbins, T. W., & Roberts, A. C. (2007). Differential regulation of fronto-executive function by the monoamines and acetylcholine. *Cerebral Cortex, 17*(Suppl. 1), 151–160.

Roca, M., Parr, A., Thompson, R., Woolgar, A., Torralva, T., Antoun, N., et al. (2010). Executive function and fluid intelligence after frontal lobe lesions. *Brain, 133*(Pt 1), 234–247.

Rolls, E. T. (2004). The functions of the orbitofrontal cortex. *Brain and Cognition, 55*(1), 11–29.

Rucklidge, J. J., & Tannock, R. (2002). Neuropsychological profiles of adolescents with ADHD: Effects of reading difficulties and gender. *Journal of Child Psychology and Psychiatry, 43*(8), 988–1003.

Schenker, N. M., Desgouttes, A. M., & Semendeferi, K. (2005). Neural connectivity and cortical substrates of cognition in hominoids. *Journal of Human Evolution, 49*(5), 547–569.

Schoenemann, P. T., Sheehan, M. J., & Glotzer, D. L. (2005). Prefrontal white matter volume is disproportionately larger in humans than in other primates. *Nature Neuroscience, 8*(2), 242–252.

Semendeferi, K., Armstrong, E., Schleicher, A., Zilles, K., & Van Hoesen, G. W. (1998). Limbic frontal cortex in hominoids: A comparative study of area 13. *American Journal of Physical Anthropology, 106*(2), 129–155.

Semendeferi, K., Armstrong, E., Schleicher, A., Zilles, K., & Van Hoesen, G. W. (2001). Prefrontal cortex in humans and apes: A comparative study of area 10. *American Journal of Physical Anthropology, 114*(3), 224–241.

Semendeferi, K., & Damasio, H. (2000). The brain and its main anatomical subdivisions in living hominoids using magnetic resonance imaging. *Journal of Human Evolution, 38*(2), 317–332.

Siddiqui, S. V., Chatterjee, U., Kumar, D., Siddiqui, A., & Goyal, N. (2008). Neuropsychology of prefrontal cortex. *Indian Journal of Psychiatry, 50*(3), 202–208.

Sowell, E. R., Delis, D., Stiles, J., & Jernigan, T. L. (2001). Improved memory functioning and frontal lobe maturation between childhood and adolescence: A structural MRI study. *Journal of International Neuropsychological Society, 7*, 312–322.

Sowell, E. R., Peterson, B. S., Thompson, P. M., Welcome, S. E., Henkenius, A. L., & Toga, A. W. (2003). Mapping cortical change across the human life span. *Nature Neuroscience, 6*(3), 309–315.

Sowell, E. R., Thompson, P. M., Holmes, C. J., Jernigan, T. L., & Toga, A. W. (1999). In vivo evidence for post-adolescent brain maturation in frontal and striatal regions. *Nature Neuroscience, 2*(10), 859–861.

Sowell, E. R., Thompson, P. M., Leonard, C. M., Welcome, S. E., Kan, E., & Toga, A. W. (2004). Longitudinal mapping of cortical thickness and brain growth in normal children. *The Journal of Neuroscience, 24*(38), 8223–8231.

Sowell, E. R., Thompson, P. M., Tessner, K. D., & Toga, A. W. (2001). Mapping continued brain growth and gray matter density reduction in dorsal frontal cortex: Inverse relationships during post-adolescent brain maturation. *The Journal of Neuroscience, 21*(22), 8819–8829.

Stevens, J., Quittner, A. L., Zuckerman, J. B., & Moore, S. (2002). Behavioral inhibition, self-regulation of motivation, and working memory in children with attention deficit hyperactivity disorder. *Developmental Neuropsychology, 21*(2), 117–139.

Stuss, D. T. (1991). Self, awareness and the frontal lobes: A neuropsychological perspective. In J. Strauss & G. R. Goethals (Eds.), *The self: Interdisciplinary approaches* (pp. 255–278). New York: Springer.

Stuss, D. T., & Alexander, M. P. (2007). Is there a dysexecutive system? *Philosophical Transactions of the Royal Society of London. Series B, Biological Sciences, 362*(1481), 901–915.

Tarullo, A. R., Milner, S., & Gunmar, M. R. (2011). Inhibition and exuberance in preschool classrooms: Associations with peer social experiences and changes in cortisol across the preschool years. *Developmental Psychology, 47*(5), 1374–1388.

Teffer, K., & Semendeferi, K. (2012). Human prefrontal cortex: Evolution, development, and pathology. In M. A. Hofman & D. Falk (Eds.), *Progress in brain research* (pp. 191–218). Amsterdam: Elsevier.

Van Snellenberg, J. X., & Wager, T. D. (2009). Cognitive and motivational functions of the human prefrontal cortex. In A. L. Christensen, D. Bougakov, & E. Goldberg (Eds.), *Luria's legacy in the 21st century* (pp. 30–61). New York: Oxford University Press.

Vijayraghavan, S., Wang, M., Birnbaum, S. G., Williams, G. V., & Arnsten, A. F. (2007). Inverted-U dopamine D1 receptor actions on prefrontal neurons engaged in working memory. *Nature Neuroscience, 10*(3), 376–384.

Wang, M., Vijayraghavan, S., & Goldman-Rakic, P. S. (2004). Selective D2 receptor actions on the functional circuitry of working memory. *Science, 303*(5659), 853–856.

Weinberger, D. R., Berman, K. F., & Zec, R. F. (1986). Physiologic dysfunction of dorsolateral prefrontal cortex in schizophrenia. I. Regional cerebral blood flow evidence. *Archives of General Psychiatry, 43*(2), 114–124.

Welsh, M. C., & Pennington, B. F. (1988). Assessing frontal lobe functioning in children: Views from developmental psychology. *Developmental Neuropsychology, 4*(3), 199–230.

Yakovlev, P. A., & Lecours, I. R. (1967). The myelogenetic cycles of regional maturation of the brain. In A. Minkowski (Ed.), *Regional development of the brain in early life* (pp. 3–70). Oxford: Blackwell.

The Development of Hot and Cool Executive Functions in Childhood and Adolescence: Are We Getting Warmer?

Eric Peterson and Marilyn C. Welsh

The domain of cognitive skills collectively referred to as executive function has intrigued and stymied researchers for the better part of a century. This chapter explores a distinction made within this general domain that has emerged only recently in the theoretical and empirical literature: hot vs. cool executive functions. Cool executive functions are defined as the goal-directed, future-oriented skills such as planning, inhibition, flexibility, working memory, and monitoring that are manifested under relatively decontextualized, nonemotional, and analytical testing conditions (e.g., Miyake, Freidman, Emerson, Witzki, & Howerter, 2000; Stuss & Benson, 1984; Welsh & Pennington, 1988). In contrast, hot executive functions are goal-directed, future-oriented cognitive processes elicited in contexts that engender emotion, motivation, and a tension between immediate gratification and long-term rewards (e.g., Zelazo & Muller, 2002; Zelazo, Qu, & Müller, 2005). Our examination of the validity of the hot/cool distinction in the context of developmental research is just one example of a burgeoning area of scientific inquiry into the intersection of cognition and emotion in the mental life and adaptive functioning of developing individuals. It is indeed startling that we are only beginning the discussion of the cognition-emotion intersection in executive functions given that arguably the most well-known case of frontal lobe damage could have served as a springboard to begin this inquiry more than 150 years ago. This story of Phineas Gage has provided a captivating opening for countless psychology chapters on the relationship between the brain and behavior; and yet until very recently, the central question that emerges from the study of frontal lobe damage—how does the prefrontal cortex contribute to cognition *and* emotion in the service of adaptive behavior—has been slighted.

In 1848, when the unfortunate railroad man, Phineas Gage, became the unwitting first case study of frontal lobe damage, the symptoms that drew the most attention were those that were the most "out of character" and disruptive to his daily life and functioning. His impulsive, profane, irresponsible, and slovenly manner was particularly difficult to fathom in the context of what seemed to be intact intellectual and language functions, albeit it is unclear to what degree his cognitive functions were actually put to the test in the way we now think about examining the neuropsychological sequelae of brain damage. As we look back on the history of this construct known as "executive function," Gage's case of frontal damage and behavioral changes illustrates two points that were largely ignored until the 1980s: the ventromedial and orbitofrontal aspects of the prefrontal cortices participate in the affective component of cognitive processes; patients with damage to these regions make very poor decisions

E. Peterson (✉) • M.C. Welsh
University of Northern Colorado, Greeley, CO, USA
e-mail: eric.peterson@unco.edu

and demonstrate other maladaptive behaviors despite relatively intact cognitive ability. Surprisingly, for several decades after Phineas Gage's story, classic descriptions of the frontal cortex, and prefrontal region in particular, focused exclusively on its role in cold cognition.

We begin this chapter with a historical perspective. First, we will review selectively the historical influences on cool executive function, which is essentially the history of theoretical formulations and research on the domain of executive function as it has been traditionally defined. We will not attempt a comprehensive review of the developmental research examining the traditional cool executive functions, as several reviews on this topic have been published in recent years (e.g., Anderson, 2002; Best & Miller, 2010; Espy & Kaufmann, 2002; Garon, Bryson, & Smith, 2008; Luciano, 2003; Romine & Reynolds, 2005; Welsh, 2001; Welsh, Friedman, & Spieker, 2006; Zelazo, Craik, & Booth, 2004). Our historical review finishes with a description of the emergence, or, should we say, reemergence, of hot executive functions, which reflects, at least in part, the influence of research in typical child development and recent adult neuropsychology. Following the historical discussion, our treatment of the current empirical literature will focus on studies comparing and contrasting hot and cool executive functions in children from the preschool period through adolescence. Our synthesis of the recent research examining these two aspects of executive functioning centers on a few issues related to construct validity. First, we examine the degree to which current research yields evidence of construct validity in the form of differential developmental trajectories and patterns of correlations, indicating the separability of hot and cool executive functions. Second, we examine an interesting methodological approach—altering the "temperature" of tasks (e.g., increasing the affective context of a traditionally cool task)—which allows researchers to hold task, and many of its demands, constant. In most current research on hot and cool executive functions, *different* tasks are used to measure each form of executive function making it difficult to isolate which of the multiple factors distinguishing the cool and hot tasks may be mediating task performance. By systematically isolating and varying a single factor presumed to underlie hot executive function, such as affective context or motivational significance, in a single task, one may be able to more effectively identify separable hot and cool processes. Third, we examine construct validity from the perspective of whether hot and cool executive functions differentially predict other aspects of the broader phenotype such as intelligence, temperament, and academic performance. Finally, we conclude with a discussion of some intriguing lines of research that indicate the complexity of teasing apart the domain of executive function into its potentially hot and cool characteristics, particularly as this relates to highly significant "real-world" behaviors of children and adolescents, moral behavior, and risk taking.

The Traditional Executive Function Framework: A Focus on the Cold

Our understanding of executive function and its evolution over the past century is rooted in the clinical and neuropsychological observations and assessments of individuals who had sustained frontal cortical damage. The focus of interest was on a set of goal-directed, future-oriented behaviors that were essential to adaptive behavior but largely independent of general intelligence. These deficits took on a decidedly cognitive flavor as the zeitgeist in the field of psychology also strongly emphasized cognition and information processing in the latter half of the twentieth century. In what follows, we selectively review research and theory as it relates to executive function, demonstrating the manner in which cold cognition has dominated the discussion.

Brain Damage Demands an Understanding of Frontal Lobe Function

As described by Welsh et al. (2006), the construct of executive functions, and particularly the cool version, evolved from cases of focal brain damage, typically from missile wounds incurred by soldiers in wartime. Decades of such case studies led pioneers in the field such as Tueber (1964) and Luria (1973) to describe the distinct differences in sequelae following damage to the frontal cortex vs. more posterior areas of the brain. The surprising dissociation in symptoms for these clinicians was not between affective and cognitive functions, as presumably observed in Phineas Gage, but between certain preserved cognitive functions and those particular cognitive skills that were irrevocably damaged. Although decades of observation and neuropsychological testing of individuals with focal frontal damage yielded what at first appeared to be a wide-ranging collection of symptoms, a unifying theme began to emerge. In the early reports, frontal patients were characterized as lacking the skills of anticipation, planning, and monitoring necessary for purposeful, self-initiated behavior (Luria, 1966; Stuss & Benson, 1984). Patients perseverated in tasks that required flexibility (i.e., a failure to shift mental set); they experienced difficulty maintaining effort over time and were unable to integrate feedback. Individuals with frontal lobe damage exhibited "supramodal" deficits that cut across specific cognitive, sensory, and motor domains (Lezak, 1995), a reduced appreciation of context (Fuster, 1989; Pribram, 1969), and clear impairments in novel problem solving (Duncan, Burgess, & Emslie, 1995). Essentially, these patients lacked the ability to marshal basic cognitive functions in service of a future goal. While it must have been obvious to the patients, their families, and the clinicians that these core deficits also manifested in social contexts requiring emotional regulation, social sensitivity, and daily adaptive functioning, this appreciation did not appear to impact the traditional cold cognition definition of prefrontal cortical functions in the literature.

The Understanding of Executive Function Becomes Even Cooler

The frameworks that emerged to explain the constellation of deficits consequent to frontal lobe damage illustrate several strong influences of the cognitive revolution that began in the 1960s. For example, one influential information processing perspective on executive function was proposed by Norman and Shallice (1986) to characterize the neuropsychological deficits typical of frontal lobe damage. In their Supervisory Attention System (SAS) model, these authors highlight the distinction between routine and nonroutine environmental contingencies when defining the essence of executive function. In this framework, SAS is recruited in novel situations requiring an analysis of the problem at hand, followed by strategy generation, monitoring, and flexible revision of these strategies based on feedback. In their view, frontal lobe and executive function represented a domain of conscious, effortful, cognitive processes that reflected the models of information flow and processing (e.g., Atkinson & Shiffrin, 1968) of the era.

This emphasis on the cognitive functions mediated by the prefrontal cortex was also a consequence of the development of neuropsychological tests during this period. For example, in the early 1960s, Milner utilized a card-sorting task originally developed by Grant and Berg (1948) to identify deficits following frontal lobe damage, and the Wisconsin Card Sorting Task quickly became the yardstick by which individual differences in frontal function were measured (e.g., Milner, 1963). This task requires inhibition and flexibility of mental set, as well as inductive reasoning. The fact that adult levels of performance were not observed until about age 10 years (e.g., Chelune & Baer, 1986; Welsh, Pennington, & Groisser, 1991) led many neuropsychologists and researchers to suggest that the prefrontal cortex did not effectively "turn on" and influence behavior until preadolescence (Golden, 1981). Such a proposition dovetailed nicely with the dominant cognitive development theory of the time in which the systematic cognitive functions of formal operations emerged at about the same age (Piaget, 1972). However, how one defines, and therefore *assesses*, executive functions as a reflection of prefrontal activity will determine when in development one is likely to observe the putative cognitive functions. Welsh and Pennington (1988) pointed out the many potential manifestations of rudimentary executive functions that are exhibited by infants and toddlers, and the relatively recent appreciation of the emotion-based hot executive functions (Zelazo et al., 2005) has created a renewed interest in the early development of this behavioral domain (e.g., Garon et al., 2008).

Unitary or Multifaceted: Cold Cognition Remains the Emphasis

New computer technologies and statistical techniques in the past 2 decades have also served to shape the definition of executive function, but again emphasizing its cognitive components. An important debate of this period in the evolution of the executive function construct concerned whether a unitary or multifactorial view on this domain was a more accurate representation. Well-known computational or connectionist models of the sequelae observed after frontal lobe damage supported a unitary view of executive function, emphasizing either a limited capacity working memory system (Kimberg & Farah, 1993) or a system that effectively represents and maintains contextual information (Cohen & Servan-Schreiber, 1992). Consistent with this "single function" perspective, Zelazo and Frye (1998) proposed the Cognitive Complexity and Control Theory of executive function. This theory likens executive function to mental representation of logical rules (if-then) that are needed to solve novel, goal-oriented problems, and indeed, it was examined in research using the non-affective, decontextualized, "cool" task known as the Dimensional Card Sorting Task.

In contrast to this univariate definition of executive function, multivariate statistical techniques have supported a multifactorial construct with independent factors that nonetheless work together depending upon the particular task or situation. Early factor-analytic studies of school-aged children found separable factors that reflected cognitive processes such as "fluency and organized responding," "planning," and "hypothesis testing" (Brookshire, Levin, Song, & Zhang, 2004; Welsh et al., 1991). One of the most influential structural models of executive function, developed from an adult sample, demonstrated both the "unity" (correlated factors) and "diversity" (three factors of working memory, shifting, and inhibition) of executive function (Miyake et al., 2000). This model has subsequently been examined in developmental samples with mixed results. Lehto, Juujärvi, Kooistra, and Pulkkinen (2003) found the three-factor Miyake model to be the best fitting model for a sample of 8–13-year-olds; however, Huizinga, Dolan, and van der Molen (2006) found the best fitting model data across the age range from 7 to 21 years included only shifting and working memory. It is important to note here that, as in the early neuropsychological studies of executive functions, the tasks one uses to measure the construct will determine one's findings and can lead to inconsistency across studies. These multivariate statistical approaches and attempts to model executive function will depend on the tasks researchers select to represent the hypothesized components of executive function. Throughout our discussion of the new method of dichotomizing executive function, hot vs. cool, we will find that the very definition of each of these concepts is inextricably connected to the instruments one uses for measurement purposes.

Recent Status of Cool Executive Function in the Developmental Literature

Current reviews of executive function development have focused on the three independent yet interrelated constructs of working memory, inhibition, and shifting identified by Miyake et al. (2000) in their structural model, despite the fact that the model has not been adequately tested in developmental samples. Garon et al. (2008) discussed evidence for an early emergence of these executive function components in infancy but with a dynamic period of development between 3 and 5 years of age. These authors suggest that development of attentional mechanisms may underlie improvements in more complex executive function tasks that require the integration of the three components and the need to resolve conflict. In their review of studies focused on school-aged children, also organized according to the Miyake et al. (2000) tripartite model, Best and Miller (2010) concluded that there is substantial improvement across this developmental period, with differences in the trajectories depending upon the component examined. In their meta-analysis of classic neuropsychological tests of cool executive functions, Romine and Reynolds (2005) identified developmental trajectories indicating the most rapid development from 5 to 8 years, moderate to strong development in the age periods of 8–14 years, and slowing development during adolescence (14–17 years). The fact remains, however, that the majority of comprehensive reviews of executive function development across childhood and adolescent have maintained a laser focus on cool, cognitive processes.

The degree to which our current conceptualization of cool executive function is task dependent is an important question that must be addressed. Although there is some consistency in the executive function tasks used for preschoolers, school-age children, and adolescents, these age-appropriate sets of tasks are generally *different* across age groups, so both cross-sectional and longitudinal research findings addressing stability and change in executive functions must be tempered with questions of task equivalency related to content and difficulty. The search for "clean" measures of core cognitive processes comprising executive function that can be used with little or no modification across development has represented the "holy grail" of executive function research. It is unclear whether the often contradictory findings regarding convergence of executive function measures, even within the cool domain, as well as the predictive associations between executive functions and "real-world" behaviors, are an indictment of the tasks currently used or the construct itself (or both). Finally, decades of clinical and experimental analysis of cool executive functions, across several levels of analysis (e.g., brain damage, computational models), have brought clear consensus that the dorsolateral prefrontal cortical system mediates this complex set of goal-oriented cognitive processes, although the precise neural mechanisms underlying these phenomena are still in question (e.g., Duncan & Owen, 2000; Miller & Cohen, 2001; O'Reilly, 2010). As we will see in what follows, these central conceptual questions are now doubled, at the minimum, with the introduction of the construct of hot executive function.

The Rise of Hot Executive Functions

The separation of cognition and emotion and the favored status of mentalistic, cognitive processes like reason and will over the "lower" emotional processes have a long history in Western thought. In 1980, Robert Zajonc offered the first serious critique of this position, arguing instead for the independence and primacy of affect over cognition (Zajonc, 1980), giving rise to a new era in emotion research. Given the historical study of frontal lobe damage beginning with Phineas Gage and the ascendance of emotion/cognition interaction across the past few decades, it is interesting to note that the clear emergence of "hot executive functions" occurred as late as the mid-2000s. Although many factors likely contributed to this recent direction, in the review below we discuss just two important influences: the developmental research involving delay of gratification (Mischel Ebbesen, & Zeiss, 1972; Mischel, Shoda, & Rodriguez, 1989) and the adult neuropsychological work examining patients with ventromedial and orbitofrontal damage (Bechara, 2004; Bechara & Damasio, 2000).

The Development of Delay of Gratification: Hot Before Its Time

Decades before the emergence of hot executive functions, Mischel and colleagues (e.g., Mischel & Metzner, 1962; Mischel et al. 1972) examined the child's ability to delay immediate reward in an affective context. In one series of experiments (e.g., Mischel & Metzner, 1962), children were tested in a paradigm that involved choosing between an immediate reward and a reward of greater value at some distant time (e.g., a week). In essence, the task assessed the relationship between the length of delay before receiving the deferred reward and the degree to which the participant discounts its value as measured by the choice of the immediate lesser award. Consistent with Mischel's (Mischel & Metzner, 1962) original observation, the task is sensitive to both development and individual differences in general intelligence (Green, Fry, & Myerson, 1994; Shamosh & Gray, 2008). It is easy to imagine real-world scenarios in which individuals may discount the value of a delayed reward in favor of some form of immediate payoff. It is not surprising, therefore, that the delay discounting paradigm has been adapted for the study of human behavior across a range of disciplines (for review, see Shamosh & Gray, 2008).

In a conceptually similar paradigm, Mischel examined young children's ability to delay gratification within reach of an immediate reward. In the classic delay of gratification paradigm, each child was offered a treat, a single marshmallow, with the opportunity to double the reward to two marshmallows if the child could resist the urge for immediate gratification. The importance of these seminal investigations was very recently demonstrated in a follow-up study in which a group of middle-aged participants, originally tested as preschoolers in the marshmallow study, was retested in hot and cool versions of a go/no-go paradigm (Casey et al., 2011). Participants who showed relatively weaker delay of gratification when tested as preschoolers, 4 decades ago, showed increased difficulty in the no-go condition involving a happy emotion face. Importantly, their relatively poor performance in a task involving inhibition was selectively impaired in an emotional context (an emotion face relative to a neutral face).

Mischel's seminal research preceded by decades the current interest in integrating hot and cool executive development. However, more recently Metcalfe and Mischel (1999) articulated an explanatory model involving hot and cool processes that has since been cited by many developmental reviews of hot and cool executive development. It should be noted that the 1999 paper did not actually reference the notion of executive function. In their framework, maturation reflects a gradual shift of dominance such that immature hot processes are regulated by later maturing cool processes. As will be clear below, this perspective stands in contrast to the more contemporary view of hot executive processes that continue to mature with age (like cool processes) and facilitate performance in more emotionally challenging contexts.

The Adult Neuropsychological Framework: A Model for the Development of Hot Executive Functions

Perhaps the strongest influence on the current goal of integrating hot and cool executive processes comes from the study of adults with brain damage in the orbitofrontal and ventromedial cortices (Bechara, 2004). These two largely overlapping brain regions are richly connected with limbic areas associated with emotional and social processing (Bechara, 2004; Beer, 2006). The systematic study of such patients with orbital and ventromedial prefrontal damage, as opposed to dorsolateral damage, has provided strong support for the notion that adaptive decision making and related goal-oriented behavior cannot be explained entirely by "cold" cognitive processes. In spite of relatively intact general cognitive abilities, such patients display a range of behaviors that can be characterized by poor social regulation and an inability to consider future consequences when making decisions. In essence, such individuals suffer from poor "social executive functioning" (Beers, 2006). In an effort to provide a neurocognitive explanation for this dissociation, Bechara & Damasio (2000) proposed the "somatic marker hypothesis"; they posited that in the process of making decisions about the future, neurotypical individuals access a positive or negative emotion-based representation from past experience, a somatic marker, that guides the selection of future-oriented choices. To test this hypothesized role of the ventromedial cortex, they created the Iowa Gambling Task (IGT). In this task, participants choose cards from across four possible decks and are either rewarded or penalized with each card. Two of the decks are disadvantageous, coupling high immediate rewards with unpredictable large losses that outweigh early gains. The other two decks are advantageous, yielding smaller initial gains but also smaller losses for a net profit across the game.

Using the IGT, these investigators have amassed a body of research demonstrating that patients with ventromedial damage have particular difficulty integrating future positive or negative consequences in the service of making adaptive decisions (Bechara, Damásio, Damásio, & Anderson, 1994). Unlike healthy adults, they remain with the disadvantageous decks even as th high reward of the initial cards has been replaced by large punishing losses. Importantly, such deficits can be observed without significant impairment in traditional cognitive control processes like working memory and planning associated with dorsolateral prefrontal cortex. As an explanation, the somatic marker hypothesis was supported by an investigation involving skin conductance response (SCR). Like neurotypical control participants, ventromedial patients generated appropriate SCRs after experiencing the reward or punishment following selection. However, unlike the control participants, they did not develop *anticipatory* SCRs ahead of the card selection, particularly from the risky deck. Presumably, the feeling experience that soon precedes card selection influences healthy controls to avoid excessive risk. In essence, the IGT is assumed to provide an adequate laboratory analogue of real-life situations in which one must perform an implicit cost/benefit analysis between immediate reward and future consequences that may involve punishment. Importantly, in real-life social contexts, choices cannot often be subjected to a precise rational analysis but rather must be assessed by "gut" feeling (Bechara, 2004).

In summary, more than 3 decades before the first use of the term "hot executive functions," Mischel and colleagues appreciated the importance of understanding a less "purely cognitive" development, the capacity to resist an impulse in a highly motivational context toward the goal of a greater long-term reward. Metcalfe and Mischel (1999) did not consider neural mechanisms that might mediate either individual differences at a given age or the gradual normative change with maturation. Instead, the primary influence on the current thinking about a neural substrate for developmental change in hot executive functioning came from the adult neuropsychological work with ventromedial patients.

The Emergence of Hot Executive Functions

The study of adult lesion patients soon provided a framework for conceptualizing a range of behavioral developments that can each be related to the interaction of neural mechanisms mediating hot and cool processes (e.g., morality: Green & Haidt, 2002; risk taking in adolescence: Steinberg, 2005; atypical development: Zelazo & Muller, 2002). In 2004, the journal *Brain and Cognition* published a special issue dedicated to placing the developing orbitofrontal region within the prefrontal cortex of the developing child (e.g., Bechara, 2004; Happaney, Zelazo, & Stuss, 2004; Kerr & Zelazo, 2004). Although hot and cool executive processes in development had been discussed earlier (Zelazo & Muller, 2002), this special issue likely influenced the burgeoning literature that would follow. Many developmental researchers (e.g., Crone, 2009; Crone, Bullens, Van der Plas, Kijkuit, & Zelazo, 2008; Crone & van der Molen, 2004) have since developed tasks modeled after the IGT.

Today, numerous tasks exist for examining the relative contributions of cooler and warmer processes across development, and many investigators are committed to such an integrative framework. Clearly, the goal of a more comprehensive explanation of development with an appreciation of context represents a positive and important direction. However, the central questions outlined at the conclusion of the review of traditional cool executive functioning pertain equally to the study of hot processes. For example, the degree to which hot and cool processes are mediated by separate dissociable systems (as suggested by the adult neuropsychological framework) parallels the cool question regarding a single unitary system vs. multiple interacting processes. Progress in such a fundamental question remains constrained by the available tasks. As will be clear in the review below, this limits our examination of the dissociation of hot and cool executive functions.

Construct Validity of Hot and Cool Executive Function: A View from Developmental Research

While the literature on the development of cool executive functions is vast, the published research on the development of hot executive functions is but a decade old. Clearly, any attempt at integration will not easily be resolved. However, evidence for the slow developmental maturational gradient of prefrontal cortex (e.g., Giedd et al., 1999; Sowell, Thompson, Tessner, & Toga, 2001) suggests that a continued examination of hot and cool processes across development should be fruitful. In this section, we explore three questions that, collectively, examine the current status of our understanding of this exciting new direction. First, we focus on the degree to which patterns of correlations and developmental differences provide evidence of construct validity. As a second approach to the issue of construct validity, we review research in which a single task is manipulated. In several studies, researchers have either amplified or attenuated the affective component of a task, effectively changing its temperature. This approach to exploring hot and cool executive processes within a single task allows researchers to minimize the messy issues associated with across-task comparisons. Finally, we examine the degree to which hot and cool executive functions differentially predict real-world developmental phenomena, another set of evidence that would suggest separability.

Do Developmental Trajectories and Correlational Patterns Support the Independence of Hot and Cool Executive Functions?

While this question may seem straightforward, our review below highlights a number of difficult and challenging issues. As echoed throughout other sections, we are limited by our tasks. While many tasks have been used to measure cool executive functions, the newer construct of hot executive functions has been probed by a relatively small set of tasks, in a relatively small number of research groups. The cool executive function tasks primarily tap the skills of working memory, inhibition, and flexibility/switching and include measures such as the Dimensional Card Sorting Task, Self-Ordered Pointing Task, and a variety of conflict tasks (e.g., Pencil Tapping, Bear/Dragon, Grass/Snow) for young children, whereas tasks such as the Wisconsin Card Sorting Task, tower tasks, and more complex working memory and inhibition tasks have been administered to older children and adolescents. The hot executive function tasks for young children involve decision making in a reward-loss context with strong motivational significance, such as the Children's Gambling Task, patterned after the IGT. A second set of tests involve delay and prohibitions when the child is faced with an attractive, desired, appetitive stimulus, as in the classic delay of gratification task (Mischel et al. 1972). As described earlier, the classic hot executive function tasks for older children and adolescents include the IGT and delay discounting paradigms. Not only do tasks differ across hot and cool executive function, they differ across age, which complicates the assessment of developmental trajectories across wider ranges of age.

Therefore, addressing the construct validity issue by inspecting patterns of correlations, as well as differential developmental trajectories, is severely compromised by this task issue. Although the adult neuropsychological evidence (e.g., Bechara, 2004) is consistent with the hypothesis of separate developmental mechanisms for hot and cool executive functions, it may not follow that tasks can be designed to tap one or the other system exclusively. Indeed, Hongwanishkul, Happaney, Lee, and Zelazo (2005) emphasized that most tasks are likely to elicit aspects of both hot and cool executive function, with the shared contributions of various genetic and environmental factors increasing the cross-domain correlations. This dichotomy would be demonstrated by a convergent-divergent validity approach in which cool and hot executive function tasks should be more highly correlated within domain than across domain. Further, if the constructs do represent different underlying neurocognitive processes, we might expect to observe different developmental trajectories, for example, one system maturing ahead of the other.

Examining the development of both hot and cool executive functions in a sample of 3–5-year-olds, Hongwanishkul et al. (2005) hypothesized that both correlational patterns and differential developmental trajectories would provide evidence for the independence of these two constructs. Using the Dimensional Card Sorting and Self-Ordered Pointing Task as measures of cool executive function and the Children's Gambling Task and delay of gratification to tap hot executive function, the authors did not find substantively different developmental trajectories across the two domains of executive function. All four tasks demonstrated relatively similar improvements after age 3. Their findings for the Children's Gambling Task replicated the earlier reports of Kerr and Zelazo (2004), however; the delay of gratification results were inconsistent with the absence of age effects reported by Peake, Hebl, and Mischel (2002). Moreover, the tasks did not covary in a way that provided strong evidence for the dissociation of the two types of executive function. With age and intelligence controlled, the two cool tasks intercorrelated; however, both cool tasks correlated with one hot task (Children's Gambling Task), and the two hot tasks *negatively* correlated, clearly contrary to predictions. The authors provide several explanations for the unexpected negative correlation between the scores on the two hot tasks; however, the finding that the hot Children's Gambling Task correlated more predictably with the two cool tasks, than with another hot task, severely weakens the argument for two distinct executive function constructs.

Recent research conducted by Carlson and colleagues (Carlson, Davis, & Leach, 2005; Carlson & Wang, 2007) examined two types of inhibitory control: delay vs. conflict. Although they did not discuss their tasks within the cool vs. hot framework, their delay tasks bear a strong resemblance to the tasks used to measure hot executive function in other laboratories. Similarly, their conflict tasks, such as Simon Says, Bear/Dragon, and Dimensional Card Sort, are the very tasks used as cool executive function measures in the Zelazo laboratory. In the Carlson and Wang (2007) study, age significantly correlated with performance on the cool executive function task (Simon Says) and two of the hot tasks (Gift Delay and Disappointing Gift) across the age range of 4–6 years. Performance on Simon Says was uncorrelated with the hot executive function tasks (Disappointing Gift, Gift Delay, Secret Keeping, and Forbidden Toy) indicating some independence of the two constructs; however, only two of the hot tasks (Gift Delay and Forbidden Toy) significantly correlated with each other. On balance, the reported patterns of correlations do not strongly support independent constructs of hot and cool executive function. Their main hot executive function task, Less is More, which requires children to select the smaller reward to gain the larger one, correlated with *both* cool executive function tasks *and* the other hot, delay task. However, it is important to note these correlations disappeared when age and verbal intelligence were controlled in the analyses.

In the studies reviewed above, there is substantial evidence for development of *both* hot and cool executive function skills in the age period of 3–5 years, although there is not clear evidence for different developmental trajectories for the two executive function domains. For example, in a secondary analysis of data collected in her own laboratory, Carlson (2005) found no evidence that tasks labeled as "cool" due to non-affective, arbitrary rules and demands could be differentiated from affective, reward-sensitive "hot" tasks in terms of difficulty levels for samples of children ages 2–6 years. Additionally, there was not a clear pattern of correlations that demonstrated separable domains of hot and cool executive function. Although one would expect a modicum of shared variance between these two sets of tasks, since they both assess goal-directed behavior, one does expect the correlations to indicate some degree of convergent and divergent validity. As stated earlier, a small number of laboratories are engaged in the investigation of hot and cool executive functions in young children, notably the Zelazo group, and thus, the constructs are defined by the particular tasks that have been selected to examine cool executive function (e.g., DCCS) and hot executive function (e.g., Children's Gambling Task). It will be a challenge for future work in this area to extricate the definition of these two forms of executive functions from the limited number of tasks currently utilized in order to develop an understanding of hot and cool processes that is relatively task independent.

Studies involving adolescent samples are important as evidence that different developmental trajectories of hot and cool executive processes may emerge beyond early childhood. Hooper, Luciana, Conklin, and Yarger (2004) compared a large sample of children and adolescents (9–17 years of age) on three tasks: two traditional cool tasks (digit span and go/no-go) that tap memory and inhibition and one hot task (IGT). Performance increases with age were observed for all three tasks. However, the IGT showed the most protracted developmental trajectory (superior performance by the oldest children tested). After controlling for age, gender, and intellectual ability, performance on the IGT was not predicted by the two cool measures. Thus, these findings are consistent with the hypothesis that, in adolescence, the IGT taps additional cognitive processes that do not mediate performance in the cool tasks. Crone and van der Molen (2004) also examined a wide age range of participants using an adaptation of the IGT. They obtained evidence of improvement through the oldest age group tested (18–25 years). Importantly, developmental change in performance did not reflect changes in either working memory as indexed by backward digit span or inductive reasoning as measured by Raven Standard Progressive Matrices. Thus, taken together, these studies support a protracted development through adolescence on the presumably hot process that mediates performance in the IGT.

Two more studies examined developmental performance changes on the IGT in comparison to traditional cool executive tasks (Lamm, Zelazo, & Lewis, 2006; Prencipe et al., 2011). Because these studies also included an additional hot task, delay discounting, their results may provide a better test of the relationship among cooler and warmer tasks. On balance, the results did not yield strong support for dissociable systems. For example, across both studies, performance on the IGT did not correlate with delay discounting. Across an age range of 7–16 years, Lamm et al. found no evidence for age-related performance change in delay discounting though performance on the IGT positively correlated with age. After partialling for age, none of the tasks, cool or warm, were correlated with each other. Prencipe et al. (2011) examined children across a similar age range. Notably, the two cool tasks (Stroop and Digit Span) correlated with each other and with the IGT, consistent with the findings of Hongwanishkul et al. (2005) in 3–5-year-olds. Prencipe et al. did obtain some evidence consistent with the hypothesis that cool executive functions matured ahead of hot executive functions; performance differences between the youngest and oldest groups were evident in delay discounting and the IGT. Of course, it is important to remember that overall task difficulty (e.g., Stroop vs. IGT) is not equated, so we should be cautious when drawing conclusions about different developmental trajectories. Of interest to the overarching question of whether or not hot and cool executive functions are mediated by separate brain functions (as suggested by the adult neuropsychological model), the authors performed an exploratory factor analysis and did not obtain evidence of dissociable hot and cool processes.

In summary, studies involving adolescence have shown that performance on the IGT develops across adolescence, consistent with a model in which cool processes mature ahead of hot processes. However, the failure to obtain evidence of a correlation between IGT and delay discounting is not suggestive of separable mechanisms. Again, it is important to consider the problem of individual task difficulty when making inferences about differing developmental trajectories. With this caveat in mind, these results suggest that future studies with adolescent samples should include cool and warm tasks in order to continue to explore the possibility that some hot processes are later developing.

Can We Manipulate the "Temperature" of Our Tasks to Probe the Nature of Hot and Cold Processes?

As discussed above, the degree to which hot and cool executive functions reflect somewhat dissociable systems remains unclear, and again, our progress on this question is constrained by our tasks. Presumably, some of the cool tasks may evoke a stronger affective response than others (and vice versa). As discussed by Garon et al. (2008), a promising methodological direction has been the manipulation of the affective context within a single task.

In a preschool study, Carlson et al. (2005) "cooled down" the Less is More task, replacing the appetitive stimulus (i.e., candy) with a symbolic representation, such as a picture. This manipulation improved the performance of 3-year-olds, particularly when the picture was further removed from the candy stimulus. Therefore, although the delay and prohibition tasks in the Carlson and colleagues work can be considered "hot," in contrast to the "cooler" conflict inhibition tasks, the consequences of the manipulation suggest that the "temperature" of an executive function task depends to a large extent on the task demands and conditions.

Lewis, Lamm, Segalowitz, Stieben, and Zelazo (2006) studied children from 5 to 16 years of age using a go/no task that included an emotion manipulation. In addition to behavioral measures, event-related potentials were collected. The paradigm was divided into three blocks. By design, in the middle block, participants steadily lost points. Therefore, the third block provided insight into how children performed immediately after a negative emotion inducement. Consistent with the slow development of adaptive regulation in frustrating circumstances, older children were relatively less impaired by the emotion inducement. Consistent with the behavioral data, electrophysiological evidence supported the hypothesis of increased inhibitory control mediated by prefrontal cortex across development. Figner, Mackinlay, Wilkening, and Weber (2009) manipulated the affective context of the Columbia Card Task using an older sample (13–19 years of age). Each trial of this task begins with a set of cards faced down such that each card's value (magnitude of win or loss) is unknown. The quantity of loss cards as well as the total loss value is indicated. Participants have the chance to turn over cards, one after the other, for points until either they elect to stop the trial to accept the current winnings or they hit a loss card which both costs points and terminates the trial. In the cool version of the task, participants indicated in advance the total number of cards they elected to turn over and did not receive any feedback until the end of the trial. Alternatively, in the hot version of the task, participants made stepwise incremental decisions and received ongoing feedback; that is, after each trial, the points gained or lost were revealed and, assuming a loss card was not encountered, the participant needed to decide again to continue or stop. Evidence that the hot/cool manipulation was successful was supported by measures of electrodermal activity in each condition (i.e., greater activity during hot tasks). The hot version of the task was associated with greater risk taking in adolescents but not in the adult comparison sample.

Crone, Bullens, Van der Plas, and Zelazo (2008) altered the temperature of a gambling task by manipulating whether the participants were playing for themselves (hot) or another (cooler). Children made less risky choices with age (8–18 years). More important, across all age groups, participants made less risky choices when they were playing for another. The asymmetry of risky choices between the self and other condition was largest for the youngest group (8–9-year-olds). Crone, Bunge, Latenstein, and van der Molen (2005) studied children between 7 and 12 years of age using a children's version of the gambling task. Across different versions, the authors manipulated the task to examine whether development is associated with increased capacity for task difficulty (two-choice vs. four-choice options), the ability to switch response set, or a decrease in the sensitivity to punishment frequency. Development was associated with an increased ability to make adaptive choices with infrequent punishment, a finding that is consistent with the notion that children continue to develop sensitivity to somatic markers such that frequent punishment is less necessary for learning from experience.

Although it has been acknowledged that any single task presumably taps both hot and cool processes to some degree, it remains a challenge for the field to determine precisely which elements of a task should be manipulated systematically to elicit one process preferentially. Several studies have demonstrated that systematic task manipulation is a promising method for supporting specific hypotheses regarding mechanisms presumed to mediate development. To date, this small emerging literature suggests that temperature manipulations yield age effects consistent with the hypothesis that hot executive functions show a protracted development.

Do Hot and Cool Executive Processes Differentially Relate to Adaptive Behaviors of "Real-World" Significance?

A longtime challenge for executive function research is to connect developments as indexed by task performance with real-world consequences. From this perspective, we assume that the recent integration of hot executive functions promises to provide a more comprehensive framework for prediction across a range of settings beyond the laboratory.

Hongwanishkul et al. (2005) did find some evidence for divergent validity. Performance on the two cool executive functions correlated significantly with measures of intelligence, whereas performance on the hot executive function tasks did not. These results are consistent with a conceptualization in which cool executive function tasks recruit cognitive functions to a greater degree than do hot executive function measures. Cool executive function performance was correlated with some aspects of temperament, such as effortful control, whereas the hot executive function scores were not related to any measures of temperament. Although the authors used this latter finding as evidence for the independence of hot and cool executive function, they originally hypothesized that hot executive function would significantly covary with the negative affect measure of temperament. Thus, although there is some evidence for a dissociation between hot and cold, the findings, particularly for hot executive findings, were not consistent with predictions.

In two somewhat similar studies, Brock et al. (2009) and Willoughby, Kupersmidt, Voegler-Lee, and Bryant (2011) examined the associations between sets of tasks measuring cool and hot executive functions and behavioral and academic outcomes. Both Brock et al. (2009) and Willoughby et al. (2011) utilized the Balance Beam and Pencil Tapping tasks as their cool executive function measures for their 3–5-year-olds and kindergarten children, respectively. These might be seen as unusual choices of cool executive function tasks, given that both rely much more on motor planning than on cold cognition executive functions such as working memory, goal planning, and flexibility. Both studies involved self-regulation tasks characterized by prohibition and delay as their measures of hot executive function. The findings of the two studies were remarkably similar as well, despite the different age groups (3–5 years vs. kindergarten) and some differences in the specific tasks utilized. In both studies, confirmatory factor analysis indicated separate factors aligning with cool and hot executive function; however, the factors were moderately correlated. Additionally, both studies reported that cool executive function performance, but not hot, predicted academic outcomes in the children studied. It is important to highlight that the cool executive function tasks represented a departure from the typical measures of this construct, and in fact, it may be the case that these cool tasks tell us less about executive function, per se, and more about overall neural integrity. It is well established that motor dysfunction symptomatology is consistent with a range of developmental disorders that would be associated with academic deficits (e.g., Piek & Dyck, 2004; Visser, 2003). Therefore, in both the Brock et al. (2009) and the Willoughby et al. (2011) studies, the separate factors representing their two sets of tasks may, or may not, reflect a dissociation between cool and hot executive functions, depending upon whether one would accept these motor-based tasks as appropriate exemplars of cool executive function.

Several studies have provided support for the assumption that tasks assumed to measure hot executive processes predict outcomes in real-world settings. In an investigation aimed at predicting academic performance among eighth graders, Duckworth and Seligman (2005) supported and extended Mischel's initial evidence (e.g., Mischel, Shoda, & Peak, 1988) that the ability to delay gratification has greater value than IQ for predicting academic outcome. These authors created a composite of self-discipline that included questionnaire data obtained from student, parent, and teacher. Most important for the present review, the measure also included performance on two versions of a basic delay discounting paradigm. Each of the delay discounting paradigms correlated with pencil and paper ratings obtained by eighth grade participant, parent, and teacher (all r's > .50). Of course, a challenge for future research is to continue to explore the variables that drive a relationship between performance on hot tasks and behavioral outcomes. Current evidence suggests that at least some of the variance associated with the relationship between delaying reward and outcome may be associated with general intelligence. For example, meta-analytical evidence has demonstrated a negative correlation between intelligence and delay discounting performance (Shamosh et al., 2008; Shamosh & Gray, 2008). That is, greater intelligence is associated with an increased ability to defer acceptance of a smaller reward in order to obtain a reward of greater value.

In summary, there are mixed results yielded by studies with regard to the extent to which hot and cool executive processes differentially predict real-world behaviors in children and adolescents. Whereas it has been hypothesized that cool executive function would better predict academic achievement and hot executive function contribute to "warmer" behaviors such as temperament and self-regulation, the findings of the few studies reviewed do not clearly align with these predictions. Moreover, the fact that two hot tasks, delay of gratification and delay discounting, predict current or later academic outcomes among adolescents serves to point out the difficulty of determining the "hotness" or "coolness" of particular real-world behaviors. For example, most would agree that achievement in high school and college demands both cold cognition and hot emotion regulation abilities. Therefore, more clarity in the definitions of hot and cool processes, in both experimental measures and real-world contexts, is needed.

Intriguing Intersections with Other Developmental Questions

Our previous review of developmental research examining hot and cool executive functions highlighted recent studies that specifically compared sets of tasks identified by authors as putatively tapping the two domains, or by examining the consequences of manipulating a single executive function task. In what follows, we examine the potential intersections between the hot and cool distinction in two areas of developmental research involving cognition/emotion interaction in which executive function is only peripherally mentioned, if at all. The children's compliance research by Kochanska and colleagues is particularly intriguing given that many of her behavioral measures have been incorporated into the hot executive function testing batteries of other researchers. The adolescent risk-taking literature is interesting to consider in light of the risky decision-making tasks (e.g., IGT) that are typically used to measure hot executive function in adolescents, as well as the emerging evidence regarding sex differences in task performance and the development and function of the orbitofrontal cortex.

Compliance and Moral Development in Young Children

A recent example of an investigation of the cognition-emotion intersection in the development of adaptive behavior is the seminal work of Kochanska and colleagues in which they trace the roots of compliance and moral behavior in young children. In the early years of this investigation, Kochanska does not explicitly connect compliance behavior to executive functions, let alone to hot vs. cool executive functions, although she acknowledges that compliance clearly is a manifestation of self-regulation, autonomy, and assertiveness that characterize early childhood development (Kuczynski & Kochanska, 1990; Kuczynski, Kochanska, Radke-Yarrow, & Girnius-Brown, 1987). The goal of her early work was to move beyond the primarily "cold cognition" perspective on compliance behavior expressed by Grusec and Goodnow (1994; as cited in Kochanska, 1994), who pointed out that internalization of parents' rules for appropriate behavior may have "executive" aspects. She aimed to include the social, emotional, and temperamental underpinnings of how young children learned right from wrong.

Of relevance to this chapter, Kochanska (2002) identified two types of compliance: (1) "don't compliance" involves following rules in emotionally charged contexts requiring delay of gratification and prohibition to touch attractive toys and (2) "do compliance" involves relatively neutral contexts, such as following the mothers' directives to clean up toys. Her findings indicated that "don't compliance" behaviors were associated with individual differences in temperament, specifically fearfulness, while the "do compliance" behaviors were, instead, associated with attention. Moreover, the "don't compliance" behaviors appeared to emerge earlier in development than the "do compliance" behaviors, and manifestations of the two types of compliance were only weakly correlated. Although never explicitly linked to the hot vs. cool distinction in executive function, there are intriguing parallels with both "hot" delay and prohibition contexts of "don't compliance" and the "cooler" contexts of "do compliance." Specifically, the contexts that elicit "don't compliance" are emotionally laden, motivationally significant situations presumed to tap hot executive function. Furthermore, performance on these "don't compliance" tasks is correlated with temperamental characteristics. The "don't" vs. "do" compliance behaviors were found to be somewhat independent, and each exhibited different developmental trajectories.

In essence, Kochanska and her colleagues have explored the "hotter" components of compliance and moral behavior of young children in much the same way that researchers have, in the past decade, begun to examine the "hotter" manifestations of executive function. Indeed, the tasks that Kochanska (1997, 2002) has used, as well as developed, for her investigations of early compliance and moral behavior of children—delay of gratification and prohibition tasks—are precisely the measures that have been used to assess hot executive functions in recent research. Interestingly, Kochanska (2002) found that "committed compliance" during the age range of 14–45 months in "don't contexts," but *not* "do contexts," predicted measures of the "moral self" and moral behavior at 56 months, and *only for boys*. Given that "don't compliance" is measured via the same tasks that are referred to as hot executive function measures in the contemporary literature, do these findings suggest that hot executive functions may predict later rule-guided, compliant, and moral behavior in male children?

To our knowledge, no one has explicitly connected the Kochanska research to the hot vs. cool executive function literature, except to incorporate several of her prohibition tasks, such as the Gift Wrap Task (Brock et al., 2009), into the current batteries of hot executive function tests. It will be of great interest to examine the degree to which particularly hot aspects of "effortful control," as measured by Kochanska, Aksan, Penney, and Doobay (2007), are similar to, or the same as, hot executive function as described in this chapter. Based on Kochanska's work, there may be substantial implications for individual differences in hot executive function as predictors of later rule-based moral behavior in children and adolescents, particularly males. This is a provocative direction for future research as it may dovetail with emerging research on the links between hot executive function and risk-taking behavior (often characterized as the breaking of rules) and the seemingly counterintuitive findings with regard to a male superiority.

Male Superiority in Hot Executive Function in the Context of Gender Differences in Adolescent Risk Taking

An examination of sex differences in hot executive function is of interest in light of the potential implications of this domain of functioning for everyday behaviors, such as moral decision-making, as explored by Kochanska, and risk-taking activities exhibited by adolescents. Whereas research has indicated associations between individual differences in executive function skills and risk taking in adolescents (e.g., Pharo, Sim, Graham, Gross, & Hayne, 2011; White et al., 1994), there is a paucity of studies examining whether hot and cool processes *differentially* predict risk behavior. Intuitively, if hot executive function involves decision making in highly charged emotional contexts, typified by reinforcement and motivational forces, real-world risk-taking contexts should provide a fertile field to observe hot executive function in action during adolescence.

As described earlier in the chapter, the orbitofrontal cortex mediates hot executive function, and the Object Reversal Task has been found to be sensitive to the integrity of this brain system in both monkeys and young children (Overman et al., 2004), whereas the IGT is the prototypical hot executive function test for adolescents and adults. In the case of both experimental measures, a male superiority has been found in young children, adolescents, and adults, with the suggestion that this reflects earlier development of the orbitofrontal cortex in males as a result of androgen activity (Overman, Bachevalier, Schuhmann, & Ryan, 1996). While young male children outperform their female counterparts on the Object Reversal Test (Overman et al., 2004), evidence for sex differences on the Child Gambling Task is less clear. Kerr and Zelazo (2004) predicted a male superiority in performance on the Child Gambling Task but only found nonsignificant statistical trends in which 3-year-old boys outperformed 3-year-old girls on two of the five blocks of the task. In a later study, Hongwanishkul et al. (2005) found no main effect of sex or sex by age interaction on Child Gambling Task performance for 3–5-year-olds.

Sex differences have been somewhat clearer in adolescent performance on the IGT with males making more advantageous choices than females (e.g., Crone et al., 2005); however, this finding was not replicated by Hooper et al. (2004). In a fascinating study reported by Overman, Graham, Redmond, Eubank, and Boettcher (2006), the researchers tested several hypotheses posed to explain the male superiority that had been found on this measure of hot executive function. The results indicated that requiring participants to consider "personal moral dilemmas" concurrently with decision making on the IGT brought female performance more in line with male performance. That is, the typical male superiority in the selection of advantageous cards (i.e., less risky decision making) disappeared when deliberation of personal moral dilemmas was coincident with the task. The authors speculate that brain regions involved in moral decision making, specifically the dorsolateral prefrontal cortex, were activated in this experimental condition. Moreover, the recruitment of this prefrontal cortical region, associated with cool executive functions, facilitated the performance on the IGT, the most commonly used measure of hot executive functions. This finding is consistent with a theme permeating this chapter: a single task associated with either hot or cool executive function will likely recruit the cognitive and neurologic mechanisms underlying *both* types of executive function to greater or lesser degrees depending on the particular testing contexts. In this case, females' use of the more "emotional" processing of the orbitofrontal cortex did not serve them well on the IGT, and instead the activation of more "cognitive" processes involved in imagining future consequences facilitated their performance (Overman et al., 2006). Essentially, it appears that the "temperature" of an experimental task not only depends on the testing contexts, but the *perception of these contexts* may likely vary with gender, as well as with other as yet un-indentified individual differences.

Finally, one must examine the degree to which the male superiority on hot executive function tasks, such as the IGT, which involve decision making in risk/reward contexts, aligns with evidence regarding gender differences in risk taking in the real world. More advanced maturation of the orbitofrontal cortex coupled with superior performance on some measures of hot executive function would suggest that males should engage in less risky decision making, or at least more calculated risky decision making, than females. However, one need only look at the statistics regarding accidental death rates by gender (e.g., Centre for Accident Research and Road Safety—Queensland) to question this assumption. In a 1999 meta-analysis of 150 studies examining gender differences in risk taking, Byrnes, Miller, and Schafer found evidence for greater risk taking in males than in females on 14 of 16 indicators. In addition, the researchers found evidence that the gender gap may be diminishing, a finding corroborated by an Australian study demonstrating higher levels of risk taking among females as compared to their mothers' generation (Abbott-Chapman, Denholm, & Wyld, 2008). In fact, contemporary research examining the nature of gender differences in risk-taking behavior has focused less on biological sex and more on the influence of sex role socialization factors (Granié, 2009) and gender-typed beliefs about the developmental tasks of emerging adulthood (Cheah, Trinder, & Govaki, 2010). Therefore, although the current evidence of a male advantage on measures of hot executive function does not appear to converge with our anecdotal or empirical evidence of higher levels of adolescent male risk taking, this picture is complicated by generational and socialization factors that undoubtedly interact in complex ways with the biological differences between males and females.

Summary

In discussing the research areas of compliance and moral development in young children and gender differences in adolescent risk taking, we have selected two lines of research that intersect in interesting ways with hot and cool executive functions, irrespective of whether the specific term "executive function" is ever mentioned. The research by Kochanska and colleagues utilizes tasks involving delay of gratification and prohibition, many of which are the very measures that have become synonymous with hot executive function in early childhood. Their studies have yielded compelling evidence that task performance of young children predicts behaviors in contexts that challenge their moral understanding and decision making. The child's sense of "right and wrong" on these moral decision-making tasks, such as cheating and rule breaking, can be linked to risk-taking behaviors of adolescents in which there is often an element of pushing boundaries, rule violation, and future negative consequences. The adolescent risk-taking research likewise incorporates hot executive function tasks, such as the IGT and delay discounting, into the methodology; however, the nature of the gender differences is so far contradictory in the two research areas. This begs the question: to what extent do hot executive functions, as defined in current research, underlie early compliance and moral development, as well as a tendency towards risky behaviors? Longitudinal studies examining the predictive relationships between both hot and cool executive functions and these important real-world behaviors will be illuminating in this regard.

Conclusions and Future Directions

In spite of the clinical evidence provided by Phineas Gage and countless other frontal patients, the long history of executive function research has stressed cool cognitive control processes that can be observed in laboratory settings with tasks that minimize emotional incentive. Today, however, psychological scientists across a wide range of subdisciplines take seriously the notion that adaptive behavior in real-world contexts involves continuous interactions between emotional and cognitive processes. While many influences likely converged to support this current zeitgeist, the neuropsychological studies highlighting the differing roles of the orbitofrontal/ventromedial and dorsolateral aspects of frontal cortex have played a critical role. The notion of a "dual route" involving a thoughtful, cognitive pathway and a more automatic, emotional pathway has been posited across a range of literatures. Given such a hypothetical neural framework, executive functions researchers have embraced the difficult challenge of examining the full range of hot and cool processes that support adaptive behavior across contexts. This more integrative approach to understanding the goal-directed skills of executive functions in more natural settings represents a new and exciting direction for research that has wide-ranging implications. However, some of the most intractable questions that have long challenged executive functions researchers remain.

Our review of current studies specifically comparing hot and cool executive functions in preschool-aged children suggests that, thus far, this research has not yielded strong behavioral evidence for dissociable constructs. The case for separable hot and cool executive functions in older children and adolescents is somewhat more compelling, as reflected by a more protracted developmental trajectory for IGT, as compared to cool tasks; however, it is unclear whether task differences other than "temperature," such as cognitive demands, may be responsible for the later maturation of the hot task. Studies in children and adolescents that have included both the IGT and a delay discounting task highlight some of the limitations of our current tasks. Across several studies, performance on these two presumably hot tasks has not been correlated. Further, they have not consistently both yielded the same degree of evidence for development.

A close examination of the presumably hot task, delay discounting, may illustrate some problems for resolution in future research. That delay discounting has important predictive value for outcome is very clear. Although our review focused on research involving children and adolescents, a review of delay discounting among participants in college or beyond makes clear that this measure relates to both academic outcome (e.g., grade point average, Kirby, Winston, & Santiesteban, 2005) and broader adaptive functioning (e.g., likelihood of having substance abuse problem, Kollins, 2003). However, meta-analytic evidence has established clearly that general intelligence contributes to individual differences in discounting behavior (Shamosh & Gray, 2008; see also, Shamosh et al., 2008). A more recent developmental study (Anokhin, Golosheykin, Grant, & Heath, 2011) demonstrated that discounting behavior is associated with personality (e.g., novelty seeking) and family socioeconomic status. Thus, it seems clear that discounting behavior involves both cool and hot processes and the relative contribution of each may differ within an individual. Presumably, other tasks like the IGT also involve hot and cool processes mediated by different neural regions, an assumption that is consistent with the appreciation that executive tasks are likely to involve both hot and cool processes to differing degrees (e.g., Hongwanishkulh, Happeney, Lee & Zelazo, 2005). Clearly, in the absence of a more precise understanding of the component processes that contribute to overall performance in each task, across-task comparisons are difficult.

The evidence for differential associations between cool and hot task performance and real-world behaviors, such as academic performance and behavioral regulation, respectively, also has not clearly aligned with predictions in childhood or adolescence (e.g., Brock et al., 2009; Duckworth & Seligman, 2005; Willoughby et al., 2011). Again, the hypotheses regarding which real-world behaviors and contexts should be predicted by hot and cool executive functions presume that we have a clear and unambiguous definition of the nature of these behaviors and contexts in terms of affective and motivational significance. It is likely that the temperature of both the experimental tasks and these adaptive behaviors will vary across individuals and situations. In light of this, the most promising direction for research may be the use of paradigms in which the specific task is held constant, and contextual features are manipulated in order to systematically turn up, or down, the temperature of the task. The small set of studies that have utilized this strategy have demonstrated that cooling down a hot executive function task improves young children's performance (Carlson et al., 2005) and heating up tasks have a more substantial negative impact on older children's performance, relative to adolescents (e.g., Crone, Bullens, van der Plas, & Zelazo, 2008). A powerful approach to addressing the hypothesized differential development of hot and cool processes would be to administer the same task (e.g., a hot task such as IGT) across a wide age span with systematic temperature manipulations. The ages at which these manipulations significantly impact performance on the task would be illuminating with regard to whether the two executive function processes demonstrate different trajectories. Moreover, examining the associations between these manipulated versions of the executive function task and an outcome, such as academic performance, that itself has identified hot and cool components, holds promise for establishing links between executive function and successful behavior in real-world contexts.

A limitation of this review is that our focus centered on behavioral studies, and not on research examining the dissociability of the processes on a neurophysiological level. As described in our historical review, the genesis of the hot vs. cool distinction was the neuropsychological examination of adults with dorsolateral vs. orbitofrontal damage. To date, some studies (e.g., Eshel, Nelson, Blair, Pine, & Ernst, 2007; Galvan et al., 2006; Perlman & Pelphrey, 2011) have examined brain mechanisms that correlate with behavior during task performance. This is an important direction for continued research. Studies in which temperature is manipulated within task may be particularly revealing with the addition of direct measures of brain activation.

While acknowledging the many theoretical and empirical impediments to research examining the validity of the hot and cool executive function constructs, the *value* of the integration of cold cognition with emotional and motivational forces in our conceptualization of this critically important ability should not be dismissed. Our review of two exemplar areas of research, young children's compliance and moral behavior and adolescent risk taking, highlights the need for a more comprehensive perspective on executive function to understand the complex interweaving of cognitive skill, emotional impetus, and motivational drive that undoubtedly comprise the developmental and individual differences observed. Current and future examinations of executive function, in all its rich complexity, will not only inform our understanding of brain function and development, but also our appreciation for the mechanisms underlying an individual's ability to consider the consequences of decisions made in natural contexts that potentially optimize or impede adaptive development.

References

Abbott-Chapman, J., Denholm, C., & Wyld, C. (2008). Gender differences in adolescent risk taking: Are they diminishing?: An Australian intergenerational study. *Youth & Society, 40*, 131–154.

Anderson, V. (2002). Executive function in children: Introduction. *Child Neuropsychology, 8*, 69–70.

Anokhin, A. P., Golosheykin, S., Grant, J. D., & Heath, A. C. (2011). Heritability of delay discounting in adolescence: A longitudinal twin study. *Behavior Genetics, 41*(2), 175–183.

Atkinson, R. C., & Shiffrin, R. M. (1968). Human memory: A proposed system and its control processes. In K. Spence & J. T. Spence (Eds.), *The psychology of learning and motivation* (Vol. 2, pp. 89–1952). New York: Academic.

Bechara, A. (2004). The role of emotion in decision making: Evidence from neurological patients with orbitofrontal damage. *Brain and Cognition, 55*, 30–40.

Bechara, A., Damasio, H., & Damasio, A. R. (2000). Emotion, decision-making, and the orbitofrontal cortex. *Cerebral Cortex, 10*, 295–307.

Bechara, A., Damásio, A. R., Damásio, H., & Anderson, S. W. (1994). Insensitivity to future consequences following damage to human prefrontal cortex. *Cognition, 50*(1–3), 7–15.

Beer, J. S. (2006). Orbitofrontal cortex and social regulation. In J. T. Cacioppa, P. S. Visser, & C. L. Picket (Eds.), *Social neuroscience: People thinking about thinking people* (pp. 153–166). Cambridge: MIT Press.

Best, J. R., & Miller, P. H. (2010). A developmental perspective on executive function. *Child Development, 81*(6), 1641–1660.

Brock, L. L., Rimm-Kaufman, S. E., Nathanson, L., & Grimm, K. J. (2009). The contributions of 'hot' and 'cool' executive function to children's academic achievement, learning-related behaviors, and engagement in kindergarten. *Early Childhood Research Quarterly, 24*, 337–349.

Brookshire, B., Levin, H. S., Song, J. X., & Zhang, L. (2004). Components of executive function in typically developing and head-injured children. *Developmental Neuropsychology, 25*, 61–83.

Carlson, S. M. (2005). Developmentally sensitive measures of executive function in preschool children. *Developmental Neuropsychology, 28*(2), 595–616.

Carlson, S. M., Davis, A. C., & Leach, J. G. (2005). Less is more: Executive function and symbolic representation in preschool children. *Psychological Science, 16*(8), 609–616.

Carlson, S. M., & Wang, T. S. (2007). Inhibitory control and emotion regulation in preschool children. *Cognitive Development, 22*, 489–510.

Casey, B. J., Somerville, L. H., Gotlib, I. H., Ayduk, O., Franklin, N. T., Askren, M. K., et al. (2011). Behavioral and neural correlates of delay of gratification 40 years later. *Proceedings of the National Academy of Sciences, 108*(36), 14998–15003.

Cheah, C. S. L., Trinder, K. M., & Govaki, T. N. (2010). Urban/rural gender differences among Canadian emerging adults. *International Journal of Behavioral Development, 34*, 339–344.

Chelune, G. J., & Baer, R. L. (1986). Developmental norms for the Wisconsin Card Sorting Test. *Journal of Clinical and Experimental Neuropsychology, 8*, 219–228.

Cohen, J. D., & Servan-Schreiber, D. (1992). Context, cortex, and dopamine: A connectionist approach to behavior and biology in schizophrenia. *Psychological Review, 99*, 45–77.

Crone, E. A. (2009). Executive functions in adolescence: Inferences from brain and behavior. *Developmental Science, 12*(6), 825–830.

Crone, E. A., Bullens, L., Van der Plas, A., Kijkuit, E. J., & Zelazo, P. D. (2008). Developmental changes and individual differences in risk and perspective taking in adolescence. *Development and Psychopathology, 20*, 1213–1229.

Crone, E. A., Bunge, S. A., Latenstein, H., & Van der Molen, M. (2005). Characterization of children's decision making: Sensitivity to punishment frequency, not task complexity. *Child Neuropsychology, 11*, 245–263.

Crone, E. A., & Van der Molen, M. (2004). Developmental changes in real life decision making: Performance on gambling task previously shown to depend on the ventromedial prefrontal cortex. *Developmental Neuropsychology, 25*(3), 251–279.

Duckworth, A. I., & Seligman, M. E. P. (2005). Self-discipline outdoes IQ in predicting academic performance of adolescents. *Psychological Science, 16*(12), 939–944.

Duncan, J., & Owen, A.M. (2000). Common regions of the human frontal lobe recruited by diverse cognitive demands. *Trends in Neurosciences, 23*(10), 475–483.

Duncan, J., Burgess, P., & Emslie, H. (1995). Fluid intelligence after frontal lobe lesions. *Neuropsychologia, 33*, 261–268.

Eshel, N., Nelson, E. E., Blair, R. J., Pine, D. S., & Ernst, M. (2007). Neural substrates of choice selection in adults and adolescent: Development of the ventrolateral prefrontal and anterior cingulate cortices. *Neuropsychologia, 45*, 1270–1279.

Espy, K. A., & Kaufmann, P. M. (2002). Individual differences in the development of executive function in children: Lessons from the delayed response and A-not-B tasks. In D. L. Molfese & V. J. Molfeses (Eds.), *Developmental variations in learning: Applications to social, executive function, language, and reading skills* (pp. 113–137). Mahwah, NJ: Lawrence Erlbaum.

Figner, B., Mackinlay, R. J., Wilkening, F., & Weber, E. U. (2009). Affective and deliberative processes in risky choice: Age differences in risk taking in the Columbia Card Task. *Journal of Experimental Psychology, 35*(3), 709–730.

Fuster, J.M. (1989). *The prefrontal cortex* (2nd Ed.). New York: Raven Pres.

Galvan, A., Hare, T. A., Parra, C. E., Penn, J., Voss, H., Glover, G., et al. (2006). Earlier development of the accumbens relative to orbitofrontal cortex might underlie risk-taking behavior in adolescents. *The Journal of Neuroscience, 26*(25), 6885–6892.

Garon, N., Bryson, S. E., & Smith, I. M. (2008). Executive function in preschoolers: A review using an integrative framework. *Psychological Bulletin, 134*(1), 31–60.

Giedd, J.N., Blumenthal, J., Jeffries, N.O., Castellanos, F.X., Liu, H., Zijdenbos, A., Paus, T., Evans, A.C., & Rapoport, J.L. (1999). Brain development during childhood and adolescence: a longitudinal MRI study. *Nature Neuroscience, 2*(10), 861–863.

Golden, C. J. (1981). The Luria-Nebraska Children's Battery: Theory and formulation. In G. W. Hynd & J. E. Obrzut (Eds.), *Neuropsychological assessment and the school-aged child* (pp. 277–302). New York: Grune & Stratton.

Granié, M.-A. (2009). Effects of gender, sex-stereotype conformity, age and internalization on risk-taking among adolescent pedestrians. *Safety Science, 47*, 1277–1283.

Grant, D. A., & Berg, E. A. (1948). A behavioral analysis of reinforcement and ease of shifting to new responses in a Weigel-type card-sorting problem. *Journal of Experimental Psychology, 38*, 404–411.

Green, J., & Haidt, J. (2002). How (and where) does moral judgment? *Trends in Cognitive Sciences, 6*(12), 517–523.

Green, L., Fry, A. F., & Myerson, J. (1994). Discounting of delayed rewards: A life-span comparison. *Psychological Science, 5*(1), 33–36.

Green, L., Fry, A. F., & Myerson, J. (1994). Discounting of delayed rewards: A life-span comparison. *Psychological Science, 5*(1), 33–36.

Happaney, K., Zelazo, P. D., & Stuss, D. T. (2004). Development of orbitofrontal function: Current themes and future directions. *Brain and Cognition, 55*, 1–10.

Hongwanishkulh, D., Happaney, K. R., Lee, W. S. C., & Zelazo, P. D. (2005). Assessment of hot and cool executive function in young children: Age-related changes and individual differences. *Developmental Neuropsychology, 28*(2), 617–644.

Hooper, C. J., Luciana, M., Conklin, H. M., & Yarger, R. S. (2004). Adolescents' performance on the Iowa Gambling Task: Implications for the development of decision making and ventromedial prefrontal cortex. *Developmental Psychology, 40*(6), 1148–1158.

Huizinga, M., Dolan, C. V., & van der Molen, M. V. (2006). Age-related change in executive function: Developmental trends and a latent variable analysis. *Neuropsychologia, 44*, 2017–2036.

Kerr, A., & Zelazo, P. D. (2004). Development of "hot" executive function: The children's gambling task. *Brain and Cognition, 55*, 148–157.

Kimberg, D. Y., & Farah, M. J. (1993). A unified account of cognitive impairments following frontal lobe damage: The role of working memory in complex, organized behavior. *Journal of Experimental Psychology. General, 122*, 411–428.

Kirby, K. N., Winston, G. C., & Santiesteban, M. (2005). Impatience and grades: Delay discount rates correlate negatively with college GPA. *Learning and Individual Differences, 15*, 213–222.

Kochanska, G. (1994). Beyond cognition: Expanding the search for the early roots of internalization and conscience. *Developmental Psychology, 30*, 20–22.

Kochanska, G. (1997). Multiple pathways to conscience for children with different temperaments: From toddlerhood to age 5. *Developmental Psychology, 33*, 228–240.

Kochanska, G. (2002). Committed compliance, moral self, and internalization: A mediational model. *Developmental Psychology, 38*, 339–351.

Kochanska, G., Aksan, N., Penney, S. J., & Doobay, A. F. (2007). Early positive emotionality as a heterogeneous trait: Implications for children's self-regulation. *Journal of Personality and Social Psychology, 93*, 1054–1066.

Kollins, S. H. (2003). Delay discounting is associated with substance use in college students. *Addictive Behaviors, 28*, 1167–1173.

Kuczynski, L., & Kochanska, G. (1990). Development of children's noncompliance strategies from toddlerhood to age 5. *Developmental Psychology, 26*, 398–408.

Kuczynski, L., Kochanska, G., Radke-Yarrow, M., & Girnius-Brown, O. (1987). A developmental interpretation of young children's noncompliance. *Developmental Psychology, 23*, 799–806.

Lamm, C., Zelazo, P. D., & Lewis, M. D. (2006). Neural correlates of cognitive control in childhood and adolescence: Disentangling the contributions of age and executive function. *Neuropsychologia, 44*, 2139–2148.

Lehto, J. E., Juujärvi, P., Kooistra, L., & Pulkkinen, L. (2003). Dimensions of executive functioning: Evidence from children. *The British Journal of Developmental Psychology, 21*, 59–80.

Lewis, M. D., Lamm, C., Segalowitz, S. J., Stieben, J., & Zelazo, P. D. (2006). Neurophysiological correlates of emotion regulation in children and adolescents. *Journal of Cognitive Neuroscience, 18*(3), 430–443.

Lezak, M. D. (1995). *Neuropsychological assessment* (3rd ed.). New York: Oxford University Press.

Luciano, M. (2003). The neural and functional development of human prefrontal cortex. In M. de Haan & M. H. Johnson (Eds.), *The cognitive neuroscience of development* (pp. 157–180). New York: Psychology Press.

Luria, A. R. (1966). *Higher cortical functions in man*. New York: Basic.

Luria, A. R. (1973). *The working brain*. New York: Basic.

Metcalfe, J., & Mischel, W. (1999). A hot/cool-system analysis of delay of gratification: Dynamics of willpower. *Psychological Review, 106*(1), 3–19.

Miller, E. K., & Cohen, J. D. (2001). An integrative theory of prefrontal cortex function. *Annual Review of Neuroscience, 24*, 167–202.

Milner, B. (1963). Effects of different brain lesions on card sorting. *Archives of Neurology, 9*, 90–100.

Mischel, W., Ebbeson, E. B., & Ziess, A. R. (1972). Cognitive and attentional mechanisms in delay of gratification. *Journal of Personality and Social Psychology, 16*, 204–218.

Mischel, W., & Metzner, R. (1962). Preference for delayed reward as a function of age, intelligence, and length of delay interval. *Journal of Abnormal and Social Psychology, 64*(6), 425–431.

Mischel, W., Shoda, Y., & Peak, P. K. (1988). The nature of adolescent competencies predicted by preschool delay of gratification. *Journal of Personality and Social Psychology, 54*, 687–696.

Mischel, W., Shoda, Y., & Rodriguez, M. L. (1989). Delay of gratification in children. *Science, 244*(4907), 933–938.

Miyake, A., Freidman, N. P., Emerson, M. J., Witzki, A. H., & Howerter, A. (2000). The unity and diversity of executive functions and their contributions to complex "frontal lobe" tasks: A latent variable analysis. *Cognitive Psychology, 41*, 49–100.

Norman, D. A., & Shallice, T. (1986). Attention to action: Willed and automatic control of behavior. In R. J. Davidson, G. E. Schwartz, & D. Shapiro (Eds.), *Consciousness and self-regulation. Advances in research and theory* (Vol. 4, pp. 1–18). New York: Plenum.

O'Reilly, R. C. (2010). The what and how of prefrontal cortical organization. *Trends in Neurosciences, 33*(8), 355–361.

Overman, W. H., Bachevalier, J., Schuhmann, E., & Ryan, P. (1996). Cognitive gender differences in very young children parallel biologically based cognitive gender differences in monkeys. *Behavioral Neuroscience, 110*, 673–684.

Overman, W., Graham, L., Redmond, A., Eubank, R., & Boettcher, L. (2006). Contemplation of moral dilemmas eliminates sex differences on the Iowa Gambling Task. *Behavioral Neuroscience, 120*, 817–825.

Overman, W. H., Frassrand, K., Ansel, S., Trawalter, S., Bies, B., & Redmond, A. (2004). Performance on the IOWA card task by adolescents and adults. *Neuropsychologia, 42*, 1838–1851.

Peake, P. K., Hebl, M., & Mischel, W. (2002). Strategic attention deployment for delay of gratification in working and waiting situations. *Developmental Psychology, 38*(2), 313–326.

Perlman, S. B., & Pelphrey, K. A. (2011). Developing connections for affective regulation: Age-related changes in emotional brain connectivity. *Journal of Experimental Child Psychology, 108*, 607–620.

Pharo, H., Sim, C., Graham, M., Gross, J., & Hayne, H. (2011). Risky business: Executive function, personality, and reckless behavior during adolescence and emerging adulthood. *Behavioral Neuroscience, 125*, 970–978.

Piaget, J. (1972). Intellectual evolution from adolescence to adulthood. *Human Development, 15*, 1–12.

Piek, J. P., & Dyck, M. J. (2004). Sensory-motor deficits in children with developmental coordination disorder, attention deficit hyperactivity disorder and autistic disorder. *Human Movement Science, 23*(3–4), 475–488.

Prencipe, A., Kesek, A., Cohen, J., Lamm, C., Lewis, M. D., & Zelazo, P. D. (2011). Development of hot and cool executive function during the transition to adolescence. *Journal of Experimental Child Psychology, 108*, 621–637.

Pribram, K. H. (1969). The amnestic syndrome: Disturbances in coding? In G. A. Talland & N. C. Waugh (Eds.), *The pathology of memory* (pp. 127–157). New York: Academic.

Romine, C. B., & Reynolds, C. R. (2005). A model of the development of frontal lobe functioning: Findings from a meta-analysis. *Applied Neuropsychology, 12*, 190–201.

Shamosh, N. A., DeYoung, C. G., Green, A. E., Reis, D. L., Johnson, M. R., Conway, A. R. A., et al. (2008). Individual differences in delay discounting: Relation to intelligence, working memory, and anterior prefrontal cortex. *Psychological Science, 19*(9), 904–911.

Shamosh, N. A., & Gray, J. R. (2008). Delay discounting and intelligence: A meta-analysis. *Intelligence, 36*(4), 289–305.

Sowell, E.R., Thompson, P.M., Tessner, K.D., & Toga, A.W. (2001). Mapping Continued Brain Growth and Gray Matter Density Reduction in Dorsal Frontal Cortex: Inverse Relationships during Post-adolescent Brain Maturation. *The Journal of Neuroscience, 21*(22), 8819.

Steinberg, L. (2005). Cognitive and affective development in adolescence. *Trends in Cognitive Science, 9*(2), 69–74.

Stuss, D. T., & Benson, D. F. (1984). Neuropsychological studies of frontal lobes. *Psychological Bulletin, 95*, 3–28.

Tueber, H. L. (1964). The riddle of frontal lobe function in man. In J. Warren & K. Akert (Eds.), *The frontal granular cortex and behavior* (pp. 410–440). New York: McGraw-Hill.

Visser, J. (2003). Developmental coordination disorder: A review of research on subtypes and comorbidities. *Human Movement Science, 22*(3–4), 475–488.

Welsh, M. C. (2001). The prefrontal cortex and the development of executive functions. In A. Kalverboer & A. Gramsbergen (Eds.), *Handbook of brain and behaviour development* (pp. 767–789). The Netherlands: Kluwer.

Welsh, M. C., Friedman, S. L., & Spieker, S. J. (2006). Executive functions in developing children: Current conceptualizations and questions for the future. In K. McCartney & D. Phillips (Eds.), *Blackwell handbook of early childhood development* (pp. 167–187). Malden: Blackwell Publishing.

Welsh, M. C., & Pennington, B. F. (1988). Assessing frontal lobe functioning in children: Views from developmental psychology. *Developmental Neuropsychology, 4*, 199–230.

Welsh, M. C., Pennington, B. F., & Groisser, D. B. (1991). A normative-developmental study of executive function: A window on prefrontal function in children. *Developmental Neuropsychology, 7*, 131–149.

White, J. L., Moffitt, T. E., Caspi, A., Bartusch, D. J., Needles, D. J., & Stouthamer-Loeber, M. (1994). Measuring impulsivity and examining its relationship to delinquency. *Journal of Abnormal Psychology, 103*(2), 192–205.

Willoughby, M., Kupersmidt, J., Voegler-Lee, M., & Bryant, D. (2011). Contributions of hot and cool self-regulation to preschool disruptive behavior and academic achievement. *Developmental Neuropsychology, 36*(2), 162–180.

Zajonc, R. B. (1980). Feeling and thinking: Preferences need no inferences. *American Psychologist, 35*(2), 151–175.

Zelazo, P. D., Craik, F. I. M., & Booth, L. (2004). Executive function across the life span. *Acta Psychologica, 115*, 167–183.

Zelazo, P. D., & Frye, D. (1998). Cognitive complexity and control: II. The development of executive function in children. *Current Directions in Psychological Science, 7*, 121–126.

Zelazo, P. D., & Muller, U. (2002). Executive function in typical and atypical development. In U. Goswami (Ed.), *Blackwell handbook of childhood cognitive development*. Oxford: Blackwell Publishers.

Zelazo, P. D., Qu, L., & Müller, U. (2005). Hot and cool aspects of executive function: Relations in early development. In W. Schneider, R. Schumann-Hengsteler, & B. Sodian (Eds.), *Young children's cognitive development: Interrelationships among executive functioning, working memory, verbal ability, and theory of mind* (pp. 71–93). Mahwah, NJ: Lawrence Erlbaum.

Part II

Practical Implications

A Review of the Use of Executive Function Tasks in Externalizing and Internalizing Disorders

Lisa L. Weyandt, W. Grant Willis, Anthony Swentosky, Kimberly Wilson, Grace M. Janusis, Hyun Jin Chung, Kyle Turcotte, and Stephanie Marshall

Executive function (EF) is a complex construct that encompasses a variety of cognitive abilities that allow for impulse control, strategic planning, cognitive flexibility, and goal-directed behavior. Executive functions have been studied in nearly every major childhood disorder including externalizing and internalizing disorders. A universally accepted definition of EF does not exist, and many have criticized the broad definitions of the construct. For example, Pennington and Ozonoff (1996) noted, "in both neuropsychology and cognitive psychology, the definition of EFs is provisional and under-specified" (p. 55). Fletcher (1996) also acknowledged that EFs are difficult to define and described EFs as "factorially complex." More recently, Jurado and Rosselli (2007) acknowledged that the fundamental question of "whether there is one single underlying ability that can explain all the components of executive functioning or whether these components constitute related but distinct processes" remains unanswered. To complicate matters, a large variety of tasks that purportedly measure executive functions have been used in the literature. What remains unclear is specifically which executive function tasks are used most often in the literature and on which executive tasks are groups most likely to differ? Hence, the purpose of this review is to conduct a systematic search of the childhood internalizing and externalizing literature to determine (a) executive function tasks that are used in the literature, (b) executive function tasks that are most commonly used, (c) executive function tasks on which clinical and control groups differ most frequently, and (d) executive function tasks on which clinical groups differ most frequently. To begin, a review will be provided regarding executive function performance of children with commonly diagnosed externalizing and internalizing disorders. Next, specific findings regarding the type, usage, and discriminant ability of executive function tasks will be presented followed by implications and suggestions for future research.

Attention-Deficit/Hyperactivity Disorder

Attention-deficit/hyperactivity disorder (ADHD) is characterized by developmentally inappropriate levels of impulsivity and hyperactivity, and attention deficits, and affects 3–7 % of the school-age population (American Psychiatric Association, 2013). Executive functions have been studied extensively in children, and in general, studies have found that children with ADHD tend to perform poorly on EF tasks relative to nondisabled peers and these deficits may begin early in life (e.g., Barkley, Edwards, Laneri, Fletcher, & Metevia, 2001; Barkley, Murphy, &

L.L. Weyandt (✉) • W.G. Willis • A. Swentosky
K. Wilson • G.M. Janusis • H.J. Chung • K. Turcotte
S. Marshall
University of Rhode Island, Kingston, RI, USA
e-mail: lisaweyandt@uri.edu

Bush, 2001; Fuggetta, 2006; Klimkeit, Mattingley, Sheppard, Lee, & Bradshaw, 2005; Nigg, Blaskey, Huang-Pollock, & Rappley, 2002; Seidman, Biederman, Faraone, Weber, & Ouellete, 1997; Weyandt, Rice, Linterman, Mitzlaff, & Emert, 1998; Willcutt, Doyle, Nigg, Faraone, & Pennington, 2005). For example, preliminary studies have found that preschoolers with ADHD demonstrated EF impairments relative to their peers (Byrne, DeWolfe, & Bawden, 1998; DuPaul, McGoey, Eckert, & VanBrakle, 2001; Mahone, Pillion, & Heimenz, 2001) and inhibition problems in preschool may be predictive of EF deficits and ADHD in later childhood (Berlin, Bohlin, & Rydell, 2003; Friedman et al., 2007). With regard to long-term outcome, Biederman et al. (2007) completed a 7-year follow-up study of 85 males with ADHD and reported that the majority (69 %) maintained EF deficits into adulthood. Others (e.g., Fischer, Barkley, Smallish, & Fletcher, 2005; Hinshaw, Carte, Fan, Jassy, & Owens, 2007; Rinsky & Hinshaw, 2011) have conducted similar longitudinal studies of children with ADHD into adolescence and adulthood, and collectively, these findings suggest that EF deficits may emerge early in life in children with ADHD and the impairments are likely to persist into adolescence and possibly adulthood.

Not all studies have found EF deficits in children with ADHD, however, and impairments are commonly found on *some* but not all EF measures (Barkley, Grodzindky, & DuPaul, 1992; Berlin, Bohlin, Nyberg, & Janols, 2004; Geurts, Verte, Oosterlaan, Roeyers, & Sergeant, 2005; Lawrence et al., 2002; Rhodes, Coghill, & Matthews, 2005; Seidman, Biederman, Weber, Hatch, & Faraone, 1998; Tsal, Shalev, & Mevorach, 2005; Weyandt, 2004; Weyandt & Willis, 1994). These findings raise questions about the specificity and sensitivity of EF tasks and collectively suggest that ADHD is not associated with global deficits in EF as has been frequently reported in the literature, but may, however, be characterized by specific EF deficits (Barkley, 2010; Pennington & Ozonoff, 1996; Sergeant, Geurts, & Oosterlaan, 2002; Weyandt, 2005; Wu, Anderson, & Castiello, 2002). The specific EF components that might be compromised in ADHD are equivocal, although response inhibition has been implicated in multiple studies (Barkley, 1997, 2010; Denckla, 1996; Mahone & Hoffman, 2007; Wu et al., 2002). As noted by Weyandt (2009), the inconsistencies across studies may be due in part to methodological factors including sample size, statistical power, inclusion and diagnostic criteria used for ADHD, subtypes of ADHD, EF tasks employed, psychometric properties of the EF tasks, age, sex, ethnicity, comorbidity, intelligence, and statistical methods used to analyze data.

In summary, the literature suggests that EF deficits are not necessarily unique to ADHD and they are not necessary or sufficient for a diagnosis of ADHD. In addition, when EF impairments are present in children with ADHD, they tend to be specific rather than global impairments.

Conduct Disorder and Oppositional Defiant Disorder

Conduct disorder (CD) is characterized by a "persistent pattern of behavior in which the basic rights of others or major age-appropriate societal norms or rules are violated" (DSM-IV-TR, APA, 2000). CD is diagnosed more frequently in males than females and is estimated to affect 1–10 % of the child population (DSM-IV-TR, APA, 2000). Oppositional defiant disorder (ODD) is defined as a "recurrent pattern of negativistic, defiant, disobedient, and hostile behavior toward authority figures." Like CD, ODD occurs more frequently in males than females and affects 2–16 % of the child population (DSM-IV-TR, APA, 2000).

Compared to the ADHD literature, fewer studies have explored EFs in children with CD or ODD. Earlier studies by Moffitt and Henry (1989) and McBurnett et al. (1993) found that children with ADHD and comorbid CD displayed EF deficits but not children with CD only. Speltz, DeKlyen, Calderon, Greenberg, and Fisher (1999) examined EF performance in preschoolers with ADHD and ODD or ODD alone and found that those with ADHD and ODD performed more poorly on two EF measures (Motor Planning Task and the Verbal Fluency subtest of the McCarthy Scales of Children's Abilities) compared to preschoolers with ODD only. Clark, Prior, and Kinsella (2000) found similar results with adolescents with ADHD relative to adolescents with CD or ODD. Giancola, Mezzich, and Tarter (1998), however, found that females with CD displayed EF deficits relative to a control group as measured by overall performance on seven neuropsychological tasks. Unfortunately, given the analyses used in the study (i.e., Principal Components Analysis), group performance on specific EF tasks was not reported.

More recently, Herba, Tranah, Rubia, and Yule (2006) found that adolescents with conduct problems demonstrated EF impairments on a motor response inhibition task (i.e., Stop Task) but not on other EF measures. It is important to note that these children were identified in schools based on rating scales and did not necessarily have diagnosed CD. In addition, similar to the ADHD literature, many of the children had comorbid attention problems, and therefore it is difficult to determine the degree to which attention and impulsivity contributed to the findings rather than conduct related issues. Kim, Kim, and Kwon (2001) also reported that adolescents with CD displayed EF deficits on an inhibition task (i.e., Wisconsin Card Sorting Test, WCST) relative to a control group, but differences were not found on additional EF measures (Visual Performance Test, Contingent Continuous Performance Test, Stroop Test, Spatial Memory Test, and Recognition Test). Toupin, Dery, Pauze, Mercier, and Fortin (2000), however, found that children with CD displayed significant impairments on four of five EF tasks (WCST number of preservative errors, WCST number of preservative responses, Rey-Osterrieth Complex Figure (ROCF) copy accuracy, and Stroop number of word colors) even after ADHD and socio-economic status were statistically controlled. Morgan and Lilienfeld (2000) conducted a meta-analysis that included 39 studies (4,589 participants) that examined EF and CDs. Studies that included one or more of six commonly used EF tasks were included in the meta-analyses (Porteus Mazes, Category Test, Stroop Test, WCST, Verbal Fluency tests, and Trail Making Test). Overall, Morgan and Lilienfeld found those with conduct problems performed worse than control participants on EF measures. The effect sizes ranged substantially, however, depending on the specific task. Lastly, Sergeant et al. (2002) conducted a selective review of CD, ODD, and ADHD studies and concluded that deficits in EF are not unique to ADHD and that children with CD and ODD often display inhibition deficits on EF measures.

In conclusion, earlier reviews reported that EF deficits were not characteristic of children and adolescents with CD after comorbid ADHD was factored out (e.g., Pennington & Ozonoff, 1996). More recent studies, however, suggest that inhibition deficits may be characteristic of both ADHD and CD but whether children with CD display impairments on additional EF measures is equivocal. Similar to the ADHD literature, methodological problems characterize many of the CD/ODD studies including differences in inclusionary criteria, diagnostic criteria, age, gender, and measurement variables.

Tourette's Disorder

Tourette's disorder (TD) is characterized by multiple motor tics and one or more vocal tics and is estimated to affect 5–30 children per 10,000 (American Psychiatric Association, 2000). The onset of TD is before age 18 and the disorder occurs more often in males than females. Compared to ADHD and CD/ODD, very few studies are available concerning EFs in children with TD. Of the studies that have been conducted, no consistent EF finding has emerged. A few studies have found that children with TD display a slower and/or more variable reaction time on continuous performance tests compared to children without TD, but some have questioned the role attention problems may have played in these findings (e.g., Harris et al., 1995; Schuerholz, Singer, & Denckla, 1998; Shucard, Benedict, Tekok-Kilic, & Lichter, 1997). In an effort to address the comorbidity issue, Harris et al. (1995) compared the performance of children with TD only, children with ADHD only, and children with both ADHD and TD on ten EF tasks (including Test of Variables of Attention

(TOVA), WCST, ROCF, Multilingual Aphasia Examination-Controlled Word Fluency subtest). Results revealed that children with ADHD and children with ADHD plus TD performed more poorly on EF tasks compared to children with TD only (although children with TD also displayed EF impairments). After controlling for IQ, scores on the ROCF were significantly worse in TD plus ADHD than TD only. A control group was not included in this study, however, precluding comparison with nondisabled children.

Channon, Pratt, and Robertson (2003) also compared EF performance in three groups—those with TD only, TD and ADHD, and TD and obsessive-compulsive disorder (OCD). Results revealed that those with TD and ADHD performed poorly on several EF measures (e.g., Six Elements Test, Hayling category A and B), whereas those with TD performed poorly on only one EF measure (e.g., Hayling Test category A) as compared to the control group. Similar results were reported by Schuerholz et al. (1998) who compared the performance of children with TD only, ADHD only, and TD and ADHD and a comparison group and found girls with TD performed lower on Letter Word Fluency than children in the other groups. Furthermore, girls with TD and ADHD had the greatest variability of reaction time on the TOVA and were slowest on the Letter Word Fluency. Mahone et al. (2002) also reported that children with TD plus ADHD and children with ADHD only demonstrated poorer performance on the five measures of the Behavior Rating Inventory of Executive Function (BRIEF). The TD only group did not differ from comparison children (or children with ADHD). Similarly, Chang, McCracken, and Piacentini (2007) administered EF tasks to children who had TD alone, OCD alone, and control participants and did not find significant differences in performance across the three groups. The authors noted that those with TD showed "trends toward impairments" in EF but the findings were not robust. Ozonoff and Strayer (2001) compared the performance of children with TD, children with autism, and comparison children on working memory tasks and also did not find group differences. Other studies have also reported no differences between adolescents with TD and control participants on working memory tasks (e.g., Crawford, Channon, & Robertson, 2005).

In an earlier study, Ozonoff and Jensen (1999) compared EF performance using the WCST, Tower of Hanoi (TOH), and the Stroop Color-Word Test in three groups of children, those with TD only, ADHD only, and autism only, and nondisabled comparison children. Results revealed that children with TD did not show impairments on any of the EF tasks, children with ADHD showed impairment on only one task (Stroop), and children with autism showed deficits on two of the EF tasks (WCST and TOH). Cirino, Chapieski, and Massman (2000) compared the WCST performance of children and adolescents with TD only to children with ADHD and comorbid TD and did not find differences between the two groups.

In summary, no clear pattern of EF deficits emerges in the literature concerning children with TD. A few studies, but not all, have reported greater response-time variability on continuous performance tasks. Preliminary studies suggest that working memory is not characteristically impaired with children with TD. Some studies suggest that children with TD may have EF deficits particularly with response time and memory search (e.g., poor performance on Hayling Test, timed continuous performance task on TOVA, and Letter Word Fluency), but they may not be as severe as EF deficits in other disorders such as ADHD. Results are equivocal with respect to the performance of children with TD on planning tasks and measures of cognitive flexibility. What is clear is that distinct and robust impairments in EF do not appear to be characteristic of children with TD. Additional, methodologically sound studies are needed to address whether subtle differences in EF may exist between children with TD and other types of childhood psychopathology.

Anxiety Disorders

Taxometric approaches in the field of developmental psychopathology typically identify a general anxious-depressed syndrome within a broader

Table 5.1 Tasks most commonly used to assess executive function

Executive function test	Number of times used	Sensitivity to group differences	Percentage of significant differences between clinical and control groups	Percentage of significant group differences between two clinical groups
Stroop Color and Word Test and variants	41	28/73 = 38 %	22/37 = 59 %	6/36 = 17 %
Wisconsin Card Sorting Test (including computerized and non-computerized versions)	34	75/226 = 33 %	60/139 = 43 %	14/88 = 16 %
Trail Making Test and variants	26	43/121 = 36 %	35/79 = 44 %	8/42 = 19 %
Continuous Performance Test and variants	19	31/72 = 43 %	26/52 = 50 %	5/15 = 33 %
BRIEF	16	177/266 = 67 %	88/104 = 85 %	24/64 = 38 %
Go/No-Go Test	14	37/81 = 46 %	23/41 = 56 %	7/17 = 41 %
Tower of London Test and variants	13	3/75 = 4 %	1/39 = 3 %	2/39 = 5 %
Rey-Osterrieth Complex Figure (ROCF) Test or Rey Complex Figure Test (RCFT)	12	31/93 = 33 %	24/56 = 43 %	7/37 = 19 %

grouping of internalizing disorders (Achenbach & Edelbrock, 1983). As such, unlike the DSM-IV-TR (American Psychiatric Association, 2000), anxiety usually is not differentiated from depression as a separate diagnostic category, and distinctions usually are not made among various types of anxiety disorders. In contrast, the DSM-IV-TR specifically identifies one kind of anxiety disorder (i.e., separation anxiety disorder) as one of the "Disorders Usually First Diagnosed in Infancy, Childhood, or Adolescence" (p. 39) and provides child and adolescent diagnostic criteria for several other anxiety disorders including panic disorder, specific phobia, social phobia, OCD, posttraumatic stress disorder (PTSD), acute stress disorder, and generalized anxiety disorder.

Toren et al. (2000) compared a group of children ($n = 19$; M age = 11.5 years) who had been diagnosed with separation anxiety disorder and overanxious disorder (based on DSM-III criteria) (American Psychiatric Association, 1987) to a group of children who were comparable in terms of age and gender ($n = 14$; M age = 11.5 years) with no history of psychopathology. Despite the common comorbidity of anxiety and depression among children, a strength of this study was that not one of the children in the clinical group met criteria for major depression. Neuropsychological and EF measures included the California Verbal Learning Test (Delis, Cullum, Butters, Cairns, & Prifitera, 1988), the ROCF Test (Osterrieth, 1944), and the WCST (Heaton, 1981; Spreen & Strauss, 1991). Thus, neurocognitive functions assessed included verbal processing, visuospatial processing, and EF.

When the ten measures on the California Verbal Learning Test were analyzed as a single composite (using multivariate analysis of variance techniques), children with anxiety disorders showed verbal-processing deficits relative to children with no history of psychopathology (Toren et al., 2000). Follow-up univariate analyses, however, failed to identify any one of those ten measures as a reliable discriminator when considered separately. The three measures on the ROCF Test failed to discriminate between children with anxiety disorders and those with no history of psychopathology, either as a composite or when considered as separate dependent variables. Similarly, no group differences were found on a composite of the five measures on the WCST. Univariate results, however, showed differences on two WCST measures: total errors and perseverative responses (see Table 5.1). Overall, then, these findings suggest possible generalized deficits in verbal or linguistic abilities and in a set-shifting or cognitive-flexibility component of EF, with sparing of visuospatial organization skills among children with anxiety disorders.

Emerson, Mollet, and Harrison (2005) also found that boys who were anxious and depressed showed EF deficits relative to boys with no psychiatric history on two tests designed to measure set shifting and concept formation. These investigators argued that although anxiety and depression each may contribute separately to neurocognitive processing, the fact that these two categories commonly co-occur highlights the importance of examining the comorbid condition. Expressed here, of course, is the traditional trade-off between internal and external validity issues: on the one hand, a concern for the clinical integrity of nosological categories vs., on the other hand, the value of representativeness and generalizability to existing populations.

A sample of boys ($n = 19$; age = 9–11 years) who scored high (albeit not necessarily at clinical levels) on measures of both depression (i.e., Child Depression Inventory; Kovacs & Beck, 1977) and anxiety (the Trait subscale of the State-Trait Anxiety Inventory for Children; Spielberger, Edwards, Lushene, Montuori, & Platzek, 1973) was compared to a similar group of boys who scored low on both of these measures on Parts A and B of the Trail Making Test and on the Concept Formation subtest of the Woodcock Johnson Test of Cognitive Abilities (Woodcock & Johnson, 1989). Both groups showed similar completion times on Part A and Part B of the Trail Making Test. In contrast, however, they differed in terms of the number of perseverative errors that they made (on both Part A and Part B). Participants in the anxious-depressed group showed lower levels of accuracy, implicating possible impairments in the ability to shift mental set. The two groups also differed in terms of accuracy on the Concept Formation subtest, with those in the anxious-depressed group showing greater degrees of difficulty on a task that requires one to solve problems based on abstract rules of categorization. Although acknowledging the lack of any data at a physiological or anatomical level of analysis, Emerson et al. (2005) speculated that these findings were consistent with evidence implicating frontal lobe EF deficits in children with symptoms of both anxiety and depression.

In a well-designed study of children with PTSD, Beers and De Bellis (2002) also found evidence of EF deficits among children with this anxiety-based disorder. Children ($n = 14$, M age = 11.4 years) who had been identified as maltreated (e.g., sexual abuse, physical abuse, witnessing domestic abuse) by a child-protective-service agency and who subsequently had been diagnosed with PTSD were compared to a similar group of children who were healthy and who had not been maltreated. An extensive battery of neuropsychological tests (described by Spreen & Strauss, 1998) was administered that included two measures of language, six measures of attention, six measures of abstract reasoning/EF, six measures of learning and memory, five measures of visuospatial functioning, and four measures of psychomotor speed. After corrections for multiple significance tests, children with PTSD showed deficits relative to children without PTSD on two of the six measures of attention (i.e., Stroop Color-Word Test: Color/Word, which is a measure of interference control; cf. Doyle et al., 2005; Digit Vigilance Tests: omission errors, which is a measure of sustained attention) and on two of the six measures of abstract reasoning/EF (i.e., WCST, categories completed, which is an EF measure of problem solving and set shifting; cf. Doyle et al., 2005; Controlled Oral Word Association Test, animal naming, which is an EF measure of verbal fluency). There were no differences found on any of the language, learning and memory, visuospatial functioning, or psychomotor-speed tests, suggesting some degree of specificity in terms of the neurocognitive domains that were assessed.

In concluding that these results supported EF differences between children with and without maltreatment-related PTSD, Beers and De Bellis (2002) acknowledged that the sample size studied was relatively small and that these children also experienced comorbid conditions such as major depressive disorder, dysthymic disorder, separation anxiety disorder, and ADHD (inattentive subtype). Moreover, the extent to which results were related to maltreatment or to the presence of psychopathology could not be assessed.

Using neuroimaging technology, Carrion, Garrett, Menon, Weems, and Reiss (2008) found differences between youth with and without post-traumatic stress symptoms when they performed a Go/No-Go task during a functional magnetic resonance imaging (fMRI) scan. The accuracy and response times for the Go/No-Go task were similar between the two groups; however, the control group displayed greater middle frontal cortex activation, whereas the group with post-traumatic stress symptoms demonstrated greater medial frontal activation.

Finally, an anxiety disorder that has received a great deal of research attention, particularly in terms of its neurological basis, is OCD. This disorder is characterized by recurring obsessions or compulsions that are time-consuming or cause significant distress or impairment (American Psychiatric Association, 2000). The perseverative and ego-dystonic nature of cognitive impulses (i.e., obsessions) and the repetitive behaviors that accompany them as an attempt to reduce anxiety (compulsions) have led many to consider the disorder from an executive-dysfunction perspective (see Friedlander & Desrocher, 2006). Once believed to be rare among children, OCD now has been shown to present lifetime prevalence rates that range from about .5 to 2.1 %, with comparable estimates for both children and adults (Evans & Leckman, 2006). The age of onset for childhood OCD ranges from about 6 to 11 years (M age = 10.3 years); Shin et al. (2008) studied children from Korea diagnosed with OCD, ADHD, tic disorder, and depressive disorder, and controls were compared using the Wechsler Intelligence Scale-Revised (WISC-R) and the WCST. The children with OCD performed the worst on the perceptual organization tasks and had significantly more errors and used fewer strategies on the WCST compared to the control group. This study also demonstrated that deficits in EF that are apparent in adults with OCD are similar in children with OCD. Both children and adults with OCD have demonstrated EF deficits in mental set shifting, which provides further evidence for the hypothesis of frontal-striatal dysfunction in individuals with OCD.

Unfortunately, little research has addressed the neuropsychological characteristics of children with OCD. For example, it would be interesting to compare the EF abilities of children with early-onset OCD to children without OCD as well as to adults with early-onset OCD, adults with late-onset OCD, and adults without OCD. Indeed, these kinds of studies could help to clarify if differences between earlier- and later-onset OCD are associated with neurodevelopmental issues, qualitative differences between potential subtypes of the disorder, or perhaps an interaction between these factors.

In summary, the neurophysiological basis of anxiety disorders has been widely studied (e.g., Gray, 1987; Zahn-Waxler, Klimes-Dougan, & Slattery, 2000), and the neuropsychological construct of EF among youth with separation anxiety disorder, overanxious disorder, PTSD, and OCD has received recent attention. Specifically, EF deficits in set shifting, cognitive flexibility, concept formation, interference control, and verbal fluency have been documented among children with separation anxiety disorder, overanxious disorder, and PTSD (Beers & De Bellis, 2002; Emerson et al., 2005; Toren et al., 2000). Little research has addressed the EFs of children with OCD, but studies of adults with early- vs. late-onset OCD subtypes implicate EF deficits associated with working memory and the ability to shift mental set for those with later-onset OCD but not for those with earlier-onset OCD. Additional research is needed to help clarify the relationship between the development of EF in childhood and adolescence and OCD subtypes.

Depression

Although several investigators convincingly have argued that depression is probably better conceptualized as a taxometric continuum vs. a discrete category of psychopathology (Hankin, Fraley, Lahey, & Waldman, 2005; Prisciandaro & Roberts, 2005), the three depressive, unipolar disorders described in the currently used DSM-IV-TR

(APA, 2000) nosology are (a) major depressive disorder, (b) dysthymia, and (c) adjustment disorder with depressed mood. The overall 30-day prevalence rate of a major depressive episode is about 5 %, with the highest prevalence rates among female teens and young-adult males (Kessler et al., 2008). The prevalence of unipolar depression among children younger than 15 years is relatively rare and ranges from .4 to 2 % (Costello et al., 2002; Hankin et al., 1998).

Channon (1996), using nonclinical samples, found that older adolescents (i.e., first- and second-year university undergraduate students) who scored relatively higher (suggesting a naturally occurring dysphoric mood) on the Beck Depression Inventory (BDI; Beck, 1978) required more trials to attain criterion-level performances and made more perseverative as well as nonperseverative errors on the standard version of the WCST than a similar group of individuals who scored relatively lower on the BDI. Thus, in this study, older adolescents in the dysphoric group showed more difficulty than those in the control group in EF shifting set correctly and in altering their behavior in response to feedback.

In summary, little research has been conducted on the EF abilities among youth who are depressed. Studies that have included late adolescents (e.g., Channon, 1996) have suggested some degree of sensitivity of EF tasks in identifying unipolar depression, but less specificity. Given the neuropsychological discontinuities that are characteristic of human development and the heterotypic nature of child, adolescent, and adult depression, however, generalizing EF findings to younger individuals with depression requires great caution.

Bipolar Disorder

There has been a history of controversy about the identification of mania, a major criterion for bipolar disorder (BD), among children and adolescents, but recent evidence suggests that much of this debate may be associated with diagnostic challenges rather than to extremely low incidence rates as once was thought (Biederman, 2003). For example, Costello et al. (2002) reported no epidemiological evidence of BD among children younger than 13 years old and only about a 1 % lifetime prevalence rate among adolescents aged 14–18 years. The reliability of these estimates, however, has been questioned because of (a) inadequate sample sizes, (b) different definitions and criteria applied to the diagnosis of BD, and (c) the putative heterotypic continuity of symptoms of this disorder across the lifespan (Costello et al., 2002). BD among children may present as irritability and aggression rather than as euphoric mood and is likely to present as continuous, chronic, and rapidly cycling (Geller & Luby, 1997).

Biederman (2003) has documented a growing consensus on the significance of serious consequences of affective dysregulation among children and adolescents and the increased scientific and clinical attention that they have received in recent years. Much of this attention has focused on the neurocognitive concomitants of BD among children, especially those involving EF. When BD has been studied among child and adolescent populations, however, results often have been complicated by its comorbidity with ADHD and other disruptive as well as internalized disorders. When reviewing the epidemiology of BD within this population, Costello et al. (2002) found that the majority of children and adolescents classified with BD (i.e., 60–90 %) showed comorbid ADHD, leading to significant diagnostic confusion.

Doyle et al. (2005) studied the EF (e.g., working memory, interference control) and other (e.g., processing speed, sustained attention, visuospatial organization) EF characteristics of a sample ($n=57$) of youth with BD between 10 and 18 years old, comparing them to same-aged youth with no history of bipolar or other mood disorders on a variety of tasks. Not surprisingly, comorbidity rates between the two samples differed markedly for several externalizing disorders (ADHD, CD, and ODD) as well as anxiety disorders. In fact, 74 % of those in the bipolar group met criteria for ADHD, whereas only 17 % in the control group met those criteria. An even larger discrepancy occurred for comorbid ODD between those with and without BD (i.e., 93 % and 20 %, respectively). Results demonstrated that youth with BD showed impairments relative to those without any history of mood disorders on selected measures within several of the EF and non-EF neurocognitive areas that were assessed. Children with BD demonstrated significantly poorer performance on Digit Span, Digit Symbol/Coding, Stroop Color, Stroop Color-Word, and the Auditory Continuous Performance Test (Doyle et al., 2005). Among other EF tasks, no differences were found on any of the three measures of abstract problem solving and set shifting; among the non-EF tasks, no differences were found on any of the two measures of visuospatial organization or one measure of verbal learning. Thus, although measures of planning (e.g., Tower of London, TOL) and verbal fluency were not administered, traditional EF measures, such as those assessed on the WCST for shifting of set, failed to discriminate between the groups. One strength of this study was the fact that the effects of ADHD were statistically controlled; therefore, results cannot be attributed to this comorbidity. On the other hand, one might question the representativeness and generalizability of the findings, given the natural coincidence of ADHD with BD among youth. That is, perhaps these are not really dissociable clusters of symptoms. It also is unclear why ADHD was controlled, whereas other forms of psychopathology (e.g., ODD) were not. This study, therefore, provided limited evidence that particular aspects of EF may be affected in youth with BD, which included one measure of interference control and two measures of working memory.

Dickstein et al. (2004), however, did include some of these more traditional EF constructs, such as an intradimensional/extradimensional set-shifting task (described as a task that "mirrors the WCST"; p. 34) and the Stockings of Cambridge (described as a "modified version of the TOL task," which is essentially a spatial planning task; p. 34). In this study, a group of children with pediatric BD (M age = 13 years) was compared with a similar group of normal controls on a variety of neurocognitive tasks including a simple pointing task, pattern recognition, and several tasks of spatial memory, in addition to the set-shifting task and the Stockings of Cambridge.

Over 20 t-tests were conducted on measures derived from these tasks in analyzing potential differences between children with and without BD. Given the preliminary nature of this study, no corrections for multiple significance tests were conducted, but only two of these tests were significant at $p<.01$: (a) the number of errors made prior to an extradimensional shift on the set-shifting task and (b) the mean correct latency of a pattern-recognition memory task. Here, children with BD made more errors and took longer to respond than those without BD. Interestingly, there were no differences between the groups on any of five measures for the Stockings of Cambridge task. Moreover, other than the number of errors made prior to an extradimensional shift, only one other of seven measures on the intradimensional/extradimensional set-shifting task discriminated between the two groups at $p<.05$. Results suggested that deficits in attentional set shifting and visuospatial memory potentially may be implicated in pediatric BD. Unfortunately, the sample sizes for each group were small ($n=21$ in each group), and at the time of testing, all children in the bipolar group were taking at least one psychotropic medication (over half were taking four or more medications) and over half of these children were diagnosed with comorbid ADHD (over 70 % met ADHD criteria during their lifetimes). Thus, Dickstein et al. (2004) cautioned that these findings are best considered as preliminary.

Meyer et al. (2004) found that 6 of 32 offspring of mothers with BD and 3 of 42 offspring of mothers with unipolar depression were diagnosed with BD as young adults (Radke-Yarrow, 1998). Records of the EF measures administered (WCST, Trail Making Test Part B [TMT-B]) to these nine individuals when they were adolescents were compared with similar measures for offspring who were later diagnosed with unipolar depression ($n=22$) and those who showed no evidence of a mood disorder as young adults ($n=64$). On WCST measures of EF obtained during adolescence, those who were diagnosed with BD during young adulthood ($n=9$) showed more preservative errors, fewer categories completed, and fewer conceptual-level responses than those who showed no evidence of mood disorders as adults (Meyers et al., 2004; Radke-Yarrow, 1998). No differences were found among the young-adult groups on their adolescent performances on the Trail Making Test. Results showed that, among those participating in this study, 67 % of the high-risk offspring who were diagnosed with BD as young adults previously had shown EF deficits as adolescents. In contrast, only 19 % of the high-risk offspring who were diagnosed with unipolar depression as young adults previously had shown EF deficits as adolescents, and only 17 % of the high-risk offspring who showed no mood-disorder symptoms as young adults previously had shown EF deficits as adolescents. Thus, this study suggests that there may be a specific profile of neurocognitive deficits that include EF (as well as visuospatial memory and sustained attention) that *precede* the adult onset of bipolar disorder but are unrelated to unipolar depression (Klimes-Dougan, Ronsaville, Wiggs, & Martinez, 2006; cf. Dickstein et al., 2004).

In summary, there is a growing consensus about the significance of BD among children, and several studies have targeted its EF concomitants. Although results often have been confounded with significant comorbidity issues, children and adolescents with BD reliably have demonstrated impairments relative to those without any history of mood disorders on several EF and non-EF neurocognitive measures: EF tasks that have been implicated include working memory, interference control, and set shifting and non-EF tasks that have been implicated include processing speed, sustained attention, and visuospatial memory (Dickstein et al., 2004; Doyle et al., 2005; Klimes-Dougan et al., 2006). Moreover, in addition to evidence of the sensitivity of these EF and non-EF neurocognitive constructs as potential risk factors for BD, there also is some evidence of their specificity. In other words, there is accumulating evidence for a specific profile of EF and non-EF neurocognitive deficits that precede the adult onset of BD but are unrelated to unipolar depression. For example, data from Meyer et al. (2004) show that individuals identified as having EF impairments during adolescence are much more likely to develop BD, but no more likely to develop unipolar depression, during young adulthood than adolescents without these deficits. Although these findings require replication with other samples, they present significant implications for prevention and early intervention.

Collectively, these studies indicate that EF deficits often accompany a variety of childhood psychopathologies; however, the nature of these deficits remains equivocal. Findings also suggest that EF tasks may be sensitive to the identification of deficits in childhood populations; however, they often lack specificity. Methodological differences across studies are problematic and may obfuscate subtle differences in EFs among children with different types of disorders. Future research is needed to elucidate further the specific types of EF deficits that co-occur with childhood externalizing and internalizing disorders and to determine whether these deficits are global or unique to each disorder.

It is the purpose of this review to examine the executive tests most commonly used within the research literature. This review will also examine the percentage of times that different executive function tests have shown significant between-group differences. In addition, this review will examine the percentage of comparisons between individual internalizing disorders and externalizing disorders, and control groups have shown significant differences.

Method

Search and Retrieval

We attempted to identify and retrieve all empirical studies and meta-analyses published after 1999 that examined executive functioning abilities in children and adolescents with specific internalizing and externalizing disorders. The search and retrieval process was conducted using the keywords: *Executive Function* +

ADHD, Executive Function + Attention Deficit/Hyperactivity Disorder, Executive Function + Conduct Disorder, Executive Function + Oppositional Defiant Disorder, Executive Function + Tourette's Disorder, Executive Function + Major Depressive Disorder, Executive Function + Dysthymia, Executive Function + Bipolar Disorder, Executive Function + Cyclothymia, Executive Function + Generalized Anxiety Disorder, Executive Function + Social Phobia, Executive Function + Obsessive Compulsive Disorder, Executive Function + OCD, Executive Function + Posttraumatic Stress Disorder, Executive Function + PTSD, Executive Function + Specific Phobia, Executive Function + Panic Disorder, Executive Functioning Deficits. The search and retrieval process included a comprehensive search of the following bibliographic databases: PsycINFO, PsycARTICLES, and MEDLINE. All studies were retrieved from the University of Rhode Island's electronic library between June and August of 2010 and then again in January of 2012.

Eligibility Criteria

Studies were selected for review based on the following criteria:
1. The study was published in English.
2. The study was published no earlier than 2000.
3. The study only included children and adolescent ages 18 and younger.
4. The study involved the comparison of executive functioning performance between individuals with at least one of the specified internalizing or externalizing psychological disorders and a control group or a comparison group characterized by the presence of a psychological or psychiatric disorder. The specified internalizing and externalizing disorders included ADHD, conduct disorder, ODD, Tourette's disorder, major depressive disorder, dysthymia, bipolar disorder, cyclothymia, generalized anxiety disorder, OCD, social phobia, PTSD, specific phobia, and panic disorder.

Results

A total of 141 studies were identified that met eligibility criteria and were included in the study. The studies are summarized in Table 5.1. Note that the "Number of Times Used" column represents the number of published articles that have used each of these tests to examine between-group comparisons. The "Sensitivity to Group Differences" column represents the total number of times each of these tests was used to examine between-group comparisons.

Across all studies that were examined, there were a total of 164 different tests used to measure executive function. As seen in Table 5.1, there were eight tests that were used to assess executive functioning in more than ten studies. The other 156 tests were used to assess executive function in less than ten studies. As expected, the percentage of times tests showed significant between-group differences varied across executive function tests. The BRIEF was associated with significant differences between clinical groups (% of comparisons), or between clinical and control groups (% of comparisons), 67 % of the times it was used. The Go/No-Go Test was associated with significant differences between clinical groups (% of comparisons), or between clinical and control groups (% of comparisons), 46 % of the times it was used.

The Continuous Performance Test and test variants were associated with significant differences between clinical groups (33 % of comparisons), or between clinical and control groups (50 % of comparisons), 43 % of the times it was used. Significant between-group differences on the Stroop Color and Word Test and test variants occurred 38 % of the total times it was used, with significant differences occurring during 17 % of comparisons between clinical groups and 59 % of comparisons between clinical and control groups. Significant between-group differences on the Trail Making Test occurred 36 % of the total times it was used, with significant between-group differences occurring during 19 % of comparisons between clinical groups and 44 % of comparisons between clinical and control groups.

Table 5.2 Proportion of comparisons showing significant differences between groups

Diagnosis	Percentage of significant between group differences	Percentage of significant differences between clinical groups	Percentage of significant differences between clinical and control groups
Conduct disorder	13/31 = 42 %	No comparisons	13/31 = 42 %
Attention deficit/hyperactivity disorder	643/1,621 = 40 %	131/669 = 20 %	512/952 = 54 %
Posttraumatic stress disorder	4/13 = 31 %	No comparison	4/13 = 31 %
Major depressive disorder	8/29 = 28 %	3/9 = 33 %	5/20 = 25 %
Obsessive compulsive disorder	12/58 = 21 %	2/25 = 8 %	10/33 = 30 %
Bipolar disorder	14/70 = 20 %	2/42 = 5 %	12/28 = 43 %
Oppositional defiant disorder	7/36 = 19 %	No comparisons	7/36 = 19 %
Tourette's disorder	12/81 = 15 %	8/50 = 16 %	4/31 = 13 %

The WCST was associated with significant between-group differences 33 % of the total times it was used, with significant differences occurring during 16 % of comparisons between clinical groups and 43 % of comparisons between clinical and control groups. The ROCF Test and Rey Complex Figure Test were associated with significant differences between groups 33 % of the total times it was used, with significant differences occurring during 19 % of comparisons between clinical groups and 43 % comparisons between clinical and control groups. Finally, the TOL test and variants of the TOL were associated with significant differences between clinical groups or between clinical groups and controls only 4 % of the total times it was used, with significant differences occurring during 5 % of comparisons between clinical groups and 3 % of comparisons between clinical and control groups.

The percentage studies showing significant between-group differences across all executive function tests varied across clinical groups. The results are summarized in Table 5.2. Note that subtypes of clinical disorders were not differentiated. Also, comparisons that included subtypes of the same disorder (e.g., ADHD-HI and ADHD-C) and comparisons between groups of individuals with the same diagnosis (e.g., ADHD and ADHD comorbid with Specific Learning Disability) were excluded. In addition, comparisons using comorbid conditions for the diagnosis of interest were also excluded.

Studies comparing individuals with conduct disorder to other clinical or control groups found significant between-group differences 42 % of the time. Studies comparing individuals with ADHD to other clinical or controls groups found significant between-group differences on executive function test performance 40 % of the time. PTSD and a comparison group found significant between-group differences 31 % of the time. Studies comparing executive function test performance in individuals with major depressive disorder and a comparison group found significant between-group differences 28 % of the time, while studies comparing executive function test performance in individuals with OCD and a comparison group found significant between-group differences 21 % of the time. Studies comparing individuals with bipolar disorder to other clinical or control groups found significant between-group differences 20 % of the time, while studies examining executive function test performance in individuals with ODD found significant between-group differences 19 % of the time. Finally, studies examining Tourette's disorder and a comparison group found significant between-group differences 15 % of the time.

Discussion

Currently there are a vast number of neuropsychological measures purported to measure executive functioning or individual isolated components of executive functioning. While some studies use standardized neuropsychological assessments such as the Development Neuropsychological Assessment (NEPSY) (Spruyt, Capdevila, Kheirandish-Gozal, & Gozal, 2009) to assess executive functioning, many others have primarily relied on tests created within research labs and that have been used exclusively for research purposes (Gohier et al., 2009). Due to the enormous variability in the type of assessments used for executive functioning assessment, the purpose of this study was to conduct a comprehensive review of the literature to determine which executive functioning tests are used most often among researchers when assessing executive functioning in samples of individuals with internalizing and externalizing forms of psychopathology. Similarly, this review also examined the frequency with which the most commonly used tests of executive function reveal significant differences in performance between clinical and control participant groups as well as between multiple clinical participant groups. In addition, as executive functioning impairments are reported to be characteristic of a number of different internalizing and externalizing forms of psychopathology (Willcutt, Pennington, Olson, Chhabildas, & Hulslander, 2005), the proportion of research studies in which individuals with specific forms of internalizing and externalizing forms of psychopathology demonstrate impairments on executive functioning test performance remains unexplored. Therefore, this review also provided a comparison of the total proportion of studies that have found significant differences between clinical and control participant groups and multiple clinical participant groups across different forms of psychopathology.

In order to identify the executive functioning tests and assessments most commonly used within the internalizing and externalizing psychopathology research literature, we conducted an extensive literature search and review of all related articles published between 2000 and 2011. We compiled a database that included all relevant studies that met our inclusion/exclusion criteria and were then able to identify the wide range of different executive functioning tests that have been used and identify the frequency of their usage. Although there were a total of 164 identified measures of executive functioning, we only provided further analyses for the executive tests that were used with a relatively higher frequency. Based on our findings, authors tend to use The Stroop Color-Word Test and variants, WCST and variants, and Trail Making Test and variants more often than other measures of executive function. This finding suggests that much of what is currently known about childhood/adolescent executive functioning abilities in internalizing and externalizing forms of psychopathology is dependent on the quality of these three assessments.

Despite their widespread usage, these tests are not necessarily the most effective at discriminating between different clinical and control groups. Indeed, results have revealed that the BRIEF, the Continuous Performance Test and variants, and the Go/No-Go Test are sensitive to between-group differences more often than each of the three more commonly used tests. These results indicate that perhaps different executive functioning assessments may be more appropriate for addressing different types of research questions. Furthermore, these results may help shed light on the inconsistent results between studies examining executive functioning impairments in individuals with psychopathology. Given that there were a total of 164 executive tests used, it is undeniable that these tests ultimately measure the executive function construct or components of the executive function construct with more or less accuracy. Similarly, since some measures are more sensitive to group differences than others, the reported findings from individual studies may be a function of the test used rather than the executive function purported to be assessed.

Findings also revealed considerable variability in the proportion of times that participant groups with specific forms of internalizing and externalizing forms of psychopathology demonstrated impaired performance on executive function tests. Based on the present findings, when compared to clinical or control groups, individuals with conduct disorder and ADHD tend to more often show performance differences on executive function measures compared to individuals with disorders such as Tourette's disorder and ODD. In addition, there was also noticeable variability in the degree to which the specified clinical groups demonstrated significant differences in executive function test performance when their performance was compared to other clinical groups or control groups. These findings suggest that when compared to other clinical and/or control groups, individuals with some forms of psychopathology, such as major depressive disorder, demonstrate executive functioning performance differences *more often* than individuals with other forms of psychopathology.

Limitations

The present study identified specific inclusion criteria and 141 studies were included. Clearly a comprehensive review of the frequencies and proportions of executive tests used across the entire body of executive functioning literature was beyond the scope of this paper. It is quite possible, however, that the inclusion of all disorders within the Diagnostic and Statistical Manual of Mental Disorders, Fourth Edition (DSM-IV-TR; American Psychiatric Association, 2000) would lead to different results. Similarly, although not a focus of the present review, these results did not disaggregate by age, which prohibits an examination of the degree to which different tests are used across different developmental levels.

For several of the clinical groups examined, there were only a very limited number of studies that have examined executive function performance. For instance, there were no between-group comparisons that included individuals with panic disorder, social phobia, generalized anxiety, specific phobia, cyclothymia, and dysthymia. This finding considerably limits the degree to which we can compare the results for these disorders to the disorders more commonly examined (e.g., ADHD, bipolar disorder).

Although we examined the proportion of studies that found significant between-group differences across different executive functioning measures, we did not use meta-analytic procedures that take into account effect sizes and the magnitude of between-group differences. However, the purpose of this review was to examine the current body of existing literature and identify the most commonly used executive functioning assessments, as well as determine the proportion of studies related to internalizing and externalizing forms of psychopathology that report significant between-group differences in executive functioning performance. Although the application of meta-analytic procedures would be quite informative, examining the magnitude of between-group differences on executive functioning tests was not the aim of this review.

Suggestions for Future Research

This selective review of literature concerning EFs in externalizing and internalizing childhood disorders has revealed a number of interesting findings. Regarding ADHD, studies collectively suggest that EF deficits are neither unique to the disorder nor are they necessary or sufficient for a diagnosis of ADHD. When EF impairments are present with ADHD, they tend to be specific rather than global impairments. Findings are inconsistent as to whether comorbid disorders are related to EFs in children with ADHD. With regard to CD, earlier studies suggested that EF deficits were not characteristic of children with CD after comorbid ADHD was statistically controlled for; however, recent findings suggest that EF deficits may indeed be characteristic of both children with CD and children with ADHD. No clear pattern of EF deficits emerges in the literature concerning children with TD. A few studies, but not all, have reported greater response-time variability on continuous performance tasks with this population. Preliminary studies suggest that working memory is not characteristically impaired with children with TD. Results are equivocal with respect to the performance of these children on planning tasks and measures of cognitive flexibility. Overall, distinct and robust impairments in EF do not appear to be characteristic of children with TD. Additional, methodologically sound studies are needed to address whether subtle differences in EF may exist between children with TD and other types of childhood psychopathology.

With regard to internalizing disorders of childhood, the current research literature on the relationship between EF deficits and internalized forms of developmental psychopathology is unequally distributed. Specifically, in contrast to research conducted with adult populations, very little research has been conducted linking EF to child and adolescent depression. This dearth of evidence seems curious, given the popularity of cognitive-behavioral interventions and the use of psychotropic medications to treat depression among youth. On the other hand, it is perhaps understandable, given the taxometric identification of a broadband, internalizing, or anxious-depressed syndrome that has been identified in empirical investigations of child and adolescent populations (Achenbach & Edelbrock, 1983), contrasted with less support for the kinds of narrowband classification distinctions made among DSM-IV-TR diagnoses (American Psychiatric Association, 2000).

In addition, several studies have been conducted in the area of childhood BD and EF (Dickstein et al., 2004; Doyle et al., 2005), and the longitudinal work that has been conducted in this area (Klimes-Dougan et al., 2006; Meyers et al., 2004) provides important implications for prevention and early intervention. Similarly, with the exception of OCD, research also has begun to accumulate that assesses potential EF deficits among children with a variety of kinds of anxiety disorders (Beers & De Bellis, 2002; Emerson et al., 2005; Toren et al., 2000). Finally, there seems to be an emerging consensus about the importance of differentiating early- vs. late-onset of OCD (e.g., Geller et al., 2001), and research has documented EF differences among adults with these two subtypes (Roth, Milovan, Baribeau, & O'Connor, 2005). Developmental studies, however, have yet to be conducted in this area. Also, future studies examining a wider range of psychological and psychiatric disorders may better reflect the true frequency with which different executive function tests are used.

Given the findings of the current review, authors of future studies examining between-group performances in executive functioning should choose their executive function measures carefully. It must be recognized that some executive function measures (e.g., the BRIEF) are more likely to reveal between-group differences than other measures (e.g., TOL and variants) despite the fact that both measures are purported to measure executive functioning. These findings are pertinent to clinicians as it appears that some neuropsychological tests of executive functioning are better than others at discriminating between clinical groups or clinical groups and controls groups.

References

Achenbach, T. M., & Edelbrock, C. (1983). *Child behavior checklist*. Burlington, VT: University Associates in Psychiatry.

American Psychiatric Association. (1987). *Diagnostic and statistical manual of mental disorders (rev.)*. Washington, DC: Author.

American Psychiatric Association. (2013). *Diagnostic and statistical manual of mental disorders* (4th ed., text revision). Washington: Author.

American Psychiatric Association. (2013). Diagnostic and statistical manual of mental disorders (5th ed.). Washington, DC: Author.

Barkley, R. A. (1997). Behavioral inhibition, sustained attention, and executive functions: Constructing a unifying theory of ADHD. *Psychological Bulletin, 121*, 65–94.

Barkley, R. A. (2010). Differential diagnosis of adults with ADHD: The role of EF and self regulation. *The Journal of Clinical Psychiatry, 71*(7), e17. doi:10.4088/JCP.9066tx1c.

Barkley, R. A., Edwards, G., Laneri, M., Fletcher, K., & Metevia, L. (2001). Executive functioning, temporal discounting, and sense of time in adolescents with attention deficit hyperactivity disorder (ADHD) and oppositional defiant disorder (ODD). *Journal of Abnormal Child Psychology, 29*, 541–556.

Barkley, R. A., Grodzindky, G., & DuPaul, G. J. (1992). Frontal lobe functions in attention deficit disorder with and without hyperactivity: A review and research report. *Journal of Abnormal Child Psychology, 20*, 163–188.

Barkley, R. A., Murphy, K. R., & Bush, T. (2001). Time perception and reproduction in young adults with attention deficit hyperactivity disorder. *Neuropsychology, 15*, 351–360.

Beck, A. T. (1978). *Depression inventory (BDI)*. Philadelphia: Center for Cognitive Therapy.

Beers, S. R., & De Bellis, M. D. (2002). Neuropsychological function in children with maltreatment-related post-traumatic stress disorder. *American Journal of Psychiatry, 159*(3), 483–486.

Berlin, L., Bohlin, G., Nyberg, L., & Janols, L. O. (2004). How well do measures of inhibition and other executive functions discriminate between children with ADHD and controls? *Child Neuropsychology, 10,* 1–13.

Berlin, L., Bohlin, G., & Rydell, A. M. (2003). Relations between inhibition, executive functioning, and ADHD symptoms: A longitudinal study from age 5 to 8 ½ years. *Child Neuropsychology, 9,* 255–266.

Biederman, J. (2003). Pharmacotherapy for attention-deficit/hyperactivity disorder (ADHD) decreases the risk for substance abuse: Findings from a longitudinal follow-up of youths with and without ADHD. *Journal of Clinical Psychiatry, 64,* 3–8.

Biederman, J., Petty, C. R., Fried, R., Doyle, A. E., Spencer, T., Seidman, L. J., et al. (2007). Stability of executive function deficits into young adult years: A prospective longitudinal follow-up study of grown up males with ADHD. *Acta Psychiatrica Scandinavica, 116,* 129–136.

Byrne, J. M., DeWolfe, N. A., & Bawden, H. N. (1998). Assessment of attention-deficit hyperactivity disorder in preschoolers. *Child Neuropsychology, 4,* 49–66.

Carrion, V. G., Garrett, A., Menon, V., Weems, C. F., & Reiss, A. L. (2008). Posttraumatic stress symptoms and brain function during a response inhibition task: An fMRI study in youth. *Depression and Anxiety, 25*(6), 514–526.

Chang, S. W., McCracken, J. T., & Piacentini, J. C. (2007). Neurocognitive correlates of child obsessive compulsive disorder and Tourette's syndrome. *Journal of Clinical and Experimental Neuropsychology, 29,* 724–733.

Channon, S. (1996). Executive dysfunction in depression: The Wisconsin card sorting test. *Journal of Affective Disorders, 39*(2), 107–114.

Channon, S., Pratt, P., & Robertson, M. M. (2003). Executive function, memory, and learning in Tourette's syndrome. *Neuropsychology, 17,* 247–254.

Cirino, P. T., Chapieski, M. L., & Massman, P. J. (2000). Card sorting performance and ADHD symptomatology in children and adolescents with Tourette syndrome. *Journal of Clinical and Experimental Neuropsychology, 22,* 245–256.

Clark, C., Prior, M., & Kinsella, G. J. (2000). Do executive function deficits differentiate between adolescents with ADHD and oppositional defiant/conduct disorder? A neuropsychological study using the Six Elements Test and Hayling Sentence Completion Test. *Journal of Abnormal Child Psychology, 28,* 403–414.

Costello, E. J., Pine, D. S., Hammen, C., March, J. S., Plotsky, P. M., Weissman, M. M., et al. (2002). Development and natural history of mood disorders. *Biological Psychiatry, 52*(6), 529–542.

Crawford, S., Channon, S., & Robertson, M. M. (2005). Tourette's syndrome: Performance on tests of behavioural inhibition, working memory and gambling. *Journal of Child Psychology and Psychiatry, 46,* 1327–1336.

Delis, D. C., Cullum, C. M., Butters, N., Cairns, P., & Prifitera, A. (1988). Wechsler memory scale-revised and california verbal learning test: Convergence and divergence. *The Clinical Neuropsychologist, 2*(2), 188–196.

Denckla, M. B. (1996). A theory and model of executive function: A neuropsychological perspective. In G. R. Lyon & N. A. Krasnegor (Eds.), *Attention, memory, and executive function* (pp. 263–277). Baltimore: Paul H. Brookes.

Dickstein, D. P., Treland, J. E., Snow, J., McClure, E. B., Mehta, M. S., Towbin, K. E., et al. (2004). Neuropsychological performance in pediatric bipolar disorder. *Biological Psychiatry, 55*(1), 32–39.

Doyle, A. E., Wilens, T. E., Kwon, A., Seidman, L. J., Faraone, S. V., Fried, R., et al. (2005). Neuropsychological functioning in youth with bipolar disorder. *Biological Psychiatry, 58*(7), 540–548.

DuPaul, G. J., McGoey, K. E., Eckert, T. L., & VanBrakle, J. (2001). Preschool children with attention-deficit/hyperactivity disorder: Impairments in behavioral, social, and school functioning. *Journal of the American Academy of Child and Adolescent Psychiatry, 40,* 508–515.

Emerson, C. S., Mollet, G. A., & Harrison, D. W. (2005). Anxious-depression in boys: An evaluation of executive functioning. *Archives of Clinical Neuropsychology, 20*(4), 539–546.

Evans, D. W., & Leckman, J. F. (2006). Definitions and clinical descriptions 405. *Developmental Psychopathology: Risk, Disorder, and Adaptation, 3,* 404.

Fischer, M., Barkley, R. A., Smallish, L., & Fletcher, K. (2005). Executive functioning in hyperactive children as young adults: Attention, inhibition, response perseveration, and the impact of comorbidity. *Developmental Neuropsychology, 27,* 107–133.

Fletcher, J. M. (1996). Executive functions in children: Introduction to the special series. *Developmental Neuropsychology, 12,* 1–3.

Friedlander, L., & Desrocher, M. (2006). Neuroimaging studies of obsessive–compulsive disorder in adults and children. *Clinical Psychology Review, 26*(1), 32–49.

Friedman, N. P., Haberstick, B. C., Willcutt, E. G., Miyake, A., Young, S. E., Corley, R. P., et al. (2007). Greater attention problems during childhood predict poorer executive functioning in late adolescence. *Psychological Science, 18,* 898–900.

Fuggetta, G. P. (2006). Impairment of executive functions in boys with attention deficit/hyperactivity disorder. *Child Neuropsychology, 12,* 1–21.

Geller, D. A., Biederman, J., Faraone, S. V., Bellordre, C. A., Kim, G. S., Hagermoser, L., et al. (2001). Disentangling chronological age from age of onset in children and adolescents with obsessive-compulsive disorder. *The International Journal of Neuropsychopharmacology, 4*(2), 169–178.

Geller, B., & Luby, J. (1997). Child and adolescent bipolar disorder: A review of the past 10 years. *Journal of the American Academy of Child and Adolescent Psychiatry, 36*(9), 1168–1176.

Geurts, H. M., Verte, S., Oosterlaan, J., Roeyers, H., & Sergeant, J. A. (2005). ADHD subtypes: Do they differ in their executive functioning profile? *Archives of Clinical Neuropsychology, 20*, 457–477.

Giancola, P. R., Mezzich, A. C., & Tarter, R. E. (1998). Executive cognitive functioning, temperament, and antisocial behavior in conduct-disordered adolescent females. *Journal of Abnormal Psychology, 107*, 629–641.

Gohier, B., Ferracci, L., Surguladze, S. A., Lawrence, E., El Hage, W., Kefi, M. Z., et al. (2009). Cognitive inhibition and working memory in unipolar depression. *Journal of Affective Disorders, 116*(1), 100–105.

Gray, J. A. (1987). Perspectives on anxiety and impulsivity: A commentary. *Journal of Research in Personality, 21*(4), 493–509.

Hankin, B. L., Abramson, L. Y., Moffitt, T. E., Silva, P. A., McGee, R., & Angell, K. E. (1998). Development of depression from preadolescence to young adulthood: Emerging gender differences in a 10-year longitudinal study. *Journal of Abnormal Psychology, 107*(1), 128–140.

Hankin, B. L., Fraley, R. C., Lahey, B. B., & Waldman, I. D. (2005). Is depression best viewed as a continuum or discrete category? A taxometric analysis of childhood and adolescent depression in a population-based sample. *Journal of Abnormal Psychology, 114*(1), 96–110.

Harris, E. L., Schuerholz, L. J., Singer, H. S., Reader, M. J., Brown, J. E., Cox, C., et al. (1995). Executive function in children with Tourette syndrome and/or attention deficit hyperactivity disorder. *Journal of the International Neuropsychological Society, 1*, 511–516.

Heaton, R. K. (1981). *A manual for the Wisconsin card sorting test*. Los Angeles: Western Psychological Services.

Herba, C. M., Tranah, T., Rubia, K., & Yule, W. (2006). Conduct problems in adolescence: Three domains of inhibition and effect of gender. *Developmental Neuropsychology, 30*, 659–695.

Hinshaw, S. P., Carte, E. T., Fan, C., Jassy, J. S., & Owens, E. B. (2007). Neuropsychological functioning of girls with attention-deficit/hyperactivity disorder followed prospectively into adolescence: Evidence for continuing deficits? *Neuropsychology, 21*, 263–273.

Jurado, M. B., & Rosselli, M. (2007). The elusive nature of executive functions: A review of our current understanding. *Neuropsychology Review, 17*, 213–233.

Kessler, R. C., Gruber, M., Hettema, J. M., Hwang, I., Sampson, N., & Yonkers, K. A. (2008). Co-morbid major depression and generalized anxiety disorders in the National Comorbidity Survey follow-up. *Psychological Medicine, 38*(3), 365–374.

Kim, M. S., Kim, J. J., & Kwon, J. S. (2001). Frontal P300 decrement and executive dysfunction in adolescents with conduct problems. *Child Psychiatry and Human Development, 32*, 93–106.

Klimes-Dougan, B., Ronsaville, D., Wiggs, E. A., & Martinez, P. E. (2006). Neuropsychological functioning in adolescent children of mothers with a history of bipolar or major depressive disorders. *Biological Psychiatry, 60*(9), 957–965.

Klimkeit, E. I., Mattingley, J. B., Sheppard, D. M., Lee, P., & Bradshaw, J. L. (2005). Motor preparation, motor execution, attention, and executive functions in attention deficit/hyperactivity disorder (ADHD). *Child Neuropsychology, 11*, 153–173.

Kovacs, M., & Beck, A. T. (1977). An empirical-clinical approach toward a definition of childhood depression. In J. G. Schulterbrandt & A. Raskin (Eds.), *Depression in childhood: Diagnosis, treatment, and conceptual models* (pp. 1–25). New York: Raven Press.

Lawrence, V., Houghton, S., Tannock, R., Douglas, G., Durkin, K., & Whiting, K. (2002). ADHD outside the laboratory: Boys' executive function performance on tasks in videogame play and on a visit to the zoo. *Journal of Abnormal Child Psychology, 30*, 447–462.

Mahone, E. M., Cirino, P. T., Cutting, L. E., Cerrone, P. M., Hagelthorn, K. M., Hiemenz, J. R., et al. (2002). Validity of the Behavior Rating Inventory of Executive Function in children with ADHD and/or Tourette syndrome. *Archives of Clinical Neuropsychology, 17*, 643–662.

Mahone, E. M., & Hoffman, J. (2007). Behavior ratings of executive function among preschoolers with ADHD. *Clinical Neuropsychology, 21*, 569–586.

Mahone, E. M., Pillion, J. P., & Heimenz, J. R. (2001). Initial development of an auditory continuous performance test for preschoolers. *Journal of Attention Disorders, 5*, 93–106.

McBurnett, K., Harris, S. M., Swanson, J. M., Pfiffner, L. J., Tamm, L., & Freeland, D. (1993). Neuropsychological and psychophysiological differentiation of inattention/overactivity and aggression/defiance symptom groups. *Journal of Clinical Child Psychology, 22*, 165–171.

Meyers, C. A., Smith, J. A., Bezjak, A., Mehta, M. P., Liebmann, J., Illidge, T., et al. (2004). Neurocognitive function and progression in patients with brain metastases treated with whole-brain radiation and motexafin gadolinium: Results of a randomized phase III trial. *Journal of Clinical Oncology, 22*(1), 157–165.

Moffitt, T. E., & Henry, B. (1989). Neuropsychological assessment of executive functions in self-reported delinquents. *Development and Psychopathology, 1*, 105–118.

Morgan, A. B., & Lilienfeld, S. O. (2000). A meta-analytic review of the relation between antisocial behavior and neuropsychological measures of executive function. *Clinical Psychology Review, 20*, 113–136.

Nigg, J. T., Blaskey, L. G., Huang-Pollock, C., & Rappley, M. D. (2002). Neuropsychological executive functions and DSM-IV ADHD subtypes. *Journal of the American Academy of Child and Adolescent Psychiatry, 41*, 59–66.

Osterrieth, P. (1944). The test of copying a complex figure: A contribution to the study of perception and memory. *Archive de Psychologie, 30*, 206–356.

Ozonoff, S., & Jensen, J. (1999). Brief report: Specific executive function profiles in three neurodevelopmental disorders. *Journal of Autism and Developmental Disorders, 29*, 171–177.

Ozonoff, S., & Strayer, D. L. (2001). Further evidence of intact working memory in autism. *Journal of Autism and Developmental Disorders, 31*, 257–263.

Pennington, B. F., & Ozonoff, S. (1996). Executive functions and developmental psychopathology. *Journal of Child Psychology and Psychiatry, 37*, 51–87.

Prisciandaro, J. J., & Roberts, J. E. (2005). A taxometric investigation of unipolar depression in the National Comorbidity Survey. *Journal of Abnormal Psychology, 114*(4), 718–728.

Radke-Yarrow, M. (1998). *Children of depressed mothers: From early childhood to maturity*. Cambridge, England: University Press.

Rhodes, S. M., Coghill, D. R., & Matthews, K. (2005). Neuropsychological functioning in stimulant-naïve boys with hyperkinetic disorder. *Psychological Medicine, 35*, 1109–1120.

Rinsky, J. R., & Hinshaw, S. P. (2011). Linkages between childhood executive functioning and adolescent social functioning and psychopathology in girls with ADHD. *Child Neuropsychology, 17*, 368–390.

Roth, R. M., Milovan, D., Baribeau, J., & O'Connor, K. (2005). Neuropsychological functioning in early-and late-onset obsessive-compulsive disorder. *The Journal of Neuropsychiatry and Clinical Neurosciences, 17*(2), 208–213.

Schuerholz, L. J., Singer, H. S., & Denckla, M. B. (1998). Gender study of neuropsychological and neuromotor function in children with Tourette syndrome with and without attention-deficit hyperactivity disorder. *Journal of Child Neurology, 13*, 277–282.

Seidman, L. J., Biederman, J., Faraone, S. V., Weber, W., & Ouellete, C. (1997). Toward defining a neuropsychology of attention deficit-hyperactivity disorder: Performance of children and adolescents from a large clinically referred sample. *Journal of Consulting and Clinical Psychology, 65*, 150–160.

Seidman, L. J., Biederman, J., Weber, W., Hatch, M., & Faraone, S. V. (1998). Neuropsychological functioning in adults with attention-deficit hyperactivity disorder. *Biological Psychiatry, 44*, 260–268.

Sergeant, J. A., Geurts, H., & Oosterlaan, J. (2002). How specific is a deficit of executive functioning for attention-deficit/hyperactivity disorder? *Behavioural Brain Research, 130*, 3–28.

Shin, M. S., Choi, H., Kim, H., Hwang, J. W., Kim, B. N., & Cho, S. C. (2008). A study of neuropsychological deficit in children with obsessive-compulsive disorder. *European Psychiatry, 23*(7), 512–520.

Shucard, D. W., Benedict, R. H., Tekok-Kilic, A., & Lichter, D. G. (1997). Slowed reaction time during a continuous performance test in children with Tourette's syndrome. *Neuropsychology, 11*, 147–155.

Speltz, M. L., DeKlyen, M., Calderon, R., Greenberg, M. T., & Fisher, P. A. (1999). Neuropsychological characteristics and test behaviors of boys with early onset conduct problems. *Journal of Abnormal Psychology, 108*, 315–325.

Spielberger, C. D., Edwards, C. D., Lushene, R. E., Montuori, J., & Platzek, D. (1973). *STAIC, state-trait anxiety inventory for children*. Redwood City, CA: Mind Garden.

Spreen, O., & Strauss, E. (1991). *A compendium of neuropsychological tests: Administration, norms and commentary*. New York: Oxford University Press.

Spreen, O., & Strauss, E. (1998). *A compendium of neuropsychological tests: Administration, norms, and commentary* (2nd ed.). New York: Oxford University Press.

Spruyt, K., Capdevila, O. S., Kheirandish-Gozal, L., & Gozal, D. (2009). Inefficient or insufficient encoding as potential primary deficit in neurodevelopmental performance among children with OSA. *Developmental Neuropsychology, 34*(5), 601–614.

Toren, P., Wolmer, L., Rosental, B., Eldar, S., Koren, S., Lask, M., et al. (2000). Case series: Brief parent–child group therapy for childhood anxiety disorders using a manual-based cognitive-behavioral technique. *Journal of the American Academy of Child and Adolescent Psychiatry, 39*(10), 1309–1312.

Toupin, J., Dery, M., Pauze, R., Mercier, H., & Fortin, L. (2000). Cognitive and familial contributions to conduct disorder in children. *Journal of Child Psychology and Psychiatry, 41*, 333–344.

Tsal, Y., Shalev, L., & Mevorach, C. (2005). The diversity of attention deficits in ADHD: The prevalence of four cognitive factors in ADHD versus controls. *Journal of Learning Disabilities, 38*, 142–157.

Weyandt, L. L. (2004). Neuropsychological performance of adults with ADHD. In D. Gozal & D. Molfese (Eds.), *Attention deficit hyperactivity disorder: From genes to animal models to patients*. Totowa, NJ: Humana Press.

Weyandt, L. L. (2005). Executive function in children, adolescents, and adults with attention deficit hyperactivity disorder: Introduction to the special issue. *Developmental Neuropsychology, 27*, 1–10.

Weyandt, L. L. (2009). Executive functions and attention deficit hyperactivity disorder. *The ADHD Report, 17*(6), 1–7.

Weyandt, L. L., Rice, J. A., Linterman, I., Mitzlaff, L., & Emert, E. (1998). Neuropsychological performance of a sample of adults with ADHD, developmental reading disorder, and controls. *Developmental Neuropsychology, 14*, 643–656.

Weyandt, L. L., & Willis, W. G. (1994). Executive functions in school-aged children: Potential efficacy of tasks in discriminating clinical groups. *Developmental Neuropsychology, 10*, 27–38.

Willcutt, E. G., Doyle, A. E., Nigg, J. T., Faraone, S. V., & Pennington, B. F. (2005). Validity of the executive function theory of attention-deficit/hyperactivity disorder: A meta-analytic review. *Biological Psychiatry, 57*, 1336–1346.

Willcutt, E. G., Pennington, B. F., Olson, R. K., Chhabildas, N., & Hulslander, J. (2005). Neuropsychological analyses of comorbidity between reading disability and attention deficit hyperactivity disorder: In search of the common deficit. *Developmental Neuropsychology, 21*, 35–78.

Woodcock, R. W., & Johnson, M. B. (1989). *WJ-R tests of cognitive ability*. Itasca, IL: Riverside.

Wu, K. K., Anderson, V., & Castiello, U. (2002). Neuropsychological evaluation of deficits in executive functioning for ADHD children with or without learning disabilities. *Developmental Neuropsychology, 22*, 501–531.

Zahn-Waxler, C., Klimes-Dougan, B., & Slattery, M. (2000). Internalizing disorders of childhood and adolescence: Progress and prospects for advances in understanding anxiety and depression. *Development and Psychopathology, 12*, 443–466.

The Neuropsychology of Executive Functioning and the DSM-5

Cecil R. Reynolds and Arthur MacNeill Horton Jr.

The Diagnostic and Statistical Manual (DSM) of the American Psychiatric Association (APA) is the primary source for diagnostic classification in the United States. The DSM provides standard criteria for the diagnosis and classification of mental disorders and is used extensively by clinicians, researchers, government agencies, managed care companies, and pharmaceutical companies in the United States and in other countries (APA, 2000).

This chapter will address the relationship of the DSM to the neuropsychology of executive functioning. Currently, the fifth edition of the DSM (DSM-5) is being developed by numerous workgroups (APA, 2012). In order to provide a framework for this chapter, first there will be a discussion of the DSM including both the previous DSM-IV and the new DSM-5 and revisions that are being made with respect to the neuropsychology of executive functioning. After laying a framework with a discussion of the DSM, there will be a discussion of the neuropsychology of executive functioning with specific attention to diagnostic classification of selective phenomena of psychopathology. Then there will be a critical discussion of the ways that the neuropsychology of executive functioning is addressed in the DSM-5. It should be noted that at the time that this chapter is being written, DSM-5 has not been completed, so, while every effort to ensure accuracy will be made, there is the possibility that some of the discussion will be rendered moot by later developments.

Historical Background of the DSM

The need for a framework for the classification and diagnosis of mental disorders has been clear since the beginning of the art and science of medicine. Exactly what disorders and how the mental disorders should be classified however have been very controversial. Various classification systems over the past 2,000 years have differed with respect to their emphasis on behavioral features and self-reported complaints and etiology and course of the mental disorder. In the United States, the need for collection of accurate statistical data regarding individuals with mental illness was a major force for the development of better classifications systems of mental disorders. Another major force was the mental health needs with respect to soldiers and sailors who fought in World War II. The US Army and later the Veterans Administration developed a system for classifying mental disorders and served as a basis for the DSM (APA, 2000). In addition, the World Health Organization (WHO) which publishes the International Classification of Diseases (ICD) added a section on mental disorders that was

C.R. Reynolds (✉)
Texas A&M University, College Station, TX, USA
e-mail: crrh@earthlink.net

A.M. Horton Jr.
Psychological Associates of Maryland,
Towson, MD, USA

based on the classification system developed by the Veterans Administration. It should be mentioned that the WHO maintains statistics with respect to medical illnesses throughout the world. In 1952 of the American Psychiatric Association Committee on Nomenclature and Statistics published the first edition of the Diagnostic and Statistical Manual: Mental Disorders (DSM-I) that was largely based on the then current version of the ICD. It should be noted that by international treaty, the current versions of the ICD and the DSM must be roughly comparable. Or in other words, it must be possible to translate from one version to the other because the statistics collected from the DSM must be integrated in the overall WHO statistics (APA, 2012).

While DSM-I and DSM-II were relatively similar, DSM-III, which is published in 1980, made major changes. The major problem confronting psychiatry was the unreliability of psychiatric diagnoses. Simply put, psychiatrists were unable to reliably diagnose similar psychiatric patients. This state of affairs contributes to a perception that psychiatry was unscientific and of little value to mainstream medicine. With DSM-III, major efforts were made to improve the reliability of psychiatric diagnoses. These included the development of explicit diagnostic criteria for classification and a descriptive approach that was not based on various theories of etiology of psychiatric illness. In other words, it was neither psychodynamic nor behavioral, but rather empirically based on scientific findings. DSM-III was a success because the reliability of psychiatric diagnoses increased somewhat. It was noteworthy the best reliabilities were obtained for diagnoses of sexual disorders. DSM-III also provided a medical nomenclature consistent for both researchers and clinicians who have facilitated the improvement of further empirical research.

And with DSM-IV the focus on methodology and empirical literature continued. A three-stage empirical process was developed that included systematic reviews of empirical literature, re-analyses of available sets of psychiatric data, and field trials to evaluate and refine proposed criteria for classification prior to adoption. As before of course, new versions of the DSM needed to be essentially comparable with the contemporary ICD (APA, 2012).

A point raised in the DSM-IV was the uncertain definition of a mental disorder. Essentially:

> ...each of the mental disorders is conceptualized as a clinical significant behavioral or psychological syndrome or pattern that occurs an individual and that is associated with the presence distress (e.g., a painful symptom) or disability (impairment in one or more important areas of functioning) or with a significantly increased risk of suffering death, pain, disability or an important loss of freedom. In addition, the syndrome or pattern must not be merely an expectable and culturally sanctioned response to a particular event, for example, the death of a loved one. Whatever its original causes, it must currently be considered a manifestation of a behavioral, psychological, or biological dysfunction in the individual. Neither deviant behavior (e.g., political, religious, or sexual) nor conflicts that are primarily between individual and society are mental disorders unless the deviance or conflict is a symptom of a dysfunction in the individual, as described above. (APA, 2000, page xxxi)

The term mental disorder suggests a difference between mental disorders and physical disorders, which increasingly is becoming illusionary. Put another way, it is suggesting that the mind is entirely separate from the brain, which is an absurd idea. Unfortunately, however, the DSM organizational system is categorical. In other words, the supposition is that each mental disorder can be defined in a way that is distinct from other mental disorders and/or physical disorders. It is worth noting that categorization is the most common way of transmitting information in society and has been the approach used in traditional systems of medical diagnosis. Categorization of course works best when members of a category are homogeneous, mutually exclusive, and distinct from other categories. Unfortunately many mental disorders do not share these characteristics. Rather many mental disorders have multiple overlapping features, and individuals sharing a diagnosis are likely to be heterogeneous with respect to the defining features of the primary diagnosis. Rather, in many ways, categorization has some significant limitations in the diagnosis of mental disorders. It is noteworthy that psychometrically, the scales of measurement are nominal, ordinal, interval, and ratio, in the order of empirical rigor. A ratio scale of measurement has a true zero point and equal intervals, interval has equal intervals, ordinal places measurement in an order of greater of the quality measures, and nominal is simply a classification. Nominal is the least sophisticated scale of measurement and the least useful (Reynolds & Livingston, 2012) and is, of course, what is used in the DSM system (Horton & Wedding, 1984).

An alternative might be a dimensional model, where individuals are seen as being placed on a continuum with respect to specific attributes and there is no sharp defining line between normal and abnormal. For example, one could categorize the dimension of height as short, medium, or tall categories, but a more telling metric is to describe someone's height in feet and inches. Or, in other words, use a more sophisticated scale of measurement (Horton & Wedding, 1984).

With respect to DSM-IV, the section of greatest interest for this chapter is that with respect to a section on Delirium, Dementia, and Amnestic and Other Cognitive Disorders. This section will be described in order to provide some perspective with respect to the changes proposed for DSM-5 and the relevance for assessing the neuropsychology of executive functioning. Essentially with respect to organic conditions and neuropsychology, the section on Delirium, Dementia, and Amnestic and Other Cognitive Disorders is intended to cover suspected organic conditions. As specifically noted in the DSM-IV manual:

> The predominant disturbance is a clinically significant deficit in cognition, which represents a clinically significant change from a previous level of functioning. For each disorder in the section, etiologies are either a general medical condition (although the specific general medical condition may not be identifiable) or substance (i.e., a drug of abuse, medication, or toxin), or a combination of these factors. (APA, 2000, p. 135)

It is noteworthy that in the previous edition DSM-III-R, there was a section titled "Organic Mental Syndromes and Disorders" which included the disorders now placed in the section on Delirium, Dementia, and Amnestic and Other Cognitive Disorders. DSM-IV expunged the term organic mental disorder because of a reluctance to consider that any mental disorder did not have a biological basis. Some might consider this a bias towards biological rather than psychological etiologies of mental disorders (APA, 2000).

The conditions of Delirium, Dementia, and Amnestic and Other Cognitive Disorders as identified in DSM-IV will be briefly described in an oversimplified fashion. This is done in order to provide context in terms of understanding proposed changes and DSM-5.

Delirium

Essentially a delirium is a change in cognition in the absence of a dementia that includes a quickly developing and fluctuating disturbance of consciousness. A delirium is usually considered a direct physiological consequence of a general medical condition, substance withdrawal or intoxication, medication effect or exposure to a neurotoxin, or multiple etiologies. Diagnostic criteria for delirium due to a general medical condition include:

A. *Disturbance of consciousness (i.e., reduced clarity of awareness of the environment) with reduced ability to focus, sustain, or shift attention.*
B. *A change in cognition (such as memory deficit, disorientation, language disturbance) or development of a perceptual disturbances not been accounted for by a preexisting, established, or evolving dementia.*
C. *The disturbance develops over a short period of time (usually hours to days) and tends to fluctuate during the course of the day.*
D. *There is evidence from history, physical examination, or laboratory findings at disturbances caused by the direct physiological consequences of the general medical condition* (APA, 2000).

Other subcategories of delirium can be from substance intoxication, substance withdrawal, and multiple etiologies. When a specific etiology for delirium cannot be determined, the subcategory is specified as Not Otherwise Specified (NOS).

Dementia

Essentially dementia is memory impairment that includes at least one other cognitive disturbance including language impairment, motor coordination difficulties, perceptual difficulties, or executive functioning impairment in the absence of a delirium. A dementia must represent a decline from a previous higher level of adaptive functioning with significant impairment in the ability to work and function in society (APA, 2000).

Diagnostic criteria for dementia, the Alzheimer's type, include:
A. *The development of multiple cognitive deficits manifested by both*
   1. *Memory impairment (impaired ability to learn new information recall previously learned information)*
   2. *One (or more) of the following cognitive disturbances*
      (a) *Aphasia (language disturbance)*
      (b) *Dyspraxia (impaired ability carry of motor activities despite intact motor function)*
      (c) *Agnosia (failure to recognize or identify objects despite intact sensory function)*
      (d) *Disturbance of executive functioning (i.e., planning, organization, sequencing, abstracting)*
B. *The cognitive deficits in criteria A1 and A2 each cause significant impairment in social or occupational functioning and represent a significant decline from a previous level of functioning*
C. *The course is characterized by gradual onset and continuing cognitive decline*
D. *The cognitive deficits in criteria A1 and A2 are not caused by the following*
   1. *Other central nervous system conditions that cause progressive deficits in memory and cognition (e. g, cerebral vascular disease, Parkinson's disease, Huntington's disease, subdural hematoma, normal pressure hydrocephalus, brain tumor)*
   2. *Systemic conditions are known to cause dementia (e.g., hypothyroidism, vitamin B12 or folic acid deficiency, niacin deficiency, hypocalcemia, neurosyphilis, HIV infection)*
   3. *Substance-induced conditions*
E. *The deficits did not occur exclusively during the course of a delirium*
F. *Disturbance is not better accounted for by another Axis I disorder (Major Depressive Disorder, Schizophrenia)* (APA, 2000)

A dementia of the Alzheimer's type can also be characterized as with or without a clinically significant people disturbance and with either early onset (at age 65 years or below) or late onset (after age 65 years).

Other subtypes of dementia in DSM-IV include Vascular Dementia, Dementia due to HIV disease, Dementia due to Head Trauma, Dementia due to Parkinson's disease, Dementia due to Huntington's disease, Dementia due to Pick's disease, Dementia due to Creutzfeldt-Jakob disease, Dementia due to General Medical Condition, Substance-Induced Persisting Dementia, Dementia due to Multiple Etiologies, and Dementia NOS (APA, 2000).

Amnestic Disorder

The essential feature of amnestic disorder is that it is not due to a dementia or delirium, but is memory impairment for either the ability to learn new information and/or to recall previously learned information or events. The memory impairment must represent a decline from a previously higher level of adaptive functioning and cause significant difficulty in terms of working and adapting to the demands of daily living (APA, 2000).

Diagnostic criteria for Amnestic Disorder due to General Medical condition include:
A. *The development of memory impairment as manifested by impairment and the ability to learn new information or the inability to recall previously learned information.*
B. *The memory disturbance causes significant impairment in social or occupational functioning and represents a significant decline from a previous level functioning.*
C. *The memory disturbance does not occur exclusively during the course of a delirium or dementia.*
D. *There is evidence from history, physical examination, or laboratory findings that*

the disturbance is the direct physiological consequence of a general medical condition (including physical trauma). (APA, 2000)

Amnestic disorders can also be classified as either transient (memory impairment lasts for 1 month or less) or chronic (if memory impairment lasts more than 1 month).

Other subtypes of amnestic disorders include Substance-Induced Persisting Amnestic Disorder and Amnestic Disorder NOS.

Other Cognitive Disorders

Essentially the other cognitive disorders are a NOS condition and that are disorders of cognitive dysfunction that do not meet the criteria for any of the previously described deliriums dementias or amnestic disorders, but are presumed to be the result of a general medical condition. Research criteria are provided for Mild Neurocognitive Disorder and Postconcussional Disorder.

Mild Neurocognitive Disorder

Essentially, mild neurocognitive disorder is the impairment of two or more areas of neurocognitive functioning of a mild nature that is due to a general medical condition and not better time accounted for by another mental disorder (APA, 2000).

Research criteria for mild neurocognitive disorder include:

A. *The presence of two (or more) of the following impairments and cognitive function, lasting most the time for at least 2 weeks (as reported by the individual or a reliable informant)*
   1. *Memory impairment as identified by reduced ability to learn or recall information*
   2. *Disturbance in executive functioning (i.e., planning, organizing, sequencing, abstracting)*
   3. *Disturbance in attention or speed of information processing*
   4. *Impairment in perceptual motor abilities*
   5. *Impairment in language (e.g., comprehension, word finding)*

B. *There is objective evidence from physical examination or laboratory findings, (including neuroimaging imaging techniques) of a neurological or general medical condition that is judged to be etiologically related to the cognitive disturbance*

C. *There is evidence from neuropsychological testing or quantified cognitive assessment of an abnormality or decline in performance*

D. *The cognitive deficits caused marked distress or impairment in social, occupational, or other important areas of functioning and represent a decline from a previous level of functioning*

E. *The cognitive disturbance does not meet the criteria for a delirium, a dementia, or amnestic disorder and is not better accounted for by another mental disorder (e.g., you. a Substance-Related Disorder, Major Depression Disorder)* (APA, 2000)

Postconcussional Disorder

Postconcussional disorder is essentially a combination of impaired cognitive functioning and multiple neurobehavioral symptoms which are a consequence of a closed head injury which has caused a cerebral concussion. Cognitive deficits would need to be documented in either attention or memory and neurobehavioral sequelae, and attention or memory deficits would have to follow a head injury for a Postconcussional disorder to be diagnosed.

Research criteria for Postconcussional Disorder are as follows:

A. *A history of head injury that has caused significant cerebral contusion*
B. *Evidence for neuropsychological testing or quantified cognitive assessment of difficulty in attention (concentrate, shifting focus of attention, performing simultaneous cognitive tasks) or memory (learning or recalling information)*
C. *Three (or more) of the following occur shortly after the trauma and last at least 3 months*
   1. *Becoming fatigued easily*
   2. *Disordered sleep*
   3. *Headache*

4. *Vertigo or dizziness*
   5. *Irritability or aggression on little or no provocation*
   6. *Anxiety, depression, or affective lability*
   7. *Changes in personality (e.g., social or sexual inappropriateness)*
   8. *Apathy or lack of spontaneity*
D. *The symptoms of criteria B and C have their onset following head injury or else represent a substantial worsening of preexisting symptoms*
E. *The disturbance causes significant impairment in social or occupational functioning and represents a significant decline from a previous level of functioning. In school-age children, the impairment may be made manifest by significant worsening in school or academic performance dating from the trauma*
F. *These symptoms do not meet the criteria for Dementia Due to Head Trauma and are not better accounted for by another mental disorder (e.g., Amnestic Disorder Due to Head Trauma, Personality Change Due to Head Trauma) (APA, 2000)*

Discussion of DSM-IV Delirium, Dementia, and Amnestic and Other Cognitive Disorders

Essentially it can be seen that under DSM-IV, cognitive disorders are conceptualized as always being related to a medical condition and limited to four categories. Basically, Delirium, which is a disturbance of consciousness and attention, Amnestic Disorder, which is memory impairment, Dementia, which is memory impairment in addition to either impairment of executive functioning, language, perceptual motor abilities or sensory perception, and Other Cognitive Disorders which are impairment in cognitive functioning, not better diagnosed in the three preceding mental disorder categories. It would be clear from the fact that Other Cognitive Disorders including Research Criteria for Mild Neurocognitive Disorder and Postconcussional Disorder that the diagnosis of mental disorders that had previously been described as Organic Mental Disorders under DSM-III was a still evolving area of research. It might also be noted that while executive functioning impairment was noted as a specific criteria in both the Dementia and Other Cognitive Disorders-Mild Cognitive Disorder categories, executive functioning was not singled out as a specific category. Similarly the discussion of executive functioning impairment was relatively superficial, but it should be mentioned that neuropsychological testing was specifically included with respect to diagnosis in the research criteria for Mild Cognitive Disorder and Postconcussional Disorder (APA, 2000).

Next, attention will be devoted to the current DSM-5 proposal for diagnosing and classifying what was labeled as Organic Mental Disorders under DSM-III and was in the section labeled Delirium, Dementia, and Amnestic and Other Cognitive Disorders under DSM-IV.

Proposed DSM-5 Organizational Structure Disorders Names

As of the time this chapter was written, the following are the proposed names for the mental disorders included in DSM-5:
Neurodevelopmental Disorders
Schizophrenia Spectrum and Other Psychotic Disorders
Bipolar and Related Disorders
Depressive Disorders
Anxiety Disorders
Obsessive Compulsive and Related Disorders
Trauma and Stressor Related Disorders
Dissociative Disorders
Somatic Symptom Disorders
Feeding and Eating Disorders
Elimination Disorders
Sleep-Wake Disorders
Sexual Dysfunctions
Gender Dysphoria
Disruptive, Impulse Control and Contact Disorders
Substance Use and Addictive Disorders
Neurocognitive Disorders

Personality Disorders
Paraphilic Disorders
Other Disorders (APA, 2012)

For the purposes of this chapter, attention will be focused on the Neurocognitive Disorders as this section would be the most relevant to the neuropsychology of executive functioning. Essentially, what had been the mental disorders covered in the Organic Mental Disorders section in DSM-III and the Delirium, Dementia, Amnestic and Other Cognitive Disorders section in DSM-IV are now assigned to the Neurocognitive Disorders section in DSM-5. The Dementia and Amnestic Disorders of DSM-IV have been removed in the DSM-5 Neurocognitive Disorders Work Group's proposal and have replaced by categories of Major Neurocognitive Disorders and Minor Neurocognitive Disorders.

Neurocognitive Disorders in DSM-5

As of the writing of this chapter, the Neurocognitive Disorders Work Group's proposal had suggested three broad syndromes of Neurocognitive disorders. These broad syndromes are Delirium, Major Neurocognitive Disorder, and Mild Neurocognitive Disorder. As might be expected Delirium is somewhat similar to what was in DSM-IV but was a greater emphasis on attention, and there are also subcategories similar to DSM-IV for Not Otherwise Specified (NOS) for where the formal diagnostic criteria cannot be met, but Mild Neurocognitive Disorder and Major Neurocognitive Disorder are new categories in DSM-5. Essentially the category of Dementia in DSM-IV was replaced by Major Neurocognitive Disorder in DSM-5 (APA, 2012).

Members of the Neurocognitive Disorders Work Group at the time of the writing of this chapter were as follows:
Jeste, Dilip, V., M.D. (Chairperson from 2007 to 2011, was elected APA president in 2011)
Blacker, Deborah, M.D., Sc.D
Blazer, Dan, M.D., Ph.D., M.P.H. (Chairperson)
Gangull, Mary, M.D., M.P.H.
Grant, Igor, M.D.
Lenze, Eric, J., M.D.
Paulsen, Jane, Ph.D.
Ronald, Petersen, Ph.D., M.D. (Co-Chairperson)
Sachdev, Perminder, M.D., Ph.D., FRAZCP (APA, 2012)

A list of the proposed Neurocognitive Disorders categories and subtypes is shown below:
Delirium
Substance-Induced Delirium
Delirium Not Elsewhere Classified
Mild Neurocognitive Disorder
Major Neurocognitive Disorder
Subtypes of Major and Mild Neurocognitive Disorders
Neurocognitive Disorder due to Alzheimer's disease
Vascular Neurocognitive Disorder
Frontotemporal Neurocognitive Disorder
Neurocognitive Disorder due to Traumatic Brain Injury
Neurocognitive Disorder due to Lewy Body Dementia
Neurocognitive Disorder due to Parkinson's Disease
Neurocognitive Disorder due to HIV Infection
Substance Induced Neurocognitive Disorder
Neurocognitive Disorder due to Huntington's Disease
Neurocognitive Disorder due to Prion Disease
Neurocognitive Disorder due to Another Medical Condition
Neurocognitive Disorder Not Elsewhere Classified (APA, 2012)

Delirium Criteria in DSM-5

A. *Disturbance of attention (i.e., reduced ability to focus, sustained and shift attention) and orientation to the environment.*
B. *The disturbance develops over a short period of time (usually hours to days) and represents an acute change from baseline that is not solely attributable to another neurocognitive disorder and tends to fluctuate in severity during the course of the day.*

C. A change in an additional cognitive domain, such as memory deficit, disorientation or language disturbance, or perceptual disturbances, that is not better accounted for by a preexisting, established, or evolving neurocognitive disorder.
D. Disturbance and in criteria A and C must not be occurring in the context of the severely reduced level of arousal such as coma. (APA, 2012)

Neurocognitive Disorder Domains

Neurocognitive Disorder in DSM-5 is conceptualized as involving either modest (Minor Neurocognitive Disorder) or substantial (Major Neurocognitive Disorder) cognitive decline in one or more of the following domains:

Complex attention (sustained attention, divided attention, selective attention processing speed)

Executive ability (planning, decision making, working memory, responding to feedback/error correction, overriding habits, mental flexibility)

Learning and memory (immediate memory, recent memory [including free recall, cute recall, and recognition memory])

Language (expressive language [including naming, fluency, grammar, and syntax] and receptive language)

Visuoconstructive-perceptual ability (construction and visual perception)

Social Cognition (recognition of emotions, theory of mind, behavioral regulation) (APA, 2012)

The diagnostic criteria in DSM-5 are based on the above Neurocognitive Disorder domains.

Minor Neurocognitive Disorder Criteria

A. Evidence of modest cognitive decline from previous levels performance in one or more of the domains outlined above based on:
  1. Concerns of the individual, a knowledgeable informant, or the clinician that there has been a modest decline in cognitive function
  2. A decline in neurocognitive performance, typically involving test performance in the range of 1 and 2 standard deviations, below appropriate norms (i.e., between the 3rd and 16th percentile) on formal testing or equivalent clinical evaluation
B. The cognitive deficits are insufficient to interfere with independence (i.e., instrumental activities of daily living [more complex tasks such as paying bills or managing medications] are preserved), but greater effort, compensatory strategies, or accommodations may require to maintain independence
C. The cognitive deficits do not occur exclusively in the context of a Delirium
D. The cognitive deficits are not primarily attributable to another mental disorder (e.g., Major Depressive Disorder, Schizophrenia) (APA, 2012)

Major Neurocognitive Disorder Criteria

A. Evidence of substantial cognitive decline from previous levels performance in one or more of the domains outlined above based on
  1. Concerns of the individual, a knowledgeable informant, or the clinician that there has been a substantial decline in cognitive function
  2. A decline in neurocognitive performance, typically involving test performance in the range of 2 or more standard deviations, below appropriate norms (i.e., below the third percentile) on formal testing or equivalent clinical evaluation
B. The cognitive deficits are sufficient to interfere with independence (i.e., requiring assistance at a minimum with instrumental activities of daily living [more complex tasks such as paying bills or managing medications])
C. The cognitive deficits do not occur exclusively in the context of a Delirium

D. *The cognitive deficits are not primarily attributable to another mental disorder (e.g., Major Depressive Disorder, Schizophrenia)* (APA, 2012)

As earlier mentioned Major Neurocognitive Disorder in DSM-5 is intended to replace Dementia as conceptualized in DSM-IV. Minor Neurocognitive Disorder in DSM-5, however, is not a replacement of Amnestic Disorder in DSM-IV. While Minor Cognitive Disorder in DSM-5 includes memory impairment, it is one of the cognitive domains that can be impaired; there are also a number of other cognitive domains that can be impaired in order to support the diagnosis of Minor Cognitive Disorder in DSM-5. According to the Neurocognitive Disorders Work Group, the category of Minor Neurocognitive Disorder was devised for the assessment of clinical needs of individuals who have mild cognitive deficits in one or more cognitive domains, but still are able to function independently with respect to instrumental activities of daily living, even though increased effort or compensation strategies are required to successfully complete activities of daily living. In other words, these are individuals with minor cognitive impairment who require assessment and treatment from a clinician, but can still function independently outside of an institutional environment. The point is that patients could be identified for early intervention efforts which could forestall the loss of independence by many individuals with brain-related disorders. Neuromedical conditions such as diabetes, early stages of cerebral vascular disease, substance-use-related brain disorders, HIV, mild brain injury, and the early prodromal stage of Alzheimer's disease (pre-diagnosis of Alzheimer's disease) are all conditions for which early identification and intervention could be extremely beneficial. Appropriate services would include treatment of associated mood symptoms, instruction in behavioral compensation strategies, identification of possible treatable causes by further investigation of brain functioning, and lifestyle modifications (APA, 2012).

Under DSM-IV, such cognitively impaired individuals would have been coded as Cognitive Disorder NOS. It is also worth noting that in many clinical settings, the syndrome has been described as Mild Cognitive Impairment (MCI) and has been an important area of clinical research in recent years. It might also be noted that the term "minor" for Minor Neurocognitive Disorder was chosen by the Neurocognitive Disorders Work Group in order to be parallel with the term "major" for Major Neurocognitive Disorder and to also be able to maintain within the Major Neurocognitive Disorder diagnosis distinctions with respect to mild, moderate, and severe versions of Major Neurocognitive Disorder. At the same time the Neurocognitive Disorders Work Group noted that specific term "minor" might be misinterpreted to suggest a lack of need for services, which is not the intention of the Neurocognitive Disorders Work Group (APA, 2012).

Both Minor Neurocognitive Disorder and Major Neurocognitive Disorder clearly specify that the cognitive deficits must be a change from a previously more adequate level of performance so that individuals who have suffered from a lifetime of cognitive impairment would not qualify for either Neurocognitive Disorder diagnosis. In other words only acquired cognitive deficits are considered as Neurocognitive Disorders. Perhaps the major dividing line between Minor Neurocognitive Disorder and Major Neurocognitive Disorder is related to functional independence. Simply put, in Major Neurocognitive Disorder, the substantial cognitive deficits interfere with independence such that patients require assistance with respect to complex daily living tasks such as paying bills or managing medications (APA, 2012).

With respect to Major Neurocognitive Disorder in DSM-5, while the clear intention is to replace Dementia as conceptualized in DSM-IV, the term Major Neurocognitive Disorder and not specifically associated with Alzheimer's disease is was to a large degree the term Dementia. Rather, Major Neurocognitive Disorder is expected to be better accepted with respect to individuals who have cognitive deficits related to head injury or HIV (APA, 2012).

Another aspect of Major Neurocognitive Disorder is that memory impairment is no longer required in every case for diagnosis. With the Dementia diagnoses in DSM-IV, in every case memory impairment was required for a diagnosis of dementia. In DSM-5, memory impairment can be evidence for cognitive impairment, but other neurocognitive domains can also be considered as evidence of cognitive impairment. While in Alzheimer's disease, often memory impairment is the initial area of cognitive decline and other cognitive domains become impaired in later stages of the progression of the disease, in other neurocognitive disorders, the progression of cognitive impairment may be different. For example, in frontotemporal degeneration, HIV-related cognitive decline and cerebral vascular disease, among other neurocognitive disorders, executive functioning, attention, or language, among other cognitive domains, could be the first cognitive domain to be impaired. In addition, it might be recalled that in the research criteria previously reviewed earlier under DSM-IV for Postconcussional Disorder, a cognitive deficit in the domain of attention alone would have been considered as evidence for the diagnosis of Postconcussional Disorder (APA, 2012).

Brief Review of the Concept of Executive Functioning

The conceptualization of executive functioning should be briefly described. Executive functioning is a multifaceted neuropsychological construct that is presumed to be supported by a widely distributed neural network in the human brain. The overall construct of executive functioning is thought to include multiple neuropsychological subdomains such as working memory, mental set-shifting, and planning/cognitive flexibility, among others (Goldberg, 2001). Executive functions also may be differentiated conceptually by contrasting them with the concept of knowledge. Executive functions, as a class of conceptual elements, primarily deal with making mental judgments and performing adaptive actions, while the concept of knowledge relates to retention and recall of an array of previously learned and organized sets of facts. Executive functioning, as a concept, denotes active decision making and behavioral outputs that are adaptive to external demands rather than the storage and reproduction of a number of varieties of organized information items (Reynolds & Horton, 1994).

Lezak (1995) states that each component of executive functioning "…involves a distinctive set of activity related behaviors" (p. 650) and are necessary for successful adaptive self-direction. Another way of understanding executive functions would contrast them with knowledge. Simply put, executive functions plan, monitor, and evaluate while performing adaptive actions, carrying out behavioral plans, and doing tasks, while knowledge relates to storing and recall of an organized set of facts which may be related to a specific cultural context and environment. Executive functioning involves active real-time decision making, action and motor outputs that are adaptive to external demands rather than the passive retention, and retrieval of varieties of factual information and events.

Initial problem-solving has been described as first matching to previous exemplars and later problem-solving has been described as following a rule-based system. The point being that in problem-solving, while initial problem-solving efforts are often subsumed by the right cerebral hemisphere, when the strategy can be verbalized, then executive functioning moves from a right hemisphere to a left hemisphere focus. This switching of problem-solving styles matches the Goldberg and Costa's (1981) conceptualization of hemispheric differences in executive functioning and might have implications for better understanding the domain of executive functioning.

Equating frontal lobe lesions with executive functioning is a common but serious error. Essentially, frontal lobe lesions are a neuroanatomical condition. On the other hand, executive functioning is a neuropsychological construct. Executive functioning certainly has some overlap with frontal lobe lesions as the frontal lobe controls motor output and virtually every way of measuring executive functioning requires some degree of motor output. At the same time, lesions in other areas of the brain other than the frontal lobe may produce impairment in executive functioning (Heaton, 1981). While often in clinical neuropsychology, executive functioning as a concept has been often associated with frontal lobe functioning, more contemporary, sophisticated understanding of brain behavior functioning avers that impairments in brain areas other than the frontal lobes may impair executive functioning (Reitan & Wolfson, 2000).

With respect to the conceptualization of executive functioning, including as measures of executive functioning what are primarily reading or psychomotor speed tests is another common error. For example, including the Trail Making Test (TMT) part A as a measure of executive functioning is a mistake. Most clinical neuropsychologists would consider part A as a measure of psychomotor processing abilities and perhaps attention. Similarly, the Word portion of the Stroop Color-Word Test (SCWT) should not be included as a measure of executive functioning, whereas most clinical neuropsychologists would consider the Word portion of the SCWT as a reading speed measure. This conceptualization of executive functioning which includes tests not thought of as measures of executive functioning is a fatal error.

Therefore the concept of executive functioning does not postulate specific localization for any particular brain area as being solely responsible for executive functioning performance. Similarly, the entire range of possible executive functions is potentially limitless, but there are a limited number of clinical neuropsychological procedures that have been accepted as measuring aspects of and objectifying this important area of mental functioning and adaptive functioning (Reynolds & Horton, 1994).

Simply put, theoretical models of executive functioning are very controversial. As one clinical neuropsychologist said, "The only thing that neuropsychologists can agree about executive functioning, is that they do not agree about executive functioning" (name withheld at request of author). Similarly, a theory of executive functioning is like a toothbrush, everyone appears to want to have their own (Reynolds & Horton, 1994).

A paradox, however, is that while there is little theoretical agreement regarding executive functioning, there is substantial agreement concerning what neuropsychological tests are appropriate measures of executive functioning. Neuropsychological procedures that have traditionally been most commonly accepted as measures of executive functioning performance include card sorting tasks, category and letter retrieval tasks, and "trail-making" tasks. This list is not exhaustive of all measures of executive functioning but rather a limited selection that is considered "gold standard" measures of executive functioning (Reynolds & Horton, 1994). Each clinical neuropsychology procedure will be briefly reviewed to provide a clinical perspective.

Card Sorting Tasks

Tests that sort cards, to assess mental abilities have been widely used in neuropsychology (Lezak, 1995). The best known card sorting test has been the Wisconsin Card Sorting Test (WCST) (Grant & Berg, 1948). The purpose of the WCST is to assess abstract thinking and the ability to set shift (Lezak, 1995). The WCST test materials are 2 decks of 64 cards, which are sorted based on the principles of color, form, and number (Heaton, 1981). The colors are red, green, yellow, and blue; the numbers are one, two, three, and four; and the forms are triangles, stars, crosses, and circles. Four stimulus cards are used as guides for sorting. The four stimulus cards are one red triangle, two green stars, three yellow crosses, and four blue circles. The person assessed is to sort the decks of cards in four piles, each pile under one of the stimulus cards (Heaton, 1981). The examinee does not know the sorting principle that is to guide the sorting but is given immediate feedback that each sort is correct or incorrect. The sorting principle is by color, form, and numbers and the sequence is repeated. After the examinee achieves ten correct sorts, the sorting principle is changed (Heaton, 1981). The test ends after the examinee completes six correct sorts or exhausting the two decks of cards (Heaton, 1981). The most commonly used current WCST scoring system is the one devised by

Heaton (1981) and later revised and expanded (Heaton, Chelune, Talley, Kay, & Curtiss, 1993).

The WCST was first advocated as a measure of the integrity of the frontal lobes by Milner (1963). Subsequent research finding was mixed as a measure of the frontal lobes (Lezak, 1995). Heaton (1981) averred that the WCST is a generalized measure of executive functioning (planning ability) rather than a measure of frontal lobe functioning. Interrater reliability of the WCST has been adequate (Axelrod, Goldman, & Woodward, 1992) and age effects (Lezak, 1995) and education effects were found to be small (Heaton, Grant, & Mathews, 1991). In patients that have successfully solved the card sorting principles and have adequate memory functioning, the WCST is less useful for repeated assessment of executive functioning (Lezak, 1995) as the patient is not naive to the sorting principles.

Category and Letter Retrieval

Verbal fluency measures are frequently impaired in brain-damaged patients (Lezak, 1995). While general language abilities are impaired by aphasic symptoms, certain aspects of verbal fluency appear to have unique advantages for the assessment of executive functioning. Estes (1974) has suggested that verbal fluency measures can be sensitive measures of executive functioning. The two most common verbal fluency assessment formats in neuropsychology have been category and letter retrieval tasks.

Letter retrieval was used by Arthur L. Benton when he developed the Controlled Oral Word Association Test (COWAT) (Benton and Hamsher (1989)). The COWAT uses three letters, "FAS," for three word naming trials of 1 min each. The examinee says words that start with either F, A, or S. The three trials are added together to compute a total score that is adjusted for age, education, and gender (Benton & Hamsher, 1989).

Benton (1968) found that letter retrieval was impaired in frontal lobe patients and the left frontal lobe patients did more poorly than right frontal lobe patients but bilaterally impaired patients did more poorly than either left or right frontal lobe patients. Subsequent research confirmed Benton's findings (Lezak, 1995). Reliability was found to be adequate (Benton & Hamsher, 1989).

The other common type of verbal fluency-category naming has been used to assess mental performance in Alzheimer's patients (Rosen, 1980). This research has focused on the use of category retrieval to assess the brain's semantic category organization. The most common category used is "animals." Patients are asked to say as many different names of animals as they can in 1 min. Other commonly used categories are "items in a super market," "fruits," and "vegetables." Animal naming has been shown to discriminate between dementia and depression (Hart, Kwentus, Taylor, & Hamer, 1988). Reliability data on category naming appears to be adequate.

Trail-Making Tasks

Trail-making tasks appear to be sensitive measures of brain function (e.g., Reitan, 1955, 1958) and are frequently used by neuropsychologists (Mitrushina, Boone, & D'Elia, 1999). Reitan (1992) noted that trail-making tasks require

> ...immediate recognition of the symbolic significance of numbers and letters, ability to scan the page continuously to identify the next number or letter in sequence, flexibility in integrating the numerical and alphabetical series, and completion of these requirements under the pressure of time. (p. ii)

Mitrushina et al. (1999) suggested that trail-making tasks appeared to be measures of attention, visual scanning, speed of eye-hand coordination, and information processing. Lezak (1995) similarly averred that trail-making tasks measured complex visual scanning, motor speed, and attention, concluding that trail making was sensitive to the effects of brain injury. Mitrushina et al. (1999) also agreed that the abilities required for trail making and their coordination require good executive function and are

> ...known to be highly vulnerable to deterioration resulting from brain pathology of different etiologies. (p. 33)

A large body of empirical research has shown that trail-making tasks are useful in assessing and documenting brain damage or dysfunction in mild cases of brain injury, early detection of dementia, and detection of attention/concentration deficits and frontal activation defects (Anderson, 1994; Mitrushina et al., 1999; Stuss, Bisschop, Alexander, Levine, Katz, & Izukawa, 2001). Research has also documented that trail-making tests are sensitive measures of executive function (Lezak, 1995; Mitrushina et al., 1999; Storandt, Botwinick, & Danziger, 1984). Reitan (1992) concluded that trail making is one of the best available measures of general brain functions, a conclusion echoed by Jarvis and Barth (1984) and Horton (1979).

The Original Trail-Making Test. The original trail-making task, now known as the TMT, parts A and B, first appears in clinical application according to Horton (1979) as part of the US Army Individual Test Battery in 1944. Spreen and Strauss (1998) note that the test was first developed in 1938 by Partington (Partington & Leiter, 1949) as a measure of divided attention. The original TMT consisted of two parts, trails A and trails B. On part A, the examinee is required to quickly draw a line to connecting circles that contain numbers going from 1 to 25. On part B, the examinee also quickly draws a line connecting circles, but the circles contain either numbers or letters. The task is to draw the line alternating between numbers and letters to be connected in the sequence, 1-A-2-B-3-C. Reitan (e.g., 1955, 1958; also see Reitan & Wolfson, 1993) found that the TMT was "… extremely sensitive to the biological condition of the brain" (Reitan & Wolfson, 1993, p. 40).

The validity of the TMT as a measure of brain function and its integrity is well supported by empirical studies (Lezak, 1995; Mitrushina et al., 1999; Reitan & Wolfson, 1993; Spreen & Strauss, 1998; Stuss et al., 2001).

The Comprehensive Trail-Making Test (CTMT). The original TMT however had a number of methodological weaknesses primarily regarding normative data. The more recently developed CTMT (Reynolds, 2003) revised and re-normed the time-honored TMT to making it more reliable and more sensitive to brain dysfunction. Data has shown that the CTMT has become a very frequently used neuropsychological test.

Test of Verbal Conceptualization and Fluency

An example of a published neuropsychological test of executive functioning that includes subtests of card sorting, verbal fluency, and trail making is the Test of Verbal Conceptualization and Fluency (TVCF).

The TVCF presents a standardized set of four subtests with a total administration time of 25–30 min for most individuals. The test is designed to measure multiple aspects of executive functions. The TVCF was designed and standardized for use with individuals ranging in age from 8 years through 89 years. Standardized (or scaled scores) are provided in the form of smoothed linear T-scores, having a mean of 50 and a standard deviation of 10, along with their accompanying percentile ranks.

The four subtests of the TVCF are as follows:
1. Category Fluency measures word retrieval by conceptual category (e. g., things to eat, wear) and fluency of ideation.
2. Classification is a verbal measure of set-shifting and rule induction that is designed as a verbal or language-based analog to the well-known WCST (Grant & Berg, 1948).
3. Letter Naming measures word retrieval by initial sound and fluency of ideation.
4. Trails C measures the ability to coordinate high attentional demands, sequencing, visual search capacity, and the ability to shift rapidly between Arabic numerals and linguistic representations of numbers. The Trails task used and conormed with the other TVCF subtests is a variation of several other "trail-making" tasks and was taken from the earlier published CTMT (Reynolds, 2003).

The materials needed to administer the TVCF are the TVCF test booklet, the Classification subtest set of cards for sorting, the Trails C subtest form, a stopwatch, and a sharpened pencil, preferably without an eraser. Chapter 16 in this volume contains a more detailed description of the TVCF, and other later chapters describe other tests of executive functioning.

Comparison of DSM-5 and the Neuropsychology of Executive Functioning

Earlier in this chapter, the proposed DSM-5 proposal, as of the time of the writing of this chapter, was compared and contrasted with the earlier DSM-IV manual. In addition, the concept of the neuropsychology of executive functioning was briefly reviewed. In the current discussion, a brief discussion of neuropsychological assessment is offered as a foundation for later comments.

Neuropsychological assessment has been largely based on the empirical research and clinical contributions of Ralph M. Reitan (1955) and other important neuropsychological contributors too numerous to mention (please see Fitzhugh-Bell, 1997, for a review of the early history of clinical neuropsychology). The importance of neuropsychological assessment techniques in diagnosis and in the design of rehabilitation and related treatments has been recognized officially by the American Academy of Neurology (AAN) in an official report of its Therapeutics and Technology Assessment Subcommittee (American Academy of Neurology, 1996). The major contribution of clinical neuropsychology has been the diagnosis of brain injury and other forms of central nervous system dysfunction. These include contributions to diagnosis (e.g., Reitan & Wolfson, 2001; Reynolds, 2003) and to rehabilitation (e.g., Bennett, 2001; Hartlage, 2001). More recently, there has been a rise in the application of neuropsychological measures for determining the functional deficits and related functional implications of CNS damage or dysfunction and greatly increased use in psychiatric, rehabilitation, and forensic settings. Neuropsychological assessment of potential brain damage deficits and an evaluation of the implications of such findings are necessary through neuropsychological testing because neuroimaging methods are limited in the assessment of subtle brain injuries and cannot specify the functional implications of any visualized abnormality or brain injury. Neuropsychological testing is required for the determination of the clinical correlates of brain abnormalities. Such large government programs as Medicare, Social Security, and CHAMPUS reimburse appropriately credentialed professionals and clinicians for clinical neuropsychological assessment, and the American Medical Association (2001) assigns unique billing codes for neuropsychological testing. Among the most useful and sensitive of neuropsychological measures have been tests of executive functioning. It is worthwhile recalling that the specific terms such as attention, language, memory, spatial relations, and executive functioning are all psychological constructs initially researched and devised by psychologists and neuropsychologists (Horton & Wedding, 1984).

What is most troubling about the role of the neuropsychology of executive functioning in the proposed DSM-5 is the dearth of a role for neuropsychological assessment in general and the neuropsychology of executive functioning in particular. While the proposed neurocognitive domains in DSM-5 is a very clear borrowing of psychological and neuropsychological constructs, there are few mentions of neuropsychological assessment and/or neuropsychological testing and even fewer mentions of the neuropsychology of executive functioning. Despite the tremendous body of empirical research on the use of neuropsychological assessment in the diagnosis of brain damage, an equivalent clinical evaluation is seen as equal to formal testing. The equivalent clinical evaluation proposed would appear to be a brief cognitive screening instrument such as the Mini-Mental State Examination (MMSE) (Folstein, Folstein, & McHugh, 1975) or the Montreal Cognitive Assessment (MoCA) (Nasreddine et al., 2005) which may take 10 min or less to administer.

The cognitive impairment assessed by the MMSE or MoCA is general cognitive functioning (i.e., both instrument initially developed to assess early and late stages of dementia), and the

MMSE and the MoCA were not designed to be specific to the neuropsychology of executive functioning. If the MMSE or MoCA is used to evaluate brain-behavior relationships, then the relationship seen is the relationship with dementia rather than executive functioning. To use the MMSE or MoCA does not add a clear evidence for or against impairment of executive functioning particularly with respect to very subtle early forms of brain damage and dysfunction that are the most likely to be amenable to early interventions and prevention of further decline.

The analogy might be offered of a patient going to his/her primary care practitioner (PCP) for an annual checkup. If the PCP said that no laboratory values of thyroid functioning or blood cell count, etc., were needed because the PCP could assess those laboratory values by looking at the patient, then the patient would know his/her physician was a quack. Neuropsychological assessment was developed to be laboratory values regarding higher cortical brain functioning for the neurologist. As earlier stated, terms such as attention, memory, language, spatial relations, and executive functioning are all psychological and neuropsychological constructs that were initiated, developed, and researched by psychologists and neuropsychologists (Horton & Wedding, 1984). A brief cognitive screening instrument is a poor substitute for a comprehensive neuropsychological evaluation that can take 5–6 hours to administer.

It might also be noted that a comprehensive neuropsychological evaluation covers a wide range of neuropsychological domains. These neuropsychological domains include intelligence and academic achievement, sensory-perceptual functioning, motor abilities, perceptual-motor abilities, language, spatial relations, executive functioning, attention and memory, processing speed, and emotional and personality functioning. The list of neuropsychological domains found to be necessary for a comprehensive neuropsychological evaluation was determined by empirical clinical research over decades. In addition, an appropriate comprehensive neuropsychological evaluation requires the neuropsychological test scores to be interpreted through multiple levels of inference. The multiple levels of inference include consideration of neuropsychological test scores based on level of performance, patterns of performance, pathognomonic signs, and left-right comparisons of the two sides of the body. Extensive clinical research has proved repeatedly that the use of all four methods of inference yielded better diagnostic results than a single method of inference. These multiple levels of inference allow for a more rigorous methodological examination of the neuropsychological test data than the use of a single method of inference. The multiple methods of inference and comprehensive research-based inclusion of all relevant neuropsychological ability domains allow for extensive considerations of interrelationships of the multiple brain-behavior components. Simply put, human brains are very complex and require complex models and extensive neuropsychological test data for an appropriate neuropsychological evaluation (Horton & Wedding, 1984). In DSM-5, the criteria also allow for only the use of a single method of inference—the level of performance (APA, 2012). This insures that suboptimal diagnostic results will be obtained for DSM-5.

While the neurocognitive domains mentioned in DSM-5 are a step in the right direction on one hand, they are a step backward on the other hand. On the plus side, there are multiple domains of neuropsychological abilities such as memory, language, visual-spatial skills, attention, and executive functioning, among others. In advanced dementia (now Major Neurocognitive Disorder), patients could have all of the neuropsychological domains impaired to a significant degree, not just executive functioning. In addition, there is a predictable pattern of impairment over the course of dementia progression with memory being impaired initially, then executive functioning, and later impairment of visual-spatial skills, language, and attention. As earlier mentioned, language would be expected to be impaired later in the process of becoming demented than executive functioning. On the minus side, defining all forms of human brain damage as neurocognitive disorders ignores the significant contribution of neuropsychological abilities in motor and sensory perceptual domains as well as intelligence, academic achievement, and other domains. Limiting the data to be considered to neurocognitive domains only handicaps the diagnostic process and will harm many vulnerable patients by failing to appropriately diagnose their brain functioning. As noted by Gestalt psychology, the whole is greater than the sum of its parts. DSM-5 has reduced the complexity of human brain functioning to five neurocognitive domains and much is lost.

An example of the importance of assessing motor functioning might be mentioned. As mentioned earlier, the frontal lobes are the external behavioral pathway by which executive functions are carried out. There is an overlap in functioning between the neuropsychological domain of executive functioning and the frontal lobes, but they are, of course, not identical constructs as the frontal lobes are a neuroanatomical area, as earlier mentioned. The frontal lobes are intimately involved in the expression of motor behavior. Reynolds and Horton (1994) specifically recommend that any assessment of executive functioning includes both motor-enhanced and motor-reduced measures.

Summary

This chapter has considered the role of the neuropsychology of executive functioning in the DSM-5 proposal, as it was at the time the chapter was written. For historical perspective, the previous versions of the DSM system were briefly reviewed with special attention given to DSM-IV, the immediate previous version of DSM-5. The current DSM-5 Neurocognitive Disorders Work Group's proposal was reviewed with special attention to Minor Neurocognitive Disorders and Major Neurocognitive Disorders and the evolution from DSM-IV to DSM-5. The neuropsychological construct of executive functioning and methods for the assessment of executive functioning were briefly reviewed in order to contrast what is known about assessing executive functioning with the current DSM-5 Neurocognitive Disorders Work Group's proposed diagnostic criteria. Finally, the neuropsychology of executive functioning was contrasted with the current DSM-5 Neurocognitive Disorders Work Group's proposal. Unfortunately, the current DSM-5 Neurocognitive Disorders Work Group's proposal was found to be severely lacking and has the potential to fail to appropriately diagnose the brain-related disorders of vulnerable patients. The major failing of the current DSM-5 Neurocognitive Disorders Work Group's proposed diagnostic criteria is a lack of appreciation and understanding of clinical neuropsychological assessment in general and the neuropsychology of executive functioning in particular. The hope and expectation is that this chapter will serve to identify flaws inherent to the current DSM-5 Neurocognitive Disorders Work Group's proposal in the hope of correcting and improving the DSM-5.

References

American Academy of Neurology. (1996). Assessment: Neuropsychological testing of adults; considerations for neurologists. *Neurology, 47*, 592–599.

American Medical Association. (2001). *Current procedural terminology*. Washington, DC: Author.

American Psychiatric Association. (2000). *Diagnostic and statistical manual of mental disorders. 4th edition, technical revision*. Washington, DC: American Psychiatric Association.

American Psychiatric Association. (2012). *Diagnostic and statistical manual of mental disorders. 5th edition. Neurocognitive disorders work group proposal*. Retrieved June 20, 2012 from http://www.dsm5.org

Anderson, R. M. (1994). *Practitioner's guide to clinical neuropsychology*. New York: Plenum Press.

Axelrod, B. N., Goldman, B. S., & Woodward, J. L. (1992). Interrater reliability in scoring the Wisconsin Card Sorting Test. *The Clinical Neuropsychologist, 6*, 143–155.

Bennett, T. L. (2001). Neuropsychological evaluation in rehabilitation planning and evaluation of functional skills. *Archives of Clinical Neuropsychology, 16*(3), 237–253.

Benton, A. (1968). Differential behavioral effects in frontal lobe disease. *Neuropsychologia, 6*, 53–60.

Benton, A., & Hamsher, K. (1989). *Multilingual Aphasia Examination*. Iowa City, IA: AJA Associates.

Estes, W. K. (1974). Learning theory and intelligence. *American Psychologist, 29*, 740–749.

Fitzhugh-Bell, K. (1997). Historical antecedents of clinical neuropsychology. In A. M. Horton, D. Wedding, &

J. Webster (Eds.), *The neuropsychology handbook* (pp. 67–90). New York: Springer.

Folstein, M. E., Folstein, S. E., & McHugh, P. R. (1975). "Mini-mental state": A practical method for grading the cognitive status of patients for the clinician. *Journal of Psychiatric Research, 12*, 228–235.

Goldberg, E. (2001). *The executive brain: Frontal lobes and the civilized mind*. Oxford: Oxford University Press.

Goldberg, E., & Costa, L. (1981). Hemispheric differences in the acquisition and use of descriptive systems. *Brain and Language, 14*, 144–173.

Grant, D. A., & Berg, E. A. (1948). A behavioral analysis of the degree of reinforcement and ease of shifting to new responses in a Weigl-type card sorting problem. *Journal of Experimental Psychology, 38*, 404–411.

Hart, R. P., Kwentus, J. A., Taylor, J. R., & Hamer, R. M. (1988). Productive naming and memory in depression and Alzheimer's type dementia. *Archives of Clinical Neuropsychology, 3*, 313–322.

Hartlage, L. C. (2001). Neuropsychological testing of adults: Further considerations for neurologists. *Archives of Clinical Neuropsychology, 16*(3), 201–213.

Heaton, R. (1981). *Wisconsin Card Sorting Test (WCST)*. Odessa, FL: Professional Assessment Resources.

Heaton, R., Chelune, G., Talley, J., Kay, G., & Curtiss, G. (1993). *Wisconsin Card Sorting Test (WCST) manual-revised and extended*. Odessa, FL: Professional Assessment Resources.

Heaton, R., Grant, I., & Mathews, C. (1991). *Comprehensive norms for an expanded Halstead-Reitan battery*. Odessa, FL: Psychological Assessment Resources.

Horton, A. M., Jr. (1979). Some suggestions regarding the clinical interpretation of the trail making test. *Clinical Neuropsychology, 1*, 20–23.

Horton, A. M., Jr., & Wedding, D. (1984). *Clinical and behavioral neuropsychology*. New York: Praeger.

Jarvis, P. E., & Barth, J. T. (1984). *Halstead-Reitan test battery: An interpretive guide*. Odessa, FL: Psychological Assessment Resources.

Lezak, M. D. (1995). *Neuropsychological assessment* (3rd ed.). Oxford: Oxford University Press.

Milner, B. (1963). Effects of different brain lesions on card sorting. *Archives of Neurology, 9*, 90–100.

Mitrushina, M., Boone, K., & D'Elia, L. (1999). *Handbook of normative data for neuropsychological assessment*. Oxford: Oxford University Press.

Nasreddine, Z., Phillips, N. A., Beditian, V., Charbonneau, S., Whitehead, V., Collin, I., et al. (2004). The Montréal cognitive assessment (MoCA); a screening tool for mild cognitive assessment. *Journal of American Geriatrics Society, 53*, 695–699.

Partington, J. E., & Leiter, R. G. (1949). Partington's pathway test. *The Psychological Service Center Bulletin, 1*, 9–20.

Reitan, R. M. (1955). The relation of the trail making test to organic brain damage. *Journal of Consulting Psychology, 19*, 393–394.

Reitan, R. M. (1958). The validity of the trail making test as an indicator of organic brain damage. *Perceptual and Motor Skills, 8*, 271–276.

Reitan, R. M. (1992). *Trail Making Test: Manual for administration and scoring*. Tucson, AZ: Reitan Neuropsychology Laboratory.

Reitan, R. M., & Wolfson, D. (1993). *The Halstead-Reitan neuropsychological test battery: Theory and clinical interpretation* (2nd ed.). Tucson, AZ: Neuropsychology Press.

Reitan, R. M., & Wolfson, D. (2000). *Mild head injury: Intellectual, cognitive, and emotional consequences*. Tucson, AZ: Neuropsychology Press.

Reitan, R. M., & Wolfson, D. (2001). Critical evaluation of assessment: Neuropsychological testing of adults. *Archives of Clinical Neuropsychology, 16*(3), 215–226.

Reynolds, C. R. (2003). *Comprehensive Trail-Making Test: Examiner's manual*. Austin, TX: PRO-ED.

Reynolds, C, & Horton, A. M., Jr., (1994). *Test of verbal conceptualization and fluency manual*. Austin, TX: PRO-ED.

Reynolds, C. R., & Livingston, R. A. (2012). *Mastering modern psychological testing*. Boston: Pearson.

Rosen, W. G. (1980). Verbal fluency in aging and dementia. *Journal of Clinical Neuropsychology, 2*, 135–146.

Spreen, O., & Strauss, E. (1998). *A compendium of neuropsychological tests* (2nd ed.). Oxford: Oxford University Press.

Storandt, M., Botwinick, J., & Danziger, W. L. (1984). Psychometric differentiation of mild senile dementia of the Alzheimer's type. *Archives of Neurology, 41*, 497–499.

Stuss, D., Bisschop, S. M., Alexander, M., Levine, B., Katz, D., & Izukawa, D. (2001). The Trail Making Test: A study in focal lesion patients. *Psychological Assessment, 13*, 230–239.

Executive Functioning Theory and ADHD

Kevin M. Antshel, Bridget O. Hier, and Russell A. Barkley

Attention-deficit/hyperactivity disorder (ADHD) is the current diagnostic label for individuals presenting with significant problems with attention and/or impulsiveness and hyperactivity. While the disorder has not always been called ADHD, the history of the clinical syndrome of inattention and overactivity dates back over 200 years. Across the last 200+ years, different aspects of the disorder (hyperkinesis, inattention, etc.) have been emphasized yet there has been an increasing recognition of the heterogeneity of the disorder. ADHD remains among the most common reasons that a child is referred for mental health treatment and is increasingly a common reason that adults are referred for treatment. Individuals with ADHD display considerable variation in the degree of symptoms, functional impairments from these symptoms, domains of impairment, age of diagnosis, response to treatment, and psychiatric comorbidity. While not currently a symptom of ADHD, there is evidence that executive functioning (EF) deficits may be a defining aspect of the disorder and even that its two symptom dimensions actually represent dimensions of EF. This chapter presents an overview of EF theory and ADHD.

Executive Functions

While described far more completely in previous chapters of this book, the term executive function is a rather ambiguous one that refers to a set of various interrelated cognitive abilities that operate metaphorically as a company "executive" (Denckla, 1996) and are considered to be largely mediated by prefrontal cortical/subcortical circuits (Goldman-Rakic, 1995). Yet there remains no consensus definition of the term nor has an operational definition been provided that could easily serve to segregate executive from nonexecutive mental abilities (Barkley, 2012). The term executive function also has been used to encompass the actions of planning, inhibiting responses, strategy development and use, flexible sequencing of actions, maintenance of behavioral set, and resistance to interference (Denckla, 1996). Even more globally, Lezak defined executive functions as "those capacities that enable a person to engage successfully in independent, purposive, self-serving behavior" (p. 42) (Lezak, Howieson, Loring, & Hannay, 2004). Nevertheless, it is generally agreed that these multifaceted abilities all pertain to goal-directed behaviors.

Executive functions seem to be mediated, at least in part, by the frontal cortex (particularly the

K.M. Antshel (✉)
SUNY—Upstate Medical University, Syracuse, NY, USA
e-mail: AntshelK@upstate.edu

B.O. Hier
Syracuse University, Syracuse, NY, USA

R.A. Barkley
Medical University of South Carolina, Charleston, SC, USA

prefrontal cortex) and are implicated in the neuropsychology of ADHD (Castellanos, Sonuga-Barke, Milham, & Tannock, 2006; Nigg & Casey, 2005; Sagvolden, Johansen, Aase, & Russell, 2005). Among others, Barkley (1997b) has theorized that problems with executive functioning (EF) specifically and self-regulation more generally are central to ADHD and give rise to the more obvious surface behavioral symptoms represented in the DSM-5 diagnostic criteria.

Barkley's EF Theory of ADHD

The Centrality of Response Inhibition and Self-Regulation

Over fifteen years ago, Barkley proposed his EF theory of ADHD (Barkley, 1997b) which assumes that behavioral inhibition, self-control, and executive functioning are overlapping and interacting human abilities. In his theory, EF is self-regulation and behavioral inhibition is essential to its performance. The purpose of self-control and EF are inherently social—humans engage in reciprocal social exchanges and cooperative ventures as a means to their survival and must both track prior such exchanges with others and prepare for such future interactions with others. In Barkley's original theory, response inhibition was seen as a central feature of EF because it provided the delay in automatic responding that was essential to permitting the executive functions to monitor, interrupt, and otherwise guide behavior toward goals. Inhibition referred to three overlapping yet somewhat distinct and separately measurable processes: (a) inhibiting the initial prepotent (dominant) response to an event so as to create a delay in responding, (b) interrupting an ongoing response that is proving ineffective thereby permitting a delay in and reevaluation of the decision to continue responding, and (c) protecting the self-directed (executive) responses that will occur within these delays as well as the goal-directed behavior they generate from disruption by competing events and responses (interference control or resistance to distraction) (Barkley, 1997b).

From Barkley's perspective, without the initial inhibition of the dominant response, thinking and related goal-directed actions are impossible (Bronowski, 1977). The ability to inhibit the dominant response and subsequently engage in self-change for the sake of attaining a goal requires self-control. Self-control is a response made by the individual that alters the probability of their subsequent response to an event and in so doing thereby changes the likelihood of a later or delayed consequence related to that event (Skinner, 1953). Self-control has been defined as generally involving the choice of a larger, later reward over a smaller, sooner (Ainslie, 1974). However, this general definition does not consider the self-directed actions in which the individual must engage so as to value the delayed over the immediate reward and then to pursue that delayed consequence. Four steps appear involved in adequate self-control: (1) the inhibition of the prepotent response directed toward some environmental event and (2) the directing of actions (both cognitive and motoric) toward oneself, (3) which will result in the alteration of the subsequent response from what it would have been had none of these self-directed actions been, and (4) that leads to the change in the likelihood of a delayed (future) consequence that arises as a function of this change in the behavior that will be employed.

Goal-directed behaviors require the ability to contemplate a future time point so as to juxtapose the "now" against the "later" and to evaluate the value or desirability of that later state vs. the current one. This capacity to consider delayed or future events requires some mental capacities for understanding time and the temporal ordering of events, for holding this information actively in mind, and for using this information to order and execute timely responses to them (Shimamura, Janowsky, & Squire, 1990). To accomplish the long chains of behavior that will be needed to bridge the delay in time between now and later, behavior must be hierarchically organized using smaller units nested within larger goals that are themselves nested within even larger goals (Badre, 2008). In Barkley's theory of EF and ADHD, executive functions represent classes of self-directed behavior or actions that we employ for purposes of self-regulation (changing our future) (Barkley, 1997b). The key then to operationalizing EF is that all EFs are self-directed actions, the central requirement for distinguishing an executive from nonexecutive mental ability. Any executive act then achieves the requirements for self-stopping, self-management to time, self-organization and problem solving across time, self-activation to initiate them, self-motivation to sustain them toward the goal, and emotional self-regulation. Such actions may be covert yet are volitional, effortful, conscious, and self-initiated actions. Neuroimaging research suggests that this covert behavior is measurable (D'Esposito et al., 1997; Ryding, Bradvik, & Ingvar, 1996).

Response inhibition is a requirement for self-regulation because one cannot direct actions or behavior toward one's self if one is automatically responding impulsively to an immediate event. In Barkley's theory, the EFs represent the general classes of self-directed actions that humans use in self-regulation *following this delay in the immediate response* (Barkley, 1997b). Without this initial response delay, however, the EFs are poorly accessed, arise after the fact, or even fail to be utilized at all.

The EFs depend on the individual being capable of perceiving and valuing future over immediate outcomes. If there is no sense of the future, there is no self-control, and there is no point in engaging in socially cooperative behavior that requires the subordination of one's immediate self-interests to those of others to attain greater longer-term self-interests. As we develop, we become far more capable of showing a preference for larger delayed rewards over smaller more immediate ones. This development requires the capacity to sense the future, that is, to construct hypothetical futures, particularly for social consequences. It also simultaneously involves the weighing of alternative responses and their temporally proximal and distal outcomes—a calculation of risk/benefit ratios over time.

Barkley originally theorized that humans have at least five classes of action that they direct toward themselves to change themselves to improve their future. In his model, these five classes are (a) self-stopping (response inhibition), (b) sensing to the self, (c) self-speech, (d) emoting and motivating to the self, and (e) self-play (Barkley, 1997b). Barkley's model relies heavily on the work of others (Bronowski, 1977; Damasio, 1995; Fuster, 1997; Goldman-Rakic, 1995; Vygotsky, 1987) and is therefore a hybrid model.

In his original hybrid model, response inhibition sets the occasion for the occurrence of the other EFs. The other four EFs are interactive and share a common purpose: to "internalize" or more accurately privatize certain self-directed behavior so as to anticipate and prepare for the future and maximize net long-term vs. short-term outcomes.

Sensing to the Self (Nonverbal Working Memory). The second executive function is the nonverbal working memory, or the visual–spatial sketchpad (Baddeley, 1986, 2003; Baddeley & Hitch, 1994). In Barkley's theory, this concept represents the privatization of sensory–motor actions—sensing to the self (literally, resensing to the self). The most important of the senses to humans are vision and hearing and so this EF is chiefly comprised of visual imagery and covert audition—re-seeing and re-hearing to the self. This EF has both retrospective (sensory or resensing) and prospective (preparatory motor) elements (Goldman-Rakic, 1995) and is the mental module for sensing the hypothetical future from the experienced past. By generating the private or mental representations (images, auditions, etc.) that bridge the cross-temporal elements within a contingency arrangement (event–response–outcome), humans are able to use such mental representations to guide and sustain actions over time and manage themselves relative to time (or time management) to attain the contemplated goal.

Speech to the Self (Verbal Working Memory). Barkley posits that the internalization of speech (Diaz & Berk, 1992) serves as the basis for the verbal working memory system of EF (Baddeley, 1993) and transitions outer-directed behavior toward the self as a means to control one's own behavior. In Barkley's theory, Vygotsky's model of the developmental internalization of speech

(Vygotsky, 1987) figures prominently and represents what other neuropsychologists call verbal working memory. Self-speech permits self-description and reflection, self-instruction, self-questioning, and problem solving, as well as the invention of rules and meta-rules to be applied to oneself (Diaz & Berk, 1992). Self-speech contributes to self-control and provides the basis for private verbal reasoning, strategy (rule) development, verbal problem solving, and moral conduct (internalizing socially prescribed rules of conduct).

Emotion to the Self (Self-Regulation of Affect/Motivation/Arousal Emotion). This EF may occur initially as a consequence of the first three (inhibition, private sensing, and self-speech). These mentally represented events have associated affective and motivational properties or valences (Damasio, 1995). Initially these affective valences have publicly visible counterparts—emotional displays—as when we laugh out loud in response to a mentally visualized incident. Eventually, however, these emotional displays are kept private or covert in form. In Barkley's model, private self-directed affect and its motivational properties—feeling (emoting/motivating) to the self—develop from using the other EFs above to generate the mental representations that provoke such secondary emotional states in the absence of such provocative events in the environment. The development of emotion to the self provides the self-restraint of emotion so important to cooperative social interactions. Yet it also provides the intrinsic motivation (willpower) so necessary to support future-directed behavior, especially across large delays in schedules of reinforcement or where external consequences for such future-directed action are otherwise not available in the immediate context. It also provides the motivational basis for persistence (sustained attention) toward future goals (Barkley, 1997b). These two functions of this EF, emotional self-control and self-motivation, may be partially separable or measurable given that they may be mediated by different zones of the anterior cingulate cortex.

Self-Play (Reconstitution). The last EF in the original theory is self-directed private (covert) play, or reconstitution. Fluency, flexibility, and generativity are other terms by which this EF is known in neuropsychology. This EF is the source of self-organization and innovation (problem solving) during goal-directed actions. In Barkley's model, reconstitution occurs through a two-step process: analysis and synthesis (Barkley, 1997b). Both are applied to the mental contents being held in the working memory systems (self-sensing and self-speech systems). In analysis, old behavior contingencies (stimulus–response arrangements and sequences) are broken down into smaller units. These units are then recombined (synthesized) into new sequences that can be tested against the requirements of the problem to be solved (Fuster, 1997). Reconstitution, or private planning and problem solving, arises from the internalization of play (both sensory–motor and symbolic) and creates the source for generating novel future-directed actions. Such novel actions will be needed when obstacles to a goal are encountered (problems) in order to overcome them and successfully attain the goal.

Theoretical Amendment. More recently, Barkley (2012) has amended this original model in three important ways. First, he has argued that there is a sixth self-directed action, or EF, which arises in conjunction with inhibition during developing and that is self-awareness (self-monitoring). Self-awareness arises like the other EFs through a process of self-directing a pre-executive function, in this case attention. Thus, self-directed attention codevelops with inhibition as neither makes any sense in the absence of the other. One cannot inhibit an automatic action if one is not aware of or attending to his or her own behavior, and there is little point to self-monitoring more automatic behavior if it cannot be inhibited or interrupted so as to make it more consistent with a longer-term goal. Second, neither self-monitoring nor self-inhibition provides much of any benefit if there is not a mental capacity to contemplate the future generally or goals specifically and alternative means to attain it. This requires the nonverbal working memory or self-directed sensory–motor

EF above in which the future is being contemplated. Therefore, in the more recent iteration of this theory, Barkley now argues that inhibition, self-awareness, and self-directed sensory–motor actions likely codevelop as a unity to form the initial level of EF as it likely evolved in human evolution. It has subsequently been expanded to include self-directed speech, emotion/motivation, and play (planning/problem solving).

Third, Barkley now explicitly shows how this initial instrumental self-directed or cognitive level of EF expands over time to link up with methodical–self-reliant behavior and the next higher level in a hierarchy of EF functioning in daily life. Cognitive EFs extend their effects outward to guide self-regulatory behavior and daily adaptive functioning more generally creating executive actions. EF = executive cognition (EC) + executive actions or behavior (EA or EB). This linkage between instrumental and self-reliant levels of EF is part of what Barkley argues as an extended phenotype of EF into daily and lifelong effective social functioning. He then goes on to elaborate to additional levels of this extended phenotype model of EF in which the effects of EF lead upward (outward) to socially reciprocal actions with others to accomplish goals (the tactical–reciprocal level) and eventually to socially cooperative ventures (the strategic–cooperative level). At the latter level, groups form to accomplish mutual goals that no single individual can attain alone or through mere reciprocity or exchange. With each new level in this hierarchy, additional mental faculties may be needed, such as theory of mind and vicarious learning through the mirror-neuronal system of the prefrontal cortex (PFC). Larger social networks are also required as is a greater reliance on cultural devices and methods to form the external scaffolding needed to boost EF upward to these higher levels of human social life across major domains of activities. In this way, EF can now be linked not only to daily adaptive or self-reliant activities but also outward to ethics and morality, social exchange and economics, law and criminality, and even government and politics through such extended phenotypic effects. All have in common the essential requirement to be capable of contemplating the future, or the later versus the now. EF disorders like ADHD therefore result in a weakening or even collapsing of this hierarchy downward resulting in serious social and adaptive impairments.

Further Implications. Each of the EFs noted above is also hypothesized to contribute to developmental shifts in the sources of control over human behavior from (a) external events to mental representations related to those events, (b) control by others to control by the self, (c) immediate reinforcement to delayed gratification, and (d) the temporal now to the conjectured social future. Across development, individuals progressively become more guided by covert representations that permit self-control, deferred gratification, and goal- and future-directed actions. In Barkley's model, these five EFs provide a powerful set of mind tools that greatly facilitate adaptive functioning in anticipation of the future (Barkley, 1997b). These EFs permit the private simulation of actions that can be tested out mentally for their probable consequences before a response is selected for public execution. This constitutes a form of mental trial and error learning that lacks real-world consequences for one's mistakes.

The Impact of ADHD on Self-Regulation

Behavioral inhibition is a central problem for those with ADHD (Nigg, 2001). Barkley's theory originally asserted that a deficit in inhibition associated with ADHD would result in a cascading of secondary deficits into the remaining EFs. Behavioral disinhibition leads to nonverbal working memory deficiencies and therefore (1) forgetfulness (forgetting to do things at certain critical points in time), (2) impaired ability to organize and execute their actions relative to time (e.g., time management), and (3) reduced hindsight and forethought, (4) leading to a reduction in the creation of anticipatory action toward future events. As a result, the capacity for the cross-temporal organization of behavior in those with ADHD is diminished, disrupting the ability to sequence together complex chains of actions directed, over time, to a future goal. In its most recent iteration, however, Barkley now concedes that since self-awareness, self-restraint (inhibition), and self-sensory–motor actions (nonverbal working memory) may codevelop as a unity, all become the primary deficits in ADHD and the starting point for understanding the symptoms associated with the disorder (and its two-dimensional structure). Research does show that not only inhibition but also nonverbal working memory, timing, and forethought are deficient in ADHD (Frazier, Demaree, & Youngstrom, 2004; Rapport et al., 2008). The greater the degree to which time separates the components of the behavioral contingency (event, response, consequence), the more difficult the task will prove for those with ADHD. Thus, Barkley now argues that working memory, especially nonverbal, may be as much a primary deficit in ADHD alongside that of poor inhibition and diminished self-awareness rather than the latter being secondary effects of the inhibitory problem, consistent with more recent research on this issue (Rapport et al., 2008). In sum, inhibition, self-monitoring, and working memory are interactive with deficits in each being likely to adversely affect the others. Indeed, Barkley now suggests that the very process of self-directing and eventually internalizing (privatizing) the instrumental EFs may be a more general developmental deficit in ADHD.

In addition to deficiencies in working memory, the privatization of speech should also be delayed in ADHD, resulting in greater public speech (excessive talking), less verbal reflection before acting, less organized and rule-oriented self-speech, a diminished influence of self-directed speech in organizing and controlling one's own behavior, and difficulties following the rules and instructions given by others. Research supports this hypothesis (Berk & Potts, 1991; Winsler, Diaz, Atencio, McCarthy, & Chabay, 2000). Verbal working memory tasks such as digit span backward, mental arithmetic, paced auditory serial addition, paired associated learning, and other tasks believed to reflect verbal working memory are difficult for those with ADHD (Frazier et al., 2004; Hervey, Epstein, & Curry, 2004).

These deficits lead to a third problem—impaired emotional/motivational self-regulation. Those with ADHD will display (1) greater impulsive emotional expressions in their reactions to events, (2) less objectivity in the selection of a response to an event, (3) diminished social perspective taking as the individual does not delay his or her initial emotional reaction long enough to take the view of others and their own needs into account, (4) greater difficulties in self-soothing the initially strong emotional reaction, (5) greater problems with self-distracting and otherwise modifying their attention to the emotionally provocative event so as to diminish its ongoing impact, and (6) a diminished ability to construct more socially appropriate and moderate emotions in place of the original emotion that are more supportive of their long-term welfare or social interests. ADHD should also impair the capacity to induce drive and motivational states in the service of goal-directed behavior. Those with ADHD remain more dependent upon the environmental contingencies within a situation or task to determine their motivation than do others (Barkley, 1997a).

Barkley's EF model further predicts that ADHD will be associated with impaired reconstitution, or self-directed play, evident in a diminished use of analysis and synthesis in the formation of both verbal and nonverbal responses to events. The capacity to mentally visualize or verbalize, manipulate, and then generate multiple plans of action (options) in the service of goal-directed behavior and to select from among them those with the greatest likelihood of succeeding should, therefore, be reduced. This impairment in reconstitution will be evident in everyday verbal fluency when the person with ADHD is required by a task or situation to assemble rapidly, accurately, and efficiently the parts of speech into messages (sentences) so as to accomplish the goal or requirements of the task. It will also be evident in tasks where visual information must be held in mind and manipulated to generate diverse scenarios to help solve problems (Barkley, 1997a). In general poorer self-organization and problem solving in support of one's goals or assigned tasks should result.

Evidence for a deficiency in verbal and nonverbal fluency, planning, problem solving, and strategy development more generally in ADHD is limited, but what exists is consistent with Barkley's theory (Clark, Prior, & Kinsella, 2000; Klorman et al., 1999).

In general, individuals with ADHD will be more under the control of external events than mental representations about time and the future, under the influence of others rather than acting to control one's self, pursuing immediate gratification over deferred gratification, and under the influence of the temporal now more than of the probable social futures that lie before them. From this vantage point, ADHD is not a disorder of attention, at least not to the moment or the external environment, but more of a disorder of intention—that is, attention to the future and what one needs to do to prepare for its arrival. It is also a disorder of time—time management specifically—in that the individual manifests an inability to regulate their behavior relative to time as well as others of their developmental level. This creates a sort of temporal myopia in which the individual responds to or prepares for only events that are relatively imminent rather than ones that lie further ahead in time yet which others of their age are preparing for so as to be ready for their eventual arrival (Barkley, 1997a).

Other Perspectives on ADHD and EF

In viewing ADHD as a disorder of self-regulation (and its underlying executive functioning), Barkley's theory has proposed a model of how ADHD disrupts the normal structure and processes of self-regulation to produce what is known about the disorder. Barkley's theory also suggests new hypotheses about what may be disrupted by the disorder. In this way, Barkley's theory has been very influential in spurring research into the relationship between EF and ADHD.

However, some researchers have suggested that EF deficits may not be specific to the disorder and that it is more likely that only a subpopulation of individuals with ADHD experience clinically significant executive dysfunction. For example, although neuropsychological theories have implicated executive dysfunction as a main characteristic of ADHD, some researchers have questioned the role of EF as a core deficit of the disorder. In fact, some (e.g., Trani et al., 2010) have posited that evidence from neuropsychological studies suggests that only a subpopulation of individuals with ADHD experience clinically significant EF deficits. Thus, some believe that executive dysfunction is only a partial explanation of a comprehensive model of ADHD. The potential flaw in such arguments is the premise that "cold" cognitive EF psychometric measures as collected in clinical or lab settings are the sole or gold standard for evaluating EF. Barkley's recent extended phenotype model of EF shows why this is not the case and why EF tests have low or no ecological validity (Barkley, 2012).

Theories implicating executive dysfunction as a causal mechanism underlying ADHD have been tested by comparing groups of individuals with and without DSM-defined ADHD. However, it has been suggested based on reviews of the results only of psychometric EF tests that no one neuropsychological model, including Barkley's model of executive dysfunction, currently provides a complete account of ADHD (Nigg & Casey, 2005). For example, Nigg et al. (2005) reported on neuropsychological data gathered from 600 children without ADHD and 287 children with ADHD combined subtype. Of the administered neuropsychological measures, the Response Suppression Task: Stop Task was the most discriminative with approximately 50 % of the children with ADHD demonstrating clinically significant impairment. However, this suggests that nearly half of the children with ADHD were not impaired on tasks of response inhibition. Furthermore, although approximately 80 % of children with ADHD exhibited impairment on at least one EF task, so did nearly half of the control participants. Thus, because only some individuals with ADHD experienced executive dysfunction across the measured tasks, it was concluded that EF is not the only causal pathway leading to ADHD (Nigg et al., 2005).

Similarly, results of several meta-analyses have also indicated that EF deficits are likely experienced by some, but not all, individuals with ADHD. One such meta-analysis (Boonstra, Oosterlaan, Sergeant, & Buitelaar, 2005) quantitatively examined the difference between adults with ADHD and control participants across EF measures and non-EF variables. Thirteen studies were reviewed, and data from five tests of EF were analyzed. Specifically, data on participants' verbal fluency, attention and response inhibition, working memory, and mental inhibition were analyzed. Participants' performance on non-EF variables (e.g., processing speed, verbal memory) and EF tasks was also compared. Results indicated that adults with ADHD tended to demonstrate greater difficulty than control participants on measures of both EF and non-EF. Further, because average effect sizes were similar for both the EF and non-EF domain ($d=.40$ and .43, respectively), Boonstra et al. (2005) concluded that EF is not a specific deficit for adults with ADHD. Rather, the authors suggested that adults with ADHD demonstrated poorer performance than adult control participants in a variety of cognitive domains, including EF.

In another meta-analysis, Willcutt, Doyle, Nigg, Faraone, and Pennington (2005) included 83 studies that measured EF psychometrically among groups of individuals with ADHD (total $N=3,734$) and without ADHD ($N=2,969$). Although ADHD groups demonstrated significant EF impairment across all measured EF domains (i.e., response inhibition, vigilance, set-shifting, planning, organization, verbal working memory, spatial memory), effect sizes were moderate in size ($d=.46-.69$) (Willcutt et al., 2005). Others (Nigg et al., 2005) have criticized the interpretation that moderate effect sizes are evidence that a unified, core EF deficit is characteristic of all children with ADHD. Specifically, Nigg et al. (2005) argued that such effect size magnitudes suggest distributional overlap between ADHD and non-ADHD samples on EF performance and that the performance of some with ADHD falls within the normal range.

Some evidence suggests that tests of EF are sensitive but not specific to the diagnosis of ADHD (e.g., Doyle, Biederman, Seidman, Weber, & Faraone, 2000). Clinically, this means that poor scores on these tests may indicate that an individual has ADHD, but average or above-average scores cannot be used in isolation to rule out the possibility of ADHD. For example, one study (Wodka et al., 2008) examined the predictive ability of four subtests of the Delis–Kaplan Executive Function System (D-KEFS) toward an ADHD diagnosis in 69 children without ADHD and 54 children with ADHD. Results indicated that those without ADHD performed significantly better than those with ADHD on only two of the four selected D-KEFS measures. The authors concluded that this measure of EF lacks specificity in ADHD diagnosis (Wodka et al., 2008).

While these data suggest that EF tests are not specific to ADHD diagnosis, there are also data that suggest otherwise (Clark et al., 2000; Holmes et al., 2010). For example, Clark et al. argued that EF impairment is specific to ADHD. They found that two groups of adolescents with ADHD (e.g., ADHD only and ADHD with comorbid ODD/CD) demonstrated a significant deficit on measures of EF in comparison to adolescents with ODD/CD only and a typically developing control group. Thus, the relationship between ADHD and EF continues to evolve and may depend on how EF is being defined and assessed.

As noted above, the argument has been made (Barkley, 2012; Barkley & Fischer, 2011; Barkley & Murphy, 2010, 2011) that all such conclusions about the nature of EF in ADHD are undermined by the exclusive reliance of such research on psychometric approaches to measuring EF. When other methods, such as rating scales of EF, have been used, the vast majority (86–98 %) of individuals with ADHD are found to place in the deficient range ($\leq$ 7th percentile). The fact that EF ratings are only weakly related if at all to EF tests scores further undermines the credibility of the latter as the exclusive approach to studying EF in disordered populations such as ADHD.

Barkley's EF Theoretical Expansion

In the 15 years since Barkley first proposed his theory of EF and ADHD, much research, often conflicting, has considered the relationship between EF and ADHD (see above). As briefly noted above, Barkley (2012) recently expanded his EF theory upward to involve four additional levels beyond the level of basic cognitive EFs. A main goal of this theoretical expansion was to integrate a more traditional view of EF at the neurocognitive level with how it plays out in everyday life, referred to in Biology as the extended phenotype.

Barkley's Expanded Phenotypic Theory of EF

In Barkley's expanded theory, EF and EF subcomponents arise out of two developmental processes, the *self-direction of actions* and their *internalization*. It is the self-direction of human actions that makes an act, function, or component executive in nature. The self-directed action is being done to alter subsequent behavior from what it would have otherwise been (it is a means to an end) and that is done to alter the likelihood of future consequences for the individual (ends or goals). This constitutes the definition of self-regulation. Therefore, the initial definition of EF was clarified and made more specific as follows: *the use of self-directed actions so as to choose goals and to select, enact, and sustain actions across time toward those goals*. Although the cross-temporal nature of EF is implied in the definition of self-regulation, Barkley believed it was important to make it explicit. Humans bind current status, intermediate means, and future ends together into cross-temporal structures that are mentally represented and serve to guide goal-directed actions (Fuster, 1997).

In Barkley's expanded theory of EF, as noted above, there are now six self-directed actions rather than five that are identified as being used for self-regulation and as being self-evident in any human's existence: (1) self-directed attention to create self-awareness, (2) self-directed inhibition to create self-restraint, (3) self-directed sensory–motor actions to create mental representations and simulations (ideation), (4) self-directed speech to create verbal thinking, (5) self-directed emotion and motivation to create conscious appraisal, and (6) self-directed play (nonverbal and verbal reconstitution) to create problem solving, fluency, or innovation. Humans are using at least six forms of self-regulation in directing and sustaining action toward a goal, and each is an EF in Barkley's expanded theory.

Using Barkley's extended phenotype viewpoint, EF has radiating effects outside of and at considerable spatial and temporal distances from the organism. This leads to an appreciation for the important role of EF in the initial zone of that phenotype that respects the group living niche in which humans exist and so includes other humans as self-interested competitors and manipulators. This expansion of the EF phenotype identified fellow humans as likely to be engaging in the manipulation of others as a means to attain ends at this methodical–self-reliant level of the EF phenotype. Thus, Barkley's definition of EF was broadened to incorporate this initial social context and became the use of self-directed actions so as to choose goals and to select, enact, and sustain actions across time toward those goals usually *in the context of others*.

EF is not just indispensable for social self-defense. Rather, by adopting a longer view of one's self-interests, others can be construed as a means to goals that are symbiotically beneficial to both parties. This connected EF to the practice of social reciprocity and exchange as well as the larger field of economic behavior and formed the tactical–reciprocal level of the EF phenotype. Extending the time horizon over which one is contemplating the longer-term even further ahead, it becomes possible to demonstrate that reciprocity may be improved through cooperation (acting in unison) which itself often results in division of labor with trade. This forms the strategic–cooperative level of the EF phenotype. Understanding these extended phenotypic effects of EF led Barkley to a further expansion of the definition of EF: the use of self-directed actions so as to choose goals and to select, enact, and sustain actions across time toward those goals usually in the context of others *often relying on social means*.

Barkley's extended phenotype model of EF also argues for an increasing use of cultural scaffolding to ratchet up the human capacities for goal-directed actions. Humans create and use culture (stored and shared information)—its knowledge, inventions, devices, and products—to bootstrap their EF capacities upward for the attaining of larger goals, extending over longer time spans, spatial distances, and social networks. Thus, the definition of EF was further expanded to recognize this fact: the use of self-directed actions so as to choose goals and to select, enact, and sustain actions across time toward those goals usually in the context of others often relying on social *and cultural means*.

Finally, to contrast forms of cultural scaffolding (principles, policies, and governments) that do and do not promote this upward ratcheting of the human ability for goal-directed action, Barkley emphasized that human EF is motivated out of self-interest, albeit over the longer term. Such self-interest can only be determined by the individual using reason. EF is motivated by subjective appraisal of longer-term self-interest and is essentially self-determination. Forms of cultural scaffolding that accept and promote these basic features of human nature allow EF to succeed, extend outward to have wider phenotypic effects, and permit human life to thrive and prosper as individuals pursue their longer-term self-interests. EF was therefore concluded to be the use of self-directed actions so as to choose goals and to select, enact, and sustain actions across time toward those goals usually in the context of others often relying on social and cultural means *for the maximization of one's longer-term welfare as the person defines that to be*.

A number of features distinguish Barkley's extended phenotypic model of EF including Barkley's belief that EF and its components have arisen out of two psychological processes: self-direction of action and internalization or privatization. EF consists of private self-directed actions (self-regulation) and is viewed as active effortful behavior-to-the-self. Barkley's theory posits that much of this initial instrumental level of self-directed activity gradually becomes private in form across development such that by adulthood it gives rise to a private and cognitive domain of behavior as distinguished from behavior that is readily observed. Humans therefore possess both a private and a public self (Bronowski, 1977).

Barkley's model of EF is presently the only model which employs the biological concept of an extended phenotype. The goal is to demonstrate how EF radiates outward to produce effects on the physical and social environment at a distance from the genotype and conventional phenotypic levels to give rise to self-reliance, reciprocity, cooperation, and social mutualism. In this way, Barkley's EF model links EF to the social interactive behavior of individuals as well as their social self-defense, reciprocity, cooperation, mutualism, and communalism. In doing so it shows how EF is essential to functioning in most major life activities (occupational, educational, financial–economic, cohabiting/marital, parental, etc.). All of these domains of human activity are predicated on a capacity to sense the future – to contemplate the likely consequences for the various actions one may choose to do in order to attain a goal.

Emphasizing and Predicting Real-World Functioning

Barkley's expanded theory of EF was developed to revise his original theory and go beyond the neurocognitive level and emphasize how EF affects everyday life and functioning, not simply performance on a laboratory measure of EF. This theoretical expansion follows a move in the fields of psychiatry and psychology which emphasizes collecting real-world information about the deployment of EF rather than relying on laboratory tests (Gioia, Isquith, Guy, & Kenworthy, 2000; Roth, Isquith, & Gioia, 2005).

As a hierarchically organized model, Barkley's theory makes obvious how impairments at lower EF levels may radiate upward to affect higher levels; yet Barkley's theory also shows that deficits at higher levels need not always radiate downward to the detriment of functioning at the lower level. For instance, individuals may not be capable of sustained cooperative ventures (acting in unison to attain a common goal in which all share) but may still be able to engage in social reciprocity and exchange. The radiating effects of disturbances at lower levels outward to later, higher levels of human functioning can show how ADHD can have adverse effects on many fields or domains of human functioning, such as marriage and parenting, education, health maintenance, economic behavior (occupational functioning, financial management), transportation (driving), and community participation (politics and government). The impact that EF deficits may have on traditional neuropsychological tests may be trivial in comparison to those occurring at higher levels of the extended EF phenotype (Barkley & Fischer, 2011).

Implications of Barkley's Theory for Managing and Treating ADHD

Barkley's extended phenotype model of EF views EF as conscious, effortful, self-initiated, and self-directed activities that strive to modify otherwise automatic behavior so as to alter the likelihood of future consequences (longer-term goals and desires). Barkley's theory views these self-directed activities as consisting of self-directed attention, self-restraint, sensory–motor action to the self using visual imagery, speech to the self, emotion to the self, self-motivation, and self-directed play. Barkley's theory encourages those who wish to develop or rehabilitate their EF to repeatedly practice: self-monitoring, self-stopping, seeing the future, saying the future, feeling the future, and playing with the future so as to effectively "plan and go" toward that future.

The extended phenotype view of EF argues that the problems posed for those with EF deficits in major life activities have more to do with not using what they know at critical points of performance in their natural environments than with not knowing what to do. In short, information is not self-regulation. The extended phenotype model views EF as self-regulation and impairment in EF poses more of a problem with doing what one knows rather than one of knowing what to do—it is a performance vs. knowledge (skills) distinction.

With the performance vs. knowledge distinction in mind, interventions are most helpful when they assist with the performance of a particular behavior at the *point of performance* in the natural environments where and when such behavior should be performed. The further away in space and time a treatment is from this point of performance, the less effective it is likely to be in assisting with the management of EF deficits. Conveying more knowledge does not prove as helpful as altering the parameters associated with the performance of that behavior at its appropriate point of performance.

If the process of regulating behavior by internally represented forms of information (working memory or the internalization of self-directed behavior) is impaired or delayed in those with ADHD, then they will be best assisted by "externalizing" those forms of information; the provision of physical representations of that information will be needed in the setting at the point of performance. Since covert or private information is weak as a source of stimulus control, making that information overt and public may assist with strengthening control of behavior by that information. Consequently, those with ADHD will require the provision of externalized sources of motivation. For instance, the provision of artificial rewards, such as tokens, may be needed throughout the performance of a task or other goal-directed behavior when there is otherwise little or no such immediate consequence associated with that performance. Such artificial reward programs become for the person with ADHD what prosthetic devices such as mechanical limbs are to the physically disabled, allowing them to perform more effectively in some tasks and settings with which they otherwise would have considerable difficulty.

Several EF-based cognitive behavioral therapy (CBT) approaches related to Barkley's model have been recently developed, researched, and published in manual form for clinicians

(Ramsay & Rostain, 2008; Safren et al., 2005, 2010; Solanto et al., 2010). All of these CBT protocols go beyond simply conceptualizing ADHD as a disorder of purely cognitive underpinnings. Rather, these protocols also consider the adaptive or self-reliant and higher levels of EF in Barkley's extended phenotype model (e.g., time management, self-organization, problem solving, emotional self-control, self-motivation). These CBT protocols also consider different EF levels of dysfunction. For example, deficits at the basic instrumental level of EF might be dealt with by training in self-directed inhibition, imagery, audition, and speech, among others, that is often the focus of cognitive rehabilitation (often computer based) training programs. Medications may also serve to temporarily improve or even normalize some or all of these instrumental EFs and thus be valuable supplements to such CBT programs. Adverse effects at the self-reliant level may need to focus more on helping the individual to reorganize their external environment to facilitate performance of EF, self-care, and general adaptive functioning at this level. This could also be facilitated and amplified by artificial devices such as digital memory recorders, computers, personal data assistants, or cell phones to which periodic prompts and reminders are sent, and other such environmental prostheses. Deficits at the strategic levels will likely require training and ongoing assistance with social skills, etiquette, emotional self-regulation in social settings, and other therapies aimed at the social nature of these levels (reciprocity, cooperation, mutualism).

Conclusions/Future Directions

The expanded model explains why EF tests may be insufficient to capture deficits in EF, even at the instrumental level, because their window of ascertainment of cognition may be too brief for how humans deploy EF in daily life. Such tests also focus on "cold" cognition rather than on the social purposes of the EF system, fail to evaluate the self-regulation of emotion and motivation, and do not capture the reciprocal relationship between EF and cultural scaffolding needed to operate at higher levels of EF as it occurs in human daily life activities. For centuries, individuals with disorders of the PFC have been noted to have marked changes in their personality, ethics and morality, capacity for effective occupational and educational functioning, a preference for immediate gratification, emotional dysregulation, and an adverse impact on social reciprocity and cooperation none of which are the focus of the "cold" psychometric approach to evaluating EF. Barkley's latest model integrates EF with these larger important human endeavors attempting to demonstrate why disorders of EF produce profound disturbances in human adaptive functioning across numerous major domains of daily life activities while being only partially detectable by lab tests of the EFs.

Based on such a model, it is also evident that ADHD has to be EFDD, not only because the neural networks of the PFC that give rise to the executive brain are deficient in ADHD but also because the behavioral symptoms of ADHD are dimensions of EF in daily life (behavioral regulation and metacognition) listed under other names. Moreover, ADHD has to equal EFDD given the profound deficits evident in EF in daily life activities as captured by rating scales of EF even if such deficits are only evident in a minority of cases on "cold" cognitive test batteries that only partially evaluate the instrumental level of EF cognition.

There is considerable room for future research based on this extended phenotypic view of EF and its outward extension into daily human activities, especially over time, and as applied to understanding ADHD (and other disorders of EF). New tests could be developed to improve laboratory evaluation of EF provided that they integrate social motives into their content, use more extended time intervals, evaluate self-regulation including that of emotion and motivation, and are combined with other measures of higher level EF functioning, such as rating scales of EF and direct behavioral observations across time in natural settings. New measures of EF also need to be developed to more directly capture the tactical–reciprocal and strategic–cooperative levels of the EF phenotype beyond the value that adaptive behavior inventories, social skills ratings, general impairment rating scales, and even archival records (e.g., driving, education, work history) may have at the moment to partially detect such impairments. As Barkley emphasized 15 years ago (Barkley, 1997b), such theories of EF are always imperfect when first proposed yet they serve to provide a time-limited tool for better understanding, evaluating, and managing EF until better models can be designed based on research and experience with the earlier theory.

References

Ainslie, G. W. (1974). Impulse control in pigeons. *Journal of Experimental Analysis of Behavior, 21*(3), 485–489.

Baddeley, A. D. (1986). *Working Memory*. New York: Oxford University Press.

Baddeley, A. D. (1993). Verbal and visual subsystems of working memory. *Current Biology, 3*(8), 563–565.

Baddeley, A. D., & Hitch, G. J. (1994). Developments in the concept of working memory. *Neuropsychology, 8*(4), 485–493.

Baddeley, A. D. (2003). Working memory and language: An overview. *Journal of Communication Disorders, 36*(3), 189–208.

Badre, D. (2008). Cognitive control, hierarchy, and the rostro-caudal organization of the frontal lobes. *Trends in Cognitive Sciences, 12*(5), 193–200.

Barkley, R. A. (1997a). *ADHD and the nature of self-control*. New York: Guilford.

Barkley, R. A. (1997b). Behavioral inhibition, sustained attention, and executive functions: Constructing a unifying theory of ADHD. *Psychological Bulletin, 121*(1), 65–94.

Barkley, R. A. (2012). *Executive functioning and self-regulation: Extended phenotype, synthesis, and clinical implications*. New York: Guilford.

Barkley, R. A., & Fischer, M. (2011). Predicting impairment in major life activities and occupational functioning in hyperactive children as adults: Self-reported executive function (EF) deficits vs EF tests. *Developmental Neuropsychology, 36*(2), 137–161.

Barkley, R. A., & Murphy, K. R. (2010). Impairment in occupational functioning and adult ADHD: The predictive utility of executive function (EF) ratings versus EF tests. *Archives of Clinical Neuropsychology, 25*(3), 157–173.

Barkley, R. A., & Murphy, K. R. (2011). The nature of executive function (EF) deficits in daily life activities in adults with ADHD and their relationship to EF tests. *Journal of Psychopathology and Behavioral Assessment, 33*, 137–158.

Berk, L. E., & Potts, M. K. (1991). Development and functional significance of private speech among attention-deficit hyperactivity disordered and normal boys. *Journal of Abnormal Child Psychology, 19*(3), 357–377.

Boonstra, A. M., Oosterlaan, J., Sergeant, J. A., & Buitelaar, J. K. (2005). Executive functioning in adult ADHD: A meta-analytic review. *Psychological Medicine, 35*(8), 1097–1108.

Bronowski, J. (1977). *Human and animal languages. A sense of the future*. Cambridge, MA: MIT Press.

Castellanos, F. X., Sonuga-Barke, E. J., Milham, M. P., & Tannock, R. (2006). Characterizing cognition in ADHD: Beyond executive dysfunction. *Trends in Cognitive Sciences, 10*(3), 117–123.

Clark, C., Prior, M., & Kinsella, G. J. (2000). Do executive function deficits differentiate between adolescents with ADHD and oppositional defiant/conduct disorder? A neuropsychological study using the Six Elements Test and Hayling Sentence Completion Test. *Journal of Abnormal Child Psychology, 28*(5), 403–414.

D'Esposito, M., Detre, J. A., Aguirre, G. K., Stallcup, M., Alsop, D. C., Tippet, L. J., et al. (1997). A functional MRI study of mental image generation. *Neuropsychologia, 35*(5), 725–730.

Damasio, A. R. (1995). On some functions of the human prefrontal cortex. In K. J. H. J. Grafma & F. Boller (Eds.), *Structure and functions of the human prefrontal cortex* (Vol. 769, pp. 241–251). New York: New York Academy of Sciences.

Denckla, M. B. (1996). A theory and model of executive function: A neuropsychological perspective. In G. R. Lyon & N. A. Krasnegor (Eds.), *Attention, memory and executive function* (pp. 263–278). Baltimore, MD: Paul H. Brookes.

Diaz, R. M., & Berk, L. E. (1992). *Private speech: From social interaction to self-regulation*. Mahwah, NJ: Lawrence Erlbaum Associates, Inc.

Doyle, A. E., Biederman, J., Seidman, L. J., Weber, W., & Faraone, S. V. (2000). Diagnostic efficiency of neuropsychological test scores for discriminating boys with and without attention deficit-hyperactivity disorder. *Journal of Consulting and Clinical Psychology, 68*, 477–488.

Frazier, T. W., Demaree, H. A., & Youngstrom, E. A. (2004). Meta-analysis of intellectual and neuropsychological test performance in attention-deficit/hyperactivity disorder. *Neuropsychology, 18*(3), 543–555.

Fuster, J. M. (1997). *The prefrontal cortex*. New York: Raven.

Gioia, G. A., Isquith, P. K., Guy, S. C., & Kenworthy, L. (2000). Behavior rating inventory of executive function. *Child Neuropsychology, 6*(3), 235–238.

Goldman-Rakic, P. S. (1995). Architecture of the prefrontal cortex and the central executive. *Annals of the New York Academy of Sciences, 769*, 71–83.

Hervey, A. S., Epstein, J. N., & Curry, J. F. (2004). Neuropsychology of adults with attention-deficit/hyperactivity disorder: A meta-analytic review. *Neuropsychology, 18*(3), 485–503.

Holmes, J., Gathercole, S. E., Place, M., Alloway, T. P., Elliott, J. G., & Hilton, K. A. (2010). The diagnostic utility of executive function assessments in the identification of ADHD in children. *Child and Adolescent Mental Health, 15*(1), 37–43.

Klorman, R., Hazel-Fernandez, L. A., Shaywitz, S. E., Fletcher, J. M., Marchione, K. E., Holahan, J. M., et al. (1999). Executive functioning deficits in attention-deficit/hyperactivity disorder are independent of oppositional defiant or reading disorder. *Journal of the American Academy of Child and Adolescent Psychiatry, 38*(9), 1148–1155.

Lezak, M. D., Howieson, D. B., Loring, D. W., & Hannay, H. J. (2004). *Neuropsychological assessment* (4th ed.). London: Oxford University Press.

Nigg, J. T. (2001). Is ADHD a disinhibitory disorder? *Psychological Bulletin, 127*(5), 571–598.

Nigg, J. T., & Casey, B. J. (2005). An integrative theory of attention-deficit/ hyperactivity disorder based on the cognitive and affective neurosciences. *Development and Psychopathology, 17*(3), 785–806.

Nigg, J. T., Willcutt, E. G., Doyle, A. E., & Sonuga-Barke, E. J. (2005). Causal heterogeneity in attention-deficit/hyperactivity disorder: Do we need neuropsychologically impaired subtypes? *Biological Psychiatry, 57*(11), 1224–1230.

Ramsay, J. R., & Rostain, A. L. (2008). *Cognitive-behavioral therapy for adult ADHD: An integrative psychosocial and medical approach.* New York: Routledge.

Rapport, M. D., Alderson, R. M., Kofler, M. J., Sarver, D. E., Bolden, J., & Sims, V. (2008). Working memory deficits in boys with attention-deficit/hyperactivity disorder (ADHD): The contribution of central executive and subsystem processes. *Journal of Abnormal Child Psychology, 36*(6), 825–837.

Roth, R. M., Isquith, P. K., & Gioia, G. A. (2005). *Behavior rating inventory of executive function–adult version (BRIEF-A) Lutz.* FL: Psychological Assessment Resources.

Ryding, E., Bradvik, B., & Ingvar, D. H. (1996). Silent speech activates prefrontal cortical regions asymmetrically, as well as speech-related areas in the dominant hemisphere. *Brain and Language, 52*(3), 435–451.

Safren, S. A., Otto, M. W., Sprich, S., Winett, C. L., Wilens, T. E., & Biederman, J. (2005). Cognitive-behavioral therapy for ADHD in medication-treated adults with continued symptoms. *Behaviour Research and Therapy, 43*(7), 831–842.

Safren, S. A., Sprich, S., Mimiaga, M. J., Surman, C., Knouse, L., Groves, M., et al. (2010). Cognitive behavioral therapy vs relaxation with educational support for medication-treated adults with ADHD and persistent symptoms: A randomized controlled trial. *Journal of the American Medical Association, 304*(8), 875–880.

Sagvolden, T., Johansen, E. B., Aase, H., & Russell, V. A. (2005). A dynamic developmental theory of attention-deficit/hyperactivity disorder (ADHD) predominantly hyperactive/impulsive and combined subtypes. *The Behavioral and Brain Sciences, 28*(3), 397–419. discussion 419–368.

Shimamura, A. P., Janowsky, J. S., & Squire, L. R. (1990). Memory for the temporal order of events in patients with frontal lobe lesions and amnesic patients. *Neuropsychologia, 28*(8), 803–813.

Skinner, B. F. (1953). *Science and human behavior.* New York: Macmillan.

Solanto, M. V., Marks, D. J., Wasserstein, J., Mitchell, K., Abikoff, H., Alvir, J. M., et al. (2010). Efficacy of meta-cognitive therapy for adult ADHD. *The American Journal of Psychiatry, 167*(8), 958–968.

Trani, M. D., Casini, M. P., Capuzzo, F., Gentile, S., Bianco, G., Menghini, D., et al. (2010). Executive and intellectual functions in attention-deficit/hyperactivity disorder with and without comorbidity. *Brain & Development, 33*(6), 462–469.

Vygotsky, L. S. (1987). *Thinking and speech (Vol. 1— Problems in general psychology).* New York: Plenum.

Willcutt, E. G., Doyle, A. E., Nigg, J. T., Faraone, S. V., & Pennington, B. F. (2005). Validity of the executive function theory of attention-deficit/hyperactivity disorder: A meta-analytic review. *Biological Psychiatry, 57*(11), 1336–1346.

Winsler, A., Diaz, R. M., Atencio, D. J., McCarthy, E. M., & Chabay, L. A. (2000). Verbal self-regulation over time in preschool children at risk for attention and behavior problems. *Journal of Child Psychology and Psychiatry, 41*(7), 875–886.

Wodka, E. L., Loftis, C., Mostofsky, S. H., Prahme, C., Larson, J. C., Denckla, M. B., et al. (2008). Prediction of ADHD in boys and girls using the D-KEFS. *Archives of Clinical Neuropsychology, 23*(3), 283–293.

Executive Functioning Theory and Autism

Hilde M. Geurts, Marieke de Vries, and Sanne F.W.M. van den Bergh

Autism Spectrum Disorders

The triad of characteristics that defines the autistic disorder includes the following: social and communication impairments, and restricted, stereotypical patterns of behavior and interests (American Psychiatric Association [APA], 1994, 2000, for all symptoms see Table 8.1). There are different classic autism-like conditions, and these other pervasive developmental disorders (PDD), such as Asperger syndrome and PDD not otherwise specified (PDDNOS), are part of the broader phenotype of autism. In the current classification system, DSM-IV (APA, 1994, 2000), also Rett syndrome and the Disintegration disorder are considered autism-like conditions. However, in the current chapter we will focus solely on classic autism, Asperger syndrome, and PDDNOS. The combination of these three disorders is referred to as an autism spectrum disorder (ASD), which is the term we will use throughout this chapter.

H.M. Geurts (✉) • S.F.W.M. van den Bergh
Department of Psychology, Dutch Autism & ADHD Research Center (d'Arc), University of Amsterdam, Amsterdam, The Netherlands

Dr Leo Kannerhuis, Amsterdam/Doorwerth, The Netherlands
e-mail: h.m.geurts@uva.nl

M. de Vries
Department of Psychology, Dutch Autism & ADHD Research Center (d'Arc), University of Amsterdam, Amsterdam, The Netherlands

ASD is a heterogeneous, lifelong neurobiological disorder, with an enormous impact on all developmental domains of which the prevalence is estimated between 60 and 100 cases per 10,000 (Baird et al., 2006; Brugha et al., 2011; Gezondheidsraad, 2009). ASD can be diagnosed as early as 18 months of age and leads to a wide array of affective, behavioral, and cognitive problems that are waxing and waning across the lifespan (Rapin & Tuchman, 2008; Volkmar, Lord, Bailey, Schultz, & Klin, 2004). Approximately 70 % of the individuals with an ASD diagnosis have an IQ below 80, indicating an intellectual disability (Fombonne, 2005; Matson & Boisjoli, 2008). ASD also commonly co-occurs with other disorders such as attention-deficit hyperactivity disorder (ADHD), anxiety, and mood disorders (Hofvander et al., 2009; Leyfer et al., 2006; Matson & Nebel-Schwalm, 2007). In children with ASD 71 % has at least one comorbid disorder, and 41 % at least two (Simonoff et al., 2008). This high prevalence of comorbid disorders is persistent into adulthood (Geurts & Jansen, 2012; Hofvander et al., 2009) and has probably a large impact on the observed cognitive problems of individuals with ASD.

An influential cognitive theory of ASD purports that the symptoms observed in individuals with ASD arise from executive function (EF) deficits (Damasio & Maurer, 1978; Hill, 2004; Maurer & Damasio, 1982; Pennington & Ozonoff, 1996; Russell, 1997; Russo et al., 2007). As described in the previous chapters, executive functions (EFs) encompass the ability to suppress responses (inhibition), to keep and

Table 8.1 ASD symptoms

DSM-IV-TR criteria autistic disorder

A   1. Qualitative impairment in social interaction
    (a) Marked impairments in the use of multiple nonverbal behaviors such as eye-to-eye gaze, facial expression, body posture, and gestures to regulate social interaction
    (b) Failure to develop peer relationships appropriate to developmental level
    (c) A lack of spontaneous seeking to share enjoyment, interests, or achievements with other people (e.g., by a lack of showing, bringing, or pointing out objects of interest to other people)
    (d) Lack of social or emotional reciprocity (e.g., not actively participating in simple social play or games, preferring solitary activities, or involving others in activities only as tools or "mechanical" aids)

2. Qualitative impairments in communication
    (a) Delay in, or total lack of, the development of spoken language (not accompanied by an attempt to compensate through alternative modes of communication such as gesture or mime)
    (b) In individuals with adequate speech, marked impairment in the ability to initiate or sustain a conversation with others
    (c) Stereotyped and repetitive use of language or idiosyncratic language
    (d) Lack of varied, spontaneous make-believe play or social imitative play appropriate to developmental level

3. Restricted repetitive and stereotyped patterns of behavior, interests and activities
    (a) Encompassing preoccupation with one or more stereotyped and restricted patterns of interest that is abnormal either in intensity or focus
    (b) Apparently inflexible adherence to specific, nonfunctional routines or rituals
    (c) Stereotyped and repetitive motor mannerisms (e.g., hand or finger flapping or twisting, or complex whole body movements)
    (d) Persistent preoccupation with parts of objects

B   Delays or abnormal functioning in at least one of the following areas, with onset prior to age 3 years: in social interaction, language as used in social communication, symbolic or imaginative play
C   The disturbance is not better accounted for by Rett's disorder or childhood disintegrative disorder

Note: In the DSM-IV-TR, one needs a total of six (or more) items from (1), (2), and (3), with at least two from (1), and one each from (2) and (3) one to meet criteria for the autistic disorders, for Asperger's disorder domain (2) is not part of the criteria. Georgiades and colleagues (2007) showed that the three categorical DSM-IV ASD domains, social relationships, communication, and restrictive repetitive and stereotyped behavior are very heterogeneous. For example, communication includes behavior that regulates social interaction, but also includes flexible use of language. Also repetitive behavior consists of both repetitive stereotyped movements and inflexible behavior. They suggested three new factors (1) social communication; (2) inflexible language and behavior; and (3) repetitive sensory and motor behavior. Especially the last two might be related to inflexibility, respectively to cognitive and to motor inflexibility. In the proposal for the DSM-5 two domains are included, Persistent deficits in social communication and social interaction across contexts, not accounted for by general developmental delays and restricted, repetitive patterns of behavior, interests, or activities including Hyper- or hyporeactivity to sensory input or unusual interest in sensory aspects of environment

manipulate information online (working memory), to change strategies (cognitive flexibility), and to plan ahead (planning). Individuals with ASD seem to encounter deficits in each of these domains. Some even argued (Damasio & Maurer, 1978) that EF deficits might be at the core of ASD as individuals with ASD have problems with exerting effortful control when they need to deal with novel, complex, or ambiguous situations in everyday life. Moreover, it seems that these deficits in ASD are associated with structural and functional abnormalities in the underlying frontostriatal network (Amaral, Schumann, & Nordahl, 2008; Gilbert, Bird, Brindley, Frith, & Burgess, 2008; Kana, Keller, Minshew, & Just, 2007; Luna et al., 2002; Schmitz et al., 2006; Shafritz, Dichter, Baranek, & Belger, 2008).

In this chapter we will first address the origin of this theory, followed by a short overview of the literature focusing on the (dis)functioning of the frontostriatal network in ASD. Hereafter, we will describe how several ASD symptoms might arise from an EF deficit for the following executive functioning domains: inhibition, working memory, cognitive flexibility, and planning. For each of these four EF domains, a short overview of the most recent findings in ASD will be provided.

The Analogy with Patients with Frontal Lobe Damage

The first who postulated an executive dysfunction account of ASD were Damasio and Maurer (1978). In their influential paper they described how individuals with frontal lobe damage show specific behavior which is also typical for people with ASD (Eslinger & Damasio, 1985). The observation that patients with frontal lobe lesions have social difficulties led to the hypothesis that ASD might be a frontal lobe disorder. This idea has inspired various research groups across the world to determine whether or not individuals with ASD indeed encounter EF deficits, whether individuals with ASD show deficits in all or just in some EF domains, and whether there is evidence for a disruption of the frontal network.

The Involvement of the Frontostriatal Network in ASD

Brain imaging studies of ASD demonstrate abnormalities in both structure and function of several brain regions including the prefrontal cortex (Agam, Huang, & Sekuler, 2010; Amaral et al., 2008; Mcalonan et al., 2005; Stanfield et al., 2008). Other studies have suggested that rather than deficits in localized activity, ASD may be better conceptualized as dysfunctions in activity of distributed brain network, or deficient synchronization within those networks (Courchesne & Pierce, 2005). According to Courchesne and Pierce (2005) the "autistic brain" is characterized by local over-connectivity and long-range under-connectivity of the *frontal* cortex. Just, Cherkassky, Keller, and Minshew (2004) postulated that ASD arises from reduced synchronization between *frontal and posterior regions* of the cortex. This reduced synchronization has been observed during performance on a broad range EF tasks (Agam, Huang, et al., 2010; Just et al., 2004; Kana, Keller, Cherkassky, Minshew, & Just, 2006; Kleinhans et al., 2008; Mason, Williams, Kana, Minshew, & Just, 2008; Solomon et al., 2009) but, for example, also during social processing (Kana, Keller, Cherkassky, Minshew, & Just, 2009; Welchew et al., 2005). Moreover, this connectivity has been related to the presence of repetitive behavior in individuals with ASD, which is one of the key aspects of the ASD diagnosis (e.g., Agam, Huang, et al., 2010; Langen, Durston, Kas, Van Engeland, & Staal, 2011). Imaging studies revealed that, while performing EF tasks, people with ASD show activation abnormalities in the frontostriatal circuitry (Gilbert et al., 2008; Kana et al., 2007; Luna et al., 2002; Schmitz et al., 2006; Shafritz et al., 2008). They often recruit *more* brain areas when performing these tasks as compared to healthy people, but both over- and under-activation have been observed in individuals with ASD as compared to controls (Gilbert et al., 2008; Kana et al., 2007; Luna et al., 2002; Schmitz et al., 2006; Shafritz et al., 2008).

With respect to the different EF domains, several ASD studies focused on the frontostriatal and frontoparietal network. For example, imaging studies focusing on inhibition reported more frontal and less parietal activation (Kana et al., 2007; Schmitz et al., 2006). In working memory studies (e.g., Belmonte & Yurgelun Todd, 2003; Gomarus, Wijers, Minderaa, & Althaus, 2009; Koshino et al., 2005; Luna et al., 2002), less task-related activation has been observed, for example, in the dorsolateral prefrontal cortex and the posterior cingulate (Luna et al., 2002), the left inferior frontal area (Koshino et al., 2008), and in the middle frontal gyrus and superior parietal lobe activation (Belmonte & Yurgelun Todd, 2003). In line with these findings, anterior-posterior coherence in brain connectivity is higher in children with ASD, which is associated with worse working memory performance (Chan et al., 2011). Moreover, reduced connectivity in the prefrontal regions is not just related to working memory but also to ASD severity (Poustka et al., 2012). In a meta-analysis (Di Martino et al., 2009) it was shown that when performing the so-called nonsocial tasks (these were mainly EF tasks), the pre-supplementary motor area and the dorsal anterior cingulate cortex (ACC) were hypo-activated in individuals with ASD, while in social tasks (including facial processing tasks and theory of mind [ToM] tasks), the perilingual wall of the ACC and right anterior insula were hypo-activated. Hence, currently ASD is seen as a brain connectivity disorder (see Schipul, Williams, Keller, Minshew, & Just, 2012; Vissers, Cohen, & Geurts, 2012; Wass, 2011), and the observed EF deficits have been related to the increased connectivity within the prefrontal cortex and the decreased connectivity of the frontal cortex with more posterior regions of the brain.

Do People with ASD Have Specific Executive Functioning Deficits?

Even though the EF dysfunction account does have an intuitive appeal to explain the observed behavior in individuals with ASD, there are some difficulties with this idea. A complication for an executive dysfunction account of ASD is that various other disorders (e.g., ADHD, see Chap. 10) are also associated with EF deficits. Hence, the specificity of the EF hypothesis has been widely disputed (Pennington & Ozonoff, 1996; Sergeant, Geurts, & Oosterlaan, 2002) as, for example, working memory deficits seem to be present in a wide range of disorders (Willcutt, Sonuga-Barke, Nigg, & Sergeant, 2008). Even though executive dysfunctions are not specific for ASD, this does not imply that it is not worthwhile studying EF in relation to ASD as the pattern of EF deficits might gain insight in the day-to-day deficits people with ASD encounter.

Another complication is that recent reviews and meta-analyses suggest that findings regarding EF in ASD are rather inconsistent across studies (Geurts, Corbett, & Solomon, 2009; Hill, 2004; Russo et al., 2007). For example, some argue that there is a clear deficit in cognitive flexibility (Hill, 2004; Russo et al., 2007) while this is doubted by others (Geurts, Corbett, et al., 2009). To explain these different findings, various arguments have been proposed. First, it has been noted that the participants included in the ASD groups may differ in their clinical diagnosis (i.e., autism, Asperger syndrome, PDDNOS). Even though these subgroups seem to have similar patterns of EF deficits (Verté, Geurts, Roeyers, Oosterlaan, & Sergeant, 2006b), this is often used as an explanation for the different pattern of findings. Second, there are differences among studies in the IQ range of the included participants and in how IQ differences between groups are handled. However, even studies in which the IQs of the participants were similar have shown inconsistent results (e.g., Geurts, Verté, Oosterlaan, Roeyers, & Sergeant, 2004). Third, some studies focus on children (e.g., Corbett, Constantine, Hendren, Rocke, & Ozonoff, 2009; Goldberg et al., 2005; Happe, Ronald, & Plomin, 2006; Luna, Doll, Hegedus, Minshew, & Sweeney, 2007), others on adults (e.g., Bramham et al., 2009; Lopez, Lincoln, Ozonoff, & Lai, 2005; Luna et al., 2007) or even the elderly (Geurts & Vissers, 2012), and some even include individuals from a very broad age range (Ambery, Russell, Perry, Morris, & Murphy, 2006; Hill & Bird, 2006). Hence, the inconsistencies might be due to the deviant developmental trajectory of EFs in people with ASD (see Happe et al., 2006; Luna et al., 2007). Fourth, both the types of task used to measure the different EF domains and the reported dependent measures of these tasks vary widely (Sergeant et al., 2002). There is some truth in each of these four arguments, but the inconsistent findings also reflect the genuine heterogeneity in the cognitive deficits of the ASD population. A recent study by Pellicano (2010b) seems to support this argument; considerable individual differences were found in EF abilities in very young children with ASD (see also Geurts, Sinzig, Booth, & Happé, submitted [children]; Johnston, Madden, Bramham, & Russell, 2011 [adults]).

An important discussion in the EF literature regarding ASD is how the EF theory relates to other dominant ASD theories. Two other dominant cognitive theories about ASD are the central coherence theory (Frith, 1989; Frith & Happé, 1994; Happé, 1999) and the ToM (e.g., Baron Cohen, 2001; Frith, Morton, & Leslie, 1991). Central coherence refers to an information processing style in which one processes information in its specific context, and a weak central coherence would result in piecemeal processing which is often observed in individuals with ASD (Happé, 1999; Pellicano, 2007, 2010b). The relationship between EF and central coherence is hardly studied as the assumption is that these theories explain different aspects of the autism spectrum (see also Happe et al., 2006). ToM refers to the ability to attribute mental states to oneself and to others and to the ability to understand how mental states influence human behavior.

A well-developed ToM is crucial for making social inferences and guiding social behavior in everyday life communicative interactions. Importantly, people with ASD may have impaired ToM abilities (e.g., Baron Cohen, 1995; Happé, 1994), even when their performance on inference tasks that do not require understanding of mental states is unimpaired (Baron Cohen, 1995, 2001; Charman & Baron Cohen, 1992; Happé, 1994; Ozonoff, Pennington, & Rogers, 1991). The relationship between EF and ToM has received a great deal of attention (Fisher & Happé, 2005; Hughes & Ensor, 2007; Ozonoff et al., 1991; Pellicano, 2007) as these constructs seem to be highly interlinked (Frye, Zelazo, Brooks, & Samuels, 1996; Hala, Hug, & Henderson, 2003; Hughes, 1998; Perner & Lang, 1999; Sabbagh, Xu, Carlson, Moses, & Lee, 2006). For example, cognitive flexibility involves the ability to switch rapidly between multiple tasks and may be crucial to change strategies or perspective in ToM tasks or during everyday conversation. Moreover, ToM tasks require working memory (Mckinnon & Moscovitch, 2007) as intermediate steps are needed to perform well on various complex ToM tasks. The intermediate steps need to be kept in mind, evaluated, and perhaps adjusted, so more intermediate steps may require a larger involvement of working memory. In EF tasks such as classical cognitive flexibility tasks, ToM may play a role as participants have to adjust their behavior based on feedback given by the assessor of the task (Ozonoff & Miller, 1995). In most EF tasks the participants need to conceptualize (i.e., infer) what the experimenter wants them to do (see for details also Pellicano, 2007) which is an important aspect of ToM. To put it differently, to perform adequately on these EF tasks, one needs to have a representational understanding of mind (Perner & Lang, 2000). Hence, it is no surprise that EF and ToM deficits often go hand in hand.

Interestingly, the development of ToM seems intricately intertwined with the development of EF (e.g., Carlson, Mandell, & Williams, 2004; Hughes & Ensor, 2007). In fact, some researchers have argued that EF ability is necessary to perform adequately on many ToM tasks (Frye et al., 1996) and, more generally, that development of EF is a prerequisite for the development of ToM (e.g., Hughes, 1998; Russell, 1997). In contrast, some researchers have proposed a reverse relationship (e.g., Perner & Lang, 1999, 2000), namely, that the metarepresentational capacity that underlies ToM, the understanding that behavior is guided by internal states, is required for the development of EF. The results from longitudinal studies in typically developing children thus far support the notion that EF competence is important for the acquisition of ToM (Carlson et al., 2004; Flynn, O'Malley, & Wood, 2004; Hughes, 1998; Hughes & Ensor, 2007; Muller, Liebermann-Finestone, Carpendale, Hammond, & Bibok, 2012).

The results from studies that have focused on EF and ToM in ASD also underline the strong relationship between these two constructs (e.g., Pellicano, 2007, 2010a; Zelazo, Jacques, Burack, & Frye, 2002). However, the precise nature of the EF-ToM relationship remains unclear. On the one hand, the correlation analyses typical of most research studies do not allow for any causal inferences (but see Pellicano, 2007). On the other hand, however, training EF abilities in children with ASD seems to improve their performance on ToM tasks, whereas training on ToM does not result in improved EF (Fisher & Happé, 2005). These findings hint at the possibility that EF deficits are primary in ASD. Pellicano (2007) hypothesized that ToM and EF are crucially linked at an early stage of development when both abilities begin to emerge, but do not influence one another when children are older and conceptual understanding has been developed. To be able to establish whether this is indeed the case, more longitudinal studies are needed (see for an example Pellicano, 2010a). In all, the consensus thus far seems that EF and ToM abilities can interact with one another, share a developmental timetable, and are both impaired in people with ASD. In the current chapter we will focus on EF, but if relevant for interpreting the EF findings in relation to ASD, we will also discuss ToM studies.

Inhibition

Inhibition problems are often observed in day-to-day behavior in people with ASD. For example, the ability to generate appropriate responses during social interactions involves selecting the most fitting response while inhibiting those responses deemed inappropriate. Also in language use it is necessary to inhibit one (frequently used) meaning of a word (e.g., a *bank* to sit on vs. a *bank* to withdraw money from) if you need to use the other (less frequently used) meaning of a word. Taking language literally is one of the often observed behavior characteristics in people with ASD. Repetitive behavior in individuals with ASD might also be due to difficulties in suppressing behavior even when the consequences are negative (e.g., Langen et al., 2012). In several studies (Geurts & De Wit, in press; Solomon, Ozonoff, Cummings, & Carter, 2008), the observed ASD behavior (as measured with parent reports, diagnostic interviews, or observational schedules) correlated with performance on inhibitory control tasks (but see Happé, Booth, Charlton, & Hughes, 2006). Hence, several key characteristics of ASD might be related to deficits in inhibitory control.

Since the first series of studies by Ozonoff and Russell and colleagues in the 1990s (Hughes & Russell, 1993; Hughes, Russell, & Robbins, 1994; Ozonoff & Strayer, 1997; Ozonoff, Strayer, Mcmahon, & Filloux, 1994), various research groups around the globe focused on inhibitory control in children and adolescents (Adams & Jarrold, 2009; Christ, Holt, White, & Green, 2007; Christ, White, Mandernach, & Keys, 2001; Corbett et al., 2009; Eskes, Bryson, & Mccormick, 1990; Geurts, Begeer, & Stockmann, 2009; Geurts, Luman, & Meel, 2008; Geurts et al., 2004; Goldberg et al., 2005; Happé & Frith, 2006; Johnson et al., 2007; Kilincaslan, Mukaddes, Kucukyazici, & Gurvit, 2010; Lee et al., 2009; Lemon, Gargaro, Enticott, & Rinehart, 2011; Mahone et al., 2006; Pellicano, 2007; Raymaekers, Van Der Meere, & Roeyers, 2006; Russo et al., 2007; Semrud-Clikeman, Walkowiak, Wilkinson, & Butcher, 2010; Sinzig, Morsch, Bruning, Schmidt, & Lehmkuhl, 2008) and adults (e.g., Agam, Joseph, Barton, & Manoach, 2010; Barnard, Muldoon, Hasan, O'Brien, & Stewart, 2008; Johnston et al., 2011; Kana et al., 2007; Langen et al., 2012; Mosconi et al., 2009; Nydén et al., 2010; Raymaekers, Antrop, Van Der Meere, Wiersema, & Roeyers, 2007; Schmitz et al., 2006) with ASD.

The findings across these studies seem, at first sight, not very consistent as inhibitory control deficits in ASD did not come to the fore in various studies. This inconsistency of findings seems to be independent of the age of the participants. According to Luna et al. (2007), inhibitory control seems to be deficient in ASD throughout development even though there are developmental improvements in the capacity to inhibit in ASD.

Inhibition can be divided into prepotent response inhibition, resistance to distractor interference, and resistance to proactive interference (Friedman & Miyake, 2004), and a broad range of measures has been used to measure these three inhibitory control constructs. The inhibitory control impairments in people with ASD seem to be most prominent in resistance to distractor interference tasks (e.g., Adams & Jarrold, 2012; Christ, Kester, Bodner, & Miles, 2011; Geurts et al., 2008; but see Henderson et al., 2006; Johnston et al., 2011; Solomon et al., 2008), for example, called flanker (Eriksen & Eriksen, 1974) tasks, while proactive interference seems to be relatively intact (Bennetto, Pennington, & Rogers, 1996; Christ et al., 2011). On typical prepotent response inhibition tasks such as the Go-NoGo task (Casey et al., 1997) and the Stop task (Logan, 1994), findings seem to be inconsistent as various studies reported null findings (Adams & Jarrold, 2012; Christ et al., 2007, 2011; Eskes et al., 1990; Geurts, Begeer, et al., 2009; Goldberg et al., 2005; Happé & Frith, 2006; Kana et al., 2007; Kilincaslan et al., 2010; Ozonoff & Jensen, 1999; Ozonoff & Strayer, 1997; Raymaekers et al., 2007; Russell, Jarrold, & Hood, 1999; Schmitz et al., 2006; Semrud-Clikeman et al., 2010), while some others do report deficits in individuals with ASD (Adams & Jarrold, 2009; Corbett et al., 2009; Geurts et al., 2004; Johnston et al., 2011; Ozonoff et al., 1994; Raymaekers, Van Der Meere, & Roeyers, 2004). In a recent study Christ et al. (2011) suggest that the observed impairment in resistance to distractor interference might be due to a developmental delay which resolves with aging. This would suggest that adults with ASD will probably not have this type of inhibitory control impairments, but so far this has not been tested in sufficiently powered studies.

In most interference control tasks, but also in the Stop task (Logan, 1994) and the Stroop task (Macleod, 1991), the participants need to inhibit a formerly learned response to a specific stimulus. Yet, this is in contrast with a typical Go-NoGo task, in which a NoGo stimulus is typically not associated with a response. Hence, children with ASD might mainly have difficulties with inhibiting a learned response instead of having difficulties in just not responding. However, the null findings on Stroop like tasks contradict this interpretation (Christ et al., 2007; Goldberg et al., 2005; Kilincaslan et al., 2010; Semrud-Clikeman et al., 2010). An alternative explanation might lie in the role of working memory in most inhibitory control tasks. Previous studies have demonstrated that children with ASD are especially challenged by inhibitory control tasks with a heavy working memory load (Hughes & Russell, 1993; Joseph, Steele, Meyer, & Tager-Flusberg, 2005; Kana et al., 2007; Luna et al., 2007; Ozonoff & Strayer, 1997; Ozonoff et al., 1994; Russell, 1997). It could well be that working memory deficits are partly underlying the reported difficulties with inhibitory control (but see Christ et al., 2011). Nonetheless, so far the evidence suggests that people with ASD have difficulties in their ability to ignore and/or suppress irrelevant (interfering) information and there is no convincing evidence for an ASD-related impairment in prepotent response inhibition.

Working Memory

Individuals with ASD also seem to experience working memory (WM) problems, as, for example, a common complaint by parents is that their child with ASD is not able to execute instructions or commands. Even children with well-developed hearing and verbal understanding seem to demonstrate such difficulties in implementation of instructions. Especially when more than one instruction is given at once, individuals with ASD have difficulties to follow them all. This seems to be a WM problem; although information seems to be understood, and possibly stored, the transmission to actually manipulate and use the information subsequently seems to be disturbed (Baddeley, 1992). In everyday life WM is necessary in various situations, e.g., remembering directions while driving or remembering the name of someone who introduced himself. For children, WM is necessary when a teacher at school explains a future assignment or when parents instruct their children. Apart from this obvious role of WM, WM deficits might also influence social behavior as in social situations WM plays an important role. When meeting new people, it is necessary to introduce oneself, remember not just the name of the person you meet, but also the subject of the conversation. Moreover, for a smooth social interaction, it is important to remember, process, and interpret information like a person's face, facial expression, tone of voice, and body language. To be able to interact appropriately with others, new information needs to be stored and combined with familiar information and needs to be interpreted fast and accurately. These different aspects of social interaction require WM (Causton-Theoharis, Ashby, & Cosier, 2009). Hence, when individuals with ASD indeed encounter WM deficits, this is of crucial importance for their day-to-day functioning.

In the WM literature a distinction is often made between (1) the central executive, (2) the visual-spatial sketch pad, and (3) the phonological loop (Baddeley, 1992; Gathercole & Alloway, 2006; Gathercole, Pickering, Ambridge, & Wearing, 2004). However, in the ASD literature, the main distinction made is whether verbal or visual information needs to be processed. Therefore, the latter distinction will be discussed in this section. Overall, it seems that individuals with ASD do show deficits in both verbal and visual-spatial WM (Willcutt et al., 2008), but some argue that the deficits in visual-spatial WM are the most prominent (Williams, Goldstein, Carpenter, & Minshew, 2005; Williams, Goldstein, & Minshew, 2006).

Memory span tasks are often used to measure verbal WM; a list of stimuli (e.g., digits, letters, or sentences) has to be remembered and reproduced (Bennetto et al., 1996; Cui, Gao, Chen, Zou, & Wang, 2010; Gabig, 2008; Minshew &

Goldstein, 2001; Williams et al., 2005, 2006). Another verbal WM measurement, which is commonly used in ASD research, is the *n*-back task (Kana et al., 2007; Koshino et al., 2005, 2008; Williams et al., 2005). Verbal stimuli are visually displayed and participants have to alternately point out if a certain stimulus is similar to a target stimulus (0-back), the previous stimulus (1-back), or two stimuli earlier (2-back). The *n*-back task is thought to be mainly verbal as even pictures are mostly remembered in words (Williams et al., 2005). When WM load is minimal, individuals with ASD seem to have no impairment in verbal WM (Cui et al., 2010; Williams et al., 2005), but when a large amount of complex information has to be processed, individuals with ASD do show verbal WM deficits (Williams et al., 2006). More specifically, increasing WM load seems to impair children with ASD more than typically developing children (Cui et al., 2010). These deficits are reported in various age groups (Bennetto et al., 1996; Gabig, 2008; Minshew & Goldstein, 2001) and a similar pattern is seen in everyday life. Children with ASD seem particularly disabled when several complex or ambiguous tasks have to be performed consecutively, thus when WM load is high. When performing or finishing a relatively difficult task, WM seems to get overloaded. When given one task at a time, with clear instructions, or step-by-step guidance—hence low WM load—individuals with ASD are indeed able to perform one or more tasks.

Classical visual-spatial WM tasks widely used in ASD research are the Corsi Block-Tapping Task (Berch, Krikorian, & Huha, 1998; Corsi, 1972) and the highly similar CANTAB spatial WM task (Cambridge, 2002; Corbett et al., 2009; Goldberg et al., 2005; Happé et al., 2006; Landa & Goldberg, 2005; Sinzig et al., 2008; Steele, Minshew, Luna, & Sweeney, 2007) and the CANTAB spatial span task (Barnard et al., 2008; Cambridge, 2002; Corbett et al., 2009). Visual-spatial WM seems to be impaired in ASD when measured with the aforementioned tasks. Although not all studies are confirmative (Ozonoff & Strayer, 2001; Yerys, Hepburn, Pennington, & Rogers, 2007), evidence that there are actual problems in this area is increasingly convincing. Children with ASD show difficulty in storing, maintaining, and retrieving visual-spatial information (Corbett et al., 2009; Goldberg et al., 2005; Happé et al., 2006; Landa & Goldberg, 2005; Sinzig et al., 2008; Williams et al., 2005). Moreover, visual-spatial WM deficits seem to correlate with ASD symptoms (Verté, Geurts, Roeyers, Oosterlaan, & Sergeant, 2006a). Also adults with ASD do still show similar WM deficits (Barnard et al., 2008; Gomarus et al., 2009; Luna et al., 2007; Steele et al., 2007; Williams et al., 2005). In everyday life, individuals with ASD often use pictures, symbols, or icons to represent tasks that have to be performed and events that will happen during a certain day or period of time (Ganz, Davis, Lund, Goodwyn, & Simpson, 2011). It might be that this visually offered information supports the less well-developed visual-spatial WM. By displaying the pictures externally, WM load will be reduced, which might, in turn, increase self-reliance, by helping individuals with ASD to keep up with daily routines.

WM interacts with inhibitory control (see para. 2.1) but also with the other EF domains; to be able to execute a task, one needs to keep a certain rule in mind that needs to be followed (Barnard et al., 2008). WM and other EFs are thought to be mutual influential (Stoet & López, 2010). Not only does WM influence executive functioning per se, but under certain conditions, WM itself is used or triggered by other EFs. WM is influenced by, and influences, attention, inhibition, flexibility, and planning. In executive functioning, firstly, an individual has to pay attention to certain information. If information does not get proper attention, it will not be processed sufficiently and as a result, will not be stored and enter the WM process. Secondly, one can only attend to certain information, when other information will simultaneously be ignored (i.e., a response towards this information needs to be inhibited) as it is impossible to pay attention to, and process, all available information (Chun, Golomb, & Turk-Browne, 2011). Thirdly, one can only focus on one aspect of incoming information and ignore other information, when one can flexibly switch between a variety of available information. Fourthly, to plan an action, WM is needed to trace, scan, and choose what information to use and react to. Especially when more complex tasks are used to measure EF constructs in individuals with ASD, such as cognitive flexibility and planning, it is important to determine the role of WM abilities on task performance as WM in itself is already impaired in individuals with ASD (Willcutt et al., 2008).

Cognitive Flexibility

In the diagnostic criteria of ASD (APA, 1994, 2000), stereotypical and repetitive behavior is the third domain of the ASD triad of symptoms. Also in the social and communication domains, inflexible behavior is part of the ASD criteria (see for a detailed review Geurts, Corbett, et al., 2009). This is one of the reasons why especially the EF construct of cognitive flexibility has an immediate appeal when one tries to explain ASD-related behavior. Cognitive flexibility involves the ability to rapidly switch between multiple tasks (Monsell, 2003) and may therefore be crucial for the ability to change strategies or perspective during everyday conversation. The difficulties of people with ASD to respond to unexpected events might also be related to an inability to flexibly adjust one's behavior to the changing environment. However, the face validity of this relationship between ASD symptoms and cognitive flexibility is difficult to reveal in experimental studies (see Geurts, Corbett, et al., 2009).

In a wide range of cognitive flexibility studies in ASD, the classical neuropsychological task, the Wisconsin Card Sorting Task (WCST; Berg, 1948), has been used (e.g., Bennetto et al., 1996; Griebling et al., 2010; Liss et al., 2001; Maes, Eling, Wezenberg, Vissers, & Kan, 2011; Minshew, Goldstein, Muenz, & Payton, 1992; Ozonoff et al., 1991; Prior & Hoffmann, 1990; Robinson, Goddard, Dritschel, Wisley, & Howlin, 2009; Rumsey, 1985; Sumiyoshi, Kawakubo, Suga, Sumiyoshi, & Kasai, 2011). In most of these studies, children and adults with ASD indeed seem to have cognitive flexibility deficits as they perform worse on the WCST compared to typically developing controls (Geurts, Corbett, et al., 2009). However, not just cognitive flexibility is of importance to perform well on the WCST. Difficulties with learning from feedback, keeping a goal of in mind (i.e., WM), noticing that a change in strategy is necessary, inhibiting a previous motor response, switching to another response, and sustaining responding over time can lead to a decreased WCST performance (Barcelo, 1999; Geurts, Corbett, et al., 2009; Ozonoff, 1995). As we discussed in the previous paragraphs, individuals with ASD seem to have deficits in specific aspects of, for example, inhibitory control and WM, and deficits in these EF domains might already decrease the WCST performance.

However, cognitive flexibility can be measured with a wide range of tasks and not just with the WCST. The difficulty with the ASD cognitive flexibility literature is that the findings of studies using other clinical neuropsychological tasks, or of studies using more experimental tasks, are rather inconsistent (Geurts, Corbett, et al., 2009). Some studies do report cognitive flexibility deficits (Hughes et al., 1994; Ozonoff et al., 2004; Reed & Mccarthy, 2012; Reed, Watts, & Truzoli, 2013; Yerys et al., 2007, 2009), while other do not report any deficits (Corbett et al., 2009; Goldberg et al., 2005; Happé & Frith, 2006; Poljac et al., 2009; Schmitz et al., 2006; Shafritz et al., 2008; Sinzig et al., 2008; Stahl & Pry, 2002; Whitehouse, Maybery, & Durkin, 2006). Studies differ, of course, in methodology (like choice of dependent measures, age, and diagnosis of participants), but this does not seem to be the main reason for the observed inconsistency in findings. In our earlier work (Geurts, Corbett, et al., 2009) we hypothesized that the failure to find cognitive flexibility deficits in ASD in relatively pure cognitive flexibility measurements (such as switch tasks) is due to the predictability of the switches in most of these tasks, while in day-to-day life switches are often unpredictable. In recent studies it was indeed shown that children with ASD are relatively cognitive inflexible when switches occur random and unpredictable (Maes et al., 2011; Stoet & López, 2010).

Various other alternative explanations for the inconsistency in findings have been explored resulting in a series of new studies focusing on cognitive flexibility in children and adults with ASD (e.g., Dichter et al., 2010; Geurts & Vissers, 2012; Griebling et al., 2010; Pellicano, 2010b; Poljac et al., 2009; Reed & Mccarthy, 2012; Robinson et al., 2009; Van Eylen et al., 2011; Yerys et al., 2009). For example, WM load varied largely in different studies and tasks, as a task cue can be available continuously (Schmitz et al., 2006), at the beginning of a task run (Poljac et al., 2009; Shafritz et al., 2008), or only when the task starts (Maes et al., 2011). On switch tasks with minimal WM demand, children with ASD do not show difficulties (Schmitz et al., 2006; Stoet & López, 2010), but when WM demand is higher, the results are inconclusive; some studies report difficulties in ASD (Maes et al., 2011; Shafritz et al., 2008; Stoet & López, 2010) and some do not (Poljac et al., 2009; Whitehouse et al., 2006). Moreover, performance on switch tasks is more influenced by WM demand in children with ASD than in typically developing children (Dichter et al., 2010; Stoet & López, 2010). While WM has probably a large influence on task performance (but see Russo et al., 2007), two alternative hypotheses (Maes et al., 2011; Van Eylen et al., 2011) might also shed some new light on the circumstances in which individuals with ASD do encounter cognitive flexibility deficits.

The first hypothesis is that the possibility to observe flexibility impairments is determined by the degree of explicitly provided task instructions (Van Eylen et al., 2011; see for similar ideas White, Burgess, & Hill, 2009). Van Eylen et al. (2011) classified cognitive flexibility tasks based on the explicitness of task instructions and concluded that the WCST (on which individuals with ASD generally fail) is the task with the lowest degree of explicit task instructions and typical experimental task switch paradigms (on which individuals with ASD generally succeed) have the highest degree of explicit task instructions. In a task switch paradigm where explicitness of task instructions was also low (Van Eylen et al., 2011), children with ASD indeed showed cognitive flexibility problems.

The second hypothesis is that novelty processing might be impaired in individuals with ASD (Maes et al., 2011), resulting in the perseverative behavior observed on WCST-like tasks in people with ASD. The idea is that individuals with ASD are less prone to respond to novel stimuli (see also Anckarsater et al., 2006) and, therefore, keep responding to familiar stimuli. Indeed when a paradigm was used in which the tendency to pay attention to novel or familiar stimuli could be disentangled, children with ASD seemed to favor familiar stimuli (Maes et al., 2011) suggesting reduced novelty processing in individuals with ASD.

In sum, unpredictability, high WM load, the lack of explicit task instructions, and reduced novelty processing might all contribute to the observed day-to-day difficulties in cognitive flexibility. Which account is the most plausible explanation for the inconsistent findings in past cognitive flexibility studies needs to be tested, but for now it seems that especially those cognitive flexibility tasks that are complex in various aspects are those that individuals with ASD cannot succeed on.

Planning

Besides cognitive flexibility problems, in daily life people with ASD often experience planning problems. For example, difficulties are encountered when making homework assignments, organizing morning activities in order to get to work in time, or when running a household. Impairments in communication and social interaction, key characteristics of ASD, might also partly be influenced by planning deficits. Parents or partners, for example, often organize all social appointments of their relative or partner with ASD in order to keep the social relations active. It is also known that at least in children with ASD, planning skills and ToM abilities are strongly related (Pellicano, 2007), for example, performance of children with ASD on a planning task predicts ToM abilities 1 year later, independent from age and verbal ability (Pellicano, 2010b). The fact that day-to-day difficulties in planning are observed in individuals with ASD is not surprising as planning is a complex process of working towards a desired goal and various skills are needed, such as monitoring, reevaluating, and updating actions (Hill, 2004; Shallice, 1982). Hence, like cognitive flexibility, planning is a complex cognitive process in which both inhibitory control and WM are of importance (Newman, Carpenter, Varma, & Just, 2003; Welsh, Satterlee-Cartmell, & Stine, 1999; Zinke et al., 2010). It is simply not possible to plan and perform an action, without using information that is already stored, combining this information with new information, and ignoring irrelevant information. Planning usually consists of several steps, and each of these steps has to be stored but also adjusted to the changing context meaning that also cognitive flexibility is of importance for planning.

Planning is one of the EF domains that is most consistently found to be impaired in people with ASD as compared to typical developing groups (e.g., Bennetto et al., 1996; Booth, Charlton, Hughes, & Happé, 2003; Griebling et al., 2010; Lopez et al., 2005; Ozonoff & Jensen, 1999; Ozonoff & Mcevoy, 1994; Ozonoff et al., 1991; Pellicano, 2010a; Prior & Hoffmann, 1990). Moreover, planning seems even more impaired in individuals with ASD than in individuals with ADHD (Bramham et al., 2009; Geurts et al., 2004; Semrud-Clikeman et al., 2010), although these differences between ASD and ADHD groups are not confirmed in all studies (Booth et al., 2003). Moreover, some studies are not able to differentiate individuals with and without ASD with respect to planning (e.g., Boucher et al., 2005; Corbett et al., 2009; Happé & Frith, 2006; Liss et al., 2001). These null findings might challenge the idea of a general impairment in planning in people with ASD, but based on a meta-analysis of 21 ASD planning studies (Geurts & Bringmann, 2011), it seems that planning difficulties in ASD clearly exist, independent of which ASD diagnosis an individual has and which age. Not just children (e.g., Landa & Goldberg, 2005; Ozonoff et al., 2004; Pellicano, 2010a; Semrud-Clikeman et al., 2010; Verté et al., 2006a; Zinke et al., 2010) but also adults with ASD are impaired in planning compared to typically developed adults (e.g., Bramham et al., 2009; Hill & Bird, 2006; Just, Cherkassky, Keller, Kana, & Minshew, 2007; Lopez et al., 2005). Normally, planning performance improves across development, but this improvement does not seem to be quite evident in individuals with ASD (Keary et al., 2009; Ozonoff & Mcevoy, 1994). Although in a recent longitudinal study (across a 3 year period) young children with ASD showed a steeper increase in planning performance as compared to typically developing peers, the control group still outperformed the ASD group. This suggests that the developmental delay in planning abilities in children with ASD indeed remains present across development (Pellicano, 2010a). Future research is needed to gain insight in the developmental trajectories of planning skills in people with ASD.

As the planning tasks and reported outcomes in planning studies are often very different, it is challenging to make a comparison across studies (see for recent reviews Geurts & Bringmann, 2011; Hill, 2004). The inconsistencies are partly due to the fact that tasks used to measure planning often only correlate moderately (e.g., Tower of Hanoi and Tower of London-Revised, Welsh et al., 1999). Also, the large range of IQ levels in most studies makes it difficult to determine whether ASD or learning disabilities is the main contributing factor to the observed planning problems (Hill, 2004). For example, planning problems in one study (Mari, Castiello, Marks, Marraffa, & Prior, 2003) seem to be due to IQ level, while in another study (Hughes et al., 1994) planning impairments seem to be ASD specific as people with ASD are also impaired when compared to a group with moderate learning disabilities. Across the used tasks it seems that especially the so-called Tower tasks, except the "Stocking of Cambridge" (SoC), are the most effective in determining planning problems in people with ASD (Geurts & Bringmann, 2011). The SoC is a computerized task and might therefore be less related to planning in daily life, because for individuals with ASD it seems easier to perform on a computerized task than a task requiring more social interaction (Ozonoff, 1995).

Hence, the task choice does influence the outcome of ASD planning studies.

There is a cognitive neuroarchitecture model (called 4CAPS) of problem solving with a Tower task (Just & Varma, 2007). The basic idea of this model is that functional connectivity is crucial for efficient problem solving; multiple cortical networks perform multiple cognitive functions in specialized, as well as dynamic, ways (Just & Varma, 2007). This is of interest for understanding the ASD-related planning impairments, as ASD is more and more considered a brain connectivity disorder (Courchesne & Pierce, 2005; Just et al., 2004; Schipul et al., 2012; Vissers et al., 2012; Wass, 2011). The difficulty levels of Tower assignments are correlated with activation in the right and left dorsolateral prefrontal cortex and with the left, but not right superior parietal regions (Newman et al., 2003). In the 4CAPS model, four collaborating centers are proposed: the left and right hemisphere executive centers (in the dorsolateral prefrontal cortex) and the left and right spatial centers (in the superior parietal regions). The right hemisphere executive center is hypothesized to be important in strategic control by selecting and planning the moves, while the left hemisphere executive center is involved in controlling and executing the planning process. The left spatial center is proposed to spatially transform the Tower image by imagining moving the objects and, thereby, controlling the execution of planning. The right spatial center would only generate perceptual moves and, therefore, is not linked to planning difficulty (Newman et al., 2003). Fitting such a model to data obtained from individuals with ASD (see, e.g., Griebling et al., 2010 for a Tower related imaging study with ASD) could potentially inform us what brain network deficiencies underlie the observed difficulties with planning tasks in ASD.

More work involving mathematical models of cognitive functioning in planning tasks but also regarding other EF tasks would be of importance to unravel how and when individuals with ASD do or do not encounter EF deficits. This is especially important as not just in planning but also in other EF domains, the needed cognitive processes to perform well on the tasks are highly intertwined. It is not clear whether difficulties on a wide range of EF tasks are due to just one underlying deficient cognitive process or whether especially performing multiple cognitive processes at the same time is the reason for the observed failures on EF tasks.

Is ASD an Executive Function Disorder?

ASD cannot be described as an EF disorder as (1) many individuals with ASD do not encounter EF deficits (Geurts et al., submitted; Johnston et al., 2011; Pellicano, 2010b) and (2) EF deficits are not specific for people with ASD (Pennington & Ozonoff, 1996; Sergeant et al., 2002; Willcutt et al., 2008). However, as EF deficits are more common in individuals with ASD as compared to typical developing individuals, it is important to study EF in relation to ASD. That is, the idea that get its feet on the ground more and more is that ASD results from an interacting compound of cognitive deficits (and/or styles such as EF, ToM, and weak central coherence) and no single deficit might be sufficient or even necessary for the diagnostic symptom profile to arise (e.g., Happé & Ronald, 2008; Happe et al., 2006).

Various challenges for the field have been discussed in the different sections within this chapter as there is currently no consensus regarding the type of tasks which has the highest validity to measure EF in ASD, there is no consensus regarding the dependent variables that need to be reported, nor is there consensus regarding the variables we need to control for when choosing appropriate control group. However, the field is progressing as more and more EF studies include mathematical models (Just & Varma, 2007) and experimental paradigms (see, e.g., Christ et al., 2011; Maes et al., 2011; Solomon et al., 2009) in which different cognitive processes that might be affected in ASD can be disentangled.

We feel that there are two other major challenges for EF researchers. First of all, for people with ASD it could be helpful if we establish which individuals with ASD do encounter EF deficits and which persons do not show deficits. This is important as this might have implications for the determining which interventions are the most suitable for a specific individual. For example, children with ASD with a cognitive flexibility deficit (as measured with the WCST) might not benefit from social skills training, while children without such a deficit do (Berger, Aerts, Spaendonck, Cools, & Teunisse, 2003). Whether one does or does not have certain EF deficits could also be of importance when one wants to train EF. Given that EF deficits are so widely studied in ASD, it is surprising that, as far as we know, only one study focused on training EF (Fisher & Happé, 2005). This is especially startling given that EF training (mainly WM training) seems an effective intervention for ADHD (Beck, Hanson, Puffenberger, Benninger, & Benninger, 2010; Holmes et al., 2009; Klingberg, Forssberg, & Westerberg, 2002; Klingberg et al., 2005; White & Shah, 2006), a neurodevelopmental disorder which is often comorbid with ASD (Rommelse, Franke, Geurts, Hartman, & Buitelaar, 2010). A first pilot study by Fisher and Happé (2005) indeed showed that training EF (focusing on cognitive flexibility) improved the performances of children with ASD on ToM tasks. In this study children with ASD received ToM training, EF training, or no intervention. Both training programs had a strategy-based approach which was adjusted for each participant. In each training session different rules were learnt and the trainer used objects and illustrative stories to explain the rules. Directly after the ToM training sessions, the children improved on ToM tasks, but not on EF tasks. In contrast, directly after the EF training sessions, the children did neither improve at the ToM nor at the EF tasks. However, at follow-up (6–12 weeks later) all children who received a ToM training or an EF training improved in their ToM performance. These results suggest that children with ASD could pass ToM tasks after both ToM and EF training. However, the EF did not improve at all. Both the ToM and the EF training failed to reduce the EF difficulties in these children with ASD. This might suggest that in ASD EF cannot be trained, while in the ADHD literature, game-like computerized training programs seem to be successful, especially when the focus is on WM (Beck et al., 2010; Holmes et al., 2009; Klingberg et al., 2002, 2005). The first preliminary findings of our study in which we compare a game-like training of WM and cognitive flexibility, with a non-EF computer training in ASD, do suggest that children with ASD improve in their day-to-day EF skills (De Vries, Prins, Schmand, & Geurts, 2011). Hence, to determine whether or not EF training will indeed be a candidate intervention for people with ASD, more research is needed. However, when studying this type of interventions, one needs to take into account that there are individual differences in EF deficits in individuals with ASD, as the profile of EF deficits and strengths might be of great importance for the failure or success of such an intervention (Berger et al., 2003).

The second challenge for ASD researchers is to incorporate a developmental perspective. ASD is a neurodevelopmental disorder, but studies often focus on one specific age range when studying ASD. As described in both childhood and adulthood, individuals with ASD show a broad range of EF deficits, but these findings are not unambiguous. The developmental pattern of EF in children and adolescents with ASD appears to be atypical (Happé et al., 2006; Luna et al., 2007; Pellicano, 2010a). For example, children with autism between 8 and 11 years of age showed several EF deficits, while these deficits did not emerge in children with autism aged 11–16 years (Happé et al., 2006). Also, in a recent longitudinal 3 year follow-up, planning capacity in children with autism improved at a faster rate than that of typically developing children (Pellicano, 2010a). Hence, these findings indicate that at least some EF deficits decline when aging. However, this idea of abating deficits might be in contrast with adult studies in which executive dysfunctions are still present in individuals with ASD above 16 years of age (e.g., Ambery et al., 2006; Bramham et al., 2009; Geurts & Vissers, 2012; Goldstein, Johnson, & Minshew, 2001; Hill & Bird, 2006; Lopez et al., 2005; Minshew, Meyer, & Goldstein, 2002). In a cross-sectional developmental study (Luna et al., 2007) executive dysfunctions were present in people with ASD of different ages (8–12, 13–17, and 18–33 year). Across the three age groups, the autism group encountered inhibitory control deficits as well as WM deficits. However, developmental improvements in inhibitory control were similar in both groups (i.e., parallel development), while the development of WM was impaired in the autism group (Luna et al., 2007). The smaller extent of improvements in EF in children has also been reported in two longitudinal studies (Griffith, Pennington, Wehner, & Rogers, 1999; Ozonoff & Mcevoy, 1994). The combined findings from cross-sectional and longitudinal studies suggest that there are different developmental patterns for different aspects of EF. So studies focusing on both *how* and *when* EFs are disturbed are needed to fully grasp the EF impairments of individuals with ASD.

Although the aforementioned avenues for research might make it even more challenging to make clear statements about the ASD group as a whole, the large individual differences by themselves do give information as the ASD group is apparently a heterogeneous group. In sum, individuals with ASD do seem to experience EF problems, but (1) not all individuals with ASD do so, (2) not in the same areas, (3) not with similar severity, and (4) not all individuals have similar compensatory skills.

References

Adams, N. C., & Jarrold, C. (2009). Inhibition and the validity of the Stroop task for children with autism. *Journal of Autism and Developmental Disorders, 39*(8), 1112–1121.

Adams, N. C., & Jarrold, C. (2012). Inhibition in autism: Children with autism have difficulty inhibiting irrelevant distractors but not prepotent responses. *Journal of Autism and Developmental Disorders, 42*(6), 1052–1063.

Agam, Y., Huang, J., & Sekuler, R. (2010). Neural correlates of sequence encoding in visuomotor learning. *Journal of Neurophysiology, 103*(3), 1418–1424.

Agam, Y., Joseph, R. M., Barton, J. J., & Manoach, D. S. (2010). Reduced cognitive control of response inhibition by the anterior cingulate cortex in autism spectrum disorders. *NeuroImage, 52*(1), 336–347.

Amaral, D. G., Schumann, C. M., & Nordahl, C. W. (2008). Neuroanatomy of autism. *Trends in Neurosciences, 31*(3), 137–145.

Ambery, F. Z., Russell, A. J., Perry, K., Morris, R., & Murphy, D. G. (2006). Neuropsychological functioning in adults with Asperger syndrome. *Autism, 10*(6), 551–564.

American Psychiatric Association. (1994). *Diagnostic and statistical manual of mental disorders*. Washington, DC: American Psychiatric Association.

American Psychiatric Association. (2000). *Diagnostic and statistical manual of mental disorders*. Washington, DC: American Psychiatric Association.

Anckarsater, H., Stahlberg, O., Larson, T., Hakansson, C., Jutblad, S. B., Niklasson, L., et al. (2006). The impact of ADHD and autism spectrum disorders on temperament, character, and personality development. *The American Journal of Psychiatry, 163*(7), 1239–1244.

Baddeley, A. (1992). Working memory. *Science, 255*(5044), 556–559.

Baird, G., Simonoff, E., Pickles, A., Chandler, S., Loucas, T., Meldrum, D., et al. (2006). Prevalence of disorders of the autism spectrum in a population cohort of children in South Thames: The Special Needs and Autism Project (SNAP). *Lancet, 368*(9531), 210–215.

Barcelo, F. (1999). Electrophysiological evidence of two different types of error in the Wisconsin Card Sorting Test. *Neuroreport, 10*(6), 1299–1303.

Barnard, L., Muldoon, K., Hasan, R., O'Brien, G., & Stewart, M. (2008). Profiling executive dysfunction in adults with autism and comorbid learning disability. *Autism, 12*(2), 125–141.

Baron Cohen, S. (1995). *Mindblindness: An essay on autism and theory of mind*. Cambridge, MA: MIT Press.

Baron Cohen, S. (2001). Theory of mind and autism: A review. In L. M. Glidden (Ed.), *International review of research in mental retardation: Autism* (p. 312). San Diego, CA: Academic.

Beck, S. J., Hanson, C. A., Puffenberger, S. S., Benninger, K. L., & Benninger, W. B. (2010). A controlled trial of working memory training for children and adolescents with ADHD. *Journal of Clinical Child and Adolescent Psychology, 39*(6), 825–836.

Belmonte, M. K., & Yurgelun Todd, D. A. (2003). Functional anatomy of impaired selective attention and compensatory processing in autism. *Cognitive Brain Research, 17*(3), 651–664.

Bennetto, L., Pennington, B. F., & Rogers, S. J. (1996). Intact and impaired memory functions in autism. *Child Development, 67*(4), 1816–1835.

Berch, D. B., Krikorian, R., & Huha, E. M. (1998). The Corsi block-tapping task: Methodological and theoretical considerations. *Brain and Cognition, 38*(3), 317–338.

Berg, E. A. (1948). A simple objective technique for measuring flexibility in thinking. *The Journal of General Psychology, 39*, 15–22.

Berger, H. J. C., Aerts, F. H. T. M., Spaendonck, K. P. M. V., Cools, A. R., & Teunisse, J.-P. (2003). Central coherence and cognitive shifting in relation to social improvement in high-functioning young adults with autism. *Journal of Clinical and Experimental Neuropsychology, 25*(4), 502–511.

Booth, J. R., Burman, D. D., Meyer, J. R., Lei, Z., Trommer, B. L., Davenport, N. D., et al. (2003). Neural development of selective attention and response inhibition. *NeuroImage, 20*(2), 737–751.

Booth, R., Charlton, R., Hughes, C., & Happé, F. (2003). Disentangling weak coherence and executive dysfunction: Planning drawing in autism and attention-deficit/hyperactivity disorder. *Philosophical Transactions of the Royal Society of London. Series B, Biological Sciences, 358*(1430), 387–392.

Boucher, J., Cowell, P., Howard, M., Broks, P., Farrant, A., Roberts, N., et al. (2005). A combined clinical, neuropsychological, and neuroanatomical study of adults with high functioning autism. *Cognitive Neuropsychiatry, 10*(3), 165–213.

Bramham, J., Ambery, F., Young, S., Morris, R., Russell, A., Xenitidis, K., et al. (2009). Executive functioning differences between adults with attention deficit hyperactivity disorder and autistic spectrum disorder in initiation, planning and strategy formation. *Autism, 13*(3), 245–264.

Brugha, T. S., Mcmanus, S., Bankart, J., Scott, F., Purdon, S., Smith, J., et al. (2011). Epidemiology of autism spectrum disorders in adults in the community in England. *Archives of General Psychiatry, 68*(5), 459–465.

Cambridge. (2002). *CANTAB*. Cambridge, England: Cambridge Cognition.

Carlson, S. M., Mandell, D. J., & Williams, L. (2004). Executive function and theory of mind: Stability and prediction from ages 2 to 3. *Developmental Psychology, 40*(6), 1105–1122.

Casey, B. J., Castellanos, F. X., Giedd, J. N., Marsh, W. L., Hamburger, S. D., Schubert, A. B., et al. (1997). Implication of right frontostriatal circuitry in response inhibition and attention-deficit/hyperactivity disorder. *Journal of the American Academy of Child and Adolescent Psychiatry, 36*(3), 374–383.

Causton-Theoharis, J., Ashby, C., & Cosier, M. (2009). Islands of loneliness: Exploring social interaction through the autobiographies of individuals with autism. *Intellectual and Developmental Disabilities, 47*(2), 84–96.

Chan, A. S., Han, Y. M. Y., Sze, S. L., Cheung, M.-C., Leung, W. W.-M., Chan, R. C. K., et al. (2011). Disordered connectivity associated with memory deficits in children with autism spectrum disorders. *Research in Autism Spectrum Disorders, 5*(1), 237–245.

Charman, T., & Baron Cohen, S. (1992). Understanding drawings and beliefs: A further test of the metarepresentation theory of autism: A research note. *Journal of Child Psychology and Psychiatry, 33*(6), 1105–1112.

Christ, S. E., Holt, D. D., White, D. A., & Green, L. (2007). Inhibitory control in children with autism spectrum disorder. *Journal of Autism and Developmental Disorders, 37*(6), 1155–1165.

Christ, S. E., Kester, L. E., Bodner, K. E., & Miles, J. H. (2011). Evidence for selective inhibitory control impairment in individuals with autism spectrum disorder. *Neuropsychology, 25*(6), 690–701.

Christ, S. E., White, D. A., Mandernach, T., & Keys, B. A. (2001). Inhibitory control across the life span. *Developmental Neuropsychology, 20*(3), 653–669.

Chun, M. M., Golomb, J. D., & Turk-Browne, N. B. (2011). A taxonomy of external and internal attention. *Annual Review of Psychology, 62*, 73–101.

Corbett, B. A., Constantine, L. J., Hendren, R., Rocke, D., & Ozonoff, S. (2009). Examining executive functioning in children with autism spectrum disorder, attention deficit hyperactivity disorder and typical development. *Psychiatry Research, 166*(2–3), 210–222.

Corsi, P. M. (1972). Human memory and the medial temporal region of the brain [Dissertation]. *Dissertation Abstracts International, 34*(2), 819B.

Courchesne, E., & Pierce, K. (2005). Why the frontal cortex in autism might be talking only to itself: Local over-connectivity but long-distance disconnection. *Current Opinion in Neurobiology, 15*(2), 225–230.

Cui, J., Gao, D., Chen, Y., Zou, X., & Wang, Y. (2010). Working memory in early-school-age children with Asperger's syndrome. *Journal of Autism and Developmental Disorders, 40*(8), 958–967.

Damasio, A. R., & Maurer, R. G. (1978). A neurological model for childhood autism. *Archives of Neurology, 35*(12), 777–786.

De Vries, M., Prins, P., Schmand, B., & Geurts, H. M. (2011). *Executieve functietraining met een computergame bij kinderen met ASS*. Paper presented at the Even-Waardig: Autisme, praktijk en onderzoek.

Di Martino, A., Ross, K., Uddin, L. Q., Sklar, A. B., Castellanos, F. X., & Milham, M. P. (2009). Functional brain correlates of social and nonsocial processes in autism spectrum disorders: An activation likelihood estimation meta-analysis. *Biological Psychiatry, 65*(1), 63–74.

Dichter, G., Radonovich, K., Turner-Brown, L., Lam, K., Holtzclaw, T., & Bodfish, J. (2010). Performance of children with autism spectrum disorders on the Dimension-Change Card Sort Task. *Journal of Autism and Developmental Disorders, 40*(4), 448–456.

Eriksen, B. A., & Eriksen, C. W. (1974). Effects of noise letters upon the identification of a target letter in a nonsearch task. *Perception & Psychophysics, 16*(1), 143–149.

Eskes, G., Bryson, S., & Mccormick, T. (1990). Comprehension of concrete and abstract words in autistic children. *Journal of Autism and Developmental Disorders, 20*(1), 61–73.

Eslinger, P. J., & Damasio, A. R. (1985). Severe disturbance of higher cognition after bilateral frontal lobe ablation: Patient EVR. *Neurology, 35*(12), 1731–1741.

Fisher, N., & Happé, F. (2005). A training study of theory of mind and executive function in children with autistic spectrum disorders. *Journal of Autism and Developmental Disorders, 35*(6), 757–771.

Flynn, E., O'Malley, C., & Wood, D. (2004). A longitudinal, microgenetic study of the emergence of false belief understanding and inhibition skills. *Developmental Science, 7*(1), 103–115.

Fombonne, E. (2005). Epidemiology of autistic disorder and other pervasive developmental disorders. *The Journal of Clinical Psychiatry, 66*(Suppl. 10), 3–8.

Friedman, N. P., & Miyake, A. (2004). The relations among inhibition and interference control functions: A latent-variable analysis. *Journal of Experimental Psychology. General, 133*(1), 101–135.

Frith, U. (1989). A new look at language and communication in autism. *British Journal of Disorders of Communication, 24*(2), 123–150.

Frith, U., & Happé, F. (1994). Autism: Beyond "theory of mind". *Cognition, 50*(1–3), 115–132.

Frith, U., Morton, J., & Leslie, A. M. (1991). The cognitive basis of a biological disorder: Autism. *Trends in Neurosciences, 14*(10), 433–438.

Frye, D., Zelazo, P. D., Brooks, P. J., & Samuels, M. C. (1996). Inference and action in early causal reasoning. *Developmental Psychology, 32*(1), 120–131.

Gabig, C. S. (2008). Verbal working memory and story retelling in school-age children with autism. *Language, Speech, and Hearing Services in Schools, 39*(4), 498–511.

Ganz, J. B., Davis, J. L., Lund, E. M., Goodwyn, F. D., & Simpson, R. L. (2011). Meta-analysis of PECS with individuals with ASD: Investigation of targeted versus non-targeted outcomes, participant characteristics, and implementation phase. *Research in Developmental Disabilities, 33*(2), 406–418.

Gathercole, S. E., & Alloway, T. P. (2006). Practitioner review: Short-term and working memory impairments in neurodevelopmental disorders: Diagnosis and remedial support. *Journal of Child Psychology and Psychiatry, 47*(1), 4–15.

Gathercole, S. E., Pickering, S. J., Ambridge, B., & Wearing, H. (2004). The structure of working memory from 4 to 15 years of age [Article]. *Developmental Psychology, 40*(2), 177–190.

Georgiades, S., Szatmari, P., Zwaigenbaum, L., Duku, E., Bryson, S., Roberts, W., et al. (2007). Structure of the autism symptom phenotype: A proposed multidimensional model. *J Am Acad Child Adolesc Psychiatry, 46*(2), 188–196.

Geurts, H. M., Begeer, S., & Stockmann, L. (2009). Brief report: Inhibitory control of socially relevant stimuli in children with high functioning autism. *Journal of Autism and Developmental Disorders, 39*(11), 1603–1607.

Geurts, H.M., & Bringmann, L. (2010). Planning in autisme: Een kwalitatieve en kwantitatieve analyse. *Wetenschappelijk Tijdschrift Autisme, 1*, 4–17.

Geurts, H. M., Corbett, B., & Solomon, M. (2009). The paradox of cognitive flexibility in autism. *Trends in Cognitive Science, 13*(2), 74–82.

Geurts, H.M. & De Wit (in press). Goal-directed action control in children with autism spectrum disorders. *Autism.* No update on Geurts, Sinzig, Booth & Happé.

Geurts, H. M., & Jansen, M. D. (2012). Short report: A retrospective chart study: The pathway to a diagnosis for adults referred for ASD assessment. *Autism, 16*, 299–305.

Geurts, H. M., Luman, M., & Meel, C. S. V. (2008). What's in a game: The effect of social motivation on interference control in boys with ADHD and autism spectrum disorders. *Journal of Child Psychology and Psychiatry, 49*(8), 848–857.

Geurts, H. M., Sinzig, J., Booth, R., & Happé, F. (submitted). Neuropsychological heterogeneity in autism spectrum disorders: A comparison with ADHD. *Autism Research.*

Geurts, H. M., Verté, S., Oosterlaan, J., Roeyers, H., & Sergeant, J. A. (2004). How specific are executive functioning deficits in attention deficit hyperactivity disorder and autism? *Journal of Child Psychology and Psychiatry, 45*(4), 836–854.

Geurts, H. M., & Vissers, M. E. (2012). Elderly with autism: Executive functions and memory. *Journal of Autism and Developmental Disorders, 42*(5), 665–675.

Gezondheidsraad. (2009). *Autismespectrumstoornissen: Een leven lang anders.* Den Haag: Gezondheidsraad.

Gilbert, S. J., Bird, G., Brindley, R., Frith, C. D., & Burgess, P. W. (2008). Atypical recruitment of medial prefrontal cortex in autism spectrum disorders: An fMRI study of two executive function tasks. *Neuropsychologia, 46*(9), 2281–2291.

Goldberg, M. C., Mostofsky, S. H., Cutting, L. E., Mahone, E. M., Astor, B. C., Denckla, M. B., et al. (2005). Subtle executive impairment in children with autism and children with ADHD. *Journal of Autism and Developmental Disorders, 35*(3), 279–293.

Goldstein, G., Johnson, C. R., & Minshew, N. J. (2001). Attentional processes in autism. *Journal of Autism and Developmental Disorders, 31*(4), 433–440.

Gomarus, H. K., Wijers, A. A., Minderaa, R. B., & Althaus, M. (2009). ERP correlates of selective attention and working memory capacities in children with ADHD and/or PDD-NOS. *Clinical Neurophysiology, 120*(1), 60–72.

Griebling, J., Minshew, N. J., Bodner, K., Libove, R., Bansal, R., Konasale, P., et al. (2010). Dorsolateral prefrontal cortex magnetic resonance imaging measurements and cognitive performance in autism. *Journal of Child Neurology, 25*(7), 856–863.

Griffith, E. M., Pennington, B. F., Wehner, E. A., & Rogers, S. J. (1999). Executive functions in young children with autism. *Child Development, 70*(4), 817–832.

Hala, S., Hug, S., & Henderson, A. (2003). Executive function and false-belief understanding in preschool children: Two tasks are harder than one. *Journal of Cognition and Development, 4*(3), 275–298.

Happé, F. (1994). An advanced test of theory of mind: Understanding of story characters' thoughts and feelings by able autistic, mentally handicapped, and normal children and adults. *Journal of Autism and Developmental Disorders, 24*(2), 129–154.

Happé, F. (1999). Autism: Cognitive deficit or cognitive style? *Trends in Cognitive Sciences, 3*(6), 216–222.

Happé, F., Booth, R., Charlton, R., & Hughes, C. (2006). Executive function deficits in autism spectrum disorders and attention-deficit/hyperactivity disorder: Examining profiles across domains and ages. *Brain and Cognition, 61*(1), 25–39.

Happé, F., & Frith, U. (2006). The weak coherence account: Detail-focused cognitive style in autism spectrum disorders. *Journal of Autism and Developmental Disorders, 36*(1), 5–25.

Happé, F., & Ronald, A. (2008). The fractionable autism triad: A review of evidence from behavioural, genetic, cognitive and neural research. *Neuropsychology Review, 18*(4), 287–304.

Happe, F., Ronald, A., & Plomin, R. (2006). Time to give up on a single explanation for autism. *Nature Neuroscience, 9*(10), 1218–1220.

Henderson, H., Schwartz, C., Mundy, P., Burnette, C., Sutton, S., Zahka, N., et al. (2006). Response monitoring, the error-related negativity, and differences in social behavior in autism. *Brain and Cognition, 61*(1), 96–109.

Hill, E. L. (2004). Evaluating the theory of executive dysfunction in autism. *Developmental Review, 24*(2), 189–233.

Hill, E. L., & Bird, C. M. (2006). Executive processes in Asperger syndrome: Patterns of performance in a multiple case series. *Neuropsychologia, 44*(14), 2822–2835.

Hofvander, B., Delorme, R., Chaste, P., Nyden, A., Wentz, E., Stahlberg, O., et al. (2009). Psychiatric and psychosocial problems in adults with normal-intelligence autism spectrum disorders. *BMC Psychiatry, 9*, 35.

Holmes, J., Gathercole, S. E., Place, M., Dunning, D. L., Hilton, K. A., & Elliott, J. G. (2009). Working memory deficits can be overcome: Impacts of training and medication on working memory in children with ADHD. *Applied Cognitive Psychology, 24*(6), 827–836.

Hughes, C. (1998). Executive function in preschoolers: Links with theory of mind and verbal ability. *British Journal of Developmental Psychology, 16*(2), 233–253.

Hughes, C., & Ensor, R. (2007). Executive function and theory of mind: Predictive relations from ages 2 to 4. *Developmental Psychology, 43*(6), 1447–1459.

Hughes, C., & Russell, J. (1993). Autistic children's difficulty with mental disengagement from an object: Its implications for theories of autism. *Developmental Psychology, 29*(3), 498–510.

Hughes, C., Russell, J., & Robbins, T. W. (1994). Evidence for executive dysfunction in autism. *Neuropsychologia, 32*(4), 477–492.

Johnson, K. A., Robertson, I. H., Kelly, S. P., Silk, T. J., Barry, E., Daibhis, A., et al. (2007). Dissociation in performance of children with ADHD and high-functioning autism on a task of sustained attention. *Neuropsychologia, 45*(10), 2234–2245.

Johnston, K., Madden, A. K., Bramham, J., & Russell, A. J. (2011). Response inhibition in adults with autism spectrum disorder compared to attention deficit/hyperactivity disorder. *Journal of Autism and Developmental Disorders, 41*(7), 903–912.

Joseph, R. M., Steele, S. D., Meyer, E., & Tager-Flusberg, H. (2005). Self-ordered pointing in children with autism: Failure to use verbal mediation in the service of working memory? *Neuropsychologia, 43*(10), 1400–1411.

Just, M. A., Cherkassky, V. L., Keller, T. A., Kana, R. K., & Minshew, N. J. (2007). Functional and anatomical cortical underconnectivity in autism: Evidence from an fMRI study of an executive function task and corpus callosum morphometry. *Cerebral Cortex, 17*(4), 951–961.

Just, M. A., Cherkassky, V. L., Keller, T. A., & Minshew, N. J. (2004). Cortical activation and synchronization during sentence comprehension in high-functioning autism: Evidence of underconnectivity. *Brain, 127*(Pt 8), 1811–1821.

Just, M., & Varma, S. (2007). The organization of thinking: What functional brain imaging reveals about the neuroarchitecture of complex cognition. *Cognitive, Affective, & Behavioral Neuroscience, 7*(3), 153–191.

Kana, R. K., Keller, T. A., Cherkassky, V. L., Minshew, N. J., & Just, M. A. (2006). Sentence comprehension in autism: Thinking in pictures with decreased functional connectivity. *Brain, 129*(Pt 9), 2484–2493.

Kana, R. K., Keller, T. A., Cherkassky, V. L., Minshew, N. J., & Just, M. A. (2009). Atypical frontal-posterior synchronization of theory of mind regions in autism during mental state attribution. *Social Neuroscience, 4*(2), 135–152.

Kana, R. K., Keller, T. A., Minshew, N. J., & Just, M. A. (2007). Inhibitory control in high-functioning autism: Decreased activation and underconnectivity in inhibition networks. *Biological Psychiatry, 62*(3), 198–206.

Keary, C. J., Minshew, N. J., Bansal, R., Goradia, D., Fedorov, S., Keshavan, M. S., et al. (2009). Corpus callosum volume and neurocognition in autism. *Journal of Autism and Developmental Disorders, 39*(6), 834–841.

Kilincaslan, A., Mukaddes, N. M., Kucukyazici, G. S., & Gurvit, H. (2010). Assessment of executive/attentional performance in Asperger's disorder. *Türk Psikiyatri Dergisi, 21*(4), 289–299.

Kleinhans, N. M., Richards, T., Sterling, L., Stegbauer, K. C., Mahurin, R., Johnson, L. C., et al. (2008). Abnormal functional connectivity in autism spectrum disorders during face processing. *Brain, 131*(Pt 4), 1000–1012.

Klingberg, T., Fernell, E., Olesen, P. J., Johnson, M., Gustafsson, P., Dahlstrom, K., et al. (2005). Computerized training of working memory in children with ADHD—A randomized, controlled trial. *Journal of the American Academy of Child and Adolescent Psychiatry, 44*(2), 177–186.

Klingberg, T., Forssberg, H., & Westerberg, H. (2002). Training of working memory in children with ADHD. *Journal of Clinical and Experimental Neuropsychology, 24*(6), 781–791.

Koshino, H., Carpenter, P. A., Minshew, N. J., Cherkassky, V. L., Keller, T. A., & Just, M. A. (2005). Functional connectivity in an fMRI working memory task in high-functioning autism. *NeuroImage, 24*(3), 810–821.

Koshino, H., Kana, R. K., Keller, T. A., Cherkassky, V. L., Minshew, N. J., & Just, M. A. (2008). fMRI investigation of working memory for faces in autism: Visual coding and underconnectivity with frontal areas. *Cerebral Cortex, 18*(2), 289–300.

Landa, R., & Goldberg, M. (2005). Language, social, and executive functions in high functioning autism: A continuum of performance. *Journal of Autism and Developmental Disorders, 35*(5), 557–573.

Langen, M., Durston, S., Kas, M. J., Van Engeland, H., & Staal, W. G. (2011). The neurobiology of repetitive behavior: …and men. *Neuroscience and Biobehavioral Reviews, 35*(3), 356–365.

Langen, M., Leemans, A., Johnston, P., Ecker, C., Daly, E., Murphy, C. M., et al. (2012). Fronto-striatal circuitry and inhibitory control in autism: Findings from diffusion tensor imaging tractography. *Cortex, 48*, 183–193.

Lee, P. S., Yerys, B. E., Della Rosa, A., Foss-Feig, J., Barnes, K. A., James, J. D., et al. (2009). Functional connectivity of the inferior frontal cortex changes with age in children with autism spectrum disorders: A fcMRI study of response inhibition. *Cerebral Cortex, 19*(8), 1787–1794.

Lemon, J. M., Gargaro, B., Enticott, P. G., & Rinehart, N. J. (2011). Brief report: Executive functioning in autism spectrum disorders: A gender comparison of response inhibition. *Journal of Autism and Developmental Disorders, 41*(3), 352–356.

Leyfer, O. T., Folstein, S. E., Bacalman, S., Davis, N. O., Dinh, E., Morgan, J., et al. (2006). Comorbid psychiatric disorders in children with autism: Interview development and rates of disorders. *Journal of Autism and Developmental Disorders, 36*(7), 849–861.

Liss, M., Fein, D., Allen, D., Dunn, M., Feinstein, C., Morris, R., et al. (2001). Executive functioning in high-functioning children with autism. *Journal of Child Psychology and Psychiatry, 42*(2), 261–270.

Logan, G. D. (1994). On the ability to inhibit thought and action: A users' guide to the stop signal paradigm. Inhibitory processes in attention, memory, and language. In D. Dagenbach (Ed.), *Inhibitory processes in attention, memory, and language* (Vol. 14). San Diego, CA: Academic.

Lopez, B. R., Lincoln, A. J., Ozonoff, S., & Lai, Z. (2005). Examining the relationship between executive functions and restricted, repetitive symptoms of autistic disorder. *Journal of Autism and Developmental Disorders, 35*(4), 445–460.

Luna, B., Doll, S. K., Hegedus, S. J., Minshew, N. J., & Sweeney, J. A. (2007). Maturation of executive function in autism. *Biological Psychiatry, 61*(4), 474–481.

Luna, B., Minshew, N. J., Garver, K. E., Lazar, N. A., Thulborn, K. R., Eddy, W. F., et al. (2002). Neocortical system abnormalities in autism: An fMRI study of spatial working memory. *Neurology, 59*(6), 834–840.

Macleod, C. M. (1991). Half a century of research on the Stroop effect: An integrative review. *Psychological Bulletin, 109*(2), 163–203.

Maes, J. H., Eling, P. A., Wezenberg, E., Vissers, C. T., & Kan, C. C. (2011). Attentional set shifting in autism spectrum disorder: Differentiating between the role of perseveration, learned irrelevance, and novelty processing. *Journal of Clinical and Experimental Neuropsychology, 33*(2), 210–217.

Mahone, E. M., Powell, S. K., Loftis, C. W., Goldberg, M. C., Denckla, M. B., & Mostofsky, S. H. (2006). Motor persistence and inhibition in autism and ADHD. *Journal of the International Neuropsychological Society, 12*(5), 622–631.

Mari, M., Castiello, U., Marks, D., Marraffa, C., & Prior, M. (2003). The reach-to-grasp movement in children with autism spectrum disorder. *Philosophical Transactions of the Royal Society of London. Series B, Biological Sciences, 358*(1430), 393–403.

Mason, R. A., Williams, D. L., Kana, R. K., Minshew, N., & Just, M. A. (2008). Theory of mind disruption and recruitment of the right hemisphere during narrative comprehension in autism. *Neuropsychologia, 46*(1), 269–280.

Matson, J. L., & Boisjoli, J. A. (2008). The token economy for children with intellectual disability and/or autism: A review. *Research in Developmental Disabilities, 30*(2), 240–248.

Matson, J. L., & Nebel-Schwalm, M. S. (2007). Comorbid psychopathology with autism spectrum disorder in children: An overview. *Research in Developmental Disabilities, 28*(4), 341–352.

Maurer, R. G., & Damasio, A. R. (1982). Childhood autism from the point of view of behavioral neurology. *Journal of Autism and Developmental Disorders, 12*(2), 195–205.

Mcalonan, G. M., Cheung, V., Cheung, C., Suckling, J., Lam, G. Y., Tai, K. S., et al. (2005). Mapping the brain in autism. A voxel-based MRI study of volumetric differences and intercorrelations in autism. *Brain, 128*(Pt 2), 268–276.

Mckinnon, M. C., & Moscovitch, M. (2007). Domain-general contributions to social reasoning: Theory of mind and deontic reasoning re-explored. *Cognition, 102*(2), 179–218.

Minshew, N. J., & Goldstein, G. (2001). The pattern of intact and impaired memory functions in autism. *Journal of Child Psychology and Psychiatry, 42*(8), 1095–1101.

Minshew, N. J., Goldstein, G., Muenz, L. R., & Payton, J. B. (1992). Neuropsychological functioning nonmentally retarded autistic individuals. *Journal of Clinical and Experimental Neuropsychology, 14*, 749–761.

Minshew, N. J., Meyer, J., & Goldstein, G. (2002). Abstract reasoning in autism: A dissociation between concept formation and concept identification. *Neuropsychology, 16*(3), 327–334.

Monsell, S. (2003). Task switching. *Trends in Cognitive Sciences, 7*(3), 134–140.

Mosconi, M. W., Kay, M., D'Cruz, A. M., Seidenfeld, A., Guter, S., Stanford, L. D., et al. (2009). Impaired inhibitory control is associated with higher-order repetitive behaviors in autism spectrum disorders. *Psychological Medicine, 39*(9), 1559–1566.

Muller, U., Liebermann-Finestone, D. P., Carpendale, J. I., Hammond, S. I., & Bibok, M. B. (2012). Knowing minds, controlling actions: The developmental relations between theory of mind and executive function from 2 to 4 years of age. *Journal of Experimental Child Psychology, 111*(2), 331–348.

Newman, S. D., Carpenter, P. A., Varma, S., & Just, M. A. (2003). Frontal and parietal participation in problem solving in the Tower of London: fMRI and computational modeling of planning and high-level perception. *Neuropsychologia, 41*(12), 1668–1682.

Nydén, A., Niklasson, L., Stahlberg, O., Anckarsater, H., Wentz, E., Rastam, M., et al. (2010). Adults with autism spectrum disorders and ADHD neuropsychological aspects. *Research in Developmental Disabilities, 31*(6), 1659–1668.

Ozonoff, S. (1995). Reliability and validity of the Wisconsin Card Sorting Test in studies of autism. *Neuropsychology, 9*(4), 491–500.

Ozonoff, S., Cook, I., Coon, H., Dawson, G., Joseph, R., Klin, A., et al. (2004). Performance on Cambridge neuropsychological test automated battery subtests sensitive to frontal lobe function in people with autistic disorder: Evidence from the collaborative programs of excellence in autism network. *Journal of Autism and Developmental Disorders, 34*(2), 139–150.

Ozonoff, S., & Jensen, J. (1999). Brief report: Specific executive function profiles in three neurodevelopmental disorders. *Journal of Autism and Developmental Disorders, 29*(2), 171–177.

Ozonoff, S., & Mcevoy, R. E. (1994). A longitudinal study of executive function and theory of mind development in autism. *Development and Psychopathology, 6*(3), 415–431.

Ozonoff, S., & Miller, J. N. (1995). Teaching theory of mind: A new approach to social skills training for individuals with autism. *Journal of Autism and Developmental Disorders, 25*(4), 415–433.

Ozonoff, S., Pennington, B. F., & Rogers, S. J. (1991). Executive function deficits in high-functioning autistic individuals: Relationship to theory of mind. *Journal of Child Psychology and Psychiatry, 32*(7), 1081–1105.

Ozonoff, S., & Strayer, D. L. (1997). Inhibitory function in nonretarded children with autism. *Journal of Autism and Developmental Disorders, 27*(1), 59–77.

Ozonoff, S., & Strayer, D. L. (2001). Further evidence of intact working memory in autism. *Journal of Autism and Developmental Disorders, 31*(3), 257–263.

Ozonoff, S., Strayer, D. L., Mcmahon, W. M., & Filloux, F. (1994). Executive function abilities in autism and Tourette syndrome: An information processing approach. *Journal of Child Psychology and Psychiatry, 35*(6), 1015–1032.

Pellicano, E. (2007). Links between theory of mind and executive function in young children with autism: Clues to developmental primacy. *Developmental Psychology, 43*(4), 974–990.

Pellicano, E. (2010a). The development of core cognitive skills in autism: A 3-year prospective study. *Child Development, 81*(5), 1400–1416.

Pellicano, E. (2010b). Individual differences in executive function and central coherence predict developmental changes in theory of mind in autism. *Developmental Psychology, 46*(2), 530–544.

Pennington, B. F., & Ozonoff, S. (1996). Executive functions and developmental psychopathology. *Journal of Child Psychology and Psychiatry, 37*(1), 51–87.

Perner, J., & Lang, B. (1999). Development of theory of mind and executive control. *Trends in Cognitive Sciences, 3*(9), 337–344.

Perner, J., & Lang, B. (2000). Theory of mind and executive function: Is there a developmental relationship? In D. J. Cohen, S. Baron Cohen, & H. Tager Flusberg (Eds.), *Understanding other minds: Perspectives from developmental cognitive neuroscience* (2nd ed., pp. 150–181). New York: Oxford University Press.

Poljac, E., Simon, S., Ringlever, L., Kalcik, D., Groen, W. B., Buitelaar, J. K., et al. (2009). Impaired task switching performance in children with dyslexia but not in children with autism. *The Quarterly Journal of Experimental Psychology, 63*(2), 401–416.

Poustka, L., Jennen-Steinmetz, C., Henze, R., Vomstein, K., Haffner, J., & Sieltjes, B. (2012). Fronto-temporal disconnectivity and symptom severity in children with autism spectrum disorder. *The World Journal of Biological Psychiatry, 13*(4), 269–280.

Prior, M., & Hoffmann, W. (1990). Brief report: Neuropsychological testing of autistic children through an exploration with frontal lobe tests. *Journal of Autism and Developmental Disorders, 20*(4), 581–590.

Rapin, I., & Tuchman, R. F. (2008). Autism: Definition, neurobiology, screening, diagnosis. *Pediatric Clinics of North America, 55*(5), 1129–1146. viii.

Raymaekers, R., Antrop, I., Van Der Meere, J. J., Wiersema, J. R., & Roeyers, H. (2007). HFA and ADHD: A direct comparison on state regulation and response inhibition. *Journal of Clinical and Experimental Neuropsychology, 29*(4), 418–427.

Raymaekers, R., Van Der Meere, J., & Roeyers, H. (2004). Event-rate manipulation and its effect on arousal modulation and response inhibition in adults with high functioning autism. *Journal of Clinical and Experimental Neuropsychology, 26*(1), 74–82.

Raymaekers, R., Van Der Meere, J., & Roeyers, H. (2006). Response inhibition and immediate arousal in children with high-functioning autism. *Child Neuropsychology, 12*(4–5), 349–359.

Reed, P., & Mccarthy, J. (2012). Cross-modal attention-switching is impaired in autism spectrum disorders.

Journal of Autism and Developmental Disorders, 42(6), 947–953.

Reed, P., Watts, H., & Truzoli, R. (2013). Flexibility in young people with autism spectrum disorders on a card sort task. *Autism, 17*, 162–171.

Robinson, S., Goddard, L., Dritschel, B., Wisley, M., & Howlin, P. (2009). Executive functions in children with autism spectrum disorders. *Brain and Cognition, 71*, 362–368.

Rommelse, N. J., Franke, B., Geurts, H. M., Hartman, C. A., & Buitelaar, J. (2010). Shared heritability of attention-deficit/hyperactivity disorder and autism spectrum disorder. *European Child & Adolescent Psychiatry, 19*(3), 281–295.

Rumsey, J. M. (1985). Conceptual problem-solving in highly verbal, nonretarded autistic men. *Journal of Autism and Developmental Disorders, 15*(1), 23–36.

Russell, J. (1997). *Autism as an executive disorder*. New York: Oxford University Press.

Russell, J., Jarrold, C., & Hood, B. (1999). Two intact executive capacities in children with autism: Implications for the core executive dysfunctions in the disorder. *Journal of Autism and Developmental Disorders, 29*(2), 103–112.

Russo, N., Flanagan, T., Iarocci, G., Berringer, D., Zelazo, P. D., & Burack, J. A. (2007). Deconstructing executive deficits among persons with autism: Implications for cognitive neuroscience. *Brain and Cognition, 65*(1), 77–86.

Sabbagh, M. A., Xu, F., Carlson, S. M., Moses, L. J., & Lee, K. (2006). The development of executive functioning and theory of mind: A comparison of Chinese and U.S. preschoolers. *Psychological Science, 17*(1), 74–81.

Schipul, S. E., Williams, D. L., Keller, T. A., Minshew, N. J., & Just, M. A. (2012). Distinctive neural processes during learning in autism. *Cerebral Cortex, 22*, 937–950.

Schmitz, N., Rubia, K., Daly, E., Smith, A., Williams, S., & Murphy, D. G. (2006). Neural correlates of executive function in autistic spectrum disorders. *Biological Psychiatry, 59*(1), 7–16.

Semrud-Clikeman, M., Walkowiak, J., Wilkinson, A., & Butcher, B. (2010). Executive functioning in children with Asperger syndrome, ADHD-combined type, ADHD-predominately inattentive type, and controls. *Journal of Autism and Developmental Disorders, 40*(8), 1017–1027.

Sergeant, J. A., Geurts, H., & Oosterlaan, J. (2002). How specific is a deficit of executive functioning for attention-deficit/hyperactivity disorder? *Behavioural Brain Research, 130*(1–2), 3–28.

Shafritz, K. M., Dichter, G. S., Baranek, G. T., & Belger, A. (2008). The neural circuitry mediating shifts in behavioral response and cognitive set in autism. *Biological Psychiatry, 63*(10), 974–980.

Shallice, T. (1982). Specific impairments of planning. *Philosophical Transactions of the Royal Society of London. Series B, Biological Sciences, 298*(1089), 199–209.

Simonoff, E., Pickles, A., Charman, T., Chandler, S., Loucas, T., & Baird, G. (2008). Psychiatric disorders in children with autism spectrum disorders: Prevalence, comorbidity, and associated factors in a population-derived sample. *Journal of the American Academy of Child and Adolescent Psychiatry, 47*(8), 921–929.

Sinzig, J., Morsch, D., Bruning, N., Schmidt, M., & Lehmkuhl, G. (2008). Inhibition, flexibility, working memory and planning in autism spectrum disorders with and without comorbid ADHD-symptoms. *Child and Adolescent Psychiatry and Mental Health, 2*(1), 4.

Solomon, M., Ozonoff, S. J., Cummings, N., & Carter, C. S. (2008). Cognitive control in autism spectrum disorders. *International Journal of Developmental Neuroscience, 26*(2), 239–247.

Solomon, M., Ozonoff, S. J., Ursu, S., Ravizza, S., Cummings, N., Ly, S., et al. (2009). The neural substrates of cognitive control deficits in autism spectrum disorders. *Neuropsychologia, 47*(12), 2515–2526.

Stahl, L., & Pry, R. (2002). Joint attention and set-shifting in young children with autism. *Autism, 6*(4), 383–396.

Stanfield, A. C., Mcintosh, A. M., Spencer, M. D., Philip, R., Gaur, S., & Lawrie, S. M. (2008). Towards a neuroanatomy of autism: A systematic review and meta-analysis of structural magnetic resonance imaging studies. *European Psychiatry, 23*(4), 289–299.

Steele, S. D., Minshew, N. J., Luna, B., & Sweeney, J. A. (2007). Spatial working memory deficits in autism. *Journal of Autism and Developmental Disorders, 37*(4), 605–612.

Stoet, G., & López, B. (2010). Task-switching abilities in children with autism spectrum disorder. *The European Journal of Developmental Psychology, 8*(2), 244–260.

Sumiyoshi, C., Kawakubo, Y., Suga, M., Sumiyoshi, T., & Kasai, K. (2011). Impaired ability to organize information in individuals with autism spectrum disorders and their siblings. *Neuroscience Research, 69*(3), 252–257.

Van Eylen, L., Boets, B., Steyaert, J., Evers, K., Wagemans, J., & Noens, I. (2011). Cognitive flexibility in autism spectrum disorder: Explaining the inconsistencies? *Research in Autism Spectrum Disorders, 5*(4), 1390–1401.

Verté, S., Geurts, H. M., Roeyers, H., Oosterlaan, J., & Sergeant, J. A. (2006a). Executive functioning in children with an autism spectrum disorder: Can we differentiate within the spectrum? *Journal of Autism and Developmental Disorders, 36*(3), 351–372.

Verté, S., Geurts, H. M., Roeyers, H., Oosterlaan, J., & Sergeant, J. A. (2006b). The relationship of working memory, inhibition, and response variability in child psychopathology. *Journal of Neuroscience Methods, 151*(1), 5–14.

Vissers, M. E., Cohen, M., & Geurts, H. M. (2012). Brain connectivity and high functioning autism: A promising path of research that needs refined models, methodological convergence, and stronger behavioral links. *Neuroscience and Biobehavioral Reviews, 36*, 604–625.

Volkmar, F. R., Lord, C., Bailey, A., Schultz, R. T., & Klin, A. (2004). Autism and pervasive developmental disorders. *Journal of Child Psychology and Psychiatry, 45*(1), 135–170.

Wass, S. (2011). Distortions and disconnections: Disrupted brain connectivity in autism. *Brain and Cognition, 75*(1), 18–28.

Welchew, D. E., Ashwin, C., Berkouk, K., Salvador, R., Suckling, J., Baron-Cohen, S., et al. (2005). Functional disconnectivity of the medial temporal lobe in Asperger's syndrome. *Biological Psychiatry, 57*(9), 991–998.

Welsh, M. C., Satterlee-Cartmell, T., & Stine, M. (1999). Towers of Hanoi and London: Contribution of working memory and inhibition to performance. *Brain and Cognition, 41*(2), 231–242.

White, H. A., & Shah, P. (2006). Training attention-switching ability in adults with AD/HD. *Journal of Attention Disorders, 10*, 44–53.

White, S. J., Burgess, P. W., & Hill, E. L. (2009). Impairments on "open-ended" executive function tests in autism. *Autism Research, 2*(3), 138–147.

Whitehouse, A. J. O., Maybery, M. T., & Durkin, K. (2006). Inner speech impairments in autism. *Journal of Child Psychology and Psychiatry, 47*(8), 857–865.

Willcutt, E. G., Sonuga-Barke, E. J. S., Nigg, J. T., & Sergeant, J. A. (2008). Recent developments in neuropsychological models of childhood psychiatric disorders. *Advances in Biological Psychiatry, 24*, 195–226.

Williams, D., Goldstein, G., Carpenter, P., & Minshew, N. (2005). Verbal and spatial working memory in autism. *Journal of Autism and Developmental Disorders, 35*(6), 747–756.

Williams, D. L., Goldstein, G., & Minshew, N. J. (2006). The profile of memory function in children with autism. *Neuropsychology, 20*(1), 21–29.

Yerys, B. E., Hepburn, S. L., Pennington, B. F., & Rogers, S. J. (2007). Executive function in preschoolers with autism: Evidence consistent with a secondary deficit. *Journal of Autism and Developmental Disorders, 37*(6), 1068–1079.

Yerys, B. E., Wallace, G. L., Harrison, B., Celano, M. J., Giedd, J. N., & Kenworthy, L. E. (2009). Set-shifting in children with autism spectrum disorders. *Autism, 13*(5), 523–538.

Zelazo, P. D., Jacques, S., Burack, J. A., & Frye, D. (2002). The relation between theory of mind and rule use: Evidence from persons with autism-spectrum disorders. *Infant and Child Development, 11*(2), 171–195.

Zinke, K., Fries, E., Altgassen, M., Kirschbaum, C., Dettenborn, L., & Kliegel, M. (2010). Visuospatial short-term memory explains deficits in tower task planning in high-functioning children with autism spectrum disorder. *Child Neuropsychology, 16*(3), 229–241.

Executive Functioning as a Mediator of Age-Related Cognitive Decline in Adults

Dana Princiotta, Melissa DeVries, and Sam Goldstein

It is broadly accepted that even absent pathology, changes occur in cognitive functioning with advancing age. Decreased cognitive performance is noted among a variety of cognitive ability domains (Smith & Rush, 2006) and they are the result of age-related changes to brain structure and function. In this regard, the World Health Organization developed the term *aging-associated cognitive decline* which is defined as "(1) performance on a standardized cognitive test that is at least one standard deviation below age-adjusted norms in at least one of any of the following cognitive domains: learning and memory, attention and cognitive speed, language, or visuoconstructive abilities; (2) exclusion of any medical, psychiatric, or neurological disorder that could cause cognitive impairment; (3) normal activities of daily living and exclusion of dementia" (as cited in Levy, 1994).

Advancing age does not appear to impact all cognitive abilities equally, however. Certain cognitive abilities appear to be more affected than others; for example, significant declines have been reported in attention and memory (for review, see Glisky, 2007), general ability, and processing speed (Tucker-Drob, 2011) as a result of age-related processes. Though, the pattern does follow a somewhat linear trend, with increasing decline in very old age (e.g., Schaie, 2005). For example, the greatest declines have been noted past age 85 years (Baltes & Lindenberger, 1997), but the rate of decline varies dependent upon the cognitive domain and a number of other variables (Glisky, 2007). For this reason, it is useful to examine individual cognitive abilities and the interaction among groups of abilities thought to be closely related or those that interact with each other to have a mediating impact on performance. Overall it appears that tasks that are familiar and rely on existing knowledge are less impacted by advancing age as compared with tasks that involve the acquisition of new knowledge or novel problem solving (Zimprich et al., 2008).

Age-related changes in cognitive abilities also show substantial interindividual variability, although it has been suggested that individuals declining in one area are likely to be declining in other areas when compared to same age peers (Salthouse, 2004, 2009; Salthouse & Ferrer-Caja, 2003). One reason for individual differences could be *cognitive reserve* defined by Stern (2002) as the maximization of normal performance, as opposed to compensation for deficits, an efficient use of brain networks, or recruitment of alternative networks when needed to complete a task. This reserve may act as a protective factor against age-related cognitive decline. Corral, Rodriguez, Amenedo, Sanchez, and Diaz (2006) observed that individuals with higher cognitive

D. Princiotta • M. DeVries (✉) • S. Goldstein, Ph.D.
Neurology, Learning and Behavior Center,
University of Utah School of Medicine, South 500
East, Suite 100 230, Salt Lake City, UT 84102, USA
e-mail: Melissa@samgoldstein.com; info@samgoldstein.com

reserve were 6 times less likely to demonstrate deficits on neuropsychological testing, which can begin to occur as early as the third to fourth decade of life.

One variable that confounds researchers' ability to study age-related cognitive change is that the abilities demonstrating the most significant decline (e.g., memory and general ability) are complex and may in fact be mediated by other factors, for example, processing speed, attention, or executive functioning. Minimal research has attempted to examine the amount of variance in age-related change that is accounted for by executive functioning, in particular, perhaps because of its elusiveness as an independent domain of cognitive functioning (Salthouse, 2005). As discussed in Chap. 1, there exist some 33 definitions of executive functioning and those tools purported to measure executive functioning, while significantly correlated with each other, do also correlate with other cognitive ability measures (Hedden & Yoon, 2006). It appears that many researchers agree that executive functioning has a mediating effect on several, if not all, other cognitive abilities. For example, Glisky (2007) proposed that executive processes involve planning, organizing, coordinating, implementing, and evaluating nonroutine activities and are particularly crucial to performance on novel tasks.

Existing Theories of Executive Functioning

Depending on your theoretical stance, the areas subsumed under the umbrella of executive functioning vary from researcher to researcher. You may recall from Chap. 1 that the definition included as many as 33 different components by the middle of the 1990s (Barkley, 2011).

The literature is thus inconclusive regarding the operational definition of executive functioning. Salthouse and colleagues indicated that EF should be conceptualized as a single construct and that there is a statistical association between executive functioning and age-related effects on measures of cognitive functioning (Salthouse, Atkinson, & Berish, 2003). EF may be a metaconstruct. Executive functions are not independent, but interactive and probably hierarchically organized in development (BDEFS Manual, p. 14, 2011).

The heterogeneity of executive functioning is as evident from its multiple definitions as it is from the number of diverse measures purported to assess it. Salthouse (2005) examined a variety of executive functioning measures in comparison with other nonexecutive cognitive ability measures and observed that executive functioning was closely related to reasoning and processing speed and that age-related influence of executive functioning was rarely statistically independent of age-related influence on the other cognitive abilities measures, suggesting that executive functioning is not an independent construct.

Our Working Definition of Executive Functioning

In alliance with Salthouse and colleagues, with varying views and definitions of executive functioning thus presented, modified, and altered over the last 3 decades, we are attempting to define executive functioning as efficiency. Congruent with the executive decline hypothesis, our proposed working definition of executive functioning focuses on a single entity premise rather than multiple factors and can be considered a mediator of other cognitive functions; executive functioning is the efficiency with which an individual applies his or her ability and knowledge in order to deal with everyday life.

Why Have a Chapter on Aging in a Book About Executive Functioning?

Revolving around this single construct, we could hypothesize that executive functioning, or efficiency, would demonstrate a relationship with aging as the other previous constructs presently had. Efficiency may represent the parsimonious explanation, replacing the overwhelmingly diverse group of previously proposed definitions. Summarizing the existent literature on age-related changes in executive functioning could prove useful not only in informing aging research but also in better defining and understanding executive functioning. Thus, our main goal of exploration in this chapter is to review what is known about age-related cognitive decline across a number of areas of functioning including executive functioning. Facet(s) of executive functioning typically emerges after 5 years of age, notably for working memory, shifting, and planning (e.g., Best, Miller, & Jones, 2009). Much less research exists to document when these abilities begin to decline.

The outline of the chapter will be guided by a review of cognitive changes in the brain with age and a discussion of the literature that exists to explore changes specifically occurring in executive functioning with age, a discussion that would be incomplete without attention paid to the physical changes in the brain over time.

Brief Review of Physical Changes in the Brain Related to Executive Functioning (EF)

Although a plethora of research exists outlining the physical changes that occur in the human body over the life span, less is known about the brain and executive functioning, namely. Magnetic resonance imaging (MRI) and other neuroanatomical studies have, however, documented that the brain undergoes approximately a 50 % volume loss as a result of normal aging (Brickman, Habeck, Zarahn, Flynn, & Stern, 2007; DeCarli et al., 2005; Good et al., 2001; Resnick, Pham, Kraut, Zonderman, & Davatzikos, 2003), with accelerated change occurring after 50 years of age (DeCarli et al., 2005). Some specific noted changes include damage to white matter (often predicted by hypertension) or infarcts (Buckner, 2005). For example, white matter lesions have been found to correlate highly with memory deficits (Buckner, 2005); in fact, they are proposed to account for all age-related variance in speed and executive processing (Rabbitt et al., 2007). Reductions in myelination have also been significantly correlated with cognitive processing speed (Lu et al., 2011). The prefrontal cortex (PFC), cerebellum, and basal ganglia are the most common areas associated with executive functioning (Salthouse et al., 2003).

The PFC, specifically, has been cited as controlling these executive processes (see Chap. 1 for review). Neuroimaging of changes in frontal lobe structures has buttressed demonstrated declines in performance for neuropsychological tasks measuring executive functioning (Daniels, Toth, & Jacoby, 2006). Individuals diagnosed with Alzheimer's disease as well as individuals within the confines of "normal brain aging" are both "predisposed" to PFC impairment (Royall et al., 2004). Some researchers claim, "these findings reveal the cellular basis of age-related cognitive decline in dorsolateral PFC" (Wang et al., 2011).

Even without physical damage, the PFC has been found to be especially vulnerable to the aging process. The PFC is one of the first areas of the brain to demonstrate the results of degeneration (Ferrer-Caja, Crawford, & Bryan, 2002), showing a loss of function and volume (Raz, 2000). This degeneration has been identified as the cause of much age-related cognitive decline and has been coined the *frontal hypothesis of aging* (Andres & van der Linden, 2001; Crawford, Bryan, Luszcz, Obonsawin, & Stewart, 2000; Raz, 2000; Souchay, Isingrini, & Espagnet, 2000). Schretlen et al. (2000) evaluated whether age-related changes in fluid intelligence were related to declines in processing speed, the adverse impact of atrophy of frontal lobe structures on executive functioning abilities, or some combination of these two theories. Their data supported that declines in processing speed and changes in executive abilities, as supported by MRI data demonstrating declines in frontal lobe volume, accounted for 75 % of the variance in declining fluid abilities. PFC deficits progress over time, even in the absence of an abnormal mini mental status examination (Royall et al., 2004).

Recalling the case of Phineas Gage from Chap. 1, the destruction of much of Mr. Gage's PFC leads to behavioral/personality changes. Bekhterev (1905) reported that damage to the frontal lobes would result in a disruption of goal-directed behavior (Barkley, 2011). Along these lines, PFC impairment was added by the American Psychiatric Association to its working definition of dementia in 1994 (Royall et al., 2004). In summary, changes are evident in the "normal" aging brain with regard to physical structure. These include a loss of total volume, most often due to reduction in myelination of neurons. It has been proposed that when this occurs in the PFC, cerebellum, and basal ganglia, changes are observed in executive functioning.

A Brief Review of Changes that Occur in Cognitive Functioning as a Result of Normal Aging

Intelligence. Kaufman (2001) noted that verbal comprehension remains relatively stable in adulthood until the eighth decade, whereas perceptual reasoning peaks earlier in adulthood and declines steadily with age. Declines in full-scale intelligence with increasing age therefore are more likely due to slowed processing speed and reductions in perceptual reasoning because verbal comprehension and working memory tend to be better preserved despite increasing age (Miller, Myers, Prinzi, & Mittenberg, 2009). Ribot's (1882) law (as cited in Miller et al., 2009) argued that this was because cognitive abilities learned earliest and those that are more rehearsed were more resistant to aging. Several other studies have supported the verbal versus performance difference in age-related cognitive decline. Ryan, Sattler, and Lopez (2000, see Ardila refs) noted that aged individuals demonstrated worse performance on tasks measuring processing speed and perceptual reasoning when compared to younger adults, but virtually no age-related changes were observed on tasks measuring verbal comprehension. Lee et al. (2005) proposed that perceptual reasoning declines occurred linearly from the sixth decade on and were most closely related to cerebellar atrophy.

Memory. Research suggests that age correlates inversely with measures of memory (i.e., increases in age yield lower memory scores at a certain point in the life span) (Crawford et al., 2000). The correlation between memory decline and age has been proposed as either a single-factor model or a multiple-factor model. In the past decade, more emphasis has been directed to age-related decline in memory as it relates to declines in executive functioning (Crawford et al., 2000). Memory may be supported by attention as well as executive abilities. One theory is that tasks requiring more "mental effort" (e.g., working memory or free recall) will demonstrate age-related changes more so than activities requiring lower levels of mental effort (Connor, 2001).

Accelerated declines in memory and other cognitive functions are often associated with Alzheimer's disease. Under Dementia of the Alzheimer's Type in the DSM-IV-TR, memory impairment subsumes "the impaired ability to learn new information or to recall previously learned information" (APA, 2000). Alzheimer's disease may represent the acceleration of typical progressive memory loss related to aging or it may be a separate entity (Buckner, 2005). Through research with patients with Alzheimer's disease, the conceptualization of *cognitive reserve* suggests that certain factors like educational level or premorbid intelligence may result in decreased decline with age. Some factors are amenable to "retraining" to a degree, with cognitive training and better nutrition recommended as methods to decrease intellectual decline (Gunstad et al., 2006).

Although knowledge of pathology may direct understanding of the "atypical" population, what does this typical progressive memory loss look like in the "normal" population? In an effort to better comprehend the acceleration of memory loss, understanding the "typical" rate of memory loss is integral in discussing age-related functioning (Buckner, 2005).

Regarding different facets of memory, memory for content fairs much better than memory for context (Connor, 2001). Aging negatively affects memory retrieval more so than encoding or storage (Connor, 2001). However, data suggests that long-term retrieval demonstrates less developmental change than other cognitive abilities

(McGrew & Woodcock, 2001). Some memory abilities demonstrate peaks within the life span with declines thereafter; short-term memory peaks at approximately 25–30 (McGrew & Woodcock, 2001).

Popular neuropsychological measures of executive functioning (e.g., Modified Card Sorting Test, the Stroop Test, and the Tower of London Test) suggest that speed of processing declines with age (Crawford et al., 2000). In older individuals the trend for preserved performance with highly practiced skills (e.g., semantic and autobiographical information) (Buckner, 2005) is recognized. However, memory for newly acquired facts or recall of recent autobiographical experiences is more likely to be impaired.

In Gunstad and colleagues' work with adults with memory deficits, three clusters were discovered: a group with poor executive function, persons with reduced speed performance (attention, executive function, motor), and those with global cognitive decline (Gunstad et al., 2006). More notably, diminished executive functioning was noted compared to younger adults. In other words, memory deficits were never isolated, even in healthy subjects; rather they were mediated by other factors (Gunstad et al., 2006). In this particular study, researchers insisted that age-related memory impairment is not part of healthy aging (Gunstad et al., 2006). These results are somewhat consistent with the work of Naveh-Benjamin, Craik, Guez, and Kreuger (2005) who demonstrated that older adults did not demonstrate significant declines in memory performance so long as they could rely on semantic knowledge. Those adults who did appear to have declining memory performance were utilizing increased attentional resources at encoding and retrieval stages and were therefore likely using less effective strategies for memorization.

Working Memory. Working memory is the temporary holding of information in order to carry out a mental task (Huizinga, Dolan, & van der Molen, 2006). Working memory allows for the manipulation of information that is immediately available. According to Huizinga et al. (2006), working memory is still developing into young adulthood. More specifically an "adult level" of working memory is not readily established prior to 12 years of age (Huizinga et al., 2006). In their study, working memory was the strongest predictor of performance on the Wisconsin Card Sorting Task for youth 7–21 years of age, compared with inhibition and shifting (Huizinga et al., 2006). Goral (2004) supported these claims in establishing that working memory tasks demonstrate age-related differences with the exception of short-term memory.

Supported by declines in executive functioning, Wang et al. (2011) reported a marked loss of PFC in monkeys with age, specifically a decline in the firing rate of specific memory-related neurons. These memory-related neurons were purported to be associated with aging. It is unknown if these changes with normal aging alter the physiological substance of the PFC or not, although some research would suggest this is the case. Charlton et al. (2008) noted that reduced white matter integrity was correlated with age-related declines in working memory, although indirectly, as changes in working memory were mediated by reduced processing speed efficiency.

Self-Regulation/Attention. Attention abilities mediate all other cognitive domains, perhaps except for those tasks that are habitual or automatic. Therefore, one could assume that age-related declines in attention could significantly impact the functioning of older adults across a variety of domains (Glisky, 2007). Like other cognitive domains, however, all subcomponents of attention processes do not appear to be uniformly affected by normal cognitive aging. Glisky noted that older adults demonstrated the most significant impairments on attention tasks requiring switching or divided attention or flexibility of attention. Older adults were not differentially affected by distractibility although on the whole they were slower than younger adults on tasks measuring selective attention. Older adults also appeared generally similar to young adults on tasks measuring sustained attention or vigilance. Hartman and Stratton-Salib (2007) also noted that older adults struggled with selective attention under conditions where there were many stimulus items to choose from, which adversely affected their performance on tasks measuring concept formation. Thus, attention may serve as a mediating factor in age-related change among more complex ability areas such as concept formation and decision-making in healthy older adults (Isella et al., 2008).

Processing Speed. Processing speed declines with advanced age (Smith & Rush, 2006) and a plethora of research has demonstrated that age-related changes in processing speed are the catalyst for age-related changes in other cognitive domains (Finkel & Pedersen, 2000; Hertzog & Bleckley, 2001; Keys & White, 2000; Parkin & Java, 2000; Zimprich, 2002). The biological basis for processing speed changes is likely a reduction in myelination in areas such as the PFC and genu of the corpus callosum (Lu et al., 2011; Posthuma et al., 2003).

Ghisletta and Lindenberger (2003) noted that the influence of processing speed on changes in knowledge with increasing age was stronger than the influence of knowledge on changes in processing speed over time. Finkel, Reynolds, McArdle, and Pedersen (2007) indicated that declines in processing speed were responsible for changes in memory and spatial ability, but not verbal abilities. Lemke and Zimprich (2005) also noted that speed changes accounted for a portion of the variance in memory changes with age but cautioned that their data suggested no single-factor theory should be used to explain cognitive changes with age. Others have further challenged the notion that processing speed accounts for such a large portion of cognitive aging on the premise that processing speed itself was not clearly defined and measures of processing speed may, in fact, tap other abilities such as fluid reasoning (Parkin & Java, 1999).

Along these lines, Zimprich and Martin (2002) explored the relationship between speed of processing and fluid reasoning in 417 adults aged approximately 62 years at baseline. Adults were tested over a 4-year period. The results of this study indicated that changes in processing speed and fluid reasoning were correlated at .53 over the 4-year period and shared only 28 % common variance. Tucker-Drob, Johnson, and Jones (2009) measured reasoning and speed abilities in individuals ages 65–89 to evaluate the *cognitive reserve hypothesis*; that individuals higher in experiential resources would exhibit higher levels of cognitive function later in life due to either (a) these resources playing a protective role against age-related declines or (b) the persistence of high functioning present in earlier life. They demonstrated that individuals demonstrating higher levels of cognitive functioning later in life were merely reflecting earlier life differences (their cognitive abilities were higher to begin with) rather than differences in rates of age-related decline. The findings of Deary, Johnson, and Starr (2010) further supported this claim by noting that individual differences in processing speed evident at age 70 years were consistent with individual differences present at age 11 and thus processing speed was not a statistically strong biomarker of age-related cognitive change. Genetic factors that impact cognitive abilities at older ages are the same factors impacting processing speed at younger ages. A combination of these factors and environmental influences is the likely cause of individual variation in later adulthood (Finkel, Reynolds, McArdel, Hamagami, & Pedersen, 2009). It has been suggested that at least 50 % of the variance in later life cognitive ability is accounted for childhood cognitive ability (Gow et al., 2011).

Although it appears that processing speed may be an important mediator of cognitive functioning in childhood, it may not be the case that this relationship remains stable throughout development and aging. Nettelbeck and Burns (2010) examined cognitive performance in children and adults of increasing age to see if the aging process and cognitive changes associated with that were a mirror opposite of the increases in cognitive functioning associated with development through younger to older childhood. They found that for children, increasing processing speed positively impacted working memory, which allowed for greater reasoning abilities. However, the reverse was not the case for aging adults. In that case, adults over 55 years of age demonstrated declining reasoning abilities as a result of slower processing speed, but also changes in working memory that appeared to be independent of processing speed.

To summarize the findings related to the impact of processing speed on cognitive decline among other cognitive abilities, Salthouse (1996) coined the term *processing speed hypothesis*. This idea was developed to explain a common etiology for various cognitive declines noted with advancing age. Salthouse (1996) based this idea on a noted trend in the literature whereby several cross-sectional studies, after statistically controlling for simple processing speed, demonstrated reduced significance of changes among other cognitive variables as a result of increasing age.

To further evaluate this phenomenon, Salthouse (1996) conducted a longitudinal study examining the cognitive performance of 303 participants, with a mean age of 77.2 years, over a 3-year period. The results were indicative of significant declines in episodic memory, verbal fluency, verbal comprehension, vocabulary, memory span, and attention. Examining the data from the point of cross-sectional age effects, Salthouse noted that individual differences in processing speed accounted for a large portion of the variance among changes in other cognitive domains. However, within-person longitudinal effects revealed that while processing speed was still noted to play a role in age-related change, it accounted for a far smaller portion of the variance in age-related cognitive change thus suggesting that cross-sectional data likely ovserestimates the role of processing speed as a major factor in age-related cognitive decline.

Visual-Spatial Ability. Jenkins, Myerson, Joerding, and Hale (2000) evaluated differences between younger and older adults on tasks of visually based processing speed, working memory, and paired associates learning versus verbally based processing speed, working memory, and paired associates learning tasks. Their results indicated that older adults' performance was worse than that of younger adults across all tasks, but the greatest decrements in performance were noted on visually based tasks, suggesting that verbal performance is less compromised with age. These results are consistent with other research demonstrating the resistance of verbal abilities (versus visual/performance abilities) to age-related changes (Miller et al., 2009; Ryan et al., 2000).

Language. Language is purported to be affected by age, though the problem appears to be related more to search and retrieval rather than linguistic information (Goral, 2004). More specifically, of the verbal abilities, word finding and spelling appear to demonstrate the greatest age-related declines (Burke & MacKay, 1997). Difficulties are noted for lexical retrieval during word production and comprehension of complex materials (Goral, 2004). Goral demonstrated this by comparing individuals of different age brackets (30s, 40s, 50s, 60s, and 70s). Findings suggested age was a significant predictor of overall scores on the Norton Naming Tasks and the Action Naming Tests (Goral, 2004). The steepest drop in scores was observed in individuals 70 and above, though decline can be observed as early as 50 years of age (Goral, 2004). Supporting this finding, Barresi et al. (2000) investigated impaired lexical access (defined as failure to name before and after cues were given) and semantic degradation (defined here as early successful naming with later failures), both explanations for age-related declines in naming. Adults in their 70s demonstrated more difficulty with semantic degradation for object naming than other age brackets (Barresi et al., 2000).

Analyses of speech have demonstrated declines in later life in grammatical complexity, attributed to working memory decline (Kemper, Thompson, & Marquis, 2001). Vocabulary and syntax remain intact across the life span of adults. Similarly, Barresi et al. (2000) proposed that vocabulary definitions remain stable over the life span while naming pictures begins to decline after age 30 (Barresi et al., 2000). Education (e.g., professional language use, professional reading, and reduced television viewing) and age both influence naming ability (Barresi et al., 2000). The question is why some language and memory factors are more vulnerable to aging than others (Burke & MacKay, 1997). Difficulties arise for researchers and clinicians through the span of the human life in attempting to understand abilities that are preserved versus impaired and the typical patterns that emerge (Burke & MacKay, 1997).

Achievement. In 2004, Grissom demonstrated a negative relationship between academic achievement and age that remains constant over time

(Duncan, Brooks-Gunn, & Klebanov, 1994; Walker, Greenwood, Hart, & Carta, 1994). Achievement is predicted in neuropsychology by measured executive skills. The Woodcock-Johnson tests provide further insight into the developmental trajectories that occur over the life span in terms of cognitive functioning and achievement. The norming sample of both the Woodcock–Johnson Tests of Cognitive Abilities and Tests of Achievement includes data for individuals aged 2–90 (Mather & Woodcock, 2001). Subtests for both measures demonstrate growth and decline across the life span (McGrew & Woodcock, 2001). Divergent growth curves suggest unique abilities have different developmental trajectories over the life span (McGrew & Woodcock, 2001).

To name a few, both cognitive and academic fluency clusters seem to follow different developmental trajectories; cognitive fluency demonstrates less growth overall than academic fluency (McGrew & Woodcock, 2001). In contrast, the change noted from age 5 to 90 for comprehension-knowledge suggests a different story, with greater change observed (McGrew & Woodcock, 2001). In terms of academic achievement, reading and writing abilities are increasing during formal schooling (ages 5–18) and do not demonstrate a dramatic decline with age. The Oral Expression cluster does not peak until age 50–60 (McGrew & Woodcock, 2001). Single-word reading/decoding, or phonics abilities, is highly correlated with overall cognitive functioning but is also relatively insensitive to aging and early dementias (McGurn et al., 2004).

What Does the Current Literature Tell Us About What Happens to Executive Functioning with Age?

It is known that many conditions involve difficulties with executive functioning: autism spectrum disorders, attention deficit-hyperactivity disorder, conduct disorder, oppositional defiant disorder, obsessive-compulsive disorder, depression, and anxiety. Developmental issues at each point in the life span can be better addressed with an understanding of executive functioning from young childhood to late adulthood (Best et al., 2009). Often times, the research that explains executive functioning is primarily aimed towards school age children and discussions surrounding attention deficits. While this may better explain the prevalence rates for ADHD among school age children, in more recent years, the importance of examining executive functioning across the life span has become a focus, not only in this chapter (e.g., Best et al., 2009). For example, older adults do not perform as well as their younger counterparts for tasks measuring executive functioning (e.g., Wisconsin Card Sorting Task, Trails B, Tower of London tasks, and Stroop tasks) (Daniels et al., 2006). In the 1980s, Hasher and Zacks suggested that decreased performance in older adults was the outcome of inefficient inhibition, thus increasing the effort necessary to concentrate and avoid distraction (Goral, 2004).

In the past, working memory, attention, and inhibition have been linked to playing major roles in cognitive aging (Salthouse et al., 2003). On one side of the argument, a general reduction in resources for cognitive processing has been proposed as a viable explanation rather than the view that declines exist for *specific* components of executive functioning. One hypothesis explaining the foundation for *general cognitive slowing* explains that deficiencies may rest within impaired processing ability (see earlier discussion of processing speed). This view that coined a *general reduction hypothesis* has been supported by many researchers during the past 30 years (e.g., Daigneault & Braun, 1993; Daigneault, Braun, & Whitaker, 1992; Mittenberg, Seidenberg, O'Leary, & DiGiulio, 1989; Crawford et al., 2000), and so too may processing speed mitigate changes in executive abilities. Stewart, Scarisbrick, and Golden (2011, August) noted that executive functioning increases with age, while processing speed shows age-related declines and noted that processing speed may mediate the relationship between age-related changes in executive functioning. "Problematic, however, is the possibility that performance on executive tasks may be determined by more than one executive process" (Daniels et al., 2006).

Beginning in middle adulthood to expiration, executive functioning, memory, and attention begin to decline (Gunstad et al., 2006). As early as the mid-twenties, deterioration may occur in the "biological framework of thinking and reasoning" (Gunstad et al., 2006). Crawford et al. (2000) investigated the *executive decline hypothesis of cognitive aging*. They conducted two studies with adults aged 18–75 years and 60–89 years, respectively, to examine whether aging was associated with changes in executive functioning independent of performance changes in other cognitive domains. Although the studies were cross-sectional in nature, which have been criticized as resulting in overly robust findings (see discussion of processing speed above), Crawford and colleagues noted that there was no apparent differential decline in executive functioning as compared with measures of general cognitive ability (as measured by the WAIS-R; Wechsler, 1981). Although executive functioning was linked to age-related declines in measures of memory, processing speed appeared to be the mediating variable. Furthermore, executive functioning and general cognitive ability accounted for similar amounts of variance in age-related memory declines, with executive functioning making only a small unique contribution.

In summary, it appears that executive functioning, as it is currently measured (by tasks such as the Wisconsin Card Sorting Test and Tower of London Test), may demonstrate declines with age (Daniels et al., 2006), although it may only make a few independent contribution to changes in other cognitive functions as a result of normal aging (Crawford et al., 2000).

Implications

Earlier we discussed executive functioning, or *efficiency*, not only as a single entity but also as a mediator of other cognitive functions (i.e., the efficiency with which an individual applies his or her knowledge or ability to deal with everyday life). Given what we know about age-related changes in other areas of cognitive functioning, where might we posit that executive functioning is playing a role? Is it the case that true declines are observed in intelligence (specifically perceptual reasoning; Ryan et al., 2000), memory for context and retrieval of that contextual information more so than encoding or storage (Connor, 2001), working memory (Goral, 2004), selective and divided attention (Glisky, 2007; Hartman & Stratton-Salib, 2007), processing speed (Salthouse, 1996; Finkel & Pedersen, 2000; Hertzog & Bleckley, 2001; Keys & White, 2000; Parkin & Java, 2000; Zimprich, 2002), and language abilities, specifically naming (Barresi et al., 2000; Goral, 2004)? Or is it that executive functioning changes as the consequence of normal aging has reduced our capacity to utilize our innate abilities to effectively deal with everyday life, solve novel problems, and perform in alliance with our younger counterparts? Let us consider, for example, the work of Gunstad et al. (2006) who demonstrated that memory performance declines with age was mediated by others factors (one of which was executive functioning). Age-related changes in nonverbal reasoning (evident from research demonstrating reductions in perceptual reasoning on measures of intelligence; Ryan et al., 2000) could very well be the result of reduced *efficiency*, or executive functioning, in utilizing what we have to solve these problems, whereas tasks that are more knowledge based (tasks measuring verbal reasoning) remain more stable over time. The research we have reviewed overall appears to generally support that much of our knowledge remains intact over time, but our abilities are altered for the worse. We utilize innate abilities to acquire knowledge and we coordinate these two areas, via executive functioning, to solve everyday problems and perform higher order tasks. Once this performance becomes more habitual, or overlearned, our abilities are not as necessary for performance, and the tasks become more knowledge based. Thus, we could propose that the role of executive functioning, as we have defined it, is only necessary for tasks that are newly learned and that therefore, declines in executive functioning with age would result in difficulty performing tasks that are newly learned or still required the integration of knowledge and abilities, which is consistent with some of the existing literature (e.g., Buckner, 2005; Naveh-Benjamin et al., 2005).

Despite all of this, the database of literature examining age-related changes in executive functioning is still very much in its infancy and is clearly not without its own limitations.

Limitations to Current Knowledge

Virtually the same factors that limit the study of executive functioning as an independent construct also limit our ability to understand the impact of age-related change. The number of variables proposed (Salthouse, 2005) to be subsumed under the heading of executive functioning is astounding and creates a substantial deficit in the existing literature because so many of these variables have gone unexplored as they relate to normal aging. Additionally, the measurement of executive functioning is elusive because of the correlation that exists between these measures and those tapping other cognitive abilities (Hedden & Yoon, 2006; Salthouse et al., 2003; Salthouse, 2005).

Summary and Conclusions

A multitude of research studies support that cognitive functioning does not remain stable over time and that a number of declines are observed in light of advancing age, particularly in older adulthood. This research also supports the notion that different domains of functioning decline at varying rates and acquisition of new knowledge and novel problem solving appear to decline at faster rates than existing knowledge (Zimprich et al., 2008). If this were true, it would be congruent with our proposed definition of executive functioning, the efficiency with which we utilize our knowledge and abilities to deal with everyday life. This may be a more simplistic yet more sensible way to define and understand executive functioning and its impact on other cognitive abilities, which we use to acquire new knowledge as opposed to the vast number of other definitions that have been proposed. Studies of biological changes correlated with age-related decline have demonstrated volume loss in the brain due to reduced myelination. The brain areas in which this occurs most often have been proposed as those responsible for controlling executive functioning. However, as previously discussed, the concept of executive functioning and the tools suggested to measure it are poorly defined and in some ways based in circular logic. Any definition proposed in the future, including our own, will need to be subjected to rigorous research demonstrating its convergent and divergent validity (Salthouse et al., 2003).

References

American Psychiatric Association. (2000). *Diagnostic and statistical manual of mental disorders* (4th ed., text rev.). Washington, DC: American Psychiatric Association.

Andres, P., & van der Linden, M. (2001). Age related differences in supervisory attentional system functions. *Journal of Gerontology: Psychological Science, 55B*, P373–P380.

Ardila, A. (2007). Normal aging increases cognitive heterogeneity: Analysis of dispersion in WAIS-III scores across age. *Archives of Clinical Neuropsychology, 22*, 1003–1011.

Baltes, P. B., & Lindenberger, U. (1997). Emergence of a powerful connection between sensory and cognitive functions across the adult life span: A new window to the study of cognitive aging? *Psychology and Aging, 12*, 12–21.

Barkley, R. A. (2011). *Barkley deficits in executive functioning scale (BDEFS)*. New York, NY: Guilford Press.

Barresi, B. A., Nicholas, M., Connor, L. T., Obler, L. K., & Albert, M. L. (2000). Semantic degradation and lexical access in age-related naming failures. *Aging, Neuropsychology, and Cognition, 7*(3), 169–178.

Bekhterev, V. M. (1905). Fundamentals of brain function (7 vols.). St. Petersburg.

Best, J. R., Miller, P. H., & Jones, L. L. (2009). Executive functions after age 5: Changes and correlates. *Developmental Review, 29*(3), 180–200.

Brickman, A. M., Habeck, C., Zarahn, E., Flynn, J., & Stern, Y. (2007). Structural MRI covariance patterns associated with normal aging and neuropsychological functioning. *Neurobiology of Aging, 28*, 284–295.

Buckner, R. L. (2005). Memory and executive function review in aging and AD: Multiple factors that cause decline and reserve factors that compensate. *Neuron, 44*, 195–208.

Burke, D. M., & MacKay, D. G. (1997). Memory, language, and aging. *Philosophical Transactions of the Royall Society, Biological Sciences, 352*, 1845–1856.

Charlton, R. A., Landau, S., Shiavone, F., Barrick, T. R., Clark, C. A., Markus, H. S., et al. (2008). A structural equation modeling investigation of age-related variance in executive function and DTI measured white matter change. *Neurobiology of Aging, 29,* 1547–1555.

Connor, L. T. (2001). Memory in old age: Patterns of decline and preservation. *Seminars in Speech and Language, 22,* 117–125.

Corral, M., Rodriguez, M., Amenedo, E., Sanchez, J. L., & Diaz, F. (2006). Cognitive reserve, age, and neuropsychological performance in healthy participants. *Developmental Neuropsychology, 29,* 479–491.

Crawford, J. R., Bryan, J., Luszcz, M. A., Obonsawin, M. C., & Stewart, L. (2000). The executive decline hypothesis of cognitive aging: Do executive deficits qualify as differential deficits and do they mediate age-related memory decline? *Aging, Neuropsychology, and Cognition, 7,* 9–31.

Daigneault, S., Braun, C. M. J., & Whitaker, H. A. (1992). Early effects of normal aging on perseverative and non-perseverative prefrontal measures. *Developmental Neuropsychology, 8,* 99–114.

Daigneault, S., & Braun, C. M. J. (1993). Working memory and the self-ordered pointing task: Further evidence of early prefrontal decline in normal aging. *Journal of Clinical and Experimental Neuropsychology, 15,* 881–895.

Daniels, K. A., Toth, J. P., & Jacoby, L. L. (2006). The aging of executive functions. In E. Bialystok & F. I. M. Craik (Eds.), *Lifespan cognition: Mechanisms of change* (pp. 96–111). New York: Oxford University Press.

Deary, I. J., Johnson, W., & Starr, J. M. (2010). Are processing speed tasks biomarkers of cognitive aging. *Psychology and Aging, 1,* 219–228.

DeCarli, C., Massaro, J., Harvey, D., Hald, J., Tullberg, M., Auf, R., et al. (2005). Measures of brain morphology and infarction in the Framingham Heart Study: Establishing what is normal. *Neurobiology of Aging, 26,* 491–510.

Duncan, G. J., Brooks-Gunn, J., & Klebanov, P. K. (1994). Economic deprivation and early-childhood development. *Child Development, 65,* 296–318.

Ferrer-Caja, E., Crawford, J. R., & Bryan, J. (2002). A structural modeling examination of the executive decline hypothesis of cognitive aging through reanalysis of Crawford et al.'s (2000) data. *Aging, Neuropsychology, and Cognition, 9,* 231–249.

Finkel, D., & Pedersen, N. L. (2000). Contribution of age, genes, and environment to the relationship between perceptual speed and cognitive ability. *Psychology and Aging, 15,* 56–64.

Finkel, D., Reynolds, C. A., McArdel, J. J., Hamagami, F., & Pedersen, N. L. (2009). Genetic variance in processing speed drives variation in aging of spatial and memory abilities. *Developmental Psychology, 45,* 820–834.

Finkel, D., Reynolds, C. A., McArdle, J. J., & Pedersen, N. L. (2007). Age changes in processing speed as a leading indicator of cognitive aging. *Psychology and Aging, 22,* 558–568.

Ghisletta, P., & Lindenberger, U. (2003). Age-based structural dynamics between perceptual speed and knowledge in the Berlin Aging Study: Direct evidence for ability dedifferentiation in old age. *Psychology and Aging, 18,* 696–713.

Glisky, E. L. (2007). Changes in cognitive function in human aging. In D. R. Riddle (Ed.), *Brain aging: Models, methods, mechanisms* (pp. 1–15). Boca Raton, FL: CRC Press. Retrieved July 30, 2009, from http://www.ncbi.nlm.nih.gov/bookshelf/br.fcgi?book=frbrainage&part=ch1

Good, C. D., Johnsrude, I. S., Ashburner, J., Henson, R. N., Friston, K. J., & Frackowiak, R. S. (2001). A voxel-based morphometric study of ageing in 465 normal adult human brains. *NeuroImage, 14,* 21–36.

Goral, M. (2004). First-language decline in healthy aging: Implications for attrition in bilingualism. *Journal of Neurolinguistics, 17*(1), 31–52.

Gow, A. J., Johnson, W., Pattie, A., Brett, C. E., Robersts, B., Starr, J. M., et al. (2011). Stability and change in intelligence from age 11 to ages 70, 79, and 87: The Lothian birth cohorts of 1921 and 1936. *Psychology and Aging, 26,* 232–240.

Gunstad, J., Paul, R. H., Brickman, A. M., Cohen, R. A., Arns, M., Roe, D., et al. (2006). Patterns of cognitive performance in middle-aged and older adults: A cluster analytic examination. *Journal of Geriatric Psychiatry and Neurology, 19,* 59–64.

Hartman, M., & Stratton-Salib, B. C. (2007). Age differences in concept formation. *Journal of Clinical and Experimental Neuropsychology, 29,* 198–214.

Hedden, T., & Yoon, C. (2006). Individual differences in executive processing predict susceptibility to interference in verbal working memory. *Neuropsychology, 20,* 511–528.

Hertzog, C., & Bleckley, M. K. (2001). Age differences in the structure of intelligence: Influences of information processing speed. *Intelligence, 29*(3), 191–217.

Huizinga, M., Dolan, C. V., & van der Molen, M. W. (2006). Age-related change in executive function: Developmental trends and a latent variable analysis. *Neuropsychologia, 44*(11), 2017–2036.

Isella, V., Mapelli, C., Morielli, N., Pelati, O., Franceschi, M., & Appollonio, I. M. (2008). Age-related quantitative and qualitative changes in decision making ability. *Behavioural Neuropsychology, 19,* 59–63.

Jenkins, L., Myerson, J., Joerding, J. A., & Hale, S. (2000). Converging evidence that visuospatial cognition is more age-sensitive than verbal cognition. *Psychology and Aging, 15,* 157–175.

Kaufman, A. S. (2001). WAIS-III Iqs, Horn's theory and generational changes from young adulthood to old age. *Intelligence, 29,* 131–167.

Kemper, S., Thompson, M., & Marquis, J. (2001). Longitudinal change in language production: Effects of aging and dementia on grammatical complexity and propositional content. *Psychology and Aging, 16,* 600–614.

Keys, B. A., & White, D. A. (2000). Exploring the relationship between age, executive abilities, and psychomotor speed. *Journal of the International Neuropsychological Society, 6*, 76–82.

Lee, J. Y., Lyoo, I. K., Kim, S. U., Jang, H. S., Lee, D. W., Jeon, H. J., et al. (2005). Intellect declines in healthy elderly subjects and cerebellum. *Psychiatry and Clinical Neurosciences, 59*, 45–51.

Lemke, U., & Zimprich, D. (2005). Longitudinal changes in memory performance and processing speed in old age. *Aging, Neuropsychology, and Cognition, 12*, 57–77.

Levy, R. (1994). Aging-associated cognitive decline. *International Psychogeriatrics, 6*, 63–68.

Lu, P. H., Lee, G. J., Raven, E. P., Tingus, K., Khoo, T., Thompson, P. M., et al. (2011). Age-related slowing in cognitive processing speed is associated with myelin integrity in a very healthy elderly sample. *Journal of Clinical and Experimental Neuropsychology, 33*(10), 1059–1068.

Mather, N., & Woodcock, R. W. (2001). *Examiner's manual. Woodcock-Johnson III tests of achievement.* Itasca, IL: Riverside Publishing.

McGrew, K. S., & Woodcock, R. W. (2001). *Technical Manual. Woodcock-Johnson III.* Itasca, IL: Riverside Publishing.

McGurn, B., Starr, J. M., Topfer, J. A., Pattie, A., Whiteman, M. C., Lemmon, H. A., et al. (2004). Pronunciation of irregular words is preserved in dementia, validating premorbid IQ estimation. *Neurology, 62*, 1184–1186.

Miller, L. J., Myers, A., Prinzi, L., & Mittenberg, W. (2009). Changes in intellectual functioning associated with normal aging. *Archives of Clinical Neuropsychology, 24*, 681–688.

Mittenberg, W., Seidenberg, M., O'Leary, D., & DiGiulio, D. (1989). Changes in cerebral functioning associated with normal aging. *Journal of Clinical and Experimental Neuropsychology, 11*, 918–932.

Naveh-Benjamin, M., Craik, F. I. M., Guez, J., & Kreuger, S. (2005). Divided attention in younger and older adults: Effects of strategy and relatedness on memory performance and secondary task costs. *Journal of Experimental Psychology: Learning, Memory, and Cognition, 31*, 520–537.

Nettelbeck, T., & Burns, N. R. (2010). Processing speed, working memory and reasoning ability from childhood to old age. *Personality and Individual Differences, 48*, 379–384.

Parkin, A. J., & Java, R. I. (1999). Deterioration of frontal lobe function in normal aging: Influences on fluid intelligence versus perceptual speed. *Neuropsychology, 13*, 539–545.

Parkin, A. J., & Java, R. I. (2000). Determinants of age-related memory loss. In T. J. Perfect & E. A. Maylor (Eds.), Models of cognitive aging (pp. 188–203). Oxford, UK: Oxford University Press.

Posthuma, D., Baaré, W. F. C., Pol, H. E. H., Kahn, R. S., Boomsma, D. I., & de Geus, E. J. C. (2003). Genetic correlations between brain volumes and the WAIS-III dimensions of verbal comprehension, working memory, perceptual organization, and processing speed. *Twin Research, 6*, 131–139.

Rabbitt, P., Scott, M., Lunn, M., Thacker, N., Lowe, C., Pendleton, N., et al. (2007). White matter lesions account for all age-related declines in speed but not intelligence. *Neuropsychology, 21*, 363–370.

Raz, N. (2000). Aging of the brain and its impact on cognitive performance: Integration of structural and behavioural findings. In F. I. M. Craik & T. Salthouse (Eds.), *Handbook of aging and cognition* (2nd ed., pp. 1–90). Mahwah, NJ: Erlbaum.

Resnick, S. M., Pham, D. L., Kraut, M. A., Zonderman, A. B., & Davatzikos, C. (2003). Longitudinal magnetic resonance imaging studies of older adults: A shrinking brain. *Journal of Neuroscience, 23*, 3295–3301.

Ribot, T. (1882). *Diseases of memory.* New York: Appleton-Century-Crofts.

Royall, D. R., Palmer, R., Chiodo, L. K., & Polk, M. J. (2004). Declining executive control in normal aging predicts change in functional status: the Freedom House Study. *Journal of the American Geriatrics Society, 52*(3), 346–352.

Ryan, J. J., Sattler, J., & Lopez, S. J. (2000). Age effects on Wechsler adult intelligence scale-III subtests. *Archives of Clinical Neuropsychology, 15*, 311–317.

Salthouse, T. A. (1996). The processing-speed theory of adult age differences in cognition. *Psychological Review, 103*, 403–428.

Salthouse, T. A. (2004). Localizing age-related individual differences in a hierarchical structure. *Intelligence, 32*, 541–561.

Salthouse, T. A. (2005). Relations between cognitive abilities and measures of executive functioning. *Neuropsychology, 19*, 532–545.

Salthouse, T. A. (2009). When does age-related cognitive decline begin? *Neurobiology of Aging, 30*, 507–514.

Salthouse, T. A., Atkinson, T. M., & Berish, D. E. (2003). Executive functioning as a potential mediator of age-related cognitive decline in normal adults. *Journal of Experimental Psychology, 132*(4), 566–594.

Salthouse, T. A., & Ferrer-Caja, E. (2003). What needs to be explained to account for age-related effects on multiple cognitive variables? *Psychology and Aging, 18*, 91–110.

Schaie, K. W. (2005). What can we learn from longitudinal studies of adult development? *Research in Human Development, 2*, 133–158.

Schretlen, D., Pearlson, G. D., Anthony, J. C., Aylward, E. H., Augustine, A. M., Davis, A., et al. (2000). Elucidating the contributions of processing speed, executive ability, and frontal lobe volume to normal age-related differences in fluid intelligence. *Journal of the International Neuropsychological Society, 6*, 52–61.

Smith, G., & Rush, B. K. (2006). Normal aging and mild cognitive impairment. In D. K. Attix & K. A. Welsh-Bohmer (Eds.), *Geriatric neuropsychology: Assessment and intervention* (pp. 27–55). New York: Guilford Press.

Souchay, C., Isingrini, M., & Espagnet, L. (2000). Aging, episodic memory feeling-of-knowing, and frontal functioning. *Neuropsychology, 14*, 299–309.

Stern, Y. (2002). What is cognitive reserve? Theory and research application of the reserve concept. *Journal of the International Neuropsychological Society, 8*, 448–460.

Stewart, J., Scarisbrick, D., & Golden, C. J. (2011). *Mediating role of processing speed on age-related declines in executive functioning.* Presented at Annual Conference of the American Psychological Association (August, 2011), Washington, DC.

Tucker-Drob, E. M. (2011). Global and domain-specific changes in cognition throughout adulthood. *Developmental Psychology, 47*, 331–343.

Tucker-Drob, E. M., Johnson, K. E., & Jones, R. N. (2009). The cognitive reserve hypothesis: A longitudinal examination of age-associated declines in reasoning and processing speed. *Developmental Psychology, 45*, 431–446.

Walker, D., Greenwood, C., Hart, B., & Carta, J. (1994). Prediction of school outcomes based on early language production and socioeconomic factors. *Child Development, 65*, 606–621.

Wang, M., Gamo, N. J., Jin, L. E., Wang, X. J., Laubach, M., Mazer, J. A., et al. (2011). Neuronal basis of age-related working memory decline. *Nature, 476*, 210–213.

Wechsler, D. (1981). *Wechsler adult intelligence scale-revised*. San Antonio, TX: Psychological Corporation.

Zimprich, D. (2002). Cross-sectionally and longitudinally balanced effects of processing speed on intellectual abilities. *Experimental Aging Research, 28*, 231–251.

Zimprich, D., & Martin, M. (2002). Can longitudinal changes in processing speed explain longitudinal age change in fluid intelligence? *Psychology and Aging, 17*, 690–695.

Zimprich, D., Martin, M., Kliegel, M., Dellenback, M., Rast, P., & Zeintl, M. (2008). Cognitive abilities in old age: Results from the Zurich longitudinal study on cognitive aging. *Swiss Journal of Psychology, 67*, 177–195.

Part III

Assessment of Executive Functioning

Assessment of Executive Function Using Rating Scales: Psychometric Considerations

10

Jack A. Naglieri and Sam Goldstein

Introduction

In any field of scientific study the information we obtain from research is directly related to the quality of the information we obtain from the tools we use. The better the tool, the more accurate and reliable the information that is obtained. Ultimately, the validity of the tools used in science will be proportionate to the quality of the concepts being evaluated. Ultimately, better tools are more effective for researchers and clinicians. The better the tools used in research and clinical practice, the more valid and reliable the decisions will be, the useful the information obtained will be, and ultimately, the better the services that will be provided. In this chapter, the rating scales used for assessment of executive function will be examined.

There are two goals of this chapter. First, to illustrate the relevance reliability and validity have on the decisions made by clinicians and researchers, review of essential psychometric qualities of test reliability and validity will be provided. The practical implication these psychometric issues have for the assessment and the implications for interpretation of results will be emphasized. Special attention will be paid to scale development procedures, particularly methods used to develop derived scores. The second section of this chapter will focus on rating scales used to assess behaviors considered indicative of executive function. The overall aim is to provide an examination of the relevant psychometric issues and the extent to which researchers and clinicians can have confidence in the tools they may use to assess executive function.

Reliability

Good reliability is critical for any test used for clinical practice as well as research purposes. It is essential that clinicians and researchers know the reliability of a test so that the amount of accuracy and the amount of error in the measurement of the construct are known. The higher the reliability, the smaller the error and the smaller range of scores used to build the confidence interval around the estimated true score. The smaller the range, the more precision and confidence practitioners can have in their interpretation of the results.

Bracken (1987) provided suggestions for the evaluation of test reliability (evaluated using some internal reliability estimate). He stated that individual scales from a test (e.g., a subtest or subscale) should have a reliability of .80 or greater and total tests should have an internal consistency of .90 or greater. The level of precision required should be determined in relations to the reason for

J.A. Naglieri (✉)
University of Virginia, Curry School of Education, White Post Road 6622, Centreville, VA 20121, USA
e-mail: jnaglieri@gmail.com

S. Goldstein
Neurology, Learning and Behavior Center, University of Utah School of Medicine, South 500 East, Suite 100 230, Salt Lake City, UT 84102, USA
e-mail: info@samgoldstein.com

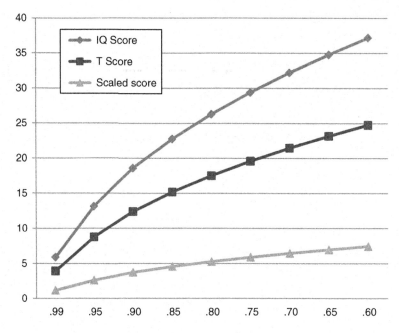

Fig. 10.1 Range of a 95 % confidence intervals as a function of reliability

testing and the importance of the decisions that may be made. If a score is to be used for screening purposes where over identification is preferred to under identification, a .80 reliability standard for a total score may be acceptable. If, however, high-stakes decisions are made, for example about special educational placement, then a higher reliability (e.g., .95) is more appropriate (Nunnally & Bernstein, 1994).

Details About Reliability

Researchers and practitioners must be aware of the reliability of any score they use. High reliability is essential for all test scores used in research and applied settings. Reliability is important to the practitioner because it reflects the amount of error in the measurement. That is, reliability describes the amount of variability to expect around the true score, assuming that any obtained score comprises the true score plus error (Crocker & Algina, 1986). Because we can never directly determine the true score, we use the reliability coefficient to describe a range of values within which the person's score likely falls with a particular level of probability. The size of the range is determined by the reliability of the measurement with higher reliability resulting in smaller ranges. This provides a way to describe an IQ score as a number and a range. For example, 105 (±5), meaning that there is a 90 % likelihood that the child's true IQ score falls within the range of 100–110 (105±5). The range of scores (called the confidence interval) is computed by first obtaining the standard error of measurement (SEM) from the reliability coefficient and the standard deviation (SD) of the score in the following formula (Crocker & Algina, 1986):

$$SEM = SD \times \sqrt{1 - \text{reliabillity}}$$

The SEM, which is the average standard deviation of a person's scores around the true score, is used to compute the confidence interval. To obtain a confidence interval, the SEM is multiplied by a z value of, for example 1.64 or 1.96 at the 90 % or 95 % levels, respectively. The resulting value is added to and subtracted from the obtained score to yield the confidence interval. In the example provided above, the confidence interval for an obtained score of 100 is 95 (100−5) to 105 (100+5). Figure 10.1 provides the range of confidence intervals (95 % level of

Table 10.1 Relationships between obtained scores, estimated true scores, and confidence intervals

Obtained standard score	Estimated true score	True score minus obtained score	Lower confidence band	Upper confidence band	Confidence interval range
55	60	5	52	67	16
70	73	3	65	81	16
85	87	2	79	94	16
100	100	0	92	108	16
115	114	−2	106	121	16
130	127	−3	119	135	16
145	141	−5	133	148	16

Note: This assumes a reliability coefficient of .90 and a 90 % confidence interval

confidence), that is, the values to be added and subtracted from an obtained score to calculate confidence intervals for a typical IQ score (Mn=100; SD=15), *T*-score (Mn=50; SD=10), and IQ test scaled score (Mn=10; SD=3) for measures with a reliability of .60 through .99. The range within which the true score is expected to fall varies as the reliability coefficient changes—the lower the reliability, the wider the range of scores that can be expected to include the true score.

It is important to know, however, that the confidence interval (and SEM) should be centered around the estimated true score rather than the obtained score (Nunnally & Bernstein, 1994). In many published tests (e.g., Wechsler Intelligence Scale for Children Fourth Edition (Wechsler, 2003) and the Cognitive Assessment System (Naglieri & Das, 1997)), the confidence intervals provided in the norms tables are centered on the estimated true score. Table 10.1 illustrates the relationships between obtained and estimated true scores, the lower and upper range of the confidence intervals in relation to the obtained scores, and the actual range of the confidence intervals for a hypothetical test (mean of 100, *SD* of 15) with a reliability of .90 at the 90 % level of confidence.

Examination of the scores in Table 10.1 shows that the confidence interval is equally distributed around a score of 100 (92 and 108 are both 8 points from the obtained score) but the interval becomes less symmetrical as the obtained score deviates from the mean. Ranges for standard scores that are below the mean are *higher* than the obtained score. As shown in Table 10.1, the range for a standard score of 70 is 65–81 (5 points below 70 and 11 points above 60). In contrast scores for standard scores that are above the mean are *lower* than the obtained score. The range for a standard score of 130 is 119–135 (11 points below 130 and 5 points above 130). This asymmetry is the result of centering the range of scores on the estimated true score rather than the obtained score even though the size of the confidence interval is constant (±8 points).

Whether confidence intervals are constructed using obtained or estimated true score methods, measurement error must be considered and communicated when scores from any test are used and particularly when results are explained to consumers. Confidence intervals, especially those that are based on the estimated true score, should be provided for all test scores including rating scales.

The importance of the SEM must be considered when two scores are compared. The lower the reliability (the larger the SEM), the more likely two scores will differ on the basis of chance. In order to account for reliability's influence on the difference between scores, a formula for determining how different two scores need to be can be applied. This formula is based on the SEM of each score and the *z* score associated with a specified level of significance. The difference needed for significance can be computed using the following formula:

$$\text{Difference} = z \times \sqrt{\text{SEM1}^2 + \text{SEM2}^2}$$

The relationship between reliability and the differences needed for significance when comparing two scores is provided in Table 10.2.

Table 10.2 Differences required for significance when comparing standard scores with a mean of 100 and SD of 15 ($p=.05$)

Reliability	.99	.95	.90	.85	.80	.75	.70	.65	.60
.99	4	7	10	12	13	15	16	18	19
.95	7	9	11	13	15	16	17	19	20
.90	10	11	13	15	16	17	19	20	21
.85	12	13	15	16	17	19	20	21	22
.80	13	15	16	17	19	20	21	22	23
.75	15	16	17	19	20	21	22	23	24
.70	16	17	19	20	21	22	23	24	25
.65	18	19	20	21	22	23	24	25	25
.60	19	20	21	22	23	24	25	25	26
.55	20	21	22	23	24	25	25	26	27
.50	21	22	23	24	25	25	26	27	28

To use this table, find the row that corresponds to the reliability of one score and the column that corresponds to the second. Read into the table for the difference required for significant. The significance level is based on the assumption that one pair is compared. The values in Table 10.2 can be used to compare more than one pair of scores; however, doing so changes the actual level of significance in proportion to the number of comparisons made. For example, using a .05 level of significance 6 times makes the experiment-wise error rate actually .265, not .05, because six pair wise increases error (the chance of a type I error is obtained using the formula 1—$(1-.05) \times 6$). One way to control for inflation in the level of significance is by using the Bonferroni correction method. This procedure controls for the number of comparisons by setting the experiment-wise error rate on the basis of making all six comparisons simultaneously (e.g., .05/6 = .008).

The differences needed for significance when comparing two scores with reliability coefficients that range from .55 to .99 are shown in Fig. 10.2 for scores that have an SD of 15 (a typical IQ test), 10 (a T-score used by many rating scales), and 3 (an IQ test subtest scaled score) calculated using the formula above. These findings demonstrate that in research and most importantly in clinical settings, test scores with high reliability are desired.

Researchers and clinicians assessing behaviors associated with executive function should use test scores possessing a reliability coefficient of .80 or higher and any total or composite score should have reliability of at least .90. If a rating scale does not meet these requirements, then their inclusion in research studies and particularly in clinical settings should be questioned. Clinicians are advised not to use measures that do not meet reliability standards because there will be too much error in the obtained scores to allow for reliable interpretation and especially comparison with other scores. This is particularly important when the decisions clinicians are making could have substantial and long-lasting impact on a child, adolescent, or adult.

Validity

While having a measure with good reliability is essential, reliable measurement of a construct that has limited validity has little use to the clinician and researcher. Validity is described as the degree to which empirical evidence supports an interpretation of scores representing a construct of interest. For example, a rating scale for evaluation of behaviors associated with executive function should include questions that accurately reflect the concept. Authors striving to produce a measure of executive function are especially burdened with the responsibility to define the concept carefully and the observable behaviors associated with it. When the behaviors and characteristics

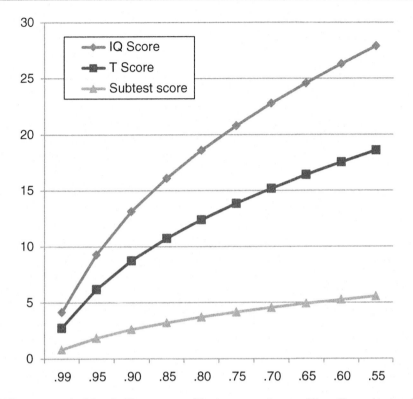

Fig. 10.2 Differences required for significance at $p = .05$ when comparing two IQ, or T, or subtest scaled scores at various levels of test reliability

associated with the disorder are thoroughly operationalized, then the way the concept has been measured by the observations of, for example, a parent or teacher can be tested and the task of establishing validity begun. Evidence of validity will be dependent upon the extent to which the items have adequate reliability.

The concept of executive function has been defined in multiple ways. Additionally, there are many different methods researchers and practitioners have utilized in attempting to measure this concept. Given that conceptualizations and methods vary and are evolving, we have particular responsibility to provide validity evidence of the effectiveness of any method we choose (rating scales, tests, interviews, etc.). Examining the validity of the concept of executive function is much hard than establishing its reliability because of the many different ways the concept has been defined in the literature (see Chap. 1 in this volume). Thus, the author of any measure of executive function defines the concept by the questions that are included. This can provide a broad or truncated view of how executive function could be measured.

Establishing concurrent validity is especially difficult for a rating scale of executive function because of the variability in the way the concept has been conceptualized and measured. That is, the author has to decide what marker test should new tests be compared to? The variability in conceptual and measurement approaches used by different authors will have direct influence on any research findings that may be found as well as the psychometric quality of the tests and methods used. Researchers and clinicians should be mindful that until there is sufficient maturity in the concept and tools used to measure executive function, any and all method should be examined carefully and high psychometric quality demanded.

Development of Scales to Assess Executive Function

There is a need for a number of well-standardized measures of executive function with demonstrated reliability and validity. At this time, there are only a few published behavior rating scales for clinical use possessing varying degrees of reliability and validity (a detailed analysis of these will be provided later in this chapter). Given the relatively small number of options, there is a need for practitioners and researchers to have an understanding of the psychometric qualities of these tools. It is particularly important to pay close attention to the development methods used by the authors of any scale intended to be used to evaluate executive function. Development of a rating scale should follow a series of steps to ensure the highest quality and validity. The development of such a scale is a task that demands well-known procedures amply described by Crocker and Algina (1986) and Nunnally and Bernstein (1994) that are summarized in the following section.

Initial development of a rating scale for executive function should begin with a clear definition of the concept or concepts and the behaviors that can be used to assess them. The items used to evaluate these behaviors must be written with sufficient clarity that they can be answered reliability over time and across raters. Items should be included that represent the author's carefully defined view of executive function.

The first test development step is to prepare an initial pool of items. The main goal at this initial stage is to evaluate the clarity of the directions and items and manage other logistical issues. For example, it is important to determine raters' reactions to the size of the fonts, clarity of the directions, colors used on the form, and position of the items on the paper.

Once initial testing is completed, a larger study of the items can be conducted. This research effort helps determine if there is confidence that the items have been adequately operationalized and the following information is obtained:

- Means and SDs and difficulty of each item should be obtained.
- The contribution each item has to the reliability of the scale(s) on which it is placed should be evaluated.
- Items designed to measure the same construct should correlate with other items designed to measure that same construct higher than items designed to measure different constructs. If this is not found, the item may be eliminated.
- The internal reliability of those items organized to measure each construct should be computed, as should the reliability of a composite score.
- The factor structure of the set of items may be examined to test the extent to which items or scales form groups, or factors, whose validity can be examined.

The number of times preliminary research studies are conducted depends upon the results of the statistical analysis which in turn is dependent upon the quality of the (a) original concepts, (b) initial pool of items, and (c) the sampling used to study the instrument. The results of these efforts should be used to develop an experimental version of an instrument that is ready to be used in a larger national standardization study. This would include sufficient data to establish quality norms and also to conduct a research program to examine the reliability and validity of the final scale.

Standardization and norms development requires that a sample represents the population of the country in which the scale will be used. Standardization samples are designed to be representative of the normal population so that those that differ from normality can be identified and the extent to which they differ from the norm (50th percentile) can be calibrated as a standard score. Dispersion from the mean should also be calculated. Development of norms is an art as much as a science. There are several ways in which this task can be accurately accomplished (see Crocker & Algina, 1986; Guilford & Fruchter, 1978; Nunnally & Bernstein, 1994; Thorndike, 1982).

The second task of national standardization efforts includes analysis of data for establishing reliability (internal, test retest, inter-rater, intra-rater) and validity (e.g., construct and content).

Of these two, validity is more difficult to establish and should be examined using a number of different methodologies and to assess the extent to which the scores the scale yields is valid for the purposes for which it is intended. The many different types of validity studies needed to fully evaluate any scale make it impossible to establish validity by a single study. According to the Standards for Educational and Psychologist Testing (AERA, APA, NCME, 1999), evidence for validity "integrates various strands of evidence into a coherent account of the degree to which existing evidence and theory support the intended interpretation of test scores for specific uses" (p. 17). There are 24 standards that relate to validity issues that should be addressed by authors and test development companies. Some of the more salient issues include the need to provide evidence that supports:

- The interpretations based on the scores the instrument yields
- The utility of the measure across a wide variety of demographic groups or its limitations based on race, ethnicity, language, culture, and so forth
- The appropriate relationships between the scores of the instrument with one or more relevant criterion variables
- The expectation that the scores provided differentiate between groups as intended
- The alignment of the factorial structure of the items or subtests with the scale configuration provided by the authors

Documentation in test manuals of scale development often focuses on construction, standardization, reliability, and validity. Reporting this information is important, but authors also have the responsibility to inform users about how the scores should be interpreted (AERA, APA, NCME, 1999). This includes how test scores should be compared. It is especially important to provide the values needed for significance when the various scores a rating scale provides are compared, for example, across raters. This information is critically important if clinicians are to be expected to interpret the scores from any instrument in a manner that is psychometrically defensible.

Professionals have a responsibility to choose scales that have been developed using the *highest* standards available because important decisions will be made on the basis of the information these measures provide. This includes ample documentation of methods used to develop the measure as well as ample evidence of validity and explicit instructions for interpretation of the scores that are obtained. Because of the impact score interpretation has on those individuals who seek help from professionals in clinical practice, in addition to being reliable, tools used to evaluate any condition must have been standardized and the scores based on norms developed from a large sample that represents the country in which the scale is used.

Obtaining information about the psychometric characteristics of psychological and educational tests is a time consuming and sometimes confusing task. Some test manual information is clear and concise, and at other times it is hard to ascertain enough details to fully evaluate the results being presented. Comparisons across instruments are complicated by this inconsistency and the logistical task of collecting the information. In the remainder of this chapter, a systematic examination of the scales used to assess the behaviors associated with executive function will be provided. The goal is to be informative of the specific details associated with important issues such as reliability, validity, standardization samples, and norming procedures. The information provided is intended to include essential topics such as description of the scale and standardization characteristics provided by the authors in their respective test manuals. Following this summary, a commentary of the relative advantages of the scales is provided.

Descriptions of Rating Scales Used to Assess Executive Function

BRIEF Parent and Teacher Reports

The Behavior Rating Inventory of Executive Function (BRIEF; Gioia, Isquith, Guy, & Kenworthy, 2000) was designed to assess the behavioral manifestations of executive functions in children aged 5–18 years as rated by parents or teachers. The 86-item rating scale evaluates two general domains—Behavioral Regulation (Inhibit, Shift, Emotional Control) and Metacognitive Problem-Solving (Initiation, Task Organization/Planning, Environmental Organization, Self-Monitoring, Working Memory) across the eight interrelated sub-domains.

The normative group for the BRIEF-Parent and BRIEF-Teacher ratings was based on data obtained from 25 schools in the State of Maryland (12 elementary, 9 middle, and 4 high schools). The sample description in the manual is very limited, mainly focused on sex (approximately equal percentages of males and females) and race/ethnicity. Table 10.3 shows that the distribution of the normative sample by race/ethnicity is quite different from that in the US population. The sample was dominated by Whites and considerably underrepresented by Hispanics. Even after statistically weighting the samples that were obtained by race/ethnicity, the values for Hispanics are still considerably lower than the US population values based on the 2011 Census.

BRIEF-Self-Report

The Behavior Rating Inventory of Executive Function—Self-Report (BRIEF-SR; Guy, Isquith, & Gioia, 2004) was designed to assess the behavioral manifestations of executive functions from self-reports of individuals aged 11–18 years. The 80-item rating scale evaluates two general domains—Behavioral Shift (Inhibit, Shift, Emotional Control, Monitor) and Cognitive Shift (Working Memory, Plan Organize, Organization of Materials, Task Completion) and eight sub-domains. Items are scored 1 (Never), 2 (Sometimes), and 3 (Often). Raw scores are converted to T-scores (mean of 50 and standard deviation of 15) and scaled so that scores above 70 are termed clinically significant. That is, the higher the score, the more difficulty with executive function is indicated. Two validity scales are also included an Inconsistency and Negativity Scale.

The BRIEF-SR norms were based on 1,000 11–18 year olds who completed the 80-item rating scale. The authors report in the manual that the sample was obtained through public and private school recruitment in the states of Maryland, Ohio, Vermont, New Hampshire, Florida, and Washington. These states do comprise the four regions of the country; however, the percentages of cases from each of these locations were not reported. Table 10.3 also illustrates that the distribution of the normative sample for African Americans was slightly underrepresented, Hispanics were underrepresented by about 40 %, and Whites were overrepresented by about 20 % in relation to the US population based on the 2011 Census. All five parental education levels deviated from the US population figures ranging from about 20 % for those without college experience to 65 % for those with bachelor's degree. These characteristics suggest that the sample characteristics are quite dissimilar to the US population based on the 2011 Census.

BDEFS-CA

The Barkley Deficits in Executive Function Scale—Children and Adolescents (BDEFS-CA, Barkley, 2012) was designed to assess the behaviors associated with executive functions as rated by parents of their children aged 6–17 years. The xx-item rating scale provides scores for five scales: Self-Management to Time, Self-Organization/Problem Solving, Self-Restraint, Self-Motivation, and Self-Regulation. Items are scored on a five-point Likert scale. Raw scores are converted to percentile scores for each sub-scale and an EF Summary Score. The scores are scaled such that the higher the score, the more the deficit in executive function.

The normative sample for the BDEFS-CA was obtained from parent raters distributed across the four regions of the USA with fairly equal proportions to the overall population. As shown in Table 10.3, the distribution of the normative sample by race/ethnicity, however, is substantially different from that in the US population. The sample was dominated by Whites and

Table 10.3 Number of items, age range, normative sample size, and percentages of normative sample by region, race/ethnicity, and educational level for the BRIEF, BDEFS, D-REF, and CEFI

	BRIEF-Parent	BRIEF-Teacher	BRIEF-Self-report	BDEFS-CA (parent)	D-REF parent	D-REF teacher	D-REF self report	CEFI-Parent	CEFI-Teacher	CEFI-Self report	US Pop %
Scale description											
No. of items	86	86	80	70	36	36	36	100	100	100	
Age range	5–18	5–18	11–18	6–17	5–18	5–18	11–18	5–18	5–18	12–18	
Standardization											
Sample size	1,416	720	1,000	1,922	500	342	220	1,400	1,400	700	
Region											
Northeast	0	0	–	18	16.1	12.2	5.4	16.0	16.2	16.0	17.0
Midwest	0	0	–	28	15.6	19.3	13.9	22.1	22.0	22.0	21.7
South	100	100	–	31	58.6	57.2	77.8	37.9	38.0	38.0	37.2
West	0	0	–	23	9.8	11.3	2.9	24.1	24.0	24.0	24.1
Race/ethnic											
Asian	3.8	6.1	(In other)	–	–	–	–	4.0	3.8	4.0	4.2
Black	11.9	13.5	14.7	7.7	16.5	19.8	5.4	14.0	14.0	14.0	13.9
Hispanic	3.1	4.2	12.5	12.4	19.2	15.8	13.9	22.0	22.0	22.0	21.2
White	80.5	72.1	67.3	73.0	58.0	56.4	77.8	56.0	56.5	56.0	56.5
Other	.5	.4	5.5	–	6.2	8.1	2.9	4.0	3.7	4.0	4.2
Parental education level											
<High school	–	–	12.1	4.1	9.2	10.0	7.9	14.1	–	13.9	14.7
High school grad	–	–	33.6	28.1	26.0	28.9	26.6	28.0	–	28.0	28.5
Some college	–	–	12.4	29.8				30.0	–	30.0	28.9
Bachelor's degree	–	–	29.0	22.6				18.0	–	18.1	17.6
Graduate degree	–	–	12.9	15.4	64.9	61.1	65.5	10.1	–	10.0	10.3

Notes: US population percentages based on 2009 Census. Percentages by race/ethnicity for BRIEF-Teacher reported in the manual do not sum to 100 %. D-REF values were averaged across age groups

considerably underrepresented by Hispanics and Blacks. Importantly, the parental education levels are not consistent with the US population figures. There are too few cases with parental education levels less than high school and too many from the top two educational attainment categories. This disparity on these two important demographic variables indicates that the characteristics of the DBEFS-CA normative sample are quite dissimilar to the US population based on the 2011 Census.

D-REF

The Delis Rating of Executive Functions (D-REF; Delis, 2012) is a set of rating forms designed to assess executive functioning in individuals ages 5–18. The scale has three forms: Parent, Teacher, and Self, each comprises 36 items. The D-REF is designed to evaluate a child or adolescent's behavioral, emotional, and executive functioning in four specific areas of executive functioning: Attention/Working Memory, Activity Level/Impulse Control, Compliance/Anger Management, and Abstract Thinking/Problem-Solving. Raw scores are converted to T-scores ($Mn=50$; $SD=10$) for each of the four index scores (low scores suggest better executive function).

The normative samples for the parent, teacher, and self scales of the D-REF were distributed across the four regions of the USA with varying correspondence to the overall population. For example, the parent, teacher, and self-rating samples underrepresented cases in the West and overrepresented cases from the South considerably. The cases from the Northeast were also very underrepresented (see Table 10.3). The inclusion of cases by race/ethnicity was also problematic. For example, Blacks were very underrepresented in the self-rating sample. Hispanic groups underrepresented in the parent, teacher, and self-rating samples, and Whites were very overrepresented in the self-rating sample. The sample by parental education was underrepresented for those with less than high school education and overrepresented for those with greater than a high school education. These differences in demographic variables indicate that the characteristics of the D-REF normative samples are substantially inconsistent with the characteristics of the US population based on the 2011 Census.

Comprehensive Executive Function Inventory

The Comprehensive Executive Function Inventory (CEFI, Naglieri & Goldstein, 2013) is a rating scale designed to evaluate observable behaviors that are related to executive function. The CEFI is completed by parents (or similar caregiver) or teachers (or similar professional) who rate behaviors of children ages 5–18 years. There is also a self-report version for 12–18 year olds. The 100 items of the CEFI items are organized on the basis of their content into nine scales (Attention, Emotion Regulation, Flexibility, Inhibitory Control, Initiation, Organization, Planning, Self-Monitoring, and Working Memory). A total (Full Scale) is also included. In addition, three scales that evaluate the quality of the ratings are provided: one that examines the consistency of the ratings (Consistency Index), one that is designed to assess the likelihood that the rater's scores are overly negative, and one that suggests an overly positive view of the person being evaluated (Negative and Positive Impression Scales, respectively). Each of these scales is scaled to have a normative mean of 100 and SD of 15 where higher scores indicate better executive function.

The norms for the CEFI were based on the standardization sample including cases from all 50 states in the USA. The normative samples included 1,400 for the Parent Form, 1,400 for the Teacher Form, and 700 for the Self-Report Form. The stratifications by region, race/ethnicity, and parental education were within one percentage point of the values for the US population. The report of demographic variables indicates that the characteristics of the CEFI normative samples are very consistent to the US population based on the 2011 Census.

Normative Sample Disparities

In this chapter we have discussed various psychometric characteristics of rating scales and the samples upon which their norms were based. It is clear from this summary that some of the demographic characteristics of the samples upon which the derived scores were based vary considerably. It is reasonable to ask, does this matter? In order to examine the impact a variable such as parental education can have on the resulting normative scores, norms for parent ratings from the CEFI (Naglieri & Goldstein, 2013) were prepared for four different groups by parent educational levels (PEL). This study began with the calculation of means and standard deviations for the standardization data ($N = 1,400$) for PEL levels reported in the manual. The mean raw scores were 252.1 (no high school diploma), 269.2 (high school diploma), 280.3 (some college), and 285.6 (bachelor's degree or higher). Using these raw scores, standard scores were computed using the formula ((raw score – *mean* raw score)/raw score SD)*15 + 100. The scores were calibrated so that high standard scores indicated better executive functioning.

The resulting values presented in Table 10.4 illustrate how much differences in CEFI total scale scores vary across parental education. The raw scores associated with a standard score of 100 vary from 250 to 285 (35 points) across the four PEL levels. The difference between the score of 100 based on the total sample and the lowest PEL level is 6 standard score points which is nearly half a standard deviation. Of particular importance are the differences that are found at the standard score of 85, which indicates a very poor score on this scale of executive function. The same raw score of 210 yields a score of 85 when based on the total sample, but a standard score of 92 (which falls in the average range) when the reference group is those with less than a high school education and a score of 81 when the highest education level is used as a reference group. The 11 point difference between the 81 and 92 represents the range of scores that can be expected due to the influence of PEL on CEFI scores.

Table 10.4 Calibration of standard scores ($Mn = 100$; $SD = 15$) across parental education levels for CEFI parent ratings

	Standard scores				
Raw score	Less than high school	High school graduate	Some college	College graduate	Total sample
180	**85**	80	76	74	79
185	86	81	77	75	80
190	87	82	79	76	81
195	88	83	80	77	82
200	90	**85**	81	79	83
205	91	86	82	80	84
210	92	87	83	81	**85**
215	93	88	**85**	82	86
220	94	89	86	84	88
225	95	90	87	**85**	89
230	96	91	88	86	90
235	97	92	89	87	91
240	98	93	90	89	92
245	99	95	92	90	93
250	**100**	96	93	91	94
255	101	97	94	92	95
260	102	98	95	94	97
265	103	99	96	95	98
270	104	**100**	98	96	99
275	105	101	99	97	**100**
280	106	102	**100**	99	101
285	107	103	101	**100**	102
290	108	105	102	101	103
295	109	106	103	102	105
300	110	107	105	104	106
305	111	108	106	105	107
310	112	109	107	106	108
315	113	110	108	107	109
320	114	111	109	109	110
325	115	112	111	110	111
330	116	114	112	111	112

Note: Standard scores of 100 (at the normative mean) and 85 (one standard deviation below the mean) are in bold text

The variability of standard scores obtained across levels of parental education illustrates the importance of having a normative sample that closely represents the US population. Of course, the results presented here represent only one variable. In those standardization samples that are not representative of the US population on more than one variable, the potential impact on the resulting scores, and the decisions made by practitioners when evaluating executive function, cannot be ignored. For this reason, it is advised that only measures that have been normed on a nationally representative sample that closely corresponds to the US population should be used in professional practice as well as research to ensure accurate calibration of an individual's executive function.

Conclusions

The information summarized in this chapter provides clinicians and researchers with information about the psychometric characteristics of rating scales used to assess behaviors associated with the concept of executive function. Special attention was paid to the quality of the standardization samples used to create norms. The information provided here illustrates very different approaches to test development and the quality of the standardization samples used to create the norms. For example, some of the scales are short (the D-REF has 36 items) while others such as the CEFI contain many items (100). Some authors provide only percentile scores (BDEFS-CA) which make use of the scores in any mathematical formula difficult, others (e.g., BRIEF) provide *T*-scores scaled so that high scores indicate more deficits in executive function, and others (CEFI) use the familiar mean of 100 and standard deviation of 15 where high scores indicate better executive function. Although these rating scales of behaviors related to executive function all strive to evaluate essentially the same concept, the characteristics of the samples upon which their derived scores are based reflect a fundamental difference in test development. That is, some normative samples are more closely matched to the US population characteristics than others. The closer the samples are to the US population, the more confidence users can have with the obtained scores and the greater likelihood that accurate and valid information can be obtained.

References

American Educational Research Association, American Psychological Association, National Council on Measurement in Education. (1999). *Standards for educational and psychological testing*. Washington, DC: American Educational Research Association, American Psychological Association, National Council on Measurement in Education.

Barkley, R. A. (2012). *Barkley deficits in executive function scale-children and adolescents*. New York: Guilford Press.

Bracken, B. A. (1987). Limitations of preschool instruments and standards for minimal levels of technical adequacy. *Journal of Psychoeducational Assessment, 5*, 313–326.

Crocker, L., & Algina, J. (1986). *Introduction to classical and modern test theory*. New York: Hold, Rinehart and Winston.

Delis, D. C. (2012). *Delis rating of executive functions*. Bloomington, MN: Pearson.

Gioia, G. A., Isquith, P. K., Guy, S. C., & Kenworthy, L. (2000). *Behavior rating inventory of executive function*. Lutz, FL: Psychological Assessment Resources, Inc.

Guilford, J. P., & Fruchter, B. (1978). *Fundamental statistics in psychology and education*. New York: McGraw Hill.

Guy, S. C., Isquith, P. K., & Gioia, G. A. (2004). *Behavior rating inventory of executive function—Self-report version*. Lutz, FL: Psychological Assessment Resources, Inc.

Naglieri, J. A., & Das, J. P. (1997). *Cognitive assessment system interpretive handbook*. Austin, TX: ProEd.

Naglieri, J. A., & Goldstein, S. (2013). *Comprehensive executive functioning index*. Toronto: Multi Health Systems.

Nunnally, J. C., & Bernstein, I. H. (1994). *Psychometric theory*. New York: McGraw-Hill.

Thorndike, R. L. (1982). *Applied psychometrics*. Boston: Houghton-Mifflin.

Wechsler, D. (2003). *Wechsler intelligence scale for children fourth edition*. San Antonio: Pearson.

# The Cambridge Neuropsychological Test Automated Battery in the Assessment of Executive Functioning

Katherine V. Wild and Erica D. Musser

Computerized administration of clinical instruments is not an entirely new phenomenon. The first personal computers were introduced into wide use in the 1970s. Rapid adoption of computer-based testing paralleled this development. By the 1980s, the research literature was replete with considerations of the inherent advantages and limitations of automated assessment of a myriad of clinical domains. In particular, the application of computers to the evaluation of cognition has been widely studied. This body of research has generally fallen into one of two categories: (1) the translation of existing standardized tests to computerized administration and (2) the development of new computer tests and batteries for the assessment of cognitive function. Somewhere between these two categories are approaches that have adapted existing tests in a new way using computer administration. The Cambridge Neuro-psychological Test Automated Battery (CANTAB) is an example of a battery that has successfully combined standard cognitive test paradigms with novel formats.

The transition from paper-and-pencil- to computer-based assessment is not necessarily straightforward, and both methods of administration have distinct advantages and drawbacks. Included among the multiple benefits of computerized tests that have been cited are their ability to cover a wider range of abilities while minimizing floor and ceiling effects, potential for more standardized administration, multiple versions applicable to repeated testing, and precisely record accuracy and speed of response with a level of sensitivity not possible in standard administrations. Another potential advantage of computerized test batteries over traditional paper-and-pencil assessments is their flexibility in terms of immediate adjustment to performance levels. Many batteries have the capability of automatically altering test order, presentation rate, and level of difficulty in response to ongoing test performance. Such characteristics can be critical both in early detection and also in extending the range of a test to be useful across the full range of cognitive performance in a given patient population.

In comparison with traditional neuropsychological assessment instruments, computerized tests may also represent a potential cost savings not only with regard to materials and supplies but also in the time required of the test administrator. Moreover, the nature of the computerized instruments may allow administration by health-care clinicians other than neuropsychologists, allowing greater scheduling flexibility in the reduced need for administration by trained personnel.

In the initial excitement of this new application of technology, however, some basic aspects of test development may have been sacrificed. One of the more persistent criticisms of computerized test batteries has been the general lack of adequately established psychometric standards (Schlegel & Gilliland, 2007). Whether included

K.V. Wild (✉) • E.D. Musser
Oregon Health and Science University, 3181 SW Sam Jackson Park Road, Portland, OR 97239, USA
e-mail: wildk@ohsu.edu

in the test development phase or as post hoc analyses, basic indices of psychometric properties are essential to the widespread acceptance of new cognitive test batteries. Schlegel and Gilliland (2007) have outlined the necessary elements of quality assurance assessments for computer-based batteries. They caution against the acceptance of computerized adaptations of paper-and-pencil tests based purely on face validity. Others have also warned that equivalence across these media cannot be assumed (Buchanan, 2002; Butcher, Perry, & Atlis, 2000; Doniger et al., 2006). At a minimum, differences in communication of instructions, stimulus presentation, and response format may yield significant differences in test performance, particularly in an older population. Differences in computer experience as an intervening variable in performance cannot be ignored. Failure to demonstrate equivalence between the examinee's experience of computer versus traditional test administration, limited—and for the elderly in particular, perhaps unfamiliar—response modality, and poorly designed computer-person interface has been problematic in early iterations of this new methodology.

CANTAB

The CANTAB was developed initially by adapting animal paradigms for cognitive testing (Robbins et al., 1994; Sahakian & Owen, 1992). At the same time, careful analysis of the processes underlying each cognitive domain yielded means for independent assessment of these processes in a systematic and controlled fashion. For example, in an adaptation of the widely used Tower of London test of planning, the CANTAB Stockings of Cambridge tasks involve two stages. In the first, the subject works out and executes a series of steps to replicate a presented configuration; in the second, the subject must arrive at the correct response by mentally solving the series of moves without actually moving the stimuli. The multiple measures obtained during these tasks are then related to discrete cognitive functions that in turn are associated with activation of specific neural networks. The capability of a computerized test battery with a well-defined theoretical foundation to dissect performance on a test of executive function, for example, into multiple related but independent factors is a unique asset in furthering our knowledge of brain–behavior relationships.

In its original format first presented over 20 years ago (Sahakian, 1990), the CANTAB consisted of three batteries of tests designed to measure visual memory, attention, and planning. Over the years the battery has expanded to include tests designed to assess visual and verbal memory, executive function, attention, decision-making and response control, and social cognition. In addition two short "induction" tests can be used to familiarize participants with the general testing milieu and response format. The publishers of the CANTAB have also assembled "core batteries" for various diagnostic applications such as attention-deficit/hyperactivity disorder (ADHD), mild cognitive impairment (MCI), and schizophrenia, to provide focused assessments of the cognitive domains relevant to each disorder.

The nature of automated tests such as the CANTAB makes possible investigations of executive function across the life span. With minimal reliance on language, and continuous and immediate adjustment to level of performance, the CANTAB is well suited to help clarify age-related changes in specific executive abilities. DeLuca et al. (2003) selected CANTAB tests thought to tap working memory, strategic planning, organization of goal-directed behavior, and set-shifting in a study of a normative sample between the ages of 8 and 64. After administration of the Spatial Span (SSP), Spatial Working Memory (SWM), Tower of London, and Intra-/Extra-Dimensional Set-Shifting (IED) tests to 93 males and 101 females, the authors concluded that the CANTAB was sensitive to age and gender effects on executive function.

Another important parameter of test batteries purported to be well suited to repeat administration, and tracking of change over time or with intervention, is test-retest reliability. Lowe and Rabbitt (1998) administered all tests of the

CANTAB on two occasions separated by 4 weeks. The authors hypothesized that practice effects may be a more significant issue for tests that assess frontal or executive functions than tests of temporal function, as the former tend to rely on identification of strategies for successful performance. Further, as task novelty decreases with repetition, practice effects and individual variability in improvement with repeat testing may be amplified. Participants, ages 60–80, were selected to represent a range of ability as measured by a test of fluid intelligence. Practice effects were found to vary with test, task difficulty, level of intellectual ability, and outcome parameters. In general, test-retest correlations were higher for tests of memory than for tests of planning or working memory. More specifically, measures such as number of moves to solution on the Tower of London had poor test-retest correlations. On some tests higher intelligence was related to greater improvement with practice, while on others the opposite was true. Speed versus accuracy measures also differed in their sensitivity to repeat testing. The authors recommend the use of correction factors for practice effects where available, or at a minimum, obtaining good baseline data by allowing participants' practice trials dependent on the task.

The CANTAB has been used extensively in research settings as well as in clinical trials and has a published bibliography of over 700 articles assessing over 100 disorders. For the purposes of this review, discussion will be limited to those publications addressing deficits in executive function. Further, we focus on a selection of diagnostic groups that represent some of the main targets of research with this instrument. Thus, the bulk of the discussion will review studies of ADHDs and autism spectrum disorders (ASD) in childhood, followed by an overview of work in age-related cognitive decline, disorders of the frontal lobe, and finally Huntington's disease as an exemplar of a progressive neurodegenerative disorder with executive function implications. While it is beyond the scope of this chapter to summarize all relevant research efforts even in these diagnostic categories, every effort has been made to include those articles describing the most rigorous scientific methodology.

CANTAB Tests

The CANTAB can be administered by a trained assistant without reliance on verbal instruction. Responses are by touch screen or with a response button, depending on the task demands. Each task begins with practice items at a basic level. The design and interface of the CANTAB tests thus attempt to minimize effects of computer experience and computer-related anxiety. The tests described below are those that assess executive function and related domains; additional tests of verbal and visual memory and social cognition that are part of the complete battery are beyond the scope of this chapter but are described on their website (http://www.cambridgecognition.com).

Attention Switching Task (Goldberg et al., 2005). The participant is initially instructed to press a left or right button in response to the direction in which an arrow in the center of the screen is pointing. The second phase requires the participant to attend to a cue at the top of the screen that will determine whether the response reflects the direction in which the arrow is pointing or the side of the screen on which the arrow is located. Outcome measures include speed, accuracy, and types of errors (commission and omission), as well as switch cost and congruency cost.

Intra-/Extra-Dimensional Set Shift (*IED*). This test, described as a computer-based analog of the Wisconsin Card Sorting Test, assesses set formation and maintenance, shifting, and attentional flexibility. The task includes nine stages of increasing difficulty. The test initially presents two simple colored shapes and the participant must determine which one is correct in response to feedback. In successive stages when criterion is reached (i.e., six correct responses), the rules and/or stimuli change, moving from intra-dimensional, in which colored shapes remain the only relevant dimension, to extra-dimensional,

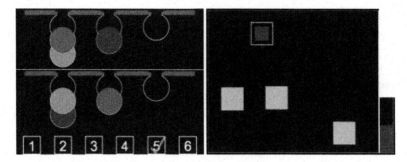

Fig. 11.1 One-Touch Stockings of Cambridge (*left*) and Spatial Working Memory (*right*) tasks

in which the participant needs to shift between white lines and colored shapes as relevant or irrelevant dimensions. Among the multiple outcome measures are errors to criterion, number of trials and stages completed, and response latencies.

Stockings of Cambridge (SOC). Based on the Tower of London test, the SOC is a measure of spatial planning in which the participant attempts to move colored balls to match a displayed pattern in the least possible number of moves. The time taken to complete the pattern, number of moves taken, and trials performed in minimum number of moves are measured.

One-Touch Stockings of Cambridge (OTS). This test relies on working memory in addition to spatial planning. In this task, two arrays of colored balls are presented. The participant's task is to choose from a series of numbered boxes, the minimum number of moves required to achieve the upper display by rearranging the lower array. The response is based on working out the solution without actually moving any balls. Outcome measures are based on speed and accuracy of response and include problems solved on first choice, mean choices to correct response, latency to first choice, and latency to correct choice.

Spatial Span (SSP). Described as a visual analog of the Digit Span Test, in this task white squares in an irregular array change color briefly in random sequences. The participant touches the boxes in the same order, or in reverse order, for varying sequence spans. The initial span consists of two boxes changing color up to a maximum of nine. The test is discontinued after three consecutive errors at a given span. Span length, errors, number of attempts, and latency are recorded.

Spatial Working Memory (SWM). Participants are asked to search through randomly arrayed boxes to locate colored tokens within the boxes, in order to fill a column on the side of the display. The number of boxes is increased over trials to a maximum of eight boxes. Measures of latency, errors, and a measure of search strategy are among the main outcome measures for this task. Errors can be further analyzed as "between-search" errors in which the subject returns to a box which has already been found to contain a token and "within-search" errors in which a box already opened and found to be empty earlier in the same trial is touched (Fig. 11.1).

Rapid Visual Information Processing (RVP). In this test of sustained attention, a white box appears in the center of the screen, containing single digits that appear in random order. The participant's task is to monitor the sequence of digits to match a target three-digit sequence. Measures of latency, hits and misses, false alarms, and rejections are tabulated.

Cambridge Gambling Task (CGT). Used to assess decision-making and risk-taking behavior, this task asks the participant to guess whether a yellow token is hidden in a red box or a blue box of ten boxes displayed across the screen. The proportion of red to blue boxes varies, as does the percent of "points" the participant chooses to gamble on the correctness of their choice. The test aims to separate out risk-taking from impulsivity, as the point percents are presented in ascending or descending order, forcing the participant to wait to make a high-risk bet. Outcome measures can include quality of decision-making (i.e., whether the subject chose the more likely outcome), time taken to choose, "bet" size, and risk adjustment.

Stop Signal Task (SST). This is a response inhibition test of two parts. In the first, participants are instructed to press a left- or right-hand button in response to an arrow pointing in that direction. In the second set of trials arrows continue to appear but the response is to be withheld if an auditory signal precedes the arrow presentation. Outcome measures include direction errors, proportion of successful stops, go-trial reaction time, stop-trial errors, and stop signal reaction time.

Assessing Executive Functioning in Childhood

Until recently, little research had been conducted on executive functioning in childhood, as it was believed that these cognitive skills did not develop until adolescence (DeLuca et al., 2003; Golden, 1981; Hughes & Graham, 2002). This paucity of research on executive functioning in children has been tied to three major factors (DeLuca et al., 2003). First, it was thought that the prefrontal cortex only became functionally mature late in development (i.e., late in adolescence or early adulthood) (Golden, 1981; Stuss, 1992). Second, several primate studies and early research on traumatic brain injuries suggested that juvenile prefrontal lesions had little or no consequence until later in adulthood (Walker, Husain, Hodgson, Harrison, & Kennard, 1998). Third, as standardized measures of executive functioning were designed to be difficult, these measures were often inappropriate for use with children, leaving this group of individuals without formal assessments of these domains of cognitive functioning (Anderson, Anderson, Northam, Jacobs, & Catroppa, 2001; Hughes & Graham, 2002; Kempton et al., 1999).

However, more recent investigations into executive functioning in children have suggested that these processes develop much earlier than once thought, beginning around 12 months of age, with a burst in development around 8 years of age (Ardila & Roselli, 1994; Case, 1992; DeLuca et al., 2003; Luciana & Nelson, 1998). Additionally, these early increases in executive functioning have been correlated with increased myelination and synaptogenesis in the frontal regions during these periods of growth (DeLuca et al., 2003; Espy, 1997; Kempton et al., 1999; Klingberg, 1997). Furthermore, life-span studies of executive functioning show that a number of domains, including short-term memory and sequencing, working memory, strategic planning and organization, and attentional set-shifting, appear to be present and measurable between ages 8 and 10 years, with the greatest functional gains in performance in each of these domains appearing between 15 and 30 years of age, followed by a gradual decline in performance over time with aging (DeLuca et al., 2003). Thus, there is much evidence to support the development executive functioning in children, and as such, there is an increased need to develop standardized executive functioning assessment measures that are appropriate for use with children.

The increased interest in the childhood development of executive functioning has been sparked by the study of clinical populations, as well as the development of several new assessment methods, which have been shown to be appropriate for use with children (Hughes & Graham, 2002). Of particular interest, several neurodevelopmental disorders have been shown to be associated with specific impairments in executive functioning, with the greatest body of literature providing evidence for these deficits in ADHD and ASD (Hughes & Graham, 2002; Ozonoff, 1997). Specifically, children with ADHD have been shown to have deficits primarily in inhibitory control, which is likely rooted in frontostriatal circuitry and decreased volume in dorsolateral prefrontal cortex, caudate, and cerebellum, as well as difficulty with strategic flexibility, planning, working memory, monitoring, and sustained attention (Chamberlain et al., 2011). In contrast, executive functioning deficits in ASD have been more often characterized by high-level cognitive, nonspatial problems, such as deficits in planning and set-shifting, as well as more secondary deficits in working memory (Hill, 2006). Furthermore, numerous studies of preschoolers have demonstrated that individual differences in executive functioning are correlated with individual differences in theory of mind, or the ability to attribute mental states to the self and others, which is a well-established deficit among individuals with ASD (Hughes & Graham, 2002; Perner & Lang, 1999).

In addition to the growth of research in the domain of executive functioning deficits in neurodevelopmental disorders, the study of normative development of executive functioning in childhood has also received increased attention. This increased focus has resulted in the development of numerous executive functioning assessment measures, which propose to be appropriate for broad age groups (Hughes & Graham, 2002; Hughes, Plumet, & Leboyer, 1998; Manly et al., 2001). However, children differ from adults in a number of important ways, and to fully assess executive functioning in children, executive functioning assessment measures must make accommodations for these differences. Specifically, children have limited language abilities and tend to have poorer motivation than adults; thus, child-appropriate measures of executive functioning need to be easy to understand, relatively independent of language skills, somewhat simplified, and in order to maintain motivation, somewhat fun (DeLuca et al., 2003; Hughes & Graham, 2002). Furthermore, in order to assess the development of executive functioning over time in both typically developing and neurodevelopmentally delayed populations, executive functioning assessment tools must be standardized across broad age ranges and populations, allowing for more reliable and valid longitudinal assessments (Hughes & Graham, 2002).

Computerized testing batteries of executive functioning may help to address several of the issues associated with the assessment of executive functioning both in children and across development. Specifically, these testing batteries can tap into a wide range of abilities, reduce floor and ceiling effects, reduce human error with more standardized formats, more precisely record speed and accuracy, reduce reliance on verbal instructions and feedback, and increase the potential for widespread screening efforts across broad age ranges (DeLuca et al., 2003; Wild, Howieson, Webbe, Seelye, & Kaye, 2008). In particular, the CANTAB may be uniquely suited to the assessment of executive functioning in children (Chamberlain et al., 2011; DeLuca et al., 2003). The CANTAB has the benefit of being widely used in academic studies with children from age 4 years and upwards (Chamberlain et al., 2011), with standardized scores and age-normative data available for individuals aged 8–80 years (Chamberlain et al., 2011; DeLuca et al., 2003; Hughes & Graham, 2002). Furthermore, the CANTAB has been shown to be sensitive to impairments in school-aged children with ADHD, ASD, and other neurodevelopmental disorders (Chamberlain et al., 2011; Hughes & Graham, 2002). In fact, recent reviews suggest that over 30 studies have utilized the CANTAB to examine executive functioning deficits in individuals with ADHD, and 20 studies have done so in individuals with ASD. We now turn our attention to a review of these literatures beginning with ADHD.

Evaluating Executive Functioning Using CANTAB in ADHD

According to DSM-IV criteria, ADHD includes the symptom domains of inattention/distraction and hyperactivity/impulsivity (American Psychological Association, 2000). Impairments in executive functioning are suggested both by the core diagnostic criteria of ADHD, with dysregulation of attention, behavior, and impulse control at the heart of the disorder, and disruptions to the frontostriatal circuitry, as supported by neuroimaging studies, which have revealed reduced volumes in the dorsolateral prefrontal cortex, caudate, and cerebellum (Seidman, Biederman, Monuteaux, & Doyle, 2005; Valera, Faraone,

Murray, & Seidman, 2007). Specifically, children with ADHD have been shown to have deficits in primarily inhibitory control, as well as difficulty with strategic flexibility, planning, working memory, self-monitoring, and sustained attention (Chamberlain et al., 2011).

The CANTAB has been widely used in the study of executive functioning deficits associated with ADHD, as well as in the study of the effects of pharmaceutical treatment on both improving executive functioning and reducing symptoms among individuals with ADHD. In fact, in a recent meta-analysis by Chamberlain et al. (2011), 13 studies examining performance on CANTAB subtests compared participants with ADHD and typically developing controls. According to this meta-analysis, medium to large effects were observed in participants with ADHD when compared to typically developing controls on the CANTAB sub-domains of response inhibition, working memory, and executive planning, with smaller effects observed in attentional set-shifting (Chamberlain et al., 2011). An additional review of the literature on the effects of salient drugs on executive functioning, as assessed by the CANTAB, in individuals with ADHD revealed that methylphenidate (Ritalin) improved working memory, modafinil improved planning, and methylphenidate, modafinil, and atomoxetine all improved inhibition (Chamberlain et al., 2011). We now turn our attention to a more in-depth examination of these literatures beginning with an examination of the literature on inhibition as assessed by the CANTAB in individuals with ADHD.

Inhibition as Assessed by the CANTAB in ADHD. Inhibition is assessed with the CANTAB using the SST, which is a classic stop signal response inhibition test. As described above, the meta-analysis by Chamberlain et al. (2011) found a large deficit in stop signal reaction time when individuals with ADHD were compared to typically developing individuals across four studies. Specifically, individuals with ADHD were found to have longer stop signal reaction times; however, individuals with ADHD did not differ from controls with respect to their reaction times on go trials, suggesting that the deficit is specific to inhibition. These findings are similar to those that have been reported in other meta-analyses using other computerized SSTs, as well as other measures of inhibition, such as the Stroop task (Boonstra, Oosterlaan, Sergeant, & Buitelaar, 2005; Lijffijt, Kenemans, Leon, Verbaten, & van Engeland, 2005; Willcutt, Doyle, Nigg, Faraone, & Pennington, 2005). Additionally, this deficit has been observed in both children (Brophy, Taylor, & Hughes, 2002; DeVito et al., 2008; Goldberg et al., 2005; Rhodes, Coghill, & Matthews, 2005) and adults (Aron, Dowson, Sahakian, & Robbins, 2003; Chamberlain et al., 2007; Clark et al., 2007; McLean et al., 2004; Turner, Clark, Dowson, Robbins, & Sahakian, 2004). Thus, the deficit in inhibition appears to be rather robust among individuals with ADHD (Garcia-Villamisar & Hughes, 2007).

These deficits have been associated with disrupted neural networks, including right inferior frontal gyrus, bilateral anterior cingulate cortex, and the superior motor region (Clark et al., 2007; Goldberg et al., 2005). Additionally, inhibition is thought to be under the control of catecholamines, specifically dopamine and norepinephrine, as several studies have shown that medications which alter dopaminergic and noradrenergic functioning improve inhibition and stop signal reaction times among individuals with ADHD (Aron et al., 2003; Chamberlain et al., 2007; DeVito et al., 2008; Turner et al., 2004). Specifically, in an acute, double-blind, placebo-controlled crossover study, Aron et al. (2003) found that methylphenidate ameliorated the deficit in stop signal reaction time among adults with ADHD, and DeVito et al. (2008) replicated these results in children. Additionally, Coghill, Rhodes, and Matthews (2007) found that chronic treatment of ADHD with methylphenidate improved performance on the SST among children. In contrast, Rhodes, Coghill, and Matthews (2006) did not observe the amelioration of the slowed stop signal reaction time among children with ADHD in a randomized, double-blind, placebo-controlled study with methylphenidate; however, it should be noted that this study utilized a low-dose design. Thus, it may be that dosing and duration of treatment with methylphenidate play a role in the treatment of disinhibition among individuals with ADHD.

It should also be noted that similar results were observed in acute, double-blind, placebo-controlled crossover studies with atomoxetine in children (Gau & Shang, 2010b) and adults with ADHD (Chamberlain et al., 2007), and modafinil has been shown to have similar effects in adults with ADHD (Turner et al., 2004). Therefore, the treatment effects do not appear to be specific to a particular medication; however, these effects do appear to be limited to those medications that target both norepinephrine and dopamine.

Interestingly, the results observed among individuals with ADHD were also shown to be relevant to healthy controls. In two acute, double-blind, placebo-controlled studies with atomoxetine, one parallel and one crossover design, significant increases in stop signal reaction times were observed among healthy adult volunteers (Chamberlain et al., 2006, 2009). In an fMRI study, the use of atomoxetine was found to increase right, frontal inferior cortex activation during the CANTAB stop signal task among healthy adult volunteers (Chamberlain et al., 2009). However, mixed results have been observed in similar trials using both methylphenidate and modafinil with some studies finding increases in stop signal reaction time and some studies finding no treatment effects (Turner et al., 2003, 2004; Winder-Rhodes et al., 2009). Finally, in a single acute, double-blind, placebo-controlled, parallel study with guanfacine in healthy adult volunteers, an overall global slowing of reaction time on both go and stop trials was observed, suggesting that guanfacine may act as a sedative, resulting in increased inhibition (Muller et al., 2005).

Working Memory as Assessed by the CANTAB in ADHD. In addition to deficits in inhibition, deficits in working memory among individuals with ADHD have also been reported in a meta-analysis of ten studies with a large effect size (Chamberlain et al., 2011). In particular, working memory on the CANTAB is assessed with the SWM Task, which assesses an individual's ability to retain spatial information and manipulate remembered items in working memory. Meta-analytic results suggest that the greatest impairments are in the areas of between-search errors and strategy, with individuals with ADHD having more errors and worse strategy scores than typically developing controls. Furthermore, this is consistent with previous meta-analytic work examining other SWM tasks revealing SWM deficits in ADHD (Martinussen, Hayden, Hogg-Johnson, & Tannock, 2005; Willcutt et al., 2005). As with the observed effect on inhibition, the working memory deficit associated with ADHD has also been observed in both children (Barnett et al., 2001; Gau, Chiu, Shang, Cheng, & Soong, 2009; Gau & Shang, 2010a; Goldberg et al., 2005; Kempton et al., 1999; Klingberg, Forssberg, & Westerberg, 2002; Rhodes, Coghill, & Matthews, 2004; Vance, Maruff, & Barnett, 2003) and adults (Chamberlain et al., 2007; Clark et al., 2007; Dowson et al., 2004; Gropper & Tannock, 2009; McLean et al., 2004). The SWM deficit may serve as an endophenotype of ADHD, as typically developing siblings of children with ADHD have also been shown to display deficits on this measure (Gau & Shang, 2010a). Additionally, poor performance on SWM in individuals with ADHD has also been linked to poor academic achievement among young adults and poor performance on progressive matrices tasks, increased motor activity, and poor inhibition among children (Clark et al., 2007; Gropper & Tannock, 2009; Klingberg et al., 2002).

Again, the catecholamines, dopamine and norepinephrine, have been implicated in the deficits in working memory observed among individuals with ADHD, as the medications which have been found to improve SWM in both children and adults with ADHD have all been shown to target the production or to block the reuptake of these neurotransmitters. Specifically, methylphenidate has been shown to improve performance on the CANTAB spatial working memory test in both children (Barnett et al., 2001; Bedard, Martinussen, Ickowicz, & Tannock, 2004; Brophy et al., 2002; Hoare & Sevar, 2007; Kempton et al., 1999; Mehta, Goodyear, & Sahakian, 2004) and adults (Turner, Blackwell, Dowson, McLean, & Sahakian, 2005). In contrast, two studies found no effects of methylphenidate on SWM among children, though both of these were low-dose trials (Coghill et al., 2007; Rhodes et al., 2004). Methylphenidate was also observed to increase accuracy on the SWM task in healthy adults, with PET imaging results suggesting that improved performance was associated with increased binding of dopamine in the striatum (Mehta, Calloway, & Sahakian, 2000).

The effect of medications other than methylphenidate on SWM performance in individuals with ADHD is less conclusive with studies reporting mixed results. In particular, atomoxetine has been shown to improve spatial short-term memory among children in an acute, double-blind, placebo-controlled study (Gau & Shang, 2010b), but no effects were observed in a similar trial with adults (Chamberlain et al., 2011). Similarly, modafinil improved short-term spatial memory span in one acute, double-blind, placebo-controlled trial (Turner et al., 2004), but not another (Turner et al., 2005). Finally, in an acute, double-blind, placebo-controlled crossover study with healthy volunteers, guanfacine improved accuracy, but not strategy scores on the CANTAB spatial working memory task, though no studies of the effects of this medication on SWM among patients with ADHD were found (Jakala et al., 1999).

Executive Planning as Assessed by the CANTAB in ADHD. Another major deficit observed among individuals with ADHD is in domain of executive planning. With respect to the CANTAB, executive planning is assessed with the Stockings of Cambridge subtest. In a meta-analysis of six studies using the CANTAB Stockings of Cambridge task in ADHD, individuals with the disorder were showed to have deficits falling in the medium effect size range on this executive planning task (Chamberlain et al., 2011), which is similar to the effects sizes reported in previous meta-analyses of executive planning deficits in ADHD (Willcutt et al., 2005). The majority of research examining executive planning deficits associated with ADHD using the Stockings of Cambridge subtest has been conducted with children (Brophy et al., 2002; Gau et al., 2009; Gau & Shang, 2010a; Kempton et al., 1999; Rhodes et al., 2005) with all but one study showing significant deficits in accuracy among children with ADHD (Goldberg et al., 2005). However, a single study of executive planning in adults with ADHD, using the Stockings of Cambridge subtest, found significant deficits in executive planning among adults with ADHD when compared to typically developing adults (McLean et al., 2004).

Among individuals with ADHD, performance on the Stockings of Cambridge test appears to be relatively unaffected by medication with the majority of studies being conducted with methylphenidate (Bedard et al., 2004; Coghill et al., 2007; Mehta et al., 2004; Rhodes et al., 2006). However, a single acute, double-blind, placebo-controlled crossover study of adults did report increased accuracy on the Stockings of Cambridge test following treatment with 200 mg of modafinil (Turner et al., 2004). Medication also appears to have an effect on accuracy on this task among healthy adults. Specifically, guanfacine and modafinil have both been shown to enhance performance accuracy and planning on the Stockings of Cambridge in healthy adults in acute, double-blind, placebo-controlled crossover and parallel designed studies (Jakala et al., 1999; Muller et al., 2005; Turner et al., 2003; Winder-Rhodes et al., 2009).

Attentional Set-Shifting as Assessed by the CANTAB in ADHD. The CANTAB assesses attentional set-shifting using the IED task. According to a recent meta-analysis of eight studies, individuals with ADHD perform significantly worse on the IED task than typically developing controls with respect to overall number of errors (Chamberlain et al., 2011). This was a somewhat smaller effect than others that have been reported previously, with deficits in attentional set-shifting in the medium effect size range for individuals with ADHD when compared to typically developing controls on the Wisconsin Card Sorting Task (Willcutt et al., 2005).

Five studies comparing children with ADHD with typically developing controls reported reduced accuracy and stages completed among the ADHD group with the greatest differences observed in the final stages of the task, particularly the extra-dimensional shifting stages (Gau et al., 2009; Gau & Shang, 2010a; Kempton et al., 1999; Rhodes et al., 2005; Vance et al., 2003). Additionally, Brophy et al. (2002) reported that while hard-to-manage children showed intact set-shifting, compared to typically developing children, they made more perseverative and rule-based errors, suggesting that they performed qualitatively differently from typically developing youth. In contrast, one study reported no such difference among children with and without ADHD (Goldberg et al., 2005). Similarly, two such studies have been conducted with adults with one reporting significant differences between ADHD and controls on accuracy during the extra-dimensional shifting stage (McLean et al., 2004) and one reporting no such differences (Chamberlain et al., 2007). It should also be noted that healthy, unaffected siblings of children with ADHD also display impaired set-shifting abilities, making more extra-dimensional shift errors, suggesting that there may be a genetic component to this specific measure of executive functioning (Gau & Shang, 2010a). Thus, it may be that while overall set-shifting ability remains somewhat intact among individuals with ADHD, there may be both qualitative and quantitative differences in performance on this task at greater levels of difficulty, and these differences may serve as an endophenotype of ADHD.

This set-shifting deficit observed among individuals with ADHD also appears to be relatively unaltered by salient mediations. Only one acute, placebo-controlled parallel study of children with ADHD treated with methylphenidate reported improved accuracy and stages completed following treatment (Mehta et al., 2004), while several others reported no such effect using similar study designs using both children and adults treated with atomoxetine, methylphenidate, and modafinil (Chamberlain et al., 2007; Coghill et al., 2007; Rhodes et al., 2006; Turner et al., 2005). Additionally, no studies have reported significant improvements among healthy volunteers treated with atomoxetine, guanfacine, methylphenidate, or modafinil in any of the nine published placebo-controlled studies reviewed (Elliot et al., 1997; Garcia-Villamisar & Hughes, 2007; Jakala et al., 1999; Muller et al., 2005; Randall et al., 2005; Randall, Fleck, Shneerson, & File, 2004; Randall, Shneerson, Plaha, & File, 2003; Rogers, 1999; Turner et al., 2003, 2004), and in fact, both methylphenidate and modafinil have been shown to impair extra-dimensional set-shifting in healthy adult volunteers (Randall et al., 2004; Rogers, 1999). As such, it may be that this domain of executive functioning is under control of a unique system of neurotransmitter control other than the noradrenergic or dopaminergic systems that these drugs typically target.

Other Domains of Cognitive Functioning Assessed by the CANTAB in ADHD. In addition to those aspects of executive functioning assessed by the CANTAB described above, assessments of sustained attention/vigilance are also relevant. In particular, the CANTAB rapid visual information processing test assesses sustained attention and vigilance, which is similar to several other continuous performance tasks. Two studies found that individuals with ADHD were less accurate at identifying targets (more commission errors) than typically developing controls (Bedard et al., 2004; Turner et al., 2004). Furthermore, it was also reported that both methylphenidate and modafinil reduced these errors in individuals with ADHD, while atomoxetine did not (Bedard et al., 2004; Chamberlain et al., 2006; Turner et al., 2004). Additionally, modafinil has been shown to enhance performance, by increasing accuracy of identifying targets, among healthy adult volunteers (Randall et al., 2005).

Summary of Executive Functioning Assessed by the CANTAB in ADHD. When assessed with the CANTAB, children with ADHD have been shown to have deficits in inhibitory control and working memory, as well as more secondary deficits in executive planning, strategic flexibility/attentional set-shifting, and sustained attention (Chamberlain et al., 2011). Additionally, these findings are congruent with neuroimaging studies which have reported disruptions to the frontostriatal circuitry and specifically reduced volumes in the dorsolateral prefrontal cortex, caudate, and cerebellum (Seidman et al., 2005; Valera et al., 2007). Finally, it has been shown that specific salient drugs have an ameliorating effect on executive functioning deficits in ADHD as assessed by the CANTAB. Specifically, methylphenidate (Ritalin) has been demonstrated to improve working memory, modafinil improves planning, and methylphenidate, modafinil, and atomoxetine all improve performance on tests of inhibition (Chamberlain et al., 2011).

Evaluating Executive Functioning Using CANTAB in Autism Spectrum Disorders

Turning our attention to autism and autism spectrum disorders, these are a class of neurodevelopmental disorders characterized by impaired social interaction and communication, as well as restricted interests and repetitive behaviors (American Psychological Association, 2000). The autism spectrum disorders include autism, Asperger's syndrome, which lacks the delays in cognitive development and language often observed in autism, and pervasive developmental disorder-not otherwise specified (American Psychological Association, 2000). Overt symptoms of these disorders gradually begin around age 6 months and become well established by age 2 or 3 years (American Psychological Association, 2000). Autism is associated with significant impairment and present in less than 1 % of all youths with a 4:1 male-to-female ratio. However, the prevalence of Asperger's syndrome is somewhat debated, as it is difficult to distinguish from high-functioning autism, though it is also believed to be less than 1 % of all youths with a male-to-female ratio ranging from 1.6:1 to 4:1 (Mattila et al., 2007).

Executive functioning deficits in these disorders have been more often characterized by high-level cognitive, nonspatial problems, such as deficits planning and set-shifting, as well as secondary deficits in inhibition and working memory. Additionally, milder versions of these deficits have also been observed among first-degree, healthy relatives of individuals with ASDs, and these deficits have been linked to several ASD-specific behaviors, including perseverative focus on details and the display of highly specific interests. Furthermore, numerous studies of preschoolers have demonstrated that individual differences in executive functioning are correlated with individual differences in theory of mind, or the ability to attribute mental states to the self and others, which is a well-established deficit in individuals with ASD (Hughes & Graham, 2002; Perner & Lang, 1999). Taken together, these findings suggest that executive dysfunction in these domains may serve as an endophenotype of ASD. We now turn our attention to specific domains of executive functioning which have been assessed using CANTAB in individuals with autism and autism spectrum disorders.

Executive Planning as Assessed by the CANTAB in ASD. Executive planning refers to a complex, dynamic sequence of planned actions that must be constantly monitored, reviewed, and updated. The CANTAB assesses executive planning with the Stockings of Cambridge test. Individuals with autism tend to perform worse than control groups on this task, including groups composed of individuals with ADHD, Tourette's syndrome, cognitive age-matched individuals with other developmental disabilities, and typically developing controls (Happe, Booth, Charlton, & Hughes, 2005; Hill, 2006; Hughes, Russell, & Robbins, 1994; Ozonoff et al., 2004; Sinzig, Morsch, Bruning, Schmidt, & Lehmkuhl, 2008; Witwer & Lecavalier, 2008), suggesting that there may be unique deficits in executive planning associated with autism spectrum disorders that are not present in other forms of neurodevelopmental disorders or psychopathology (Ozonoff et al., 2004; Sinzig et al., 2008). Furthermore, this impairment has been shown to be present in children and adolescents with autism and to be sustained over time in both cross-sectional and longitudinal studies of individuals with autism (Bramham et al., 2009; Garcia-Villamisar & Hughes, 2007). Additionally, impaired performance on the Stockings of Cambridge task has been observed among first-degree relatives of individuals with autism spectrum disorders, suggesting that deficits in executive planning may serve as an endophenotype of the disorder (Hughes & Graham, 2002; Hughes, Plumet, & Leboyer, 1999).

Attentional Set-Shifting as Assessed by the CANTAB in ASD. The CANTAB assesses attentional set-shifting using the IED task. In general, individuals with autism spectrum disorders tend to show deficits in attentional set-shifting and cognitive flexibility, as illustrated by difficulties they have with perseverative and stereotyped behaviors and interests (Hill, 2006), and individuals with autism spectrum disorders tend to respond to the IED and other tasks like it, such as the Wisconsin Card Sorting Task, with perseverative responding especially when shifting to a new rule or demand (Geurts, Corbett, & Solomon, 2009; Hill, 2006; Landa & Goldberg, 2005; Ozonoff et al., 2004). This deficit in attentional set-shifting among individuals with autism spectrum disorders has been observed in both children (Geurts et al., 2009; Landa & Goldberg, 2005; Ozonoff et al., 2004) and adults (Berger, Aerts, van Spaendonck, Cools, & Teunisse, 2003). The deficit has also been observed when individuals with ASD are compared to individuals with other neurodevelopmental disorders and typically developing controls (Hughes et al., 1994) and when matched according to age, the presence or absence of a learning disorder (Hughes et al., 1994), and/or verbal and nonverbal developmental age (Sinzig et al., 2008; Teunisse, Cools, van Spaendonck, Aerts, & Berger, 2001). As with executive planning, deficits in attentional set-shifting have also been observed as assessed on the Stockings of Cambridge task among healthy parents and siblings of individuals with autism spectrum disorders (Hughes et al., 1999; Hughes & Graham, 2002).

Inhibition as Assessed by the CANTAB in ASD. While there is a great deal of evidence to support deficits in inhibition among individuals with ADHD as assessed by the CANTAB using the SST (Chamberlain et al., 2011), as well as a number of other assessments of inhibition (Boonstra et al., 2005; Lijffijt et al., 2005; Willcutt et al., 2005), the picture is somewhat less clear in individuals with autism spectrum disorders. Specifically, inhibition (i.e., stop signal reaction time) on the SST has been shown to be relatively intact among children with autism spectrum disorders when compared both to children with ADHD and typically developing children (Corbett, Constantine, Hendren, Rocke, & Ozonoff, 2009; Edgin & Pennington, 2005; Geurts et al., 2009). However, children with autism spectrum disorders have been shown to display impaired vigilance and faster overall reaction times (i.e., both on go and stop trials) than typically developing children (Corbett et al., 2009; Edgin & Pennington, 2005). Finally, qualitative data suggest that children with autism spectrum disorders tend to view the rules of this task as somewhat arbitrary and have been reported to develop maladaptive strategies when self-reported understanding of the goals of this task has been assessed (Hill, 2006).

Working Memory as Assessed by the CANTAB in ASD. As with inhibition, there is evidence that the primary deficits observed in tasks of working memory as assessed by the CANTAB may be due to the use of maladaptive strategies or poor understanding of the rules (Steele, Minshew, Luna, & Sweeney, 2007). However, there is evidence of individual differences and heterogeneity in working memory skills both among individuals with autism spectrum disorder and their first-degree relatives (Garcia-Villamisar & Hughes, 2007). For example, healthy siblings of individuals with autism spectrum disorders have been shown to have superior SSPs, but there were no observed differences when compared to typically developing individuals with respect to working memory performance per se (Hughes et al., 1999). Others have shown improved visuospatial functioning among all autism spectrum disorder probands (Lajiness & Menard, 2008).

Summary of Executive Functioning Assessed by the CANTAB in ASD. Autism and autism spectrum disorders have been characterized by high-level cognitive, nonspatial problems, such as deficits in planning and set-shifting, as well as secondary deficits in inhibition and working memory. These findings are congruent with neuroimaging studies that have reported both structural abnormalities in the prefrontal cortexes of individuals with autism and reduced dorsolateral and ventromedial prefrontal cortex activity during these tasks among individuals with autism when compared to typically developing controls (Berger et al., 2003; Hill, 2006; Ozonoff et al., 2004). Additionally, milder versions of these deficits have been observed among first-degree, healthy relatives of individuals with ASDs, suggesting that executive dysfunction in these domains may serve as an endophenotype of ASD. Finally, it should be noted that additional research into the roles of executive functioning in distinguishing specific subtypes of autism spectrum disorders is needed.

Assessing Executive Functioning in Older Adults

Despite its application to assessment of cognitive function in patients with disorders ranging from neurologic and psychiatric to metabolic and cardiac, the CANTAB was originally developed for use with older adults and those with dementia (Robbins et al., 1994). Robbins et al. (1994) administered the CANTAB as it existed in 1994 to a large sample of healthy older participants between the ages of 55 and 80, to begin to describe the effects of age, gender, and intelligence on performance. Scores on those subtests were found to successfully differentiate between different age groups and levels of intellectual ability. Eleven performance variables (e.g., accuracy and latency scores, learning trials, and error scores) were included in a factor analysis, yielding four factors interpreted as representing general learning and memory, speed of responding, executive processes, and visual perceptual ability. The factor structure was found to remain consistent across age groups and IQ test scores, but with differential loadings across the four factors based on IQ test scores.

In a later study focused on CANTAB tests of executive function, Robbins et al. (1998) reported results from a sample of healthy older adults. Three hundred forty-one participants were administered the Spatial Working Memory, Stockings of Cambridge, and Intra/Extra Dimensional Set Shift tests as well as CANTAB tests known to demonstrate age-related decline (i.e., tests of visual memory and learning). Greatest age-related declines were seen in attentional set-shifting, where the oldest age group (75–79) made significantly more errors than the rest of the group on extra-dimensional set shifts. On the Stockings of Cambridge planning task, older adults solved fewer problems in the minimum possible steps and had significantly longer response latencies than the youngest groups. The authors conclude that their findings are consistent with neuroimaging findings that have demonstrated age-associated changes in prefrontal cortex and striatum in addition to regions of the temporal lobes.

In an attempt to replicate these findings, Rabbitt and Lowe (2000) administered the CANTAB to 162 healthy older adults between the ages of 60 and 80 years. Unlike the earlier study, they found that tests of the CANTAB that are established measures of temporal lobe function (e.g., paired associates learning) were more age-sensitive than the frontal tasks. For example, scores on the IED and Stockings of Cambridge tests did not correlate with age, while in a linear regression analysis, age predicted performance on tests of visual memory. The authors conclude that the so-called frontal tests of the CANTAB are less sensitive to changes of normal aging than the tests that assess memory functions.

Disorders of Frontal Lobe Function and the CANTAB

In describing aspects of executive function that are sensitive to prefrontal cortical dysfunction, Robbins (1996) cites psychometric and neuroimaging evidence to demonstrate the ability of relevant CANTAB subtests to further characterize deficits in planning, working memory, and attentional set-shifting. For example, patients with frontal lobe deficits exhibit impaired performance with extra-dimensional set-shifting on the CANTAB Intra-/Extra-Dimensional Set Shift test. Specifically, the impairment has been attributed to a specific failure of response inhibition based on manipulation of test instructions to induce perseveration or learned irrelevance conditions, with frontal patients making more perseverative errors (Owen et al., 1993).

The two variations of the Tower of London test, i.e., Stockings of Cambridge and OTS, are thought to rely on different aspects of planning (actual motor sequencing vs. mental imagery). Functional neuroimaging studies in healthy controls have lent support for the different demands placed by these two versions of a planning task; while both activated dorsolateral prefrontal cortical areas, the "mental" task activations were greater on the right, while the "motor" format placed greater demands on the left frontal regions (Owen, Doyon, Petrides, & Evans, 1996). These findings have been posited to be a result of differential demands of the tasks on SWM and/or memory sequencing (Robbins, 1996).

In an earlier study of planning and working memory, 26 patients with frontal lobe excisions were compared with age-matched controls on a subset of CANTAB tests (Owen, Downes, Sahakian, Polkey, & Robbins, 1990). An average of 3 years following surgery, patients were found to make significantly more search errors both within and between trials and had less successful strategies on a test of SWM. On the Stockings of Cambridge planning task, they took more moves and solved fewer problems in the minimum possible moves than their healthy counterparts. Further, they were significantly slower to execute a response after a first move, raising the possibility of impulsivity in initiating response prior to constructing a successful solution. These same patients were unimpaired relative to the healthy controls on a test of short-term spatial memory. The authors identify "strategy deficits" as a key component of performance on both the working memory and planning tasks.

An examination of cognitive test performance of patients with mild frontotemporal dementia by Rahman, Sahakian, Hodges, Rogers, and Robbins (1999) further elucidates the heterogeneity of executive functions in the prefrontal cortex. Eight patients with FTD were compared with age-matched healthy controls on a range of tests of memory and executive function. They found that even in the relatively mild stages of the disease, impairments were revealed that might not be demonstrated by more traditional neuropsychological test batteries. Specifically, even in this small sample patients showed selective deficits as illustrated by performance on a decision-making task (CGT), in which they made poorer risk adjustments in response to changing odds of success. Further, there was some evidence of impairment, in attentional set-shifting, although the findings of Owen et al. (1990) with regard to increased errors at the extra-dimensional shift stage were not replicated. However, the FTD patients in this study did not differ from healthy controls in their performance on tests of pattern or spatial recognition memory, spatial span, working memory (SWM), or planning (OTS). The authors conclude that these findings are consistent with evidence from neuroimaging studies which suggests a progression of pathology in frontotemporal dementia from early orbitofrontal or ventromedial to more lateral prefrontal regions.

Evaluating Executive Functioning Using the CANTAB in Huntington's Disease

Since the discovery of the Huntington's disease mutation, accurate determination of the genetic status of at-risk individuals has made possible the study of cognitive function in preclinical HD patients. Further, tracking of cognitive decline from the earliest stages can be related to the known pattern of progression of neuropathology, from early involvement of the dorsal caudate nucleus to gradual deterioration throughout the frontostriatal system in a dorsal to ventral, anterior to posterior, and medial to lateral direction (Watkins et al., 2000). Tests of the CANTAB have been widely used to trace the progression of cognitive deficits in HD, adding to our understanding of the neural underpinnings of specific executive functions. Lawrence et al. (1998) compared HD mutation carriers with no movement disorders, with noncarriers on a battery of tests known to be sensitive to the early changes of the disease. As hypothesized, they found mutation carriers were more impaired in extra-dimensional shifting on a test of attentional set-shifting (IED), which they attribute to a deficit in inhibitory control as demonstrated by increased perseverative responses. Performance on tests of spatial span, spatial working memory, and spatial planning was no different between groups, suggesting a specific pattern of cognitive impairment in preclinical HD which is related to early basal ganglia dysfunction. The authors recommend the attentional set-shifting task as particularly sensitive to the earliest cognitive changes in HD, with implications for initiation of therapeutic interventions.

Watkins et al. have further delineated specific patterns of executive dysfunction in HD (Watkins et al., 2000). In a comparison of patients with mild HD versus age-matched healthy controls, patients had longer response latencies and made more errors on the One-Touch Stockings of Cambridge task, while on a decision-making test (CGT), they were slower to respond but were no different than controls on size of bet or impulsivity in response to changing risks and rewards. The findings that HD patients were impaired in planning but not in decision-making are in keeping with the known progression of pathology from early dorsal to later ventral caudate involvement. Previous work has shown the One-Touch Tower of London test to be sensitive to dorsolateral prefrontal cortical damage, while impaired decision-making has been associated with orbitofrontal cortical lesions (Watkins et al., 2000). The authors relate their findings to those of Rahman et al. (1999) with FTD patients to illustrate the dissociation between deficits in decision-making and planning in these patient groups, consistent with the involvement of dorsolateral PFC and relative preservation of orbitofrontal circuitry in early HD and the reverse in FTD.

In a similar effort to demonstrate qualitative differences in cognitive decline consistent with known neuropathology, patients with Huntington's disease and Alzheimer's disease were matched for level of dementia and compared on tests of visual memory and executive function (Lange, Sahakian, Quinn, & Robbins, 1995). Predictably, patients with HD had worse performance on tests of executive function including Spatial Working Memory, planning, and set-shifting. However they were also significantly more impaired relative to patients with AD on tests less clearly dependent on executive function, including tests of visual pattern recognition, spatial span, and visuospatial paired associates learning. While these patients with later-stage HD demonstrated fairly wide ranging cognitive impairment; the pattern of deficits was nevertheless qualitatively different from that of AD patients at a similar stage of disease progression.

In one of the few longitudinal studies of cognitive decline in HD, Ho et al. (2003) followed a sample of patients with mild to moderate disease for at least 3 years. While general cognitive ability remained unchanged, patterns of decline in executive function were identified, such that planning and set-shifting deteriorated over time. Specifically, measures of errors on the One-Touch Tower of London task and response latencies on the IED task were sensitive in detecting progression of cognitive impairment in this sample. Interestingly, similar decline in performance on the Wisconsin Card Sorting Test, a widely used test of executive function, was not found, leading the authors to suggest that the practice effects of learning a strategy make the WCST less useful in longitudinal assessment. Finally, they note that delineating the component features of executive processes relies on tests capable of finer gradations of measurement that are sensitive to change over time. Certainly, the growing body of evidence describing the progression of cognitive decline has been consistent with the known frontostriatal pathology of Huntington's disease.

Summary and Conclusions

The tests of the CANTAB have been used extensively to assess cognitive performance in a wide range of neurodegenerative and neurodevelopmental, psychiatric, and metabolic disorders, among others. The CANTAB has been shown to be useful in identifying and evaluating discrete components of important cognitive domains including those of memory, attention, and executive function. In doing so, it has provided substantial evidence for the neural underpinnings of specific cognitive functions (Fray, Robbins, & Sahakian, 1996); these relationships will no doubt be further elucidated by advances in neuroimaging techniques. By comparing performances of different diagnostic groups, dissociations among specific cognitive abilities have been described which help characterize the relationship between neurological structure and function and the effects of different pathologies on those relationships.

The theoretical basis for the development of the CANTAB, set as it was in the context of animal models of neuropsychological function, has served to yield a battery that can be used across the life span and across all levels of ability. It is sensitive to early changes in cognition, to cognitive decline, and to differences among distinct neuropathological conditions. Further, extensive published research attests to its utility in describing normal cognitive development as well as in measuring pharmacological treatment effects in multiple disease states. The potential for widespread application of the CANTAB to clinical rather than research settings has not, however, been fully investigated. It is possible that as familiarity with computers becomes truly universal across all age cohorts, current barriers, both perceived and actual, to their use with older populations will dissipate. As computerized tests become more ubiquitous in the clinical setting, batteries such as the CANTAB that are based in theory and extensively researched will offer a viable option for comprehensive assessment of cognitive function.

References

American Psychological Association. (2000). *Diagnostic and statistical manual of mental disorders* (4th ed.). Washington, DC: American Psychological Association.

Anderson, V., Anderson, P., Northam, E., Jacobs, R., & Catroppa, C. (2001). Development of executive functions through late childhood and adolescence: An Australian sample. *Developmental Neuropsychology, 20*, 385–406.

Ardila, A., & Roselli, M. (1994). Development of language, memory, and visuospatial abilities in 5- to 12-year-old children using a neuropsychological battery. *Developmental Neuropsychology, 10*, 97–120.

Aron, A. R., Dowson, J. H., Sahakian, B. J., & Robbins, T. W. (2003). Methylphenidate improves response inhibition in adults with attention-deficit/hyperactivity disorder. *Biological Psychiatry, 54*, 1465–1468.

Barnett, R., Maruff, P., Vance, A., Luk, E. S. L., Costin, J., Wood, C., et al. (2001). Abnormal executive function in attention deficit hyperactivity disorder: The effect of stimulant medication and age on spatial working memory. *Psychological Medicine, 31*, 1107–1115.

Bedard, A. C., Martinussen, R., Ickowicz, A., & Tannock, R. (2004). Methylphenidate improves visual-spatial memory in children with attention-deficit/hyperactivity disorder. *Journal of the American Academy of Child & Adolescent Psychiatry, 43*, 260–268.

Berger, H. J. C., Aerts, F. H. T., van Spaendonck, K. P. M., Cools, A. R., & Teunisse, J. P. (2003). Central coherence and cognitive shifting in relation to social improvement in high-functioning young adults with autism. *Journal of Clinical and Experimental Neuropsychology, 25*, 502–511.

Boonstra, A. M., Oosterlaan, J., Sergeant, J. A., & Buitelaar, J. K. (2005). Executive functioning in adult ADHD: A meta-analytic review. *Psychological Medicine, 35*, 1097–1108.

Bramham, J., Ambery, F., Young, S., Morris, R., Russell, A., Xenitidis, K., et al. (2009). Executive functioning differences between adults with attention deficit hyperactivity disorder and autistic spectrum disorder in initiation, planning, and strategy formation. *Autism, 13*, 245–264.

Brophy, M., Taylor, E., & Hughes, C. (2002). To go or not to go: Inhibitory control in "hard to manage" children. *Infant and Child Development, 11*, 125–140.

Buchanan, T. (2002). Online assessment: Desirable or dangerous? *Professional Psychology: Research and Practice, 33*, 148–154.

Butcher, J. M., Perry, J. M., & Atlis, M. M. (2000). Validity and utility of computer-based test interpretation. *Psychological Assessment, 12*, 6–18.

Case, R. (1992). The role of frontal lobes in the regulation of cognitive development. *Brain and Cognition, 20*, 51–73.

Chamberlain, S. R., DelCampo, N., Dowson, J. H., Muller, U., Clark, L., Robbins, T. W., et al. (2007). Atomoxetine improved response inhibition in adults with attention deficit hyperactivity disorder. *Biological Psychiatry, 62*, 977–984.

Chamberlain, S. R., Hampshire, A., Muller, U., Rubia, K., DelCampo, N., Craig, K., et al. (2009). Atomoxetine modulates right inferior frontal activation during inhibitory control: A pharmacological functional magnetic resonance imaging study. *Biological Psychiatry, 65*, 550–555.

Chamberlain, S. R., Muller, U., Blackwell, A. D., Clark, L., Robbins, T. W., & Sahakian, B. J. (2006). Neurochemical modulation of response inhibition and probabilistic learning in humans. *Science, 311*, 861–863.

Chamberlain, S. R., Robbins, T. W., Winder-Rhodes, S. E., Muller, U., Sahakian, B. J., Blackwell, A. D., et al. (2011). Translational approaches to frontostriatal dysfunction in attention-deficit/hyperactivity disorder using a computerized neuropsychological battery. *Biological Psychiatry, 69*, 1192–1203.

Clark, L., Blackwell, A. D., Aron, A. R., Turner, D. C., Dowson, J. H., Robbins, T. W., et al. (2007). Association between response inhibition and working memory in adult ADHD: A link to right frontal cortex pathology? *Biological Psychiatry, 16*, 1395–1401.

Coghill, D. R., Rhodes, S. M., & Matthews, K. (2007). The neuropsychological effects of chronic methylphenidate on drug-naïve boys with attention-deficit/hyperactivity disorder. *Biological Psychiatry, 62*, 954–962.

Corbett, B. A., Constantine, L. J., Hendren, R., Rocke, D., & Ozonoff, S. (2009). Examining executive functioning in children with autism spectrum disorder, attention deficit hyperactivity disorder and typical development. *Psychiatry Research, 166*, 210–222.

DeLuca, C. R., Wood, S. J., Anderson, V., Buchanan, J., Proffitt, T. M., Mahony, K., et al. (2003). Normative data from the Cantab. I: Development of executive function over the lifespan. *Journal of Clinical and Experimental Neuropsychology, 25*, 242–254.

DeVito, E., Blackwell, A. D., Kent, L., Ersche, K. D., Clark, L., Salmond, C. H., et al. (2008). The effects of methylphenidate on decision making in attention-deficit/hyperactivity disorder. *Biological Psychiatry, 64*, 636–639.

Doniger, G. M., Dwolatzky, T., Zucker, D. M., Chertkow, H., Crystal, H., & Schweiger, A. (2006). Computerized cognitive testing battery identifies mild cognitive impairment and mild dementia even in the presence of depressive symptoms. *American Journal of Alzheimer's Disease and Other Dementias, 21*, 28–36.

Dowson, J. H., McLean, A., Bazanis, E., Toone, B. K., Young, S., Robbins, T. W., et al. (2004). Impaired spatial working memory in adults with attention-deficit/hyperactivity disorder: Comparisons with performance in adults with borderline personality disorder and in control subjects. *Acta Psychiatrica Scandinavica, 110*, 45–54.

Edgin, J. L., & Pennington, B. F. (2005). Spatial cognition in autism spectrum disorders: Superior, impaired, or just intact? *Journal of Autism and Developmental Disorders, 35*, 729–745.

Elliot, R., Sahakian, B. J., Matthews, K., Bannerjea, A., Rimmer, J., & Robbins, T. W. (1997). Effects of methylphenidate on spatial working memory and planning in healthy young adults. *Psychopharmacology, 131*, 196–206.

Espy, K. A. (1997). The shape of school: Assessing executive function in preschool children. *Developmental Neuropsychology, 13*, 495–499.

Fray, P. J., Robbins, T. W., & Sahakian, B. J. (1996). Neuropsychiatry applications of CANTAB. *International Journal of Geriatric Psychiatry, 11*, 329–336.

Garcia-Villamisar, D., & Hughes, C. (2007). Supported employment improves cognitive performance in adults with autism. *Journal of Intellectual Disability Research, 51*, 142–150.

Gau, S. S., Chiu, C. D., Shang, C. Y., Cheng, A. T., & Soong, W. T. (2009). Executive functions in adolescence among children with attention-deficit/hyperactivity disorder in Taiwan. *Journal of Developmental and Behavioral Pediatrics, 30*, 525–534.

Gau, S. S., & Shang, C. Y. (2010a). Executive functions as endophenotypes in ADHD: Evidence from the Cambridge Neuropsychological Test Battery (CANTAB). *Journal of Child Psychology and Psychiatry, 51*, 838–849.

Gau, S. S., & Shang, C. Y. (2010b). Improvement of executive functions in boys with attention deficit hyperactivity disorder: An open-label follow-up study with once-daily atomoxetine. *The International Journal of Neuropsychopharmacology, 13*, 243–256.

Geurts, H. M., Corbett, B. A., & Solomon, M. (2009). The paradox of cognitive flexibility in autism. *Trends in Cognitive Sciences, 13*, 74–82.

Goldberg, M. C., Mostofsky, S. H., Cutting, L. E., Mahone, E. M., Astor, B. C., Denckla, M. B., et al. (2005). Subtle executive impairment in children with autism and children with ADHD. *Journal of Autism and Developmental Disorders, 35*, 279–293.

Golden, C. J. (1981). The Luria-Nebraska Children's Battery: Theory and formulation. In G. W. Hynd & G. E. Obrzut (Eds.), *Neuropsychological assessment and the school-aged child* (pp. 277–302). New York: Grune & Stratton.

Gropper, R. J., & Tannock, R. (2009). A pilot study of working memory and academic achievement in college students with ADHD. *Journal of Attention Disorders, 12*, 574–581.

Happe, F., Booth, R., Charlton, R., & Hughes, C. (2005). Executive function deficits in autism spectrum disorders and attention-deficit/hyperactivity disorder: Examining profiles across domains and ages. *Brain and Cognition, 61*, 25–39.

Hill, E. L. (2006). Evaluating the theory of executive dysfunction in autism. *Developmental Review, 24*, 189–233.

Ho, A. K., Sahakian, B. J., Brown, R. G., Barker, R. A., Hodges, J. R., Ane, M., et al. (2003). Profile of cognitive progression in early Huntington's disease. *Neurology, 61*, 1702–1706.

Hoare, P., & Sevar, K. (2007). The effect of discontinuation of methylphenidate on neuropsychological performance in children with attention deficit hyperactivity disorder. *Psychiatry Investigation, 4*, 76–83.

Hughes, C., & Graham, A. (2002). Measuring executive functions in childhood: Problems and solutions. *Adolescent Mental Health, 3*, 131–142.

Hughes, C., Plumet, M. H., & Leboyer, M. (1998). Toward a cognitive phenotype for autism: Increased prevalence of executive dysfunction and superior spatial span amongst siblings of children with autism. *Journal of Child Psychology and Psychiatry, 40*, 1–14.

Hughes, C., Plumet, M. H., & Leboyer, M. (1999). Towards a cognitive phenotype for autism: Increased prevalence of executive dysfunction and superior spatial span amongst siblings of children with autism. *Journal of Child Psychology and Psychiatry, 40*, 705–718.

Hughes, C., Russell, J., & Robbins, T. W. (1994). Evidence for executive dysfunction in autism. *Neuropsychologia, 32*, 477–492.

Jakala, P., Riekkinen, M., Sirvio, J., Koivisto, E., Kejonen, K., Vanhanen, M., et al. (1999). Guanfacine, but not clonidine, improves planning and working memory performance in humans. *Neuropsychopharmacology, 20*, 460–470.

Kempton, S., Vance, A., Maruff, P., Luk, E., Costin, J., & Pantelis, C. (1999). Executive function and attention deficit hyperactivity disorder: Stimulant medication and better executive function performance in children. *Psychological Medicine, 29*, 527–538.

Klingberg, T. (1997). Concurrent performance of two working memory tasks: Potential mechanisms of interference. *Cerebral Cortex, 8*, 593–601.

Klingberg, T., Forssberg, J., & Westerberg, H. (2002). Increased brain activity in frontal and parietal cortex underlies the development of visuospatial working memory capacity during childhood. *Journal of Cognitive Neuroscience, 14*, 1–10.

Lajiness, R., & Menard, P. (2008). Brief report: An autistic spectrum subtype revealed through familial psychopathology coupled with cognition in ASD. *Journal of Autism and Developmental Disorders, 38*, 982–987.

Landa, R. J., & Goldberg, M. C. (2005). Language, social, and executive functions in high functioning autism: A continuum of performance. *Journal of Autism and Developmental Disorders, 35*, 557–573.

Lange, K. W., Sahakian, B. J., Quinn, N. P., & Robbins, T. W. (1995). Comparison of executive and visuospatial memory function in Huntington's disease and dementia of Alzheimer type matched for degree of dementia. *Journal of Neurology, Neurosurgery, and Psychiatry, 58*, 598–606.

Lawrence, A. D., Hodges, J. R., Rosser, A. E., Kershaw, A., French-Constant, C., Rubinsztein, D. C., et al. (1998). Evidence for specific cognitive deficits in preclinical Huntington's disease. *Brain, 121*, 1329–1341.

Lijffijt, M., Kenemans, J. L., Leon, J., Verbaten, M. N., & van Engeland, J. (2005). A meta-analytic review of stopping performance in attention-deficit/hyperactivity disorder: Deficient inhibitory motor control? *Journal of Abnormal Psychology, 114*, 216–222.

Lowe, C., & Rabbitt, P. (1998). Test/re-test reliability of the CANTAB and ISPOCD neuropsychological batteries: Theoretical and practical issues. *Neuropsychologia, 36*, 915–923.

Luciana, M., & Nelson, C. A. (1998). The functional emergence of prefrontally-guided working memory systems in four- to eight-year-old children. *Neuropsychologia, 36*, 273–293.

Manly, T., Anderson, V., Nimmo-Smith, I., Turner, A., Watson, P., & Robertson, I. H. (2001). The differential assessment of children's attention: The test of everyday attention for children (TEA-Ch), normative sample and ADHD performance. *Journal of Child Psychology and Psychiatry, 42*, 1065–1081.

Martinussen, R., Hayden, J., Hogg-Johnson, S., & Tannock, R. (2005). A meta-analysis of working memory impairments in children with attention-deficit/hyperactivity disorder. *Journal of the American Academy of Child and Adolescent Psychiatry, 44*, 377–384.

Mattila, M. L., Kielinen, M., Jussila, K., Linna, S. L., Bloigu, R., Ebeling, H., et al. (2007). An epidemiological and diagnostic study of Asperger syndrome according to four sets of diagnostic criteria. *Journal of the American Academy of Child and Adolescent Psychiatry, 46*, 636–646.

McLean, A., Dowson, J. H., Toone, B. K., Young, S., Bazanis, E., Robbins, T. W., et al. (2004). Characteristic neurocognitive profile associated with adult attention-deficit/hyperactivity disorder. *Psychological Medicine, 34*, 681–692.

Mehta, M., Calloway, P., & Sahakian, B. J. (2000). Amelioration of specific working memory deficits by methylphenidate in a case of adult attention deficit hyperactivity disorder. *Journal of Psychopharmacology, 14*, 299–302.

Mehta, M., Goodyear, I. M., & Sahakian, B. J. (2004). Methylphenidate improves working memory and set-shifting in ADHD: Relationships to baseline memory capacity. *Journal of Child Psychology and Psychiatry, 45*, 293–305.

Muller, U., Clark, L., Lam, M. L., Moore, R. M., Murphy, C. L., & Richmond, N. K. (2005). Lack of effects of guanfacine on executive and memory functions in healthy male volunteers. *Psychopharmacology, 182*, 205–213.

Owen, A. M., Downes, J. J., Sahakian, B. J., Polkey, C. E., & Robbins, T. W. (1990). Planning and spatial working memory following frontal lobe lesions in man. *Neuropsychologia, 28*, 1021–1034.

Owen, A. M., Doyon, J., Petrides, M., & Evans, A. C. (1996). Planning and spatial working memory examined with positron emission tomography. *European Journal of Neuroscience, 8*, 353–364.

Owen, A. M., Roberts, A. C., Hodges, J. R., Summers, B. A., Polkey, C. E., & Robbins, T. W. (1993). Contrasting mechanisms of impaired attentional set-shifting in patients with frontal lobe damage or Parkinson's disease. *Brain, 116*, 1159–1179.

Ozonoff, S. (1997). Components of executive function in autism and other disorders. In J. Russell (Ed.), *Autism as an executive disorder* (pp. 179–211). Oxford: Oxford University Press.

Ozonoff, S., Cook, I., Coon, H., Dawson, G., Joseph, R. M., Klin, A., et al. (2004). Performance on Cambridge Neuropsychological Test Automated Battery subtests sensitive to frontal lobe function in people with autistic disorder: Evidence from the collaborative programs of excellence in autism network. *Journal of Autism and Developmental Disorders, 34*, 139–150.

Perner, J., & Lang, B. (1999). Theory of mind and executive function: Is there a developmental relationship. In S. Baron-Cohen, T. Tager-Flusberg, & D. Cohen (Eds.), *Understanding other minds: Perspectives from developmental neuroscience* (pp. 150–181). Oxford: Oxford University Press.

Rabbitt, P., & Lowe, C. (2000). Patterns of cognitive aging. *Psychological Research, 63*, 308–316.

Rahman, S., Sahakian, B. J., Hodges, J. R., Rogers, R. D., & Robbins, T. W. (1999). Specific cognitive deficits in mild frontal variant frontotemporal dementia. *Brain, 122*, 1469–1493.

Randall, D. C., Fleck, N. L., Shneerson, J. M., & File, S. E. (2004). The cognitive-enhancing properties of modafinil are limited in non-sleep-deprived middle-age volunteers. *Pharmacology and Biochemical Behavior, 77*, 547–555.

Randall, D. C., Shneerson, J. M., Plaha, K. K., & File, S. E. (2003). Modafinil affects mood but not cognitive function in healthy young volunteers. *Human Psychopharmacology, 18*, 163–173.

Randall, D. C., Viswanath, A., Bharania, P., Elsabagh, S. M., Hartely, D. E., Shneerson, J. M., et al. (2005). Does modafinil enhance cognitive performance in young volunteers who are not sleep deprived? *Journal of Clinical Psychopharmacology, 25*, 175–179.

Rhodes, S. M., Coghill, D. R., & Matthews, K. (2004). Methylphenidate restores visual memory but not working memory function in attention deficit-hyperkinetic disorder. *Psychopharmacology, 175*, 319–330.

Rhodes, S. M., Coghill, D. R., & Matthews, K. (2005). Neuropsychological effects of methylphenidate in stimulant drug-naïve boys with hyperkinetic disorder. *Psychological Medicine, 35*, 1–12.

Rhodes, S. M., Coghill, D. R., & Matthews, K. (2006). Acute neuropsychological effects of methylphenidate in stimulant drug-naïve boys with ADHD II-broader executive and nonexecutive domains. *Journal of Child Psychology and Psychiatry, 47*, 1184–1194.

Robbins, T. W. (1996). Dissociating executive functions of the prefrontal cortex. *Philosophical Transactions of the Royal Society of London, 351*, 1463–1471.

Robbins, T. W., James, M., Owen, A. M., Sahakian, B. J., Lawrence, A. D., McInnes, L., et al. (1998). A study of performance on tests from the CANTAB battery sensitive to frontal lobe dysfunction in a large sample of normal volunteers: Implications for theories of executive functioning and cognitive aging. *Journal of the International Neuropsychological Society, 4*, 474–490.

Robbins, T. W., James, M., Owen, A. M., Sahakian, B. J., McInnes, L., & Rabbitt, P. (1994). Cambridge Neuropsychological Test Automated Battery (CANTAB): A factor analytic study of a large sample of normal elderly volunteers. *Dementia, 5*, 266–281.

Rogers, S. J. (1999). Executive functions in young children with autism. *Child Development, 70*, 817–832.

Sahakian, B. J. (1990). Computerized assessment of neuropsychological function in Alzheimer's disease and Parkinson's disease. *International Journal of Geriatric Psychiatry, 5*, 211–213.

Sahakian, B. J., & Owen, A. M. (1992). Computerized assessment in neuropsychiatry using CANTAB: Discussion paper. *Journal of the Royal Society of Medicine, 85*, 399–400.

Schlegel, R. E., & Gilliland, K. (2007). Development and quality assurance of computer-based assessment batteries. *Archives of Clinical Neuropsychology, 22S*, S49–S61.

Seidman, L., Biederman, J., Monuteaux, M. C., & Doyle, A. E. (2005). Learning disabilities and executive dysfunction in boys with attention deficit hyperactivity disorder. *Neuropsychology, 15*, 544–556.

Sinzig, J., Morsch, D., Bruning, N., Schmidt, M. H., & Lehmkuhl, G. (2008). Inhibition, flexibility, working memory, and planning in autism spectrum disorders with and without comorbid ADHD-symptoms. *Child and Adolescent Psychiatry and Mental Health, 2*, 1–12.

Steele, S. D., Minshew, N., Luna, B., & Sweeney, J. A. (2007). Spatial working memory deficits in autism. *Journal of Autism and Developmental Disorders, 37*, 605–612.

Stuss, D. T. (1992). Biological and psychological development of executive functions. *Brain and Cognition, 20*, 8–23.

Teunisse, J. P., Cools, A. R., van Spaendonck, K. P. M., Aerts, F. H. T., & Berger, H. J. C. (2001). Cognitive styles in high-functioning adolescents with autistic disorder. *Journal of Autism and Developmental Disorders, 31*, 55–66.

Turner, D. C., Blackwell, A. D., Dowson, J. H., McLean, A., & Sahakian, B. J. (2005). Neurocognitive effects of methylphenidate in adults attention-deficit/hyperactivity disorder. *Psychopharmacology, 178*, 286–295.

Turner, D. C., Clark, L., Dowson, J. H., Robbins, T. W., & Sahakian, B. J. (2004). Modafinil improves cognitive and response inhibition in adult attention deficit hyperactivity disorder. *Biological Psychiatry, 55*, 1031–1040.

Turner, D. C., Robbins, T. W., Clark, L., Aron, A. R., Dowson, J. H., & Sahakian, B. J. (2003). Relative lack of cognitive effects of methylphenidate in elderly male volunteers. *Psychopharmacology, 168*, 455–464.

Valera, E. M., Faraone, S. V., Murray, K., & Seidman, L. J. (2007). Meta-analysis of structural imaging findings in attention-deficit/hyperactivity disorder. *Biological Psychiatry, 61*, 1361–1369.

Vance, A. L., Maruff, P., & Barnett, R. (2003). Attention deficit hyperactivity disorder, combined type: Better executive function performance with longer-term psychostimulant medication. *Australian New Zealand Psychiatry, 37*, 570–576.

Walker, R., Husain, M., Hodgson, T. L., Harrison, J., & Kennard, C. (1998). Saccadic eye movement and working memory deficits following damage to human prefrontal cortex. *Neuropsychologia, 36*, 1141–1159.

Watkins, L. H. A., Rogers, R. D., Lawrence, A. D., Sahakian, B. J., Rosser, A. E., & Robbins, T. W. (2000). Impaired planning but intact decision making in early Huntington's disease: Implications for specific frontostriatal pathology. *Neuropsychologia, 38*, 1112–1125.

Wild, K., Howieson, D., Webbe, F., Seelye, A., & Kaye, J. (2008). The status of computerized cognitive testing in aging: A systematic review. *Alzheimer's & Dementia, 4*, 428–437.

Willcutt, E. G., Doyle, A. E., Nigg, J. T., Faraone, S. V., & Pennington, B. F. (2005). Validity of the executive function theory of attention-deficit/hyperactivity disorder: A meta-analytic review. *Biological Psychiatry, 57*, 1336–1346.

Winder-Rhodes, S. E., Chamberlain, S. R., Idris, M., Robbins, T. W., Sahakian, B. J., & Muller, U. (2009). Effects of modafinil and prazosin on cognitive and physiological functions in healthy volunteers. *Journal of Psychopharmacology, 24*, 1649–1657.

Witwer, A. N., & Lecavalier, L. (2008). Examining the validity of autism spectrum disorder subtypes. *Journal of Autism and Developmental Disorders, 38*, 1611–1624.

The Assessment of Executive Function Using the Cognitive Assessment System: Second Edition

Jack A. Naglieri and Tulio M. Otero

The purpose of this chapter is to describe how executive function (EF) can be evaluated using the Cognitive Assessment System—Second Edition (Naglieri, Das, & Goldstein, 2013a, 2013b, 2013c). We will begin with a brief discussion of the relevant history of the concept of executive function, the ways it has been conceptualized, and how it has been measured. Next we will describe how the CAS2 can be used as part of the assessment process. Next, a case study will be provided which illustrates how the CAS2 can be integrated with other assessment data for identification and treatment planning. Finally, we will provide a discussion of an executive function intervention method that has been used for improving math and reading comprehension.

The Concept of Executive Function

One of the most remarkable capacities in the human brain is its ability to reflect and direct itself. This ability is often described using the term executive function. Without a well-developed executive function (EF), the human species probably would have remained unevolved and our earliest ancestors would never have been able to develop the tools, arts, and technologies of modern civilization. We owe this amazing ability to a particular part of the brain—the frontal lobes.

The concept of executive function is inextricably linked to the functions of our frontal lobes. Early groundwork for defining the executive function system was put forth by Luria (1966). Luria proposed the existence of a system in charge of intentionality, goal formulation, the action plans leading to these goals, the identification of goal-appropriate cognitive routines, the sequential access to the routines, the temporal ordered transition from one routine to the other, and the evaluation of our actions or the outcome. Following Luria's seminal work, two broad types of cognitive operations associated to the executive system have appeared in the literature: (a) the ability to guide one's behavior by formulating strategies and then guiding our behavior through sequential action plans and (b) the ability to change our plan when the situation requires it. In order to effectively cope with such transitions, mental flexibility is required. Mental flexibility can be conceptualized as the ability to respond efficiently to unanticipated contingencies within our environment. Some researchers refer to this as the ability to shift cognitive set. Goldberg (2009) sees the executive system as critical for planning and generative processes. Fuster (1997) expanded on Luria's original conceptualizations of EF and suggested that the EF system is in charge of both external and internal (such as

J.A. Naglieri (✉)
University of Virginia, Curry School of Education,
White Post Road 6622, 20121 Centreville, VA, USA
e-mail: jnaglieri@gmail.com

T.M. Otero
The Chicago School of Professional Psychology,
Chicago, IL, USA
e-mail: TOtero@thechicagoschool.edu

logical reasoning) actions. More recently, McCloskey and Perkins (2013) have offered a model of executive function that goes beyond the generative processes and includes the idea that "trans-self-integration" processes are part of the executive system. The trans-integrative system refers to high levels of intention fueled by the desire to seek out experiences beyond typical perception of self and to experience a subjective sense of interconnectedness with all things.

Whether EF should be conceptualized as a unitary construct or several diverse functions has been a matter of considerable debate. Conceptualizations of executive function have been proposed by many researchers and clinicians. There are now more than 30 definitions for the term EF (Barkley, 2012) and at least as many as different constructs have been placed under its umbrella. For example, the Encyclopedia of Mental Disorders defines the term executive function (EF) as a set of cognitive abilities that control and regulate other abilities and behaviors and are necessary for goal-directed behavior (http://www.minddisorders.com/Del-Fi/Executive-function.html). Lezak, Howieson, Bigler, and Tranel (2012) see EF as consisting of capacities that enable a person to successfully and independently display purposeful, self-serving, and self-directed behavior. Ellioit (2003) described EF as complex cognitive processing requiring the coordination of several subprocesses to achieve a particular goal. The American Heritage Medical Dictionary (2007) defines EF as "cognitive processes that regulate an individual's ability to organize thoughts and activities, prioritize tasks, manage time efficiently, and make decisions. Impairment of executive function is seen in a range of disorders, including some pervasive developmental disorders and nonverbal learning disabilities." Banich (2009) defines EF as "… providing resistance to information that is distracting or task irrelevant, switching behavior task goals, utilizing relevant information in support of decision making, categorizing or otherwise abstracting common elements across items, and handling novel information or situations" (p. 89), and Dawson and Guare (2010) conceptualize it as "… Executive skills allow us to organize our behavior over time and override immediate demands in favor of longer-term goals" (p. 1).

Executive function develops over the course of many years. Some aspects crest in late childhood or adolescence while others advance into early adulthood as the brain continues to mature and establish connections well into adulthood. Executive function is shaped by both physical changes in the brain and by life experiences, in the classroom, and in the world at large. While critical for academic performance (Bull, Espy, & Senn, 2004; Clark, Pritchard, & Woodward, 2010; Willoughby et al., 2012), executive function is also intimately linked to emotional, behavioral, and social functioning (Kochanska, Murray, & Harlan, 2000; Schoemaker et al., 2012). In fact, it has been proposed that the construct can be dichotomized into "cool" processes that are cognitive and tapped during abstract, decontextualized situations or "hot" processes that represent affective responses to situations that are meaningful and involve regulation of affect and motivation (Zelazo, Qu, & Muller, 2004).

Current Assessment Tools for EF

Many tests have been used to evaluate executive function, most of which assess a narrow band of related executive skills. For example it is typical to use neuropsychological tests to evaluate specific topics such as goal-directed attention, impulse control, cognitive flexibility, visual planning and/or organization, and divided attention. Examples include the continuous performance test (Connors, 2000), the Cancellation subtest of the WISC-IV (Wechsler, 2003), visual attention subtests of the NEPSY-II (Korkman, Kirk, & Kemp, 2007), trail making, and the design fluency subtests of the Delis-Kaplan Executive Function System (D-KEFS) (Delis, Kaplan, & Kramer, 2001a, 2001b). Some researchers have suggested that executive functions (EFs) are best conceptualized as distinct abilities that are only loosely related, and many neuropsychologists consider working memory to be one of several disparate EFs that control cognitive performance (Blair, Zelazo, & Greenberg, 2005; Fletcher,

1996; Pennington, Bennetto, McAleer, & Roberts, 1996) while others have argued that all EFs share a common executive attention component (Blair, 2006; Duncan, Emslie, Williams, Johnson, & Freer, 1996).

If we accept that working memory is one of the various constructs that comprise EF, the question still remains as to what role it plays and what other processes it influences. Within the scientific literature, theories of executive function often differ in regard to the role working memory plays in EF. Simply stated, working memory refers to the structures and processes underlying the temporary retention and manipulation of information in support of higher cognitive tasks (Baddeley, 1986; Miyake & Shah, 1999). One known feature of WM is its limited capacity for a reduced amount of information which can be directly accessed during cognitive tasks that require active processing (Cowan, 2005; Szmalec, Verbruggen, Vandierendonck, & Kemps, 2011). Therefore, when working memory demands exceed capacity, a person is restricted to temporarily store subsets of the information and to successively update those representations as new information becomes available. Very often, however, it then becomes hard to distinguish between the older and more recent information, a behavioral phenomenon referred to as proactive interference (e.g., Jonides & Nee, 2006). Given the litany of definitions of executive function, and disagreement within and across disciplines, it is difficult to operationally define the various cognitive constructs that comprise executive function, let alone design instruments to measure them in a standardized, reliable, and valid fashion. Due to this fact, we hypothesize that this also attributes to the shortage of neuropsychological assessment tools that have attempted to measure EF, due in part to the multifactorial nature of most assessment tools currently available. This also explains the paucity of specific test batteries that directly measure executive function. Traditionally, assessment tools for executive function have focused on the adult population and typically measure EF through tasks of fluency (both visual and verbal), trail making, interference, tower, and sorting (Anderson, Northam, Hendy, & Wrennall, 2001; Baron, 2004; Isquith, Roth, & Gioia, 2010; Strauss, Sherman, & Spreen, 2006). These assessments of working memory are basic measures that are not typical of real-life tasks and are not easily generalized across settings, such as in educational environments. This can easily be conceptualized when we consider the nature of the tasks of working memory on more comprehensive assessment batteries, such as Digits Span Backward or Letter-Number Sequencing on the WISC-IV.

Isquith et al. (2010) attempted to develop a standardized measure of executive control processes, specifically working memory and inhibitory control. Their measure, the Task of Executive Control (TEC) uses a computerized, user-friendly format and is built around a cognitive neuroscience framework based upon functional neuroimaging research and the neural basis of working memory. Despite these advances in pediatric neuropsychological assessment tools, the authors of the TEC still point out that their test does not provide scores of working memory or inhibitory control. Rather, they assess these constructs *indirectly* through response accuracy, response time, and response consistency scores by increasing working memory load and inhibitory control demands as the child progresses through the assessment.

One of the best known tests of executive function is the Wisconsin Card Sorting Test (WCST), which has been referred to as a marker of executive function assessment tools (Delis et al., 2001a, 2001b). The WCST was originally developed to measure abstract reasoning and cognitive shifting abilities in response to changing environments through the use of problem-solving strategies to achieve a future goal in neurotypical adults (Berg, 1948; Grant & Berg, 1948; Luria, 1973). Welsh and Pennington (1988) identified the following constructs necessary to successfully perform the WCST: strategic planning, organized searching, utilizing environmental feedback to shift cognitive sets, directing behavior toward achieving a goal, and modulating impulsive responding. One of the best features of the WCST is the fact that it not only provides scores of success but also takes into account areas of difficulty by providing scores that measure difficulty with initiation, concept generation, failure to maintain set, perseveration, and inefficient learning across trials (inability to accept feedback). Over time, the WCST has become an increasingly popular neuropsychological evaluation tool and has been used with both adults and children in various clinical populations to assess neurological dysfunction; traditionally, within the frontal lobes, some clinicians have even referred to the WCST as a measure of frontal lobe functioning at the most primary level.

Although the WCST measures various constructs of EF, noticeably absent is working memory, researchers have been studying the role of working memory in clinical samples using the WCST for decades. This is most evident in the clinical population of schizophrenia. Gold, Carpenter, Randolph, Goldberg, and Weinberger (1997) found that impaired performance on a novel Letter-Number Working Memory Test predicted the WCST category achieved score, whereas measures of set shifting, verbal fluency, and attention were predictive of perseveration errors, suggesting that working memory may be a critical determinant of one aspect of impaired WCST performance in this population. Performance deficits associated with aging have also been identified on the WCST, and working memory is also believed to play a role in these declines over time (Fristoe, Salthouse, & Woodward, 1997). The WCST has frequently been used to assess executive functions in children with attention-deficit/hyperactivity disorder (ADHD). Mullane and Corkum (2007) compared the performance of 15 children with ADHD to 15 children of a control group (age range 6–11) on the WCST and then examined the relationship among working memory, inhibition, age, IQ, and scores from this test. Their study showed that the ADHD group made significantly more set-loss errors, and working memory was significantly correlated with perseverative errors but was also fully mediated by age and IQ.

Another classic neuropsychological assessment of executive function is the Trail Making Test (TMT). The cognitive shifting required by Part B of the TMT is a direct reflection of EF, although other neuropsychological abilities, such as psychomotor speed, visual scanning, and planning, are also required to successfully complete the test (Lezak et al., 2012). Moll, de Oliveira-Souza, Tovar, Bramati, and Andreiuolo (2002) used fMRI to assess neuroanatomical associations of a verbal adaptation of the TMT. Their study found marked asymmetry of activation in favor of the left hemisphere (likely due to the fact that this was a verbal adaptation of TMT), most notably in dorsolateral prefrontal cortex (BA 6 lateral, 44 and 46) and supplementary motor area/cingulate sulcus (BA 6 medial and 32). The intraparietal sulcus (BA 7 and 39) was bilaterally activated. These findings supported previous functional neuroimaging data, which has indicated that the dorsolateral and medial prefrontal cortices and the intraparietal sulci are associated with the regulation of cognitive flexibility, intention, and the covert execution of saccades/antisaccades. Other classic neuropsychological assessments, such as the Stroop test, WCST, and go/no-go tasks, share similar cerebral activation patterns.

Another example of an executive function assessment that does not include measures of working memory is the D-KEFS. The D-KEFS comprises nine individual assessment instruments designed to comprehensively assess higher-order cognitive functions in children and adults. These higher-order abilities are dependent on more primary neuropsychological abilities and are dependent upon attention, language, and perception and provide a foundation for more advanced cognitive processing such as problem-solving, cognitive flexibility, and abstract reasoning (Delis et al., 2001a, 2001b). Two of the nine subtests on the D-KEFS are novel and were developed by two of the test authors, while the other seven subtests were adapted from previously established clinically valid instruments or employed in prior experimental studies, such as the TMT, Stroop test, or tower tests. The D-KEFS isolates fundamental skills that may negatively impact higher-order cognitive skills and helps delineate skill deficits more precisely by identifying more clearly why examinees cannot perform the EF tasks. Similar to the WCST, no mention of working memory or its role in EF is made mention of in the examiner manual. Without working memory, however, the majority of the tasks on the D-KEFS cannot be completed successfully.

Engle (2002) discussed how measures of working memory consistently show higher correlations with measures of higher-level cognitive functions than do simple memory span tasks. He proposed that working memory is related to general fluid intelligence and executive attention. If we follow this line of thinking, then individuals with high working memory capacity should perform better on tasks requiring the inhibition of distracting information. This has been supported through various domains. Melby-Lervåg and Hulme (2012) provided several examples of this theory. These examples included research on high working memory capacity and improved performance on antisaccade tasks (Kane, Bleckley, Conway, & Engle, 2001) and high working memory capacity leading to better inhibition in the more difficult condition on a Stroop task with adults when such incongruent trials were relatively infrequent and the participant may detract attention to the goal of the task (Kane and Engle, 2003). Marcovitch, Boseovski, and Knapp (2007) found the same results on the Stroop task with children. High working memory capacity also appears to help inhibit distractions on dichotic listening tasks (Conway, Cowan, & Bunting, 2001).

Neuroimaging studies have consistently shown widespread activations in the prefrontal cortex during various forms of working memory tasks. The anterior prefrontal cortex plays a specific role in the ability to distinguish between target and nontarget stimuli during recognition and delayed working memory tasks (Leung, Gore, & Goldman-Rakic, 2005). The prefrontal cortex is also adapted to generate persistent activity that outlasts stimuli and resists distractors and has long been presumed to be the basis of working memory (Wang et al., 2006). Lesions to the frontal cortex have been believed to result in an inability to use complex information stored in working memory (Goldman-Rakic, 1991).

Deficits in executive functioning, including working memory, have also been proposed as playing an important role in ADHD and are especially related to impairments of behavioral regulation, task planning, and selective attention within this population (Klingberg et al., 2005; Mezzacappa & Buckner, 2010). Working memory also contributes to cognitive deficits observed in children with autism spectrum disorders (Kenworthy, Yerys, Anthony, & Wallace, 2008) and specific language impairments (Archibald & Gathercole, 2006).

Despite the fact that the majority of tests of executive function do not measure working memory directly, this construct is still necessary to successfully complete the majority of these tasks. Neuroimaging data clearly indicates the association between activation in the prefrontal cortex and various executive function constructs. The prefrontal cortex is also clearly associated with working memory, and it appears likely that it influences performance on assessments measuring executive function, despite the fact that this construct is often absent from definitions of EF and the assessment tools that measure it. Impaired working memory has negative influences on performance on such tasks, as shown in the examples of the WCST described above. This indicates that as new assessment tools are developed, researchers must provide clinicians with an opportunity to assess WM when examining EF, ideally in a manner that is generalizable to real-world settings. Table 12.1 provides a sample of neuropsychological measures with and without working memory requirements.

Regardless of how executive function is conceptualized, Lezak et al. (2012) and Dawson and Guare (2010) cautioned that the way tests of executive function are constructed and the demands of the instructions can limit the degree to which findings of these measures can be generalized beyond the testing situation, and several examples have been provided above. This concern raises two important topics. First, it is important that measures of executive function evaluate behaviors relevant to real-world functioning should play a role in assessment. Second,

Table 12.1 Sample neuropsychological measures of Executive Function purporting to require or not require Working Memory

Test	Example	Executive components	Involves Working Memory
Continuous performance tests	Conners' Continuous Performance tests (Connors, 2000)	Goal directed attention	No
Cancellation Tests	Visual attention subtest of the NEPSY-II (Korkman et al., 2007)	Goal directed, attention, impulse control	No
Color-Word Interference Test	Stroop Color-Word Interference Test (Golden, 1978)	Goal directed attention, set shifting, cognitive flexibility, impulse control	No
Attention Tasks	Expressive Attention subtest from the CAS2 (Naglieri et al., 2013a, 2013b, 2013c)	Focused attention, impulse control, working memory	No
Complex Figure Drawing Test	Rey Complex Figure Tests (Meyers & Meyers, 1995)	Visual planning, organization	No
Mazes	Elithorn Mazes subtest of the WISC-IV Integrated (Kaplan et al., 2004)	Visual planning, cognitive flexibility, impulse control	No
Tower Tests	Tower subtest of the NEPS and Tower of Hanoi	Visual planning, cognitive flexibility, impulse control	No
Trail Making Tests	Trail Making Subtest from the Delis-Kaplan Executive Function System (Delis et al., 2001a, 2001b)	Visual planning, goal directed attention, divided attention, set shifting	No
Planning (Planned Connections)	Planned Connections subtest from the CAS2 (Naglieri et al., 2013a, 2013b, 2013c).	Visual planning, goal directed attention, divided attention, set shifting	No
Wisconsin Card Sorting Tests	WCST 128 and 64-card versions (WCST; Grant & Berg, 1948)	Goal directed attention, set shifting, cognitive flexibility, working memory	Yes
Tasks of Executive Control (TEC)	Isquith et al. (2010)	Working memory, inhibitory control	Yes
Simultaneous Processing Tasks	Verbal-Spatial Relations subtest from the CAS2 (Naglieri et al., 2013a, 2013b, 2013c)	Simultaneous processing, working memory	Yes
Successive Processing Tasks	Sentence Repetition or Questions subtest from the CAS2 (Naglieri et al., 2013a, 2013b, 2013c)	Maintenance of order, working memory	Yes

it is also important that formal tests of executive function *should not* be highly structured, predictable, and directed by the examiner which reduces, and sometimes even eliminates, the need for planning and organization on part of the examinee. Similarly, Naglieri (1999) proposed that tests should be designed to encourage self-directed cognition and the selection, implementation, and evaluation of strategies. This was the goal of subtests designed to measure aspects of executive function on the Cognitive Assessment System (Naglieri & Das, 1997).

Description of the CAS2

The CAS2 (Naglieri et al., 2013a, 2013b, 2013c) was specifically designed to measure four neurocognitive abilities defined by the PASS theory of intelligence. Because this test was built explicitly on a specific theory of intelligence, we will briefly describe that theory and then how the CAS2 measures the PASS theory as described by Naglieri and Otero (2011).

The PASS theory of intelligence is based on a fusion of cognitive and neuropsychological constructs originally described by A.R. Luria in works such as *Higher Cortical Functions in Man* (1966, 1980) and *The Working Brain* (1973). Luria viewed the brain as a functional mosaic, the parts of which interact in different combinations to subserve cognitive processing (Luria, 1973). There is no area of the brain that functions without input from other areas. This means that cognition and behavior result from an interaction of complex brain activity across various areas. It was Luria's (1966, 1973, 1980, 1982) understanding of the functional aspects of brain structures which formed the basis for the PASS theory (Planning, Attention, Simultaneous, Successive processing), initially described by Das, Naglieri, and Kirby (1994) and operationalized by Naglieri and Das (1997) in the first and Naglieri et al. (2013a, 2013b, 2013c) in the second editions of the *Cognitive Assessment System*. These four manifestations of intelligence are more fully described in the sections that follow.

The prefrontal cortex "plays a central role in forming goals and objectives and then in devising plans of action required to attain these goals. The cognitive processes required to implement plans, coordinate these activities, and apply them in a correct order are subserved by the prefrontal cortex. Finally, the prefrontal cortex is responsible for evaluating our actions as success or failure relative to our intentions" (Goldberg, 2009, p. 23). Planning helps one to achieve goals through the development of strategies necessary to accomplish tasks for which a solution may not be initially apparent. The broad term of Planning is seen as an essential ability to all activities that require someone to figure out how to solve a problem. This includes self-monitoring and impulse control as well as making, assessment, and implementation of a plan. Thus, Planning allows for the generation of solutions, discriminating use of knowledge and skills, as well as control of Attention, Simultaneous, and Successive processes (Das, Kar, & Parrila, 1996). The essential dimension of the construct of Planning as defined by Naglieri and Das (1997) is very similar to the description of executive function provided by others (see other chapters in this volume). For example, O'Shanick and O'Shanick (1994) describe executive functions as including the abilities to formulate and set goals, assess strengths and weaknesses, plan and/or direct activities, initiate and/or inhibit behavior, monitor current activities, and evaluate results. Executive function includes abilities to formulate a goal, plan; to carry out goal-directed behaviors effectively; and to monitor and self-correct spontaneously and reliably (Lezak, 2004).

Attention is an ability that is closely connected to the orienting response. Brain structures within Luria's first functional unit, the reticular formation, allow one to focus selective attention toward a stimulus over a period of time without the loss of attention to other competing stimuli. The longer attention is needed, the more the activity necessitates vigilance. Goals and intentions related to plans provide control of Attention, while knowledge and skills play an integral part as well, especially when a learned solution to a problem is employed. In such instances, executive function is reduced and action with less cognition results. Schneider, Dumais, and Shiffrin (1984) and the attention selectivity work of Posner and Boies (1971), which relates to deliberate discrimination between stimuli, are similar to the way that the Attention process was conceptualized and measured in the CAS and CAS2. That is, tasks were designed to require focused cognitive activity and resistance to distraction over time.

Simultaneous processing is an ability that is used for organizing information into groups to form a coherent whole and seeing patterns as interrelated elements. This ability is made possible by the parieto-occipital-temporal brain regions. There is a visual-spatial dimension to tasks that demand most Simultaneous tests, but not all. In the CAS and CAS2, Simultaneous processing is measured using tasks that have a strong visual-spatial component such as that found in progressive matrices tests like those developed by Penrose and Raven (1936) and Naglieri Nonverbal Ability Test (2008). Simultaneous processing is not, however, limited to nonverbal content, as demonstrated by the important role it plays in the grammatical components of language and comprehension of word relationships, prepositions, and inflections (Naglieri, 1999). This is most apparent in the inclusion of the Verbal-Spatial Relationship subtest in the CAS (Naglieri & Das, 1997) and CAS2 (Naglieri, Das, & Goldstein, 2013a). Similarly, visual-spatial tests that use the progressive matrix format have been included in the so-called nonverbal scales of intelligence tests such as the *Wechsler Nonverbal Scale of Ability* (Wechsler & Naglieri, 2006), the perceptual reasoning portion of the *Wechsler Intelligence Scale for Children-IV* (WISC-IV; Wechsler, 2003), the *Stanford-Binet-Fifth Edition* (SB5; Roid, 2003), the *Naglieri Nonverbal Ability Test-Second Edition* (NNAT2; Naglieri, 2008), the *Kaufman Assessment Battery for Children-Second Edition* (K-ABC2; Kaufman & Kaufman, 2004), and a Simultaneous processing test (Naglieri & Das, 1997).

Successive processing ability is used when working with stimuli arranged in a defined serial order such as remembering or completing information in compliance with a specific order. Successive processing is involved with the serial organization of sounds, such as learning sounds in sequence and early reading which is an underpinning of phonological analysis (Das et al., 1994). When serial information is grouped into a pattern, however (like the number 553669 organized into 55-3-66-9), then successful repetition of the string may be a function of Planning (i.e., using the strategy of chunking) and Simultaneous (organizing the numbers into related groups). This method is often used by older children and can be an effective strategy for those who are weak in Successive processing (see Naglieri & Pickering, 2010). The concept of Successive processing ability from PASS is similar to the concept of Sequential processing included in the K-ABC2 (Kaufman & Kaufman, 2004) and tests that require recall of serial information such as Digit Span Forward on the WISC-IV (Wechsler, 2003).

The PASS theory provided the basis upon which the CAS and CAS2 were built. These basic psychological processes are measured using 4 (CAS2:Brief; Naglieri et al., 2013a, 2013b, 2013c), 8, or 12 (CAS2; Naglieri et al., 2013a, 2013b, 2013c). Additionally, the 12-subtest Extended Battery of the CAS2 provides several supplementary scales which includes measures of Executive Function more distinctly (see Table 12.2). For this reason, the CAS2 Extended Battery will be emphasized and more fully described below.

The Planning Scale

The subtests which comprise the Planning scale evaluate a students' ability to create a plan of action, apply the plan, verify that the action taken conforms with the original goal, and modify the plan as needed. It is formed by combining the results of the Planned Number Matching, Planned Codes, and Planned Connections subtests.

- *Planned Number Matching.* Each Planned Number Matching item presents the student with a page of eight rows with six numbers on each row. The student is required to find and underline the two numbers in each row that are the same within the 180 s limit per page. The numbers were written so that they can be more efficiently examined using a strategy. For example, some of the numbers are similar at the beginning, others at the ending.
- *Planned Codes.* In Planned Codes, students are provided with a legend at the top of each page that shows a correspondence of letters to specific codes (e.g., A, B, C, D to OX, XX, OO, XO, respectively). The page contains four rows and eight columns of letters without the codes which are arranged in some systematic manner the child can use to complete the page more efficiently. The student is required to write the corresponding codes in each empty box beneath each of the letters. Students have 60 s per item to complete as many empty code boxes as possible.
- *Planned Connections.* The Planned Connections subtest requires the student to connect a series of

stimuli (numbers and then alternating numbers and letters) in an order as quickly as possible. Students have between 60 and 180 s to complete each item. Students who carefully examine the task note that the lines they draw never cross and that strategies such as looking back to the previous number or letter make completion of the task more effective.

The Attention Scale

The subtest on this scale is designed to measure a students' ability to focus their cognition to detect particular stimuli and inhibit response to irrelevant competing stimuli. The CAS2 Attention scale comprises the Expressive Attention, Number Detection, and Receptive Attention subtests.

- *Expressive Attention.* The Expressive Attention subtest consists of two age-related sets of three items. Students between age 5 and 7 are presented with three items consisting of seven rows of six pictures of common animals that are depicted as either big (1 in. by 1 in.) or small (1/2 in. by 1/2 in.) for each item. In each of three items, the student is required to identify whether the animal depicted is big or small in real life ignoring the relative size of the picture on the page. In Item 1, the pictures are all the same size. In Item 2, the pictures are sized appropriately (i.e., big animals are depicted with big pictures and small animals are depicted with small pictures). In Item 3, the realistic size of the animal often differs from its printed size. Students between age 8 and 18 are presented with three items consisting of eight rows of five pictures. In Item 1, students are asked to read four black-and-white colored words (BLUE, YELLOW, GREEN, and RED) that are presented in random order. In Item 2, students are asked to name the colors of four colored rectangles (printed in blue, yellow, green, and red) that are presented in random order. In Item 3, the four colored words are printed in a different color ink than the colored word name and are presented in random order. In this item, students are required to name color of the ink in which the word is printed rather than read the word. Completion of the task demands considerable focus of attention on the critical attributes of the stimuli and resisting distractions created by stimuli that are only partially like the targets.
- *Number Detection.* Each Number Detection item presents the student with a page of approximately 200 numbers. Students are required to underline specific numbers (ages 5–7) or specific numbers in a particular font (ages 8–18) on a page with many distractors. Completion of the task demands considerable focus of attention on the important attributes of the stimuli (the number in the correct font) and resisting distractions created by stimuli that are only partially like the targets (the correct number but the incorrect font).
- *Receptive Attention.* The Receptive Attention subtest consists of two age-related sets of four items containing 180 picture or letter pairs. Both versions require the student to underline pairs of objects or letters either that are identical in appearance or that are the same from a lexical perspective (i.e., they have the same name). Completion of the task demands considerable focus of attention on the critical attributes of the stimuli (two letters that are the same but look different such as R and r) and resisting distractions created by stimuli that are only partially like the targets (letter pairs such as B and r).

The Simultaneous Scale

The Simultaneous scale evaluates a students' ability to synthesize separate elements into a cohesive whole or interrelated group. This CAS2 scale comprises the Matrices, Verbal-Spatial Relations, and Figure Memory subtests.

- *Matrices.* Matrices is a multiple choice subtest that utilizes shapes and geometric elements that are interrelated through spatial or logical organization. Students are required to analyze the relationship among the parts of the

Table 12.2 CAS2 subtests on the Core and Extended Batteries and supplemental scales

	Core and Extended Battery composites					Supplemental scores			
CAS2 subtests	Planning	Simultaneous	Attention	Successive	Full Scale	Executive Function without Working Memory	Executive Function with Working Memory	Working Memory	Verbal content	Nonverbal content
Planned Codes	x				x					x
Planned Connections	x				x		x			
Planned Number Matching	*				*	x	x			
Matrices		x			x					x
Verbal-Spatial Relations		x			x		x	x	x	
Figure Memory		*			*					*
Expressive Attention			x		x	x	x			
Number Detection			x		x					
Receptive Attention			*		*				*	
Word Series				x	x					
Sentence Repetition or Questions				x	x		x	x	x	
Visual Digit Span				*	*					

Note. x = Core Battery subtest; * = Extended Battery subtest

item and solve for the missing part by choosing the best of six options.
- *Verbal-Spatial Relations.* Verbal-Spatial Relations is a multiple choice subtest in which each item consists of six drawings and a printed question at the bottom of each page. The examiner reads the question aloud and the child is required to select the option that matches the verbal description.
- *Figure Memory.* For each Figure Memory item, the examiner presents the student with a two- or three-dimensional geometric figure for 5 s. The picture is then removed and the student is presented with a response page that contains the original figure embedded in a large, more complex geometric pattern. The student is required to trace the original figure with a red pencil in the response book.

The Successive Scale

The Successive scale evaluates students' ability to recall or comprehend a verbal statement based upon the serial organization of information. All of the Successive subtests require the student to deal with information that is presented in a specific order. The CAS2 Successive scale is composed of the Word Series, Sentence Repetition or Sentence Questions, and Visual Digit Span subtests.
- *Word Series.* The Word Series subtest utilizes nine single-syllable, high-frequency words: book, car, cow, dog, girl, key, man, shoe, and wall. The examiner reads aloud a series of these words ranging in length from two to nine words, read at the rate of one word per second. The student is required to repeat the words in the same order as stated by the examiner.
- *Sentence Repetition.* The Sentence Repetition subtest (only administered to 4–7 year olds) requires the student to repeat syntactically correct sentences containing little meaning such as "The blue is yellowing."
- *Sentence Questions.* The Sentence Questions subtest (only administered to 8–18 year olds) requires the student to listen to sentences that are syntactically correct but contain little meaning and answer questions about the sentences. For example, the student reads the sentence "The blue is yellowing" and then is asked the following question: "Who is yellowing?"
- *Visual Digit Span.* Visual Digit Span subtest requires the student to recall a series of numbers in the order they were shown using the stimulus book. Each item with from two to five digits in length is exposed for the same number of seconds as digits. Items with six digits or more are all exposed for a maximum of 5 s.

Brief, Core, and Extended CAS2 Versions

There are three configurations of the CAS2. First is a four-subtest Brief; second is the eight-subtest Core, and third is the 12-subtest Extended Battery. Regardless of which version is administered, all yield PASS Scale and Full Scale standard scores score (mean of 100 and standard deviation of 15), but only the Extended Battery has supplemental scales which includes subtests specifically used to measure Executive Function. The configuration of the subtests and the scales for the Core and Extended CAS2 Batteries are shown in Table 12.1.

The CAS2 supplemental scores are provided to extend interpretation beyond the PASS scales to concepts such as Executive Function and Working Memory that may be especially helpful when interpreting CAS2 within the context of a comprehensive evaluation. The supplemental scales are also set to have a mean of 100 and standard deviation of 15 like all the other scales of the CAS2.

Executive Function scales. As described earlier in this chapter, views of executive function often differ in regard to whether working memory should be included or not. For this reason, the CAS2 provides two scales of executive function: Executive Function without Working Memory and Executive Function with Working Memory.

The Executive Function without Working Memory scale comprises the Planning Connections and Expressive Attention subtests. We chose these subtests because Weyandt, Williis, Swentosky, Wilson, Janusis, Chung, and Turcotte (this volume) found that the Stroop test (Expressive Attention on the CAS2) and TMT (Planned Connections on the CAS2) are among the most widely used tests utilized to evaluate executive functioning. These subtests address shifting and inhibition, two important components of executive function. These subtests fall on the Planning and Attention scales of the CAS2.

The Executive Function with Working Memory scale comprises Planning Connections, Verbal-Spatial Relations, Expressive Attention, and Sentence Repetition (ages 5–7) or Sentence Questions (ages 8–18) subtests. This addresses the working memory aspect of executive functioning that is central to the view of executive function described by Baddeley and Hitch (1974). That is, the Working Memory scale comprises the Verbal-Spatial Relations and Sentence Repetition (ages 5–7) or Sentence Questions (ages 8–18) subtests because they require the examinee to store information for a short amount of time and manipulate it using a phonological loop and visual-spatial sketchpad described by Baddeley and Hitch (1974). Engle and Conway (1998) describe the visual-spatial sketchpad as mental image of visual and spatial features. The phonological loop refers to retention of information from speech-based systems that are important when retention of the order of information is required (Engle & Conway, 1998). Because the CAS2 Verbal-Spatial Relations and the Sentence Repetition/Sentence Questions subtests have similar cognitive demands as the visual-spatial sketchpad and phonological loop, respectively, they were selected to evaluate Working Memory and added to the subtests used to evaluate Executive Function.

The values used to create the Executive Function scales included in the CAS2 were normed using a sample of 1,342 individuals aged 5–17 years who were representative of the US population on a number of important demographic variables. The sample is nationally representative, stratified sample based on gender, race, ethnicity, region, and parental education (see Naglieri et al., 2013a, 2013b, 2013c). Using those data, the internal reliability coefficients for the CAS2 Executive Function scales were reported in the Manual. The median reliability across ages 5–18 years was .87 and .90 for the CAS2 Executive Function with and Executive Function without working memory scales, respectively.

The next issue to be discussed is how to use this score within the larger context of a comprehensive evaluation. We will illustrate how the CAS2 PASS and Executive Function supplemental scales can be interpreted and combined with the Comprehensive Executive Function Inventory (CEFI; Naglieri & Goldstein, 2013). We will only present relevant aspects of this case to illustrate how the CAS2 Executive Function scales and the CEFI data can be integrated and a treatment plan selected.

Case of Dennis

Dennis was referred for a psychological evaluation by his physician to assist with planning for current and future high school needs given inconsistent academic performance and problems in school and at home with distractibility, forgetfulness, and inattention. Teachers reported that Dennis does not always follow directions, appears forgetful, and does not stay focused. At home, Dennis often misses cues and details, has trouble with his chores, is easily distracted, and does not follow through with instructions. He has particular trouble shifting from one uncompleted activity to another. Dennis loses things all the time and does not appear to learn well from experience, doing something the same way even though it did not work the first, or second, or third time. All of these concerns were first apparent when he was about 8 or 9 years old. At that time, ironically, there was discussion about possible placement in the gifted and talented program after he earned a very high score on a screening test. His grades varied but overall they were about average throughout his years in school but earned a very high score on his college entrance exam.

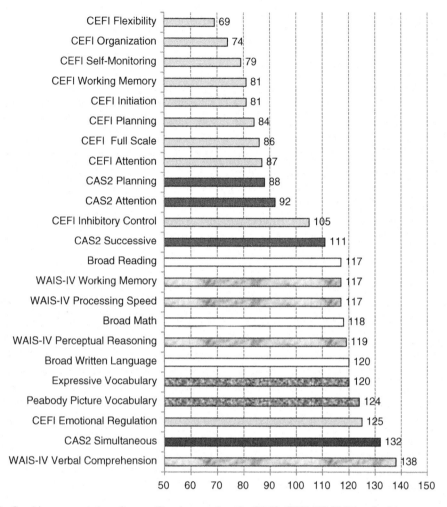

Fig. 12.1 Graphic representation of scores Dennis earned on the CAS2, CEFI, WAIS-IV, and achievement tests

Dennis earned high scores on tests that require verbal knowledge. He earned scores of 124 (95th percentile) and 120 (91st percentile) on the Peabody Picture Vocabulary Test-Fourth Edition and the Expressive Vocabulary Test-Second Edition, respectively (see Fig. 12.1). He earned an even higher score on Wechsler Adult Intelligence Scale-Fourth Edition on the Verbal Comprehension scale (138, 99th percentile) which measures general ability through tasks that require verbal knowledge and communication. Dennis' scores on the other portions of the WAIS-IV were about 20 points lower. He earned a Perceptual Reasoning score of 119 (90th percentile) and 117 (87th percentile) on both the Working Memory and Processing Speed scales. Taken as a whole, these scores suggest that Dennis earned very high scores on measures of general ability and particularly when general ability was measured using tasks that require verbal knowledge. These scores do not, however, help us understand the reported difficulties he has experienced in school and at home.

The scores Dennis earned on the CAS2 provide an understanding of this young man's abilities that are sometimes consistent and at other times very inconsistent. Dennis earned high scores on the CAS2 Simultaneous, Attention, and Successive scales. He earned a score of 132 (98th percentile) on the CAS2 Simultaneous scale of ability. This means that he is excellent at understanding interrelationships whether the task involves verbal concepts or the spatial organization of objects. This ability was also reflected in his high score on the WAIS-IV Similarities subtest on the Verbal scale which demands that he explain how seemingly different objects (e.g., a flea and a tree) are alike as well as his excellent performance on the various WAIS-IV Perceptual Reasoning score. Dennis also earned a high score on the CAS2 Attention scale (109; 73rd percentile) which demonstrates, contrary to reports of teachers and his parents, that he can focus his attention and resist distraction. In addition, Dennis is very capable of remembering sequences of words and sounds as well as working with information in order as demonstrated by his score of 111 (77th percentile) on the CAS2 Successive processing ability scale. The results of the CAS2 and the CEFI do suggest cognitive and behavioral limitations that are related to his behavioral and academic problems.

Dennis earned a score of 88 (21st percentile) on the CAS2 Planning Scale and a score of 91 (27th percentile) on the CAS2 Executive Function scale. These scores indicate that he has trouble on cognitive tests that require strategies for solving problems, control of his actions, monitoring the effectiveness of his solutions, and self-correction. The cognitive score on the CAS2 is consistent with the reports of his behavior provided by his mother's ratings on the CEFI. Her ratings yielded a total CEFI score of 86 (18th percentile). Of particular note were his very low scores on the CEFI treatment scales Flexibility, Organization, Self-monitoring, Working Memory, Initiation, and Planning and good scores on Inhibitory Control and Emotional Regulation. From both a cognitive ability (CAS2) and behavioral (CEFI) perspective, Dennis shows considerable variability and it is this pattern of strengths and weakness which explain his academic and behavioral problems at home and at school. The next important topic is what can be done to help him.

Intervention Design

There are two intervention methods that are needed for Dennis. First is to address his cognitive weakness in Planning (CAS2) and second is to improve his behaviors described by the seven low scores on the CEFI. These will require two different types of interventions. We will address the cognitive weakness in Planning from the CAS2 first, then provide a group of interventions for the seven scales of the CEFI. A brief review of the intervention research related to PASS theory will be presented first, then implementation described.

Intervention for CAS2 Planning scale. Naglieri and Otero (2011) summarized the research on Planning scores from the CAS and academic interventions in math and reading which demonstrated that students can be taught to better use their planning ability and in so doing improve in classroom work and standardized test scores. This research, which compared student with low Planning scale scores to those with high scores on the CAS, began with two published studies by Naglieri and Gottling (1995, 1997). Both studies involved students who attended a special school for those with learning disabilities. In these investigations, students completed mathematics work sheets in sessions over an approximately 2-month period. The method designed to indirectly teach skills such as strategy use, self-monitoring, and self-correction was applied in individual one-on-one tutoring sessions (Naglieri & Gottling, 1995) or within the entire classroom by the teacher (Naglieri & Gottling, 1997) about 2–3 times per week in half-hour blocks of time. During the intervention sessions, the students were encouraged to use good executive function skills when completing math work sheets. The teachers provided probes that facilitated discussion and encouraged the children to consider various ways to be more successful (see Naglieri & Gottling, 1995, 1997 for more details). The results from both these studies showed that the students with low Planning scores on the CAS improved considerably more than those with high Planning scores on the CAS.

The next study that examined PASS scores from the CAS to instruction involved students with learning disabilities and mild mental impairments (Naglieri & Johnson, 2000). They implemented the same method, called Planning Facilitation (Naglieri & Pickering, 2010), but compared students with cognitive weaknesses in each of the four PASS processes and students with no cognitive weakness. They showed that children with a cognitive weakness in Planning improved considerably over baseline rates, while those with no cognitive weakness improved only marginally. Similarly, children with cognitive weaknesses in Simultaneous, Successive, and Attention showed substantially lower rates of improvement. The importance of this study was that the five groups of children responded very differently to the same intervention. Thus, the PASS processing scores were predictive of the children's response to this math intervention (Naglieri & Johnson, 2000).

Iseman and Naglieri (2011) further demonstrated that teaching students with learning disabilities and ADHD executive function skills improves academic performance. In this randomized control study, students in the experimental group were encouraged to use good executive function skills such as planning, strategies, self-monitoring, and self-correction in math. A comparison group received additional math instruction by the regular teacher. Following the intervention, the control group outperformed significantly on math work sheets as well as standardized math test scores, illustrating the value of teaching executive skills.

Importantly, students with a Planning cognitive weakness in the experimental group improved considerably on math work sheets, but those with a Planning cognitive weakness in the comparison group did not improve. This study strongly supported the view that teaching students to better utilize Planning strategies—executive function skills, had a positive and significant influence on their academic performance in math. This finding was extended to reading comprehension by Haddad et al. (2003) with the same results. The results of these studies using academic tasks suggest that teaching cognitive skills related to Planning on the CAS should be implemented.

Interventions for low CEFI scales. Interventions that can be used to help Dennis improve on the CEFI Flexibility, Organization, Self-monitoring, Working Memory, Initiation, and Planning scales are provided in the CEFI computerized report.

Interventions for the first scale, Flexibility, should address the behaviors measured by that sale. These include how Dennis adjusts his behavior to meet circumstances, including coming up with different ways to solve problems, having many ideas about how to do things, and being able to solve problems using different approaches. These needs are also addressed by the intervention provided for the CAS2 Planning Scale.

Interventions for Organization, Self-monitoring, and Planning in the CEFI computerized report include methods from Naglieri and Pickering's (2010) *Helping Children Learn* book. The interventions are designed to provide instruction about strategies for specific instructional areas (e.g., decoding, reading comprehension, writing, math) using two basic steps. First, the teacher tells the student that a plan is a method for how to do something that involves thinking about how to solve a task. Second, using a plan means you have to (a) ask what do I want to do and what is my goal, (b) choose a plan, (c) begin work on the task using that plan, (c) see if the plan is working, (d) change the plan if necessary, and (e) evaluate the solution vis-a-vis the goal.

Interventions for Working Memory also summarized in the CEFI computerized report include an intervention handout entitled "Focusing Strategies to Improve Memory" (Naglieri & Pickering, 2010, p. 125). The goal of this intervention is to teach the student to be aware of the need to pick a good environment to work. That means being physically comfortable in a location with adequate light and temperature, working in a location with minimal visual and auditory distractions, and using self-talk strategies to control any internal distractions. Another important intervention for working memory is to use mnemonics (e.g., rhymes, acronyms, visual images) for various academic and nonacademic tasks (Naglieri & Pickering, 2010, p. 101).

Conclusions

In this chapter, we have provided a summary of the concept of Executive Function and two ways to measure the concept. The first way is to use the CAS2. This nationally standardized individually administered test can be used to measure a child or adolescent's ability to make decisions about how to complete a task, implement the solution, monitor the effectiveness, modify the solutions as needed, and recognize when the goal has been achieved adequately. These activities are the hallmark of executive function that are also evaluated by observations of the child in the natural environment using the CEFI. The case study we presented is an actual one (the name has been changed) that illustrates how the CAS2 and CEFI data can be used to understand the cognitive and behavioral manifestations of the concept of executive function and empirically supported interventions which can be applied to address the need for improvement.

References

Anderson, V., Northam, E., Hendy, J., & Wrennall, J. (2001). *Developmental neuropsychology: A clinical approach*. Philadelphia, PA: Psychology Press.
Archibald, L. M., & Gathercole, S. E. (2006). Short-term and working memory in specific language impairment. In T. P. Alloway & S. E. Gathercole (Eds.), *Working memory in neurodevelopmental conditions* (pp. 139–160). Hove: Psychology Press.
Baddeley, A. D. (1986). *Working memory*. Oxford: Clarendon Press.
Banich, M.T. (2009). Executive Function: The Search for a Integrated Account. *Current Directions in Psychological Science, 18*(2), 89–94.
Barkley, R. A. (2012). *Executive functions: What are they, how they work, and why they evolved*. New York: Guilford Press.
Baron, I. S. (2004). *Neuropsychological evaluation of the child*. New York, NY: Oxford University Press.
Berg, E. A. (1948). A simple objective technique for measuring flexibility in thinking. *Journal of General Psychology, 39*, 15–22.
Blair, C., Zelazo, P. D., & Greenberg, M. T. (2005). The measurement of executive function in early childhood. *Developmental Neuropsychology, 28*, 561–571.
Blair, C. (2006). Toward a revised theory of general intelligence: Further examination of fluid cognitive abilities as unique aspects of human cognition. *Behavioral and Brain Sciences, 29*, 145–160.
Bull, R., Espy, K. A., & Senn, T. E. (2004). A comparison of performance on the Towers of London and Hanoi in young children. *Journal of Child Psychology and Psychiatry, and Allied Disciplines, 45*, 743–754.
Clark, C. A. C., Pritchard, V. E., & Woodward, L. J. (2010). Preschool executive functioning abilities predict early mathematics achievement. *Developmental Psychology, 46*, 1176–1191.
Connors, K. (2000). *Continuous Performance Test*. Toronto: Multi-Health Systems.
Conway, A. R. A., Cowan, N., & Bunting, M. F. (2001). The cocktail party phenomenon revisited: The importance of working memory capacity. *Psychonomic Bulletin & Review, 8*, 331–335.
Cowan, N. (2005). *Working memory capacity*. Hove: Psychology Press.
Das, J. P., Kar, B. C., & Parrila, R. K. (1996). *Cognitive planning: The psychological basis of intelligent behavior*. Thousand Oaks, CA: Sage.
Das, J. P., Naglieri, J. A., & Kirby, J. R. (1994). *Assessment of cognitive processes*. Needham Heights, MA: Allyn & Bacon.
Dawson, P., & Guare, R. (2010). *Executive skills in children and adolescents: A practical guide to assessment and intervention 2nd ed*. New York: The Guilford Press.
Delis, D. C., Kaplan, E., & Kramer, J. H. (2001a). *Delis-Kaplan Executive Function System (D-KEFS)*. San Antonio, TX: The Psychological Corporation.
Delis, D. C., Kaplan, E., & Kramer, J. H. (2001b). *Delis-Kaplan Executive Function System (D-KEFS) examiner's manual* (pp. 1–218). San Antonio, TX: The Psychological Corporation.
Duncan, J., Emslie, H., Williams, P., Johnson, R., & Freer, C. (1996). Intelligence and the frontal lobe: The organization of goal-directed behavior. *Cognitive Psychology, 30*(3), 257–303.
Ellioit, R. (2003). Executive functions and their disorders: Imaging in clinical neuroscience. *British Medical Bulletin, 65*(1), 49–59. doi:10.1093/bmb/65.1.49.
Engle, R. W., & Conway, A. R. A. (1998). Comprehension and working memory. In R. H. Logie and K. J. Gilhooly (Eds.), Working memory and thinking. London: Lawrence Erlbaum.
Engle, R. W. (2002). Working memory capacity as executive attention. *Current Directions in Psychological Science, 11*, 19–23.
Fletcher, J. M. (1996). Executive functions in children: Introduction to a special series. *Developmental Neuropsychology, 12*, 1–3.
Fristoe, N. M., Salthouse, T. A., & Woodward, J. L. (1997). Examination of age-related deficits on the Wisconsin Card Sorting Test. *Neuropsychology, 11*(3), 428–436.
Fuster, J. M. (1997). Network memory. *Trends in Neurosciences, 20*, 451–459.
Gold, J. M., Carpenter, C., Randolph, C., Goldberg, T. E., & Weinberger, D. R. (1997). Auditory working memory and Wisconsin Card Sorting Test performance in

schizophrenia. *Archives of General Psychiatry, 54,* 159–165.

Golden, CJ (1978). *Stroop color and word test: A manual for clinical and experimental uses.* Chicago, IL: Skoelting.

Goldberg, E. (2009). *The executive brain: Frontal lobes and the civilized mind-second edition.* New York: Oxford University Press.

Goldman-Rakic, P. S. (1991). Prefrontal cortical dysfunction in schizophrenia: The relevance of working memory. In B. J. Carroll & J. E. Barrett (Eds.), *Psychopathology and the brain* (pp. 1–23). New York: Raven.

Grant, D. A., & Berg, E. A. (1948). A behavioral analysis of degree of reinforcement and ease of shifting to new responses in a Weigl-type card-sorting problem. *Journal of Experimental Psychology, 38,* 404–411.

Haddad, F. A., Garcia, Y. E., Naglieri, J. A., Grimditch, M., McAndrews, A., & Eubanks, J. (2003). Planning facilitation and reading comprehension: Instructional relevance of the PASS theory. *Journal of Psychoeducational Assessment, 21,* 282–289.

Iseman, J., & Naglieri, J. A. (2011). A cognitive strategy instruction to improve math calculation for children with ADHD: A randomized controlled study. *Journal of Learning Disabilities, 44,* 184–195.

Isquith, P. K., Roth, R. M., & Gioia, G. A. (2010). *Tasks of executive control. Professional manual.* Lutz, FL: PAR, Inc.

Jonides, J., & Nee, D. E. (2006). Brain mechanisms of proactive interference in working memory. *Neuroscience, 139,* 181–193.

Kane, M. J., Bleckley, M. K., Conway, A. R. A., & Engle, R. W. (2001). A controlled-attention view of working memory capacity: Individual differences in memory span and the control of visual orienting. *Journal of Experimental Psychology General, 130,* 169–183.

Kane, M. J., & Engle, R. W. (2003). Working memory capacity and the control of attention: The contributions of goal neglect, response competition, and task set to Stroop interference. *Journal of Experimental Psychology, 132,* 47–70.

Kaufman, A. S., & Kaufman, N. L. (2004). *Kaufman Assessment Battery for Children second edition.* Circle Pines, MN: American Guidance Service.

Kenworthy, L., Yerys, B. E., Anthony, L. G., & Wallace, G. L. (2008). Understanding executive control in autism spectrum disorders in the lab and in the real world. *Neuropsychological Review, 18,* 320–338.

Klingberg, T., Fernell, E., Olesen, P. J., Johnson, M., Gustafsson, P., Dahlström, K., et al. (2005). Computerized training of working memory in children with ADHD: A randomized, controlled trial. *Journal of the American Academy of Child & Adolescent Psychiatry, 44,* 177–186.

Kochanska, G., Murray, K., & Harlan, E. (2000). Effortful control in early childhood: Continuity and change, antecedents, and implications for social development. *Developmental Psychology, 36,* 220–232.

Korkman, M., Kirk, S., & Kemp, U. (2007). *NEPSY-II.* San Antonio, TX: Pearson.

Leung, H. C., Gore, J. C., & Goldman-Rakic, P. S. (2005). Differential anterior prefrontal activation during the recognition stage of a spatial working memory task. *Cerebral Cortex, 15,* 1742–1749.

Lezak, M. D., Howieson, D. B., Loring, D. W., Hannay, H. J., & Fischer, J. S. (2004). *Neuropsychological assessment* (4th ed.). New York: Oxford University Press.

Lezak, M. D., Howieson, D. B., Bigler, E. D., & Tranel, D. (2012). *Neuropsychological assessment, (5th ed.).* New York, NY: Oxford University Press.

Luria, A. R. (1966). *Human brain and psychological processes.* New York: Harper and Row.

Luria, A. R. (1973). *The working brain.* New York: Basic Books.

Luria, A. R. (1980). *Higher cortical functions in man* (2nd ed.). New York: Basic Books.

Luria, A. R. (1982). *Language and cognition.* New York: Wiley.

Marcovitch, S., Boscovski, J. J., & Knapp, R. J. (2007). Use it or lose it: Examining preschoolers' difficulty in maintaining and executing a goal. *Developmental Science, 10,* 559–564.

McCloskey, G. & Perkins, L. A. (2013). *Essentials of executive functions assessment.* Hoboken, New Jersey: Wiley.

Melby-Lervåg, M., & Hulme, C. (2012). Is working memory training effective? A meta-analytic review. *Developmental Psychology, 49,* 270–291. doi:10.1037/a0028228.

Mezzacappa, E., & Buckner, J. C. (2010). Working memory training for children with attention problems or hyperactivity: A school-based pilot study. *School Mental Health, 2*(4), 202–208.

Meyers, J. E., & Meyers, K. R. (1995). *Rey complex figure test and recognition trial: Professional manual.* Lutz, FL: Psychological Assessment Resources.

Miyake, A., & Shah, P. (Eds.). (1999). *Models of working memory: Mechanisms of active maintenance and executive control.* New York: Cambridge University Press.

Moll, J., de Oliveira-Souza, R., Tovar, F. T., Bramati, I. E., & Andreiuolo, P. A. (2002). The cerebral correlates of set-shifting. *Arquivos de Neuro-Psiquiatria, 60,* 900–905.

Mullane, J. C., & Corkum, P. V. (2007). The relationship between working memory, inhibition, and performance on the Wisconsin Card Sorting Test in children with and without ADHD. *Journal of Psychoeducational Assessment, 25,* 211–221.

Naglieri, J. A. (1999). *Essentials of CAS assessment.* New York: Wiley.

Naglieri, J. A. (2008). *Naglieri Nonverbal Ability Test—2nd edition.* San Antonio, TX: Pearson.

Naglieri, J. A., & Das, J. P. (1997). *Cognitive Assessment System.* Itasca: Riverside Publishing Company.

Naglieri, J. A., Das, J. P., & Goldstein, S. (2013a). *Cognitive Assessment System2.* Austin: ProEd.

Naglieri, J. A., Das, J. P., & Goldstein, S. (2013b). *Cognitive Assessment System2—Brief.* Austin: ProEd.

Naglieri, J. A., Das, J. P., & Goldstein, S. (2013c). *Cognitive Assessment System2—Rating sales*. Austin: ProEd.

Naglieri, J. A., & Goldstein, S. (2013). *Comprehensive Executive Functioning Index*. Toronto: Multi Health Systems.

Naglieri, J. A., & Gottling, S. H. (1995). A cognitive education approach to math instruction for the learning disabled: An individual study. *Psychological Reports, 76*, 1343–1354.

Naglieri, J. A., & Gottling, S. H. (1997). Mathematics instruction and PASS cognitive processes: An intervention study. *Journal of Learning Disabilities, 30*, 513–520.

Naglieri, J. A., & Johnson, D. (2000). Effectiveness of a cognitive strategy intervention to improve math calculation based on the PASS theory. *Journal of Learning Disabilities, 33*, 591–597.

Naglieri, J. A., & Otero, T. (2011). Cognitive Assessment System: Redefining intelligence from a neuropsychological perspective. In A. Davis (Ed.), *Handbook of pediatric neuropsychology* (pp. 320–333). New York: Springer.

Naglieri, J. A., & Pickering, E. B. (2010). *Helping children learn: Intervention handouts for use at school and home (second edition)*. Baltimore, MD: Brookes Publishing.

O'Shanick, G. J., & O'Shanick, A. M. (1994). Personality and intellectual changes. In J. M. Silver, S. C. Yudofsky, & R. E. Hales (Eds.), *Neuropsychiatry of traumatic brain injury* (pp. 163–188). Washington, DC: American Psychiatric Press.

Pennington, B. F., Bennetto, L., McAleer, O., & Roberts, R. J. (1996). Executive functions and working memory. In G. R. Lyons & N. A. Krasnegor (Eds.), *Attention, memory and executive function* (pp. 327–348). Paul H. Brookes: Baltimore.

Penrose, L. S., & Raven, J. C. (1936). A new series of perceptual tests: Preliminary communication. *British Journal of Medical Psychology, 16*, 97–104.

Posner, M. I., & Boies, S. J. (1971). Components of attention. *Psychological Review, 78*, 391–408.

Roid, G. (2003). *Stanford-Binet fifth edition*. Itasca, IL: Riverside.

Schneider, W., Dumais, S. T., & Shiffrin, R. M. (1984). Automatic and controlled processing and attention. In R. Parasuraman & D. R. Davies (Eds.), *Varieties of attention* (pp. 1–28). New York: Academic.

Schoemaker, K., Bunte, T., Wiebe, S. A., Espy, K. A., Dekovic, M., & Matthys, W. (2012). Executive function deficits in preschool children with ADHD and DBD. *Journal of Child Psychology and Psychiatry, and Allied Disciplines, 53*, 111–119.

Strauss, E., Sherman, E. M. S., & Spreen, O. (2006). *A compendium of neuropsychological tests: Administration, norms, and commentary* (3rd ed.). New York: Oxford University Press.

Szmalec, A., Verbruggen, F., Vandierendonck, A., & Kemps, E. (2011). Control of interference during working memory updating. *Journal of Experimental Psychology: Human Perception and Performance, 37*, 137–151.

The American Heritage® Medical Dictionary. (2007). *Executive function*. Retrieved December 10, 2012, from http://medical-dictionary.thefreedictionary.com/executive+function

Wang, Y., Markram, H., Goodman, P. H., Berger, T. K., Ma, J., & Goldman-Rakic, P. S. (2006). Heterogeneity in the pyramidal network of the medial prefrontal cortex. *Nature Neuroscience, 9*, 534–542.

Wechsler, D. (2003). *The Wechsler Intelligence Scale for Children—4th ed. (WISC-IV)*. San Antonio: The Psychological Corporation.

Wechsler, D., Kaplan, E., Fein, D., Morris, E., Kramer, J. H., Maerlender, A., et al. (2004). *The Wechsler Intelligence Scale for Children—Fourth edition integrated technical and interpretative manual*. San Antonio, TX: Harcourt Assessment, Inc

Wechsler, D., & Naglieri, J. A. (2006). *Wechsler Nonverbal Scale of Ability*. San Antonio: Harcourt Assessment.

Welsh, M.C. & Pennington, B.F. (1988). Assessing frontal lobe functioning in children: Views from developmental psychology. *Developmental Neuropsychology, 4*(3), 199–230.

Willoughby, M. T., Wirth, R. J., Blair, C. B., & Greenberg, M. (2012). The measurement of executive function at age 5: Psychometric properties and relationship to academic achievement. *Psychological Assessment, 24*(1), 226–239.

Zelazo, P., Qu, L., & Muller, U. (2004). Hot and cool aspects of executive function: Relations in early development. In W. Schneider, R. Schumann-Hengsteler, & B. Sodian (Eds.), *Young children's cognitive development: Interrelationships among executive functioning, working memory, verbal ability, and theory of mind* (pp. 71–93). Mahwah: Lawrence Erlbaum Associates Publishers.

The Assessment of Executive Functioning Using the Delis-Kaplan Executive Functions System (D-KEFS)

Tammy L. Stephens

The *Delis-Kaplan Executive Function System* (D-KEFS) is the first nationally standardized set of tests to evaluate higher-cognitive functions (executive functions) in both children and adults (ages 8–89 years). Published in 2001, the D-KEFS includes nine stand-alone tests that measure a wide array of verbal and nonverbal executive functions.

Theoretical Approach

A single theoretical model was not utilized within the development of the D-KEFS. Instead, the authors and developers incorporated a cognitive-process approach in the design of the tests in order to ensure that both the fundamental and higher-level components of executive functions could be quantified (Delis, Kaplan, & Kramer, 2001a, 2001b).

Constructs

- *Executive Functions* (*EF*) have been defined as mental functions associated with the ability to engage in purposeful, organized, self-regulated, and goal-directed behaviors (McCloskey, Perkins, & Divner, 2009) and integrate, synthesize, and organize other cognitive processes (Wecker, Kramer, Wisniewski, Delis, & Kaplan, 2000). EF permit a person to perform certain higher-order cognitive tasks that enable academic achievement.
- *Planning* involves the setting of short- or long-term goals and the establishment of a behavioral routine (strategy) to accomplish the set goals (McCloskey et al., 2009).
- *Inhibition* has been defined as the ability to resist or suppress urges to perceive, feel, think, or act on first impulse (McCloskey et al., 2009).
- *Attention* is a basic cognitive skill that allows for successful completion of set goals. The primary aspect of attention is the ability to focus on and respond to stimuli in the environment (Dehn, 2006).
- *Perception* involves the use of sensory and perception processes that take information in from the external environment or "inner awareness" to tune into perceptions, emotions, thoughts, or actions as they are occurring (McCloskey et al., 2009).
- *Switching* refers to a change of focus or alteration of perceptions, emotions, thoughts, or actions in reaction to what is occurring in the internal and external environments (McCloskey et al., 2009).

Description of the D-KEFS

Subtest Background and Structure

The D-KEFS consists of nine tests that measure verbal and nonverbal executive functions. Each test is designed as a stand-alone instrument that

T.L. Stephens (✉)
Pearson Clinical Assessments, San Antonio, TX, USA
e-mail: Tammy.Stephens@Pearson.com

can be administered individually or along with other D-KEFS tests, depending on the referral question. The D-KEFS is composed of the following instruments.

D-KEFS Trail Making Test

D-KEFS Trail Making Test consists of two tasks, (1) visual cancellation and (2) a series of connect-the-circle tasks. The test includes five conditions with *Condition 4: Number-Letter Switching* measuring the primary executive-function task, a means of assessing flexibility of thinking on a visual-motor sequencing task. The other four conditions allow the examiner to quantify and derive normative data for several key components necessary for performing the switching task. The five conditions are as follows:
- *Condition 1*: *Visual Scanning*
  - Task: Locate and slash through all the 3s on the page
- *Condition 2*: *Number Sequencing*
  - Task: Connect numbers in numerical order (e.g., 1-2-3-4)
- *Condition 3*: *Letter Sequencing*
  - Task: Connect letters in alphabetical order (e.g., A-B-C-D)
- *Condition 4*: *Number-Letter Sequencing*
  - Task: Switch between connecting the numbers and letters alternating between numerical order and alphabetical order (e.g., 1-A-2-B)
- *Condition 5*: *Motor Speed*
  - Task: Trace dotted lines from "Start" to "End"

Scales the Test Yields

Several types of scores are derived from the *D-KEFS Trail Making Test*: scaled scores ($M = 10$; $SD = 3$) based on completion times and one error measure, a composite scaled score, contrast scaled score, and cumulative percentile ranks for most of the error measures. An in-depth description of each score can be found within the *D-KEFS Examiner's Manual*. A brief description of each follows:

- *Completion-Time Scores*:
  - The primary scoring measure for each of the five conditions of the D-KEFS Trail Making Test is the number of seconds that the examinee takes to complete each condition. The raw score (in seconds) for each of the Conditions 1–5 is converted to a scaled score ($M = 10$; $SD = 3$). Additional combined scores can also be derived.
- *Contrast Measures*:
  - This measure allows the examiner to determine whether an impairment exists in an underlying component skill resulting in poor performance on Condition 4: Number-Letter Switching. For this reason, performance on each of the four baseline tasks is parceled out from performance on the Number-Letter Switching condition by the computation of a series of contrast measures. The contrast measures are derived by subtracting the completion-time scaled score for each component task (Conditions 1, 2, 3, or 5) or the Number Sequencing + Letter Sequencing composite from the completion-time scaled score on the switching task (Condition 4). A new scaled score, with a mean of 10 and a standard deviation of 3, is derived for each scaled score difference (Delis et al., 2001a, 2001b, p. 47).
- *Optional Error Scores*:
  - Cumulative percentile ranks can be derived for several types of errors. An in-depth description of each type of error can be found within the D-KEFS Examiner's Manual (Delis et al., 2001a, 2001b, p. 48). A brief description of each follows:
    Omission Error occurs whenever an examinee fails to mark a target 3 for Condition 1: Visual Scanning.
    Commission Error occurs whenever an examinee marks a letter or a number that is not a 3.
    Sequencing Error occurs when an examinee makes a connection within the correct set of symbols for the condition being administered (numbers or letters) but connects the wrong item within that set.

Set-Loss Error occurs when an examinee draws a line connecting an item that belongs to the wrong set of symbols (numbers or letters) for the condition being administered.

Time-Discontinue Error occurs when an examinee failed to connect one or more items because the time limit for that condition had elapsed (Delis et al., 2001a, 2001b).

D-KEFS Verbal Fluency Test

The *D-KEFS Verbal Fluency Test* evaluates an individual's ability to generate words fluently in a phonemic format (Letter Fluency), from overlearned concepts (Category Fluency), and while simultaneously shifting between overlearned concepts (Category Switching). The *D-KEFS Verbal Fluency Test* consists of a standard form and an alternate form. The three testing conditions for the *D-KEFS Verbal Fluency Test* consist of the following:
- *Condition 1: Letter Fluency* consists of three 60-s timed trials.
  - Trial 1: Examiner names as many words as possible that begin with letter F (B).
  - Trial 2: Examiner names as many words as possible that begin with letter A (H).
  - Trial 3: Examiner names as many words as possible that begin with letter S (R).
- *Category Fluency* consists of two 60-s timed trials
  - Trial 1: Examiner names as many animals (clothing) as possible in a 60-s time limit.
  - Trial 2: Examiner names as many boys' names (girls' names) as possible in a 60-s time limit.
- *Category Switching* consists of one 60-s timed trial.
  - Trial: Examiner names as many vegetables and musical instruments while switching back and forth.

A 60-s time limit for each trial of each condition is allowed (Delis et al., 2001a, 2001b).

Scales the Test Yields

Several types of scores are derived from the *D-KEFS Verbal Fluency Test*: Total Correct Scores, Primary Contrast Scores, and Optional Error Scores. An in-depth description of each score can be found within the *D-KEFS Examiner's Manual* (Delis et al., 2001a, 2001b). A brief description of each follows:
- *Total Correct Scores*:
  - This measure represents the number of correct words generated within each 60-s trial for each condition. Consult the D-KEFS Examiner's Manual for guidelines for scoring correct responses (Delis et al., 2001a, 2001b).
- *Primary Contrast Scores*:
  - These measures allow the examiner to compare the examinee's performance on one task in comparison (e.g., Letter Fluency) to another task (e.g., Category Fluency) to determine whether the examinee exhibits a disproportionate impairment in one relative to the other.
- *Optional Error Scores*:
  - Numerous optional measures can be derived for the *D-KEFS Verbal Fluency Test* (e.g., three letter-fluency trials, two category-fluency trials, and one witching trial). Refer to the D-KEFS Examiner's Manual for in-depth descriptions (Delis et al., 2001a, 2001b).

D-KEFS Design Fluency Test

The *D-KEFS Design Fluency Test* evaluates an examinee's ability to draw as many different designs as possible in a 60-s time limit. The examinee is presented with a record form containing rows of boxes, with each containing an array of dots and instructed to draw a different design in each box using only four lines to connect the dots. The *D-KEFS Design Fluency Test* consists of the following three conditions:
- *Condition 1: Filled Dots*
  - Task: Examinee is asked to draw the designs connecting filled dots.
  - Assesses basic design fluency.

- Condition 2: *Empty Dots Only*
- Condition 3: *Switching*
  - Task: Examinee is asked to draw designs by alternately connecting filled and empty dots.
  - Assesses design fluency and cognitive flexibility.

Scales the Test Yields

Several types of scores are derived from the *D-KEFS Design Fluency Test*: Primary Scores, Contrast Scores, and Optional Error Scores. An in-depth description of each score can be found within the *D-KEFS Examiner's Manual*.
- *Primary Scores*:
  - This measure represents the total number of correct designs generated in each of the three 60-s individual conditions. Consult the D-KEFS Examiner's Manual for guidelines for scoring correct responses (Delis et al., 2001a, 2001b).
- *Contrast Scores*:
  - Some examinees exhibit adequate design fluency skills, except when they must simultaneously engage in cognitive shifting to generate the designs in the switching condition. Consequently, a contrast score is derived for directly comparing the examinee's ability to generate designs with switching (Condition 3) relative to his or her ability to generate designs without switching (composite scaled score for Conditions 1 and 2). Refer to the D-KEFS examiner's Manual for additional information regarding Contrast Scores (Delis et al., 2001a, 2001b).
- *Optional Error Scores*:
  - Numerous process and error measures can be derived for the *D-KEFS Design Fluency Test*. Refer to the D-KEFS Examiner's Manual for in-depth descriptions (Delis et al., 2001a, 2001b).

D-KEFS Color-Word Interference Test

The *D-KEFS Color-Word Interference Test* evaluates the examinee's ability to inhibit an overlearned verbal response (i.e., reading the printed words) in an attempt to generate a conflicting response naming the dissonant ink colors in which the words are printed. The D-KEFS Color-Word Interference Test is based on the procedure. The test consists of the following three conditions:
- Condition 1: Basic naming of color patches.
- Condition 2: Basic reading of color words printed in black ink.
- Condition 3: (Inhibition); examinee must inhibit reading the words in order to name the dissonant ink colors in which those words are printed.
- Condition 4: (Inhibition/Switching/Cognitive Flexibility); examinee must switch back and forth between naming the dissonant ink colors and reading the words.

Scales the Test Yields

Several types of scores are derived from the *D-KEFS Color-Word Interference Test*: Completion-Time Scores, Contrast Scores, and Error Scores. An in-depth description of each score can be found within the *D-KEFS Examiner's Manual*.
- *Completion-Time Scores*:
  - This measure is the primary method used to analyze performance on the D-KEFS Color-Word Interference Test and is based on the number of seconds that the examinee takes to complete each of the four conditions (Delis et al., 2001a, 2001b).
- *Contrast Scores*:
  - This measure allows the examiner to distinguish between a deficit in higher-level abilities of inhibition or cognitive flexibility

and impairments in the fundamental skills of basic naming and reading. Refer to the D-KEFS Examiner's Manual for in-depth descriptions (Delis et al., 2001a, 2001b).
- *Error Scores*:
  - This measure categorizes errors for each of the four conditions of the D-KEFS Color-Word Interference Test. Naming Errors, Reading Errors, Inhibition Errors, and Inhibition/Switching Errors can be analyzed. Refer to the D-KEFS Examiner's Manual for in-depth descriptions (Delis et al., 2001a, 2001b).

D-KEFS Sorting Test

The *D-KEFS Sorting Test* evaluates the individual's ability to initiate problem-solving behavior in verbal and nonverbal modalities. The D-KEFS Sorting Test consists of two conditions: Free Sorting and Sort Recognition.

For Condition 1: Free Sorting—the examinee is presented six mixed-up cards that display both perceptual features and printed words. The examinee is asked to sort the cards into two groups, with three cards per group. The sorts are made according to many different concepts or rules as possible, and upon completion of the sort, the examinee must describe the concepts employed to generate each sort. Each of the two card sets has a maximum of eight target sorts: three sorts based on verbal-semantic information from the printed words and five based on visual-spatial features or patterns on each card (Delis et al., 2001a, 2001b).

For Condition 2: Sort Recognition—the examiner sorts the same sets of cards into two groups (three cards in each group) according to the eight target sorts. After the examiner completes is each sort, the examinee instructed to identify and describe the correct rules or concepts used to generate the sort. Corrective feedback is never given during administration of this test, to minimize possible adverse effects of repetitive negative feedback. Additionally, the examinee's problem-solving performance is scored in terms of both accuracy of the sorting responses and the descriptions of the sorting concepts. Finally, the formal assessment of examinees' descriptions of the sorting rules provides information about their conceptual reasoning skills.

Scales the Test Yields

Several types of scores are derived from the *D-KEFS Sorting Test*: Sorting and Descriptive Measures, Composite Scores and Contrast Scaled Scores. An in-depth description of each score can be found within the *D-KEFS Examiner's Manual*.
- *Sorting and Descriptive Measures*, raw scores are converted into scaled scores or cumulative percentile ranks.
- *Composite Scores* for combined description measures (e.g., Condition 1: Free Sorting and Condition 2: Sort Recognition) can also be derived.
- *Contrast Scaled Scores*, a percent accuracy score, and a percent description accuracy score can be determined (Delis et al., 2001a, 2001b, p. 115).

D-KEFS Twenty Questions Test

The *D-KEFS Twenty Questions Test* evaluates the examinee's ability to identify the various categories and subcategories represented in the 30 objects and to formulate abstract, yes/no questions that eliminate the maximum number of objects regardless of the examiner's answer. The task involves the examinee being presented with a stimulus page depicting pictures of 30 common objects. The examinee is instructed to ask the fewest number of questions that result in the elimination of the most objects (Delis et al., 2001a, 2001b).

Executive functions assessed by the *D-KEFS Twenty Questions Test* include:
- The ability to perceive the various categories and subcategories represented by the 30 objects

- The ability to formulate abstract, yes/no questions that eliminate the maximum number of objects regardless of the examiner's answer
- The ability to incorporate the examiner's feedback in order to formulate more efficient yes/no questions

Scales the Test Yields

Several types of scores are derived from the *D-KEFS Twenty Questions Test*: an Initial Abstraction Score, Total Questions Asked Score, and Total Weighted Achievement Score can be derived for the Primary Measure. Additionally, three optional process and error scores can also be derived: spatial questions, set-loss questions, and repeated questions. An in-depth description of each score can be found within the *D-KEFS Examiner's Manual* (Delis et al., 2001a, 2001b). A brief summary of each score follows:

- *Initial Abstraction Score*:
  - Quantifies the level of abstract thinking represented by the first question asked by an examinee on each item. The minimum number of objects eliminated by the first question is summed across items 1–4 to obtain a raw score. The raw score is then converted to a scaled score.
- *Total Questions Asked Score*:
  - The fewer yes/no questions an examinee asks, the better is his/her performance on the *D-KEFS Twenty Questions Test*. The Total Questions Asked Score quantifies this aspect of performance and serves as a global achievement measure. Further, this variable is based on the number of yes/no questions asked until the target object is identified for each item, and these scores are summed across the four items on the test to obtain a raw score. The raw score is then converted to a scaled score.
- *Total Weighted Achievement Score*:
  - This measure was developed specifically to account for those individuals who fortuitously arrive at the correct answer after asking only one or two highly concrete questions. A total weighted achievement raw score is obtained by summing the raw scores for the four items. The raw score is converted to a scaled score.
- *Spatial Questions*:
  - This measure reflects the number of yes/no questions asked that attempt to eliminate objects based on their location on the stimulus page (e.g., "Is it in the top left side of the page?"). Only 2.7% of the normative sample asked spatial questions. The total raw score is summed across all four items and is transformed into a cumulative percentile rank.
- *Repeated and Set-Loss Questions*:
  - These two error measures reflect the number of repetitive and set-loss questions, respectively, summed across all four items of the test. The total raw scores for repetition errors and set-loss questions are transformed into cumulative percentile ranks.

D-KEFS Word Context Test

The *D-KEFS Word Context Test* evaluates executive functions in the verbal modality and assesses skills such as deductive reasoning, integration of multiple bits of information, hypothesis testing, and flexibility of thinking. Specifically, the test measures the examinee's ability to discover the meaning of made-up words or mystery words based on clues given in sentences. For each mystery word, the examinee is shown five sentences (clues) that help him or her to decode the meaning of the word. With each new clue sentence for the word, previously presented sentences are also displayed. The first few sentences for each word provide vague clues about the mystery word's meaning; clues become more detailed in subsequent sentences. The examinee's task is to decode the mystery word with as few clue sentences as possible while continuing to report the correct response to the remaining clue sentences of each item (Delis et al., 2001a, 2001b).

Scales the Test Yields

Several types of scores are derived from the *D-KEFS Word Context Test*: Total Consecutively Correct, Consistently Correct Ratio, Repeated Incorrect Responses, No/Don't Know Responses, and Correct-to-Incorrect Errors. An in-depth description of each score can be found within the *D-KEFS Examiner's Manual* (Delis et al., 2001a, 2001b). A brief summary of each score follows:

- *Total Consecutively Correct*:
  - The primary achievement measure for the *D-KEFS Word Context Test*. A raw score is calculated by totaling the *Total Consecutively Correct Score* across items 1–10. The raw score is then converted to a scaled score.
- *Consistently Correct Ratio*:
  - This measure is computed by dividing the Total Consecutively Correct Raw Score/First Sentence Correct Raw Score and multiplying it by 100. The denominator (First Sentence Correct Raw Score) reflects the first sentence in which the examinee provides a correct response. The raw score obtained for this measure is converted to a scaled score.
- *Repeated Incorrect*:
  - This measure reflects the number of incorrect responses that are repeated within the same item, summed across the ten items of the test. The raw score is converted to a scaled score.
- *No/Don't Know Responses*:
  - This measure is the number of clue sentences to which the examinee provides either no response or "don't know" responses, after being prompted by the examiner to take a use, summed across the ten items of the test. The raw score is converted to a scaled score.
- *Correct-to-Incorrect Errors*:
  - The number of times an examinee provides a correct response on an early clue sentence and then lose set and report an incorrect response for the very next clue sentence presented
- *Repeated Incorrect Responses*:
  - This measure reflects the number of incorrect responses that are repeated within the same item, summed across the ten items of the test. A raw score is calculated and converted to a scaled score.
- *No/Don't Know Responses*:
  - This measure is the number of clue sentences to which the examinee provides either no response or "don't know" responses, after being prompted by the examiner to take a guess. This measure is summed across one item of the test to obtain a raw score. The raw score is converted to a scaled score.
- *Correct-to-Incorrect Errors*:
  - This measure reflects the number of times the examinee provides a correct response on an early clue sentence and then loses set and reports and incorrect response for the very next clue sentence presented. The raw score is transformed into a cumulative percentile rank.

D-KEFS Tower Test

The *D-KEFS Tower Test* evaluates the examinee's ability to move disks varying in size from small to large across three pegs to build a designated tower in the fewest number of moves possible (Delis et al., 2001a, 2001b). When completing the task, the examinee must follow the following rules:

1. Move only one disk at a time.
2. Never place a larger disk over a smaller disk.

The *D-KEFS Tower Test* assesses several key executive functions, including spatial planning, rule learning, inhibition of impulsive and perseverative responding, and the ability to establish and maintain instructional set.

Scales the Test Yields

Several types of scores are derived from the *D-KEFS Tower Test*: Total Achievement Score and five optional process scores (Mean First-Move

Time, Time-Per-Move Ratio, Move Accuracy Ratio, Total Rule Violations, and Rule-Violations-Per-Item Ratio). An in-depth description of each score can be found within the *D-KEFS Examiner's Manual* (Delis et al., 2001a, 2001b). A brief summary of each score follows:

- *Total Achievement Score*:
  - This score reflects the sum of the achievement scores, including bonus points, for all items administered. The raw score is converted to a scaled score.
- *Mean First-Move Time*:
  - This score reflects the average of the examinee's first-move times. Specifically, the score is computed by taking the sum of the examinee's first-move times (in seconds) for all the items administered divided by the number of items administered. A raw score is converted into a scaled score.
- *Time-Per-Move Ratio*:
  - This measure reflects the average time the examinee takes to make each of his or her moves. The score is computed by summing the completion times for all items administered and divided by the total number of moves made for all items administered. The raw score is then converted into a scaled score.
- *Move Accuracy Ratio*:
  - This measure reflects the efficiency with which the examinee constructed the towers. The total number of moves used by the examinee across all items administered is divided by the fewest number of moves required across all items administered. The raw score is converted to a scaled score.
- *Total Rule Violations*:
  - This measure represents the total number of rule violations committed by the examinee across all items administered. The two rule violations of the test include moving more than one disk at a time and placing a larger disk on a smaller disk. The raw score is converted into a cumulative percentile rank.
- *Rule-Violations-Per-Item Ratio*:
  - This measure reflects the average number of rule violations made by the examinee relative to the number of items administered. The obtained raw score is converted to a scaled score.

D-KEFS Proverb Test

The *D-KEFS Proverb Test* qualitatively evaluates the nature of an individual's verbal abstraction skills (Delis et al., 2001a, 2001b). The D-KEFS Proverb Test consists of eight sayings that are presented in two conditions:
- *Condition 1*: *Free Inquiry*
  - Proverbs are read individually to the examinees, who attempt to interpret them orally without assistance or cues.
- *Condition 2*: *Multiple Choice*
  - The same eight proverbs are presented individually along with four alternative interpretations from which the examinee must select the best one.
  - The set of multiple-choice response alternatives for each proverb consists of:
    A correct abstract interpretation
    A correct concrete interpretation
    An incorrect, phonemically similar response
    An unrelated saying

Scales the Test Yields

Several types of scores are derived from the *D-KEFS Proverb Test*: normative data are provided for seven measures for the Free Inquiry condition and six variables for the Multiple Choice condition. An in-depth description of each score can be found within the *D-KEFS Examiner's Manual*. A brief summary of each score follows:
Scores for Free Inquiry Condition
- *Total Achievement Score*:
  - The primary measure for the Free Inquiry condition of the D-KEFS Proverb Test. The raw score is based on the sum of the individual achievement scores of all eight items. The raw score is converted to a scaled score.
- *Common Proverb Achievement Score*:
  - This measure is based on the examinee's performance on the first five items on the D-KEFS Proverb Test, which consist of high-frequency sayings with which most people are likely to be familiar. The raw score is the sum of the raw achievement scores for the first five items. The raw score is then converted to a scaled score.

- *Uncommon Proverb Achievement Score*:
  - This measure is based on the examinee's performance on the last three items on the D-KEFS Proverb Test, which consist of low-frequency sayings with which most people are less likely to have heard. The raw score is the sum of the achievement scores for the last three items. The raw score is converted to a scaled score.
- *Accuracy Only Score*:
  - This measure reflects the extent to which examinees can provide accurate interpretations of the proverbs regardless of whether their interpretations are abstract or concrete. The raw score is the sum of the accuracy scores for the eight proverbs; the raw score is converted to a scaled score.
- *Abstraction Only Score*:
  - This measure reflects the degree to which examinees provide abstract responses to the proverbs regardless of the degree of accuracy of their interpretations. The raw score is the sum of the abstraction scores for the eight proverbs. The raw score is converted to a scaled score.
- *No/Don't Know Response and Repeated Responses*:
  - This measure reflects the frequency with which an examinee makes these error types across the eight proverbs of the Free Inquiry condition. The raw score for the no/don't know and repeated responses measures is converted into cumulative percentile ranks.

Scores for Multiple-Choice Condition
- *Total Achievement Score*:
  - This measure reflects the sum of an examinee's item achievement scores on the eight items. The raw score is converted into a cumulative percentile rank.
- *Common and Uncommon Proverb Achievement Scores*:
  - The raw score for the common proverb achievement index is the sum of an examinee's item achievement scores on the first five items of the Multiple Choice condition. Raw scores are transformed into a cumulative percentile rank.
- *Endorsement Measures*:
  - This measure reflects the number of times the examinee endorses each type of alternative response (correct abstract, correct concrete, incorrect phonemic, or incorrect unrelated) summed across the items and transformed into a cumulative percentile rank (Delis,).

Administration and Scoring

Tips on Administration

The D-KEFS tests are cognitive assessment instruments, and therefore examiners must have formal training and experience in the assessment of intellectual and cognitive functions (Delis et al., 2001a, 2001b). With one exception, the D-KEFS tests were designed for use with both children and adults (ages 8–89). The D-KEFS Proverb Test was designed for adolescents and adults (ages 16–89).

Examiners should become familiar with the standardized administration procedures for each of the nine tests that make up the D-KEFS. Each of the nine D-KEFS tests stands alone; therefore, the examiner should pick the tests that will best provide the information needed to answer the referral question.

Scoring the Tests

Scaled scores, cumulative percentile ranks, contrast measures, combined scaled scores, and various optional scores can be derived from each test. For most of the measures provided by the D-KEFS tests, the raw scores are converted to scaled scores, with a mean of 10 and a standard deviation of 3. An in-depth description of scoring procedures for each test can be found within the *D-KEFS Examiner's Manual* (Delis et al., 2001a, 2001b).

Use of Scoring Software

The *D-KEFS Scoring Assistant* software automatically computes the standardized scores for both the primary and optional measures of the standard and alternate forms of the D-KEFS and prints them in a report format. Thus, the scoring software greatly enhances the efficiency with which the D-KEFS can be used in clinical practice (p. 29, manual).

Standardization, Norms, and Psychometrics

Characteristics of the Standardization Sample

The D-KEFS was standardized on a nationally representative, stratified sample of 1,750 children, adolescents, and adults, ages 8–89 years. Stratification was based on age, sex, race/ethnicity, years of education, and geographic region. The 2000 US Census figures were used as target values for the composition of the D-KEFS normative sample (Delis et al., 2001a, 2001b, p. 1).

Sixteen age groups make up the D-KEFS normative sample: 8 years, 9 years, 10 years, 11 years, 12 years, 13 years, 14 years, 15 years, 16–19 years, 20–29 years, 30–39 years, 40–49 years, 50–59 years, 60–69 years, 70–79 years, and 80–89 years. The D-KEFS sample consisted roughly of equal proportions of men and women at each age group, with the exception of the older age groups, which had more women than men, which is consistent with the census data (Delis et al., 2001a, 2001b, pp. 1–2).

The proportion of African-American, Hispanic, white, and other racial/ethnic groups sampled were stratified to approximate the 2000 US Census population estimates. Additionally, the D-KEFS sample was divided into the five major educational groups used by the US Census: less than or equal to 8 years of education, 9–11 years, 12 years, 13–15 years, and 16 years or more. Finally, the United States was divided into four major geographical areas as specified by the US Census data: northeast, north central, south, and west. All regions are well represented in the normative sample (Delis et al., 2001a, 2001b, p. 2). A more in-depth description of the normative sample can be found in the *D-KEFS Technical Manual*.

Reliability of the Scales

Reliability indicates the consistency of measurements. Consistency has several meanings, consistent within itself (internal reliability), consistent over time (test-retest reliability), and consistent with an alternate form of the measure (alternate-form reliability) (Sattler, 2008). The psychometric properties of internal consistency, stability coefficients, and alternate-form reliability were determined for the D-KEFS instruments. These measures of reliability provide a basis for deriving the standard error of measurement and confidence intervals. An in-depth synopsis of all reliability measures can be found within the *D-KEFS Technical Manual* (Delis et al., 2001a, 2001b). A brief description of each reliability scale follows:

Internal Consistency: It assumes that all items measure the same trait or construct. It is established by dividing the test into two equivalent halves (split-half reliability). This division creates two alternative forms of the test. The most common way of dividing the test is to assign odd-numbered items to one form and even-numbered items to the other (Sattler, 2008). Internal Consistency for the nine tests that make up the D-KEFS instrument is as follows:
- *D-KEFS Trail Making Test* (Combined Number and Letter Sequencing Composite Score) ranges from .60 to .81
- *D-KEFS Verbal Fluency Test* by age group and per conditions 1–3:
  – Condition 1: .68–.90
  – Condition 2: .53–.76
  – Condition 3 (Total Correct): .37–.68
  – Condition 3 (Total Switching): .51–.76
- *D-KEFS Design Fluency Test*: Not reported due to time constraints

- **D-KEFS Color-Word Interference Test** (Combined Color Naming and Word Reading Composite Score) ranges from .62 to .86
- **D-KEFS Sorting Test** by age group and per conditions 1–3:
  - Condition 1: .55–.86
  - Condition 2: .55–.84
  - Condition 3: .62–.81
- **D-KEFS Twenty Questions Test**: Internal Consistency for this test is reported based on age group and by Initial Abstraction and Total Weighted Achievement
  - Initial Abstraction: Based on age group, Internal Consistency ranges from .72 to .87
  - Total Weighted Achievement: Based on age group, Internal Consistency ranges from .10 to .55
- **D-KEFS Word Context Test**: Internal Consistency for this test is reported based on age group and Total Consecutively Correct
  - Total Consecutively Correct: Based on age group, Internal Consistency for this test ranges from .47 to .74
- **D-KEFS Tower Test**: Internal Consistency for this test is reported based on age group and Total Achievement
  - Total Achievement: Based on age group, internal consistency for this test ranges from .43 to .84
- **D-KEFS Proverb Test**: Internal Consistency on this test is reported based on age group and Total Achievement: Free Inquiry
  - Total Achievement: Free Inquiry: Based on age group, internal consistency for this test ranges from .68 to .81

Additional forms of reliability were also established for the nine D-KEFS tests; these included Alternate-Form Reliability and Test-Retest Reliability. *Alternate-Form Reliability* is the equivalent or parallel form reliability determined by administering two equivalent tests to the same group of examinees (Sattler, 2008). Specific correlation coefficients can be found within *Chapter 2 of the D-KEFS Technical Manual* (Delis et al., 2001a, 2001b). Moderate to high correlations coefficients were found.

Test-Retest is an index of stability a measure of how consistent scores are over time (Sattler, 2008). Test-Retest Reliability coefficients for the nine tests that make up the D-KEFS instrument can be found within *Chapter 2 of the D-KEFS Technical Manual* (Delis et al., 2001a, 2001b). Test-retest correlations range from moderate to high across the D-KEFS tests.

Use of the Test

Interpretation Methods

The D-KEFS variables measure different aspects of test performance, to include accuracy of responses, error rates, and response times. Most of the D-KEFS measures provide scaled scores. The directionality of the scaled scores is used to interpret performance; specifically, the higher the scaled score, the better the performance. This rule pertains to measures reflecting (1) accuracy scores, (2) error rates (e.g., the more errors generated, the lower the scaled score), or (3) completion times (e.g., the slower the time to solve an item or complete a condition, the lower the scaled score). Moreover, there are two types of measures in which either low- or high-scaled scores reflect different types of cognitive problems (Delis et al., 2001a, 2001b).

Contrast measures may signal cognitive difficulties if the scaled score is either too low or too high. For example, on the D-KEFS Verbal Fluency Test, one of the contrast measures is letter fluency versus category fluency. For this measure, a contrast scaled score of 7 or lower may reflect greater difficulty with letter fluency than with category fluency. In contrast, a scaled score of 13 or higher may indicate greater difficulty with category fluency than with letter fluency (Delis et al., 2001a, 2001b).

Certain time variables that measure the examinee's latency to make a response might also reflect different types of cognitive problems, depending whether the scaled score is too high or too low. For example, on the D-KEFS Tower Test, the mean first-move time variable measures the average time the examinee takes to initiate the first move for each item administered. This measure offers an estimate of the examinee's initial planning time before engaging in problem-solving behavior. According to Delis et al. (2001a, 2001b) examinees that are too slow or too quick in generating their first moves both may be demonstrating cognitive problems for different reasons (e.g., those with activation problems may take longer for initial response, while those with impulsive tendencies may respond too fast).

The directionality of the cumulative percentile ranks is used to interpret examinee performance. Cumulative percentile ranks on the D-KEFS were scaled to reflect the percentage of the normative sample that obtained raw scores that were equal to or worse than the raw score obtained by the examinee. A cumulative percentage rank of 10 % on an error measure indicates that 10 % of the normative sample made the same or more errors on that measure.

Various levels of interpretation may be used to interpret the process-oriented tests that make up the D-KEFS instruments. Specifically, performance can be interpreted at four general levels: interpretation of achievement measures, interpretation of process measures, integration of D-KEFS findings with results from the entire cognitive and motor test battery, and inferences regarding risk factors for cognitive difficulties. A brief description of each interpretation method follows; however, more in-depth interpretation can be found within *Chapter 2 of the D-KEFS Examiner's Manual* (Delis et al., 2001a, 2001b).

Interpretation of the D-KEFS tests or conditions through the global achievement measures allow the examiner to gauge an examinee's overall level of performance on the task. Therefore, achievement measures usually provide an initial level of interpretation in addressing whether or not an examinee generally performed well or poorly on the test (Delis et al., 2001a, 2001b). Additionally, some D-KEFS tests provide achievement measures for baseline conditions that isolate more fundamental cognitive or motor skills that are needed to perform the higher-level conditions of the test.

The D-KEFS results can also be interpreted through process measures. According to Delis et al. (2001a, 2001b), the main intention of the process approach to cognitive assessment is the use of multiple measures that are designed to isolate and quantify specific aspects of test performance. Consequently, these measures provide pertinent information necessary to quantify the examinee's performance in comparison to normative data for a wide range of cognitive abilities, which include problem-solving strategies, ratio measures, error types, and temporal aspects of responding. Finally, process measures allow the examiner to pinpoint key areas of deficit and thereby guide appropriate intervention selection.

The integration of the D-KEFS findings with results from the entire battery given to the examinee is of utmost importance. Specifically, the cognitive deficit found through the administration of the D-KEFS test(s) should be validated by the use of several test measures that are also designed primarily for evaluating that ability area or are highly dependent on that ability for successful performance. The determination of whether the examinee's overall profile of cognitive strengths and weaknesses is typical of those seen in specific patient populations or atypical for most clinical disorders should also be determined (Delis et al., 2001a, 2001b).

The final level of interpretation relates to the examiner's understanding of risk factors for cognitive difficulties. When interpreting low scores on the D-KEFS, the examiner should not automatically associate the low scores to brain damage. Instead, the examiner should consider other reasons for cognitive difficulties. Possible neurostructural etiologies of cognitive deficits include prenatal exposure to alcohol, severe birth trauma, traumatic brain injuries (TBIs), or environmental neurotoxic exposure. Additionally, the examiner must consider possible nonneurostructural factors, which include depression, anxiety, obsessive thoughts, pain symptoms, or sleep deprivation and fatigue (Delis et al., 2001a, 2001b). The examiner should be familiar with characteristics of such factors and consider them within the interpretation of test results.

Identification of Special Populations

Results of the D-KEFS test(s) should be integrated and interpreted in consideration with results obtained from other cognitive and achievement tests. When determining eligibility for special education services under one of the 13 categories listed within the Individuals with Disability Education Act (IDEA, 2004), the D-KEFS results should be used to assist in the verification of the disability criteria. Specifically, when considering a student as having a Learning Disability (LD), evaluation personnel should use a variety of assessment results (to include the D-KEFS) to investigate patterns of strengths and weaknesses in cognitive and or achievement (IDEA, 2004). Further, by investigating the examinee's performance in the areas of higher-level executive functioning, the D-KEFS results can be utilized to verify characteristics of Attention Deficit Disorder (ADD) and/or Attention Deficit Hyperactivity Disorder (ADHD) and TBI. In-depth clinical interpretation of scores can be found within the *D-KEFS Examiner's Manual* for each of the nine tests (Delis et al., 2001a, 2001b).

Interventions Based on Test Results

Results of the examinee's performance on any of the nine D-KEFS tests can be used to guide intervention selection. Each of the nine individual D-KEFS tests targets low- and high-level executive functions, allowing for the identification of specific areas of weakness. Consequently, an in-depth clinical analysis of scores should be used to pinpoint specific areas of deficit resulting in the appropriate identification and selection of intervention practices.

Validity

Relationships to Other Similar Measures

The validity of the D-KEFS tests (e.g., the Stroop procedure, Trail Making Test, verbal and design fluency tests, tower tests, twenty questions procedure, and proverb interpretations) has been demonstrated in numerous neuropsychological studies conducted over the past 50 years or more.

The evidence of validity has been provided in terms of sensitivity of the tests in the detection of brain damage, specifically in the frontal lobe area, and in the ability of the tests to measure areas of higher-level executive functions (Delis et al., 2001a, 2001b). Numerous studies have been conducted which investigated the validity of the nine D-KEFS tests in comparison to other cognitive measures, specifically the *California Verbal Learning Test-Second Edition* (CVLT-II; Delis, Kaplan, Kramer, & Ober, 2000) and the *Wisconsin Card Sorting Test* (WCST; Heaton, Chelune, Talley, Kay, & Curtiss, 1993).

The validity study between the CVLT-II and the D-KEFS consisted of 292 adults (33 % male and 63 % female). Standardized scores from the two instruments were used in the analyses. Correlations between the CVLT-II and the D-KEFS Sorting Test measures results found the CVLT-II immediate and delayed recall measures correlated in the low positive range with the key Sorting Test measures, including the confirmed correct sorts, free sorting description score, and sort recognition description score indices (Delis et al., 2001a, 2001b). Additionally, the vast majority of the correlations were not significant, indicating little overlap between the functions assessed by the two instruments.

A small pilot study was also conducted to investigate the relationship between the WCST and the D-KEFS tests. The sample size of the study consisted of 23 adults (65 % male and 35 % female). Correlations were run between the test scores obtained on the WCST and the D-KEFS tests. According to Delis et al. (2001a, 2001b), the number of categories completed on the WCST tended to have moderate correlations with several of the primary measures of the D-KEFS tests. Additionally, findings indicated that the perseverative responses measures of the WCST tended to correlate at somewhat lower levels with key D-KEFS measures. Complete results can be found in Table 3.59 of the D-KEFS Technical Manual (Delis et al., 2001a, 2001b).

Numerous studies have been conducted over the years investigating the use of the D-KEFS tests with individuals diagnosed with fetal alcohol syndrome (FAS), Alzheimer's disease, and Huntington disease. Additionally, studies were conducted which analyzed the evidence of validity across the nine D-KEFS tests. Results for all the validity studies can be found within *Chapter 3 of the D-KEFS Technical Manual* (Delis et al., 2001a, 2001b).

Summary and Conclusions

The D-KEFS is the first nationally standardized set of tests to evaluate higher-cognitive functions (executive functions) in both children and adults. Published in 2001, the D-KEFS includes nine stand-alone tests that measure a wide array of verbal and nonverbal executive functions. The D-KEFS allows for the identification of areas of strengths and weaknesses in lower- and higher-level executive functions. Thus, results of each of the nine tests can be clinically interpreted and used with other assessment results in the validation of various disabilities (e.g., LD, ADD, ADHD, or TBI). Finally, an in-depth analysis of examinee performance can be used to pinpoint specific areas of weakness, making intervention selection more appropriate.

References

Dehn, M. (2006). *Essentials of processing assessment*. Hoboken, NJ: Wiley.

Delis, D.C., Kramer, J., Kaplan, E., & Ober, B. (2000). California Verbal Learning Test-Second Edition (CVLT-II). San Antonio, TX: The Psychological Corporation.

Delis, D., Kaplan, E., & Kramer, J. (2001a). *Delis Kaplan (D-KEFS) examiner's manual*. San Antonio, TX: NCS Pearson, Inc.

Delis, D., Kaplan, E., & Kramer, J. (2001b). *Delis Kaplan (D-KEFS) technical manual*. San Antonio, TX: NCS Pearson, Inc.

Heaton, R. K., Chelune, G. J., Talley, J. L., Kay, G. G., & Curtiss, G. (1993). *Wisconsin Card Sorting Test manual—Revised and expanded*. Lutz, FL: Psychological Assessment Resource, Inc.

Individuals with Disabilities Education Improvement Act of 2004 (IDEA), Pub. L. No. 102-446, 118 Stat. 2647 (2004).

McCloskey, G., Perkins, L., & Divner, B. (2009). *Assessment and intervention for executive function difficulties*. New York, NY: Routledge Taylor & Francis Group.

Sattler, J. M. (2008). *Assessment of children: Cognitive applications* (5th ed., pp. 102–104). La Mesa, CA: Jerome M. Sattler.

Wecker, N. S., Kramer, J. H., Wisniewski, A., Delis, D. C., & Kaplan, E. (2000). Age effects on executive ability. *Neuropsychology, 14*, 409–414.

Using the Comprehensive Executive Function Inventory (CEFI) to Assess Executive Function: From Theory to Application

Jack A. Naglieri and Sam Goldstein

Interest in executive function has grown exponentially in recent years because the concept helps us better understand the differences between what a child does and what a child can do in the classroom and in life (Naglieri & Goldstein, 2013). Interest in the mental application of human brain behavior relationships has, to a significant degree, driven interest in executive function. Nonetheless, this concept is in a relatively early stage of development (McCloskey, Perkins, & Van Divner, 2009). As in all areas of science and practice, the information we obtain from any evaluation depends upon the quality of the instruments used. As interest in executive function and its impact upon children's development has grown, so has an interest in developing valid and reliable instruments to measure all aspects of behaviors related to this important ability. Better assessment tools yield more valid and reliable decisions, more useful information, and ultimately greater benefit to the clients. This chapter includes a discussion of the psychometric qualities of measures of executive function.

J.A. Naglieri (✉)
University of Virginia, Curry School of Education, White Post Road 6622, 20121, Charlottesville, VA, USA
e-mail: jnaglieri@gmail.com

S. Goldstein
Neurology, Learning and Behavior Center,
University of Utah School of Medicine,
230 South 500 East, Suite 100, Salt Lake City, UT 84102, USA

Measuring Executive Function with the CEFI

Our main goal when developing the Comprehensive Executive Function Inventory (CEFI; Naglieri & Goldstein, 2013) was to create a rating scale to evaluate behaviors associated with executive function. We chose to measure *behaviors* related to executive function to evaluate how a child actually behaves in everyday life (e.g., vocational and academic settings). This may or may not be consistent with his or her ability to solve problems requiring executive function as measured in a formal testing session. That is, a person may have the ability to function (e.g., behave) in a particular manner, but not apply the ability to do so. Ultimately, the assessment goal is (a) to use rating scales to evaluate behaviors associated with the observable actions related to the conceptual definition of executive function *and* (b) to use tests that are administered to an examinee to evaluate executive function ability. Both rating scales and direct tests help us better understand if a person has the ability to make decisions about how and what to do (definition of EF) and how a person actually behaves to achieves goals. Assessment of the ability to and the level of functioning related to that ability demands a definition of executive function and tests to measure it.

We began the development of the CEFI with an examination of the literature on the concept of executive function which is often described as having three components: inhibitory control, set shifting (flexibility), and working memory (Davidson, Amso, Anderson, & Diamond, 2006; Miyake et al., 2000; Zelazo & Müller, 2002). Some researchers suggest that these three concepts are relatively independent (Wiebe, Espy, & Charak, 2008) and therefore propose a multidimensional model of executive function (Friedman et al., 2006; Miyake et al., 2000). Others argue that executive function is a unitary construct (e.g., Duncan & Miller, 2002; Miller & Cohen, 2001). There are those that further suggest both views may be correct. When developing the CEFI we chose not to follow any of these three approaches but rather to let the results of our research on the statistical relationships among behaviors related to executive function using a large representative sample of children and adolescents as rated by parents, teachers, and the individual himself or herself. This would allow us to build a rating scale of executive function for children and adolescents from 5 to 18 years of age (parent and teacher raters) and 12–18 (self-raters) that was supported by strong scientific evidence and addressed the question of dimensionality.

The initial content of the CEFI was determined following a comprehensive review of the literature and the authors' clinical and research experience regarding the conceptualization and assessment of executive function. A wide variety of items were written to capture key components such as time management, working memory, decision making, goal-directed behavior, planning, resistance to distraction, persistence, attention to detail, perspective taking, sustained attention, cueing, shifting, stopping and starting, motor inhibition, motivation, flexibility, regulation, and stress tolerance. These items could be organized in many ways (e.g., cognitive, behavioral, emotional) on the basis of their content, but we decided to include many aspects of executive function and let the results of our analyses determine the structure of the final scale.

We used the CEFI standardization sample which included 3,500 ratings (2,800 for the 5–18-year-olds rated by parents and teachers, and 700 for the 12–18-year-olds who completed the self-report form). These samples comprised ratings of 50 males and 50 females at each whole-year age and were representative of the US population across several demographic variables. In addition to collecting data for creation of the CEFI norms, we obtained 872 ratings of children with various diagnoses, including 432 with a primary diagnosis of Attention-Deficit/Hyperactivity Disorder (ADHD), 202 children/youths identified as having a learning disorder, 99 with a diagnosis of an Autism Spectrum Disorder, 95 with a Mood Disorder diagnosis, and 44 cases diagnosed with other disorders (e.g., Anxiety Disorders, Oppositional Defiant Disorder, and Traumatic Brain Injury). (See the CEFI Manual for more details about these samples.) All these data were used to answer the question "Is executive function a unitary or multidimensional concept?"

Executive Function or Executive Functions?

The plan for determining how many factors explained the relationships among the many items included in the initial version of the CEFI was based on two ways of examining the data—by factor analysis at the item level and using items organized into scales based on the content of the items. To do so we split the standardization samples (for ratings by parents, teachers, and self-ratings) in half and used one method on each of the paired samples. This provided a means of confirming the solution within each type of rater. Thus, six-factor analytic procedures were obtained (two per type of rater). This level of replication provided an excellent opportunity to understand the relationships among these behaviors associated with the concept of executive function.

Exploratory factor analysis at the item level (using principal axis extraction and direct oblimin rotation) was performed and a series of procedures were applied to evaluate the number of factors to retain (Velicer, Eaton, & Fava, 2000; Zwick & Velicer, 1986). A non-graphical solution to the scree plot test (Raiche & Magis, 2010) and the very simple solution (VSS) criterion

(Revelle, 2011; Revelle & Rocklin, 1979) indicated that only one factor should be retained. The ratio of the first and second eigenvalues was greater than four for all three forms, which is a common rule to support unidimensionality (Zwick & Velicer, 1986). These results supported a unidimensional factor structure for the CEFI items as rated by parents, teachers, and self-ratings. One factor clearly explained the relationships of the various behaviors included in the CEFI. These findings were further verified with the second group of factor analyses.

The second level of exploratory factor analysis (i.e., using half of the data from the normative samples) was conducted to determine if the nine CEFI Scales formed one or multiple factors. Both the Kaiser rule (eigenvalues >1) and the Eigenvalue Ratio criterion (>4) were met by the data, unequivocally supporting the conclusion that the CEFI Scales are best described as one single factor. These findings are strong evidence of unidimensionality (Slocum-Gori & Zumbo, 2011) of the CEFI at the scale level. We then performed a third series of factor analyses.

The consistency of the unidimensional factor structure of executive function as measured by the CEFI was further evaluated across gender, age group, race/ethnicity, and clinical/educational statuses. The exploratory factor analysis procedure was conducted for each demographic group to determine if the factor structure was consistent across genders (males vs. females), ages (5–11 vs. 12–18 years on the parent and teacher forms; 12–15 vs. 16–18 years on the self-report form), races/ethnicities (white vs. nonwhite), and clinical/educational statuses (nonclinical vs. clinical/educational). The factor loadings for the groups were correlated using the coefficient of congruence (Abdi, 2010); results revealed a very high degree of consistency across all groups, indicating unidimensionality. These findings, in conjunction with the item-level and scale-level exploratory factor analyses, lead to the conclusion that the concept of executive function as measured by the behaviors included in the CEFI should be considered a unidimensional construct for parent, teacher, and self-ratings. See the CEFI Manual (Naglieri & Goldstein, 2013) for more information.

Description of the CEFI

Structure of the Scales

The CEFI comprises 100 items which are combined to yield a Full Scale score set to have a mean of 100 and standard deviation of 15 scaled such that high scores imply good executive function. It is this score which represents the strongest examination of a child's or adolescent's behaviors in the natural environment that are associated with the concept of executive function. That is, the Full Scale describes how a child or adolescent does what he or she intends to do. In clinical practice, however, it is important to be able to explain the various groups of behaviors that comprise the Full Scale score, particularly for intervention planning. For this reason the CEFI provides additional scores to cover the content areas of Attention (12 items), Emotion Regulation (9 items), Flexibility (7 items), Inhibitory Control (10 items), Initiation (10 items), Organization (10 items), Planning (11 items), Self-Monitoring (10 items), and Working Memory (11 items). Items for these scales were chosen based on their content and utility for developing intervention strategies specific to each content area. Each of these nine scales is set on the same metric as the Full Scale (Fig. 14.1).

Reliability of the CEFI

Internal consistency of the scores on the CEFI, assessed using Cronbach's alpha, is summarized in Table 14.1 for the parent, teacher, and self-ratings used to standardize the scales. The results indicate that the Full Scale has excellent internal reliability. The Full Scale coefficients for the normative samples were all .97 or higher and results for the nine CEFI Scales were also high. The median CEFI reliability coefficient for the normative sample was .89 for parent raters, .93 for teacher raters, and .80 for the self-report. Naglieri and Goldstein (2013) also reported that the CEFI Scales also had good reliability estimates in the clinical and special educational

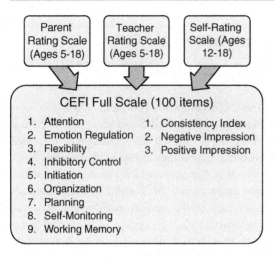

Fig. 14.1 Structure of the Comprehensive Executive Function Inventory for parent, teacher, and self-raters. Copyright (c) 2013 Jack A. Naglieri, All rights reserved

Table 14.1 CEFI internal reliability coefficients for the normative sample

	Parent raters ($N=1,396$)	Teacher raters ($N=1,400$)	Self-raters ($N=700$)
Full scale	.99	.99	.97
Attention	.93	.96	.86
Emotion regulation	.89	.93	.78
Flexibility	.85	.90	.77
Inhibitory control	.90	.94	.80
Initiation	.89	.93	.80
Organization	.91	.94	.85
Planning	.92	.96	.85
Self-monitoring	.87	.92	.78

samples (median Full scale values of .86, .91, and .81 for the parent, teacher, and self-report forms, respectively). Other types of reliability were also reported in the manual.

Naglieri and Goldstein (2013) reported that the CEFI has excellent test-retest reliability with correlations ranging from .77 to .91 (all $p<.001$) for the Full Scale and from .74 to .91 ($p<.001$) across the nine scales. Inter-rater reliability was also reported in the CEFI Manual. The Inter-rater reliability for the Full Scale was .88 between parents and .68 between teachers. Similar levels of rater agreement were found across all CEFI Scales for parent raters (correlations ranged from .73 to .86), and moderate correlations were found for teacher raters (correlations ranged from .54 to .68).

Validity of the CEFI

Several types of validity are presented in the CEFI Manual such as factor analyses across groups and inter-rater agreement. The construct validity evidence showing that the behaviors measured by the CEFI are best described as one factor presented earlier in this chapter strongly supports the use of the Full Scale score for describing a child or adolescent. Our behavioral data from the CEFI provide very strong evidence to answer the question Jurado and Rosselli (2007) asked: "whether there is one single underlying ability that can explain all the components of executive functioning or whether these components constitute related but distinct processes" (p. 214). Clearly, the application of two different ways of organizing the items (individually or in groups) across the three types of raters all shows that one factor best describes these behaviors which describe executive function. The evidence presented here and in the CEFI Manual provides very strong support for the construct validity of the scale. This finding means that users should use the Full Scale score when making decisions about how well or poorly a child or adolescent is functioning when making decisions about what and how to behave. (We will address the use of the nine treatment scales in the Interpretation section of this chapter.)

In the remainder of this section in this chapter, we will focus on the criterion-related validity of the CEFI, which includes examination of the mean differences between the general population (matched on age, gender, race/ethnicity, and parental education) and samples of children and youths previously diagnosed with specific clinical or learning disorders. We will also discuss how the CEFI scores are related to other measures of executive function, intelligence, neurocognitive abilities, and academic achievement.

CEFI Scores Across Groups

Executive function problems are typical for children and adolescents with many different psychological and educational problems (Willcutt, Doyle, Nigg, Faraone, & Pennington, 2005). Researchers have found executive function deficits in those with ADHD and Mood Disorders (e.g., Weyandt et al., in press), as well as Autism Spectrum Disorders (ASD; e.g., Gilbert, Bird, Brindley, Frith, & Burgess, 2008; Happé, Booth, Charlton, & Hughes, 2006; Solomon et al., 2009), Conduct and Oppositional Defiant Disorders (Herba, Tranah, Rubia, & Yule, 2006; Morgan & Lilienfeld, 2000), Anxiety and Depression and learning disabilities (Naglieri & Gottling, 1995). This means that CEFI scores will likely be similar across a variety of clinical and developmentally delayed groups. For this reason, the criterion-related validity of the CEFI was examined for groups of children and adolescents who were previously diagnosed with a clinical disorder of either ADHD, an ASD (parent and teacher forms only), or a Mood Disorder (including Major Depressive Disorder, Dysthymia, or Bipolar Disorder) and those who had previously been identified as having a learning disorder (LD). The LD group was not limited to one type of academic difficulty (e.g., reading decoding), and therefore we anticipated that this group would be less likely to have low executive function scores.

CEFI data obtained for the ADHD, ASD, Mood Disorder, and LD groups were used only if the following criteria were met: (a) a single primary diagnosis was indicated, (b) the diagnosis was made by a qualified professional (e.g., psychiatrist, psychologist), and (c) appropriate methods (e.g., record review, rating scales, observation, interview) were used during diagnosis. For all those in the ADHD, Mood Disorders, and ASD samples, relevant criteria were assessed using either the Diagnostic and Statistical Manual of Mental Disorders: Text Revision or the International Statistical Classification of Diseases and Health Related Problems 10th Revision. For the LD sample, cases were identified as having an LD if they met DSM-IV-TR or ICD-10 criteria, or they were identified using federal (e.g., IDEA) or state (e.g., New York's Part 200) guidelines.

CEFI standard scores on the parent, teacher, and self-report forms for each clinical and educational group were compared to scores obtained from ratings of a matched group from the standardization sample without a clinical diagnosis or a learning disorder. The demographic characteristics of the clinical and educational samples and the groups matched on sex, race/ethnicity, and parental education are described in the CEFI Manual (Tables 8.16 through 8.18). The results are provided in Fig. 14.2.

As anticipated the CEFI Full Scale standard scores for the various group and raters showed that the samples of children and adolescents with ADHD, ASD, and Mood Disorder had low scores when compared to matched samples from the standardization sample. There were moderate to large effects sizes across all forms for all three groups as anticipated. Similarly the differences between the LD group and general population samples were not expected to be as large as those between the clinical groups and their respective matched samples. The LD group differed significantly in the parent and teacher forms, but not in the self-report form. Overall, these results were consistent with expectations and suggest that the CEFI is sensitive to differences in behaviors related to executive function: *how* children and youth do what they do.

CEFI and BRIEF

The criterion-related validity of the CEFI was studied by Naglieri and Goldstein (2013) by comparing it with the Behavior Rating Inventory of Executive Functioning (BRIEF; Gioia, Isquith, Guy, & Kenworthy, 2000; Guy, Isquith, & Gioia, 2004). Ratings were obtained for the CEFI and BRIEF completed by parents ($N=108$) and teachers ($N=83$). For individuals aged 5–18 years, the BRIEF and CEFI self-report ($N=61$) for youths aged 12–18 were used. All of the participants had been diagnosed with (a) ADHD or (b) Anxiety Disorder ($N=6$), Autism Spectrum Disorder ($N=10$), Mood Disorder ($N=6$), generic learning disability ($N=15$), or other clinical disorders

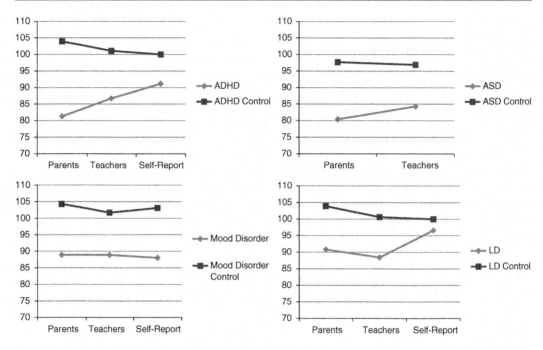

Fig. 14.2 Scores for CEFI parent, teacher, and self-reports for four groups of children and adolescents. Copyright (c) 2013 Jack A. Naglieri, All rights reserved

($N=8$) (see Table 8.23 in the CEFI Manual for more information).

The CEFI and the BRIEF are scaled differently. The CEFI scores are set to have a mean of 100 and standard deviation of 15 with high scores indicating good executive function. The BRIEF yields T-scores which have a mean of 50 and standard deviation of 10 and is configured so that higher scores indicate poor executive function. To simplify the comparison between the BRIEF and the CEFI, the BRIEF scores were converted to a mean of 100 and standard deviation of 15, and the scores were also inverted so that low scores indicate poor executive function. These mathematical modifications of the BRIEF scores have no influence on the results but they do allow for easier comparisons to the CEFI.

Means, standard deviations, sample sizes, and correlations of the BRIEF and CEFI total scores for the sample of individuals with ADHD and the mixed sample are provided in Tables 14.2 and 14.3.

Table 14.2 Means, standard deviations, sample sizes of the BRIEF and CEFI total scores for the sample of individuals with ADHD and the mixed Clinical sample

Form	CEFI			BRIEF			Effect size
	N	Mn	SD	N	Mn	SD	
ADHD							
Parent	57	81.9	11.7	57	71.8	13.7	.79
Teacher	51	87.4	11.1	51	71.2	23.7	.88
Self-rating	32	90.2	14.2	32	86.7	15.9	.23
Mixed group							
Parent	53	83.9	12.9	53	74.9	16.8	.60
Teacher	55	90.8	13.5	55	77.4	23.9	.69
Self-rating	30	96.6	19.7	30	93.8	22	.13

Note: Effect sizes of .2 are considered small, .5 medium, and .8 large

Examination of the correlations between these two rating scales for both samples indicates that the CEFI Full Scale and BRIEF Global Executive Composite scores, as rated by parents, teachers, and self-ratings, are highly correlated, providing

Table 14.3 Correlations of the BRIEF and CEFI total scores for the sample of individuals with ADHD and Mixed Clinical Group mixed sample

Form	ADHD		Mixed group	
	N	r	N	r
Parent	57	.85	53	.78
Teacher	51	.64	55	.66
Self-rating	32	.68	30	.63

Note: All correlations are significant, $p < .01$
All correlations were corrected for range instability

strong evidence for the criterion-related validity of the CEFI. Importantly, however, the scores in the ADHD and Mixed Group samples are much lower on the BRIEF for both parent (effect size = .79 and .60) and teacher (effect size = .60 and .69) raters. Ratings on the self-report, which are based on a different standardization sample from the BRIEF-parent and BRIEF-teacher forms, are lower for the BRIEF (effect size = .23 and .13). These findings can be better understood when the norming procedures for the BRIEF are examined and compared to the methods used for the CEFI as was done in Chap. X of this book.

Differences between the mean scores obtained on the CEFI and the BRIEF were expected because of the way in which the BRIEF-Parent and BRIEF-Teacher norms were normed. The BRIEF normative data was obtained from 25 schools in only one state (Maryland, which has a very high percentage of highly educated people); the sample was dominated by whites (see Table 10.4 in Chap. 10 of this volume), and no information on parental education levels were reported. Referring to Table 4 in Chap. XX, it is clear that scores on a measure of executive function (in this case the CEFI) do vary across parental education levels. Failure to control for this variability by adequately representing these groups in a standardization sample can have a substantial impact of the scores which are obtained. As shown in Table 10.4 in Chap. 10, for example, the range of scores across parental education levels at the lower end of the normative scale (e.g., 82 which is the approximate mean for Parent ratings of children with ADHD in Table 3) was 11 points (77–88). The range of scores for a score closer to the norm (e.g., 97 which is the approximate mean for Self-ratings of the Mixed Group in Table 3) is 8 points. These findings indicate that the more extreme scores obtained on the BRIEF are likely due to the imprecise methods used to create norms for that scale. This suggestion is further supported by the finding that the BRIEF-Self-Rating scores were not as discrepant from the CEFI (but still somewhat lower), because that scale is normed on a better sample than the BRIEF-Parent or BRIEF-Teacher scales (but still not as precisely normed as the CEFI).

CEFI and Measures of Intelligence

The relationship between the CEFI and the WISC-IV (Wechsler, 2003) and the CEFI and the Cognitive Assessment System (Naglieri & Das, 1997) was examined by Naglieri and Goldstein (2011). The results of the two studies they reported will be summarized here (interested readers should see the CEFI Manual for more details). These two measures of ability differ substantially. The WISC-IV is built on the concept of general intelligence measured using four scales (Verbal Comprehension, Perceptual Reasoning, Working Memory, and Processing Speed. Ratings on the CEFI (5–18 years) Teacher Form and WISC-IV scores were obtained for a predominately white sample of male and female students ($N = 43$) who varied across parental education levels, and all of which had some clinical (35 % ADHD, 21 % Anxiety Disorders) or educational diagnosis (7 % generic learning disability). The means and standard deviations of the CEFI Full Scale and the WISC-IV scales, as well as their correlations (corrected for range instability in both distributions), are provided in Table 14.4.

The correlation between the CEFI Full Scale and the WISC-IV Full Scale was .39 and the correlations with the separate WISC-IV scales ranged from a nonsignificant value of .27 (Perceptual Reasoning) to a significant value of .44 (Verbal Comprehension). The size of these correlations indicates that the two tests are not strongly related even though the CEFI and WISC-IV Full Scale means were similar, differing by only about 2 points. The correlations between the nine CEFI scales and the five

Table 14.4

CEFI	WISC-IV					CEFI	
	Full scale	Verbal comprehension	Perceptual reasoning	Working memory	Processing speed	Mn	SD
Full scale	.39*	.44**	.27	.30	.34*	93.0	11.9
Attention	.39*	.33*	.32*	.40**	.35*	91.8	11.2
Emotion regulation	.14	.25	.08	−.06	.11	97.2	14.7
Flexibility	.57**	.68**	.45**	.46**	.37*	93.8	11.0
Inhibitory control	.21	.20	.13	.08	.27	97.7	13.5
Initiation	.25	.31*	.14	.21	.25	91.2	15.1
Organization	.15	.17	.06	.14	.17	92.2	13.6
Planning	.46**	.54**	.31*	.38*	.39*	93.6	11.1
Self-monitoring	.39*	.45**	.31*	.33*	.27	92.0	11.3
Working memory	.38*	.43**	.31*	.36*	.23	92.5	13.6
WISC-IV M	95.5	96.8	101.5	92.6	90.7		
WISC-IV SD	18.1	14.7	17.5	17.5	19.4		

Note: All correlations were corrected for range instability
*p<.05; **p<.01

Table 14.5

CEFI	CAS					CEFI	
	Full scale	Planning	Simultaneous	Attention	Successive	Mn	SD
Full scale	.45**	.49**	.43**	.37**	.32*	91.4	13.2
Attention	.40**	.42**	.39**	.30*	.35**	90.3	12.8
Emotion regulation	.26*	.22	.23	.24	.13	96.9	14.7
Flexibility	.52**	.54**	.51**	.40**	.42**	92.2	13.0
Inhibitory control	.27*	.29*	.22	.18	.21	96.0	13.9
Initiation	.40**	.37**	.31*	.30*	.20	89.0	16.3
Organization	.29*	.36**	.21	.20	.23	90.5	14.3
Planning	.47**	.54**	.46**	.37**	.38**	92.5	12.4
Self-monitoring	.48**	.50**	.49**	.43**	.35**	91.2	12.4
Working memory	.48**	.46**	.45**	.38**	.30*	91.0	14.0
CAS Mn	95.8	92.4	101.6	96.5	98.0		
CAS SD	17.1	14.5	17.0	15.1	14.6		

Note: All correlations were corrected for range instability
*p<.05; **p<.01

WISC-IV scales also showed variability. The CEFI Attention, Flexibility, Planning, Self-Monitoring, and Working Memory scales showed the most relationship to the various WISC-IV scales. The Working Memory scales on the WISC-IV and CEFI were significantly correlated (.39), but the WIC-IV Working Memory scores correlated the highest with the CEFI Flexibility scores. Interestingly, the CEFI Flexibility score correlated the highest with nearly all of the WISC-IV Scales. Overall, these findings suggest that the WISC-IV and CEFI are modestly correlated and are smaller than the correlations between the CEFI and CAS.

The relationships between the scores on the CEFI and CAS are provided in Table 14.5. The CAS provides information about neurocognitive abilities defined by the Planning, Attention, Simultaneous, and Successive theory of intelligence (Naglieri & Otero, 2012). These four abilities are measured by four corresponding scales on the CAS. Briefly, Planning is a neurocognitive ability used to determine, apply, evaluate, and manage thoughts and actions so that efficient solutions to problems can be attained. Attention is an ability used to selectively focus on a particular stimulus while inhibiting responses to competing stimuli presented over time. Simultaneous ability is used to understand how separate elements fit together into conceptual whole. Successive ability is used to integrate information into a specific serial order (Naglieri, Das, & Goldstein, 2013). We anticipated that the CAS would be related to the CEFI because the concepts of PASS, and in particular the Planning scale, assess an ability thought to be related to behaviors associated with the concept of executive function.

The correlations between the CEFI Full Scale and the CAS Full Scale ($r=.45$) and the correlations with the separate PASS scales were all significant ranging from .32 (Successive) to .49 (Planning). The size and consistency of these correlations indicate that the CAS and the CEFI are strongly related. Interestingly, the CAS Planning scale not only correlated the highest with the CEFI but the mean scores were most similar (92.4 and 91.4, respectively). The correlations between the nine CEFI and the CAS PASS and Full Scale scales were higher and more often significant than those found for the WISC-IV. These findings are logical as the CAS, like the CEFI, is based on concepts related to brain function whereas the WISC-IV is based on the concept of general ability measured using verbal and nonverbal tasks mostly developed in the early 1900s (Goldstein, 2013). These findings also suggest that although the CEFI and CAS scores are related, the correlation is moderate in size and therefore, as described in the initial portion of this chapter, there will be instances where what a person can do (their ability as measured by, e.g., the Planning scale of the CAS) may not be what they actually do (their behavior as rated by the CEFI).

CEFI and Achievement

There is considerable interest in the relevance of executive function to academic achievement. Researchers have found that executive function measured using one-on-one testing procedures has been significantly related to academic achievement in children of various ages with and without specific learning disabilities (see Best, Miller, & Jones, 2009; Best, Miller, & Naglieri, 2011; Müller, Lieberman, Frye, & Zelazo, 2008, for reviews). Recently, however, Sadeh, Burns, and Sullivan (2012) have reported that behavioral ratings of executive function (based on items from the Behavior Assessment Scale for Children, Reynolds & Kamphaus, 1992) "had very week relationships with concurrent achievement scores and scores obtained 1 year later" (p. 243).

In order to investigate this aspect of validity for the CEFI, Naglieri and Goldstein (2013) reported a study of the relationship between academic achievement scores and executive function as rated by the CEFI. The relationships between the CEFI and academic achievement were examined using the reading, math, and written language scores from the Woodcock-Johnson III Tests of Achievement (WJ-III; Woodcock, McGrew, & Mather, 2001). Ratings on the CEFI (5–18 years) Teacher Form were collected for a sample of 58 students who completed the WJ-III. The means and SDs of the CEFI Full Scale and the WJ-III scales, as well as their correlations, are provided in by Naglieri and Goldstein (2013) and summarized here in Table 14.5. The results indicate that the CEFI Full Scale score was significantly and substantially correlated with the WJ-III Total Achievement test scores ($r=.51$, $p<.01$). Correlations with the separate areas of achievement ranged from .47 (Broad Written Language, $p<.01$) to .49 (Broad Math, $p<.01$). Additionally, the CEFI and WJ-III Total Achievement mean scores were similar, differing by only approximately 3 points (see Naglieri & Goldstein, 2013 Appendix H for more details). These findings suggest that the behaviors related to executive function as provided in the CEFI are strongly related to achievement test scores in reading, math, and written language from the WJ-III (Table 14.6).

Interpretation of the CEFI

Interpretation of the CEFI begins with computation of the ten scores the scale yields. The CEFI

Table 14.6 Correlations between the CEFI and achievement test scores ($N=58$)

CEFI scales	WJ-III achievement tests				
	Total	Broad reading	Broad math	Broad written language	Median
Full scale	.51**	.48**	.49**	.47**	.49**
Attention	.59**	.52**	.46**	.55**	.54**
Emotion regulation	.18	.27*	.15	.17	.18
Flexibility	.61**	.50**	.55**	.54**	.55**
Inhibitory control	.23	.32*	.15	.26	.25
Initiation	.32*	.26	.38**	.28	.30
Organization	.32*	.31*	.33*	.33*	.33*
Planning	.58**	.54**	.57**	.50**	.56**
Self-monitoring	.53**	.51**	.51**	.49**	.51**
Working memory	.57**	.48**	.60**	.47**	.53**

Note: All correlations were corrected for range instability
*$p<.05$; **$p<.01$

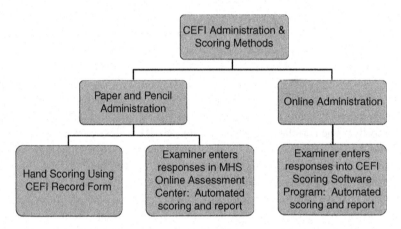

Fig. 14.3 Administration and scoring options for the CEFI

(5–18 years) Parent Form, the CEFI (5–18 years) Teacher Form, and the CEFI (12–18 years) Self-Report Form can be scored using several different options (see Fig. 14.3). All of the forms can be administered via paper-and-pencil or the MHS Online Assessment Center; completed forms can be scored via paper-and-pencil, the CEFI Scoring Software Program, or the MHS Online Assessment Center. Details about administration and scoring of the CEFI are found in the Manual (Naglieri & Goldstein, 2013). Although the CEFI can be completely administered and scored using the paper-and-pencil form administration, scoring, graphic, and narrative interpretation are provided in three types of computer-generated or online reports.

The Interpretive Report is for a single administration (this automated report also provides intervention recommendations), the Comparative Report provides a multi-rater perspective by combining results from up to five different raters, and the Progress Monitoring and Treatment Effectiveness Report combines the results from up to four ratings by the same rater to examine changes in behavior that may have occurred over time.

Once the scoring is completed interpretation of the CEFI begins by evaluating if the rater gave consistent ratings, gave an overly negative or positive impression, or neglected to respond to various items. For online administrations, interpretation also involves examining how long the rater took to complete the ratings. Once response style has been evaluated, scores on the Full Scale and the CEFI Scales are examined, and CEFI Scale scores are compared with the rated child's average standard score (not the normative mean of 100). Item-level responses may be examined, and results are compared across different raters and over time. All these procedures will be described in the sections that follow.

Rater Characteristics

The first step in the interpretation process is to determine if the rater provided usable data or if the ratings suggested inconsistent or misleading responses. CEFI results will only be accurate if the information provided by the respondent is a reasonable reflection of the individual that was rated. Standard scores on the Consistency Index provide information about whether the rater responded to very similar items differently. Inconsistent responding can occur intentionally or unintentionally and could be due to deliberate noncompliance, misunderstanding of the items or instructions, inattention, or a lack of interest in completing the form. The next two scales used to evaluate the rater's responses are the Negative and Positive Impression Scales. A negative impression response style makes the rated individual appear unrealistically negative leading to underestimation of the rated individual's functioning. Conversely, a positive impression response style makes the rated individual appear unrealistically positive, thus overestimating executive function. Additionally, there is a Time to Completion (available only for online administrations) score which identifies whether the rater completed the items so quickly that they clearly did not take adequate time to read, understand, and respond to the questions. All of these scales are designed to ensure that the results can be considered an accurate reflection of the rated individual's behaviors which reflect executive function.

Interpretation of the CEFI Scores

The CEFI Full Scale score is the most reliable and valid measure of behaviors that are indicative of executive function. The standard score this scale yields (mean of 100 and standard deviation of 15 with high scores indicating strength) is obtained from the sum of the raw scores for all the items. Because the CEFI is a one factor instrument, the Full Scale score should be used when describing a child's or adolescent's level of executive function. Standard scores that are below or above the normal range will indicate weaknesses or strengths, respectively. It is important to realize that the subscales of the CEFI (Attention, Emotion Regulation, Flexibility, Inhibitory Control, Initiation, Organization, Planning, Self-Monitoring, and Working Memory) were developed on the basis of the content of the items and for explanatory and intervention purposes. Regardless of the degree of variation among the CEFI Scales, the Full Scale score will be the most accurate description of executive function.

The nine CEFI Scales measure groups of behaviors that are indicative of executive function. These separate scales are not intended to be used in isolation to evaluate a child's or adolescent's executive function, but rather, to evaluate a specific set of behaviors that are related to the concept of executive function. That is, these specific scales, based on specific behaviors, do not constitute a complete evaluation of the executive function, but rather, they provide specific information regarding behaviors that may require intervention or suggests strengths. The scale descriptions provided below are based on the kinds of items included in each corresponding scale of the CEFI:

- The Attention scale describes how well a child or adolescent can avoid distractions, concentrate on tasks, and sustain attention.
- The Emotion Regulation scale gives information about control and management of emotions, including staying calm when handling small problems and reacting with the right level of emotion.
- The Flexibility scale reflects adjusting behavior to meet circumstances, including coming up with different ways to solve problems, having many ideas about how to do things, and being able to solve problems using different approaches.
- The Inhibitory Control scale measures control of behavior or impulses, including thinking

CEFI Scales	Standard Score	Difference From Youth's Average	Statistically Significant? (Yes/No)	Executive Function Strength/Weakness
Attention (AT)	95	-6.7	Yes	-
Emotion Regulation (ER)	82	-19.7	Yes	Weakness
Flexibility (FX)	112	10.3	Yes	Strength
Inhibitory Control (IC)	99	-2.7	No	-
Initiation (IT)	120	18.3	Yes	Strength
Organization (OG)	99	-2.7	No	-
Planning (PL)	101	-0.7	No	-
Self-Monitoring (SM)	102	0.3	No	-
Working Memory (WM)	105	3.3	No	-
Sum of Standard Scores	915 ÷9 =	101.7	Youth's Average	

Note. Differences from the Child's/Youth's Average are significant at $p < .10$.

Fig. 14.4 Illustration of comparing the nine CEFI scores to determine strengths and weaknesses

about consequences before acting, maintaining self-control, and keeping commitments.
- The Initiation scale describes how a child or adolescent starts tasks or projects on his/her own including being motivated and taking the initiative when needed.
- The Organization scale reflects management of personal effects, work, multiple tasks, behavior and thoughts, time, and working neatly.
- The Planning scale describes how well the youth can develop and implement strategies to accomplish tasks, including planning ahead and making good decisions.
- The Self-Monitoring scale describes the evaluation of one's own behavior in order to determine when a different approach is necessary, including noticing and fixing mistakes, knowing when help is required, and understanding when a task is completed.
- The Working Memory scale evaluates how well a child or adolescent can keep information in mind that is important for knowing what to do, including remembering important things, instructions, and steps.

The purpose of the nine CEFI scales is to provide a specific level of analysis of behaviors that are related to executive function. Because the items for these scales were chosen based on their content, the scores these scales yield are useful for developing intervention strategies specific to each content area. The method suggested is based on two criteria. First is a comparison of the nine CEFI Scale standard scores to the child's or adolescent's average of those scores. This determines if any of the nine scales' scores are significantly high or low in relation to the individual's own level. Comparing the CEFI Scale scores to the individual's average (and not the normative mean of 100) is called an intraindividual (or ipsative) interpretation, a technique often used in intelligence testing (see Kaufman, 1994; Naglieri, 2000). This method has the advantage of providing statistical guidelines for examining the variability of individual CEFI scale profiles. The second criteria for this process is to compare any scores that are substantially below or above the normative mean of 100 (see Naglieri, 2011). Figure 14.4 provides an example of the method used to identify strengths and weaknesses.

In this example, there is variability among the nine CEFI scales used for intervention planning as determined by the significance of the difference between the score on each scale and the child's average of 101.7. The Attention score of 82 is significantly lower than the mean of 101.7

as is the Emotional Regulation score of 95. The Attention score is considered a weakness that warrants intervention; however, the Emotion Regulation score is 95 and is in an average range (90–109). When the two rules (a) statistically different from the child's mean and (b) below what is considered the average range (e.g., 90) are applied, only the Attention scale is deemed a weakness. Inversely, the Inhibitory Control and Initiation scales' scores are significantly above the child's mean and above the average range so they are both considered strengths. Overall, this child (whose Full Scale CEFI score is 100) does not have poor executive function, but there is evidence of behaviors related to attention that warrants further investigation and may require interventions. Suggestions for interventions are provided in the CEFI Interpretive Report for scales with standard scores that are below average (i.e., less than 90).

Examination of CEFI Item Scores

To more completely understand the scores on the nine CEFI scales, individual items within each scale should be examined to determine which items contributed to a low or high score. To accomplish this goal item scores are identified that are substantially lower or higher (i.e., one standard deviation below or above the mean item score) than those from the normative group. This approach is the same technique used by Naglieri, McNeish, and Bardos (1991), Naglieri, LeBuffe, and Pfeiffer (1994), and LeBuffe and Naglieri (2003); they all suggested that an individual item score that falls in the bottom or top 15 % of the normative group distribution (e.g., exceeds the mean normative item score minus or plus one standard deviation) can be considered a weakness or a strength. In order to designate CEFI item scores as Below Average or Above Average, using the normative sample, item scores that fell at, or below, the 15th percentile rank were considered a weakness. Item scores that fell at, or above, the 85th percentile rank were considered a strength. The CEFI computerized reports provide information about which items are below or above average. When the hand-scoring method is used, the CEFI Manual provides tables that designate the classifications (i.e., Below Average, Average, and Above Average) for every item scores. Items found to be below average help pinpoint which behaviors are good targets for intervention. Items with high scores help identify behaviors that could be used as strengths within treatment programs.

Comparison of CEFI Scores Between Raters

Comparing CEFI scores obtained from parents, teachers, and the self-report can help gain a more complete perspective of the child's or adolescent's behaviors related to executive function. For example, comparing results from two or more teacher raters can help determine if there is consistency across environments or differences that may help identify where a student is doing well or poorly. Different scores may represent, for example, the student's response to unstructured or highly structured settings and can provide insight into factors that may improve an individual's functioning and those that should be avoided.

Comparing scores obtained from different raters must take measurement error into consideration by examining statistically significant differences between raters' scores. The differences needed for significance ($p<.05$) when comparing CEFI Full Scale scores range from 4 (comparing multiple teacher scores) to 7 (comparing a parent to self-report scores). Differences required for significance (at $p<.10$ and .05) when comparing different raters' scores for all nine scales and the Full Scale are provided in the CEFI Manual (Naglieri & Goldstein, 2013). For example, although a 16-year-old youth's standard score of 98 (Average) for the Emotion Regulation scale as rated by a parent is numerically higher than the score of 86 (Low Average) for the same scale as rated by a teacher, the difference is not statistically significant. A difference of 13 points is needed when the measurement error associated with these scales is taken into account. These two scores should be interpreted as similar, rather than different, despite their differing classifications. The low score provided by the teacher, however, would warrant further examination.

Intervention and the CEFI

Interventions

The purpose of the CEFI Full Scale is to provide a calibration of the child's or adolescent's behaviors related to executive function. This is the score that should be used to make decisions about the status of the individual. In contrast, the purpose of the nine content-based scales is to provide calibration of the various types of behaviors that represent executive function and comprise the entire CEFI. These groups of behaviors describe the day-to-day functioning of the child or adolescent in closer detail so that behavioral and cognitive interventions can be identified and applied. Once implemented, the effectiveness of the interventions can be determined using the methods described earlier in this chapter.

In addition to generation of the Full Scale and all nine scales' scores, narrative and graphic reporting, and item-level analysis, the CEFI automated interpretive report provides intervention suggestions for any of the nine scales that have scores less than 90. Figure 14.5 provides an example of one of the interventions that would be provided if a child's score on the Working Memory scale of the CEFI suggested that intervention was necessary. Many of the interventions which are provided come from the book *Helping Children Learn* (Naglieri & Pickering, 2010) which includes handouts for teachers, parents, and students across a wide range academic and behavioral areas. These empirically based handouts are provided in English and Spanish. Application of these handouts should include examination of the effectiveness.

Treatment Effectiveness

Whenever CEFI results suggest that specific areas of executive function require improvement, the effectiveness of interventions designed to modify behaviors related to executive function should be evaluated. The measurement of change to evaluate treatment effectiveness has been controversial (see Jensen, 2001; Ogles, Lunnen, & Bonesteel, 2001). Part of the difficulty in assessing meaningful change is controlling for measurement error, age effects, and other psychometric issues (see Naglieri & Goldstein, 2013). Many of these issues are accounted for by using the method described in the CEFI Manual. The approach requires that pre- and posttest scores are obtained from the same respondent and the interval of time should be more than 4 weeks.

The values needed for significance when addressing the statistical significance of pretest to posttest score differences are provided in the CEFI Manual. The values are based on the reliability of each of the scales and thereby take the standard error of measurement of each of the scales into account. The CEFI Progress Monitoring and Treatment Effectiveness Report provides information on the statistical significance of changes in scores over time. When hand-scoring the CEFI, the values are provided in the Manual for comparing ratings administrations at the $p<.05$ and $p<.10$ levels of significance. For example, if an 8-year-old student was found to have a low CEFI score on Attention (e.g., 80) and after interventions were implemented, the CEFI score was 87. At the $p<.10$ level, this difference is significant (a difference of 7 is needed) and suggests that interventions were effective. There is, however, an additional consideration. If the change in scores is statistically significant, as it is in this example, then evaluating the clinical importance of the change is also needed. When differences between scores are significant (e.g., a reliable difference) and the rater now describes the behaviors related to executive function as being at least in the Average range (90–109 or higher), then the pre-post change is clinically meaningful. When the difference between pre- and posttreatment scores is significant, but the posttreatment score is not 90 or higher, then the treatment appears to be effective; however, additional intervention is warranted.

Computerized Administration, Scoring, and Reporting

CEFI forms can be administered and/or scored automatically with either the CEFI Scoring Software Program or with the MHS Online

CEFI (5–18 Years) Parent Interpretive Report for Brittany   Admin Date: 05/19/2011

Intervention Strategies for Working Memory

Using Focusing Strategies to Improve Memory

Actively employing strategies that improve learning helps students remember more information. If a student's environment is not distracting, the student is more likely to be able to manipulate information in his or her mind. In turn, the student will be better able to remember the information over time. Furthermore, if the student employs strategies to self-monitor how distracted he or she is, the student is more likely to be able to focus. This strategy uses the mnemonic acronym PATS. PATS stands for:

Pick the right environment to study.
- Pick a good place to study that is comfortable. Consider how quiet the place should be, how busy it should be, and how bright it should be (bright light can be distracting and low light can make it difficult to see).
- Set aside a dedicated place to study. A student's mind might be confused and distracted by trying to study in bed, for example, because a bed is associated with sleeping.

Always reduce visual distractions.
- Find a place such as at a desk facing away from activity.
- Only have the necessary material. Other books, toys, magazines, and computers can be distracting.

Try to eliminate noise around you.
- Study in a quiet room. Lights and fans may contribute noise, so earplugs may be helpful.
- Some people like to study with music. Be sure it is not distracting. If it is, pick a quieter volume or different style of music.

Self-talk to control internal distractions.
- Some students may be distracted by internal factors such as thoughts about other things, hunger, or worry. Students should monitor their internal distractions and use positive self-talk to focus. For example, if a student is eager to e-mail a friend, the student should say to himself, "I'm distracted by wanting to e-mail, but I need to study more. I'll study for 15 more minutes and then take a break to e-mail." In this example, a timer would be a great way to help quantify study time and focus.

The student should be explicitly taught PATS and guided to use it. During class or study at home, a teacher or parent can remind the student to use PATS when he or she needs to really focus and remember information.

Naglieri, J. A., & Pickering, E. B., Helping Children Learn: Intervention Handouts for Use at School and at Home, Second Edition, 2010. Baltimore: Paul H. Brookes Publishing Co., Inc. www.brookespublishing.com. Used with the permission of the publisher.

Improving Working Memory

- Explain errorless learning techniques to the child. In errorless learning, individuals are not allowed to guess on recall tasks, but are immediately provided with the correct response, instructed to read the response, and write it down. If errors do occur they are followed by nonjudgmental corrective feedback.
- Teach study skills to help the child remember course material for tests and assignments.
- Combine the actions of seeing, saying, writing and doing when presenting information to the child, to help reinforce the child's ability to learn and remember the information.
- Teach memory mnemonic strategies (e.g., rhymes, acronyms, visual images, method of loci, catch phrases, and alliteration) to increase working memory ability.
- Use working memory tasks such as counting, spatial, word, and digit recall to help train and improve working memory.
- Start a memory log for the child that may include maps, checklists, schedules, a journal for thoughts and feelings, cues and reminders and instructions for different activities.
- Encourage the child to bring an audio tape recorder to class to help reinforce his/her learning.
- Provide a copy of in-class presentations and notes to the child.
- Use a study buddy strategy for each class subject to help the child learn course material and good study habits.
- Set up co-operative learning groups or peer tutoring for the child.

Copyright © 2013 Multi-Health Systems Inc. All rights reserved.   MHS

Fig. 14.5 Example of intervention strategies provided by the automated CEFI interpretive report

Assessment Center (http://mhsassessments.com/MAC). Both the CEFI Scoring Software Program and the MHS Online Assessment Center offer the following:

- Optional double-entry feature to verify data entry accuracy
- Optional report features that can be enabled or disabled according to the examiner's needs
- Immediate generation of reports once responses are entered
- Reports that can be printed or saved in .pdf or .rtf format allowing for using copy and paste functions into the examiner's full evaluation report
- A save option that allows results to be scored at a later time
- Ability to sort and search records by using various criteria (e.g., name/ID, test type)

In addition, the CEFI Scoring Software Program is based on a USB drive that can be easily transferred from one computer to another. When this program is used, the CEFI is first administered via paper-and-pencil, and then the rater's responses are entered into the CEFI Scoring Software Program for automatic scoring. The MHS Online Assessment Center is accessible from any computer with internet access that meets online requirements (view online requirements via the About option listed in the MHS drop down menu at http://mhsassessments.com/MAC). There are two ways to input responses online: responses from paper-and-pencil forms can be entered into the online program to generate reports or raters (i.e., parent, teacher, or youth) can take the CEFI directly online. Online administration involves sending the respondent a link to the CEFI, allowing the respondent to complete the test at a time and location (e.g., home, school) that is convenient to him/her. The MHS Online Assessment Center also allows for cases to be put into folders (e.g., school, rater type, location). Additional information about the options and features of these automated administration, scoring, and report generation are available in the CEFI Manual.

Case Study: Charles S.

In this section we will provide selected results from the CEFI-automated Interpretive Report for an 8-year-old boy we will call Charles S. The purpose here is to show how the CEFI results could be included in a larger report about a child with obvious weakness in executive function. The text and some of the figures were taken directly from the automated report as recommended by the authors. Some very minor customizing to the report, for example, selecting some but not all of the text to include, can and was done in this case.

CEFI Results

Ratings on the CEFI obtained from Mrs. S for Charles were first examined to ensure that the results could be interpreted with confidence. All the scores are set to have a mean of 100 and standard deviation of 15 and scaled so that high scores are good. Mrs. S's ratings were very consistent (score of 110), she omitted no items, and there was no suggestion that she was rating Charles too positively or too negatively. The CEFI results, summarized in Fig. 14.6 and more fully explained in the text that follows, clearly suggest that Charles has problems with behaviors related to executive function.

Charles's CEFI *Full Scale* standard score of 78 falls in the Below Average range and is ranked at the 7th percentile. This means that his score is equal to, or greater than, 7 % of those obtained by children his age in the standardization group. There is a 90 % probability that Charles's true Full Scale standard score is within the range of 75–81. The CEFI Full Scale score is made up of items that belong on nine separate scales which varied considerably, and therefore the Full Scale score will sometimes be higher, and other times lower, than scores on these Scales. The Initiation, Self-Monitoring, and Working Memory scales were found to be significant weaknesses, which means that Charles's behavior in these areas was a weakness both in relation to his average score and in relation to the norm.

Scales in the Average Range

Charles's *Emotion Regulation* scale score reflects his control and management of emotions, including

Full Scale			
Standard Score	90% Confidence Interval	Percentile Rank	Classification
78	75-81	7	Below Average

CEFI Scales							
Scale	Standard Score	90% Confidence Interval	Percentile Rank	Classification	Difference from Child's Average (80.7)	Statistically Significant? (p < .10)	Executive Function Strength/ Weakness
Attention	81	76-89	10	Low Average	0.3	No	-
Emotion Regulation	106	97-113	66	Average	25.3	Yes	-
Flexibility	74	69-87	4	Below Average	-6.7	No	-
Inhibitory Control	105	97-112	63	Average	24.3	Yes	-
Initiation	66	62-78	1	Well Below Average	-14.7	Yes	Weakness
Organization	80	75-90	9	Low Average	-0.7	No	-
Planning	73	68-82	4	Below Average	-7.7	No	-
Self-Monitoring	70	66-83	2	Below Average	-10.7	Yes	Weakness
Working Memory	71	67-82	3	Below Average	-9.7	Yes	Weakness

Fig. 14.6 Example of CEFI results for the case of Charles

staying calm when handling small problems and reacting with the right level of emotion. His standard score of 106 falls in the Average range and is ranked at the 66th percentile. There is a 90 % probability that his true Emotion Regulation standard score is within the range of 97–113.

Charles's *Inhibitory* Control scale score reflects his ability to control behavior or impulses, including thinking about consequences before acting, maintaining self-control, and keeping commitments. His standard score of 105 falls in the Average range and is ranked at the 63rd percentile. There is a 90 % probability that his true Inhibitory Control standard score is within the range of 97–112.

Scales Below the Average Range

Charles's *Attention* scale score reflects how well he can avoid distractions, concentrate on tasks, and sustain attention. His standard score of 81 falls in the Low Average range and is ranked at the 10th percentile. There is a 90 % probability that his true Attention standard score is within the range of 76–89. Item score variability suggests that ratings for Charles were high on focusing on one thing. Item score variability suggests that ratings for Charles were low on avoiding distraction, working well for a long time and reading with concentration. (See the CEFI Items by Scale section of this report for additional low item scores.)

Charles's *Organization* scale score reflects his ability to manage personal effects, work, or multiple tasks, including organizing tasks and thoughts well, managing time effectively, and working neatly. His standard score of 80 falls in the Low Average range and is ranked at the 9th percentile. There is a 90 % probability that his true Organization standard score is within the range of 75–90. Ratings for Charles were low on finishing one task before starting another, completing homework or tasks on time and handling several tasks at once. (See the CEFI Items by Scale section of this report for additional low item scores.)

Charles's *Flexibility* scale score describes how he adjusts his behavior to meet circumstances, including coming up with different ways to solve problems, having many ideas about how to do things, and being able to solve problems using different approaches. His standard score of 74 falls in the Below Average range and is ranked at the 4th percentile. There is a 90 % probability that his true Flexibility standard score is within the range of 69–87. Variability in item scores indicates that ratings for Charles were low on coming up with a new way to reach a goal, finding different ways to solve problems and solving problem in different ways. (See the CEFI Items by Scale section of this report for additional low item scores.)

Charles's *Planning* scale score reflects how well he can develop and implement strategies to accomplish tasks, including planning ahead and making good decisions. His standard score of 73 falls in the Below Average range and is ranked at the 4th percentile. There is a 90 % probability that his true Planning standard score is within the range of 68–82. Item score variability suggests that ratings for Charles were low on preparing for school or work, solving problems creatively and doing things in the right order. (See the CEFI Items by Scale section of this report for additional low item scores.)

Charles's *Working Memory* scale standard score was less than 90 and significantly lower than his average score on the CEFI Scales. This indicates that he scored especially low on how well he can keep information in mind that is important for knowing what to do and how to do it, including remembering important things, instructions, and steps. Charles's Working Memory scale standard score of 71 falls in the Below Average range and is ranked at the 3rd percentile, which means he scored as well as or better than 3 % of the children his age in the standardization group. There is a 90 % probability that his true Working Memory standard score is within the range of 67–82. Ratings for Charles were low on taking note of instructions, having many things in mind at one time and keeping track of his goals when making decisions. (See the CEFI Items by Scale section of this report for additional low item scores.)

Charles's *Self-Monitoring* scale standard score was less than 90 and significantly lower than his average score on the CEFI Scales. This indicates that he scored especially low on his ability to evaluate his own behavior in order to determine when a different approach is necessary, including noticing and fixing mistakes, knowing when help is required, and understanding when a task is completed. Charles's Self-Monitoring scale standard score of 70 falls in the Below Average range and is ranked at the 2nd percentile, which means he scored as well as or better than 2 % of the children his age in the standardization group. There is a 90 % probability that his true Self-Monitoring standard score is within the range of 66–83. Variability in item scores indicates that ratings for Charles were low on fixing his mistakes, changing a plan that isn't working and monitoring time. (See the CEFI Items by Scale section of this report for additional low item scores.)

Charles's *Initiation* scale standard score was less than 90 and significantly lower than his average score on the CEFI Scales. This indicates that he scored especially low on his skill at beginning tasks or projects on his own, including starting tasks easily, being motivated, and taking the initiative when needed. Charles's Initiation scale standard score of 66 falls in the Well Below Average range and is ranked at the 1st percentile, which means he scored as well as or better than 1 % of the children his age in the standardization group. There is a 90 % probability that his true Initiation standard score is within the range of 62–78. Item score variability suggests that ratings for Charles were low on beginning something without being asked, adopting new projects and cueing himself to get started on things. (See the CEFI Items by Scale section of this report for additional low item scores.)

Interventions Based on CEFI Results

Charles's CEFI Scores suggest that interventions are warranted for the behaviors related to executive function included in the Attention, Flexibility, Organization, and Planning Scales, especially the Initiation, Self-Monitoring, and Working Memory scales which were particularly low. These three scales are an initial priority as far as intervention need is concerned because they are the lowest. The second step in intervention should be to address the remaining low scales. The CEFI Report provides interventions for all seven scales; however, we will begin with a focus on the areas of Planning (because of its relationship with Initiation), Self-Monitoring, and Working Memory. The intervention recommendations (these recommendations are summarized from Naglieri & Pickering, 2010) for these three areas are as follows.

Intervention Strategies for Planning and Initiation

- Teach children about plans and strategy use.
- Discuss the importance of planning in class and how it helps students organize themselves so that they can be more successful and finish on time.
- Encourage children to develop, use, and evaluate their own strategies.
- Encourage verbalization of ideas and strategies.
- Explain why some methods work better than others.
- Ask questions that encourage initiation and planning, such as:
  - "How did you do the task?"
  - "Did you make a plan before you started the task?"
  - "What did you do last time? Did it work?"
  - "Is there a better way or another way to do this?"
  - "What strategy worked for you?"
  - "Do you think you will do anything differently next time?"
  - "How can you check your work to see if it is right?"
- Encourage a child to think strategically and plan ahead by giving the child training in problem solving, verbal reasoning, study skills, and task-specific skills.
- Use a calendar to map out and plan long-term goals and tasks.
- Use daily or weekly worksheets or notebooks to plan and organize short-term tasks.
- Build a list with the child that prioritizes tasks and activities. Have the child refer to this list regularly in order to plan his/her time.
- Teach the child how to tackle complex tasks by breaking them up into smaller steps.
- Provide checklists of step-by-step instructions with examples of how to accomplish a task or goal.
- Create smaller quotas or more benchmarks to increase the sense of productivity. Increase these work quotas as the child's productivity improves.
- Limit the amount of time spent on each task by setting reasonable time limits and providing the child with a means to keep track of time (e.g., a timer).
- Set up resources for the child to use when he/she needs help at home or at school. Encourage the child to use these resources and to understand that it is okay to ask for help.

Intervention Strategies for Self-Monitoring

- Provide specific description of academic accuracy and academic productivity.
- Hand out a record sheet, and explain that at the end of each session the child is to record the number of items completed with the total number of items given (productivity) and the number of items correct with the total number of items given (accuracy) in the appropriate columns.
- Explain that self-monitoring is important for on-task behavior and successful learning and demonstrate how to calculate and record the percentages for accuracy and productivity at the end of the session (10–30-min period).
- Provide a session in which the students work on a task with a specific number of items (e.g., spelling list, math problems, and question sheets related to a story). It is acceptable for students at different levels to have different activities.
- At the end of the session, have students record and calculate their progress.
- Have students keep daily logs and encourage students to compare percentages of previous sessions to recent sessions. Teachers may choose to have students graph their own progress or to post a graph in class charting the productivity and accuracy of individual students or the whole class. Reinforcement or rewards are not necessary, but some teachers do choose to reward students for certain levels of success.

Throughout these steps the teacher should model self-recording and monitoring, provide feedback, allow students to independently record their performance, encourage students to examine their performance over time, praise accurate self-reporting, and be patient—success may not come immediately.

Intervention Strategies for Working Memory

Actively employing strategies that improve learning helps students remember more information. If a student's environment is not distracting, the student is more likely to be able to manipulate information in his or her mind. In turn, the student will be better able to remember the information over time. Furthermore, if the student employs strategies to self-monitor how distracted he or she is, the student is more likely to be able to focus. This strategy uses the mnemonic acronym PATS. PATS stands for:

*P*ick the right environment to study
- Pick a good place to study that is comfortable. Consider how quiet the place should be, how busy it should be, and how bright it should be (bright light can be distracting and low light can make it difficult to see).
- Set aside a dedicated place to study. A student's mind might be confused and distracted by trying to study in bed, for example, because a bed is associated with sleeping.

*A*lways reduce visual distractions
- Find a place such as at a desk facing away from activity.
- Only have the necessary material. Other books, toys, magazines, and computers can be distracting.

*T*ry to eliminate noise around you
- Study in a quiet room. Lights and fans may contribute noise, so earplugs may be helpful.
- Some people like to study with music. Be sure it is not distracting. If it is, pick a quieter volume or different style of music.

*S*elf-talk to control internal distractions
- Some students may be distracted by internal factors such as thoughts about other things, hunger, or worry.
- Students should monitor their internal distractions and use positive self-talk to focus. For example, if a student is eager to e-mail a friend, the student should say to himself, "I'm distracted by wanting to e-mail, but I need to study more. I'll study for 15 more minutes and then take a break to e-mail." In this example, a timer would be a great way to help quantify study time and focus.
- The student should be explicitly taught PATS and guided to use it. During class or study at home, a teacher or parent can remind the student to use PATS when he or she needs to really focus and remember information.

The evaluation of treatment effectiveness is, of course, very important and should be accomplished using the guidelines presented in this chapter and in the CEFI Manual. Once there have been improvements in the scores for the Planning, Initiation, Self-Monitoring, and Working Memory scales, then additional interventions should be applied to manage the behaviors described in the Attention, Flexibility, and Organization scales. In all cases both the reliability of the pre-post intervention scores and the proximity of the post intervention scores to the Average range need to be considered.

Conclusions

Interest in executive function has grown exponentially with an appreciation and understanding that the manner by which children and adults manipulate and integrate knowledge to learn, solve problems, and function in everyday life is significantly contributed by *how* they go about doing it. The several factorial research studies using the CEFI for the national standardization sample provided strong evidence that executive function is a unitary behavioral construct. The evidence summarized here also supports the conclusion that CEFI is an instrument capable of measuring behaviors associated with executive function in a reliable and valid way. In addition, because the scores the CEFI provides are based on large samples representative of the US population, users can generalize the results with confidence. In summary, results of the CEFI can be used to better understand children's behavior related to executive function and, most importantly, to assist helping struggling children succeed in the classroom, on the playground, and at home.

References

Abdi, H. (2010). Congruence: Congruence coefficient, RV coefficient, and mantel coefficient. In N. J. Salkind, D. M. Dougherty, & B. Frey (Eds.), *Encyclopedia of research design* (pp. 222–229). Thousand Oaks, CA: Sage.

Best, J. R., Miller, P. H., & Jones, L. L. (2009). Executive functions after age 5: Changes and correlates. *Developmental Review, 29,* 180–200.

Best, J. R., Miller, P. H., & Naglieri, J. A. (2011). Relations between executive function and academic achievement from ages 5 to 17 in a large, representative national sample. *Learning and Individual Differences, 21,* 327–336.

Davidson, M. C., Amso, D., Anderson, L. C., & Diamond, A. (2006). Development of cognitive control and executive functions from 4–13 years: Evidence from manipulations of memory, inhibition, and task switching. *Neuropsychologia, 44,* 2037–2078.

Duncan, J., & Miller, E. K. (2002). Cognitive focusing through adaptive neural coding in the primate prefrontal cortex. In D. Stuss & R. T. Knight (Eds.), *Principles of frontal lobe function* (pp. 278–291). New York: Oxford University Press.

Friedman, N. P., Miyake, A., Corley, R. P., Young, S. E., DeFries, J. C., & Hewitt, J. K. (2006). Not all executive functions are related to intelligence. *Psychological Science, 17,* 172–179.

Gilbert, S. J., Bird, G., Brindley, R., Frith, C. D., & Burgess, P. W. (2008). Atypical recruitment of medial prefrontal cortex in autism spectrum disorders: An fMRI study of two executive function tasks. *Neuropsychologia, 46,* 2281–2291.

Gioia, G. A., Isquith, P. K., Guy, S. C., & Kenworthy, L. (2000). *Behavior rating inventory of executive function: Professional manual.* Lutz, FL: Psychological Assessment.

Goldstein, S. (2013). The science of intelligence testing: Commentary on the evolving nature of interpretations of the Wechsler scales. *Journal of Psychoeducaitonal Assessment, 31,* 132–137.

Guy, S. C., Isquith, P. K., & Gioia, G. A. (2004). *Behavior rating inventory of executive function: Self report version.* Odessa, FL: Psychological Assessment Resources.

Happé, F., Booth, R., Charlton, R., & Hughes, C. (2006). Executive function deficits in autism spectrum disorders and attention-deficit/hyperactivity disorder: Examining profiles across domains and ages. *Brain and Cognition, 61,* 25–39.

Herba, C. M., Tranah, T., Rubia, K., & Yule, W. (2006). Conduct problems in adolescence: Three domains of inhibition and effect of gender. *Developmental Neuropsychology, 30,* 659–695.

Jensen, P. S. (2001). Clinical equivalence: A step, a misstep, or just a misnomer? *Clinical Psychology: Science and Practice, 8,* 436–440.

Jurado, M. B., & Rosselli, M. (2007). The elusive nature of executive functions: A review of our current understanding. *Neuropsychology Review, 17,* 213–233.

Kaufman, A. S. (1994). *Intelligent testing with the WISC-III.* New York: Wiley.

LeBuffe, P. A., & Naglieri, J. A. (2003). *Devereux early childhood assessment clinical form (DECA-C).* Lewisville, NC: Kaplan Press.

McCloskey, G., Perkins, L. A., & Van Divner, B. (2009). *Assessment and intervention for executive function difficulties. School-based practice in action series.* New York, NY: Routledge Taylor & Francis.

Miller, E. K., & Cohen, J. D. (2001). An integrative theory of prefrontal cortex function. *Annual Review of Neuroscience, 24,* 167–202.

Miyake, A., Friedman, N. P., Emerson, M. J., Witzki, A. H., Howerter, A., & Wagner, T. D. (2000). The unity and diversity of executive functions and their contributions to complex "frontal lobe" tasks: A latent variable analysis. *Cognitive Psychology, 41,* 49–100.

Morgan, A. B., & Lilienfeld, S. O. (2000). A meta-analytic review of the relation between antisocial behavior and neuropsychological measures of executive function. *Clinical Psychology Review, 20,* 113–136.

Müller, U., Lieberman, D., Frye, D., & Zelazo, P. D. (2008). Executive function, school readiness, and school achievement. In S. K. Thurman & C. A. Fiorello (Eds.), *Applied cognitive research in K-3 classrooms* (pp. 41–84). New York: Routledge.

Naglieri, J. A. (2000). Can profile analysis of ability test scores work? An illustration using the PASS theory and CAS with an unselected cohort. *School Psychology Quarterly, 15,* 419–433.

Naglieri, J. A. (2011). The discrepancy/consistency approach to SLD identification using the PASS theory. In D. P. Flanagan & V. C. Alfonso (Eds.), *Essentials of specific learning disability identification* (pp. 145–172). Hoboken, NJ: Wiley.

Naglieri, J. A., & Goldstein, S. (2011). Assessment of cognitive and neuropsychological processes. In S. Goldstein & J. A. Naglieri (Eds.), *Understanding and managing learning disabilities and ADHD in late adolescence and adulthood* (2nd ed.). New York: Wiley.

Naglieri, J. A., & Goldstein, S. (2013). *Comprehensive Executive Functioning Index.* Toronto: Multi Health Systems.

Naglieri, J. A., & Das, J. P. (1997). *Cognitive assessment system.* Austin, TX: Pro-Ed.

Naglieri, J. A., Das, J. P., & Goldstein, S. (2013). *Cognitive assessment system—Second edition.* Austin, TX: ProEd.

Naglieri, J. A., & Goldstein, S. (2013). *Evaluation and treatment effectiveness in the field of autism:*

Psychometric considerations and an illustration, in S. Goldstein and J.A. Naglieri *handbook of autism treatment*. New York: Springer.

Naglieri, J. A., & Gottling, S. H. (1995). A cognitive education approach to math instruction for the learning disabled: An individual study. *Psychological Reports, 76*, 1343–1354.

Naglieri, J. A., LeBuffe, P. A., & Pfeiffer, S. I. (1994). *Devereux scales of mental disorders*. San Antonio, TX: The Psychological Corporation.

Naglieri, J. A., McNeish, T. J., & Bardos, A. N. (1991). *Draw a person: Screening procedure for emotional disturbance*. Austin, TX: ProEd.

Naglieri, J. A., & Otero, T. M. (2012). The cognitive assessment system: From theory to practice. In D. P. Flanagan & P. L. Harrison (Eds.), *Contemporary intellectual assessment, third edition: Theories, tests, and issues* (pp. 376–399). New York: Guilford.

Naglieri, J. A., & Pickering, E. B. (2010). *Helping children learn. Intervention handouts for use in school and at home* (2nd ed.). Baltimore, MD: Paul H. Brookes.

Ogles, B. M., Lunnen, K. M., & Bonesteel, K. (2001). Clinical significance: History, application, and current practice. *Clinical Psychology Review, 21*, 421–446.

Raiche, G., & Magis, D. (2010). *nFactors: Parallel analysis and non graphical solutions to the Cattell Scree Test*. R package version 2.3.2. http://CRAN.R-project.org/package=nFactors

Revelle, W. (2011). *Psych: Procedures for personality and psychological research*. Northwestern University, Evanston. R. package version 1.1.12.

Revelle, W., & Rocklin, T. (1979). Very simple structure: An alternative procedure for estimating the optimal number of interpretable factors. *Multivariate Behavioral Research, 14*, 403–414.

Reynolds, C. R., & Kamphaus, R. W. (1992). *BASC: Behavioral assessment system for children*. San Antonio: Pearson.

Sadeh, S. S., Burns, M. K., & Sullivan, A. L. (2012). Examining an Executive Function Rating Scale as a predictor of achievement in children at risk for behavior problems. *School Psychology Quarterly, 27*, 236–246.

Slocum-Gori, S. L., & Zumbo, B. D. (2011). Assessing the unidimensionality of psychological scales: Using multiple criteria from factor analysis. *Social Indicators Research, 102*, 443–461.

Solomon, M., Ozonoff, S. J., Ursu, S., Ravizza, S., Cummings, N., Ly, S., et al. (2009). The neural substrates of cognitive control deficits in autism spectrum disorders. *Neuropsychologia, 47*, 2515–2526.

Velicer, W. F., Eaton, C. A., & Fava, J. L. (2000). Construct explication through factor or component analysis: A review and evaluation of alternative procedures for determining the number of factors or components. In R. D. Goffin & E. Helmes (Eds.), *Problems and solutions in human assessment: Honoring Douglas N. Jackson at seventy* (pp. 41–72). Norwell, MA: Kluwer Academic.

Wechsler, D. (2003). *Wechsler intelligence scale for children—4th edition (WISC-IV®)*. San Antonio, TX: Harcourt Assessment.

Weyandt, L. L., Willis, W. G., Swentosky, A., Wilson, K., Janusis, G. M., & Marshall, S. (in press). A review of the use of executive function tasks in externalizing and internalizing disorders. In S. Goldstein, & J. A. Naglieri (Eds.), *Comprehensive handbook of executive functioning*. New York, NY: Springer.

Wiebe, S. A., Espy, K. A., & Charak, D. (2008). Using confirmatory factor analysis to understand executive control in preschool children: I. Latent structure. *Developmental Psychology, 44*, 575–587.

Willcutt, E. G., Doyle, A. E., Nigg, J. T., Faraone, S. V., & Pennington, B. F. (2005). Validity of the executive function theory of attention-deficit/hyperactivity disorder: A meta-analytic review. *Biological Psychiatry, 57*, 1336–1346.

Woodcock, R. W., McGrew, K. S., & Mather, N. (2001). *Woodcock-Johnson III*. Itasca, IL: Riverside Publishing.

Zelazo, P. D., & Müller, U. (2002). Executive function in typical and atypical development. In U. Goswami (Ed.), *Handbook of childhood cognitive development* (pp. 445–469). Oxford: Blackwell.

Zwick, W. R., & Velicer, W. F. (1986). Comparison of five rules for determining the number of components to retain. *Psychological Bulletin, 99*, 432–442.

The Assessment of Executive Functioning Using the Barkley Deficits in Executive Functioning Scales

Russell A. Barkley

The assessment of executive functioning (EF) has been plagued by several problems, not the least of which is the lack of any consensus definition for the term itself (Castellanos, Sonuga-Barke, Milham, & Tannock, 2006). Despite frequent use of the term in various research papers and books over the past 40 years since the term was first coined by Pribram in 1973, more than 30 definitions exist for the term (Barkley, 2012a) and at least as many different constructs have been placed under it, making it more akin to an "umbrella" term or meta-construct (Eslinger, 1996). Typically, reviews of the scientific findings on EF often sidestep the problem of definition and simply list those constructs thought to be included in the term, such as inhibition, working memory, planning, and problem-solving (Frazier, Demareem, & Youngstrom, 2004; Hervey, Epstein, & Curry, 2004; Willcutt, Doyle, Nigg, Faraone, & Pennington, 2005). The reviews then discuss findings with regard to measures of these constructs without further consideration for the rather glaring problem of specifying just what makes these constructs representative of EF while other neuropsychological functions are not so classified. What specific, operational criterion can be used to determine what mental functions are or are not EF? To my knowledge, none have previously existed. Declaring that EF is what the prefrontal cortex (PFC) does is unhelpful as this simply refers to a different level of analysis at the neurological level rather than defining the term properly at the neuropsychological one, thus conflating two distinct levels of scientific analysis (Denckla, 1996). It is also prone to circularity of argument, in that EF is stated to be what the PFC does and then declaring that what the PFC does is EF.

As a consequence of lack of definitional clarity, the second problem that arises in the assessment of EF is that of just how it is to be assessed. Many tests and measures have been declared to be EF tests without much basis for challenge. Absent any general professional consensus on the meaning of the term EF or any operational definition of it, it will prove difficult to test the validity of any measure claiming to evaluate this domain. As I have recently noted elsewhere (Barkley, 2012a), the field of neuropsychology seems to have addressed this problem by largely endorsing the use of psychometric tests of various constructs said to comprise EF as the principal or sole basis for evaluating EF and its deficits in clinical patients and in research studies. Other than convenience or tradition, why are tests given in clinical or lab settings widely considered the basis for measuring EF? Over the past 40 years, occasional voices have been raised warning that neuropsychological tests of EF were problematic (Dimond, 1980; Dodrill, 1997; Lezak, 1995, 2004). The tests were unlikely to be capturing much of what is considered to be the essence of EF or its important features as humans use it in their daily life or to be adversely affected by injuries to the PFC

R.A. Barkley (✉)
Medical University of South Carolina,
Charleston, SC, USA
e-mail: drbarkley@russellbarkley.org

(Barkley, 2001; Rabbitt, 1997; Shallice & Burgess, 1991). The warnings have largely gone unheeded as EF tests and test batteries have come to dominate the field of assessment of EF and represent an inchoate gold standard for the determination of EF and its deficits. With very few exceptions, the vast majority of studies published on the topic of EF have used EF tests or batteries of tests to determine if certain disorders impaired EF or how EF developed in normal samples. But as with the EF constructs noted above which these tests are thought to evaluate, one can rightly ask just what makes these tests measure EF while other tests, such as those of academic achievement, intelligence, or other psychological domains, are widely believed not to be so?

A further problem with the assessment of EF by tests is that of their low to moderate reliability (Lezak, 1995, 2004; Rabbitt, 1997) and rates of detection of deficits in patients with PFC (and presumed EF) disorders. As I recently discussed elsewhere (Barkley, 2012a), only a minority of patients experiencing frontal lobe injuries or those with ADHD known to have a frontal lobe disorder score in the impaired range on these measures even if mean differences between clinical and control groups are statistically significant. In contrast, consider the fact that the vast majority of such patients are placed in that range of impairment on ratings of EF in daily life activities or in direct observations of EF performance in natural settings (Alderman, Burgess, Knight, & Henman, 2003; Barkley, 2011a; Barkley & Murphy, 2011; Burgess, Alderman, Evans, Emslie, & Wilson, 1998; Gioia, Isquith, Kenworthy, & Barton, 2002; Kertesz, Nadkarni, Davidson, & Thomas, 2000; Mitchell & Miller, 2008; Wood & Liossi, 2006). Obviously people with PFC disorders and injuries have EF deficits in their daily life activities even if the EF tests do not detect them. And it is the deficits occurring in daily life, not those manifested on tests, that are the most important to understand and to clinically assess and rehabilitate or manage. Abundant research also has shown that these tests do not correlate well, if at all, with more ecologically valid means of assessing EF in everyday life circumstances. This has been evident repeatedly in studies of these tests in comparison to systematic observations, structured interviews, or ratings of daily self-care and adaptive functioning and to behavior ratings of EF in adults (Alderman et al., 2003; Bogod, Mateer, & MacDonald, 2003; Burgess et al., 1998; Chaytor, Schmitter-Edgecombe, & Burr, 2006; Mitchell & Miller, 2008; Ready, Stierman, & Paulsen, 2001; Wood & Liossi, 2006) or in children with frontal lobe lesions, traumatic brain injuries (TBI), or other neurological or developmental disorders (Anderson, Anderson, Northam, Jacobs, & Mikiewicz, 2002; Mangeot, Armstrong, Colvin, Yeates, & Taylor, 2002; Vriezen & Pigott, 2002; Zandt, Prior, & Kyrios, 2009). This is also the case in both adults with ADHD and children with ADHD followed to adulthood (Barkley & Fischer, 2011; Barkley & Murphy, 2011). If a primary clinical aim is to predict how well an individual will do with executive functioning in the real world of their daily life activities, then EF tests have very low ecological validity proving to be of minimal or no help.

Specifically, it appears that any single EF test shares just 0–10 % of its variance with EF ratings or observations of EF in daily adaptive functioning as found in the aforementioned studies. The relationships are frequently not statistically significant. Even the best combination of EF tests shares just 9–20 % of the variance with EF ratings or observations as reflected in the above and other studies (Barkley & Fischer, 2011; Barkley & Murphy, 2011; O'Shea et al., 2010; Ready et al., 2001; Stavro, Ettenhofer, & Nigg, 2007; Zandt et al., 2009). If IQ is statistically removed from the results, the few significant relationships found in these studies between EF tests and EF ratings may even become nonsignificant (Mangeot et al., 2002). There is also the related problem that tests of EF deficits are not very good at predicting impairments in various major life activities that ought to be rife with EF such as occupational functioning, educational history, driving, money management, and criminal conduct, among others (Barkley & Fischer, 2011; Barkley & Murphy, 2010, 2011).

The nearly slavish devotion to the use of EF tests as the sole or gold standard for its evaluation has also resulted in some serious logical errors in various research studies on EF and reviews of that literature, as I discussed elsewhere (Barkley, 2012a). For instance, my own area of clinical and research specialization is ADHD. The following current situation in this field represents this error:

- The PFC is the "executive" brain (Pribram, 1973).
- ADHD is a disorder arising largely from structural and functional abnormalities in the PFC (Bush, Valera, & Seidman, 2005; Valera, Faraone, Murray, & Seidman, 2007).
- ADHD is largely *not* a disorder of EF (Boonstra, Oosterlaan, Sergeant, & Buitelaar, 2005; Jonsdottir, Bouma, Sergeant, & Scherder, 2006; Marchetta, Hurks, Krabbendam, & Jolles, 2008; Nigg, Willcutt, Doyle, & Sonuga-Barke, 2005; Willcutt et al., 2005).

This conclusion was reached because the studies cited above and others (Barkley & Fischer, 2011; Barkley & Murphy, 2011; Biederman et al., 2008) demonstrated that the majority of individuals with ADHD are not impaired on EF tests, even if groups of ADHD cases differ statistically in mean scores from control groups on many such tests (Frazier et al., 2004; Hervey et al., 2004). So if EF tests are to be the sole standard for assessing the presence of EF deficits, then most cases of ADHD do not have such deficits. Ergo, ADHD is not a disorder of EF in most cases. This logical error has been repeated countless times in the literature on other disorders. Consider that a recent study concluded that the risk for developing a substance use disorder in adolescence or young adulthood is unrelated to the presence of EF deficits in childhood or adolescence; a conclusion reached solely on the basis of EF tests (Wilens et al., 2011). Such studies of disorders and the role of EF in them that relied exclusively on the psychometric approach to evaluating EF obviously must now have the generality of conclusions greatly restricted by the qualifier "as measured by EF tests." The same caveat applies as well to studies of the normal course of development of EF (Anderson, 2002; Best, Miller, & Jones, 2009). All such studies will need to be redone using other approaches to evaluating EF before any conclusions about the involvement of EF in these disorders or normal developmental course of EF and its constructs can be taken as being generally indicative of the nature of EF.

Specifying the Nature of EF

These and other problems with the definition and assessment of EF led me to develop a more specific definition of EF beginning with the idea that EF is self-regulation (Barkley, 1997a, 1997b, 2001, 2012a). After all, self-regulation is among the most commonly ascribed constructs to the umbrella term of EF (Eslinger, 1996). It also provides a gateway to a more specific definition of EF. As indicated in a separate chapter in this textbook by Antshel and I, self-regulation refers to the use of self-directed actions so as to modify one's subsequent behavior in order to alter the likelihood of a future consequence or event. EF refers to such self-directed actions and each class or type of self-directed activity can be usefully considered as an EF component. Thus, an executive function is considered EF because it is self-directed, designed to alter subsequent behavior, and so strives to change the future. EF, like self-regulation, is therefore always directed at a delayed event, or "the later" rather than "the now." I have gone on to identify six such self-directed actions that, by adulthood, are largely covert, internalized, privatized, or "cognitive" in their execution; in childhood, however, most of these self-directed activities are overt (Barkley, 2012a). These are self-directed attention (self-awareness and monitoring), self-restraint (inhibition), self-directed sensing (nonverbal working memory and especially visual imagery), self-directed speech (verbal working memory), self-directed emotions and motivations, and self-directed play (planning and problem-solving). One can now readily see how the most common constructs assigned to the term EF (in parentheses above) qualify as being EF—they are all forms of self-directed actions designed to alter subsequent behavior and so achieve a change in the probable future (delayed consequences).

I have gone on to show that EF extends outward from the initial and proximal phenotypic level into higher levels of functioning as in an extended phenotype that produces effects at considerable spatial and temporal distances from the individual (Barkley, 2012a), following the concept of the extended phenotype championed by Dawkins in biology (1982). Such effects at a distance include daily adaptive functioning, self-care, self-reliance, and social self-defense against social parasitism, onward to social functioning involving reciprocity, cooperative, and even mutualistic activities, and further on to incorporate the larger domains of economic behavior, morality and ethics, and even legal conduct (Barkley, 2012a). All of the latter fields of human endeavors are predicated on one of the most basic and essential EF components—the capacity to contemplate the future (to contrast "the later" with "the now").

Besides their self-directed nature, EF involves other characteristics. The use of EF to generate and sustain future-directed behavior spans considerable periods of time as typically used in daily life, such as over days, weeks, and even months as people pursue various goals over various durations and with various others and social networks. There is a cross-temporal, purposive, or future-directed nature to EF not seen in other mental abilities (Fuster, 1997). Also, as suggested above, EF is essential for effective social functioning (Dimond, 1980; Eslinger, 1996; Lezak, 1995; Luria, 1966). Indeed, I have argued that its social purposes are the principle reason for the evolution of EF (Barkley, 2001, 2012a). Just as important, I have argued that EF provides the means by which individuals both adopt existing culture and create new cultural devices and methods for use in their goal-directed activities, even leading to new goal-directed activities that would not be possible in earlier historical cultural epochs lacking those cultural inventions. I have therefore defined EF as: *the use of self-directed actions so as to choose goals and to select, enact, and sustain actions across time toward those goals usually in the context of others often relying on social and cultural means for the maximization of one's long-term welfare as the person defines that to be.*

Characterizing EF this way illustrates the problems inherent in using tests to assess EF in contrast to methods such as rating scales or other ethological procedures. Because EF extends over longer spans of time than can be conveniently assessed in a clinic or lab, often involves performance in major life activities and social contexts, and often involves the use of various cultural methods and devices, rating scales should prove superior to tests in capturing the various levels of EF. That is because rating scales involve much wider ascertainment windows for the behavior to be rated than do tests, can quantify qualitative features of that behavior that are more difficult to capture in tests, and capture behavior in many important major life activities outside of the clinic or lab, and can use items that involve various social activities or interactions than can tests. Moreover, given the near consensus that EF is self-regulation, it is also not obvious how current EF tests can assess such self-modification for long-term self-interestedness, self-sufficiency, and social independence.

EF tasks are not only complex but contaminated, involving multiple cognitive processes many of which are not considered part of EF, as I noted elsewhere (Barkley, 2012a). Only some of those processes are supposedly reflecting the EF construct that is intended to be sampled (Anderson, 2002; Castellanos et al., 2006). A related concern is that many EF tests are often found to be significantly influenced by overall general cognitive ability or level of intelligence (Mahone et al., 2002; Riccio, Hall, Morgan, Hynd, & Gonzalez, 1994), making their results difficult to interpret as reflecting unadulterated measures of a particular EF construct. This likely explains findings that statistically removing IQ from relationships between EF tests and observations and ratings of EF in natural settings often reduces any significant relationships to nonsignificant status (Mahone et al., 2002; Mangeot et al., 2002). And it may also account for the fact that some of the strongest relationships noted to date have been between EF tests and academic achievement scores (Biederman et al., 2008; Gropper & Tannock, 2009; Thorell, 2007) or self-ratings of academic performance (Ready et al., 2001). Given that both are significantly related to IQ, not to mention shared method (testing) where academic tests are used, this finding is not surprising. This problem of conceptual contamination is far less so for EF rating scales whose items have been intentionally selected to directly sample the various behaviors specified in EF constructs (Barkley & Murphy, 2011). Such scales or direct observations also have little or no significant relationships with intelligence (Alderman et al., 2003; Barkley & Murphy, 2011), and so the issue of contamination by general cognitive ability is far less problematic for ratings than for EF tests. Hence the conceptual or face validity of rating scales may be superior to that of EF tests merely as a consequence of their initial construction. As noted above, many EF tests were not initially designed to measure the construct of EF. These issues help to explain why EF tests and ratings are so poorly related to each other. Undoubtedly, rating scales suffer from their own set of problems, as I have detailed elsewhere (Barkley, 2011a). But despite their limitations, ratings also have many advantages over tests (see Barkley, 2011a) and they have proven superior to the tests in capturing the cross-temporal, self-regulatory, and social nature of EF as well as in predicting impairment in major life activities, as noted above.

For these and other reasons, I have spent more than a decade developing rating scales for the assessment of EF in children and adults. This chapter provides a brief summary of those scales, their development, and the evidence for the reliability and validity of their scores.

Development of the Barkley Deficits in Executive Functioning Scale

Much of what follows comes from the manuals for the adult and child versions of the Barkley Deficits in Executive Functioning Scale (BDEFS) (Barkley, 2011a, 2012b). The scale began originally as an attempt to evaluate EF deficits in daily life activities for use in studying in adults with attention deficit hyperactivity disorder (ADHD). A prototype EF scale (Barkley & Murphy, 2011) was developed for use in two large research projects on adults with ADHD (Barkley, Murphy, & Fischer, 2008). One of those projects examined clinic-referred adults diagnosed with ADHD in comparison to both clinical and community control groups. The second study was a follow-up study of hyperactive children into young adulthood (mean age 27) (see Barkley et al., 2008 for details of both studies).

The prototype was largely based on an earlier theory of EF and its five constructs and their specific adaptive purposes (Barkley, 1997a, 1997b, 2001) as well as the larger literature on the nature of EF (see Denckla, 1996; Fuster, 1997; Lyon & Krasnegor, 1996; Stuss & Benson, 1986) and the rich and lengthy history of descriptions of the symptoms of patients with PFC injuries (Luria, 1966). The original item pool consisted of 91 items. Items were developed to reflect inhibition, nonverbal working memory (self-directed sensing, especially visual imagery, sense of time, and time management), verbal working memory (self-directed private speech, verbal contemplation of one's behavior before acting, etc.), emotional-motivational self-regulation (inhibiting emotion, motivating one's self during boring activities, etc.), and reconstitution (generativity, planning, problem-solving, and goal-directed inventiveness). According to this theory, the constructs are interactive and serve the overarching purpose of self-organizing behavior across time to prepare for and attain future goals. Some additional items were also generated from a review of more than 200 charts of adults diagnosed with ADHD given that ADHD is largely a disorder of PFC functioning (Bush et al., 2005; Hutchinson, Mathias, & Banich, 2008; Mackie et al., 2007; Paloyelis, Mehta, Kuntsi, & Asherson, 2007; Valera et al., 2007), has long been construed as such (Pontius, 1973), and is characterized by many theorists as being a disorder involving EF (Barkley, 1997a; Castellanos et al., 2006; Nigg & Casey, 2005; Sagvolden, Johansen, Aase, & Russell, 2005). Each item was to be answered on a 4-point Likert scale (rarely or not at all, sometimes, often, and very often).

The scale items focused on problematic symptoms (deficit measurement) rather than on positive or normative EF functioning. The BDEFS is not intended to assess the broad variation of EF in the general population in order to identify the range of individual differences in normal functioning that may exist in that population. Scales focusing on typical EF in a general population may be quite useful in studying the range of individual differences, as in studies of behavior genetics, normal development across the lifespan, or other purposes in which normal variation in a psychological trait is of interest. The BDEFS, in contrast, was intended for clinical purposes to be used to evaluate the range of deficits in clinic-referred or high-risk adults in their EF; these are symptoms of executive dysfunctioning.

Principal components factor analysis was applied to the ratings obtained on a large sample of adults with ADHD, adults with other disorders, and a general population sample (Barkley & Murphy, 2011). Results showed five factors that had at least ten items having their highest loading on a factor and accounted for at least 2 % or more of the variance before rotation (and incidentally had Eigenvalues of 1.8 or higher). Eighty-eight items had loadings of at least .400 on any of these five factors. Three items were dropped from the scale because they did not have a loading of ≥.400 on any of the final five factors in that analysis. These factors were labeled Self-Management to Time, Self-Organization and Problem-Solving, Self-Restraint, Self-Motivation, and Self-Activation and Concentration. A factor analysis with varimax rotation was also conducted on the version of this rating scale completed by others (Barkley & Murphy, 2011). The same 5-factor solution emerged with nearly identical item content. Scores were computed for each factor simply by summing the individual item scores for items assigned to that factor. The factor scales were significantly intercorrelated, ranging from .74 to .88 for the self-ratings and .75–.88 for the other ratings. The scales therefore shared 56–77 % of their variance which may indicate the possible existence of a single underlying meta-construct of deficits in executive functioning (Barkley & Murphy, 2011).

Three of the five scales showed a low but significant correlation with participant age: Self-Management to Time ($r=-.11$, $p=.05$), Self-Motivation ($r=-.21$, $p<.001$), and Self-Activation ($r=-.11$, $p=.04$). This was true for these same three scales rated by others. Only one of the five scales was modestly but significantly correlated with participant IQ: Self-Organization/Problem-Solving scale ($r=-.15$, $p=.007$). This pattern was also the case for the other ratings ($r=-.17$, $p=.008$) (Barkley & Murphy, 2011). These results suggest that, unlike EF tests, ratings of EF in daily life activities are largely not contaminated by general intelligence. Even the single scale that was related to intelligence shared just 2.9 % of its variance with the IQ measure used in our study. The initial study groups noted above differed significantly from each other on these scales. Eighty-nine percent to 98 % of the ADHD group and 83–93 % of the clinical group fell into the clinically significant range (7th percentile of community group) compared to just 7–11 % of the community group on the five subscales. A similar pattern of results was found using other ratings of the participants as well as an interview version of the scale items (Barkley & Fischer, 2011).

A subsequent version of the BDEFS rating scale was then developed in which the items that had factor loadings of at least .500 or greater on one of the five factors in our earlier study were carried forward to the new scale. A few items on each of the prototype scales were abandoned in an attempt to reduce the scale length and hence the time needed to complete the scale without loss of information on the five subscales. A further examination of the scale items, as well as the resulting factor structure, suggested that one domain of EF appeared to be substantially underrepresented in the original scale and so did not have an opportunity to emerge in the foregoing analyses—that component was the self-regulation of emotion. The few items on the scale that dealt with emotion were primarily representing the impulsive expression of impatience, frustration, and anger and loaded on the self-restraint (inhibition) factor as might be expected. But no items pertaining to self-management of emotion were on the initial BDEFS. This is an important component of EF that is often neglected in developing EF test batteries but appears to be a commonplace observation of the deficits in clinical patients associated with PFC injuries (Fuster, 1997; Luria, 1966; Stuss & Benson, 1986). The modal model of emotional self-control developed by Gross (1998; Gross & Thompson, 2007) was used to generate an additional ten items to reflect these problems with poor self-regulation of emotional states. Two additional items reflecting self-motivation were also added to the pool to strengthen that scale.

The pool of 100 items was used to conduct a survey of 1,249 adults with nearly equal representation of males (623) and females (626) and equal representation of six age groups (18–29, 30–39, 40–49, 50–59, 60–69, 70 and older) and whose ages ranged from 18 to 92 years. These adults were representative of the US population being drawn from all regions, ethnic, educational, employment, and income groups relative to their representation in the 2000 US Census (Barkley, 2011a). Following further analyses, the final scale was reduced to 89 items. A factor analysis of this item pool on this sample revealed five factors. The five factors explained 53.2 % of the variance. Four of them were replicates of those found above. The fifth was the Self-Regulation of Emotion indicating that the addition of items reflecting this domain was successful. The factor of Self-Activation and Concentration did not replicate in the national sample, indicating that this domain may be a problem specific to individuals varying in severity of ADHD symptoms. The final published manual contains this self-report scale, a version to be completed by others, a short form of the self-rating scale consisting of 20 items (the five highest loading items from each scale), and an interview form based on these 20 items.

Psychometric Properties of the BDEFS

As in the earlier study of the prototype, in national sample age was observed to correlate to a small but significant degree with 5 of 6 scores: Self-Management to Time=−.13, $p<.001$; Self-Restraint=−.08, $p=.007$; Self-Regulation of Emotion=−.11, $p<.001$; Self-Motivation=−.126, $p<.001$; and Total EF Summary Score=−.09, $p=.002$, but not with the Self-Organization scale. There is a decline of about 10 % in EF deficits across the six age groups between the youngest and oldest. This could reflect merely cohort effect, true developmental decline in such deficits, or differential survival rates in which individuals with higher EF deficits die younger than those with lower levels. Further analyses resulted in the creation of norms for four separate age ranges: 18–34 years, 35–49, 50–64, and 65–81. These yielded sample sizes per age grouping of 305, 316, 322, and 275, respectively, for a total sample of 1,218. Though significant sex differences emerged on two scales (Self-Regulation of Emotion and Self-Motivation) ($p<.05$), the differences in mean scores for those two scales are relatively small. Even so, norms were created separately for men and women in scoring the BDEFS. Age and sex did not interact significantly in this normative sample. There were no significant differences among the ethnic groups on four of the five scale scores except for the Self-Motivation scale; even then, mean differences across groups were trivial. The geographic regions of the USA did not differ on any scales.

Reliability of the scores is quite satisfactory as evidenced by high internal consistency (Cronbach's alpha ranging from .91 to .95 scores across the five scales). Interobserver agreement using the prototype indicates acceptable reliability (ranging from .66 to .79 across scales) (Barkley & Murphy, 2011). The scale also has high test-retest reliability (ranging from .62 to .90 across scales and .84 for the Total EF Summary Score) over a 2–3-week interval using a subset of adults from the normative sample.

Validity of the scale scores was evident in numerous analyses (Barkley, 2011a). These included the factor analyses and correlations, regression analyses, and group comparisons concerning disorder discrimination. Concurrent validity was also established with various measures of functional impairment in major life activities such as educational history, occupational functioning, social relationships, marriage, driving, financial management, crime and drug use, parenting stress, and offspring psychopathology, among other domains.

Unlike other rating scales of EF deficits, this adult normative sample was not filtered to remove those with psychiatric, developmental, learning, or medical disorders or those receiving psychotropic medications (Roth, Isquith, & Gioia, 2005), a sample that does not represent the general US population. Also unique among rating scales is the provision of a limited license by the publisher to the owner of the manual to photocopy the scales and scoring sheets eliminating the inconvenience of having to reorder forms from the publisher. This permission also applies to the child and adolescent version of the scale discussed next.

Development of the BDEFS for Children and Adolescents

To create a parent report form for obtaining ratings of EF deficits in children, the 14 items that had the highest loading on each subscale of the adult BDEFS were chosen to represent that same subscale on the BDEFS for Children and Adolescents (BDEFS-CA), resulting in a 70-item scale for parents (Barkley, 2012b). This reduced the time needed to complete the scale with little if any loss of information contained on each dimension or subscale. The wording of each item was then changed from the first to the third person so parents could report about their child. Also, the phrasing of each item was examined for its appropriateness for children and adolescents. Where necessary, items were rephrased to suit these developmental stages. For instance, anytime an item referred to work, it was rephrased to refer to school work or work done at home. The ascertainment window given to parents for completing these items was the same as that used in the adult BDEFS—individuals were asked to rate the occurrence of each item based on the previous 6 months. The four anchor points for rating each item were also retained, these being Not at All or Rarely, Sometimes, Often, and Very Often. Parents were also instructed that if their child was taking any medication for a psychiatric or psychological disorder, they were to rate the child's behavior based on how the child behaved off of their medication.

The scale was then completed by a nationally representative sample of 1,922 parents of children ages 6–17 years with at least 75 children of each sex within each age level being represented in this sample. The sample was eventually reduced to 1,800 children with equal numbers of mothers and fathers represented in the sample in addition to equal numbers of boys and girls at each age level. All geographic regions of the USA were sampled in proportion to their representation in the 2000 US Census, unlike other rating scales of EF for children (Gioia, Isquith, Guy, & Kenworthy, 2000). Further analyses showed that the normative sample of parents reasonably approximated that of the US adult population as based on the US Census from the year 2000 concerning regional distribution, sex, race/ethnic group, and employment status. The normative sample has a somewhat higher representation of parents who are married and who are college graduates. As with the adult scale, the sample was not filtered to remove individuals with developmental, learning, psychiatric, or medical disorders or those children receiving psychiatric medication or special education services, as has been reportedly done with other EF rating scales (Gioia et al., 2000). Factor analysis of the scale revealed the same five-factor structure as was found on the adult BDEFS discussed above.

Age was found to be correlated to a low but significant extent to three of the five scales (range r's = −.05 to +.07), indicating only slight shared variance between the scale scores and the children's age levels. However, comparisons of the age groupings revealed no significant changes in any of the five scales with age. Even so, when age levels were clustered into 6–11- and 12–17-year-olds, significant age differences now emerged, albeit small, on two of the scales. For this reason, norms were created based on these two age groupings for scoring the scale. No differences on any scales were found between ratings of mothers and fathers. Significant differences between boys and girls, however, were found for four of the five scales, which argued for presenting the norms separately for boys and girls for scoring purposes. No significant interactions of age with sex were evident on any of the scale scores. There were no significant differences among ethnic groups. While parent education and income correlated to a very low but significant extent with the ratings on most scales (range of r's = −.05 to −.10), each accounted for less than 1 % of the variance on any scales. There were no differences across geographic regions on any of the scales. In view of the above findings, separate EF profile score sheets were created for males ages 6–11 ($N=451$), males ages 12–17 ($N=450$), females ages 6–11 ($N=451$), and females ages 12–17 ($N=448$).

Similar to the BDEFS adult form, reliability of the scale scores is quite satisfactory as evidenced by high internal consistency (Cronbach's alpha ranging from .95 to .97 scores across the five scales). There is evidence of high test-retest reliability (ranging from .73 to .82 across scales and .82 for the Total EF Summary Score) over a 3–5-week interval based on a subset of the normative sample. Validity of the scale scores was evident in numerous analyses including factor analyses and correlations with other rating scales of EF. It was also evident in correlations, regression analyses, and group comparisons concerning disorder discrimination and concurrent validity with various measures of functional impairment in major life activities such as family functioning, peer relations, education functioning, community activities, and risk for accidental injuries.

As an example of the ability of the BDEFS-CA to discriminate among various neurological, developmental, learning, and psychiatric disorders, examine the results in Table 15.1 taken from the manual (Barkley, 2012b). Here is summarized all of the comparisons among the various disorders identified in the children in the normative sample as discussed in the manual. These disorders were based on parent reports of the professional diagnoses that their children had received. It is possible to take the results from the analyses reported in the manual for each disorder and compute effect sizes (Cohen's d) to create a common metric on which to compare them. It also permits one to obtain some idea of which disorders are most likely to disrupt which domains of EF in daily life and to what degree relative to the other disorders. This gives a picture of the pattern of EF deficits across disorders and within each disorder, allowing scores on all scales to be compared directly to each other by this standard metric (effect size). An effect size here is the difference in the mean raw scores between each disorder and the control group divided by the pooled standard deviation (for both the disordered and control group). It is therefore the number of standard deviations that separate the mean scores for the disorder from the general population sample not having that disorder. Because ADHD has such a pervasive and substantial impact on EF in daily life activities in all domains, and because it can be comorbid with each of these disorders in a significant proportion of cases, those comorbid cases involving ADHD were removed from these analyses except of course the findings for ADHD itself.

Table 15.1 shows the effect size differences for each disorder on each of the five subscales of the BDEFS-CA. For each disorder except ADHD, of course, all cases having comorbid ADHD (by research criteria; see manual) were removed. So the effect sizes reflect just what would likely be associated with each disorder in the absence of empirically diagnosed cases of ADHD. Typically, an effect size of .20 is considered small, one of .50 is medium, and one of .80 is large (Cohen, 1992). Also shown in this table is whether or not the comparison of these disorder cases was significantly different from the control group when the pairwise comparisons discussed above were conducted (where $p \leq .05$) between the disorder only groups and the control groups. Again, all cases of ADHD were removed from these comparisons. Finally, and perhaps most clinically informative, this table also shows the percentage of cases of each disorder that placed in the clinically deficient range on each of the subscales. These figures are highlighted by the shaded gray rows in the table. The clinically deficient range is defined here (and traditionally) as having a score

Table 15.1 Effect sizes (Cohen's *d*), results of significance tests, and percentage of cases deemed clinically deficient for various developmental, learning, and psychiatric disorders relative to the general population sample not having that disorder on the five subscale scores and EF Summary Score for the BDEFS-CA Long Form (with ADHD removed from all but its own comparison)

Disorder[a]	Time mgnt	Self-organize	Self-restraint	Self-motivate	Emotion regulation
ADHD (parent report)					
ES	1.32	.99	1.26	1.35	1.04
Magnitude[b]	Large	Large	Large	Large	Large
Significant[c]	Yes	Yes	Yes	Yes	Yes
%Deficient[d]	43.4	32.9	43.9	42.2	34.1
ADHD (research dx)					
ES	2.48	1.93	2.48	2.86	2.16
Magnitude	X-large	X-large	X-large	X-large	X-large
Significant	Yes	Yes	Yes	Yes	Yes
%Deficient	72.8	68.0	73.6	77.6	67.2
Speech/language					
ES	.26	.63	.21	.40	.27
Magnitude	Small	Medium	Small	Small	Small
Significant	No	Yes	No	Yes	No
%Deficient	2.8	19.4	6.9	9.7	9.7
Motor disorders					
ES	.43	.87	.45	.66	.46
Magnitude	Small	Large	Small	Medium	Small
Significant	Yes	Yes	Yes	Yes	Yes
%Deficient	10.0	27.5	10.0	7.5	7.5
DD/MR					
ES	.77	1.46	.71	1.16	.91
Magnitude	Medium	Large	Medium	Large	Large
Significant	Yes	Yes	Yes	Yes	Yes
%Deficient	22.2	55.6	11.1	11.1	11.1
Seizures/epilepsy					
ES	.59	1.02	.41	.54	.34
Magnitude	Medium	Large	Small	Medium	Small
Significant	No	Yes	No	Yes	No
%Deficient	16.7	41.7	25.0	25.0	16.7
Tic disorders/TS					
ES	.73	.93	.98	.86	1.40
Magnitude	Medium	Large	Large	Large	Large
Significant	Yes	Yes	Yes	Yes	Yes
%Deficient	12.5	12.5	25.0	.0	37.5
Autism spectrum					
ES	.36	1.14	.87	.51	1.05
Magnitude	Small	Large	Large	Medium	Large
Significant	No	Yes	Yes	Yes	Yes
%Deficient	.0	31.6	36.8	15.8	21.1
Reading disorders					
ES	.55	.80	.33	.72	.37
Magnitude	Medium	Large	Small	Medium	Small
Significant	Yes	Yes	No	Yes	No
%Deficient	11.1	16.7	13.0	13.0	5.6

(continued)

Table 15.1 (continued)

Disorder[a]	Time mgnt	Self-organize	Self-restraint	Self-motivate	Emotion regulation
Spelling disorders					
ES	.69	.75	.44	.83	.54
Magnitude	Medium	Medium	Small	Large	Medium
Significant	Yes	Yes	No	Yes	Yes
%Deficient	9.1	15.2	6.1	18.2	3.0
Math disorders					
ES	.87	1.16	.38	1.07	.55
Magnitude	Large	Large	Small	Large	Medium
Significant	Yes	Yes	No	Yes	Yes
%Deficient	20.7	27.6	13.8	20.7	6.9
Writing disorders					
ES	.80	1.11	.65	1.05	.68
Magnitude	Large	Large	Medium	Large	Medium
Significant	Yes	Yes	Yes	Yes	Yes
%Deficient	14.3	32.1	14.3	25.0	7.1
Anxiety disorders					
ES	.50	.45	.37	.52	.81
Magnitude	Medium	Small	Small	Medium	Large
Significant	Yes	Yes	No	Yes	Yes
%Deficient	8.8	8.8	8.8	11.8	17.6
Depression					
ES	1.01	.64	.74	.92	1.14
Magnitude	Large	Medium	Medium	Large	Large
Significant	Yes	Yes	Yes	Yes	Yes
%Deficient	17.4	21.7	13.0	21.7	26.1
Oppositional defiant					
ES	1.34	1.25	1.80	1.10	1.49
Magnitude	Large	Large	X-large	Large	Large
Significant	Yes	Yes	Yes	Yes	Yes
%Deficient	18.2	27.3	36.4	9.1	36.4
Bipolar disorder					
ES	1.38	1.10	1.63	1.21	1.92
Magnitude	Large	Large	X-large	Large	X-large
Significant	Yes	Yes	Yes	Yes	Yes
%Deficient	14.3	14.3	28.6	28.6	57.1

From Barkley, R. A. (2012). *Barkley Deficits in Executive Functioning Scale—Children and Adolescents*. New York: Guilford Press. Copyright by Guilford Press. Reprinted with permission

ES effect size (Cohen's *d*), *BDEFS-PF* Barkley Deficits in Executive Functioning—Parent Form, *ADHD* attention deficit hyperactivity disorder, *research dx* ADHD diagnosed by the research criteria reported in this manual (93rd percentile on ADHD symptom ratings and impairment in at least one domain), *DD/MR* developmentally disabled/mental retardation, *TS* Tourette's syndrome. All diagnoses are by parent report except for ADHD which is defined by research diagnostic criteria. *Time mgmt* Self-Management to Time subscale, *self-organize* Self-Organization and Problem Solving subscale, *self-motivate* Self-Motivation subscale, *emotion regulation* Self-Regulation of Emotion subscale, *Summary Score* Total EF Summary Score

[a] All disorders are based on a parent that their child received a professional diagnosis of that disorder, except for ADHD where it was also diagnosed by research criteria presented in this manual (see below). All disorders except ADHD also have cases with comorbid ADHD removed from these analyses

[b] Magnitude is graded as .20+ = small, .50+ = medium, .80+ = large (see Cohen, 1992) and 1.5+ = X-large or extra large

[c] Significant" means that the comparison between this disorder and the control group (both without ADHD cases) was significantly different from the remainder of the same at $p < .05$

[d] %Deficient" means the percentage of children with this disorder who placed at or above the 93rd percentile for the normative sample ($N = 1,800$) for their age group (6–11, 12–17 years) and sex

at or above the 93rd percentile for the individual's age group (6–11 or 12–17 years) and sex. Just 7 % (approximately) of the population would expect to be in this range, which essentially places the participant +1.5 SD above the mean for the general population. Therefore, figures for each disorder which are higher than 7 % indicate that the disorder is associated with an elevation in risk for being clinically deficient on that subscale.

One can see very quickly from these three indicators (effect size, significance, and percentage deficient) what the impact is of each disorder on each domain of EF in daily life controlling for comorbid cases that also had ADHD. The results are quite informative and, to the author's knowledge, exist nowhere else in the scientific literature on either these disorders or on any parent rating scale of EF deficits. Bear in mind that the absolute number for the effect size here is not so important. It might well be different had a more rigorous approach to diagnosing these childhood disorders been used instead of parent reports of a professional diagnosis of their child. It is the *relative* pattern of effect sizes or deficits that is the most informative here when comparing disorders to each other as well as among the EF domains for any specific disorder.

As discussed in the manual (Barkley, 2012b), these results show that ADHD has, by far, the most adverse effect on all domains of EF in daily life activities, regardless of which indicator is used. Whether it is based on parent-reported diagnosis by a professional or by the more rigorous research criteria used in this manual, ADHD results in far more severe effects on EF domains and in far more cases being in the clinically deficient range than any other disorder. Indeed, using more rigorous criteria results in effect sizes about twice as large as those associated with parent reported diagnoses, which placed well above the "large" designation of an EF. Using research criteria to define ADHD almost doubles the percentage of cases considered clinically deficient. In virtually every EF domain, the adverse effect of ADHD (research diagnosis) is 2–3 standard deviations above the mean; this effect size isn't just large by traditional definition (Cohen, 1992) but huge. The findings essentially mean that the distributions of these scores for the ADHD group overlap minimally, if at all, with the remaining population of children. Indeed as measured by effect sizes, the impact of ADHD is 2–11 times greater in each domain than is the case for any other disorder. This cannot be attributed to the possibility that the BDEFS-CA contains many symptoms of ADHD as described in DSM-IV. That is because those symptoms were intentionally not included in the BDEFS or BDEFS-CA to permit the study of EF deficits in ADHD without such contamination of the dependent measure (EF) with the independent variable of interest (ADHD).

Not surprising, both ODD and BPD are the next most adverse disorders for their effects on EF in daily life across all domains on the BDEFS-PF as measured by effect sizes, but especially Self-Regulation of Emotion. To understand why, consider that even after removing cases of ADHD diagnosed by research criteria, some variation in ADHD symptoms is still present within the remaining sample for that other disorder. Given the strong correlation of ADHD symptom severity to that of ODD and BDP (indeed some symptoms of ADHD occur on the BPD symptom list), it is likely that the remaining variation in ADHD symptoms may be driving even these differences between ODD or BPD and the control group, or at least a portion of them. Looking at the percentage of cases in the clinically deficient range also tells us that while ODD or BPD are associated with some elevations on each subscale, the vast majority of ODD or BPD cases are not so deficient. The subscale with the greatest such deficient cases is the Self-Regulation of Emotion, which makes perfect sense when one understands that ODD and BPD involve mood dysregulation, particularly for anger, hostility, and even aggression in the case of ODD.

Now consider the various learning disabilities in Table 15.1, whether reading, spelling, math, language, or writing disorders. The vast majority of children with these disorders do not place in the clinically deficient range on the EF subscales when cases of comorbid ADHD are removed. The evidence from effect sizes tells us that these disorders are associated with a small impact on EF in daily life across these domains, and particularly so for Self-Organization and Problem-Solving. Yet when examining the proportion of cases that would likely fall in the clinically deficient range on any BDEFS-CA subscales, the vast majority of cases are unlikely to do so. It is of interest that math and writing disorders seem to have a greater adverse association with EF domains than is the case for reading, spelling, and speech/language disorders. The reason for this is unclear.

Another interesting finding has to do with anxiety disorders. Notice that once ADHD cases have been removed, anxiety disorders have only a modest adverse effect on the EF subscales as indexed by effect sizes. Yet even this is not especially impressive when translated into the percentage of cases that are likely to be clinically deficient on any subscales, where the proportion is barely above that expected from the population average of 7 % for most scales (8–11 %). The only elevation in risk here from anxiety disorders seems to be on the Self-Regulation of Emotion scale as one might expect from the very nature of this class of child psychiatric disorders. In contrast, the adverse impact of depression on EF in daily life is significantly greater resulting in an approximately twofold increase in risk of being clinically deficient in each of the EF subscale domains. Once again, as would be expected from the nature of this mood disorder, the greatest impact of depression is in the domain of Self-Regulation of Emotion. All of the above discussion of Table 15.1 is to say that the BDEFS-CA demonstrates a substantial ability to differentiate among various disorders.

Scoring and Interpretation of the BDEFS and BDEFS-CA

Both versions of the BDEFS come with a long form and a short form as well as with an interview version of the short form. Norms are not available for the interview version of these scales but its results correlate highly with the results for the rating scale versions, at least in studies of adults (see Barkley, 2011a). The long forms of the scales can be scored by simple summation to yield individual scale scores for the five main factors of Self-Management to Time, Self-Organization and Problem-Solving, Self-Restraint, Self-Motivation, and Self-Regulation of Emotion. A Total EF Summary Score can also be computed representing the sum across all items. One can also compute a Total EF Symptom Count representing the number of items answered as occurring "Often" or "Very Often." A separate ADHD-EF Index can also be computed based on the items most predictive of ADHD so as to identify cases in which the risk for a clinical diagnosis of ADHD is highly likely. The short form versions of the scales are used for a quick screening of EF and are scored to yield simply a Total EF Summary Score.

The BDEFS forms are not intended to replicate or replace the use of neuropsychological tests of EF. Indeed, the two types of measures are only modestly related if at all and so neither can replace the other nor serve as an alternative means of measuring the same EF components. Instead, EF is best viewed as a hierarchically organized series of levels or outwardly projecting concentric rings of the extended phenotype of EF into human major life activities (Barkley, 2012a). EF tests and EF ratings are likely evaluating quite different levels of this extended phenotype of EF, much like lab measures of reaction time and a parent rating scale of their teen's safe driving habits while operating a motor vehicle are measuring very different levels of driving performance. Therefore, the selection of a measure or set of measures of EF should be based largely upon the purposes of the evaluation and the levels of EF one wishes to assess. If evaluating the lowest, instrumental, moment-to-moment level of cold cognitive functioning believed to represent basic EF components is of importance, then EF tests or batteries might be preferred to EF ratings in daily life. This may be the case in neuroimaging research, molecular genetics of the instrumental EFs, or in the brain localization of specific EF functions where the EF tests and batteries may be useful. The point is arguable given the numerous problems evident in using tests to measure EF discussed above. If, however, the purpose of the evaluation is to identify EF at higher levels, of greater behavioral complexity, over longer terms, and in socially significant settings and domains of daily life activities and to predict the likelihood of impairment in various domains of major life activities, then the BDEFS would be of greater utility than would EF tests. Use of the BDEFS with populations, countries, cultures, or ethnic groups outside of the USA may be inappropriate to the extent that such groups may vary from the normative reference samples described in the manuals (Barkley, 2011a, 2012b).

As reported in the manuals, the results for the scales can be interpreted using four different approaches.

Face Validity

This approach to interpretation is aided by a complete familiarity with the items contained on the scale and the five sections (subscales) to which those items have been assigned. High scores on each scale (typically at or above the 93rd percentile and especially at or above the 98th) may signal a deficiency in this area of EF in daily life activities. One initial means of interpreting each scale in a clinical report is on the basis of the face validity of that subscale. The examiner can actually interpret the meaning of the scale score from the name given to that scale (e.g., Self-Management to Time and Self-Organization and Problem-Solving). This interpretation can be enhanced by selecting the individual items from that subscale on which the respondent has answer a 3 or 4 (a symptom) and directly quote these items in that portion of their report as to the meaning of that scale. Given that the scales are moderately to highly intercorrelated, it is most unlikely that an individual will produce deviant scores in just one or two subscales of the BDEFS. While not an impossible occurrence, such an occurrence in which only a single scale contained a clinically significant score would be the cause for further exploration of the reason for such an unusual event. One possibility may be a very discrete, focal brain legion in a small area of the PFC or cerebellum. Another would be malingering by the parent or child. Also in view of this interrelatedness of the scales, computing interscale differences is not encouraged as the meaning of disparities across the scales has not been explored in any research to date. This does not mean, however, that relative fluctuations across the scales will not be evident—only that interpreting the meaning of large disparities between scales has no empirical basis at this time. Just interpreting the meaning of each scale separately is likely to be sufficient to convey the meaning of the BDEFS scores rather than trying to torture the score comparisons to yield up more information than exists about them at the present time.

Normative Comparisons

Next, a comparison can be made of the individual's scores to those of the normative population using the EF Profiles and the resulting percentiles so obtained for them. This is then reported as the relative position of the individual in comparison to the general population of adults and specifically to the sex and age group in which they place and the norms used to obtain the percentile score. In short, using the percentile score, the individual is said to place at or above this percentage of individuals in the general population of their age group and sex in their EF in daily life activities.

Risk Analysis (Using the Evidence on Criterion Validity)

As discussed in the manuals, high scores on the BDEFS subscales were associated with a variety of psychiatric disorders, domains of impairments, risks for various treatments, and other difficulties in various domains of major life activities. Therefore, the higher the scores observed on the BDEFS, the more likely is the individual to be at risk for problems in these same domains of life activities. These risks should be noted in the clinical interpretation of the larger meaning of the BDEFS scores beyond their simply reflecting the five dimensions of EF in daily life for which the subscales are named. Problems with educational, family, peer, and community functioning, and various forms of psychological distress or psychiatric disorders (anxiety, depression, etc.) were all correlates of higher BDEFS scores that deserve notice by the examiner and probably should be conveyed to the adult patient or parent where appropriate and applicable. Such information can also be linked to what is already known about the patient from the larger evaluation in which the BDEFS should serve as simply one component when used for a clinical evaluation of a patient. The examiner will have information on the history of the patient with regard to these and other major domains of life activity that can serve to place the findings from the BDEFS in this larger context. The BDEFS results can perhaps provide additional hypotheses or even explanations as to why adverse events in these various domains may have been a consequence of or at least associated with deficits in EF. This risk analysis concerning impairments can be further aided by using the Barkley Functional Impairment Scales (Barkley, 2011b, 2012c) on which national norms are available for potential impairments in 15 major domains of adult or of children's daily life activities.

Change Resulting from Treatment

In addition to these routine clinical interpretations of the BDEFS scores, it is also possible to use the scales to assess change in patients as a consequence of treatment. Guidelines are provided in the manuals for both the adult and child scales for computing the Reliable Change Index (RCI) developed by Jacobson (Jacobson, Roberts, Berns, & McGlinchey, 1999; Jacobson & Truax, 1991) for each of the BDEFS scale scores and summary scores. Unlike arbitrarily defined thresholds for change, such as a change of 20 or 30 % from baseline to posttreatment, this statistical approach takes into consideration the test-retest reliability of the measure and the variation in change scores in the general population. Tables are provided in each manual for conveniently determining just how much change from pre- to posttreatment can be considered reliable and normalizing.

An example from the manual for the BDEFS-CA (Barkley, 2012b) illustrates the value of this information about reliable change and recovery. Consider a male child within the age range of 6–11 years with ADHD or even a closed head injury who is manifesting significant deficits in EF in daily life as reflected in an EF Summary Score from the BDEFS-CA Long Form of 260. According to the table of norms in the EF Profile for a boy of this age for this Long Form, this patient is above the 99th percentile of the population of their age group and sex. Following the introduction of treatment for a month with an ADHD medication, the parent of this child completes the BDEFS-CA Long Form again and now has an EF Summary Score of 160. Is this degree of change from the medication treatment reliable? That is, is it likely to fall outside the realm of merely measurement error or unreliability? To find out, compute the RCI as follows: Pre-test score (260) minus post-test score (160)=100 points. The table in Chapter 9 of the manual for the BDEFS-CA shows that for 6–11-year-old boys, the RCI needed to be significant for the EF Summary Score is 48.3. The patient's change score of 100 is well above this RCI threshold. Thus, the clinician can feel confident that the improvement in EF deficits is not simply due to measurement error or unreliability. It is likely a consequence of the treatment. It is reliable. The patient could be considered as a treatment responder in this instance (assuming that side effects of the medication are not so annoying as to warrant treatment discontinuation).

One can also determine if the child has been normalized by the treatment—that is, has the intervention brought them to within the normal range. A separate table in the manual shows the threshold for +2 SDs above the normal mean for boys that are aged 6–11 years (it is 200) which corresponds to approximately the 98th percentile. The patient's post-test score of 160 is well below this figure and so could be considered to place within the broadly defined normal range. However, some clinicians and researchers believe that this is too generous a definition of normalcy for determining recovery. Using the criterion of +1.5 SD above the normal mean may be preferred. In that case, this same table shows that this threshold is 179. The patient's post-test score of 160 is also below this level. And so even by this more conservative definition of normalization, the patient can be considered to place within the broadly defined normal range for their age group and sex following treatment. This may be considered "recovery."

Conclusion

As I concluded in the manuals for the BDEFS (Barkley, 2011a, 2012b), EF is a neuropsychological construct that does not depend for its existence and utility as such on some obsessively slavish linkage to a lower biological level of analysis (the PFC) or to some consensus-ordained lab test(s) heralded as the sine qua non of EF. It can be studied for its utility and contribution to understanding human behavior and affairs in its own right independent of the field of neurology and using a variety of measures, not just psychometric tests. We can do so provided that EF is operationally defined at its own neuropsychological level of analysis and that theoretical models of it at that level are testable as well as scientifically and clinically useful. EF ratings (and other observational methodologies) have as or more valuable a role to play in such conceptualizations, model building, and testing as do EF tests as well as in the clinical and scientific evaluation of EF. It should be the evidence available as well as the level of analysis and extended phenotypic domain of interest to the examiner that determines the best means by which to assess EF and not some dogmatic adherence to a historically dated obsession with laboratory testing as the gold standard for the evaluation of EF. These adult and child versions of the BDEFS provide a clinically informative means of evaluating deficits in executive functioning as reported by patients or parents in daily life activities along five interrelated dimensions of EF deficits, these being Self-Management to Time, Self-Organization/Problem-Solving, Self-Restraint, Self-Motivation, and Self-Regulation of Emotion. The evidence to date shows the two versions of the BDEFS to be reasonably reliable, valid, and useful in assessing these dimensions of EF in the daily life activities of either adult or children or teenagers, respectively. The scales can do so by a means that is efficient and cost-effective. They also yield valuable information on the potential risks that such deficits may be posing in the extent to which these deficits in EF may relate to potential impairments in a variety of domains of major life activities. The adaptive meaningfulness of the BDEFS is documented in its association with impairment in a variety of domains of major life activities beyond simply documenting the existence of such EF deficits. The scales can be used in clinical practice, research and educational/organization settings, or other such venues where the evaluation of potential deficits in EF in the daily life activities is of interest. They can also serve as a basis for evaluating change in such EF deficits as a consequence of treatment.

References

Alderman, N., Burgess, P. W., Knight, C., & Henman, C. (2003). Ecological validity of a simplified version of the multiple errands shopping test. *Journal of the International Neuropsychological Society, 9*, 31–44.

Anderson, P. (2002). Assessment and development of executive function (EF) during childhood. *Child Neuropsychology, 8*, 71–82.

Anderson, V. A., Anderson, P., Northam, E., Jacobs, R., & Mikiewicz, O. (2002). Relationships between cognitive and behavioral measures of executive function in children with brain disease. *Child Neuropsychology, 8*, 231–240.

Barkley, R. A. (1997a). Inhibition, sustained attention, and executive functions: Constructing a unifying theory of ADHD. *Psychological Bulletin, 121*, 65–94.

Barkley, R. A. (1997b). *ADHD and the nature of self-control*. New York: Guilford.

Barkley, R. A. (2001). Executive functions and self-regulation: An evolutionary neuropsychological perspective. *Neuropsychology Review, 11*, 1–29.

Barkley, R. A. (2011a). *Barkley deficits in executive functioning scale*. New York: Guilford.

Barkley, R. A. (2011b). *Barkley Functional Impairment Scale*. New York: Guilford.

Barkley, R. A. (2012a). *Executive functions: What they are, how they work, and why they evolved*. New York: Guilford.

Barkley, R. A. (2012b). *Barkley Deficits in Executive Functioning Scale—Children and Adolescents*. New York: Guilford.

Barkley, R. A. (2012c). *Barkley Functional Impairment Scale—Children and Adolescents*. New York: Guilford.

Barkley, R. A., & Fischer, M. (2011). Predicting impairment in occupational functioning in hyperactive children as adults: Self-reported executive function (EF) deficits vs. EF tests. *Developmental Neuropsychology, 36*(2), 137–161.

Barkley, R. A., & Murphy, K. R. (2010). Impairment in major life activities and adult ADHD: The predictive utility of executive function (EF) ratings vs. EF tests. *Archives of Clinical Neuropsychology, 25*, 157–173.

Barkley, R. A., & Murphy, K. R. (2011). The nature of executive function (EF) deficits in daily life activities in adults with ADHD and their relationship to EF tests. *Journal of Psychopathology and Behavioral Assessment, 33*, 137–158.

Barkley, R. A., Murphy, K. R., & Fischer, M. (2008). *ADHD in adults: What the science says*. New York: Guilford.

Best, J. R., Miller, P. H., & Jones, L. J. (2009). Executive functions after age 5: Changes and correlates. *Developmental Review, 29*, 180–200.

Biederman, J., Petty, C. R., Fried, R., Black, S., Faneuil, A., Doyle, A. E., et al. (2008). Discordance between psychometric testing and questionnaire-based definitions of executive function deficits in individuals with ADHD. *Journal of Attention Disorders, 12*, 92–102.

Bogod, N. M., Mateer, C. A., & MacDonald, S. W. S. (2003). Self-awareness after traumatic brain injury: A comparison of measures and their relationship to executive functions. *Journal of the International Neuropsychological Society, 9*, 450–458.

Boonstra, A. M., Oosterlaan, J., Sergeant, J. A., & Buitelaar, J. K. (2005). Executive functioning in adult ADHD: A meta-analytic review. *Psychological Medicine, 35*, 1097–1108.

Burgess, P. W., Alderman, N., Evans, J., Emslie, H., & Wilson, B. A. (1998). The ecological validity of tests of executive function. *Journal of the International Neuropsychological Society, 4*, 547–558.

Bush, G., Valera, E. M., & Seidman, L. J. (2005). Functional neuroimaging of attention-deficit/hyperactivity disorder: A review and suggested future directions. *Biological Psychiatry, 57*, 1273–1296.

Castellanos, X., Sonuga-Barke, E., Milham, M., & Tannock, R. (2006). Characterizing cognition in ADHD: Beyond executive dysfunction. *Trends in Cognitive Science, 10*, 117–123.

Chaytor, N., Schmitter-Edgecombe, M., & Burr, R. (2006). Improving the ecological validity of executive functioning assessment. *Archives of Clinical Neuropsychology, 21*, 217–227.

Cohen, J. (1992). A power primer. *Psychological Bulletin, 112*, 155–159.

Dawkins, R. (1982). *The extended phenotype*. New York: Oxford University Press.

Denckla, M. B. (1996). A theory and model of executive function: A neuropsychological perspective. In G. R. Lyon & N. A. Krasnegor (Eds.), *Attention, memory, and executive function* (pp. 263–278). Baltimore, MD: Paul H. Brookes.

Dimond, S. J. (1980). *Neuropsychology: A textbook of systems and psychological functions of the human brain*. London: Butterworths.

Dodrill, C. B. (1997). Myths of neuropsychology. *The Clinical Neuropsychologist, 11*, 1–17.

Eslinger, P. J. (1996). Conceptualizing, describing, and measuring components of executive function: A summary. In G. R. Lyon & N. A. Krasnegor (Eds.), *Attention, memory, and executive function* (pp. 367–395). Baltimore, MD: Paul H. Brookes.

Frazier, T. W., Demareem, H. A., & Youngstrom, E. A. (2004). Meta-analysis of intellectual and neuropsychological test performance in attention-deficit/hyperactivity disorder. *Neuropsychology, 18*, 543–555.

Fuster, J. M. (1997). *The prefrontal cortex*. New York: Raven.

Gioia, G. A., Isquith, P. K., Guy, S. C., & Kenworthy, L. (2000). *BRIEF: Behavior Rating Inventory of Executive Function—Professional manual*. Odessa, FL: Psychological Assessment Resources.

Gioia, G. A., Isquith, P. K., Kenworthy, L., & Barton, R. M. (2002). Profiles of everyday executive function in acquired and developmental disorders. *Child Neuropsychology, 8*, 121–137.

Gropper, R. J., & Tannock, R. (2009). A pilot study of working memory and academic achievement in college students with ADHD. *Journal of Attention Disorders, 12*, 574–581.

Gross, J. J. (1998). The emerging field of emotion regulation: An integrative review. *Review of General Psychology, 2*, 271–299.

Gross, J. J., & Thompson, R. A. (2007). Emotion regulation: Conceptual foundations. In J. J. Gross (Ed.), *Handbook of emotion regulation* (pp. 3–24). New York: Guilford.

Hervey, A. S., Epstein, J. N., & Curry, J. F. (2004). Neuropsychology of adults with attention-deficit/hyperactivity disorder: A meta-analytic review. *Neuropsychology, 18*, 495–503.

Hutchinson, A. D., Mathias, J. L., & Banich, M. T. (2008). Corpus callosum morphology in children and adolescents with attention deficit hyperactivity disorder: A meta-analytic review. *Neuropsychology, 22*, 341–349.

Jacobson, N. S., Roberts, L. J., Berns, S. B., & McGlinchey, J. B. (1999). Methods for defining ad determining the clinical significance of treatment effects: Description, application, and alternatives. *Journal of Consult and Clinical Psychology, 67*, 300–307.

Jacobson, N. S., & Truax, P. (1991). Clinical significance: A statistical approach to defining meaningful change in psychotherapy research. *Journal of Consulting and Clinical Psychology, 59*, 12–19.

Jonsdottir, S., Bouma, A., Sergeant, J. A., & Scherder, E. J. A. (2006). Relationship between neuropsychological measures of executive function and behavioral measures of ADHD symptoms and comorbid behavior. *Archives of Clinical Neuropsychology, 21*, 383–394.

Kertesz, A., Nadkarni, N., Davidson, W., & Thomas, A. W. (2000). The Frontal Lobe Inventory in the differential diagnosis of frontotemporal dementia. *Journal of the International Neuropsychological Society, 6*, 460–468.

Lezak, M. D. (1995). *Neuropsychological assessment* (3rd ed.). New York: Oxford University Press.

Lezak, M. D. (2004). *Neuropsychological assessment* (4th ed.). New York: Oxford University Press.

Luria, A. R. (1966). *Higher cortical functions in man.* New York: Basic Books.

Lyon, G. R., & Krasnegor, N. A. (Eds.). (1996). *Attention, memory, and executive function.* Baltimore, MD: Paul H. Brookes.

Mackie, S., Shaw, P., Lenroot, R., Greenstein, D. K., Nugent, T. F., III, Sharp, W. S., et al. (2007). Cerebellar development and clinical outcome in attention deficit hyperactivity disorder. *The American Journal of Psychiatry, 76*, 647–655.

Mahone, E. M., Hagelthora, K. M., Cutting, L. E., Schuerholz, L. J., Pelletier, S. F., Rawlins, C., et al. (2002). Effects of IQ on executive function measures in children with ADHD. *Child Neuropsychology, 8*, 52–65.

Mangeot, S., Armstrong, K., Colvin, A. N., Yeates, K. O., & Taylor, H. G. (2002). Long-term executive function deficits in children with traumatic brain injuries: Assessment using the Behavior Rating Inventory of Executive Function (BRIEF). *Child Neuropsychology, 8*, 271–284.

Marchetta, N. D. J., Hurks, P. P. M., Krabbendam, L., & Jolles, J. (2008). Interference control, working memory, concept shifting, and verbal fluency in adults with attention-deficit/hyperactivity disorder (ADHD). *Neuropsychology, 22*, 74–84.

Mitchell, M., & Miller, S. (2008). Executive functioning and observed versus self-reported measures of functional ability. *The Clinical Neuropsychologist, 22*, 471–479.

Nigg, J. T., & Casey, B. J. (2005). An integrative theory of attention-deficit/hyperactivity disorder based on the cognitive and affective neurosciences. *Development and Psychopathology, 17*, 765–806.

Nigg, J. T., Willcutt, E. G., Doyle, A. E., & Sonuga-Barke, J. S. (2005). Causal heterogeneity in attention-deficit/hyperactivity disorder: Do we need neuropsychologically impaired subtypes? *Biological Psychiatry, 57*, 1224–1230.

O'Shea, R., Poz, R., Michael, A., Berrios, G. E., Evans, J. J., & Rubinstein, J. S. (2010). Ecologically valid cognitive tests and everyday functioning in euthymic bipolar disorder patients. *Journal of Affective Disorders, 125*, 336–340.

Paloyelis, Y., Mehta, M. A., Kuntsi, J., & Asherson, P. (2007). Functional MRI in ADHD: A systematic literature review. *Expert Reviews in Neurotherapeutics, 7*, 1337–1356.

Pontius, A. A. (1973). Dysfunction patterns analogous to frontal lobe system and caudate nucleus syndromes in some groups of minimal brain dysfunction. *Journal of the American Medical Women's Association, 26*, 285–292.

Pribram, K. H. (1973). The primate frontal cortex—Executive of the brain. In K. H. Pribram & A. R. Luria (Eds.), *Psychophysiology of the frontal lobes* (pp. 293–314). New York: Academic.

Rabbitt, P. (1997). Introduction: Methodologies and models in the study of executive function. In P. Rabbitt (Ed.), *Methodology of frontal and executive function* (pp. 1–38). Hove: Psychology Press.

Ready, R. E., Stierman, L., & Paulsen, J. S. (2001). Ecological validity of neuropsychological and personality measures of executive functions. *The Clinical Neuropsychologist, 15*, 314–323.

Riccio, C. A., Hall, J., Morgan, A., Hynd, G. W., & Gonzalez, J. J. (1994). Executive function and the Wisconsin Card Sort Test: Relationship with behavioral ratings and cognitive ability. *Developmental Neuropsychology, 10*, 215–229.

Roth, R. M., Isquith, P. K., & Gioia, G. A. (2005). *Behavior Rating Inventory of Executive Function—Adult version.* Odessa, FL: Psychological Assessment Resources.

Sagvolden, T., Johansen, E. B., Aase, H., & Russell, V. A. (2005). A dynamic developmental theory of attention-deficit/hyperactivity disorder (ADHD) predominantly hyperactive/impulsive and combined subtypes. *The Behavioral and Brain Sciences, 25*, 397–468.

Shallice, T., & Burgess, P. W. (1991). Deficits in strategy application following frontal lobe damage in man. *Brain, 114*, 727–741.

Stavro, G. M., Ettenhofer, M. L., & Nigg, J. T. (2007). Executive functions and adaptive functioning in young adults with attention-deficit/hyperactivity disorder. *Journal of the International Neuropsychological Society, 13*, 324–334.

Stuss, D. T., & Benson, D. F. (1986). *The frontal lobes.* New York: Raven.

Thorell, L. B. (2007). Do delay aversion and executive function deficits make distinct contributions to the functional impact of ADHD symptoms? A study of early academic skill deficits. *Journal of Child Psychology and Psychiatry, 48*, 1061–1070.

Valera, E. M., Faraone, S. V., Murray, K. E., & Seidman, L. J. (2007). Meta-analysis of structural imaging findings in attention-deficit/hyperactivity disorder. *Biological Psychiatry, 61*, 1361–1369.

Vriezen, E. R., & Pigott, S. E. (2002). The relationship between parental report on the BRIEF and performance-based measures of executive function in children with moderate to severe traumatic brain injury. *Child Neuropsychology, 8*, 296–303.

Wilens, T. E., Martelon, M., Fried, R., Petty, C., Bateman, C., & Biederman, J. (2011). Do executive function deficits predict later substance use disorders among adolescents and young adults? *Journal of the American Academy of Child and Adolescent Psychiatry, 50*, 141–149.

Willcutt, E. G., Doyle, A. E., Nigg, J. T., Faraone, S. V., & Pennington, B. F. (2005). Validity of the executive function theory of attention-deficit/hyperactivity disorder: A meta-analytic review. *Biological Psychiatry, 57*, 1336–1346.

Wood, R. L. I., & Liossi, C. (2006). The ecological validity of executive function tests in a severely brain injured sample. *Archives of Clinical Neuropsychology, 21*, 429–437.

Zandt, F., Prior, M., & Kyrios, M. (2009). Similarities and differences between children and adolescents with autism spectrum disorder and those with obsessive compulsive disorder: Executive functioning and repetitive behavior. *Autism, 13*, 43–57.

Assessment with the Test of Verbal Conceptualization and Fluency (TVCF)

Cecil R. Reynolds and Arthur MacNeill Horton Jr.

1. Conceptualization underlying the Test of Verbal Conceptualization and Fluency (TVCF).
   (a) Historical information, definition constructs, and development of subtests

This chapter provides an introduction to the TVCF, a new measure for the assessment of executive functions. The TVCF is a standardized set of four subtests with a total administration time of 25–30 min for most individuals. The test is designed to measure multiple aspects of executive functions, through the use of several different forms of tasks commonly used by clinical neuropsychologists to assess executive functioning.

Neuropsychological Assessment

Neuropsychological assessment has grown rapidly because of the empirical work of Ralph M. Reitan (1955) and other important contributors (please see Fitzhugh-Bell, 1997, for a review of the early history of clinical neuropsychology). The value of neuropsychological assessment techniques for diagnosis and treatment has been recognized officially by the American Academy of Neurology (AAN) in an official report of its Therapeutics and Technology Assessment Subcommittee (American Academy of Neurology, 1996) and many publications detailing contributions to diagnosis (e.g., Reitan & Wolfson, 2001; Reynolds, 2001) and to rehabilitation (e.g., Bennett, 2001; Hartlage, 2001). Clinical neuropsychological testing results are used in settings such as medical, educational, and legal arenas and by professionals, primarily clinical neuropsychologists, but also neurologists; psychiatrists; pediatricians; clinical, counseling, school, educational, rehabilitation, and pediatric psychologists; occupational therapists; speech and language professionals; physical therapists; life care planners; and vocational rehabilitation experts among others. Neuropsychological testing requires the comprehensive assessment of multiple neuropsychological domains, including intelligence, academic achievement, motor abilities, sensory-perceptual functioning, language, visual spatial skills, processing speed, executive functioning, attention and memory, and emotional and personality functioning (Horton & Wedding, 1984). Among the most important of all neuropsychological domains has been executive functioning.

Executive Functioning as a Construct

The TVCF is composed of tasks related to executive functions. Lezak (1995) postulated that executive functions fall into four components, and these are "volition, planning, purposive action,

C.R. Reynolds (✉)
Texas A&M University, College Station, TX, USA
e-mail: crrh@earthlink.net

A.M. Horton Jr.
Psychological Associates of Maryland,
Towson, MD, USA

effective performance" and that each component "...involves a distinctive set of activity related behaviors" (p. 650) necessary for successful adaptive self-direction. Executive functions can be contrasted with knowledge that relates to retention of an organized set of facts, and executive functions deal with problem-solving and performing adaptive actions. Knowledge relates to the passive retention of information, while executive functioning involves motor outputs that are adaptive to external demands. A crucial insight is that while executive functioning impairment can be caused by frontal lobe brain injury, executive functioning impairment can also be caused by impairments in brain areas other than the frontal lobes (Reitan & Wolfson, 2000, 2004).

Simply put, executive functioning, although it can be related to impairment in the frontal lobes, is not synonymous with frontal lobe impairment. The frontal lobes are a neuroanatomical area of human brain, while executive functioning is a neuropsychological concept/domain similar to attention and memory. A paradox is that while there is great controversy and a lack of agreement among neuropsychologists regarding the conceptualization of executive functioning, there is relative agreement regarding what clinical neuropsychological tests are considered measures of executive functioning (Reynolds & Horton, 2006).

(b) Subtest development

Based on the understanding that executive functioning is a neuropsychological concept, the TVCF was developed based on a review of selected previously developed neuropsychological tests accepted as measures of executive functioning. While recognizing that frontal lobe functions are important for the assessment of executive functioning because all motor outputs go through the frontal lobes, the TVCF does not postulate specific localization for any particular brain area as being solely responsible for performance on the TVCF. The TVCF also does not attempt to include every possible aspect of executive functioning but rather has selected a limited number of constructs that sample executive functioning—assessing all aspects of executive functioning is a nearly impossible task since nearly all aspects of cognition are in some way linked to executive functions, necessitating the need to sample key aspects of EF. The aim of the TVCF is to provide such a standardized assessment of executive functioning. Clinicians often pull together tasks from different normative bases in assessing EF, and it is hoped that the standardized assessment and the use of co-normed tasks (i.e., tasks with a common normative base) will allow more accurate interpretation of discrepancies in executive functioning abilities.

Traditional clinical neuropsychological measures of executive functioning that were selected for inclusion and modification in the TVCF include card sorting tasks, category and letter retrieval tasks, and "trail-making" tasks. These neuropsychological measures will be briefly reviewed.

Card Sorting Tasks

Sorting objects, such as blocks and cards, into categories to assess mental abilities has had a long history in neuropsychology (Lezak, 1995). The most well-known card sorting test has been the Wisconsin Card Sorting Test (WCST) (Grant & Berg, 1948). The test assesses abstract thinking and the ability to set shift (Lezak, 1995). The test materials consist of decks of cards, which are sorted based on abstract principles exemplified by four stimulus cards (Heaton, 1981). The person being tested sorts the cards into piles under one of the four stimulus cards (Heaton, 1981). The examinee is provided feedback but not informed of the sorting principle which is varied after a specific number of correct sorts (Heaton, 1981). The most commonly used current scoring system for the WCST is the one devised by Heaton (1981) and later revised and expanded (Heaton, Chelune, Talley, Kay, & Curtiss, 1993).

Milner (1963) first proposed the WCST as measure of frontal lobe abilities. Later research was mixed (Lezak, 1995). Heaton (1981) suggests that the WCST is a measure of planning ability rather than a measure of frontal lobe functioning. The TVCF Classification subtest was designed as a verbal- or language-based analog to the nonlanguage-based WCST in order to assess set-shifting and rule induction when verbally related stimuli are involved.

Category and Letter Retrieval

Verbal fluency measures have been recognized as frequently impaired in brain damaged patients (Lezak, 1995). Estes (1974) proposed verbal fluency measures can assess how thinking processes are organized. The two most common formats in neuropsychology for assessing verbal fluency have been category and letter retrieval tasks.

Benton and Hamsher (1989) developed the Controlled Oral Word Association Test (COWAT) to assess letter fluency. The COWAT uses a set of three letters for three word naming trials. The person-assessed examinee says as many words that have the designated letter as the first letter in the words generated in 1 min during each trial. Benton (1968) found that letter retrieval was impaired in frontal lobe patients. Left frontal lobe patients did more poorly than right frontal lobe patients but bilaterally impaired patients did more poorly than either left or right frontal lobe patients (Benton, 1968).

Category naming has used to assess mental performance in Alzheimer's patients (Rosen, 1980). Category retrieval assesses semantic organization in the brain. The most common category used naming as many "animals" as possible in 1 min. Other categories of words that have been included are "super market items," "fruits," and "vegetables" among others. Animal naming has been shown to discriminate between dementia and depression (Hart, Kwentus, Taylor, & Hamer, 1988).

Trail-Making Tasks

Although used extensively in research as early as the 1930s, and subsequently as part of the US Army Individual Test Battery in 1944, trail-making tasks were first incorporated in clinical neuropsychology by Reitan (1955) and are sensitive measures of brain function (e.g., Reitan, 1955, 1958). Trail-making tasks are also among the most frequently administered of all neuropsychological tests (Mitrushina, Boone, & D'Elia, 1999). Reitan (1992) postulated trail-making tasks require "immediate recognition of the symbolic significance of numbers and letters, ability to scan the page continuously to identify the next number or letter in sequence, flexibility in integrating the numerical and alphabetical series, and completion of these requirements under the pressure of time" (p. ii). Mitrushina et al. (1999) agreed with Reitan and suggested the abilities required for trail making are "known to be highly vulnerable to deterioration resulting from brain pathology of different etiologies" (p. 33). Multiple researchers have averred that trail-making tasks are sensitive measures of executive function (Anderson, 1994; Mitrushina et al., 1999; Storandt, Botwinick, & Danziger, 1984; Stuss et al., 2001).

The Original Trail-Making Test: The original trail-making task, currently known as the trail-making test (TMT), parts A and B, was first used in psychological assessment as part of the US Army Individual Test Battery in 1944 (Horton, 1979; Mitrushina et al., 1999). Spreen and Strauss (1998) note that the TMT was developed in 1938 by Partington (Partington & Leiter, 1949) as an experimental psychology measure of divided attention. The original TMT consisted of two parts, known as Trails part A and Trails part B. On part A, the person draws a line to connect numbered circles in an ascending sequence (1-2-3-4-5…) as quickly as possible. In part B, the task is similar except the circles each contain either a number or a letter and the ascending sequence required is to connect alternate numbers and letters (1-A-2-B-3-C…) as quickly as possible.

Normative Status of the Original Trail-Making Test: Published normative information for the original TMT are limited and do not reflect population proportionate stratified random sampling reflecting United States Bureau of the Census statistics. In reviews, Mitrushina et al. (1999), Spreen and Strauss (1998) and Soukup, Ingram, Grady, and Schiess (1998) identified serious psychometric difficulties with available normative data for the TMT.

The Comprehensive Trail-Making Test: The original TMT had a number of methodological weaknesses such as unreliable normative data. The Comprehensive Trail-Making Test (CTMT) (Reynolds, 2003) was developed to improve the original TMT by using a stratified random sample and by making the test more reliable and more sensitive to brain dysfunction.

2. Description of the TVCF
   (a) Subtest background and structure

The subtests of the TVCF emphasize multiple aspects of verbal fluency, set-shifting, and rule induction, along with sequencing and visual search skills. The TVCF was designed and standardized for use with individuals ranging in age from 8 years through 89 years. Standardized (or scaled scores) are provided in the form of smoothed linear T-scores, having a mean of 50 and a standard deviation of 10, along with their accompanying percentile ranks.

The four subtests of the TVCF are as follows:
1. Category Fluency measures word retrieval by conceptual category (e. g., things to eat, wear) and fluency of ideation.
2. Classification is a verbal measure of set-shifting and rule induction that is designed as a verbal- or language-based analog to the well-known Wisconsin Card Sorting Test (WCST; Grant & Berg, 1948).
3. Letter Naming measures word retrieval by initial sound and fluency of ideation.
4. Trails C measures the ability to coordinate high attentional demands, sequencing, visual search capacity, and the ability to shift rapidly between Arabic numerals and linguistic representations of numbers. The trails task used and co-normed with the other TVCF subtests is a variation of several other "trail-making" tasks and was taken from the earlier published CTMT (Reynolds, 2003).

The T-scores for the subtests are age-corrected deviation scaled scores based on the cumulative frequency distributions of the raw scores at varying ages. The T-scores were computed directly from the percentiles of raw score means and standard deviations calculated at intervals from age 8 to 89 and the raw scores were converted to normalized T-scores at each interval. The resulting data were smoothed across age levels to allow for a consistent progression. Normative data for the subtests are provided in the tables in the TVCF Test Manual (Reynolds & Horton, 2006).

The subtests of the TVCF are interpretable from several theories of brain functioning, particularly that of Luria (see Luria, 1966; Reynolds & French, 2003) as well as more recent empirical and theoretical models of functional neuroanatomy in humans (e.g., Joseph, 1996).

The four subtests of the TVCF measure aspects of verbal fluency, set-shifting and rule induction, sequencing, inhibition, and visual search skills. From the four subtests, six scores are derived that include five age-adjusted T-scores and an age-adjusted percentile score.

Category Fluency measures the examinee's ability to retrieve words that fit within a conceptual category (e.g., animals) and fluency of ideation. Examinees are required to retrieve words from five conceptual categories (i.e., animals, things to eat, things to go in a house, things you can ride on, things you wear) during 30 s trials. An age-adjusted T-score is calculated for the total number of correct words.

Letter Naming is a measure of word retrieval by initial sound and fluency of ideation. Examinees are required to retrieve words starting with four letters (i.e., S, P, T, and D) during 30 s trials. An age-adjusted T-score is calculated for the total number of correct words produced.

Classification is a verbal measure of set-shifting and rule induction that was designed as a language-based analog to the Wisconsin card sorting task (Grant & Berg, 1948). There are four conceptual categories: animals, food, means of transportation, and clothing. Three scores are obtained for this measure: number of items correct, number of perseverative errors, and number of categories attained. Age-adjusted T-scores are calculated for the total number of correct items and total number of perseverative errors. No T-score is calculated for the number of categories attained score because the distribution of number of categories attained was not normal but a percentile score is available for interpretation.

Trails C measures the ability to coordinate high attentional demands, sequencing, and visual search capacity, and the ability to shift rapidly between Arabic numerals and linguistic representations of numbers. Trails C is a variation of other trail-making tasks; it is included as one of the subtests of the CTMT (Reynolds, 2003). An age-adjusted *T*-score is calculated for the total time to complete this task.

3. Administration and scoring

   (a) Tips on administration

   Before testing in a quiet room with no distractions and/or third-party observers, check that the stopwatch works and there are sharpened pencils without erasers for use by the examinee on Trails C. Examiners should also verify that no major interruption occurs or exogenous distractions disturb the examinee. In addition, language (primary language?) and/or cultural (recent immigrant?) factors need to be considered to determine the meaning of TVCF subtest performance.

4. Standardization, norms, and psychometrics

   (a) Characteristics of the standardization sample

   One thousand seven hundred eighty-eight individuals from the USA served as the normative sample for norming the TVCF. Examiners located in each of the four major US geographic regions, as defined by the US Bureau of the Census, tested volunteers with the TVCF based on specified age range and sample characteristics. The percentages reported by the Bureau of the Census (1998) are estimates of the US population demographic characteristics. The normative sample was selected to be representative of the US population as a whole. Demographic data (i.e., age, gender, race, ethnicity, education, and geographical region of residence) are reported as percentages which compare the sample with data reported in the *Statistical Abstracts of the United States* (U.S. Bureau of the Census, 1998), for the school age sample and the US population age distribution. See the TVCF Examiner's Manual for the data (Reynolds & Horton, 2006). While the western region of the USA and nonminority status are overrepresented in the TVCF normative sample, analysis of responses across region and ethnicity showed no significant effects on performance on the TVCF subtests. It should be noted that Reitan and Wolfson (2001) reported that these variables have minimal effects on measures of neuropsychological ability and/or sensitivity to neuropsychological deficits.

   (b) Reliability of the TVCF scales

   Tests can be either speed or power tests: with speeded tests the individuals' scores depend on how quickly an individual can complete the task. With power tests, individuals are not timed as they complete the task. The TVCF contains both speed and power tests. The Classification subtest of the TVCF is a power test as the score depends on the number of right and wrong answers but the subtest is not timed. The Category Fluency and Letter Naming subtests include elements of both power and speed tests.

   The Trails C subtest is a speed test. Coefficient alpha, perhaps the best measure of test score reliability, was computed for each TVCF subtest for selected subgroups in the TVCF normative sample. These groups include both genders and the largest ethnic groups in the USA as well as clinical groups of learning disabled and brain-injured subjects. See the TVCF Examiner's Manual for the data (Reynolds & Horton, 2006).

   The majority of the TVCF subtest reliability scores reach at least .70 or higher for all of the selected subgroups. For each gender and for the four major ethnic groups included in the sampling plan (European-Americans, African-Americans, Hispanic-Americans, and Asian-Americans) the majority of the coefficient alphas were excellent with almost all subtests but Classification-Number Correct showing values of .90 or better. For the Trails C scores, values of .70 or higher were obtained for all of the gender and ethnic subsamples.

   The standard errors of measurement for the Category Fluency, Classification and Letter Naming TVCF subtests are grouped by the age intervals used to stratify the TVCF normative sample. See the TVCF Examiner's Manual for the data (Reynolds & Horton, 2006).

   For the Trails subtest, the standard error of measurement values was computed from the earlier standardization of the CTMT (Reynolds, 2003). The overall standard error of measurement for Trails C was 6 and further information can be found in the CTMT manual (Reynolds, 2003). Smaller SEMs indicate better reliability.

Reliability estimates may vary by subgroups within a population. The test scores need to be reliable for each tested subgroup that might experience test bias because of racial, ethnic, disability, or linguistic differences (Reynolds, 2000). The TVCF subtest scores are essentially equally reliable for all the subgroups investigated and suggest minimal bias relative to those subgroups. Using the Feldt (see Reynolds, 2000) technique, no significant differences among the reliability values of the groups were found. TVCF subtest score reliabilities were similar for males and females. When European-Americans, African-Americans, Hispanic-Americans, and Asian-Americans were considered, reliability estimates suggested comparable reliability across racial-ethnic groups.

The stability of the Trails C scores over time was investigated using the test–retest method. Thirty adults (i.e., ages 20–57 years) in the Southwestern USA were tested twice, with a 1-week period between testings. Age-corrected T-scores were calculated to control for any effects of age and the values were of sufficient magnitude to allow confidence in the Trail C test scores' stability over time.

Test–retest studies also allow for assessment of practice effects. Novel tasks and nonverbal tasks such as Trails C tend to show the greatest practice effect (Reynolds & Horton, 2006). An average improvement of .3 SDs was calculated across the two testings of the Trails C. Future studies to examine longer-term practice effects and the effects of more than two administrations are needed but a smaller practice effect might be anticipated with brain-injured subjects as well since new learning is often impaired in such groups. Test–retest data with the Classification subtest was considered problematic because once the subtest is administered, the novel concept could have been discovered and recalled by many subjects and subsequent re-administration may be compromised due to reduction of the novel quality of the test stimuli. At the same time, however, clinical observations of some brain-damaged patients who were re-administered with the TVCF and demonstrated similarly poor or worse Classification scores had provided clear evidence of brain impairment and suggest that in cases of brain damaged subjects practice effects may not be seen on Classification scores. (See the "Case Study" in this chapter for an example.) The Categorical Fluency subtest showed a small practice effect (.5 SD) and the Letter Naming subtest showed a minimal practice effect.

5. Use of the test
   (a) Interpretation methods

Just as the validation of the meaning of test scores is an ongoing process (AERA et al., 1999), so is making the correct interpretation of performance on a neuropsychological test. The processes of validation and interpretation are reciprocal interdependent processes and are complementary. As additional clinical research with TVCF is conducted and reported in the clinical research literature, interpretations will change in response to evolving knowledge. This chapter, of course, contains only the currently available research findings and future research findings are expected to provide additional information that will assist in interpretation of the TVCF.

Steps to Interpreting TVCF Results

Interpreting Individual TVCF Subtests

The TVCF does not yield an overall or total score for reasons discussed in the test manual (Reynolds & Horton, 2006) so the individual subtest level is the first level of interpretation of performance on the TVCF. The TVCF subtest scores should be assigned to a qualitative category. Qualitative descriptions of performance on the TVCF as a function of T-score ranges are based upon deviations from the mean T-score of 50 of the standardization sample and consistent with similar descriptive systems (e.g., Heaton, Miller, Taylor & Grant, 2004; Kaufman & Kaufman, 1983; Reynolds & Bigler, 1994; Wechsler, 1991, 1997). The T-scores obtained for each TVCF subtest can be described with their appropriate percentile ranks as well. See the TVCF Examiner's Manual for the data (Reynolds & Horton, 2006). Please recall that the number of categories attained score only has a percentile score. Percentile ranks are an alternate to interpretation of *T*-scores since the percentiles indicate the percentage of the population scoring at or below the designated score level. *T*-scores at 30 are 2 SDs from the estimated population mean of 50 on a TVCF subtest and have a percentile rank of 2. Thus, only 2 % or 2 of 100 people at this age level earn lower scores while 98 % or 98 out of 100 (100 − 2) earn this or a higher score. Scores ≥ 1.5 SDs below the mean *T* of 50 imply mildly/moderately impairment.

Scores in an impaired range suggest neuropsychological impairment but in some cases the person being tested was not motivated to perform the test task and therefore earned a very poor score because of poor effort rather than neuropsychological impairment. In addition, variables such as mental retardation and/or visual or physical disabilities need to be ruled out. Prior to determining the interpretation of TVCF test scores, the examiner should review all available interview observations; medical, occupational, and educational records; other neuropsychological test data (including tests of symptom validity); behaviors of the examinee before, during, and after the neuropsychological testing with the TVCF; and the examinee's history.

The examiner should also evaluate differences in the performance of an individual across all of subtest scores of the TVCF. The profile of *T*-scores for these visual comparisons is on the TVCF Profile/Examiner Record Form. The first step is to average the five *T*-scores and then to subtract each *T*-score from the mean of all five TVCF subtest *T*-scores. See the TVCF test manual (Reynolds & Horton, 2006) for the values for significant differences between each individual TVCF subtest score and the mean of all TVCF subtest scores. If a difference between the individual TVCF *T*-score and the mean of all TVCF subtests is significantly different from chance, then it may be interpreted as a strength or weakness (Kaufman, 1990).

Another level of interpretation is to consider the TVCF test scores relative to extra-test variables such as intelligence, academic achievement, vocational attainment, previous educational test scores, and other ability measures. Mental abilities tend to cluster together in normal individuals and wide variation in neuropsychological performance need to be explained. TVCF scores need to be considered in relationship to other past and present measures of human abilities and achievements. While other normative comparisons compare the person's TVCF test results to an "average person" model, a more sensitive measure is to compare the person's TVCF test results to other past and present evidence of the individual person's mental abilities (Reitan & Wolfson, 1993).

Specific Interpretation of TVCF Subtests (See Case Study Table 16.1 for data interpreted)

For the first level of interpretation, please note that many TVCF scores were impaired. For the second level of interpretation, please note that when TVCF scores are averaged and differences from the mean of the *T*-scores were computed for all three administrations, the Letter Naming score was a weakness on all three administrations and the only strength to emerge was Number Correct at the post-surgery-1 administration (Table 16.1).

All of the TVCF were low and when re-administered the TVCF *T*-scores and percentile scores (CN) did not show pronounced practice effects, most likely because of the severity of brain impairment. Indeed the CN score was 0 % for the 3rd administration (PE was NA as the patient needed to complete at least one category for there to be any perseverative errors) as the patient was unable to complete a single category despite being given the test for the third time. Regarding the third level of interpretation, intelligence for the patient in the "Case Study" was previously evaluated as falling into the high average range and the patient had a college degree and held a job position that required considerable technical computer skills. Expectations would have been that the majority of his TVCF scores would have been at least in the average range but the patient had a number of TVCF scores over the three administrations that were much lower than might have been expected. The point is that other extra-TVCF information regarding the patient's mental abilities, educational achievement, social

Table 16.1 Comparison of TVCF scores over three neuropsychological evaluations

Selected tests	Pre-surgery	Post-surgery-1	Post-surgery-2
Test of Verbal Conceptualization and Fluency (TVCF)***			
TVCF-CF	43t (24 %)	39t (14 %)	37t (10 %)
TVCF-NC	45t (31 %)	50t (50 %)	36t (8 %)
TVCF-PE	41t (18 %)	44t (27 %)	No categories achieved
TVCF-CN*	3 (39 %)	3 (39 %)	0 (5 %)
TVCF-LN	33t (5 %)	29t (2 %)	30t (2 %)
TVCF-TC	38t (12 %)	49t (46 %)	39t (14 %)

These TVCF scores are from the Case Study that is described at the end of this chapter. Please refer to the "Case Study" for detail of this case. TVCF Subtests titles are: *CF* Category Fluency, *NC* Number Correct, *PE* preservation errors, *CN* categories number, *LN* letter number, *TC* Trails C. Where: Unless indicated scores are *T*-scores; *=raw scores, ***=PE score not computed for psychometric reason

status, occupational functioning, and pre-morbid functioning can be considered in evaluating TVCF scores. Obviously, a great degree of clinical judgment and caution will be required in drawing any conclusions.

6. Validity

Validity as a Psychometric Concept

According to the AERA, APA, and NCME (1999) *Standards for educational and psychological testing*, validity refers to "… the degree to which *evidence* and *theory* support the interpretations of test scores entailed by proposed users of tests" (emphasis added, p. 9). Reynolds (1998) also noted that validity refers to the appropriateness and accuracy of the interpretation of performance on a test, usually expressed as a test score. Validation of the meaning of test scores is an ongoing process and requires continuing efforts to provide evidence to increase the scientific basis for proposed test score interpretations (AERA et al., 1999; Reynolds, 1998). Validity is an evolving concept as validity (i.e., evidence to support an interpretation of test performance) of an interpretation will vary according to why test scores are being used, who is being examined, and the setting in which specific interpretations are to be made.

Test manuals present evidence for espoused interpretations of test scores, but that evidence summary is time limited. Validation continues because new research studies continue to be completed and test users should base their interpretations of test scores on the most current test validity research literature. The *Standards* (AERA et al., 1999) note that "validation is the joint responsibility of the test developer and test user" (p. 11).

The *Standards* proposed five categories of validity evidence. These categories include how test scores are used, the consequences of test results interpretive errors, the consequences of not using an objective psychological test, and the population to which the test is applied. The five areas suggested in the 1999 *Standards* are as follows:

1. *Evidence based on test content* (i.e., themes, wording, and format of the items, questions, guidelines for test administration and test scoring, and the like);
2. *Evidence based on response processes* (i.e., the fit between the latent constructs of the test and the detailed nature of behavioral performance by the examinee and behavioral conduct of the examiner);
3. *Evidence based on internal structure* (i.e., the degree to which the relationships among the component parts of the test conform to the hypothesized mental and physical constructs);
4. *Evidence based on relations to other (external) variables* (i.e., the relationships between test

scores and variables external to the test scores including developmental variables and scores on other tests of similar and dissimilar mental and physical constructs); and

5. *Evidence based on consequences of testing* (i.e., the intended and unintended outcomes of the clinical use or educational application of a test) (Reynolds & Fletcher-Janzen, 2006, p. 44).

Theory is important in guiding test interpretation and theories of brain function may be referenced in support of the validity of interpretations of test performance. The *Standards* also note that particularly with regard to evidence based on content and response processes that evidence may be logical.

Theory-Based Evidence

Luria's (1966) brain theory postulated as a systemic network composed of an integrated series of three complex functional systems emphasizes that the brain functions as a whole, as a system of neural networks that perceive, integrate, monitor, and evaluate information in order to perform behavioral actions adaptive to the environment (also see Joseph, 1996; Reynolds, 1981; Reynolds & French, 2003). Luria (1969) conceived of tertiary areas in the frontal lobes and other brain areas as subserving executive functioning, with additional contributions from other brain areas (Goldberg, 2001; Reynolds, 1981; Reynolds & French, 2003; Stuss et al., 2001). The various tasks of the TVCF are consistent with Luria's (1966) model of brain function and other contemporary conceptualizations of executive function in particular as noted in Joseph (1996) and in Zillmer and Spiers (2001). The TVCF was not intended to assess every possible theoretical aspect of executive functioning, but rather selectively samples what past research has suggested are the most important aspects of the executive functioning.

Evidence Based on Test Content

The *Standards* (AERA et al., 1999) define test content as the "…themes, wording, and format of the items, tasks, or questions…" as well as guidelines… for administration and scoring" (p. 11). The *Standards* concluded that evidence related to test content could be logical or empirical. Similarly, expert judgments of the relationships between the test scores and the proposed ability constructs assessed by the test are also appropriate. Comparison of the stimuli and the format of the TVCF subtests to other very well-researched tasks widely considered to assess executive functioning (i.e., "card sorting," various forms of verbal fluency and "trail making") provide positive evidence of validity based on the TVCF test content. The recently revised Heaton norms (Heaton et al., 2004) also include these well-researched tasks (i.e., "card sorting," various forms of verbal fluency and "trail making") under the grouping identified as "Executive Abilities" (Heaton et al., 2004) providing additional support from expert judges (Heaton et al., 2004).

Evidence Based on Response Processes

Evidence based on the response processes deals with the correspondence between the nature of the motor performance the test requires from the examinee and the constructs being assessed. Put another way, what type of motor actions is the examinee to perform and what is the relationship of the motor actions to what the test is intended to measure. The content demands of the TVCF include verbal responding and motor reduced and motor enhanced methods of responding which are consistent with the proposed construct of executive functioning. With Classification, the TVCF subtest was based on the well-researched and time-honored WCST and the major visual and motor performance measures are similar for both tests. The Category Fluency and Letter Naming subtests of the TVCF share a common response mode of speeded verbal utterances that are consistent with other executive functioning verbal measures of fluency.

The Trails C subtest of the TVCF requires rapid perceptual-motor and should be considered as requiring enhanced motor responses as adequate abilities in vision, attention, concentration, set-shifting, and resistance to distraction are required for adequate performance. The Trails C subtest is similar to other "trail-making" tasks used to assess executive functioning. In the case of examinees with intact perceptual-motor skills, the response processes of Trails C appear clearly linked to the hypothesized latent processes of executive functioning. Ruling out peripheral perceptual and/or motor confounds to response processes is necessary to ensure validity of the TVCF tasks as measures of executive functioning. Subtests with common executive functioning requirements but different response formats can be used to elucidate possible peripheral perceptual and motor confounds (Reynolds & Horton, 2006).

Evidence Based on Internal Structure

Analyses of the internal structure of a test can elucidate the interrelationships of the subtests and how the interrelationships conform to the hypothesized construct(s) being assessed (AERA et al., 1999). The internal consistency studies of the TVCF subtests demonstrated alpha coefficient values which approached or exceed .70 across age groups and a variety of demographic and clinical samples, with majority of the values in the .80 and .90 ranges. Because most of the coefficients were well above .70, the data suggested common dimensions among the TVCF subtests. An intercorrelation matrix of the scores for the subtests was calculated to assess the latent structure of the TVCF subtests. Given the need for large sample sizes to achieve stability (e.g., Cattell, 1978), it was decided to base the correlation matrix on the entire sample but because of statistical requirements related to missing data for the correlation analysis, the sample size of the analysis was very slightly reduced from the original standardization sample. Two factors appeared to represent the most appropriate solution of the correlation matrix. See the TVCF Examiner's Manual for the data (Reynolds & Horton, 2006).

All of the subtests intercorrelations were statistically significant at the .0001 level suggesting a strong general factor as is a common occurrence in tests involving cognitive processes (Jensen, 1980). The levels of correlations would all be considered to be in at least the moderate range and support the notion that the subtests of the TVCF are tapping a common cognitive construct. The correlation matrix also reveals a secondary dimension related to the two measures of verbal fluency: Category Fluency and Letter Naming. Therefore, the correlation matrix suggests a significant general factor previously identified as "Complex Sequencing" in the previous work done on the CTMT (Reynolds, 2003). The "Complex Sequencing" factor appears to load strongly on the Trails C and Classification-Number Correct subtests. The second factor appears to load on the two primary verbal subtests, Category Fluency and Letter Naming, and might be labeled as "Verbal Processing."

Age Effects on the TVCF

TVCF performance is anticipated to decline as executive functions overall decline with aging over the life span (Coffey & Cummings, 2000). Executive functioning tasks are important components of the clinical neuropsychological examination of elderly (Podell & Lovell, 2000; Reynolds, 2001b) and TVCF performance needs to be interpreted in relation to an individual's age in order to be clinically meaningful. Examination of age effects on the TVCF was done by calculating the means and standard deviations for the Classification, Category Fluency, and Letter Naming subtests of the TVCF at 12 age intervals from 8 to 89 years. See the TVCF Examiner's Manual for the data for the Classification, Category Fluency and Letter Naming subtests (Reynolds & Horton, 2006).

As can be seen, scores initially improve through early adulthood but begin to decline in middle age and the decline increases with greater age. Childhood, adolescence, and young adulthood are a period of rapid growth in the neocortex and the maturation of human brain development and well-known brain aging effects on human beings begin in middle age. Other major human brain functions, such as memory and fluid intelligence, peak around age 16 years on most major psychological tests (e.g., Kaufman, 1990, 1994; Reynolds & Bigler, 1994).

Internal Structure of the TVCF Among Selected Gender, Ethnic, and Clinical Groups

The *Standards* (AERA et al., 1999) strongly suggest that an analysis of the internal structure segregated by various subgroups is important. Separate analyses of demographic and clinical samples suggest that the latent structure of the component subtests of the TVCF is constant across the demographic and clinical groups.

Alpha coefficients were calculated for gender (e.g., male–female), ethnic status (e.g., European-American, Hispanic-American, and African-American), and clinical groups (e.g., learning disabled and brain injured). See the TVCF Examiner's Manual for the data (Reynolds & Horton, 2006).

The alpha coefficients are comparable across gender, ethnic, and clinical groups. The lack of evidence for a need for differential interpretation of score as a function of gender, ethnicity, or clinical status suggests a common set of latent constructs for the TVCF variables.

Evidence Based on Relations to Other (External) Variables

An important aspect of the validation process is the evaluation of the relationship of scores on the test to variables that are external to the test itself. External variables can include tests that may measure similar or dissimilar constructs, diagnostic categorizations, educational or occupational achievement, and relationships with developmental constructs. The *Standards* (AERA et al., 1999) note that a wide range of variables are of potential interest. These different relationships can have different degrees of importance to examiners who function in different settings or who use the test for multiple different purposes. The test user must evaluate the evidence of relationships with external variables and determine its implications for the intended test use.

Relationships with Other Tests

Tests that measure related constructs are important for understanding what functions are assessed by the TVCF. These include tests of intelligence, academic achievement, memory, and differential abilities. See the TVCF Examiner's Manual for the data (Reynolds & Horton, 2006)

Intelligence

Intelligence is related to neuropsychological tests but the relationship may vary by specific tasks with visual-motor and motor tasks showing lesser effects (Horton, 1999). Samples of TVCF standardization participants completed the appropriate age version of the Wechsler Intelligence Scales (Wechsler Intelligence Scales for Children-Third Edition (WISC-III) ($N=29$) and Wechsler Adult Intelligence Scales-Revised (WAIS-R) ($N=21$). For children, the Categorical Fluency subtest scores were related to the WISC-III Verbal and Full Scale IQs but not the Performance IQ. By contrast, the Classification-Number Correct and the Trails C subtest scores were related to the WISC-III Performance IQ but not the Verbal IQ or Full Scale IQ. Letter Naming and Classification-Perseveration Errors scores were unrelated to IQs. The results appear to support a two factor model of executive functioning as previously suggested.

With adults on the WAIS-R, Categorical Fluency and Letter Naming were related to the all three WAIS-R IQs as might be expected given the verbal loadings on all three WAIS-R IQs. On the other hand, Classification-Number Correct was related to the WAIS-R Performance IQs and Trails C was related to all three WAIS-R IQs. Similarly to the WISC-III results, Classification-

Perseveration Errors scores were unrelated to WAIS-R IQs. The pattern of correlations can be seen as providing some additional support for the earlier described internal structure of the TVCF. See the TVCF Examiner's Manual for the data (Reynolds & Horton, 2006).

An additional research study correlated the TVCF and a new measure of intelligence, the Reynolds Intellectual Assessment Scale (RIAS), in a sample of 14 children referred by neurologists and psychiatrists for outpatient neuropsychological evaluation at a private practice office who were administered full neuropsychological batteries that included the RIAS, a measure of intelligence, and the TVCF, a measure of executive functioning; and Symptom Validity Tests (SVT) such as the Word Memory Test (WMT), Amsterdam Short-Term Memory Test (ASTMT), and Test of Malingered Memory (TOMM) (Horton & Reynolds, 2012a). The RIAS is composed of subtests designed to assess components of intelligence such as Verbal Long-Term Memory (Guess What, GWH) and Nonverbal Abstract Reasoning (Odd Item Out, OIO) and Nonverbal Long-Term Memory (What's Missing, WHM). The TVCF includes subtests that assess Categorical Fluency (CF), Number Correct (NC), Perseverative Errors (PE), Number of Categories Achieved (NC), Letter Naming (LN), and Trails C (TC) which are measures of verbal fluency, card sorting, and trail-making tasks considered to be traditional executive functioning measures. The patients included 10 males, 4 females, 7 Caucasians, 5 African-Americans, and 2 Hispanics and 12 children were right handed. Diagnoses include head trauma—12, Aspergers—1, Epilepsy—1. Ages ranged from 8 to 17 (mean—11.8, standard deviation—2.7) and education ranged from 2 to 11 years (mean—6.4, standard deviation—2.5). All subjects' parents signed inform consent documents and the subjects passed SVTs. Correlations among subtests of intelligence and executive functioning were moderate to low (see Table 16.2).

Intelligence and executive functioning subtests in children had moderate to low correlations. These results are consistent with other research that found similar results with different measures of intelligence and executive functioning. As executive functioning subtests had variances unexplained by intelligence test subtests, this finding supports the construct of executive functioning in children.

Table 16.2 Correlation among intelligence and executive functioning subtests

RIAS Intelligence subtests	TVCF Executive functioning subtests					
	CF	NC	PE	CN	LN	TC
GWH	.52	.09	−.36	−.04	.50	.33
OIO	.02	−.03	−.29	.07	.50	.45
VRZ	.32	.38	.09	.17	.03	.11
WHM	.20	−.07	−.21	.01	.30	.46

Academic Achievement

On the WRAT-III Arithmetic and Reading subtests, the correlations of the TVCF subtests demonstrate that the Category Fluency subtest was related to Arithmetic and the Letter Naming subtest was related to Reading. While Trails C was related to Arithmetic but not Reading, the Classification-Number Correct and Classification-Perseveration Errors subtests were unrelated to either Reading or Arithmetic. The non-association of the Classification-Number Correct and Classification-Perseveration Errors subtests with academic achievement measures supports the notion that the Classification-Number Correct and Classification-Perseveration Errors scores are measuring an executive functioning dimension unrelated to academic achievement measures. See the TVCF Examiner's Manual for the data (Reynolds & Horton, 2006).

Memory

The Test of Memory and Learning (TOMAL) and TVCF were administered to a sample of age-appropriate subjects ($N=35$) but there were few strong relationships among subtests. Such data appear to support the conceptualization of the

TVCF as a measure of executive functioning that is distinct from memory abilities. See the TVCF Examiner's Manual for the data (Reynolds & Horton, 2006).

Differential Abilities Scale

The Differential Abilities Scale (DAS) and TVCF were administered to a sample of age-appropriate subjects ($N=46$) but again similar to the TOMAL there were few strong relationships among subtests. Such data suggest the construct of the TVCF as a measure of executive functioning rather than differential abilities. The independence of the TVCF subtests from the DAS supports the notion that the TVCF contributes new information to the clinical neuropsychological assessment of human beings. See the TVCF Examiner's Manual for the data (Reynolds & Horton, 2006).

Traditional Measures of Executive Functioning

Developmental research suggests that executive functioning abilities develop over time as children mature but much remains to be learned about how executive functioning abilities develop in children. A research study examined the relationships between traditional measures of executive functioning in older children, the Trail-Making Test for Older Children-Part B (TMT-OC-B) and Stroop Color Word Test-Color Word (CW)) and TVCF in order to provide evidence for concurrent validation of the TVCF in older children between the ages of 9 and 14 (Horton & Reynolds, 2012b). Eleven older children referred by neurologists and psychiatrists for outpatient neuropsychological evaluation at a private practice office were administered full neuropsychological batteries that included the TMT-OC-B, CW, and TVCF and SVT such as the WMT and TOMM. The TVCF consists of card sorting and verbal fluency tasks and a trail-making task and includes subtests such as Category Fluency (CF), Number Correct (NC), Perseverative Errors (PE), Number of Categories (CN), Letter Naming (LN), and Trails C (TC). The TMT-OC-B and CW measures are frequently administered and well-known neuropsychological tests for this age group and traditionally considered appropriate measures of executive functioning in older children. The patients included 6 males and 5 females, 5 Caucasians, 4 African-American, and 2 Hispanic/Latino-Americans; 10 patients were right handed. Diagnoses include head trauma—9, Epilepsy—1, and Asperger's disorder—1. Ages ranged from 9 to 14 (mean—11.9, standard deviation—2.1) and education ranged from 4 to 9 years (mean—6.5, standard deviation—1.9). All subjects' parents signed informed consent documents and all of the subjects passed SVTs. Correlations were moderate to low and ranged for TMT-OC-B from −.68 (TMT-OC-B/LN) to .02 (TMT-OC-B/CN) and for CW from −.50 (CW/CN) to −.03 (CW/NC) (see Table 16.3).

Correlations between traditional and new measures of executive functioning were moderate for at least one traditional measure of executive functioning and one new measure of executive functioning for both comparisons supporting the concurrent validity of the TVCF as a measure of executive functioning and the theoretical construct of executive functioning. The fact that some of the correlations were low suggests that there are different mental processes underlying the theoretical construct of executive functioning in older children. Further research is needed to understand the relationships between child mental maturation and the development of executive functioning abilities in children. It is clear that much additional research is needed to further elucidate the complex concept of executive functioning in children.

Table 16.3 Correlation among traditional and new measures of executive functioning

Traditional executive functioning measures	New executive functioning measures					
	CF	NC	PE	CN	LN	TC
TMT-OC-B	.07	.19	.49	.02	−.68	−.05
CW	−.04	−.03	−.26	−.50	.09	.30

Evidence Based on the Consequences of Testing

The most controversial of all aspects of the validation process as presented in the *Standards* (1999) is that of evidence based on the consequences of testing. The most negative consequences of testing are applicable to tests designed for selection for advancement opportunities. For example, if a scholastic aptitude test was shown to be biased against a particular ethnic group or gender then that would be evidence of negative consequences of using the test. For a clinical diagnostic test such as the TVCF, accurate diagnosis would appear to be the anticipated consequence of testing that should be the key to evaluating the "consequential" validity. The evidence previously reviewed suggests that the TVCF assists in the accurate diagnosis of brain dysfunction and to does so accurately across demographic groupings as gender and ethnicity. Ample evidence of a lack of TVCF cultural bias for males, females, European-Americans, African-Americans, and Hispanic-Americans has been provided previously in this chapter. Evidence based on a lack of negative consequences thus far has been positive but additional work is needed, particularly with ethnic groups, such as Asian-Americans and Native Americans.

Also, normative studies in other countries than the USA are needed.

Summary

The TVCF appears to be an appropriate measure of executive functioning with a strong theoretical and empirical basis. Validation of the meaning of test scores, however, is an ongoing clinical process and continuing investigation of the TVCF and relationships of the TVCF to various types of neuropsychological impairment and their relative levels of severity and sensitivity across various demographic groups and geographical locations is warranted. As emphasized throughout this chapter (AERA et al., 1999; Reynolds, 1998), test validation is a continuing process and test users should follow the accumulated and contemporary empirical clinical research literature of the ongoing process of validation of test score interpretations. As noted in the *Standards*, "validation is the joint responsibility of the test developer and test user" (AERA et al., 1999, p. 11). The hope and expectation is that the TVCF will greatly assist in the neuropsychological assessment of executive functioning and selection of the most appropriate and helpful treatment options for children, adolescents, adults, and elderly individuals.

Case Study ("xxx" indicates information deleted to protect patient confidentiality)

(The Case Study has been condensed due to page limitations)

Background

This 50-year-old Caucasian right-handed, married male, who had completed 16 years of education and was employed in Computer Network Security Support, was referred for neuropsychological evaluation to determine the current extent of neuropsychological deficits and to make recommendations regarding treatment options and was tested three times with neuropsychological tests that included the TVCF.

The patient had been neuropsychologically tested prior to his undergoing neurosurgery to correct his hydrocephalus that was diagnosed 10 years ago after he had pain in his arm and an MRI scan of his brain showed fluid on his brain. The patient had been significantly neuropsychologically impaired when neuropsychologically evaluated on xxx.

From the xxx First Neuropsychological Evaluation Report

This 52-year-old, married, Caucasian right-handed male, who has obtained 16 years of education, is functioning in the high average range of measured intellectual abilities (Composite Intelligence Index = 112—high average, Verbal Intelligence Index = 113—high average, Nonverbal Intelligence Index = 109—average, Composite Memory Index = 78—borderline). There was a statistically significant difference between the Composite Intelligence Index and Composite Memory Index to the .01 level, and such a difference was found at less than 1.4 % of the normative sample, a very rare occurrence.

Academic achievement was in the average range for Word Reading (90, 25 %, 11.2 years) and Math Computation (98, 45 %, 11.2 years).

Evidence for effort was adequate and neuropsychological testing results were considered valid.

Major areas of neuropsychological deficits include nonverbal abstract concept-formation skills, creative problem-solving skills, short-term verbal memory, short-term nonverbal memory, simple attention and concentration skills, sustained attention abilities, bilateral tactual perceptual discrimination skills (including bilateral very simple tactual perception abilities), phonemic verbal fluency, motor strength difficulties with the right hand (the dominant upper extremity), visual-spatial/organizational skills, and simple and complex set-shifting and cognitive flexibility. The impaired creative problem-solving strategies are related to difficulties in adapting to new and ambiguous situations. The patient had difficulty generating creative alternate strategies to solve problems when the strategy he was implementing was unsuccessful. This sort of neuropsychological impairment has been associated with frontal lobe impairment. The neuropsychological profile was moderately impaired and diffusely impaired. The pattern and severity of neuropsychological impairment was consistent with a progressive neurological disorder such as hydrocephalus. Given the level of neuropsychological impairment, it is remarkable that the patient is able to maintain competitive employment.

Subsequently, Dr. XXX, a neurosurgeon, operated on the patient on XXX at the XXX to correct the hydrocephalus. Subsequently the patient was again neuropsychologically evaluated on xxx to assess his post-neurosurgical neuropsychological status.

From the xxx Second Neuropsychological Evaluation Report

This 53-year-old, married, Caucasian, right-handed male, who has obtained 16 years of education, was previously assessed as functioning in the high average range of measured intellectual abilities. Evidence for effort was adequate and the neuropsychological test results are considered valid. Relative to the earlier neuropsychological evaluation, the patient demonstrated substantial improvement in the neuropsychological abilities of attention, immediate memory, delayed memory, language, and visual/constructive skills over previously neuropsychological test scores even when practice effects were considered. While there had been clear improvement at the same time, a number of important neuropsychological abilities remain impaired.

Major areas of residual neuropsychological deficits include nonverbal abstract concept-formation skills, short-term verbal memory, short-term nonverbal memory, complex tactual perceptual discrimination skills, categorical and phonemic verbal fluency, motor strength difficulties with the right hand (the dominant upper extremity), and complex set-shifting and cognitive flexibility. Relative to the previous evaluation, the left-hand motor speed appeared to have decreased.

The neuropsychological profile was mildly and diffusely impaired. The pattern of impairment was consistent with the patient's neurological history. It might be noted that prognostic indices do suggest some potential for additional recovery of neuropsychological functioning.

For an opinion to be rendered regarding the patient's possible permanent degree of neuropsychological deficit, it will be necessary to reevaluate the patient on the 2-year anniversary of the neurosurgery.

At this time there have been 6 months since the patient's injury and the patient might be seen as still in the initial phase of recovery from the neurosurgery.

Chief Complaints of xxx at the Time of Third Interview

The patient self-reported current cognitive symptoms which included problems with short-term memory, difficulty with organizing (paperwork), and focusing (completing tasks). The date of onset of problems was reported by the patient to have been prior to the neurosurgery (xxx), but in the previous neuropsychological evaluation the patient had only reported the short-term memory problem. Antecedents include hydrocephalus that was diagnosed 10 years ago after he had pain in his arm and an MRI scan of his brain showed fluid on his brain and the more recent (xxx) neurosurgical procedure. Recent consequences since the neurosurgery have included difficulties at work, organizing paperwork, and learning new procedures. The patient had attempted to return to work but had to be put on medical leave again. The patient reported his gait, however, had substantially improved since the neurosurgery.

The patient denied any prior head injuries or strokes.

Interview with the Patient's Spouse

The spouse came to the interview and was also interviewed in the patient's presence. The spouse indicated the patient's cognitive deficits were as the patient had described them and also noted that since the neurosurgery and also she reported that now outside noises would upset the patient and he would yell at their children. She also reported that she thought the patient's gait had improved, but cognitive deficits had persisted. The patient's wife appeared to be very supportive and concerned over her husband's cognitive deficits.

Analysis of Neuropsychological Test Results

A selected battery of neuropsychological tests was administered including measures of adaptive ability and language screening test measures of memory, and tactile perceptual functioning as well as measures of motor skills and emotional status (Table 16.1).

Summary/Formulation

This 50-year-old, married, Caucasian, right-handed male, who has obtained 16 years of education, was previously assessed (xxx) as functioning in the high average range of measured intellectual abilities (RIAS-CIX-112). Evidence for effort was adequate and the neuropsychological test results are considered valid.

Relative to the two earlier neuropsychological evaluations, pre-neurosurgery and post-neurosurgery, the patient continued to demonstrate improvement in selected neuropsychological abilities (see Table 16.1 for selected neuropsychological test scores for the three neuropsychological evaluations), but there are still a number of neuropsychological abilities that have not recovered completely.

For example, to greatly over simplify, as a rule of thumb, for an individual with high average intellectual abilities, scores below the 16 % percentile would appear to be due to still impaired neuropsychological abilities and the patient still, on the third neuropsychological evaluation, almost 15 months after the neurosurgery on xxx, has a greater than expected number of neuropsychological test scores below the 16 percentile.

While there had been improvement since the first postsurgical neuropsychological evaluation, at the same time a number of important neuropsychological abilities appear to still remain impaired. In many cases, however, it should be noted that neuropsychological test scores from the third neuropsychological evaluation are better than the test scores from the pre-neurosurgical first neuropsychological evaluation, demonstrating clear benefit from the neurosurgery.

Major areas of neuropsychological deficits include nonverbal abstract concept-formation skills, short-term verbal memory, short-term nonverbal memory, complex tactual perceptual discrimination skills, categorical and phonemic verbal fluency, motor strength difficulties with the right hand (the dominant upper-extremity), and complex set-shifting and cognitive flexibility. Relative to the previous second neuropsychological evaluation, there appear to have been improvements in abstract nonverbal concept formation and complex set-shifting abilities.

The neuropsychological profile was mildly and diffusely impaired. The pattern of impairment was consistent with the patient's neurological history. It might be noted that there is still some potential for additional recovery of neuropsychological functioning. At the time of the third neuropsychological evaluation it had been almost 15 months since the patient's neurosurgery and the patient may be seen as in the late phase of recovery from the neurosurgery.

For an opinion to be rendered regarding the patient's possible permanent degree of neuropsychological deficit, it will be necessary to reevaluate the patient after the 2-year anniversary of the neurosurgery (xxx).

Disclosure statement: Drs. Reynolds and Horton have a financial interest in the TVCF.

Sections of this chapter have been adapted from the TVCF Examiner's Manual.

References

American Educational Research Association, American Psychological Association, and the National Council on Measurement in Education. (1999). *Standards for educational and psychological testing*. Washington, DC: AERA.

American Academy of Neurology. (1996). Assessment: Neuropsychological testing of adults; Considerations for neurologists. *Neurology, 47*, 592–599.

Anderson, R. M. (1994). *Practitioner's guide to clinical neuropsychology*. New York: Plenum Press.

Bennett, T. L. (2001). Neuropsychological evaluation in rehabilitation planning and evaluation of functional skills. *Archives of Clinical Neuropsychology, 16*(3), 237–253.

Benton, A. (1968). Differential behavioral effects in frontal lobe disease. *Neuropsychologia, 6*, 53–60.

Benton, A., & Hamsher, K. (1989). *Multilingual Aphasia Examination*. Iowa City, IA: AJA.

Cattell, R. (1978). *The scientific use of factor analysis in the behavioral and life sciences*. New York: Plenum Press.

Coffey, C. E., & Cummings, J. L. (Eds.). (2000). *Textbook of geriatric neuropsychiatry* (2nd ed.). Washington, DC: American Psychiatric Association.

Estes, W. K. (1974). Learning theory and intelligence. *American Psychologist, 29*, 740–749.

Fitzhugh-Bell, K. (1997). Historical antecedents of clinical neuropsychology. In A. M. Horton, D. Wedding, & J. Webster (Eds.), *The neuropsychology handbook* (pp. 67–90). New York: Springer.

Goldberg, E. (2001). *The executive brain: Frontal lobes and the civilized mind*. Oxford: Oxford University Press.

Grant, D. A., & Berg, E. A. (1948). A behavioral analysis of the degree of reinforcement and ease of shifting to new responses in a Weigl-type card sorting problem. *Journal of Experimental Psychology, 38*, 404–411.

Hart, R. P., Kwentus, J. A., Taylor, J. R., & Hamer, R. M. (1988). Productive naming and memory in depression and Alzheimer's type dementia. *Archives of Clinical Neuropsychology, 3*, 313–322.

Hartlage, L. C. (2001). Neuropsychological testing of adults: Further considerations for neurologists. *Archives of Clinical Neuropsychology, 16*(3), 201–213.

Heaton, R. (1981). *Wisconsin Card Sorting Test (WCST)*. Odessa, FL: Professional Assessment Resources.

Heaton, R., Chelune, G., Talley, J., Kay, G., & Curtiss, G. (1993). *Wisconsin Card Sorting Test (WCST) manual-revised and extended*. Odessa, FL: Professional Assessment Resources.

Heaton, R., Miller, S. W., Taylor, M. J., & Grant, I. (2004). *Revised comprehensive norms for an expanded Halstead-Reitan Battery*. Odessa, FL: Psychological Assessment Resources.

Horton, A. M., Jr. (1979). Behavioral neuropsychology: Rationale and research. *Clinical Neuropsychology, 1*, 20–23.

Horton, A. M., Jr. (1999). Above average intelligence and neuropsychological test score performance. *The International Journal of Neuroscience, 99*, 221–231.

Horton A. M. Jr., & Reynolds, C. R. (2012a, August). *Executive functioning and intelligence in children*. Poster session presented at the annual meeting of the American Psychological Association, Orlando, FL.

Horton A. M. Jr., & Reynolds, C. R. (2012b, August). *Concurrent validation of a new measure of executive functioning in older children*. Poster session presented at the annual meeting of the American Psychological Association, Orlando, FL.

Horton, A. M., Jr., & Wedding, D. (1984). *Clinical and behavioral neuropsychology*. New York: Praeger.

Jensen, A. R. (1980). *Bias in mental testing*. New York: The Free Press.

Joseph, R. (1996). *Neuropsychiatry, neuropsychology, and clinical neuroscience* (2nd ed.). Baltimore: Williams & Wilkins.

Kaufman, A. S. (1990). *Assessing adolescent and adult intelligence*. Needham, MA: Allyn & Bacon.

Kaufman, A. S. (1994). *Intelligent testing with the WISC-III*. New York: Wiley.

Kaufman, A. S., & Kaufman, N. L. (1983). *Kaufman assessment battery for children*. Circle Pines, MN: American Guidance Service.

Lezak, M. D. (1995). *Neuropsychological assessment* (3rd ed.). Oxford: Oxford University Press.

Luria, A. R. (1966). *Higher cortical functions in man*. New York: Basic Books.

Luria, A. R. (1969). *Cerebral organization of conscious acts: A frontal lobe function*. Invited address to the 19th International Congress of Psychology, London, England.

Milner, B. (1963). Effects of different brain lesions on card sorting. *Archives of Neurology, 9*, 90–100.

Mitrushina, M., Boone, K., & D'Elia, L. (1999). *Handbook of normative data for neuropsychological assessment*. Oxford: Oxford University Press.

Partington, J. E., & Leiter, R. G. (1949). Partington's pathway test. *The Psychological Service Center Bulletin, 1*, 9–20.

Podell, K., & Lovell, M. R. (2000). Neuropsychological assessment. In C. E. Coffey & J. L. Cummings (Eds.), *Textbook of geriatric neuropsychiatry* (pp. 143–164). Washington, DC: American Psychiatric Press.

Reitan, R. M. (1955). The relation of the Trail Making Test to organic brain damage. *Journal of Consulting Psychology, 19,* 393–394.

Reitan, R. M. (1958). The validity of the Trail Making Test as an indicator of organic brain damage. *Perceptual and Motor Skills, 8,* 271–276.

Reitan, R. M. (1992). *Trail Making Test: Manual for administration and scoring.* Tucson, AZ: Reitan Neuropsychology Laboratory.

Reitan, R. M., & Wolfson, D. (1993). *The Halstead-Reitan Neuropsychological Test Battery: Theory and clinical interpretation* (2nd ed.). Tucson, AZ: Neuropsychology Press.

Reitan, R. M., & Wolfson, D. (2000). *Mild head injury: Intellectual, cognitive, and emotional consequences.* Tucson, AZ: Neuropsychology Press..

Reitan, R. M., & Wolfson, D. (2001). Critical evaluation of "Assessment: Neuropsychological testing of adults". *Archives of Clinical Neuropsychology, 16*(3), 215–226.

Reitan, R. M., & Wolfson, D. (2004). The category test and the trail making test as measures of frontal lobe functioning. *The Clinical Neuropsychologist, 9,* 50–56.

Reynolds, C. R. (1981). The neuropsychological basis of intelligence. In G. Hynd & J. Obrzut (Eds.), *Neuropsychological assessment of the school-aged child* (pp. 87–124). New York: Grune & Stratton.

Reynolds, C. R. (1998). Fundamentals of measurement and assessment in psychology. In C. R. Reynolds (Series Ed.) & A. Bellack & M. Hersen (Vol. Eds.), *Comprehensive clinical psychology: Vol. 4.* Assessment (pp. 33–56). Oxford: Elsevier.

Reynolds, C. R. (2000). Methods for detecting and evaluating cultural bias in neuropsychological tests. In C. R. Reynolds & E. Fletcher-Janzen (Eds.), *Handbook of clinical child neuropsychology* (pp. 249–286). New York: Plenum Press.

Reynolds, C. R. (2001). Commentary on the American Academy of Neurology report on neuropsychological assessment. *Archives of Clinical Neuropsychology, 16*(3), 199–200.

Reynolds, C. R. (2001a). Commentary on the American Academy of Neurology report on neuropsychological assessment. *Archives of Clinical Neuropsychology,* 16(3), 199–200.

Reynolds, C. R. (2001b). *Clinical assessment scales for the elderly: Professional manual.* Odessa, FL: Psychological Assessment Resources.

Reynolds, C. R. (2003). *Comprehensive Trail Making Test: Professional manual.* Austin, TX: Pro Ed.

Reynolds, C. R., & Bigler, E. D. (1994). *Test of memory and learning.* Austin, TX: PRO-ED.

Reynolds, C. R., & Fletcher-Janzen, E. (2000). *Encyclopedia of special education* (2nd ed.). New York: Wiley.

Reynolds, C. R., & French, C. L. (2003). The neuropsychological basis of intelligence revisited: Some false starts and a clinical model. In A. M. Horton Jr. & L. C. Hartlage (Eds.), *Handbook of forensic neuropsychology* (pp. 249–286). New York: Springer.

Reynolds, C. R., & Horton, A. M., Jr. (2006). *Test of Verbal Conceptualization and Fluency: Examiner's manual.* Austin, TX: PRO-ED.

Rosen, W. G. (1980). Verbal fluency in aging and dementia. *Journal of Clinical Neuropsychology, 2*(2), 135–146.

Soukup, V. M., Ingram, F., Grady, J., & Schiess, M. (1998). Trail Making Test: Issues in normative data selection. *Applied Neuropsychology, 5,* 65–73.

Spreen, O., & Strauss, E. (1998). *A compendium of neuropsychological tests* (2nd ed.). Oxford: Oxford University Press.

Storandt, M., Botwinick, J., & Danziger, W. L. (1984). Psychometric differentiation of mild senile dementia of the Alzheimer's type. *Archives of Neurology, 41,* 497–499.

Stuss, D., Bisschop, S. M., Alexander, M., Levine, B., Katz, D., & Izukawa, D. (2001). The Trail Making Test: A study in focal lesion patients. *Psychological Assessment, 13,* 230–239.

U.S., Bureau of the Census. (1998). *Statistical abstract of the United States.* Washington, DC: Author.

Wechsler, D. (1991). *Wechsler Intelligence Scale for Children* (3rd ed.). San Antonio, TX: Psychological Corporation.

Wechsler, D. (1997). *Wechsler Adult Intelligence Scale* (3rd ed.). San Antonio, TX: Psychological.

Zillmer, E. A., & Spiers, M. V. (2001). *Principles of neuropsychology.* Belmont, CA: Wadsworth.

Examining Executive Functioning Using the Behavior Assessment System for Children (BASC)

17

Mauricio A. Garcia-Barrera, Emily C. Duggan, Justin E. Karr, and Cecil R. Reynolds

As it was suggested by Strayhorn (1993) 2 decades ago, one of the most limiting constraints of our scientific progress in measuring human behavior has been our inability to produce valid and reliable instruments, despite our technological developments. These difficulties have been made readily apparent when trying to measure psychological constructs. In the philosophy of science, a construct has been defined as a product of the mind rather than independent of it. In other words, a construct is an "ideal object" of science as oppose to a "real" one. We have agreed in psychology that as constructs represent our ideas, they are not directly observable, and it is via our observations of their outcomes that we attempt to estimate and quantify them. The self-regulatory ability we call *executive function* in neuropsychology is one of the best examples of such elusive constructs (Jurado & Rosselli, 2007). Thus, an underlying assumption of this chapter is that executive functions emerge as outcomes of multiple interactions between cognitive and emotional control processes. These processes are mediated by basic connections within and between key neural nodes, involved in rule setting and organization of internal representations. The ultimate goal of these interactions is to produce volitional, purposeful, and efficient guided behavior (Garcia-Barrera, 2013).

Due to their complexity, executive functions are best measured using a multimethod approach. In fact, there is a broad range of neuropsychological tests that have been used to examine the cognitive performance derived from executive function components. These tests employ several methodologies such as classic paper and pencil techniques (e.g., Design Fluency, Rey-Osterrieth Complex Figure), manipulation of objects and tools (e.g., Tower of London/Tower of Hanoi, Wisconsin Card Sorting Test, Behavioural Assessment of the Dysexecutive Syndrome), 2D digital media (e.g., Stroop Test, Continuous Performance Tests, Go/No-Go Task, N-back task), and 3D virtual reality (e.g., the Virtual Classroom, Rizzo et al., 2000, 2006). A growing area of interest has been the creation of behavioral ratings scales (e.g., Behavior Rating Inventory of Executive Function (BRIEF), Frontal Systems Behavior Scale (FrSBe), Behavioural Assessment of the Dysexecutive Syndrome-Dysexecutive Questionnaire (BADS-DEX)) to allow a behavioral dimension to be included in the multimethod-based assessment of executive functioning.

Although rating scales have been traditional tools in neuropsychological assessment, the resurgence of executive function rating scales is better understood in the context of a growing concern for the ecological validity of our traditional instruments (Burgess et al., 2006; Garcia-Barrera, Kamphaus, & Bandalos, 2011). That is, despite

M.A. Garcia-Barrera (✉) • E.C. Duggan • J.E. Karr
University of Victoria, Victoria, BC, Canada
e-mail: mgarcia@uvic.ca

C.R. Reynolds
Texas A&M University, College Station, TX, USA

our progress in elucidating the computational interactions that give rise to goal-oriented executive behavior, there is no gold standard test that can facilitate the translation of those estimated interactions to observable everyday executive behaviors. Behavioral ratings scales may be the kind of tool that, in lieu of direct observations, best aids the examination of executive behavior in the natural context or at least as reported by those familiar with the environment of our examinees. This chapter presents evidence of the validity of one of such rating scales, the *Behavior Assessment System for Children* (BASC; Reynolds & Kamphaus, 1992, 2004), for the assessment of executive functions.

Conceptualization of the BASC

The original *Behavior Assessment System for Children* (BASC; Reynolds & Kamphaus, 1992), now in its second edition (BASC-2; Reynolds & Kamphaus, 2004), was conceptualized as an omnibus system comprising of, and integrating, two approaches: the multimethod approach and the multidimensional approach for the evaluation of behaviors and self-perceptions of children (Reynolds & Kamphaus, 1992). The BASC has been designed successfully to improve the quality of technology available for the assessment of child behavior and emotion in order to facilitate differential diagnosis of pertinent childhood disorders and to assist with treatment plan design. The wide-scale acceptance and adoption of the BASC has generally been considered as resulting from a number of strengths including its *multimethod* and *multidimensional* conceptualization, standardization, normative information, simplicity of scoring, usefulness of composite scores and scales, and ease of interpretation. The use of individual instruments provides the clinician an assortment of valuable data that is reliable and psychometrically sophisticated. Furthermore, when used as a comprehensive system, the BASC provides the clinician a more integrated and complete understanding of a child.

As a *multimethod* system, the BASC gathers information from three sources: the self (child), teacher, and parent. This triangulation method is advantageous because it allows for the assessment of behaviors and emotions from multiple viewpoints and in the context of multiple settings (e.g., school, home). Further, the BASC includes multiple components (see Table 17.1), which can be used either individually or in any combination:

- The *Self-Report of Personality* (*SRP*)
- The *Teacher and Parent Ratings Scales* (*TRS and PRS*)
- The *Structured Developmental History* (*SDH*)
- The *Student Observation System* (*SOS*)
- The *Parenting Relationship Questionnaire* (*PRQ*)

By compiling information from different sources using a variety of methods, results from the BASC are more generalizable, and ensuing diagnoses have increased validation.

The BASC is *multidimensional* in that it gathers information regarding numerous aspects of behavior and personality. Historically, behavior rating scales and systems have focused on negative dimensions of behavior while failing to consider or assess positive dimensions (e.g., see Haynes & Heiby, 2004; Kratochwill, Sheridan, Carlson, & Lasecki, 1999). In response to this significant limitation, the BASC was developed to collect information regarding both negative (clinical) *and* positive (adaptive) dimensions of behavior, including

Table 17.1 BASC-2 instruments

Teacher perspectives	Parent perspectives	Self-perspective
• Teacher Rating Scales (TRS)	• Parent Rating Scales (PRS)	• Self-Report of Personality (SRP)
• Student Observation System (SOS)	• Structured Developmental History (SDH)	• Self-Report of Personality-Interview (SRP-I)
• BASC-2 Portable Observation Program (POP)	• Parenting Relationship Questionnaire (PRQ)	

internalizing and externalizing problems. "The BASC also assesses overt and covert behavior along with attitudes, feelings, and cognitions as well as certain affective states—for example, anxiety, depressed mood, and attributional states—giving a range of dimensions heretofore unavailable except via the use of many different scales, developed over the span of many years, and/or with disparate samples" (Reynolds & Kamphaus, 2002, p. 2).

Considering the widespread recognition and use of the BASC over time, its revision (the BASC-2) has served as a response to prior BASC criticisms and as a means to introduce new and desired assessment enhancements, while retaining the unique conceptualization and key features of the original BASC. Major changes introduced by the BASC-2 include new scale and item content; a new item response format for the SRP; an expansion of the assessable age range (from ages 2.5–18 years to 2–25 years) and improved normative samples and psychometric properties; expansion of Spanish-language forms; and the introduction of the SRP-I, PRQ, and new and improved software/technology features (e.g., Parent Feedback Reports and the BASC Portable Observation Program; *POP*). For the purpose of executive function assessment, it is worth noting the addition of an executive functioning scale to the BASC-2, which will be discussed later in this chapter.

Structure and Components of the BASC-2

As discussed above, the BASC-2 gathers information about a child's behaviors from three perspectives (self, teacher, and parent) using a multitude of separate components that can be used individually or in any combination. An array of resources providing a more extensive review of the BASC and the BASC-2 are available for professionals in clinical, research, and education settings (e.g., Reynolds & Kamphaus, 2004; Tan, 2007).

The *Self-Report of Personality* (*SRP*) is a self-report scale that a child uses to describe self-perceptions and emotions. It contains both dichotomous items to be rated either *true* or *false* and items to be rated on a four-point Likert scale of frequency anchored by 1 (*never*) and 4 (*almost always*). The SRP is available in English and Spanish and has versions (which significantly overlap in content, structure, and scales) at three age levels: child, ages 8–11; adolescent, ages 12–21; and college, ages 18–25 (new to the BASC-2). A new, adjunct instrument to this component, the *Self-Report of Personality-Interview* (*SRP-I*), was developed to gather self-report information through an interview format for younger children (ages 6–7), for which administration of a written SRP would be inappropriate. The SRP is estimated to take approximately 20–30 min to complete.

The *Teacher Rating Scales* (*TRS*) is a measure providing comprehensive information regarding a child's observable problematic and adaptive behaviors in the school setting. It contains behavior descriptors to be rated on a four-point Likert scale of frequency anchored by 1 (*never*) and 4 (*almost always*). The TRS is available in English and Spanish and has three age level forms: preschool, ages 2–5; child, ages 6–11; and adolescent to college-level adults, ages 12–21. Teachers (or individuals who fill a similar role) rate behavior descriptors, which sample broad negative domains of behaviors, such as externalizing, internalizing, and school problems, as well as positive domains of behavior like adaptive skills. Completion of the TRS takes approximately 10–15 min.

The *Parent Rating Scale* (*PRS*) is a measure providing extensive information regarding a child's observable problem and adaptive behaviors in home and community settings. Like the TRS, this scale also contains behavior descriptors to be rated on a four-point Likert scale of frequency anchored by 1 (*never*) and 4 (*almost always*). The same age level forms are available (preschool, ages 2–5; child, ages 6–11; and adolescent to college-level adults, ages 12–21) and are similar in structure, content, and time to complete to the corresponding TRS forms.

The *Structured Developmental History* (*SDH*) is a comprehensive background information (e.g., demographic, biographic, developmental, and historical information) collection tool that can be used either to conduct a structured interview with a parent/caregiver or can be filled out as a questionnaire in a clinical setting or at home. Considering that many medical problems and developmental events may impact a child's behavior, the information gathered from the SDH can provide crucial diagnostic and treatment process information. The SDH is available in English and Spanish.

The *Student Observation System* (*SOS*) is a form used to record direct observations of a child's classroom behaviors. A teacher, or clinician, can use the SOS in order to systematically sample a wide range of a child's behaviors including negative behaviors (e.g., inappropriate inattention or hyperactivity) in conjunction with positive behaviors (e.g., beneficial student-teacher interaction). Sampling consists of 3-s coding intervals spaced 30 s apart over a period of 15 min. A new electronic version, the *BASC Portable Observation Program* (*BASC POP*), is available with the BASC-2 and can be used on a laptop or personal digital assistant (PDA).

The *Parenting Relationship Questionnaire* (*PRQ*) is a new form (available independently and as part of the BASC-2) used to examine and gather more detailed information about the relationship between a child and parent. The PRQ has two forms: preschool (ages 2–5) and child and adolescent (ages 6–18). It is expected to take a parent/caregiver approximately 10–15 min to complete and is available in English and Spanish.

Data Resulting from the BASC-2

The BASC-2 yields a variety of valuable information that can be categorized into the following groups: primary scales, composites, content scales, normative scores (*T*-scores and percentiles), and indexes of validity (consistency, response pattern, *F*, *L*, and *V* Indexes).

Primary Scales. The SRP, TRS, and PRS consist of items structured to collect ratings of behaviors that have each been conceptualized a priori as belonging to a particular and meaningful scale. Scales are generally consistent across the SRP, TRS, and PRS; however, some differences exist as a result of using questions and scales that examine behavior differences that are unique to particular contexts or settings.

Broadly, the BASC-2 consists of two categories of scales: clinical and adaptive. Overall, the clinical scales measure behaviors deemed as maladaptive, with higher scores representing more negative or disadvantageous characteristics that may impact functioning in one or more settings (e.g., home, school, and relationships). Specifically, *T*-scores between 60 and 69 are considered to be in the "at-risk" range, while *T*-scores 70 and above are considered to be in the clinically significant range. BASC-2 scales categorized as clinical include aggression, alcohol abuse, anxiety, attention problems, attitude to school, attitude to teachers, atypicality, conduct problems, depression, hyperactivity, learning problems, locus of control, school maladjustment, sensation seeking, sense of inadequacy, social stress, somatization, and withdrawal. Adaptive scales examine positive behaviors, with higher scores indicating more positive or desirable attributes and lower scores indicating possible sources of behavior problems. For the adaptive scales, *T*-scores between 31 and 40 are considered to be in the "at-risk" range, while *T*-scores 30 or lower are considered to be clinically significant. Scales categorized as adaptive include activities of daily living, adaptability, interpersonal relations, functional communication, leadership, relations with parents, self-esteem, self-reliance, social skills, and study skills. The BASC-2 scales are particularly useful for identifying specific syndromes and/or behavioral strengths. A summary of the primary scales, along with the composite scores and content scales, included in the SRP, TRP, and PRS is provided in Tables 17.2 and 17.3.

Composites. Composites can be generally conceptualized as behavior dimensions and, while they lack the precision of the primary scales, composites are advantageous in formulating performance summaries and general impressions as well as drawing broad conclusions regarding both maladaptive and adaptive behaviors or personality

Table 17.2 BASC and BASC-2 composites and scales: Teacher Rating Scales and Parent Ratings Scales

Version	Teacher Rating Scales			Parent Rating Scales		
	Preschool	Child	Adolescent	Preschool	Child	Adolescent
Age range	2–5	6–11	12–21	2–5	6–11	12–21
Composite	•	•	•	•	•	•
Adaptive skills	•	•	•	•	•	•
Behavioral Symptoms Index	•	•	•	•	•	•
Externalizing problems	•	•	•	•	•	•
Internalizing problems	•	•	•	•	•	•
School problems		•	•			
Primary scale						
Activities of daily living$_a$				■	■	■
Adaptability$_a$	•	•	■	•	•	■
Aggression$_c$	•	•	•	•	•	•
Anxiety$_c$	•	•	•	•	•	•
Attention problems$_c$	•	•	•	•	•	•
Atypicality$_c$	•	•	•	•	•	•
Conduct problems$_c$		•	•		•	•
Depression$_c$	•	•	•	•	•	•
Functional communication$_a$	■	•	•	•	•	■
Hyperactivity$_c$	•	•	•	•	•	•
Leadership$_a$		•	•		•	•
Learning problems$_c$		•	•			
Social skills$_c$	•	•	•	•	•	•
Somatization	•	•	•	•	•	•
Study skills$_a$		•	•			
Withdrawal$_c$	•	•	•	•	•	•
Content scale						
Anger control	■	■	■	■	■	■
Developmental social disorders	■	■	■	■	■	■
Emotional self-control	■	■	■	■	■	■
Executive functioning	■	■	■	■	■	■
Negative emotionality	■	■	■	■	■	■
Resiliency	■	■	■	■	■	■
Item total	100	139	139	134	160	150

Note: • = scales on both the BASC and BASC-2; ■ = new scales added to the BASC-2; a = adaptive scales; c = clinical scales

tendencies. The literature reviewing the performance of overall composite scores versus partial scores (e.g., in the instance of the BASC—primary and adaptive scales) has shown us that full composites have a large amount of scientific and theoretical support, including the fact that they tend to be more stable over time (Canivez & Watkins, 2001; Neisser et al., 1996; Raguet, Campbell, Berry, Schmitt, & Smith, 1996) and have higher levels of predictive validity (Hunter & Hunter, 1984). However, the literature has also demonstrated that part scores (e.g., indexes, scales, factors) allow clinicians to perform profile analysis, in which patterns of subtest scores designed to assess individual strengths and weaknesses are interpreted. Having access to partial scores facilitates diagnosis and appropriate intervention(s) selection (Glutting, Mcdermott, Konold, Snelbaker, & Watkins, 1998), which in turn encompassed the goals of the BASC as a clinical instrument.

Table 17.3 BASC and BASC-2 composites and scales: Self-Report of Personality

	Self-Report of Personality			
Version	Interview	Child	Adolescent	College
Age range	6–7	8–11	12–21	18–25
Composite	•	•	•	•
Emotional symptoms index		•	•	•
Inattention/hyperactivity		•	•	•
Internalizing problems		•	•	•
Personal adjustment		•	•	
School problems		•	•	
Primary scale				
Alcohol abuse$_c$				■
Anxiety$_c$	■	•	•	■
Attention problems$_c$		■	■	■
Attitude to school$_c$	■	•	•	
Attitude to teachers$_c$	■	•	•	
Atypicality$_c$	■	•	•	■
Depression$_c$	■	•	•	■
Hyperactivity$_c$		■	■	■
Interpersonal relations$_a$	■	•	•	■
Locus of control$_c$		•	•	■
Relations with parents$_a$		•	•	■
School maladjustment$_c$				■
Self-esteem$_a$		•	•	■
Self-reliance$_a$		•	•	■
Sensation seeking$_c$			•	■
Sense of inadequacy$_c$		•	•	■
Social stress$_c$	■	•	•	■
Somatization$_c$			•	■
Content scale			■	■
Anger control			■	■
Ego strength			■	■
Mania			■	■
Test anxiety			■	■
Item total	65	139	176	185

Note: • = scales on both the BASC and BASC-2; ■ = new scales added to the BASC-2; a = adaptive scales; c = clinical scales

As such, the TRS and PRS have been designed to yield the four same composites: externalizing problems (e.g., disruptive behavior problems), internalizing problems (e.g., acting-out behaviors), the behavior symptoms index (i.e., an overall measure of problem behavior), and adaptive skills (e.g., positive, adaptive behaviors). The TRS has a fifth composite, school problems (e.g., problems that result in academic difficulties). The SRP produces five composite scales: school problems (unavailable for the SRP; college, ages 18–25 form), internalizing problems, inattention/hyperactivity, personal adjustment, and the emotional symptoms index (a global indicator of serious emotional disturbance, the only composite available for the SRP-I).

Content Scales. Content scales, new to the BASC-2, provide supplementary interpretations of other BASC-2 scales and assist with the detection of specific behavior patterns not captured by the other scales. They include some items belonging to the primary scales of the SRP, TRS, and PRS and some not included in any primary scale. The content scales include anger control, bullying, developmental social disorders, ego strength, emotional self-control, executive functioning, mania, negative emotionality, resiliency, and test anxiety.

Indexes of Validity. The F index, sometimes referred to as the "fake bad" index (available on the SRP, TRS, and PRS), is a measure of overly negative response patterns. A high F index score could represent an attempt to overexaggerate behavioral problems *or* an extreme behavioral and/or emotional problem. The consistency index (available on the TRS and PRS) identifies instances in which a respondent rates similar items differently. The response pattern index (available on the TRS and PRS) identifies possible instances of inattentive responding to item content by the respondent. The SRP provides two additional validity measures: the L index, sometimes referred to as the "fake good" index, which identifies instances of overly positives self-portrayal, and the V index, which serves as a basic check of validity.

Administration and Scoring

Overall, administration of individual BASC-2 components is relatively simple and quick (ranging from 10 to 30 min). Three different forms of the SRP, TRS, and PRS for each of the different age levels are available: hand scored, computer scored, and scannable. The BASC-2 offers two different software scoring and reporting programs: the BASC-2 ASSIST and the BASC-2 ASSIST PLUS. The BASC-2 ASSIST is ideal for basic scoring and reporting. It calculates scale and composite scores, displays results in profile and table formats, and generates score summaries, validity indexes, basic score narratives, and standard, progress, and multi-rater reports. The BASC-2 ASSIST PLUS software expands on the scoring and reporting features of the BASC-2 ASSIST software by including additional report sections, identifying strengths and weaknesses, illustrating relationships to DSM-IV-TR diagnostic criteria, and presenting target behaviors for intervention. Audio recordings of the PRS and SRP scales are available for use with individuals with reading problems. Comprehensive descriptions of the numerous scores, indexes, and critical items yielded from the BASC-2, as well as guidance regarding interpretation and feedback reports, are available in the BASC-2 manual (Reynolds & Kamphaus, 2004).

Standardization, Norms, and Psychometrics of the BASC-2

Standardization of the BASC-2 TRS, PRS, and SRP took place over a 2-year time period (August 2002 to May 2004) and included a total sample of more than 13,000 measures (TRS, PRS, and SRP combined) collected from 375 sites representing various settings (e.g., schools, day cares, and clinics). General norm samples were developed to reflect the US population (as of 2001) on the variables of race/ethnicity, geographic region, US population, and special-education classification. Clinical norms were developed using samples of children diagnosed with, or classified as having, one or more behavioral, emotional, or physical problem. Clinical norm samples are not demographically matched to the US population. General norms include age group and separate-sex and combined-sex norms. Scale and composite norms include T-scores and percentiles. Please refer to the BASC-2 manual (Reynolds & Kamphaus, 2004) for a more extended discussion on standardization, norms, and psychometric properties of the BASC-2.

Reliability and Validity

The reliability and validity for the BASC-2 has been extensively established (see Lett & Kamphaus, 1997; Merenda, 1996; Nowinski, Furlong, Roxanna, & Smith, 2008; Pineda et al., 2005; Reynolds & Kamphaus, 2004; Mahone, Cirino,

Table 17.4 BASC-2 summary of reliability: mean alpha correlations for composites and scales (Reynolds & Kamphaus, 2004)

	TRS	PRS	SRP
Internal consistency			
By sex and age level	.84–.89	.80–.87	.79–.83
By sex and norm group	.84–.88	.84–.87	.75–.86
Test-Retest			
Raw	.79–.88	.76–.84	.73–.83
Adjusted	.81–.86	.77–.86	.71–.84
Interrater reliability			
Raw	.52–.69	.76–.79	.76–.79
Adjusted	.53–.65	.69–.77	.69–.77

et al., 2002; Reynolds, 2010). The TRS, PRS, and SRP all show high reliability in terms of internal consistency, test-retest reliability, and interrater reliability. Scale intercorrelations, factor structure, correlations with related measures behavior, and profiles within clinical groups have all established positive validity for the TRS, PRS, and SRP, and the BASC-2 as a whole has shown strong multitrait-multimethod validity. Table 17.4 includes a summary of main reliability and validity indicators. Refer to the BASC-2 manual for a detailed overview and discussion (Reynolds & Kamphaus, 2004).

Using the BASC in the Assessment of Psychopathology Involving Executive Dysfunction

As presented above, the BASC-2 rating scales are omnibus assessment tools, collecting information on a variety of adaptive behaviors (e.g., leadership, social, and study skills) and internalizing (e.g., anxiety, depression, and withdrawal) and externalizing (e.g., conduct, hyperactivity, and aggression) problems often included in psychopathology evaluations. As such, the validity of this behavioral assessment system for the assessment of symptoms often associated to executive dysfunction has been the focus of some studies.

In particular, the utility of the BASC as a tool for the diagnosis of attention-deficit/hyperactivity disorder (ADHD) has been largely studied. Since Barkley's (1997) early postulation that an inhibitory control dysregulation was the core deficit in ADHD, a voluminous series of studies have been launched to examine this hypothesis. Although the discussion continues, it is now generally accepted that one of the underlying core neuropsychological features of ADHD is impairment in several executive domains (Willcutt, Doyle, Nigg, Faraone, & Pennington, 2005). This is particularly evident in performance on tests tapping into the "cold" aspects of executive functioning (i.e., cognitive components associated with dorsolateral prefrontal cortex; Zelazo & Mueller, 2002) but is also relevant to the more affective or "hot" aspects of executive functioning, evident in hyperactive and impulsive behavior (involving orbital and ventromedial prefrontal regions; Castellanos, Sonuga-Barke, Milham, & Tannock, 2006). Recognition of the multidimensionality of ADHD symptomatology has been met with the acknowledgement of a need for implementing a multimethod approach for its diagnostic assessment (Garcia-Barrera & Kamphaus, 2006). Behavioral rating scales such as the BASC appear to facilitate such comprehensive assessment.

As a broadband scale, the BASC has demonstrated sensitivity in identifying ADHD cases in both American (e.g., Ostrander, Weinfurt, Yarnold, & August, 1998) and other cultures (e.g., Pineda et al., 2005). Interestingly, some researchers have found the BASC scales to be informative regarding the relationship between internalizing problems and ADHD symptomatology. Specifically, children with ADHD and co-demonstrating internalizing symptoms have little to no executive impairments (Graziano, McNamara, Geffken, & Reid, 2013), unless other clinical comorbid symptomatology is present (e.g., high scores on the atypicality scale correlating with poor executive function performance on the Tower of London test; Jonsdottir, Bouma, Sergeant, & Scherder, 2006).

Furthermore, despite its status as a broadband instrument, most of the BASC-2 PRS scales correlate moderately to high with the BRIEF (Reynolds & Kamphaus, 2004), which as it was mentioned before, it is a more narrowband instrument specifically designed to assess executive dysfunction (Gioia, Isquith, Guy, & Kenworthy, 2000). For instance, the BRIEF's Global Executive Composite (GEC) correlates with the externalizing problems composite ($r=.58$ for the children 6–11-year-old sample and $r=.84$ for the adolescent sample 12–18 years old) and the Behavioral Symptoms Index ($r=.67$ for the children sample, $r=.80$ for the adolescents); at the clinical scales level, the GEC correlates moderately to highly with the hyperactivity and the attention problems scales in both children ($r=.58$, $r=.71$) and in adolescents ($r=.80$, $r=.59$). An independent study demonstrated that the BASC clinical scales derived from an ADHD sample significantly correlated with most of the scales included in the BRIEF (Jarratt, Riccio, & Siekierski, 2005). Although these associations could not be generalized to all clinical samples (e.g., Mahone, Zabel, Levey, Verda, & Kinsman, 2002), these findings largely support the utility of the BASC as an omnibus instrument for assessing behavioral problems associated with executive dysfunction-based psychopathologies such as ADHD. In general, broadband scales are superior at initial diagnosis of ADHD and most other forms of childhood psychopathology, while narrowband scales are more efficacious in evaluating treatment effects. Narrowband scales tend to overdiagnose single disorders such as ADHD and fail to identify comorbidities, whereas broadband scales provide a clearer, more comprehensive look at behavior and affect.

The BASC Frontal Lobe/Executive Control and the BASC-2 Executive Function Scales: Performance and Discussion

Among the studies examining the BASC as an assessment tool for executive function, some have specifically investigated the actual executive function measure of the BASC—referred to as the Frontal Lobe/Executive Control (FLEC) scale (obtained from the BASC PRS; Reynolds & Kamphaus, 2002) or as the Executive Function scale (obtained from the BASC-2 PRS and TRS; Reynolds & Kamphaus, 2004). The FLEC was originally developed by Barringer and Reynolds (1995) as an 18-item supplemental scale to assist in the identification of problems in frontal lobe functioning and executive control by examining behaviors typically associated with executive dysfunction, such as those often observed after traumatic brain injury (Reynolds & Kamphaus, 2002). The 18 items of the FLEC were selected from the pool of items included in the PRS-child and the PRS-adolescent forms (for a list and norms, see Reynolds & Kamphaus, 2002, pp. 236–237). Based on the FLEC scale of the BASC, the executive functioning scale (EF scale) is one of the seven content scales of the BASC-2 (Reynolds & Kamphaus, 2004) available for the PRS and TRS and is automatically obtained using the BASC-2 ASSIST PLUS software. T-scores can be obtained for both the FLEC and the EF scales using the appropriate normative information, and it should be reiterated that T-scores between 60 and 69 are considered in the at-risk range and T-scores 70 and above are considered clinically significant.

In the most widely known FLEC/EF scale study, Sullivan and Riccio (2006) administered the BASC PRS (and obtained the FLEC scale), the BRIEF (parent form; Gioia et al., 2000), and the Conners' Parent Rating Scales-Revised: Short Form (Conners, 1997) to a community sample of 92 children who were divided into one of three groups: (1) children who met the criteria for attention-deficit/hyperactivity disorder ("ADHD group"), (2) children without ADHD but who met the diagnostic criteria for some other clinical disorder (e.g., learning disabilities, adjustment, mood, substance use, conduct, and oppositional defiant disorders; "other clinical group"), and (3) children with no clinical diagnosis ("control group"). Overall, Sullivan and Riccio found the BASC FLEC scale to be sensitive to the identification of behaviors associated with executive dysfunctions in children with ADHD and clinical disorders. Mean T-scores on the FLEC scale were in the at-risk range for both the ADHD group and the other clinical group (67.18 ± 8.87 and 66.52 ± 15.00, respectively), while the control group had a mean T-score in the average range (47.50 ± 7.70). Further, Sullivan and Riccio reported significant correlations with scores on all the scales of the BRIEF and Conners' Parent Rating Scales-Revised: Short Form. Specifically, correlations with the BRIEF scales ranged from .45 (Organization of Materials) to .83 (Global Executive Composite), and correlations with the Conners' Parent Rating Scales-Revised: Short Form scales ranged from .63 (ADHD index) to .77 (Oppositional scale). According to Sullivan and Riccio's report (p. 499), all correlations were significant at the $p < .001$ level.

Recently, Reck and Hund (2011) conducted a study designed to evaluate the value of using sustained attention and age as predictors of inhibitory control—an identified component of executive function. In this study, 103 children were administered laboratory tasks of sustained attention (Picture Deletion Task for Preschoolers- Revised-PDTP-R; Byrne, Bawden, DeWolfe, & Beattie, 1998) and inhibitory control ("bear/dragon-" Carlson, Moses, & Breton, 2002; Reed, Pien, & Rothbart, 1984; "day/night-" Carlson, 2005; Gerstadt, Hong, & Diamond, 1994; "whisper-" Carlson et al., 2002; Kochanska, Murray, Jacques, Koenig, & Vandergeest, 1996; and "gift delay-" Carlson et al., 2002; Kochanska et al., 1996). Scores from these tests were used to create the "observational inhibitory control composite" measure. Additionally, a BASC-2 PRS and a Child Behavior Questionnaire—Short Form (CBQ; Rothbart, Ahadi, Hershey, & Fisher, 2001; Putnam & Rothbart, 2006) were obtained for each child participant. The EF scale and the attention problems scale from the BASC-2 (T-scores) were combined using z-score transformations with the inhibitory control subscale from the CBQ (reversed) to create a parent-rated attention problems composite score. Overall, Reck and Hund reported that both observational and parent-rated inhibitory control scores were predicted by sustained attention, with less inhibitory control predicted by increasing attention problems. Of note, the parent-rated attention problems composite, when combined with age, was correlated with the observed inhibitory control composite ($r=-.31$, $p<.01$). Additionally, the parent-rated attention problems composite was correlated with omissions on the sustained attention measure (PDTP-R) when combined with age ($r=.36$, $p<.01$) and with the CBQ attentional focusing measure ($r=-.81$, $p<.05$). Although not explicitly discussed, the authors reported significant correlations between the BASC-2 EF scale and the following inhibitory control and sustained attention measures: day/night-total ($r=-.30$, $p<.01$), gift delay-peek resistance ($r=-.22$, $p<.05$), CBQ-inhibitory control ($r=-.56$, $p<.01$), PDTP-R omissions ($r=.36$, $p<.01$), and the BASC-2 attention problems scale ($r=.73$, $p<.01$).

Two studies (Hass, Brown, Brady, & Boehm Johnson, 2012 and Volker et al., 2010) have used the BASC-2, including the EF scale, to examine behavior ratings (including executive behaviors) in children diagnosed with autism spectrum disorders. Volker and colleagues (2010) investigated clinical and adaptive features of children diagnosed with high-functioning autism spectrum disorders (HFASDs; i.e., high-functioning autism, Asperger's disorder, and pervasive developmental disorder not otherwise specified). In this study, a measure of general intelligence (WISC-IV—Short Form (vocabulary, similarities, block design, and matrix reasoning); Wechsler, 2003) and a BASC-2 PRS was obtained for 124 children (62 HFASD and 62 typically developing). Overall, IQ was not found to significantly impact PRS scores between groups, and the PRS composite scores yielded significant differences between their HFASD and control groups, with the Behavioral Symptoms Index (Cohen's $d=2.537$) and adaptive skills score ($d=-2.379$) having the largest effect sizes. For the content scales, all seven (including the EF scale) were reported to yield significant differences between the HFASD and control groups ($p<.001$). While the developmental social disorders scale (reported in the clinically significant range for the HFASD group; mean $T=74.10 \pm 8.15$) was found to produce the greatest effect size ($d=3.184$), the EF scale still differentiated well between the groups with an effect size d of 1.983. Of particular interest, Volker and colleagues reported a significant difference between the mean EF scale T-scores for the HFASD group (66.55 ± 9.73, in the at-risk range) and the control group (48.79 ± 8.12, in the normal range; $F(1,122) = 121.85, p < .001$). Mean T-scores for the hyperactivity and attention problems scales (in the at-risk range) were also found for the HFASD group (66.13 ± 11.64 and 64.50 ± 7.13).

In a similar line of research, Hass et al. (2012) examined the effectiveness of using the BASC-2 TRS to assess higher functioning in a sample of 60 students with an educational diagnosis of autism. A TRS was obtained for the autism (comprised of child and adolescent groups) and control groups. They found that the TRS discriminated between students who had received an education diagnosis of autism and typically developing controls, with significant group differences attained on almost all scales (child autistic sample: ranging from $d = .58$ (aggression) to $d = 2.18$ (atypicality) and adolescent autistic sample: $d = .72$ (depression) to $d = 2.32$ (withdrawal)) and generally greater effect sizes for the child autistic group compared to the adolescent autistic group. Moreover, these authors found significant age group differences on 10 of the 27 TRS scales, which they concluded suggests notable differences between domain scores for children and adolescents with autism. Mean T-scores for the EF scale were 6.71 ± 9.58 for the child autistic group compared to 48.97 ± 8.97 for controls and 54.27 ± 7.74 for the adolescent autistic group compared to 46.97 ± 6.42 for controls. The age group comparison on the EF scale showed a significant difference between children with autism and adolescents with autism, $F(1, 56) = 7.80$, $p = .01$.

Based on the current, though limited state of the FLEC/EF scale literature, it appears that FLEC/EF scales have consistently distinguished between controls and samples of children/adolescents with clinical disorders involving dysfunction of at least one key component of executive function (e.g., inhibitory control, attentional control). Samples of children diagnosed with ADHD and autism spectrum disorders consistently yielded EF scale mean T-scores in the at-risk range, and these scores were persistently and significantly different than controls (Hass et al., 2012; Sullivan & Riccio, 2006; Volker et al., 2010). Thus far, this suggests that the FLEC/EF scales' greatest utility lies in assisting with the identification of some types of executive behaviors indicative of executive dysfunction in clinical populations. However, the extent to which the FLEC/EF scales can be generalizable or differentiate executive dysfunction differences in various clinical pathologies remains unclear. In general, current FLEC/EF scale research has been limited to populations with ADHD and autism spectrum disorders and, therefore, has focused on the attentional and inhibitory components of executive function, with minimal consideration given to examining the relationship between the FLEC/EF scales and other components of executive function (e.g., problem solving). Considering the relative ease with which they can be obtained, and the widespread popularity of the BASC (included in approximately 45 % of behavioral assessments; Reynolds & Kamphaus, 2004), the FLEC/EF scales represent a measure of executive dysfunction that possesses substantial clinical and research utility.

Alternative Utility of the BASC in Executive Functioning Assessment

A series of studies within our lab has established the validity of an executive functioning screener derived from 25 items on the original BASC-TRS (Garcia-Barrera et al., 2011). The studies have followed three sequential goals: (a) confirming a latent four-factor model of executive functioning measurable through behavioral ratings (Garcia-Barrera et al., 2011); (b) establishing consistent measurement of this model across gender, age, and time (Garcia-Barrera et al., 2011; Garcia-Barrera, Karr, & Kamphaus, in press); and (c) identifying the shared validity of this model within cross-cultural and clinical samples (i.e., Colombian children, ADHD; Direnfeld, Karr, Garcia-Barrera, & Pineda, 2013). Through research achieving these three goals, the executive screener sits on a robust body of empirical support, serving as an accurate and efficient metric for the assessment of executive functions. The following sections will describe in greater detail this line of research.

Theoretical Four-Factor Model of Executive Functioning

For practical purposes, we will begin by discussing the four latent executive constructs measured by the screener, as explaining these factors will elucidate the value of this measure in psychological assessment. The first factor is labeled "problem solving," and it accounts for the temporal organization of behavior towards goal attainment. A lengthy history of neuropsychological theory has established the importance of this construct. Alexander Luria (1973) originally proposed problem solving as a cognitive ability derived from intention formation, planning, and programming. Since Luria, multiple neuropsychologists have provided more comprehensive but similar definitions of this executive-related ability. Philip Zelazo and colleagues (1997) established a model of executive functions developed through a problem-solving framework, emphasizing a cognitive sequence of problem representation, planning, execution, and evaluation. Our model incorporates these past theories into a relatively broad construct, defining problem solving as an ability to formulate an effective response to a novel situation. Utilizing the BASC screener, we measure this construct through the behavioral assessment of planning, decision-making, conflict resolution, and the orientation of behavior towards goal achievement (Garcia-Barrera et al., 2011).

The second construct within our model, "attentional control," involves focusing, sustaining, and shifting attention in response to current task demands. It relates closely to the executive attention theorized by Posner and colleagues (Posner & Rothbart, 1998; Rueda, Posner, & Rothbart, 2005), with an emphasis on the individual's voluntary control over attentional resources (Garcia-Barrera et al., 2011). The third factor represents "behavioral control," a construct related to inhibition and impulse control. This factor aligns with Barkley's (1997) ideas of self-regulation, where children inhibit their behaviors in response to environmental cues and adjust the likelihood of their ensuing responses. The last construct, "emotional control," relates closely to behavioral inhibition but differs in the emotional saliency of the inhibited responses. For this factor, self-regulation deals entirely with controlling or delaying emotional reactions (Barkley, 1997). As Damasio (1995) originally proposed the importance of emotions in the decision-making process, their regulation underlies an important factor within the executive system.

Latent Variable Approach

A quick explanation of the methods for deriving this screener will clarify both its quality and validity. The initial derivation process involved five phases: item selection, item distribution, data screening, item screening, and reliability and validity analysis (Garcia-Barrera et al., 2011). The item selection phase involved the identification of 28 executive-related items that fit within the four theorized executive constructs. We then distributed the items across the specific factors, operationalizing the latent variables through their assigned indicators. The next two phases (i.e., data and item screening) involved controlling for missing data and outliers to ensure the accuracy of our statistical conclusions. Lastly, a group of expert neuropsychologists rated the classification of the selected items, ultimately dropping three of them and repositioning another to a different factor. One item ultimately loaded on both problem solving and behavioral control, while all other items loaded on only one of the factors (for a full list of items per factor, see Garcia-Barrera et al., 2011).

Model Validation. Confirmatory factor analysis (CFA) allows researchers to assess how well their data matches their proposed statistical model, using model fit indices to gauge the convergence between the hypothesized model and the observed data. For our research, we used the comparative fit index (CFI) to determine model fit. The CFI has been shown to be an acceptable fit index for polytomous data (Yu, 2002), facilitating its use in our analysis of the Likert-type categorical responses of the BASC forms. Values for this index range from 0 to 1, with a .95 value serving as a cutoff for optimal model fit (Hu & Bentler, 1999) and .90 serving as a cutoff for acceptable fit (Cordon & Finney, 2008).

All analyses explained hereafter were conducted using MPlus v.6.12 or an earlier version of this software (Muthén & Muthén, 2011). For the four-factor model, the initial derivation reached a CFI of .948 (Garcia-Barrera et al., 2011), while the first replication obtained a CFI of .972 (Garcia-Barrera et al., in press). In this replication, the loadings of each item appeared highly similar to those of the initial derivation study, and again a four-factor model presented the greatest statistical fit. The second replication of the four-factor model with Colombian children achieved a CFI of .942 using the Spanish version of the BASC (Direnfeld et al., 2013), demonstrating the cross-cultural utility of the screener.

Factorial Invariance. Using a latent variable approach requires a reliable measurement beyond the indicator level, meaning each factor must measure the same construct across different groups and times of measurement. At an indicator level, measurement remains invariant as long as researchers maintain standard administration using reliable tests or scales; however, at a factor level, invariance becomes more complex, as by definition, factors cannot be directly measured. Factorial invariance requires establishing model fit and maintaining equivalent fit across progressively constrained models. The metric for equivalent fit is $\Delta CFI \leq .01$ when comparing the CFI of each model to the previous model in the analytical sequence (Cheung & Rensvold, 2002). Researchers have extensively explained invariance analyses elsewhere (e.g., Bontempo, Grouzet, & Hofer, 2012; Bontempo & Hofer, 2007), with special consideration for analysis with polytomous indicators like that of the BASC (Finney & DiStefano, 2006; Millsap & Yun-Tein, 2004).

The first model, labeled as configural, simply requires the same indicators loading on each factor across groups or time. If the configural model presents acceptable fit, the weak model involves holding the factor loadings equal across groups or waves of measurement. Thereafter, the strong model traditionally restricts the intercepts for each indicator to equal across groups or time; however, the BASC involves polytomous indicators, which results in the constraining of thresholds rather than intercepts. These "threshold" parameters are continuous values serving as thresholds that differentiate between ordinal categorical responses. Since the BASC has four responses (i.e., 1 through 4 Likert-type ratings), each item includes three thresholds (i.e., between 1 and 2, between 2 and 3, and between 3 and 4). Lastly, the strict model involves constraining the residual variances of each item across groups or time. For group invariance analyses, strong and strict invariance are equivalent, as indicator variances are constrained at baseline for the model to identify (Muthén & Muthén, 2010, p. 77).

To ensure the consistent reliability of a model, factorial invariance remains statistically important across both groups and time. In turn, we established both group and longitudinal invariances for our four-factor model. For both the model derivation study and its two replications, the model achieved strict invariance across gender and age groups (i.e., young = 6–8 years, old = 9–11; Direnfeld et al., 2013; Garcia-Barrera et al., 2011, in press). Further, the model came very close to strict invariance across control and ADHD groups, just slightly exceeding the cutoff for minimal change with added model constraints (i.e., $\Delta CFI = -.012$).

Aside from group invariance, the model presented impressive consistency within a longitudinal analysis. The invariance analyses indicated nonsignificant change between the configural and strict models for each factor ($\Delta CFI = .001-.003$). Considering the stability of these factors across time, the model lends itself to latent growth analyses in addition to its clinical applications, with clear utility in longitudinal research on executive-related development (Garcia-Barrera et al., in press). A famous statistician, George Box, once wrote, "all models are wrong, but some are useful" (Box & Draper, 1987, p. 424); thus, through this series of studies, we identified a reliable model with clear clinical "usefulness."

Summary and Conclusions

One of the most interesting new approaches to the assessment of executive functions has arisen from researchers' concerns about the ecological validity of executive function tasks. Ecological validity refers to the extent to which results obtained on a controlled standardized test generalize onto performance in naturalistic settings (Chamberlaine, 2003). Within this more *ecologically valid* direction, yet another psychometric approach has been focused on the analysis of the everyday-behavioral components of executive functions and is recognized for the development of behavioral ratings of frontal and executive functions. In this chapter, we reviewed the utility of the parent and TRS included in the two editions of the BASC, to the assessment of executive functioning. For this purpose, we first reviewed the main components of this behavioral system, administration and scoring main tenets, and its psychometric properties as a reliable and valid omnibus instrument for psychopathology assessment. Second, we summarized the literature examining the utility of the BASC as an instrument for the assessment of specific executive dysfunction-related syndromes, using ADHD as a scaffold for the discussion. Third, we examined studies using two available executive function scales, the FLEC (BASC) and the EF (BASC-2), in order to demonstrate these scales' utility in the identification of behaviors that can be considered to be outcomes of executive-based impairments often observed in developmental disorders such as ADHD and high-functioning autism. Finally, we synthesized a collection of our own studies examining an innovative approach to the examination of executive behaviors using the BASC, including an alternative model to the behavioral assessment of executive functioning. Our analyses have demonstrated the robustness of a four-factor model screener for the assessment of executive control, including top-down "cold" (problem solving, attentional control) and bottom-up "hot" (behavioral control and emotional control) executive functions.

Overall, as a multidimensional and multi-method system, the BASC has been demonstrated to be an instrument broad enough to capture sufficient data for the identification of a set of executive-based behaviors within our four categories of executive abilities and does so consistently across genders, age, clinical populations, as well as healthy individuals. It can also be used to reliably examine executive functioning during development. Our research is moving forward to evaluate the psychometric properties of the BASC executive functioning four-factor screener in the two developmental extremes it covers: a set of preschool samples (healthy, born preterm, and premature birth) and in healthy college-aged students (18–25 years old).

References

Barkley, R. A. (1997). *ADHD and the nature of self-control*. New York: Guilford Press.

Barringer, M. S., & Reynolds, C. R. (1995). Behavior ratings of frontal lobe dysfunction. Paper presented at the annual meeting of the National Academy of Neuropsychology.

Bontempo, D. E., Grouzet, F. E., & Hofer, S. M. (2012). Measurement issues in the analysis of within-person change. In J. T. Newsom, R. N. Jones, & S. M. Hofer (Eds.), *Longitudinal data analysis: A practical guide for researchers in aging, health, and social sciences* (pp. 97–142). New York: Routledge/Taylor & Francis Group.

Bontempo, D. E., & Hofer, S. M. (2007). Assessing factorial invariance in cross-sectional and longitudinal studies. In A. D. Ong & M. M. van Dulmen (Eds.), *Oxford handbook of methods in positive psychology* (pp. 153–175). New York: Oxford University Press.

Box, G. P., & Draper, N. R. (1987). *Empirical model-building and response surfaces*. Oxford: Wiley.

Burgess, P. W., Alderman, N., Forbes, C., Costello, A., Coates, L. M., Dawson, D. R., et al. (2006). The case for the development and use of "ecologically valid" measures of executive function in experimental and clinical neuropsychology. *Journal of the International Neuropsychological Society, 12*(2), 194–209.

Byrne, J. M., Bawden, H. N., DeWolfe, N. A., & Beattie, T. L. (1998). Clinical assessment of psychopharmacological treatment of preschoolers with ADHD. *Journal of Clinical and Experimental Neuropsychology, 20*, 613–627.

Canivez, G. L., & Watkins, M. W. (2001). Long-term stability of the Wechsler intelligence scale for children-third edition among students with disabilities. *School Psychology Review, 30*(2), 438–453.

Carlson, S. M. (2005). Developmentally sensitive measures of executive function in preschool children. *Developmental Neuropsychology, 28*, 595–616.

Carlson, S. M., Moses, L. J., & Breton, C. (2002). How specific is the relation between executive function and theory of mind? Contributions of inhibitory control and working memory. *Infant and Child Development, 11*, 73–92.

Castellanos, F. X., Sonuga-Barke, E. J. S., Milham, M. P., & Tannock, R. (2006). Characterizing cognition in ADHD: Beyond executive dysfunction. *Trends in Cognitive Sciences, 10*, 117–123.

Chamberlaine, E. (2003). Behavioural assessment of the dysexecutive syndrome (BADS): Test review. *Journal of Occupational Psychology, Employment and Disability, 5*(2), 33–37.

Cheung, G. W., & Rensvold, R. B. (2002). Evaluating goodness-of-fit indexes for testing measurement invariance. *Structural Equation Modeling, 9*, 233–255. doi:10.1207/S15328007SEM0902_5.

Conners, C. K. (1997). *Conners' rating scales revised*. North Tonawanda, NY: Multi-Health Systems.

Cordon, S. L., & Finney, S. J. (2008). Measurement invariance of the mindful attention awareness scale across adult attachment style. *Measurement and Evaluation in Counseling and Development, 40*(4), 228–245.

Damasio, A. R. (1995). On some functions of the human prefrontal cortex. *Annals of the New York Academy of Sciences, 769*, 241–251. doi:10.1111/j.1749-6632.1995.tb38142.x.

Direnfeld, E., Karr, J. E., Garcia-Barrera, M. A., & Pineda, D. A. (2013). Cross-cultural validation of a screener for executive functions [Abstract]. *Journal of the International Neuropsychological Society, 19*(S1), 51–52. doi:10.1017/s1355617713000362.

Finney, S. J., & DiStefano, C. (2006). Non-normal and categorical data in structural equation modeling. In G. R. Hancock & R. O. Mueller (Eds.), *Structural equation modeling: A second course* (pp. 269–314). Greenwich, CT: Information Age.

Garcia-Barrera, M. A. (2013). *The integrative neuropsychological theory of executive-related abilities and component transactions (INTERACT)*. Manuscript in preparation.

Garcia-Barrera, M. A., & Kamphaus, R. W. (2006). Diagnosis of attention-deficit/hyperactivity disorder and its subtypes. In R. W. Kamphaus & J. M. Campbell (Eds.), Wiley Press, New York. *Psychodiagnostic assessment of children: Dimensional and categorical approaches* (pp. 319–355).

Garcia-Barrera, M. A., Kamphaus, R. W., & Bandalos, D. (2011). Theoretical and statistical derivation of a screener for the behavioral assessment of executive functions in children. *Psychological Assessment, 23*, 64–79.

Garcia-Barrera, M. A., Karr, J. E., & Kamphaus, R. W. (In Press). Longitudinal applications of a behavioral screener of executive functioning: Assessing factorial invariance and exploring latent growth. *Psychological Assessment*.

Gerstadt, C. L., Hong, Y. J., & Diamond, A. (1994). The relationship between cognition and action: Performance of children 3 1/2–7 years old on a Stroop-like day–night test. *Cognition, 53*, 129–153.

Gioia, G. A., Isquith, P. K., Guy, S. C., & Kenworthy, L. (2000). *Behavior rating inventory of executive function*. Odessa, FL: Psychological Assessment Resources.

Glutting, J. J., Mcdermott, P. A., Konold, T. R., Snelbaker, A. J., & Watkins, M. W. (1998). More ups and downs of subtest analysis: Criterion validity of the DAS with an unselected cohort. *School Psychology Review, 27*, 599–612.

Graziano, P. A., McNamara, J. P., Geffken, G. R., & Reid, A. M. (2013). Differentiating co-occurring behavior problems in children with ADHD: Patterns of emotional reactivity and executive functioning. *Journal of Attention Disorders, 17*(3), 249–260. doi:10.1177/1087054711428741. Epub 2011 Dec 29.

Hass, M. R., Brown, R. S., Brady, J., & Boehm Johnson, D. (2012). Validating the BASC-TRS for use with children and adolescents with an educational diagnosis of autism. *Remedial and Special Education, 33*(3), 173–183. doi:10.1177/0741932510383160.

Haynes, S. N., & Heiby, E. M. (Eds.). (2004). *Comprehensive handbook of psychological assessment, Vol. 3: Behavioral assessment*. Hoboken, NJ: Wiley.

Hu, L., & Bentler, P. M. (1999). Cutoff criteria for fit indexes in covariance structure analysis: Conventional criteria versus new alternatives. *Structural Equation Modeling, 6*(1), 1–55. doi:10.1080/10705519909540118.

Hunter, J. E., & Hunter, R. V. (1984). Validity and utility of alternate predictors of job performance. *Psychological Bulletin, 96*(72–98).

Jarratt, K. P., Riccio, C. A., & Siekierski, B. M. (2005). Assessment of attention deficit hyperactivity disorder (ADHD) using the BASC and BRIEF. *Applied Neuropsychology, 12*, 83–93.

Jonsdottir, S., Bouma, A., Sergeant, J. A., & Scherder, E. J. (2006). Relationships between neuropsychological measures of executive function and behavioral measures of ADHD symptoms and comorbid behavior. *Archives of Clinical Neuropsychology, 21*(5), 383–394.

Jurado, M., & Rosselli, M. (2007). The elusive nature of executive functions: A review of our current understanding. *Neuropsychology Review, 17*(3), 213–233.

Kochanska, G., Murray, K., Jacques, T. Y., Koenig, A. L., & Vandergeest, K. A. (1996). Inhibitory control in young children and its role in emerging internalization. *Child Development, 67*, 490–507.

Kratochwill, T. R., Sheridan, S. M., Carlson, J., & Lasecki, K. L. (1999). Advances in behavioral assessment.

In C. R. Reynolds & T. B. Gutkin (Eds.), *Handbook of school psychology* (3rd ed., pp. 350–382). New York: Wiley.

Lett, N. J., & Kamphaus, R. W. (1997). Differential validity of the BASC student observation system and the BASC Teacher Rating Scale. *Canadian Journal of School Psychology, 13*(1), 1–14.

Luria, A. R. (1973). *The working brain: An introduction to neuropsychology*. New York: Basic Books.

Mahone, E. M., Cirino, P. T., Cutting, L. E., Cerrone, P. M., Hagelthorn, K. M., Hiemenz, J. R., et al. (2002). Validity of the behavior rating inventory of executive function in children with ADHD and/or Tourette syndrome. *Archives of Clinical Neuropsychology, 17*(7), 643–662.

Mahone, E. M., Zabel, T. A., Levey, E., Verda, M., & Kinsman, S. (2002). Parent and self-report ratings of executive function in adolescents with myelomeningocele and hydrocephalus. *Child Neuropsychology, 8*(4), 258–270.

Merenda, P. F. (1996). BASC: Behavior assessment system for children. *Measurement and Evaluation in Counseling and Development, 28*(4), 229–232.

Millsap, R. E., & Yun-Tein, J. (2004). Assessing factorial invariance in ordered-categorical measures. *Multivariate Behavioral Research, 39*(3), 479–515.

Muthén, L. K., & Muthén, B. O. (2010). *Mplus user's guide* (6th ed.). Los Angeles, CA: Muthén & Muthén.

Muthén, L., & Muthén, B. (2011). *MPlus (Version 6.12) [Computer software]*. Los Angeles, CA: Muthén & Muthén.

Neisser, U., Boodoo, G., Bouchard, T. J., Boykin, A. W., Ceci, S. J., & Al, E. (1996). Intelligence: Knowns and unknowns. *American Psychologist, 51*(2), 77–101.

Nowinski, L. A., Furlong, M. J., Roxanna, R., & Smith, S. R. (2008). Initial reliability and validity of the BASC-2, SRP, college version. *Journal of Psychoeducational Assessment, 26*(2), 156–167.

Ostrander, R., Weinfurt, K. P., Yarnold, P. R., & August, G. J. (1998). Diagnosing attention deficit disorders with the behavioral assessment system for children and the child behavior checklist: Test and construct validity analyses using optimal discriminant classification trees. *Journal of Consulting and Clinical Psychology, 66*, 660–672.

Pineda, D. A., Aguirre, D. C., Garcia, M. A., Lopera, F. J., Palacio, L. G., & Kamphaus, R. W. (2005). Validation of two rating scales for attention-deficit/hyperactivity disorder diagnosis in Colombian children. *Pediatric Neurology, 33*(1), 15–25.

Posner, M. I., & Rothbart, M. K. (1998). Attention, self-regulation and consciousness. *Philosophical Transactions of the Royal Society of London. Series B, Biological Sciences, 353*, 1915–1927. doi:10.1098/rstb.1998.0344.

Putnam, S. P., & Rothbart, M. K. (2006). Development of short and very short forms of the children's behavior questionnaire. *Journal of Personality Assessment, 87*, 103–113.

Raguet, M. L., Campbell, D. A., Berry, D. T. R., Schmitt, F. A., & Smith, G. T. (1996). Stability of intelligence and intellectual predictors in older persons. *Psychological Assessment, 8*(2), 154–160.

Reck, S., & Hund, A. (2011). Sustained attention and age predict inhibitory control during early childhood. *Journal of Experimental Child Psychology, 108*(3), 504–512.

Reed, M., Pien, D. L., & Rothbart, M. K. (1984). Inhibitory self-control in preschool children. *Merrill-Palmer Quarterly, 30*, 131–147.

Reynolds, C. R. (2010). *Behavior assessment system for children*. Hoboken, NJ: Wiley.

Reynolds, C. R., & Kamphaus, R. W. (1992). *Behavior assessment system for children*. Circle Pines, MN: American Guidance Service.

Reynolds, C. R., & Kamphaus, R. W. (2002). *The clinician's guide to the behavior assessment system for children*. New York: Guilford.

Reynolds, C. R., & Kamphaus, R. W. (2004). *Behavior assessment system for children* (2nd ed.). Circle Pines, MN: American Guidance Service.

Rizzo, A. A., Bowerly, T., Buckwalter, J. G., Klimchuk, D., Mitura, R., & Parsons, T. D. (2006). A virtual reality scenario for all seasons: The virtual classroom. *CNS Spectrums, 11*(1), 35–44.

Rizzo, A. A., Buckwalter, J. G., Bowerly, T. T., van der Zaag, C. C., Humphrey, L. L., Neumann, U. U., et al. (2000). The virtual classroom: A virtual reality environment for the assessment and rehabilitation of attention deficits. *Cyberpsychology & Behavior, 3*(3), 483–499. doi:10.1089/10949310050078940.

Rothbart, M. K., Ahadi, S. A., Hershey, K. L., & Fisher, P. (2001). Investigations of temperament at 3–7 years: The children's behavior questionnaire. *Child Development, 72*, 1394–1408.

Rueda, M. R., Posner, M. I., & Rothbart, M. K. (2005). The development of executive attention: Contributions to the emergence of self-regulation. *Developmental Neuropsychology, 28*, 573–594. doi:10.1207/s15326942dn2802_2.

Strayhorn, J. (1993). The case of the agreeable raters. *Journal of the American Academy of Child and Adolescent Psychiatry, 32*(6), 1302–1303.

Sullivan, J. R., & Riccio, C. A. (2006). An empirical analysis of the BASC frontal lobe/executive control scale with a clinical sample. *Archives of Clinical Neuropsychology, 21*, 295–501. doi:10.1016/j.acn.2006.05.008.

Tan, C. S. (2007). Test review: Behavior assessment system for children (2nd ed.). *Assessment for Effective Intervention, 32*(2), 121–124. doi:10.1177/15345084070320020301.

Volker, M. A., Lopata, C., Smerbeck, A. M., Knoll, V. A., Thomeer, M. L., Toomey, J. A., et al. (2010). BASC-2 PRS profiles for students with high-functioning autism spectrum disorders. *Journal of Autism and Developmental Disorders, 40*, 188–199. doi:10.1007/s10803-009-0849-6.

Wechsler, D. (2003). *Wechsler intelligence scale for children* (4th ed.). San Antonio, TX: The Psychological Corporation.

Willcutt, E. G., Doyle, A. E., Nigg, J. T., Faraone, S. V., & Pennington, B. F. (2005). Validity of the executive function theory of attention-deficit/hyperactivity disorder: A meta-analytic review. *Biological Psychiatry, 57*, 1336–1346.

Yu, C. Y. (2002). *Evaluating cutoff criteria of model fit indices for latent variable models with binary and continuous outcomes.* Doctoral dissertation, University of California, Los Angeles, CA.

Zelazo, P. D., Carter, A., Reznick, J. S., & Frye, D. (1997). Early development of executive function: A problem-solving framework. Review of General Psychology, 1, 198–226. doi:10.1037/1089-2680.1.2.198.

Zelazo, P. D., Carter, A., Reznick, J. S., & Frye, D. (1997). Early development of executive function: A problem-solving framework. *Review of General Psychology, 1*, 198–226. doi:10.1037/1089-2680.1.2.198.

Zelazo, P. D., & Mueller, U. (2002). Executive function in typical and atypical development. In U. Goswami (Ed.), *Blackwell handbook of childhood cognitive development* (pp. 445–469). Malden, MA: Blackwell.

Assessment of Executive Functioning Using the Behavior Rating Inventory of Executive Function (BRIEF)

Robert M. Roth, Peter K. Isquith, and Gerard A. Gioia

A Brief History of the BRIEF

The Behavior Rating Inventory of Executive Function (BRIEF) was one of the first attempts to measure executive function via self- and informant reports of everyday functioning in the real-world environment and was the first published measure of these self-regulatory capabilities in children and adolescents (Gioia, Isquith, Guy & Kenworthy, 2000a). The impetus for the BRIEF arose among the authors in 1994 while trying to reconcile the often discrepant parent and teacher reports of children's everyday functioning at home and in school with their performance on putative performance measures (i.e., "tests") of executive function. At that time, there were few such performance measures of executive function developed for children and adolescents, no rating scales or structured observational methods for evaluating executive functions, and very few published articles on executive function in children (Bernstein & Waber, 2007).

While deficits in executive functions are important features of many developmental and acquired neurological disorders, challenges in measurement have long been recognized (Denckla, 1994; Kaplan, 1988). Given the central importance of the executive functions to the direction and control of dynamic "real world" behavior, reliance on traditional performance measures potentially can yield a limited, incomplete assessment (Gioia & Isquith, 2004; Gioia, Kenworthy & Isquith, 2010; Silver, 2000). While performance tests attempt to tap executive functions in explicit and specific ways, multiple confounds can limit their ecological validity and generalizability. It has been argued that neuropsychological tests alone are inadequate for assessing executive function because they artificially and ambiguously fractionate an integrated system (Burgess, 1997). Performance-based measures tap individual components of the executive function system over a short time frame and not the integrated, multidimensional, relativistic, priority-based decision making that is often demanded in real-world situations (Goldberg & Podell, 2000).

Trained in a developmental neuropsychological assessment model articulated by Holmes-Bernstein and Waber (1990) that views executive function as a broad umbrella term within which a set of interrelated subdomains could be defined via behavioral manifestations, the BRIEF authors recognized the potential efficacy of gathering structured observations of children's everyday self-regulatory functioning from parents and teachers. This behavioral assessment approach was intended as a complement to, rather than in lieu of, traditional performance measures and as

R.M. Roth (✉) • P.K. Isquith
Geisel School of Medicine at Dartmouth School,
Lebanon, NH, USA
e-mail: Peter.Isquith@Dartmouth.edu

G.A. Gioia
Children's National Medical Center, George
Washington University School of Medicine,
Rockville, MD, USA

an index of ecological validity for findings in the clinic or laboratory setting. The guiding framework for developing the BRIEF was based on a review of the literature on executive functions across the lifespan, with particular attention to developmental models. The resulting model defined executive functions as a collection of interrelated functions, or processes, responsible for goal-directed behavior and cognitive activity, or as the "conductor of the orchestra" that controls, organizes, and directs cognitive activity, behavior, and emotional responses (Gioia, Isquith & Guy, 2001). While authors vary in which functions are viewed within an executive function framework, most models include variants of *inhibition* of prepotent responses, competing actions, and interfering stimuli; flexible *shifting* of cognitive set or problem-solving strategies when necessary; *initiation* of goal-directed behavior; *planning* and *organization* of information and behavior; and *monitoring* one's own social and problem-solving behavior. In support of these behaviors, *working memory* capacity plays a fundamental role in holding information actively "on-line" in the service of problem-solving (Pennington & Ozonoff, 1996). Importantly, the executive functions are not exclusive to cognition, or so-called "cool" executive processes, but are reflected in behavior and emotional control ("hot") executive processes (Zelazo, Qu & Muller, 2004).

Following a traditional test development pathway, items for the BRIEF were extracted from clinical interviews with parents and teachers, generated within commonly agreed upon domains of executive function while minimizing overlap with commonly employed behavior rating scales (e.g., CBCL, BASC), reviewed for readability and fit within those domains by experts in the field, and the measure was developed, refined, studied, and validated over the following 6 years until first publication in 2000 (Gioia et al., 2000a). Since publication of the original BRIEF, the instrument family has expanded to include several versions covering the span from 2 to 90 years of age. Each of the versions has been accompanied by interpretive report software and, more recently, smartphone apps and electronic manuals and are now available for web-based administration.

A BRIEF Description

The BRIEF is a family of rating scale instruments that were developed to capture the behavioral manifestations of executive dysfunction across the lifespan from the age of 2–90 years. Four different versions are available: the BRIEF-Preschool Version (BRIEF-P) for ages 2–5 years with one report form for parents and teachers/caregivers (Gioia, Espy & Isquith, 2003), the original BRIEF (BRIEF) for ages 5–18 years with separate parent and teacher report forms (Gioia et al., 2000a), the BRIEF-Self Report Version (BRIEF-SR) for adolescents aged 11–18 years (Guy, Isquith & Gioia, 2004), and the BRIEF-Adult Version (BRIEF-A) for ages 18–90 years with separate self- and informant report forms (Roth, Isquith & Gioia, 2005).

The BRIEF contains problem-oriented rating scales that ask respondents to indicate if each specific behavior is *never*, *sometimes*, or *often* a problem. Although each version of the BRIEF varies in scale composition to some degree, the general domains assessed and scale names (in parentheses) include inhibitory control (Inhibit), cognitive and behavioral flexibility (Shift), emotional regulation (Emotional Control), self-monitoring in the social context (Self-Monitor), ability to initiate activity (Initiate), ability to sustain working memory (Working Memory), planning and organization of cognition and problem-solving (Plan/Organize), organization of materials and environment (Organization of Materials), and monitoring of problem-solving and task performance for accuracy (Task Monitor). Initiation was not included in the BRIEF-SR as it was not supported by the data, but a Task Completion scale emerged. While not considered an executive function per se, the Task Completion scale captures the end result of executive difficulties, for example, getting started on tasks and following them through to completion. There was also insufficient resolution of the Initiate, Plan/Organize, Organization of Materials, and Monitor scales on the BRIEF-P resulting in retention of the Inhibit, Shift, Emotional Control, and Working Memory scales but collapsing of the remaining scales into a

Plan/Organize scale. Finally, the original BRIEF and BRIEF-SR have a unitary Monitor scale that was subsequently subdivided into Self-Monitor and Task Monitor scales based on further factor analytic research. Scales and definitions are described as follows:

- *Inhibit* measures the individual's ability to stop one's own behavior at the appropriate time (i.e., the ability to inhibit, resist, or not act on an impulse).
- *Shift* measures the ability to move freely from one situation, activity, or aspect of a problem to another, as the circumstances demand. Key aspects of shifting include the ability to make transitions, problem-solve flexibly, switch or alternate attention, and change focus from one mindset or topic to another.
- *Emotional Control* addresses the manifestation of executive functions within the emotional realm and measures the ability to modulate emotional responses. Poor emotional control can be expressed as emotional lability or emotional explosiveness. Individuals with difficulties in this domain may have overblown emotional reactions to seemingly minor events.
- *Initiate*, included in the parent, teacher, and adult forms, contains items relating to independently beginning a task or activity and generate ideas, responses, or problem-solving strategies. Poor initiation typically does not reflect noncompliance or disinterest in a specific task. Individuals with initiation problems typically want to succeed at a task, but they cannot get started. Individuals frequently report difficulties with getting started on tasks or chores, along with a need for extensive prompts or cues in order to begin a task or activity.
- *Working Memory* captures the capacity to actively hold information in mind for the purpose of completing a task or generating a response. Working memory is essential for a variety of everyday cognitive activities including carrying out multistep activities, implementing a sequence of actions, or following complex instructions. Individuals with weak working memory may have trouble remembering things (e.g., directions) even for a few minutes, lose track of what they are doing as they work, or forget what they are supposed to retrieve when instructed.
- *Plan/Organize* measures the ability to manage current and future-oriented task demands within the situational context. The *Plan* component of this scale relates to the ability to anticipate future events, implement instructions or goals, and develop appropriate steps ahead of time to carry out a task or activity. Planning often requires sequencing or stringing together a series of actions or responses. Planning is often described in terms of ability to start tasks in a timely fashion or to obtain, in advance, the correct tools or materials necessary to complete the activity. The *Organize* component of this scale relates to the ability to bring order to information, actions, or materials to achieve an objective. Individuals with organizational problems often approach tasks in a haphazard fashion or become easily overwhelmed by large amounts of information. They may have difficulty maintaining order in their environment or among their personal belongings.
- *Monitor* includes two functions, a self-monitoring and a task monitoring function. These are subsumed within one scale for the parent, teacher, and adolescent self-report forms but separated for the adult forms. Neither appears separately on the preschool forms. *Self-Monitor* measures a personal or social self-monitoring function or the extent to which one keeps track of his or her own behavior and its effect on others. Problems with monitoring are described in terms of failing to appreciate or have an awareness of one's own social behavior and the effect this might have on others. *Task Monitor* measures a problem-solving task-oriented, monitoring function. That is, the extent to which one keeps track of his or her own problem-solving success or failure. Problems with task-oriented monitoring are described in terms of failing to appreciate or have an awareness of one's own errors during such activities as problem-solving.
- *Organization of Materials* is included in all versions, except the Preschool Version, and measures one's ability to maintain organization

in his or her everyday environment, such as orderliness of work, play, living, or storage spaces such as desks, closets, and bedrooms. While this scale is not capture an executive function subdomain directly, the ability to keep ones environment organized is thought to reflect at least partly executive function abilities.
- *Task Completion* replaces the *Initiate* scale on the adolescent self-report version. It asks adolescents about their ability to complete work appropriately and in a timely manner. While this scale does not attempt to capture an executive function subdomain directly, the ability to complete tasks is an outcome of well-regulated problem-solving. And may reflect executive function difficulties.

In addition to the clinical scales, all versions of the BRIEF provide validity scales. The Negativity scale assesses the extent to which certain BRIEF items are answered in an unusually negative manner. A high Negativity score raises the possibility that the respondent had an unusually negative response style that skewed the results, though it is also possible that results represent the accurate perception of an individual with severe executive dysfunction. The Inconsistency scale indicates the extent to which the respondent answered a set of item pairs of similar content in an inconsistent manner. The BRIEF-A, but not the other versions, also includes an Infrequency scale measuring the extent to which adults endorse items in an atypical fashion. The scale includes items that are likely to be endorsed only in one direction by most people. For example, marking *often* to "I forget my name" is highly unusual, even for adults with severe cognitive impairment. An elevated Infrequency score raises the possibility of haphazard responding and/or the possibility that the respondent may have been biased toward endorsing items in an extreme manner. An elevated Infrequency scale score may raise the possibility of a purposeful attempt to portray the rated individual in a more positive or negative light than may actually be the case.

Structure of the BRIEF

Each version of the BRIEF summarizes individual scales within indexes based on theoretical considerations and the factor structure of the measures, as well as providing an overall executive function score across scales labeled the Global Executive Composite (GEC). Exploratory factor analyses suggested two factors for the BRIEF, BRIEF-SR, and BRIEF-A. The Behavior Regulation Index (BRI) summarizes the Inhibit, Shift, and Emotional Control scales for the parent/teacher forms; includes the Self-Monitor scale on BRIEF-A; and includes the Monitor scale on the BRIEF-SR. The BRI is interpreted as reflecting an individual's general ability to regulate or control his or her behavior and emotional responses, including appropriate inhibition of thoughts and actions, flexibility in shifting problem-solving set and adjusting to change, regulation of emotional responses, and, for adults and adolescents, monitoring of their own behavioral output.

The Metacognition Index (MI) summarizes the Initiate (Task Completion for BRIEF-SR), Working Memory, Plan/Organize, Organization of Materials, and Monitor scales for the parent/teacher forms. On the BRIEF-A this index also includes a Task Monitor scale. The MI can be interpreted as reflecting one's ability to get started on activity, to hold information in active working memory, to plan and organize problem-solving approaches, to complete tasks (adolescent self-report), and to maintain organization in the environment.

The BRIEF-P has three factor-based indexes: the Inhibit and Emotional Control scales forming an Inhibitory Self-Control Index, the Shift and Emotional Control scales forming a Flexibility Index, and the Working Memory and Plan/Organize scales forming an Emergent Metacognition Index.

More recently, confirmatory factor analyses have suggested that a three-factor model may more accurately reflect the underlying structure of the BRIEF. Please see the section on "Empirical Support for Test Structure" below for details.

Table 18.1 Translations of the Behavior Rating Inventory of Executive Function (BRIEF)

BRIEF-P	BRIEF	BRIEF-SR	BRIEF-A
Arabic	Afrikaans	Brazilian Portuguese	Afrikaans
Castellano	Bahasa	Chinese (simplified and traditional)	Chinese (simplified and traditional)
Catalan	Bemba (in-process)	Danish	Danish
Chichewa (in-process)	Castellano	Dutch	Dutch
Chinese (simplified)	Chichewa (in-process)	French	Dutch for Belgium
Danish	Chinese (simplified and traditional)	French for Canada	English for Australia
Dutch	Czech	German (in-process)	English for Canada
Finnish	Danish	Hebrew	English for South Africa
French	Dutch	Kannada	English for the United Kingdom
French for Canada	Finnish	Korean	Filipino
German	French	Norwegian	Finnish
Hebrew	French for Canada	Polish	French
Hungarian	German	Portuguese	French for Belgium
Italian (in-process)	Hebrew	Swedish	French for Canada
Japadhola	Icelandic		German
Kannada	Italian		German for Austria
Korean	Japadhola		German for Belgium
Latvian	Japanese		German for Switzerland
Luganda	Korean (in-process)		Hebrew
Lusoga	Luganda		Icelandic
Norwegian	Norwegian		Italian
Polish	Nyanja (in-process)		Japanese
Portuguese	Polish		Korean
Portuguese for Brazil	Portuguese (in-process)		Norwegian
Russian	Portuguese for Brazil		Portuguese
	Romanian		Portuguese for Brazil
Spanish	Russian		Romanian
Swahili	Sesotho (in-process)		Russian
Swedish	Shona (in-process)		Slovene
Teso	Slovakian (in-process)		Spanish for the USA
Thai	Slovene		Spanish for Spain
Turkish	Spanish		Spanish for Argentina
	Spanish for Puerto Rico		Spanish for Mexico
	Swedish		Spanish for Puerto Rico
	Teso		Swedish
	Thai (in-process)		
	Turkish		
	Xhosa		

Translations

At the time of publication of this book, translations of the BRIEF approved by the publisher were available in numerous languages and dialects on six continents, with additional translations in development (Table 18.1). For example, it recently was included in a multinational study of children's development in sub-Saharan Africa and has been used in national and international pharmaceutical studies. The majority of translations were undertaken to facilitate research and were created through a process of translation and back translation with author review and input to ensure that the meaning of test items was retained. Normative data developed using the translated version is available for some of the translations and several are published as standardized instruments.

Administration and Scoring

Administration

All versions of the BRIEF typically take between 10 and 15 min to complete and have very similar instructions for administration and for respondents. Standardized instructions for administration of the BRIEF are available in the published Professional Manuals. Each of the BRIEF versions also has instructions for respondents printed directly on the rating form. Additionally, it is helpful to ensure that respondents understand the time frame for which they are rating behaviors (e.g., within the past month) and to encourage them to complete all of the items. Items are rated on a 3-point scale (1 = "never," 2 = "sometimes," 3 = "often"). A minimum fourth to fifth grade reading level is recommended for respondents.

Scoring

The BRIEF may be scored by hand or through the use of published scoring software. Both methods yield raw scores, T-scores, percentiles, and 90 % confidence intervals for each of the indexes and scales, as well as validity scale scores. The BRIEF should not be scored if more than a set number of responses are missing (e.g., 12 items for the preschool version). If responses are missing for more than two items on a given scale (or one item on the BRIEF-A Self- and Task Monitor scales), then the raw score should not be calculated for that scale. Missing responses for one or two items on a given scale can be assigned a score of 1 to permit calculation of raw and standardized scores for that scale.

Hand scoring is done by users first tearing off a perforated strip and peeling away the BRIEF report form to reveal a carbonless scoring sheet behind it on which the demographic information and responses are reproduced. The scoring sheet facilitates calculation of raw scores for each of the clinical scales and the three validity scales. Raw scores are transferred to a Scoring Summary sheet that includes detailed instructions for obtaining standardized scores (using the published Professional Manual) and gauging validity. The reserve side of the Scoring Summary sheet has a graph on which one may plot T-scores for scales and index scores.

Scoring software is available separately for the BRIEF-A and for the other versions combined (BRIEF, BRIEF-SR, and BRIEF-P). Examiners first enter the client's demographic details and information about the respondent (e.g., relationship to client) as appropriate. Next, item-level responses are entered by clicking with a mouse on the corresponding score circled on the report form (i.e., N, S, O) or by entering 1 for N, 2 for S, 3 for O, and 4 or ? for a missing response. Both software packages can then generate several types of reports. These vary in the level of detail from presenting scores (raw and standardized), item-level responses, and a plot of T-scores to presenting scores along with an extensive report explaining the measure, discussing validity and each clinical scale as well as providing a number of intervention recommendations tied to scale elevations. The reports for children and adolescents also offer language appropriate for IEP and 504 plan documentation. The software can also produce a "protocol summary" report that shows scores from up to four respondents (e.g., parent and one or more teachers) in a table and figure, thus facilitating comparison of ratings and providing a ready visual presentation of findings to clients, parents, or teachers. Demographic information and scores, both at the item level and all scores, from one or more clients can be exported from the software as delimited text files to facilitate importing into database, spreadsheet, and statistical software packages.

In addition to scoring software, the publisher of the BRIEF family of measures has made available online applications (Apps) that convert

BRIEF (all versions) raw scores into *T*-scores and percentiles and graph the standardized scores of the rated individual. Online administration and scoring of the BRIEF is now available via the publisher's web site.

Standardization, Norms, and Psychometrics

Characteristics of the Standardization Sample

Tables 18.2, 18.3, 18.4, and 18.5 present the characteristics of the standardization samples for the BRIEF versions. The standardization samples were collected with the goal of approximating the population of the United States according to key demographic variables. These included age, gender, ethnicity, and geographical population density. Socioeconomic status (SES) and parental education were also considered for some of the scales. The samples were weighted as needed to reflect estimated proportions for ethnicity and gender in the US population. Of note, while the standardization sample for the original BRIEF (Gioia, Isquith, Guy & Kenworthy, 2000b) was largely drawn from the State of Maryland, studies including typically developing children from around the world over the past decade have yielded scores consistent with the normative sample.

Table 18.2 Characteristics of the BRIEF standardization sample

	Parent report	Teacher report
Sample	25 private and public schools in urban, suburban, and rural areas of Maryland and 18 adolescents from a study in Cleveland, Ohio	Same as parent report
Exclusion criteria	History of special education or psychotropic medication use; maximum 10 % missing items on BRIEF	Same as parent report
N	1,419	720
Parental education (years)	Mean = 14.2 (SD = 2.57)	Not specified
Age groupings (years)	5–6, 7–8, 9–13, 14–18	Same as parent report
Gender (%)		
Boys	43	44
Girls	57	56
Race/ethnicity (%, actual/weighted)		
White	80.5/71.7	72.1/71.7
African-American	11.9/12.2	13.5/12.2
Hispanic	3.1/11.6	4.2/11.6
Asian/Pacific Islander	3.8/3.8	6.1/3.8
Native American/Inuit	.5/.7	.4/.7
Socioeconomic status (%)		
Upper	3.0	7.4
Upper middle	21.8	20.0
Middle middle	36.1	28.0
Lower middle	31.8	21.0
Lower	6.2	2.5
Unassigned	1.2	21.0

Table 18.3 Characteristics of the BRIEF-P standardization sample

	Parent report	Teacher report
Sample	20 preschool programs including private and public schools, as well as pediatric well-child visits, in urban, suburban, and rural areas of Maryland, Illinois, Vermont, New Hampshire, Florida, and Texas	Same as parent report
Exclusion criteria	History of special education, attention problems, developmental or cognitive difficulties, or psychotropic medication use; maximum 10 % missing items on BRIEF-P	Same as parent report
N	460	302
Parental education (years)	Mean = 15.7 (SD = 2.84)	Not specified
Age groupings (years)	2–3, 4–5	Same as parent report
Gender (%)		
Boys	53.5	54.3
Girls	46.5	45.7
Race/ethnicity (% actual)		
White	73.0	71.9
African-American	13.9	12.3
Hispanic	4.8	4.6
Asian/Pacific Islander	3.0	2.0
Native American/Inuit	.7	.7
Not Specified	4.6	8.6
Socioeconomic status (%)		
Upper	18.9	23.5
Upper middle	28.9	29.1
Middle middle	26.3	24.2
Lower middle	15.7	12.3
Lower	10.0	6.6
Unassigned	.2	4.3

Table 18.4 Characteristics of the BRIEF-SR standardization sample

	Parent report
Sample	Private and public schools, in urban, suburban, and rural areas of Maryland, Ohio, Vermont, New Hampshire, Florida, and Washington state
Exclusion criteria	History of special education or psychotropic medication use; maximum 10 % missing items on BRIEF-P
N	1,000
Parental education (years)	
Mothers	Mean = 13.55 (SD = 3.31)
Fathers	Mean = 13.88 (SD = 3.50)
Age groupings (years)	11–14, 15–18
Gender (%)	
Boys	44.8
Girls	55.2
Race/ethnicity (% actual)	
White	67.3
African-American	14.7
Hispanic	12.5
Other (Asian/Pacific Islander, Native American/Inuit)	5.5

Table 18.5 Characteristics of the BRIEF-A standardization sample

	Self-report	Informant report
Sample	Internet sampling throughout the USA	Same as self-report
Exclusion criteria	History of diagnosis or treatment of psychiatric illness, learning disorder, neurological disorder, serious medical illness (e.g., cancer), history of psychotropic medication use; all items had to be completed	Same as self-report
N	1,050	1,200
Education years (%)		
≤11	15.0	11.2
12	30.9	37.7
13–15	28.3	25.4
≥16	25.8	25.8
Age groupings (years)	18–29, 30–39, 40–49, 50–59, 60–69, 70–79, 80–90	Same as self-report
Gender (%)		
Male	50	45.2
Female	50	54.8
Race/ethnicity (% actual)		
White	72.6	71.8
African-American	9.3	13.0
Hispanic	12.0	8.5
Other (Asian/Pacific Islander, Native American/Inuit)	6.1	6.8
Geographic region (%)		
Northeast	20.0	19.4
Midwest	22.0	26.7
South	35.4	34.9
West	22.6	19.0

Reliability of the Scales

Internal Consistency

Internal consistency, the degree to which items on a single scale are measuring the same construct, has been reported using Cronbach's alpha (Cronbach, 1951). Alpha coefficients in the normative samples for both the BRIEF-P (range = .80–.97) and the BRIEF (range = .80–.98) parent and teacher reports are high. This was also seen in clinical samples for the BRIEF parent ($n = 852$) and teacher ($n = 475$) reports. Internal consistency was also reported to be high for the scales (.78–.90) and index scores (.93–.96) in a sample of 847 typically developing children using a Dutch translation of the BRIEF (Huizinga & Smidts, 2011). Alphas for the BRIEF-SR normative sample are moderate to high, ranging from .72 for scales with fewer items to .87 for scales with a larger number of items. Similarly, moderate to high alphas were obtained for both the BRIEF-A normative and mixed clinical samples. Coefficients tended to be higher for the informant than the self-report form, ranging for clinical scales from .80 to .93 in the normative samples and from .85 to .95 in mixed clinical samples. All but one index score across the versions were above .90 (BRIEF-P parent report Flexibility Index alpha = .89), with most at .93 or higher. These findings indicate that all BRIEF versions have strong internal consistency.

Table 18.6 Test-retest reliability and mean T-score difference for the BRIEF

	n	Test-retest interval (mean weeks)	Reliability coefficient mean (range)	T-score difference mean (range)
BRIEF-P				
Parent normative	52	4.5	.86 (.78–.90)	1.2 (.34–2.93)
Teacher normative	67	4.2	.83 (.65–.94)	1.6 (.04–3.03)
BRIEF				
Parent clinical	40	3	.79 (.72–.84)	3.1 (1.8–7.5)
Parent normative	54	2	.81 (.76–.88)	.8 (.0–3.0)
Teacher normative	41	3.5	.87 (.83–.92)	1.2 (.0–3.1)
BRIEF-SR				
Self-report	59	4.91	.78 (.59–.89)	4.5 (.37–6.19)
BRIEF-A				
Self	50	4.22	.89 (.82–.94)	2.5 (1.88–3.26)
Informant	44	4.21	.94 (.91–.96)	1.78 (1.32–2.18)

Test-Retest Reliability

Table 18.6 presents the test-retest reliability coefficients and the mean T-scores difference in the test-retest samples for all versions of the BRIEF. Between 41 and 67 participants were retested, with a mean retest interval ranging from 2 weeks to close to 5 weeks. Stability was observed to be adequate to high across nearly all scales, index scores, and versions ($r=.59–.96$), with the most coefficients being above .80. Furthermore, stability coefficients in the normative samples were similar across raters for the BRIEF-P, BRIEF, and BRIEF-A. Slightly higher average stability was seen for the teacher (mean $r=.87$) relative to the parent (mean $r=.81$) report on the BRIEF, as well as for the informant report (mean $r=.94$) relative to the self-report (mean $r=.89$) in adults. Furthermore, a recent study using the Dutch translation of the BRIEF observed high (.73–.94) test-retest reliability over a 2-week interval (Huizinga & Smidts, 2011).

Knowledge of the degree of expected change in T-scores over repeated administrations in normative samples, where little change should be seen over modest time frames, is important. This is particularly relevant for clinical purposes such as monitoring recovery (e.g., traumatic brain injury (TBI), stroke) and evaluating treatment effects (e.g., medication, behavioral, surgery). In the test-retest samples, the change in T-scores across versions was generally less than 2–3 points for scale and index scores. Perhaps not surprisingly, the largest discrepancies were observed for the BRIEF clinical sample and the adolescent self-reports.

Together, these findings indicate little variability in scores due to the instrument itself. This supports repeated administration and provides a basis for interpreting changes over time.

Inter-Rater Reliability

The test manuals report on the inter-rater reliability for the different versions of the BRIEF. Table 18.7 provides a summary of the correlations (mean and range) between scores for different raters. In preschool children, correlations between parent and teacher ratings are modest (mean=.19). Greater inter-rater agreement was reported for the Inhibit, Shift, and Emotional Control scales than the Working Memory and Plan/Organize scales. The BRIEF manual also reports data on the correspondence between the ratings of parents and teachers, noting a modest overall correlation of .32. The lowest correlations were observed for the Initiate ($r=.18$) and Organization of Materials ($r=.15$) subscales. Parents tended to rate their children, both boys and girls, as having more problems on the BRIEF-P and BRIEF as compared to teachers, which is consistent with other literature on

Table 18.7 BRIEF inter-rater reliability

	n	Sample	Raters	Reliability coefficient mean (range)
BRIEF-P	302	Normative	Parent–teacher	.19 (.06–.28)
BRIEF	296	Normative	Parent–teacher	.32 (.15–.50)
BRIEF-SR	243	Mixed control and clinical	Self–parent	.47 (.36–.57)
BRIEF-SR	148	Mixed control and clinical	Self–teacher	.28 (.20–.41)
BRIEF-A	180	Mixed control and clinical	Self–informant	.57 (.44–.68)

parent-teacher discrepancies (Offord et al., 1996). Jarratt et al. reported an average correlation of .58 (range = .46–.72) in a sample of 40 children (Jarratt, Riccio & Siekierski, 2005). Discrepancies between parent and teacher ratings, and differences in the inter-rater reliability of specific scales, may partly reflect the consistency with which behavioral and emotional difficulties are expressed in disparate environments. For example, problems with disinhibition may be more readily observed across settings than some other aspects of executive dysfunction. Differences with respect to degrees of environmental structure and demand in home and school settings may also contribute to inconsistencies between raters.

In children and adolescents, examination of the relationship between self-ratings and other's ratings (e.g., parent) is subject to a number of methodological and developmental considerations (Surber, 1984). In particular, adolescence is commonly a period of growing self-awareness (Lyons & Zelazo, 2011) that in part parallels maturation of gray and white matter in frontal and other brain regions (Khundrakpam et al., 2012; Peters et al., 2012). Another important consideration in adolescents, as well as in other age groups, is that self-awareness may be compromised in a variety of clinical conditions such as acquired brain injuries (Beardmore, Tate & Liddle, 1999; Ciurli et al., 2010; Flashman, 2002). A study of 98 adolescents with TBI and 98 healthy teens found moderate agreement between the BRIEF-SR and BRIEF parent index scores in both the clinical ($r = .43–.54$) and control samples ($r = .40–.42$). While adolescents in both groups generally reported having fewer problems with executive function than observed by their parents, the difference was greater in the patient than healthy group for the Metacognitive Index but not the Behavioral Regulation Index (Wilson, Donders & Nguyen, 2011). In the BRIEF-SR manual, adolescent self-ratings were moderately to highly correlated with parents ratings (mean $r = .47$), while a significant but lower association was observed in relation to teacher ratings (mean $r = .28$). The lower correlation with teacher as opposed to parent ratings may reflect differences in demand characteristics of the school vs. home setting. It also possible that the greater percentage of subjects from clinical populations relative to typically developing adolescents in the self-teacher than the self-parent samples contributed to the discrepancy.

Inter-rater agreement for the BRIEF-A was examined in a mixed clinical and healthy adult sample. Correlations across the scales and indices were moderate ($r = .44–.68$), with the lowest correlation being seen for the Shift scale. Importantly, approximately 50–70 % of the T-scores for self- and informant reports were within a standard deviation. Thus, while there is often good agreement between adult raters, there is a nontrivial subset of cases in which disagreement is present. There are several possible explanations for discrepancies between self- and informant ratings such as awareness deficits in the person being rated, rater bias (e.g., the informant having overly positive or negative view of the individual being rated), and a variety of contextual factors (e.g., whether the rated individual is observed in more or less structured environments).

Use of the Rating Scale

Interpretation Methods: Case Example

Matthew was a 16-year-old tenth-grade student with a history of declining school performance since middle school and increasing attention, behavior, and mood difficulties in the context of loss of his closest friend; two unmanaged concussions in close proximity; onset of seizures that are well controlled with medication; and mononucleosis within the past year. Early history was unremarkable. In middle school, Matthew began having difficulty getting started on tasks and putting forth effort toward schoolwork unless tasks were inherently motivating, such that he earned As in classes he enjoys but failing grades in other areas such that he was at risk for not graduating. He was referred for neuropsychological consultation to assist in developing a more comprehensive understanding of his strengths and needs.

The evaluation protocol documented very superior verbal knowledge and reasoning along with superior nonverbal problem-solving, learning, memory, and motor function. Academic skills were high average. He was pleasant, social, straightforward, and aware of his own difficulties and strengths. On a variety of performance-based measures of executive function, Matthew performed well, suggesting intact fundamental "cold" executive functions when engaged in novel, short-term problem-solving. On the Tasks of Executive Control (TEC), a lengthy measure of the ability to sustain attention and vigilance when confronted with increasing working memory load and inhibitory demand, he demonstrated good performance in all respects. He was appropriately focused and vigilant for both novel and frequent information, was not impulsive, and demonstrated rapid and consistent response speed. There were no meaningful changes in his performance as the task became more demanding, suggesting good ability to hold complex information in working memory despite increasing challenges. His performance on a range of standard executive function measures (including Tower of London, Trail Making, and Stroop tasks) was similarly average to high average.

Matthew's mother and two teachers (English, Social Studies) completed the appropriate forms of the BRIEF and Matthew completed the BRIEF-SR. The protocols were scored using the BRIEF Software Portfolio to facilitate speed and accuracy of scoring, allow for comparison of profiles between raters, and generate suggestions for feedback appropriate for Matthew and his parents and teachers as well as recommendations for areas of difficulty, along with language appropriate for an IEP or 504 plan. Figure 18.1 shows T-scores and percentiles. A review of the Validity scales (Negativity and Inconsistency) at the bottom of Fig. 18.1 shows that all protocols had acceptable validity indicators with the exception of rater 3, a teacher, who obtained an Inconsistency score that was "Questionable." The raw score of 8 reflects a difference of 8 points between ten pairs of similar items on the BRIEF protocol, such as marking *never* in response to "Gets out of control more than friends" but marking *often* in response to "Acts too wild or 'out of control.'" A raw score difference of 8 is at the 99th percentile relative to other teachers' consistency scores in the standardization sample. This does not invalidate Rater 3s profile per se, but the protocol should be viewed with caution. Of note, Rater 3s scores on the clinical scales are generally higher than that seen for the other raters.

Figure 18.2 shows each rater's scores graphically in the BRIEF Software Portfolio Profile Report view. T-scores between 60 and 65 are typically considered mildly elevated, while T-scores above 65, shown in the shaded area on the profile view, are considered clinically elevated. Reading from left to right, the scales comprising the BRI are presented first: Inhibit, Shift, and Emotional Control. Rater 3s scores were on these three scales were all clinically elevated, though the protocol was "Questionable" due to highly inconsistent ratings. By comparison, Matthew, his mother, and his Social Studies teacher's ratings were clinically elevated only for the Shift scale. In viewing the profile regardless of T-scores, all four raters had the greatest

BRIEF® Protocol Summary

Index/Scale	R1 03/17/2011 T (%ile) Parent	R2 04/27/2011 T (%ile) Teacher	R3 03/31/2011 T (%ile) Teacher	R4* 3/20/2011 T (%ile) Self
R1 =Mother R2 =Social Studies(Teacher); R3 =English (Teacher) R4 Self				
Inhibit	60 (89)	56 (83)	83 (96)	57 (77)
Shift	65 (91)	79 (97)	88 (97)	70 (98)
Emotional Control	63 (89)	62 (88)	73 (93)	54 (69)
Behavioral Regulation Index (BRI)	64 (89)	66 (88)	84 (97)	60 (84)
Initiate	53 (63)	80 (97)	88 ($\geq$ 99)	--
Working Memory	71 (97)	101 ($\geq$ 99)	97 ($\geq$ 99)	77 ($\geq$ 99)
Plan/Organize	66 (93)	95 ($\geq$ 99)	98 ($\geq$ 99)	60 (86)
Organization of Materials	72 ($\geq$ 99)	130 ($\geq$99)	116 ($\geq$ 99)	62 (88)
Monitor	65 (96)	71 (91)	93 ($\geq$ 99)	52 (58)
Task Completion (BRIEF-SR Only)	--	--	--	73 (97)
Metacognition Index (MI)	68 (96)	97 ($\geq$ 99)	104 ($\geq$ 99)	70 (95)
Global Executive Composite (GEC)	68 (92)	89 ($\geq$ 99)	102 ($\geq$ 99)	67 (93)

Validity Scale	R1 Raw Score (Protocol Classification)	R2 Raw Score (Protocol Classification)	R3 Raw Score (Protocol Classification)	R4 Raw Score (Protocol Classification)
Negativity	0 (Acceptable)	2 (Acceptable)	2 (Acceptable)	2 (Acceptable)
Inconsistency	4 (Acceptable)	4 (Acceptable)	8 (Questionable)	5 (Acceptable)

Fig. 18.1 BRIEF software portfolio protocol summary report multi-rater score table. *Note*: BRIEF-SP does not currently print BRIEF-SR scores with Parent and Teacher Form scores in the same table. They are presented in the same table here for convenience

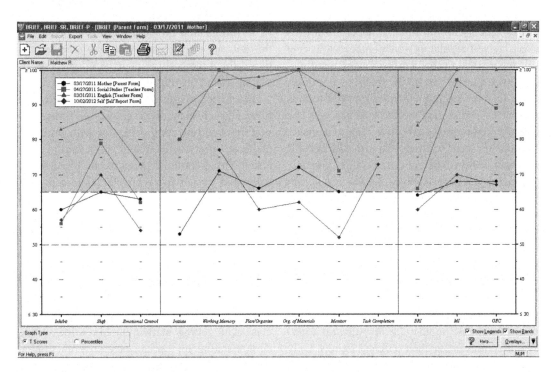

Fig. 18.2 BRIEF software portfolio parent, teacher, self-report profiles

concerns about Matthew's ability to adapt to change behaviorally and cognitively as reflected on the Shift scale. While there were some concerns with his emotional regulation (parent and teacher ratings above 60), these were of less concern. This pattern suggests that Matthew dislikes change and becomes upset when his plans or expectations are altered.

Continuing to interpret the profile from left to right, the next set of scales composes the Metacognition Index, reflecting an individual's ability to cognitively self-manage tasks. Matthew's teachers rated him as having substantial difficulty in all domains. He was described as having marked difficulty initiating, planning, organizing, and monitoring his problem-solving efforts while holding information and goals in working memory. His mother's ratings were clinically elevated for most scales with the exception of the Initiate scale. Follow-up interview revealed that Matthew initiated activities of his choosing at home, such as composing music and going out with friends, but avoided bringing any schoolwork home or letting his parents know what school-related tasks were due. Thus, his mother had little opportunity to observe the initiation difficulties noted by his teachers in the school setting.

While parent and teacher ratings on the BRIEF can provide useful information about a student's everyday self-regulatory functioning, adolescent self-reports add to the complete assessment picture by capturing the individual's own experiences and perspectives. Adolescent self-reports present a special case because adolescents are actively developing executive self-awareness. While adolescent self-reports on the BRIEF-SR correlate well with parent ratings (approximately .50 overall) and reasonably well with teacher ratings (approximately .25 overall) on the BRIEF, this does not mean that they have the same scores but instead means that the ratings tend to be parallel; that is, raters tend to describe similar peaks and valleys in their ratings though *T*-scores may be higher or lower. Indeed, some 56 % of adolescents' ratings on the BRIEF-SR were within 10 *T*-score points above or below their parents' ratings, another 38 % were up to 20 *T*-score points *below* their parents' ratings. Only 5 % of adolescents rated themselves as having over 10 *T*-score points *greater* difficulty than described by their parents. A similar pattern is seen for adolescent ratings in comparison to teacher ratings. Thus, parent, teacher, and adolescent ratings tend to show similar profiles, and adolescents' ratings are often within one standard deviation of parent and teacher ratings, many are more than one standard deviation *lower*, and it is unusual for an adolescent to report substantially *greater* difficulty than their parent or teacher.

In this example, Matthew's self-report was indeed parallel to his mother's and teachers' ratings, and within 10 *T*-score points of his mother's ratings with the exception of the Monitor scale where his ratings were much lower. This suggested good agreement between raters as to the primary areas of concern, including ability to adjust to change, to hold information in working memory, and to plan, organize, and monitor his problem-solving approaches. While Matthew's ratings on these latter scales were not "clinically elevated," his ratings on the Task Completion scale were high. This scale assesses the adolescent's ability to complete tasks appropriately and/or in a timely manner. While the scale does not attempt to capture a primary executive function, the ability to complete tasks is an outcome of well-regulated problem-solving. Problems in this area are often closely linked to other executive difficulties such as poor working memory, planning, and organization.

In essence, the evaluation profile as a whole highlights a history, observations, and formal test performance reflecting difficulty with initiating and sustaining cognitive and behavioral activity absent strong external incentive in an academically and cognitively capable adolescent boy with several risk factors upon entry to high school. In the context of superior to very superior overall current cognitive functioning, above-average learning and memory, and strong academic skills, Matthew demonstrated a pattern of difficulty initiating and sustaining cognitive activity and behavioral output that is consistent with his history of increasing problems with focusing in class and getting started on his work. While demonstrating appropriate executive function on a range of performance-based measures, ratings of Matthew's executive functioning in the real-world context indicated problems with initiating attention and behavior, sustaining working memory, and planning, organizing, and monitoring his own output, accompanied by resistance to change. This is consistent with his historical pattern of good early academic performance, but decline beginning in middle school, and teacher observations of a "drifty" inattentive style and trouble getting started on tasks in school. His difficulty initiating activity was context dependent, as he brought minimal schoolwork home thus did not exhibit initiation difficulties in that setting. Students with similar profiles tend to exhibit a pattern of good performance on tasks or activities for which they feel highly motivated vs. problems doing the work on tasks that they find less interesting. The good agreement between all raters about Matthew's resistance to change may reflect his anxiety and distress regarding perceived inability to accomplish tasks. Matthew wants to do well, to graduate, and to attend college and study music composition and production but does not know how to correct his current path.

While these everyday executive difficulties were emerging during the middle school years, the pattern was complicated by a number of risk factors. As he was making the transition into adolescence and high school, he sustained a mild TBI with significant post-injury symptoms but returned to school without rest or management. He sustained a second injury several months later, notably followed by sudden onset of depression and longer-term post-concussion symptoms. More recently, he experienced the onset of seizures and began medication. Matthew's best friend also moved away, and he had a difficult time adjusting to the loss. While none of these risk factors fully explains Matthew's functional difficulties, they may be exacerbating factors and add complexity to the clinical presentation. Concussion or mild TBI effects typically resolve within a few weeks to months of the injury. This can be prolonged, however, in students with vulnerable neurological systems such as those with attention problems or seizures. Repeated concussion can have compounding effects as well.

Adolescents who have or develop chronic illness (e.g., seizure disorders) often experience a difficult adolescent period, as the normal processes of developing competencies, self-confidence, and separating and becoming an independent individual are complicated by the illness and its effects and limitations.

Interventions Based on Test Results

The software packages for the BRIEF, BRIEF-SR, BRIEF-P, and BRIEF-A were designed to provide the user with the option of generating reports containing brief or detailed descriptions of the individual's profile and recommendations for each of the scales with an elevated T-score or, alternatively, recommendations whatever scales the user selects. The recommendations are organized according to the guiding model of executive function (e.g., Inhibit, Shift, Emotional Control, Initiate) and were compiled by the authors based on clinical experience and a review of literature on managing executive dysfunction in clinical populations (e.g., Cicerone et al., 2000; Mateer, 1999; Ylvisaker & Feeney, 1996, 1998; Ylvisaker, Szekeres & Feeney, 1998). These are similar to several recently described approaches to managing executive dysfunction (Cooper-Kahn & Dietzel, 2008; Dawson & Guare, 2010, 2012; McCloskey, Perkins & Divner, 2009).

The majority of recommendations are designed to serve as compensatory strategies that circumvent, rather than directly improve, deficits (e.g., learning cognitive strategies such as verbalization, development of an organized plan, goal setting, and strategies for aiding monitoring of behavior). Such strategies have demonstrated effectiveness in a number of patient populations (Dirette, 2002; Velligan et al., 2000; Wexler & Bell, 2005). Other recommendations emphasize the interaction of the individual and their environment, suggesting environmental modifications or accommodations (e.g., keeping work space clutter-free) that could facilitate executive functions (Ylvisaker, Hanks & Johnson-Greene, 2002; Ylvisaker, Jacobs & Feeney, 2003).

It is important to note that the offered recommendations are generic in nature and should be tailored on a case-by-case basis to conform to individual needs based on severity of deficit, preserved strengths, and environmental demands. Furthermore, the decision to use any given strategy to address executive dysfunction should be based on an appropriate assessment of the individual and tailored accordingly. The interpretive reports generated by the BRIEF software portfolio also include language appropriate for inclusion in writing an IEP or 504 plan.

Validity

Relationships to Other Similar Measures

Until very recently, the BRIEF was the only published rating scale for use with children. Thus, the majority of research on the convergence of the BRIEF with other rating scales has focused on behavioral measures with which it should theoretically correlate. For example, both the test manuals and several other studies have examined the relationship between BRIEF scores and parent ratings of attention-deficit/hyperactivity disorder (ADHD) symptoms. In the test manual, BRIEF-P scale and index scores were reported to show significant correlations with the Inattention ($r=.66-.90$) and Hyperactivity/Impulsivity ($r=.49-.87$) scales of the ADHD-IV-P. Similarly, BRIEF scale and index scores had moderate to high correlations with the Inattention ($r=.39-.67$) and Hyperactivity/Impulsivity ($r=.15-.73$) scales of the ADHD-IV. Significant correlations with the CBCL and BASC Attention Problem scales, the BASC Hyperactivity scale, and relevant scales on the Conner's Parent Rating Scale were also reported in the test manuals. Other investigators have also generally observed moderate to high correlations between the BRIEF and ADHD symptoms as measured using rating scales such as the ADHD Rating Scale (Mahone et al., 2002), BASC (Jarratt et al., 2005; McCandless & O' Laughlin, 2007), and Conner's Parent Rating Scale—Revised (Mahone & Hoffman, 2007).

In contrast to the pediatric literature, a number of other rating scales have been developed to assess executive functions in adults. The Dysexecutive Questionnaire (DEX) was designed to provide a single overall score reflecting executive functioning (Wilson, Alderman, Burgess, Emslie & Evans, 1996). All of the BRIEF-A scale and index scores correlated with the DEX in the expected direction. This was seen for both the self- and the informant ratings on the two instruments. The Frontal Systems Behavior Scale (FrSBe) was designed to assess three domains of frontal lobe deficits including Apathy, Disinhibition, and Executive Dysfunction (Grace & Malloy, 2002). Significant associations with FrSBe scores were reported in the BRIEF-A manual for both self- and informant ratings, the strongest correlations being observed with the Executive Dysfunction scale. Interestingly, the Shift and Emotional Control scales on the BRIEF-A showed some of the lowest correlations with the FrSBe scales, likely due to differences in the aspects of executive function tapped by the item content in the two measures. Nonetheless, the pattern of correlations with both the DEX and the FrSBe provides strong evidence for the convergent validity of the BRIEF.

The relationship between subjective ratings and scores on performance-based measures (e.g., Wisconsin Card Sorting Task, Trail Making) of executive function is another source of data pertaining to validity. A commonly raised concern is that rating scales of executive function do not always correlate in predictable ways with performance measures of the same or similar constructs. Indeed, there is inconsistency in the literature with respect to the presence of significant correlations between the BRIEF and performance-based tests. A recent review noted, however, that there are many measureable correlations between the BRIEF and such tests (McAuley, Chen, Goos, Schachar & Crosbie, 2010), though when such correlations are observed they tend to be small to moderate (e.g., Mahone, Cirino et al., 2002; Shimoni, Engel-Yeger & Tirosh, 2012; Toplak,

Bucciarelli, Jain & Tannock, 2009). For example, in a sample of children with TBI, a subset of BRIEF scores were reported to show correlations ($r=.24–.38$) with performance on the TEC, a computerized instrument emphasizing working memory and inhibitory control (Isquith, Roth & Gioia, 2010). Many authors have discussed explanations for the modest relationship between subjective ratings of cognitive functioning, including executive functions, and performance on tests in the clinic or laboratory setting (Gioia & Isquith, 2004; Isquith, Roth & Gioia 2013; McAuley et al., 2010; Sbordone, 1996).

Fairness, Sex, Race, and Ethnic Differences

The influence on demographic characteristics on BRIEF scores is reported in the test manuals. In addition, although the majority of respondents on the parent form were mothers (BRIEF, 83.2 %; BRIEF-P, 88.7 %), there was no significant difference between mother and father reports with respect to level of scale scores. Furthermore, for both the BRIEF-P and BRIEF, how well and for how long the teacher has known the rated child had little effect on scores, accounting for no more than 3 % of variance.

A significant effect of age was observed in the BRIEF standardization sample, for both parent and teacher report forms, with executive functions noted to improve with increasing age. This was reflected in both the index and scale scores. A small but significant effect of age was also seen in the BRIEF-P standardization sample, with younger children being reported by parents and teachers as having greater difficulty with some aspects of their executive functions. Similarly, on the BRIEF-SR younger adolescents reported themselves as having more difficulty with executive functions than older teens. Together, these findings are consistent with other evidence for developmental changes in executive functions from early childhood through adolescents (Anderson, Anderson, Jacobs & Smith, 2008; Best & Miller, 2010). Analysis of the BRIEF-A standardization sample revealed slightly greater difficulty with executive functions on several scales in younger (especially the 18–39-year-old group) than older adults for both self- and informant reports. Between 1 and 7 % of variance on the scales was accounted for by age in the adult samples.

Gender of the rated child has been found to affect scores. In the BRIEF-P standardization sample, parents tended to rate boys as having greater problems than girls on the Inhibit scale, while teachers rated boys as having more difficulty on the Inhibit, Working Memory, and Plan/Organize scales. Gender differences were also noted for the ISCI, EMI, and GEC scores. For most of these scales, however, no more than 3 % of variance was accounted for by gender. No interaction between age and gender was noted. On the BRIEF, both the parent and teacher forms show differences between boys and girls as well as there being an interaction between age and gender with boys showing dramatic improvements in executive function with increasing age. On the BRIEF-SR, boys tended to rate themselves on most scales as having slightly greater difficulty than girls, but girls reporting more problems on the Emotional Control scale. No interaction between gender and age was noted for the adolescent respondents. On the BRIEF-A, minimal gender differences were noted. Men reported more difficulty on the Initiate scale, while women reporting more problems with Emotional Control. The informant report was only found to show a gender difference for the Organization of Materials. Less than 2 % of variance was accounted for by gender on the BRIEF-A.

Significant but low correlations have been reported indicating that lower parental education is associated with report of greater problems with executive function on the BRIEF-P, BRIEF, and BRIEF-SR. These findings are consistent with that observed for parent behavioral ratings of their children's social-emotional functioning (Achenbach, McConaughy & Howell, 1987). Similarly, for both the self- and informant report forms of the BRIEF-A, several small but significant correlations were noted with years of education, indicating that lower education of the rater individual is associated with greater difficulty with executive function as reflected by some of the scales. Importantly, across the scales no more than 5 % of variance was accounted for by education in the child and adolescent versions of the BRIEF (BRIEF-P, BRIEF, BRIEF-SR) and less than 2 % of variance for the BRIEF-A. Therefore, education makes relatively small contribution to scores and is not considered a major factor when interpreting the measures.

SES has a small but significant relationship to BRIEF scores. Children from lower household SES are likely to be rated as having greater problems with executive function on both the BRIEF-P and the BRIEF, though SES accounts for no more than 2 % of variance in the former and 5 % of variance in the latter measure.

Race/ethnic group membership of the rated child does not have a significant impact on BRIEF, BRIEF-P, or BRIEF-SR scores. The BRIEF-A self-report form showed only a single significant effect indicating that African-American respondents obtained slightly higher scores than Hispanic respondents on the Organization of Materials scale. Analysis of the informant report form indicated that the race/ethnic group of the rated individual was associated with slightly different scores on the Inhibit, Self-Monitor, and Task Monitor scales. These differences on the BRIEF-A accounted for only about 1 % of variance, thus indicating that they are of little clinical significance

Profiles of Abilities and Their Relationship to Diagnosis

There is considerable evidence for the value of the BRIEF in assessing executive functions. Literature referencing the BRIEF has grown to over 300 scholarly articles and book chapters, the BRIEF-P and BRIEF-SR to over 100 articles each, and the more recently published BRIEF-A to over 50 articles, and countless conference presentations and posters include the BRIEF. The instrument has been included in many treatment trials, both pharmacological and behavioral, as well as longitudinal, multicenter studies of development (Waber et al., 2007; Waber, Forbes, Almli & Blood, 2012). The BRIEF has been demonstrated to be sensitive to executive functioning in a wide range of clinical populations. Table 18.8 presents a selection of this literature. The two diagnostic groups in which the BRIEF has been most frequently studied are ADHD and TBI. Here, we provide a brief overview of the BRIEF in relation to these groups.

ADHD

Numerous studies have examined the diagnostic utility of the BRIEF in children with ADHD. The authors presented the first such data (Gioia et al., 2000b), showing that the Working Memory scale predicted ADHD-Inattentive type (ADHD-I) and ADHD-Combined type (ADHD-C) with approximately 80 % accuracy for the parent form (81 %) and teacher form (83 %), while the Inhibit scale did so to a lesser degree (78 %, 70 %, respectively). Consistent with the requirement of impulsivity for a diagnosis of ADHD-C, the Inhibit scale was useful in distinguishing between children diagnosed with ADHD-I and ADHD-C. Reddy, Hale and Brodzinsky (2011) found similar rates of correct classification (77–86 %) in a well-controlled study comparing children diagnosed with ADHD vs. controls, providing convergent evidence for the clinical utility of the BRIEF for identifying children with attention disorders. Other ADHD subtypes may potentially also be distinguished by the BRIEF. Children with the sluggish cognitive tempo (SCT) variant of ADHD were rated as having greater problems on the Initiate (Isquith, McQuade, Crawford & Roth, 2005) or Monitoring scales of the BRIEF than were groups of children with ADHD-C or ADHD-I without SCT (Capdevila-Brophy et al., 2012). In addition to distinct profiles on the BRIEF between children diagnosed with ADHD-I vs. ADHD-C, each of these conditions was distinguishable from children with reading disorders and those with autism spectrum disorders (ASDs; Gioia, Isquith, Kenworthy & Barton, 2002). Several other studies have shown independent support for the utility of the BRIEF in identifying individuals with ADHD and subtypes (Mahone, Cirino et al., 2002; Mares, McLuckie, Schwartz & Saini,

Table 18.8 A sample of studies using the BRIEF in clinical populations

Population	Authors
ADHD	Biederman et al. (2011), Findling, Ginsberg, Jain and Gao (2009), Jarratt et al. (2005), Mahone, Cirino et al. (2002), Mahone and Hoffman (2007), Mares et al. (2007), McCandless and O' Laughlin (2007), Qian, Shuai, Cao, Chan and Wang (2010), Rotenberg-Shpigelman et al. (2008), Shimoni et al. (2012), Toplak et al. (2009), Yang et al. (2012)
Autism spectrum disorders	Chan et al. (2009), Christ et al. (2010), Semrud-Clikeman, Walkowiak, Wilkinson and Butcher (2010)
Bipolar disorder	Shear, DelBello, Lee Rosenberg and Strakowski (2002), Stange et al. (2011)
Brain tumor	Hocking et al. (2011)
Cancer survivors	Christ et al. (2010)
Cerebral palsy	Tervo, Symons, Stout and Novacheck (2006)
CHARGE syndrome	Hartshorne, Nicholas, Grialou and Russ (2007)
Cochlear implant	Holt, Beer, Kronenberger, Pisoni and Lalonde (2012)
Conduct disorder	Tobon, Puerta and Pineda (2008)
Down's syndrome	Edgin et al. (2010), Lee et al. (2011)
Epilepsy	Parrish et al. (2007), Sherman, Slick and Eyrl (2006), Slick et al. (2006)
Fetal alcohol syndrome	Chasnoff, Wells, Telford, Schmidt and Messer (2010), Wells, Chasnoff, Schmidt, Telford and Schwartz (2012)
Focal frontal lesions	Anderson et al. (2002), Lovstad et al. (2012)
Galactosemia	Antshel, Epstein and Waisbren (2004)
Hepatitis C	Rodrigue et al. (2011)
Hydrocephalus	Anderson et al. (2002), Mahone, Zabel, Levey, Verda and Kinsman (2002)
Hypersexuality	Reid, Karim, McCrory and Carpenter (2010)
Hypertension	Lande et al. (2009)
Juvenile myoclonic epilepsy	Gilotty et al. (2002), Pulsipher et al. (2009)
Lead exposure	Roy et al. (2009)
Leukemia	Campbell et al. (2009)
Liver transplant	Sorensen et al. (2011), Varni et al. (2011)
Mild cognitive impairment	Rabin et al. (2006)
Multiple sclerosis	Till et al. (2012)
Neurofibromatosis I	Payne, Hyman, Shores and North (2011)
Obstructive sleep apnea	Beebe et al. (2004)
Pathological gambling	Reid, McKittrick, Davtian and Fong (2012)
Phenylketonuria	Anderson et al. (2002), Waisbren and White (2010)
Prenatal phthalate exposure	Engel, Miodovnik, et al. (2010)
Psychosis prodrome	Niendam, Horwitz, Bearden and Cannon (2007)
Schizophrenia	Garlinghouse et al. (2010), Kumbhani, Roth, Kuck, Flashman and McAllister (2010), Power, Dragovic and Rock (2012)
Sickle cell anemia	Berg, Edwards and King (2012), Hollocks et al. (2012)
Sleep disordered breathing	Bourke et al. (2011)
Specific language impairment	Hughes, Turkstra and Wulfeck (2009)
Spina bifida	Burmeister et al. (2005), Kelly et al. (2012), Tarazi, Zabel and Mahone (2008), Zabel et al. (2011)
Subcortical band heterotopia	Jacobs, Anderson and Harvey (2001)
Substance abuse	Hadjiefthyvoulou, Fisk, Montgomery and Bridges (2012)
Tourette's	Cummings, Singer, Krieger, Miller and Mahone (2002), Mahone, Cirino et al. (2002)
Traumatic brain injury	Anderson et al. (2002), Anderson et al. (2005), Chapman et al. (2010), Conklin et al. (2008), Donders et al. (2010), Garcia-Molina et al. (2012), Maillard-Wermelinger et al. (2009), Mangeot et al. (2002), Matheson (2010), Toglia et al. (2010), Wilde et al. (2012), Wilson et al. (2011)

(continued)

Table 18.8 (continued)

Population	Authors
Turner syndrome	Lepage, Dunkin, Hong and Reiss (2013)
Type 1 diabetes	McNally, Rohan, Pendley, Delamater and Drotar (2010)
Velocardiofacial syndrome (22q11.2 deletion syndrome)	Antshel et al. (2006), Antshel, Conchelos, Lanzetta, Fremont and Kates (2005), Kiley-Brabeck and Sobin (2006)
Very low birth weight	Anderson et al. (2011), Verkerk et al. (2012)
Williams syndrome	John and Mervis (2010)

2007; Sullivan & Riccio, 2007). For example, McCandless and O'Laughlin reported that discriminant function analysis using the Inhibit scale and Metacognition Index correctly classified 77.8 % of children as having ADHD and 76 % as not having the disorder (McCandless & O' Laughlin, 2007). Jarratt et al., 2005 found that the BRIEF and BASC (Reynolds & Kamphaus, 1992) worked well complementarily in identifying children with ADHD subtypes, with the former measure providing a detailed picture of executive functions and the latter measure capturing a range of associated behavioral, social, and mood difficulties.

As with older children, preschoolers (Isquith, Gioia & Espy, 2004; Mahone & Hoffman, 2007) and adolescents (Toplak et al., 2009) with ADHD have also been found to have elevations on the BRIEF. Adults with ADHD show similar difficulties on the BRIEF-A to those seen in pediatric samples (Rotenberg-Shpigelman, Rapaport, Stern & Hartmen-Maeir, 2008), with deficits across a number of scales, most prominently Working Memory. Difficulty with monitoring on the BRIEF-A has also been reported (Chang, Davies & Gavin, 2009), consistent with other work demonstrating that individuals with ADHD have more difficulty monitoring their performance on tasks for accuracy (Herrmann et al., 2010)

The BRIEF has also been noted to be sensitive to interventions for ADHD. For example, children with the disorder show improvements in executive function with medication (Turgay et al., 2010), cognitive/behavioral intervention (Hahn-Markowitz, Manor & Maeir, 2011), and computerized working memory training (Beck, Hanson, Puffenberger, Benninger & Benninger, 2010). Interestingly, a recent study reported that treatment with a stimulant improved BRIEF-A scores in adults with ADHD, but a subgroup of patients continued to show residual executive dysfunction on the measure (Biederman et al., 2011).

Traumatic Brain Injury

Between publication of the BRIEF and 2010, 16 studies were published showing ratings for 1,248 children and adolescents with TBI (Gioia et al., 2010). In a review of ecologically valid assessment of cognition in TBI, the BRIEF was noted to stand out as the preferred caregiver rating measure in children and adolescents with acquired brain injury (Chevignard, Soo, Galvin, Catroppa & Eren, 2012). The BRIEF has also been included as a supplemental measure in the interagency Pediatric TBI Outcomes Workgroup (McCauley et al., 2012).

Studies of the BRIEF in TBI have been distributed over a wide range of time post-injury, from the first year (Maillard-Wermelinger et al., 2009; Sesma, Slomine, Ding, McCarthy & Children's Health After Trauma Study Group, 2008), through 1–5 years (Chapman et al., 2010) (Conklin, Salorio & Slomine, 2008; Donders, DenBraber & Vos, 2010; Vriezen & Pigott, 2002), and from 5 to 10 years post-injury (Kurowski et al., 2011; Mangeot, Armstrong, Colvin, Yeates & Taylor, 2002; Muscara, Catroppa & Anderson, 2008; Nadebaum, Anderson & Catroppa, 2007). All studies found deficits in at least one aspect of executive function as measured on the BRIEF. Ten studies documented global deficits on the GEC, seven found specific deficits on the Metacognition Index, seven on the BRI, and several reported elevations on specific scales, most notably Working Memory. While all studies reported elevated ratings relative to control samples in parent or teacher reports on the BRIEF, only some studies also observed poorer scores on performance tests of executive function in the TBI group (Conklin et al., 2008; Muscara et al., 2008; Vriezen & Pigott, 2002), and ratings and performance measures correlated in only one study (Mangeot et al., 2002). As a whole, these studies suggest that executive function deficits in children with TBI may be captured on the BRIEF, even when not observed in test performance, and that injury severity is reflected in greater everyday executive dysfunction as reflected on the BRIEF.

A smaller number of studies have examined the BRIEF-A in adults with TBI. A study of 32 adults with moderate to severe TBI, evaluated on average 7 months after injury, reported clinical impairment ($T \geq 65$) in a subset of patients as rated by themselves and by a family member (Garcia-Molina, Tormos, Bernabeu, Junque & Roig-Rovira, 2012). Furthermore, greater executive dysfunction as rated on the BRIEF-A was associated with poorer ability to complete activities of daily living, interpersonal behavior, emotional behavior, and cognition on the Patient Competency Rating Scale. A study of 62 adults of mixed severity TBI observed that patients classified as having executive dysfunction based on performance-based tests were rated as having greater problems on the BRIEF-A informant but not self-report form (Matheson, 2010). A small study of four patients with TBI showed sensitivity of the BRIEF-A to the effects of cognitive strategy use training (Toglia, Johnston, Goverover & Dain, 2010).

Autism Spectrum Disorders

Executive function difficulties are commonly seen in children with ASDs (Pennington & Ozonoff, 1996). While not causally linked, executive functions contribute to symptom clusters, such as repetitive behaviors and restricted interests. In the first study of BRIEF profiles in children with ASDs, the profiles were unique in their spike on the Shift scale, differentiating them from children with ADHD-I, ADHD-C, and TBI (Gioia, Isquith, Kenworthy & Barton, 2002). In a second group study of children with ASDs, children showed similar elevations on the Shift scale but also the Plan/Organize scale, consistent with broader deficits in flexibility and organization of complex information in this population (Kenworthy et al., 2005). Kenworthy and colleagues explored these relationships in 89 high-functioning children with ASDs. They found that more ecologically oriented performance measures and parent ratings on the BRIEF (Kenworthy, Yerys, Anthony & Wallace, 2008), particularly the BRI scales, were significant predictors of autism symptoms (Kenworthy, Black, Harrison, della Rosa & Wallace, 2009). Executive functioning in children with ASDs was related to real-world adaptive functioning, particularly communication competence (Gilotty, Kenworthy, Sirian, Black & Wagner, 2002). Children with ASDs who also exhibited characteristics of ADHD showed greater deficits on the Emotional Control scale and the Working Memory scale of the BRIEF, suggesting that the presence of ADHD may moderate the expression of components of the ASD cognitive and behavioral phenotype, but the combination of ASD and ADHD may not represent an ideologically distinct phenotype (Yerys et al., 2009).

One study has investigated executive functions using the BRIEF-A in adults with autistic traits. In a large sample of undergraduate students ($n = 1,847$), those with high as opposed to low self-report autistic traits endorsed significantly greater problems with executive function on several BRIEF-A scales (Christ, Kanne & Reiersen, 2010). Consistent with studies of children with ASD, the largest effect size was observed for the Shift scale, with group accounting for 33.6 % of the variance even with statistical control for the presence of ADHD symptoms.

Others

In addition to studies of clinical populations, the BRIEF has been used in a variety of investigations as a predictor or outcome measure. With respect to academic performance, several studies have observed associations between BRIEF scores and reading (Locascio, Mahone, Eason & Cutting, 2010; McAuley et al., 2010) and math (Clark, Pritchard & Woodward, 2010) abilities. Clark et al. (2010)

reported that captured teacher ratings of 4-year-old preschool children's everyday executive functions on the BRIEF-P were predictive of math achievement at age six, over and above performance-based measures of executive function, particularly for working memory, planning, and inhibitory control. Waber, Gerber, Turcios, Wagner and Forbes (2006) found that teacher ratings on the BRIEF were the best predictor of performance on statewide academic testing.

Parent and teacher ratings on the BRIEF have also been studied in relation to certain genetic polymorphisms (Acevedo, Piper, Craytor, Benice & Raber, 2010) and sensitivity to relational aggression among adolescent girls (Baird, Silver & Veague, 2010). Scores on the BRIEF-A have been reported to be associated with alexithymia (Koven & Thomas, 2010), academic procrastination (Rabin, Fogel & Nutter-Upham, 2010), social functioning (Dawson, Shear & Strakowski, 2012), disordered eating (Salmon & Figueredo, 2009), family history of alcoholism (Schroeder & Kelley, 2008), and was recently reported to be a good predictor of alcohol-induced aggression (Giancola, Godlaski & Roth, 2012). Together, these investigations provide further support for the usefulness of the BRIEF for a wide range of clinical and experimental purposes.

Empirical Support for the Test Structure

The Professional Manuals for the BRIEF, BRIEF-SR, and BRIEF-A report exploratory factor analyses that support a two-factor solution in the normative samples and in mixed clinical samples. The factors correspond to the Behavioral Regulation Index and the Metacognitive Index described above. Exploratory factor analysis of the BRIEF-P, in contrast, yielded a three-factor solution, labeled as the Inhibitory Self-Control Index, Flexibility Index, and Emergent Metacognition Index.

Gioia and Isquith (Gioia, Isquith, Retzlaff & Espy, 2002) reexamined the BRIEF and found that the item content of the Monitor scale reflected two different types of monitoring. They therefore separated the scale into a Task Monitor scale reflecting the monitoring of task-related activities and a Self-Monitor scale reflecting monitoring of the effects of one's behavior on others. These two monitoring scales were only modestly correlated with one another ($r=.47$) in a sample of children (Gioia, Isquith, Retzlaff & Espy, 2002), a finding also noted in the normative sample for the BRIEF-A (self-report form, $r=.58$; informant report form, $r=.68$) (Roth, Isquith & Gioia, 2006), and they loaded on different factors (Self-Monitor on the Behavioral Regulation Index, Task Monitor on the Metacognition Index).

Gioia et al. subsequently conducted a confirmatory factor analysis using a nine-scale version of the BRIEF parent report, with separate Task Monitor and Self-Monitor scales, in a large mixed clinical sample (Gioia & Isquith, 2002). A three-factor solution was found to best fit the data. The Metacognition Index remained unchanged, retaining the same scales. The Behavioral Regulation Index, however, separated into a Behavioral Regulation factor, consisting of the Inhibit and Self-Monitor scales, and an Emotional Regulation factor composed of the Emotional Control and Shift scales. A recent confirmatory factor analyses of the BRIEF parent and teacher forms in a mixed healthy and clinical sample supported the nine-scale, three-factor model of the BRIEF including separate Behavioral Regulation and Emotional Regulation factors (Egeland & Fallmyr, 2010). An unpublished confirmatory factor analysis in a sample of adults was also consistent with this three-factor solution (Roth, Lance, Isquith & Giancola, 2013). Two other factor analyses of the BRIEF have been reported, one in children with epilepsy (Slick, Lautzenhiser, Sherman & Eyrl, 2006) and the other TBI (Donders et al., 2010). Although both of these reported that the best fit was for a two-factor solution, they both examined factor structure based on an eight rather than nine-scale model of the BRIEF (i.e., using only the unitary Monitor scale).

Together, factor analytic studies support the argument that executive function, at least as measured by the BRIEF, is not a unitary construct. Furthermore, while research using the three-factor solution at present is limited, there is a wealth of evidence indicating that the Behavioral Regulation Index and the Metacognitive Index show differential associations with a variety of clinical populations and outcome measures (Giancola et al., 2012; Shimoni et al., 2012; Wilson et al., 2011). It should be noted, however, that at present the Professional Manuals for the BRIEF only provide normative data for the two-factor solution.

Other Evidence of Validity

Additional support for the validity of the BRIEF may be found in an extensive literature relating scores on the inventory with a variety of outcome measures. In particular, several studies have found associations between neural substrate and everyday executive function as measured by the BRIEF. In one of the first such studies, Anderson, Anderson, Northam, Jacobs and Mikiewicz (2002) found that BRIEF profiles of children with frontal focal lesions were substantially more elevated than those of children with early treated phenylketonuria (PKU) and early treated hydrocephalus who, in turn, were more elevated than controls. Children with prefrontal lesions had greater problems with inhibitory control, shifting set, and monitoring with highest BRIEF ratings for children with right prefrontal lesions (Anderson, Jacobs & Harvey, 2005). Several investigations have examined the relationship BRIEF scores and brain integrity as measured using magnetic resonance imaging (MRI) scans. Poorer working memory on the BRIEF has been reported to be associated with smaller volume of the frontal lobe on MRI in typically developing children (Mahone, Martin, Kates, Hay & Horska, 2009) and adults with schizophrenia (Garlinghouse, Roth, Isquith, Flashman & Saykin, 2010). BRIEF scores have also been found to be significantly correlated with cortical thickness (Merkley et al., 2008; Wilde et al., 2012) and white matter integrity (Wozniak et al., 2007) in children with TBI. A prospective study found that shorter corpus callosum length in infancy was found to be related to poorer executive functions on the BRIEF at age four (Ghassabian et al., 2013). Using functional MRI, a recent study observed that poorer self-report executive function was correlated with brain activation during a card sorting task in women with breast cancer who had not received chemotherapy (Kesler, Kent & O'Hara, 2011).

Summary and Conclusions

The BRIEF family of instruments began as a practical means for its authors to measure executive function in children at a time when the domain was little known outside of neuropsychology, it was rarely applied to children and adolescents, and there were no rating scale approaches to measuring everyday executive functions. As the first instrument of its kind, the BRIEF enjoyed a grassroots beginning and has enjoyed wide acceptance and use in the USA in clinical and research settings since first publication in 2000, with global presence increasing rapidly. The BRIEF has demonstrated its usefulness for capturing executive functions in the everyday lives of children, adolescents, and adults. It is sensitive to a plethora of behavioral, emotional, and social characteristics and clinical conditions across a variety of settings. The reliability data is strong, and there is a wealth of evidence supporting its validity for measuring executive function.

It is important to appreciate that, in general, behavioral assessment methods that rely on self- and/or informant reports present their own limitations. While providing a more global or molar view of behavior, they also provide less process-specific information. For example, while performance measures attempt to fractionate and measure specific subdomains of executive function such as working memory vs. planning or organization, specific components of the executive functions in the everyday context may be less separable. Indeed, much of everyday behavior may depend on the integration of executive function subdomains, such as the ability to inhibit distraction that protects working memory which, in turn, may enable planning and organization of goal-directed behavior. As such, it may be more difficult to parse deficits in specific executive functions via self- and informant reports. Evaluators have more limited control over environmental factors that may affect ratings. An individual who struggles in a highly demanding, multitasking work environment may describe his or her executive functions as inadequate, while another individual in a simpler, more routine environment may report that his or her self-regulatory capacity is up to the challenge. Rater perspective or bias must also be considered when interpreting rating scales. An individual's emotional state or personality characteristics may influence his or her ratings or simply whether or not he or she like or dislike the rated individual. For example, one teacher may dislike a child or, as is sometimes the case, simply dislike completing rating scales, while another teacher may enjoy the same child and appreciate the opportunity to provide input into the evaluation process.

With respect to the BRIEF itself, there have been several reviews since its inception. The first published review (Baron, 2000) offered no criticisms. The second review (Goldstein, 2001) was complementary of the data provided about the new instrument but noted that some of the sample sizes were small. A review of the BRIEF-SR stated that is a "theoretically and psychometrically sound self-report measure of executive functioning for adolescents" (Walker & D'Amato, 2006). The most recent edition of *A Compendium of Neuropsychological Tests* (Strauss, Sherman & Spreen, 2006) provided a thorough review and noted that there were few limitations but offered important comments, primarily that the norms for the BRIEF parent and teacher forms were not fully aligned with the US Census data, some age and gender groups were small, and the standardization samples for all versions excluded cases with any developmental or acquired disorders. The review indicated that it would be helpful to have more information about how the BRIEF behaves in specific populations (e.g., individuals of Hispanic ethnicity) and how the BRIEF scales align with diagnoses or symptoms as found in the Diagnostic and Statistical Manual of Mental Disorders (American Psychiatric Association, 2000). Otherwise, there have been very few criticisms of items, the format and layout of forms, or the use of the scoring software. Indeed, many comment that the BRIEF is easy to use, that the software is helpful in offering interventions, and that it captures aspects of an individual's functioning in the everyday environment that no other measures detect. Strauss et al. (2006) described the BRIEF as "a well-designed, psychometrically sound instrument that, despite its recent publication, already boasts a substantial body of literature on its application in clinical contexts" and noted that "the manual is well written and thorough, and it includes detailed examination of psychometric properties in clinical groups, a feature that is sometimes overlooked in other tests and questionnaires." Goldstein, 2001 concluded that the "reasonable cost relative to many of the measures we utilize during assessment, the theory-based format and the focus on executive function make this an attractive instrument."

References

Acevedo, S. F., Piper, B. J., Craytor, M. J., Benice, T. S., & Raber, J. (2010). Apolipoprotein E4 and sex affect neurobehavioral performance in primary school children. *Pediatric Research, 67*(3), 293–299.

Achenbach, T. M., McConaughy, S. H., & Howell, C. T. (1987). Child/adolescent behavioral and emotional problems: Implications of cross-informant correlations for situational specificity. *Psychological Bulletin, 101*(2), 213–232.

American Psychiatric Association. (2000). *Diagnostic and Statistical Manual of Mental Disorders* (Revised, 4th Ed.). Washington, DC: American Psychiatric Press.

Anderson, V., Anderson, P. J., Jacobs, R., & Smith, M. S. (2008). Development and assessment of executive function: From preschool to adolescence. In V. Anderson, R. Jacobs, & P. J. Anderson (Eds.), *Executive functions and the frontal lobes: A lifespan perspective* (pp. 123–154). New York: Psychology Press.

Anderson, V. A., Anderson, P., Northam, E., Jacobs, R., & Mikiewicz, O. (2002). Relationships between cognitive and behavioral measures of executive function in children with brain disease. *Child Neuropsychology, 8*(4), 231–240.

Anderson, P. J., De Luca, C. R., Hutchinson, E., Spencer-Smith, M. M., Roberts, G., & Doyle, L. W. (2011).

Attention problems in a representative sample of extremely preterm/extremely low birth weight children. *Developmental Neuropsychology, 36*(1), 57–73.

Anderson, V., Jacobs, R., & Harvey, A. S. (2005). Prefrontal lesions and attentional skills in childhood. *Journal of the International Neuropsychological Society, 11*(7), 817–831.

Antshel, K. M., Conchelos, J., Lanzetta, G., Fremont, W., & Kates, W. R. (2005). Behavior and corpus callosum morphology relationships in velocardiofacial syndrome (22q11.2 deletion syndrome). *Psychiatry Research, 138*(3), 235–245.

Antshel, K. M., Epstein, I. O., & Waisbren, S. E. (2004). Cognitive strengths and weaknesses in children and adolescents homozygous for the galactosemia Q188R mutation: A descriptive study. *Neuropsychology, 18*(4), 658–664.

Antshel, K. M., Fremont, W., Roizen, N. J., Shprintzen, R., Higgins, A. M., Dhamoon, A., et al. (2006). ADHD, major depressive disorder, and simple phobias are prevalent psychiatric conditions in youth with velocardiofacial syndrome. *Journal of the American Academy of Child and Adolescent Psychiatry, 45*(5), 596–603.

Baird, A. A., Silver, S. H., & Veague, H. B. (2010). Cognitive control reduces sensitivity to relational aggression among adolescent girls. *Social Neuroscience, 5*(5–6), 519–532.

Baron, I. S. (2000). Test review: Behavior Rating Inventory of Executive Function. *Child Neuropsychology, 6*(3), 235–238.

Beardmore, S., Tate, R., & Liddle, B. (1999). Does information and feedback improve children's knowledge and awareness of deficits after traumatic brain injury? *Neuropsychological Rehabilitation, 9*(1), 45–62.

Beck, S. J., Hanson, C. A., Puffenberger, S. S., Benninger, K. L., & Benninger, W. B. (2010). A controlled trial of working memory training for children and adolescents with ADHD. *Journal of Clinical Child and Adolescent Psychology, 39*(6), 825–836.

Beebe, D. W., Wells, C. T., Jeffries, J., Chini, B., Kalra, M., & Amin, R. (2004). Neuropsychological effects of pediatric obstructive sleep apnea. *Journal of the International Neuropsychological Society, 10*, 962–975.

Berg, C., Edwards, D. F., & King, A. (2012). Executive function performance on the children's kitchen task assessment with children with sickle cell disease and matched controls. *Child Neuropsychology, 18*(5), 432–448.

Bernstein, J. H., & Waber, D. P. (2007). Executive capacities from a developmental perspective. In L. Meltzer (Ed.), *Executive function in education: From theory to practice* (pp. 39–54). New York: The Guilford Press.

Best, J. R., & Miller, P. H. (2010). A developmental perspective on executive function. *Child Development, 81*(6), 1641–1660.

Biederman, J., Mick, E., Fried, R., Wilner, N., Spencer, T. J., & Faraone, S. V. (2011). Are stimulants effective in the treatment of executive function deficits? Results from a randomized double blind study of OROS-methylphenidate in adults with ADHD. *European Neuropsychopharmacology, 21*(7), 508–515.

Bourke, R. S., Anderson, V., Yang, J. S., Jackman, A. R., Killedar, A., Nixon, G. M., et al. (2011). Neurobehavioral function is impaired in children with all severities of sleep disordered breathing. *Sleep Medicine, 12*(3), 222–229.

Burgess, P. W. (1997). Theory and methodology in executive function research. In P. Rabbitt (Ed.), *Methodology of frontal and executive function* (pp. 81–116). Hove: Psychology Press.

Burmeister, R., Hannay, H. J., Copeland, K., Fletcher, J. M., Boudousquie, A., & Dennis, M. (2005). Attention problems and executive functions in children with spina bifida and hydrocephalus. *Child Neuropsychology, 11*(3), 265–283.

Campbell, L. K., Scaduto, M., Van Slyke, D., Niarhos, F., Whitlock, J. A., & Compas, B. E. (2009). Executive function, coping, and behavior in survivors of childhood acute lymphocytic leukemia. *Journal of Pediatric Psychology, 34*(3), 317–327.

Capdevila-Brophy, C., Artigas-Pallares, J., Navarro-Pastor, J. B., Gracia-Nonell, K., Rigau-Ratera, E., Obiols, J. E. (2012). ADHD Predominantly inattentive subtype with high sluggish cognitive tempo: A new clinical entity? Journal of Attention Disorders, doi: 10.1177/1087054712445483.

Chan, A. S., Cheung, M. C., Han, Y. M., Sze, S. L., Leung, W. W., Man, H. S., et al. (2009). Executive function deficits and neural discordance in children with autism spectrum disorders. *Clinical Neurophysiology, 120*(6), 1107–1115.

Chang, W.-P., Davies, P. L., & Gavin, W. J. (2009). Error monitoring in college students with attention-deficit/hyperactivity disorder. *Journal of Psychophysiology, 23*, 113–125.

Chapman, L. A., Wade, S. L., Walz, N. C., Taylor, H. G., Stancin, T., & Yeates, K. O. (2010). Clinically significant behavior problems during the initial 18 months following early childhood traumatic brain injury. *Rehabilitation Psychology, 55*(1), 48–57.

Chasnoff, I. J., Wells, A. M., Telford, E., Schmidt, C., & Messer, G. (2010). Neurodevelopmental functioning in children with FAS, pFAS, and ARND. *Journal of Developmental and Behavioral Pediatrics, 31*(3), 192–201.

Chevignard, M. P., Soo, C., Galvin, J., Catroppa, C., & Eren, S. (2012). Ecological assessment of cognitive functions in children with acquired brain injury: A systematic review. *Brain Injury, 26*(9), 1033–1057.

Christ, S. E., Kanne, S. M., & Reiersen, A. M. (2010). Executive function in individuals with subthreshold autism traits. *Neuropsychology, 24*(5), 590–598.

Cicerone, K. D., Dahlberg, C., Kalmar, K., Langenbahn, D. M., Malec, J. F., Bergquist, T. F., et al. (2000). Evidence-based cognitive rehabilitation: Recommendations for clinical practice. *Archives of Physical Medicine and Rehabilitation, 81*(12), 1596–1615.

Ciurli, P., Bivona, U., Barba, C., Onder, G., Silvestro, D., Azicnuda, E., et al. (2010). Metacognitive unawareness correlates with executive function impairment after severe traumatic brain injury. *Journal of the International Neuropsychological Society, 16*(2), 360–368.

Clark, C. A., Pritchard, V. E., & Woodward, L. J. (2010). Preschool executive functioning abilities predict early mathematics achievement. *Developmental Psychology, 46*(5), 1176–1191.

Conklin, H. M., Salorio, C. F., & Slomine, B. S. (2008). Working memory performance following paediatric traumatic brain injury. *Brain Injury, 22*(11), 847–857.

Cooper-Kahn, J., & Dietzel, L. (2008). *Late, lost and unprepared*. Bethesda, MD: Woodbine House.

Cronbach, L. J. (1951). Coefficient alpha and the internal structure of tests. *Psychometrika, 16*, 297–334.

Cummings, D. D., Singer, H. S., Krieger, M., Miller, T. L., & Mahone, E. M. (2002). Neuropsychiatric effects of guanfacine in children with mild Tourette syndrome: A pilot study. *Clinical Neuropharmacology, 25*, 325–332.

Dawson, P., & Guare, R. (2010). *Executive skills in children and adolescents: A practical guide to assessment and intervention* (2nd ed.). New York: Guilford Press.

Dawson, P., & Guare, R. (2012). *Coaching students with executive skills deficits*. New York: Guilford Press.

Dawson, E. L., Shear, P. K., & Strakowski, S. M. (2012). Behavior regulation and mood predict social functioning among healthy young adults. *Journal of Clinical and Experimental Neuropsychology, 34*(3), 297–305.

Denckla, M. B. (1994). Measurement of executive function. In G. R. Lyon (Ed.), *Frames of reference for the assessment of learning disabilities: New views on measurement issues* (pp. 117–142). Baltimore: Paul Brookes.

Dirette, D. (2002). The development of awareness and the use of compensatory strategies for cognitive deficits. *Brain Injury, 16*(10), 861–871.

Donders, J., DenBraber, D., & Vos, L. (2010). Construct and criterion validity of the Behaviour Rating Inventory of Executive Function (BRIEF) in children referred for neuropsychological assessment after paediatric traumatic brain injury. *Journal of Neuropsychology, 4*(Pt 2), 197–209.

Edgin, J. O., Mason, G. M., Allman, M. J., Capone, G. T., Deleon, I., Maslen, C., et al. (2010). Development and validation of the Arizona Cognitive Test Battery for Down syndrome. *Journal of Neurodevelopmental Disorders, 2*(3), 149–164.

Egeland, J., & Fallmyr, O. (2010). Confirmatory Factor Analysis of the Behavior Rating Inventory of Executive Function (BRIEF): Support for a distinction between emotional and behavioral regulation. *Child Neuropsychology, 16*(4), 326–337.

Engel, S. M., Miodovnik, A., Canfield, R. L., Zhu, C., Silva, M. J., Calafat, A. M., & Wolff, M. S. (2010). Prenatal phthalate exposure is associated with childhood behavior and executive functioning. *Environmental Health Perspectives, 18*, 656–571.

Findling, R. L., Ginsberg, L. D., Jain, R., & Gao, J. (2009). Effectiveness, safety, and tolerability of lisdexamfetamine dimesylate in children with attention-deficit/hyperactivity disorder: An open-label, dose-optimization study. *Journal of Child and Adolescent Psychopharmacology, 19*(6), 649–662.

Flashman, L. A. (2002). Disorders of awareness in neuropsychiatric syndromes: An update. *Current Psychiatry Reports, 4*(5), 346–353.

Garcia-Molina, A., Tormos, J. M., Bernabeu, M., Junque, C., & Roig-Rovira, T. (2012). Do traditional executive measures tell us anything about daily-life functioning after traumatic brain injury in Spanish-speaking individuals? *Brain Injury, 26*(6), 864–874.

Garlinghouse, M. A., Roth, R. M., Isquith, P. K., Flashman, L. A., & Saykin, A. J. (2010). Subjective rating of working memory is associated with frontal lobe volume in schizophrenia. *Schizophrenia Research, 120*, 71–75.

Ghassabian, A., Herba, C. M., Roza, S. J., Govaert, P., Schenk, J. J., Jaddoe, V. W., et al. (2013). Infant brain structures, executive function, and attention deficit/hyperactivity problems at preschool age: A prospective study. *Journal of Child Psychology and Psychiatry, 54*(1), 96–104.

Giancola, P. R., Godlaski, A. J., & Roth, R. M. (2012). Identifying component-processes of executive functioning that serve as risk factors for the alcohol-aggression relation. *Psychology of Addictive Behaviors, 26*(2), 201–211.

Gilotty, L., Kenworthy, L., Sirian, L., Black, D. O., & Wagner, A. E. (2002). Adaptive skills and executive function in autism spectrum disorders. *Child Neuropsychology, 8*(4), 241–248.

Gioia, G. A., Espy, K. A., & Isquith, P. K. (2003). *BRIEF-P: Behavior Rating Inventory of Executive Function—Preschool Version*. Lutz, FL: Psychological Assessment Resources.

Gioia, G. A., & Isquith, P. K. (2002). Two faces of monitor: They self and thy task. *Journal of the International Neuropsychological Society, 8*, 229.

Gioia, G. A., & Isquith, P. K. (2004). Ecological assessment of executive function in traumatic brain injury. *Developmental Neuropsychology, 25*(1–2), 135–158.

Gioia, G. A., Isquith, P. K., & Guy, S. C. (2001). Assessment of executive function in children with neurological impairments. In R. Simeonsson & S. Rosenthal (Eds.), *Psychological and developmental assessment* (pp. 317–356). New York: The Guilford Press.

Gioia, G. A., Isquith, P. K., Guy, S. C., & Kenworthy, L. (2000a). Behavior Rating Inventory of Executive Function. *Child Neuropsychology, 6*(3), 235–238.

Gioia, G. A., Isquith, P. K., Guy, S. C., & Kenworthy, L. (2000b). *BRIEF: Behavior Rating Inventory of Executive Function*. Lutz, FL: Psychological Assessment Resources.

Gioia, G. A., Isquith, P. K., Kenworthy, L., & Barton, R. M. (2002). Profiles of everyday executive function in acquired and developmental disorders. *Child Neuropsychology, 8*(2), 121–137.

Gioia, G. A., Isquith, P. K., Retzlaff, P. D., & Espy, K. A. (2002). Confirmatory factor analysis of the Behavior Rating Inventory of Executive Function (BRIEF) in a clinical sample. *Child Neuropsychology, 8*(4), 249–257.

Gioia, G. A., Kenworthy, L., & Isquith, P. K. (2010). Executive function in the real world: BRIEF lessons from Mark Ylvisaker. *The Journal of Head Trauma Rehabilitation, 25*(6), 433–439.

Goldberg, E., & Podell, K. (2000). Adaptive decision making, ecological validity, and the frontal lobes. *Journal of Clinical and Experimental Neuropsychology, 22*(1), 56–68.

Goldstein, S. (2001). Test review: Behavior Rating Inventory of Executive Function. *Applied Neuropsychology, 8*(4), 255–261.

Grace, J., & Malloy, P. F. (2002). *Frontal Systems Behavior Scale (FrSBe).* Lutz, FL: Psychological Assessment Resources.

Guy, S. C., Isquith, P. K., & Gioia, G. A. (2004). *Behavior Rating Inventory of Executive Function—Self Report Version.* Lutz, FL: Psychological Assessment Resources.

Hadjiefthyvoulou, F., Fisk, J. E., Montgomery, C., & Bridges, N. (2012). Self-reports of executive dysfunction in current ecstasy/polydrug users. *Cognitive and Behavioral Neurology, 25*(3), 128–138.

Hahn-Markowitz, J., Manor, I., & Maeir, A. (2011). Effectiveness of cognitive-functional (Cog-Fun) intervention with children with attention deficit hyperactivity disorder: A pilot study. *American Journal of Occupational Therapy, 65*(4), 384–392.

Hartshorne, T. S., Nicholas, J., Grialou, T. L., & Russ, J. M. (2007). Executive function in CHARGE syndrome. *Child Neuropsychology, 13*(4), 333–344.

Herrmann, M. J., Mader, K., Schreppel, T., Jacob, C., Heine, M., Boreatti-Hummer, A., et al. (2010). Neural correlates of performance monitoring in adult patients with attention deficit hyperactivity disorder (ADHD). *The World Journal of Biological Psychiatry, 11*(2 Pt 2), 457–464.

Hocking, M. C., Hobbie, W. L., Deatrick, J. A., Lucas, M. S., Szabo, M. M., Volpe, E. M., et al. (2011). Neurocognitive and family functioning and quality of life among young adult survivors of childhood brain tumors. *Clinical Neuropsychology, 25*(6), 942–962.

Hollocks, M. J., Kok, T. B., Kirkham, F. J., Gavlak, J., Inusa, B. P., DeBaun, M. R., et al. (2012). Nocturnal oxygen desaturation and disordered sleep as a potential factor in executive dysfunction in sickle cell anemia. *Journal of the International Neuropsychological Society, 18*(01), 168–173.

Holmes-Bernstein, J., & Waber, D. P. (1990). Developmental neuropsychological assessment: The systemic approach. In J. Holmes-Bernstein & D. P. Waber (Eds.), *Neuropsychology* (pp. 311–371). Totowa, NJ: Humana Press.

Holt, R. F., Beer, J., Kronenberger, W. G., Pisoni, D. B., & Lalonde, K. (2012). Contribution of family environment to pediatric cochlear implant users' speech and language outcomes: Some preliminary findings. *Journal of Speech, Language, and Hearing Research, 55*(3), 848–864.

Hughes, D. M., Turkstra, L. S., & Wulfeck, B. B. (2009). Parent and self-ratings of executive function in adolescents with specific language impairment. *International Journal of Language & Communication Disorders, 44*(6), 901–916.

Huizinga, M., & Smidts, D. P. (2011). Age-related changes in executive function: A normative study with the Dutch version of the Behavior Rating Inventory of Executive Function (BRIEF). *Child Neuropsychology, 17*(1), 51–66.

Isquith, P. K., Gioia, G. A., & Espy, K. A. (2004). Executive function in preschool children: Examination through everyday behavior. *Developmental Neuropsychology, 26*(1), 403–422.

Isquith, P. K., McQuade, D. V., Crawford, J. S., & Roth, R. M. (2005, October). *Executive function profiles of children with sluggish cognitive tempo.* Poster presented at the 25th Annual Conference of the National Academy of Neuropsychology, Tampa.

Isquith, P. K., Roth, R. M., & Gioia, G. A. (2010). *Tasks of Executive Control (TEC).* Lutz, FL: Psychological Assessment Resources.

Isquith, P. K., Roth., R. M., & Gioia, G. A. (2013). Contributions of rating scales to the assessment of executive functions. *Applied Neuropsychology: Child, 2*, 125–132.

Jacobs, R., Anderson, V., & Harvey, A. S. (2001). Neuropsychological profile of a 9-year-old child with subcortical band heterotopia or 'double cortex'. *Developmental Medicine and Child Neurology, 43*(9), 628–633.

Jarratt, K. P., Riccio, C. A., & Siekierski, B. M. (2005). Assessment of attention deficit hyperactivity disorder (ADHD) using the BASC and BRIEF. *Applied Neuropsychology, 12*(2), 83–93.

John, A. E., & Mervis, C. B. (2010). Sensory modulation impairments in children with Williams syndrome. *American Journal of Medical Genetics. Part C, Seminars in Medical Genetics, 154C*(2), 266–276.

Kaplan, E. (1988). A process approach to neuropsychological assessment. In T. Boll & B. K. Bryant (Eds.), *Clinical neuropsychology and brain function: Research, measurement and practice* (pp. 129–167). Washington, DC: American Psychological Association.

Kelly, N. C., Ammerman, R. T., Rausch, J. R., Ris, M. D., Yeates, K. O., Oppenheimer, S. G., et al. (2012). Executive functioning and psychological adjustment in children and youth with spina bifida. *Child Neuropsychology, 18*(5), 417–431.

Kenworthy, L., Black, D. O., Harrison, B., della Rosa, A., & Wallace, G. L. (2009). Are executive control functions related to autism symptoms in high-functioning children? *Child Neuropsychology, 15*(5), 425–440.

Kenworthy, L. E., Black, D. O., Wallace, G. L., Ahluvalia, T., Wagner, A. E., & Sirian, L. M. (2005). Disorganization: The forgotten executive dysfunction in high-functioning autism (HFA) spectrum disorders. *Developmental Neuropsychology, 28*(3), 809–827.

Kenworthy, L., Yerys, B. E., Anthony, L. G., & Wallace, G. L. (2008). Understanding executive control in autism spectrum disorders in the lab and in the real world. *Neuropsychology Review, 18*(4), 320–338.

Kesler, S. R., Kent, J. S., & O'Hara, R. (2011). Prefrontal cortex and executive function impairments in primary breast cancer. *Archives of Neurology, 68*(11), 1447–1453.

Khundrakpam, B. S., Reid, A., Brauer, J., Carbonell, F., Lewis, J., Ameis, S., Karama, S., Lee, J., Chen, Z., Das, S., Evans, A. C., the Brain Development Cooperative Group. (2012). Developmental changes in organization of structural brain networks. Cerebral Cortex. doi: 10.1093/cercor/bhs187.

Kiley-Brabeck, K., & Sobin, C. (2006). Social skills and executive function deficits in children with the 22q11 Deletion Syndrome. *Applied Neuropsychology, 13*(4), 258–268.

Koven, N. S., & Thomas, W. (2010). Mapping facets of alexithymia to executive dysfunction in daily life. *Personality and Individual Differences, 49*, 24–28.

Kumbhani, S., Roth, R. M., Kuck, C. L., Flashman, L. A., & McAllister, T. W. (2010). Non-clinical obsessive compulsive symptoms and executive functions in schizophrenia. *The Journal of Neuropsychiatry and Clinical Neurosciences, 22*, 304–312.

Kurowski, B. G., Taylor, H. G., Yeates, K. O., Walz, N. C., Stancin, T., & Wade, S. L. (2011). Caregiver ratings of long-term executive dysfunction and attention problems after early childhood traumatic brain injury: Family functioning is important. *PM & R: The journal of injury, function, and rehabilitation, 3*(9), 836–845.

Lande, M. B., Adams, H., Falkner, B., Waldstein, S. R., Schwartz, G. J., Szilagyi, P. G., et al. (2009). Parental assessments of internalizing and externalizing behavior and executive function in children with primary hypertension. *Journal of Pediatrics, 154*(2), 207–212.

Lee, N. R., Fidler, D. J., Blakeley-Smith, A., Daunhauer, L., Robinson, C., & Hepburn, S. L. (2011). Caregiver report of executive functioning in a population-based sample of young children with Down syndrome. *American Journal on Intellectual and Developmental Disabilities, 116*(4), 290–304.

Lepage, J. F., Dunkin, B., Hong, D. S., & Reiss, A. L. (2013). Impact of cognitive profile on social functioning in prepubescent females with Turner syndrome. *Child Neuropsychology, 19*(2), 161–172.

Locascio, G., Mahone, E. M., Eason, S. H., & Cutting, L. E. (2010). Executive dysfunction among children with reading comprehension deficits. *Journal of Learning Disabilities, 43*(5), 441–454.

Lovstad, M., Funderud, I., Endestad, T., Due-Tonnessen, P., Meling, T. R., Lindgren, M., et al. (2012). Executive functions after orbital or lateral prefrontal lesions: Neuropsychological profiles and self-reported executive functions in everyday living. *Brain Injury, 26*, 1586–1598.

Lyons, K. E., & Zelazo, P. D. (2011). Monitoring, metacognition, and executive function: Elucidating the role of self-reflection in the development of self-regulation. *Advances in Child Development and Behavior, 40*, 379–412.

Mahone, E. M., Cirino, P. T., Cutting, L. E., Cerrone, P. M., Hagelthorn, K. M., Hiemenz, J. R., et al. (2002). Validity of the Behavior Rating Inventory of executive function in children with ADHD and/or Tourette syndrome. *Archives of Clinical Neuropsychology, 17*, 643–662.

Mahone, E. M., & Hoffman, J. (2007). Behavior ratings of executive function among preschoolers with ADHD. *Clinical Neuropsychology, 21*(4), 569–586.

Mahone, E. M., Martin, R., Kates, W. R., Hay, T., & Horska, A. (2009). Neuroimaging correlates of parent ratings of working memory in typically developing children. *Journal of the International Neuropsychological Society, 15*, 31–41.

Mahone, E. M., Zabel, T. A., Levey, E., Verda, M., & Kinsman, S. (2002). Parent and self-report ratings of executive function in adolescents with myelomeningocele and hydrocephalus. *Child Neuropsychology, 8*(4), 258–270.

Maillard-Wermelinger, A., Yeates, K. O., Gerry Taylor, H., Rusin, J., Bangert, B., Dietrich, A., et al. (2009). Mild traumatic brain injury and executive functions in school-aged children. *Developmental Neurorehabilitation, 12*(5), 330–341.

Mangeot, S., Armstrong, K., Colvin, A. N., Yeates, K. O., & Taylor, H. G. (2002). Long-term executive function deficits in children with traumatic brain injuries: Assessment using the Behavior Rating Inventory of Executive Function (BRIEF). *Child Neuropsychology, 8*(4), 271–284.

Mares, D., McLuckie, A., Schwartz, M., & Saini, M. (2007). Executive function impairments in children with attention-deficit hyperactivity disorder: Do they differ between school and home environments? *Canadian Journal of Psychiatry, 52*(8), 527–534.

Mateer, C. A. (1999). Executive function disorders: Rehabilitation challenges and strategies. *Seminars in Clinical Neuropsychiatry, 4*(1), 50–59.

Matheson, L. (2010). Executive dysfunction, severity of traumatic brain injury, and IQ in workers with disabilities. *Work, 36*(4), 413–422.

McAuley, T., Chen, S., Goos, L., Schachar, R., & Crosbie, J. (2010). Is the behavior rating inventory of executive function more strongly associated with measures of impairment or executive function? *Journal of the International Neuropsychological Society, 16*(3), 495–505.

McCandless, S., & O' Laughlin, L. (2007). The Clinical Utility of the Behavior Rating Inventory of Executive Function (BRIEF) in the diagnosis of ADHD. *Journal of Attention Disorders, 10*(4), 381–389.

McCauley, S. R., Wilde, E. A., Anderson, V. A., Bedell, G., Beers, S. R., Campbell, T. F., et al. (2012). Recommendations for the use of common outcome measures in pediatric traumatic brain injury research. *Journal of Neurotrauma, 29*(4), 678–705.

McCloskey, G., Perkins, L., & Divner, B. (2009). *Assessment and intervention for executive function difficulties*. New York: Routledge.

McNally, K., Rohan, J., Pendley, J. S., Delamater, A., & Drotar, D. (2010). Executive functioning, treatment adherence, and glycemic control in children with type 1 diabetes. *Diabetes Care, 33*(6), 1159–1162.

Merkley, T. L., Bigler, E. D., Wilde, E. A., McCauley, S. R., Hunter, J. V., & Levin, H. S. (2008). Diffuse changes in cortical thickness in pediatric moderate-to-severe traumatic brain injury. *Journal of Neurotrauma, 25*(11), 1343–1345.

Muscara, F., Catroppa, C., & Anderson, V. (2008). The impact of injury severity on executive function 7–10 years following pediatric traumatic brain injury. *Developmental Neuropsychology, 33*(5), 623–636.

Nadebaum, C., Anderson, V., & Catroppa, C. (2007). Executive function outcomes following traumatic brain injury in young children: A five year follow-up. *Developmental Neuropsychology, 32*(2), 703–728.

Niendam, T. A., Horwitz, J., Bearden, C. E., & Cannon, T. D. (2007). Ecological assessment of executive dysfunction in the psychosis prodrome: A pilot study. *Schizophrenia Research, 93*(1–3), 350–354.

Offord, D. R., Boyle, M. H., Racine, Y., Szatmari, P., Fleming, J. E., Sanford, M., et al. (1996). Integrating assessment data from multiple informants. *Journal of the American Academy of Child and Adolescent Psychiatry, 35*(8), 1078–1085.

Parrish, J., Geary, E., Jones, J., Seth, R., Hermann, B., & Seidenberg, M. (2007). Executive functioning in childhood epilepsy: Parent-report and cognitive assessment. *Developmental Medicine and Child Neurology, 49*(6), 412–416.

Payne, J. M., Hyman, S. L., Shores, E. A., & North, K. N. (2011). Assessment of executive function and attention in children with neurofibromatosis type 1: Relationships between cognitive measures and real-world behavior. *Child Neuropsychology, 17*(4), 313–329.

Pennington, B. F., & Ozonoff, S. (1996). Executive functions and developmental psychopathology. *Journal of Child Psychology and Psychiatry, and Allied Disciplines, 37*(1), 51–87.

Peters, B. D., Szeszko, P. R., Radua, J., Ikuta, T., Gruner, P., Derosse, P., et al. (2012). White matter development in adolescence: Diffusion tensor imaging and meta-analytic results. *Schizophrenia Bulletin, 38*(6), 1308–1317.

Power, B. D., Dragovic, M., & Rock, D. (2012). Brief screening for executive dysfunction in schizophrenia in a rehabilitation hospital. *The Journal of Neuropsychiatry and Clinical Neurosciences, 24*(2), 215–222.

Pulsipher, D. T., Seidenberg, M., Guidotti, L., Tuchscherer, V. N., Morton, J., Sheth, R. D., et al. (2009). Thalamofrontal circuitry and executive dysfunction in recent-onset juvenile myoclonic epilepsy. *Epilepsia, 50*(5), 1210–1219.

Qian, Y., Shuai, L., Cao, Q., Chan, R. C., & Wang, Y. (2010). Do executive function deficits differentiate between children with attention deficit hyperactivity disorder (ADHD) and ADHD comorbid with oppositional defiant disorder? A cross-cultural study using performance-based tests and the behavior rating inventory of executive function. *Clinical Neuropsychology, 24*(5), 793–810.

Rabin, L. A., Fogel, J., & Nutter-Upham, K. E. (2010). Academic procrastination in college students: The role of self-reported executive function. *Journal of Clinical and Experimental Neuropsychology, 33*(3), 344–357.

Rabin, L. A., Roth, R. M., Isquith, P. K., Wishart, H. A., Nutter-Upham, K. E., Pare, N., et al. (2006). Self and informant reports of executive function in mild cognitive impairment and older adults with cognitive complaints. *Archives of Clinical Neuropsychology, 21*, 721–732.

Reddy, L. A., Hale, J. B., & Brodzinsky, L. K. (2011). Discriminant validity of the Behavior Rating Inventory of Executive Function Parent Form for children with attention-deficit/hyperactivity disorder. *School Psychology Quarterly, 26*(1), 45–55.

Reid, R. C., Karim, R., McCrory, E., & Carpenter, B. N. (2010). Self-reported differences on measures of executive function and hypersexual behavior in a patient and community sample of men. *International Journal of Neuroscience, 120*, 120–127.

Reid, R. C., McKittrick, H. L., Davtian, M., & Fong, T. W. (2012). Self-reported differences on measures of executive function in a patient sample of pathological gamblers. *International Journal of Neuroscience, 122*(9), 500–505.

Reynolds, C. R., & Kamphaus, R. W. (1992). *The behavior assessment system for children*. Circle Pines, MN: American Guidance Services.

Rodrigue, J. R., Balistreri, W., Haber, B., Jonas, M. M., Mohan, P., Molleston, J. P., et al. (2011). Peginterferon with or without ribavirin has minimal effect on quality of life, behavioral/emotional, and cognitive outcomes in children. *Hepatology, 53*(5), 1468–1475.

Rotenberg-Shpigelman, S., Rapaport, R., Stern, A., & Hartmen-Maeir, A. (2008). Content validity and internal consistency reliability of the Behavior Rating Inventory of Executive Function—Adult Version (BRIEF-A) in Israeli adults with attention-deficit/hyperactivity disorder. *Israeli Journal of Occupational Therapy, 17*(2), 77–96.

Roth, R. M., Isquith, P. K., & Gioia, G. A. (2005). *Behavior Rating Inventory of Executive Function—Adult Version (BRIEF-A)*. Lutz, FL: Psychological Assessment Resources.

Roth, R. M., Isquith, P. K., & Gioia, G. A. (2006, February). *The duality of action monitoring in healthy adults*. Poster presented at the 34th Annual Meeting of the International Neuropsychological Society, Boston.

Roth, R. M., Lance, C. E., Isquith, P. K., Fischer, A.S. & Giancola, P. R. (2013). Confirmatory factor analysis of the Behavior Rating Inventory of Executive

Function-Adult Version in healthy adults and application to Attention-Deficit/Hyperactivity Disorder. *Archives of Clinical Neuropsychology*. doi: 10.1093/arclin/act031.

Roy, A., Bellinger, D., Hu, H., Schwartz, J., Ettinger, A. S., Wright, R. O., et al. (2009). Lead exposure and behavior among young children in Chennai, India. *Environmental Health Perspectives, 117*(10), 1607–1611.

Salmon, C., & Figueredo, A. J. (2009). Life history strategy and disordered eating behavior. *Evolutionary Psychology, 7*(4), 585–600.

Sbordone, R. J. (1996). Ecological validity: Some critical issues for the neuropsychologist. In R. J. Sbordone & C. J. Long (Eds.), *Ecological validity of neuropsychological testing* (pp. 91–112). Boca Raton, FL: St. Lucie Press.

Schroeder, V. M., & Kelley, M. L. (2008). The influence of family factors on the executive functioning of adult children of alcoholics in college. *Family Relations, 57*, 404–414.

Semrud-Clikeman, M., Walkowiak, J., Wilkinson, A., & Butcher, B. (2010). Executive functioning in children with Asperger syndrome, ADHD-combined type, ADHD-predominately inattentive type, and controls. *Journal of Autism and Developmental Disorders, 40*(8), 1017–1027.

Sesma, H. W., Slomine, B. S., Ding, R., McCarthy, M. L., & Children's Health After Trauma Study Group. (2008). Executive functioning in the first year after pediatric traumatic brain injury. *Pediatrics, 121*(6), e1686–e1695.

Shear, P. K., DelBello, M. P., Lee Rosenberg, H., & Strakowski, S. M. (2002). Parental reports of executive dysfunction in adolescents with bipolar disorder. *Child Neuropsychology, 8*(4), 285–295.

Sherman, E. M., Slick, D. J., & Eyrl, K. L. (2006). Executive dysfunction is a significant predictor of poor quality of life in children with epilepsy. *Epilepsia, 47*(11), 1936–1942.

Shimoni, M., Engel-Yeger, B., & Tirosh, E. (2012). Executive dysfunctions among boys with Attention Deficit Hyperactivity Disorder (ADHD): Performance-based test and parents report. *Research in Developmental Disabilities, 33*(3), 858–865.

Silver, C. H. (2000). Ecological validity of neuropsychological assessment in childhood traumatic brain injury. *The Journal of Head Trauma Rehabilitation, 15*, 973–988.

Slick, D. J., Lautzenhiser, A., Sherman, E. M., & Eyrl, K. (2006). Frequency of scale elevations and factor structure of the Behavior Rating Inventory of Executive Function (BRIEF) in children and adolescents with intractable epilepsy. *Child Neuropsychology, 12*(3), 181–189.

Sorensen, L. G., Neighbors, K., Martz, K., Zelko, F., Bucuvalas, J. C., & Alonso, E. M. (2011). Cognitive and academic outcomes after pediatric liver transplantation: Functional Outcomes Group (FOG) results. *American Journal of Transplantation, 11*(2), 303–311.

Stange, J. P., Eisner, L. R., Holzel, B. K., Peckham, A. D., Dougherty, D. D., Rauch, S. L., et al. (2011). Mindfulness-based cognitive therapy for bipolar disorder: Effects on cognitive functioning. *Journal of Psychiatric Practice, 17*(6), 410–419.

Strauss, E., Sherman, E. M. S., & Spreen, O. (2006). *A compendium of neuropsychological tests* (3rd ed.). New York: Oxford University Press.

Sullivan, J. R., & Riccio, C. A. (2007). Diagnostic group differences in parent and teacher ratings on the BRIEF and Conners' Scales. *Journal of Attention Disorders, 11*(3), 398–406.

Surber, C. F. (1984). Issues in using quantitative rating scales in developmental research. *Psychological Bulletin, 95*, 226–246.

Tarazi, R. A., Zabel, T. A., & Mahone, E. M. (2008). Age-related differences in executive function among children with spina bifida/hydrocephalus based on parent behavior ratings. *The Clinical Neuropsychologist, 22*(4), 585–602.

Tervo, R. C., Symons, F., Stout, J., & Novacheck, T. (2006). Parental report of pain and associated limitations in ambulatory children with cerebral palsy. *Archives of Physical Medicine and Rehabilitation, 87*(7), 928–934.

Till, C., Ho, C., Dudani, A., Garcia-Lorenzo, D., Collins, D. L., & Banwell, B. L. (2012). Magnetic resonance imaging predictors of executive functioning in patients with pediatric-onset multiple sclerosis. *Archives of Clinical Neuropsychology, 27*(5), 495–509.

Tobon, O. E. A., Puerta, I. C., & Pineda, D. A. (2008). Factorial structure of the executive function from the behavioral domain. *Perspectivas En Psicologia, 4*(1), 63–77.

Toglia, J., Johnston, M. V., Goverover, Y., & Dain, B. (2010). A multicontext approach to promoting transfer of strategy use and self regulation after brain injury: An exploratory study. *Brain Injury, 24*, 664–677.

Toplak, M. E., Bucciarelli, S. M., Jain, U., & Tannock, R. (2009). Executive functions: Performance-based measures and the behavior rating inventory of executive function (BRIEF) in adolescents with attention deficit/hyperactivity disorder (ADHD). *Child Neuropsychology, 15*(1), 53–72.

Turgay, A., Ginsberg, L., Sarkis, E., Jain, R., Adeyi, B., Gao, J., et al. (2010). Executive function deficits in children with attention-deficit/hyperactivity disorder and improvement with lisdexamfetamine dimesylate in an open-label study. *Journal of Child and Adolescent Psychopharmacology, 20*(6), 503–511.

Varni, J. W., Limbers, C. A., Sorensen, L. G., Neighbors, K., Martz, K., Bucuvalas, J. C., et al. (2011). PedsQL Cognitive Functioning Scale in pediatric liver transplant recipients: Feasibility, reliability, and validity. *Quality of Life Research, 20*(6), 913–921.

Velligan, D. I., Bow-Thomas, C. C., Huntzinger, C., Ritch, J., Ledbetter, N., Prihoda, T. J., et al. (2000). Randomized controlled trial of the use of compensatory strategies to enhance adaptive functioning in outpatients with schizophrenia. *The American Journal of Psychiatry, 157*(8), 1317–1323.

Verkerk, G., Jeukens-Visser, M., Houtzager, B., Koldewijn, K., van Wassenaer, A., Nollet, F., et al. (2012). The infant behavioral assessment and intervention program in very low birth weight infants: Outcome on executive functioning, behaviour and cognition at preschool age. *Early Human Development, 88*(8), 699–705.

Vriezen, E. R., & Pigott, S. E. (2002). The relationship between parental report on the BRIEF and performance-based measures of executive function in children with moderate to severe traumatic brain injury. *Child Neuropsychology, 8*, 296–303.

Waber, D. P., De Moor, C., Forbes, P. W., Almli, C. R., Botteron, K. N., Leonard, G., et al. (2007). The NIH MRI study of normal brain development: Performance of a population based sample of healthy children aged 6 to 18 years on a neuropsychological battery. *Journal of the International Neuropsychological Society, 13*(5), 729–746.

Waber, D. P., Forbes, P. W., Almli, C. R., & Blood, E. A. (2012). Four-year longitudinal performance of a population-based sample of healthy children on a neuropsychological battery: The NIH MRI study of normal brain development. *Journal of the International Neuropsychological Society, 18*(2), 179–190.

Waber, D. P., Gerber, E. B., Turcios, V. Y., Wagner, E. R., & Forbes, P. W. (2006). Executive functions and performance on high-stakes testing in children from urban schools. *Developmental Neuropsychology, 29*(3), 459–477.

Waisbren, S., & White, D. A. (2010). Screening for cognitive and social-emotional problems in individuals with PKU: Tools for use in the metabolic clinic. *Molecular Genetics and Metabolism, 99*(Suppl 1), S96–S99.

Walker, J. M., & D'Amato, R. C. (2006). Test review: Behavior Rating Inventory of Executive Function-Self-Report version. *Journal of Psychoeducational Assessment, 24*, 394–398.

Wells, A. M., Chasnoff, I. J., Schmidt, C. A., Telford, E., & Schwartz, L. D. (2012). Neurocognitive habilitation therapy for children with fetal alcohol spectrum disorders: An adaptation of the Alert Program(R). *American Journal of Occupational Therapy, 66*(1), 24–34.

Wexler, B. E., & Bell, M. D. (2005). Cognitive remediation and vocational rehabilitation for schizophrenia. *Schizophrenia Bulletin, 31*(4), 931–941.

Wilde, E. A., Merkley, T. L., Bigler, E. D., Max, J. E., Schmidt, A. T., Ayoub, K. W., et al. (2012). Longitudinal changes in cortical thickness in children after traumatic brain injury and their relation to behavioral regulation and emotional control. *International Journal of Developmental Neuroscience, 30*(3), 267–276.

Wilson, B. A., Alderman, N., Burgess, P., Emslie, H., & Evans, J. J. (1996). *Behavioral assessment of the dysexecutive syndrome*. Bury St. Edmunds, Suffolk: The Thames Valley Test.

Wilson, K. R., Donders, J., & Nguyen, L. (2011). Self and parent ratings of executive functioning after adolescent traumatic brain injury. *Rehabilitation Psychology, 56*(2), 100–106.

Wozniak, J. R., Krach, L., Ward, E., Mueller, B. A., Muetzel, R., Schnoebelen, S., et al. (2007). Neurocognitive and neuroimaging correlates of pediatric traumatic brain injury: A diffusion tensor imaging (DTI) study. *Archives of Clinical Neuropsychology, 22*(5), 555–568.

Yang, L., Cao, Q., Shuai, L., Li H., Chan, R .C. K., Wang, Y. (2012). Comparative study of OROS-MPH and atomoxetine on executive function improvement in ADHD: A randomized controlled trial. *International Journal of Neuropsychopharmacology, 15*, 15–26.

Yerys, B. E., Wallace, G. L., Sokoloff, J. L., Shook, D. A., James, J. D., & Kenworthy, L. (2009). Attention deficit/hyperactivity disorder symptoms moderate cognition and behavior in children with autism spectrum disorders. *Autism Research, 2*(6), 322–333.

Ylvisaker, M., & Feeney, T. J. (1996). Executive functions after traumatic brain injury: Supported cognition and self-advocacy. *Seminars in Speech and Language, 17*(3), 217–232.

Ylvisaker, M., & Feeney, T. (1998). *Collaborative brain injury intervention: Positive everyday routines*. San Diego, CA: Singular.

Ylvisaker, M., Hanks, R., & Johnson-Greene, D. (2002). Perspectives on rehabilitation of individuals with cognitive impairment after brain injury: Rationale for reconsideration of theoretical paradigms. *The Journal of Head Trauma Rehabilitation, 17*(3), 191–209.

Ylvisaker, M., Jacobs, H. E., & Feeney, T. (2003). Positive supports for people who experience behavioral and cognitive disability after brain injury: A review. *The Journal of Head Trauma Rehabilitation, 18*(1), 7–32.

Ylvisaker, M., Szekeres, S., & Feeney, T. (1998). Cognitive Rehabilitation: Executive Functions. In M. Ylvisaker (Ed.), *Traumatic brain injury rehabilitation: Children and adolescents* (pp. 221–269). Boston: Butterworth-Heinemann.

Zabel, T. A., Jacobson, L. A., Zachik, C., Levey, E., Kinsman, S., & Mahone, E. M. (2011). Parent- and self-ratings of executive functions in adolescents and young adults with spina bifida. *The Clinical Neuropsychologist, 25*(6), 926–941.

Zelazo, P. D., Qu, L., & Muller, U. (2004). Hot and cool aspects of executive function: Relations in early development. In W. Schneider, R. Schumann-Hengsteler, & B. Sodian (Eds.), *Young children's cognitive development: Interrelationships among executive functioning, working memory, verbal ability, and theory of mind* (pp. 71–93). Mahwah, NJ: Lawrence Erlbaum.

Assessment of Executive Functioning Using Tasks of Executive Control

Peter K. Isquith, Robert M. Roth, and Gerard A. Gioia

Introduction

The Tasks of Executive Control (TEC; Isquith, Roth, & Gioia, 2010) is a standardized computer-administered measure of two fundamental aspects of executive control processes or functions, namely, working memory and inhibitory control. It is designed for use with children and adolescents between the ages of 5 and 18 years, including those with a wide variety of developmental and acquired neurological disorders including attention disorders, learning disabilities, autism spectrum disorders, and traumatic brain injuries, as well as those with psychiatric and other behavioral health concerns. Published in 2010, it represents a novel contribution to clinical assessment of executive function by adapting, integrating, and standardizing two of the most commonly used neuroscience methods designed to tap working memory and inhibitory control: an *n*-back paradigm that parametrically increases working memory load and a go/no-go task to manipulate inhibitory control demand.

P.K. Isquith (✉) • R.M. Roth
Geisel School of Medicine at Dartmouth School,
Lebanon, NH, USA
e-mail: Peter.Isquith@Dartmouth.edu

G.A. Gioia
Children's National Medical Center, George
Washington University School of Medicine,
Washington, DC, USA

The combination of three levels of working memory load (0-, 1-, 2-back) and two levels of inhibitory control demand (no inhibit, inhibit) yields four sequential tasks for 5–7-year-old children and six tasks for 8–18-year-old children and adolescents. The TEC calculates multiple performance accuracy (e.g., omissions, commissions), response time, and response time variability scores within each task condition, across the instrument as a whole, and over time as demands increase. Three equivalent forms and two additional research forms are included, along with reliable change scores that facilitate interpretation of performance changes on repeated assessments. In addition to scores for each task and for the measure as a whole, the TEC captures accuracy and timing variables for each individual response and by quartile within each task. All data may be exported to spreadsheet and statistical software with labels for detailed analysis and research.

History and Rationale for the TEC

The TEC was conceived as a means of attempting to bridge the gap between functional neuroimaging research on executive functions, standard performance-based measures (i.e., "tests") of executive function, and the real-world implementation of executive function in the everyday lives of individuals. The authors have spent their careers evaluating, studying, and teaching about executive functions across the developmental spectrum and were familiar with the strengths and weaknesses of the extant assessment toolkit. While that toolkit includes a range of "tried and true" measures, they are not consistently sensitive or specific to disruptions in underlying neurological bases of executive function, to diagnostic group differences, or to functioning in the everyday real-world lives of individuals, particularly children and adolescents (Welsh & Pennington, 1988; Welsh, Pennington, & Grossier, 1991). In the first review of the then limited literature on executive functions in clinical groups of children and adolescents, inconsistencies in how these "putative" executive function measures relate to underlying neurology, particularly of the prefrontal cortex (PFC) thought to be central to executive functions and to behavior in the everyday context, have long been noted (Pennington & Ozonoff, 1996). In the first meta-analytic review of children's executive function test performance in diagnostic groups, Pennington and Ozonoff (1996) proposed that the measurement properties of many executive function tasks contribute to a problem of discriminant validity or why some children with known injury to frontal systems did well on executive function tests (sensitivity), while some children with no injury or injury to non-frontal systems did poorly on such tests (specificity). They offered that executive function tasks are complex "molar" tasks that assess many interacting component processes, both executive and nonexecutive, and not a single executive function per se. As such, they can be disrupted by multiple means and in multiple ways. For example, children may do poorly on a verbal fluency task because of a language deficit rather than an executive function deficit or might do well on a verbal fluency task despite an executive function deficit but because of a strong vocabulary. Similarly, a very bright child might "over-think" a card-sorting task and do poorly for nonexecutive reasons, while a more limited child might do well on the same task despite having executive function deficits in the everyday real-world context. Pennington and Ozonoff concluded that more discrete, or less molar, executive function tasks were necessary to confirm that there may be executive profile differences between disorders and, further, that paradigms that use *within-task* manipulations were necessary to deal with the theoretical issue of how working memory and inhibition are related and how they interact.

The need for less "molar" and more uniquely executive tasks that manipulate executive demand *within* a single task rather than *between* different tasks has been echoed as essential toward developing a better understanding of the relationships between different executive functions, those functions and clinical presentations, and those functions and everyday behaviors. In their review of 83 studies using executive function measures to compare over 2,900 typically developing controls with over 3,700 children and adolescents diagnosed with ADHD, arguably the most clear diagnosis of a developmental disorder of executive functions (e.g., Barkley, 1997; Castellanos & Tannock, 2002), Willcutt, Doyle, Nigg, Faraone, and Pennington (2005) found that less than half of all children diagnosed with ADHD exhibited impairment on any such measure and that correlations between ADHD symptoms and executive function tests were significant but smaller than expected (Willcutt et al., 2001). At the same time, the meta-analysis found the strongest and most consistent effects for measures of response inhibition, vigilance, and working memory. These authors pointed again to the need to examine the measurement characteristics of traditional, "molar" measures of executive function including weaknesses in the construct for which tasks are named (e.g., is a test of "planning" truly tapping planning or something else or a multitude of functions) and to develop appropriate *within-task* and *between-task* controls (Willcutt et al., 2005).

Given the charge to develop measures that are less molar and that more directly reflect specific executive functions, the TEC authors first reviewed models of executive function in search of agreement on fundamental functions. While theoretical models vary between authors (Jurado & Rosselli, 2007; Lezak, 1995; Stuss & Benson, 1984), most share a common fundamental position, that tasks or behaviors sensitive to "frontal" dysfunction (i.e., executive function deficits) require holding plans or programs on line in *working memory* and *inhibiting* irrelevant actions. As Pennington and Ozonoff presaged, "tasks that are high in both their working memory and inhibition demands are more likely to tax the PFC, although tasks that have a very high demand for either are also hypothesized to be prefrontal tasks" (Pennington & Ozonoff, 1996, p. 56).

Inhibitory control involves the ability to withhold or defer responding to stimuli whether they are internal to the person, such as task-irrelevant impulses, or external such as distractions in the environment. This is commonly assessed with tasks that require the suppression of a prepotent or habitual response or tasks that require resistance to interference from distractions during performance on a primary task. Inhibitory control has been reported to be an important contributor to basic cognitive, emotional, social, and adaptive functions (Anderson, Barrash, Bechara, & Tranel, 2006; Clark, Prior, & Kinsella, 2002; Dennis, 1989; Gilotty, Kenworthy, Sirian, Black, & Wagner, 2002; Jarratt, Riccio, & Siekierski, 2005; Levin & Hanten, 2005; Tarazi, Mahone, & Zabel, 2007; Wozniak, Block, White, Jensen, & Schulz, 2008) as well as academic achievement (St Clair-Thompson & Gathercole, 2006).

Working memory has been defined as "a limited capacity system allowing the temporary storage and manipulation of information needed for such complex tasks as comprehension, learning and reasoning" (Baddeley, 2000). It may also be thought of as the capacity to actively hold information in mind (or "online") for the purpose of completing a task or generating a response. It is considered by many to be a central aspect of executive function (Baddeley, 2003; Goldman-Rakic, 1987), and extensive empirical evidence supports its salience for social and emotional functioning (Anderson & Catroppa, 2005; Clark et al., 2002; Jarratt et al., 2005), adaptive behavior (Gilotty et al., 2002; Tarazi et al., 2007; Wozniak et al., 2008), and academic achievement (Alloway & Alloway, 2010; Andersson, 2008; Camos, 2008; Espy et al., 2004; Gathercole, Alloway, Willis, & Adams, 2006; Gropper & Tannock, 2009; St Clair-Thompson & Gathercole, 2006).

With at least some degree of agreement as to the fundamental executive functions, we set about finding non-molar tasks that would more uniquely tap working memory and inhibitory control in isolation. After reviewing the extant toolkit of executive function performance tasks, we turned to the neuroscience and neuroimaging literatures to find tasks that have known relationships with brain function. This review yielded two such tasks that are commonly employed in cognitive neuroscience research and in functional neuroimaging research to investigate the neural bases of working memory and inhibitory control: the *n*-back for working memory and the go/no-go for inhibitory control.

Measurement of Working Memory

There are numerous paradigms employed in the measurement of working memory. In the pediatric literature, span tasks are predominant (Alloway, Gathercole, Willis, & Adams, 2004; Karatekin & Asarnow, 1998; Wang, Deater-Deckard, Cutting, Thompson, & Petrill, 2012; Wechsler, 2003). These tasks require holding information such as a series of numbers or letters in immediate memory, manipulating the information in working memory such as reversing or reordering the information, then repeating them. An important consideration in the design of the TEC, however, was the desire to challenge children's working memory over a more naturalistic time frame, such as what one would see in a classroom setting. This led to consideration of the *n*-back paradigm, which typically requires deployment of working memory capacities continuously over several minutes. The *n*-back is widely used to evaluate working memory in neuroimaging studies of children and adults (Chen et al., 2012; Owen, McMillan, Laird, & Bullmore, 2005; Stollstorff et al., 2010). In this paradigm, a series of stimuli are presented, one at a time (e.g., objects, letters), and the child is asked to press a particular button each time he or she sees a frequent, or *standard*, stimulus but a different button when he or she sees an infrequent, or *target*, stimulus. The level of difficulty varies within the task. In the 0-back condition, the target stimulus is specified at the outset (e.g., the image of a

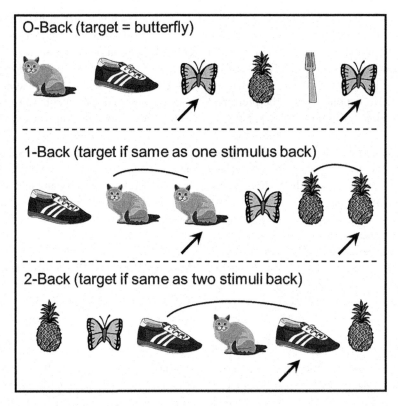

Fig. 19.1 Illustration of the TEC *n*-back paradigm. N.B., *Arrow* points to Target stimuli in each task. All other stimuli are considered Standard stimuli

butterfly), and the individual presses the target button each time a butterfly appears but presses the standard button in response to any other image. The 0-back serves as a low-level vigilance or sustained attention condition with minimal working memory demand, only requiring that the examinee remembers and detects a particular stimulus that appears periodically. In the 1-back condition, a series of stimuli are presented as in the 0-back, but no single target stimulus is specified. Instead, the child presses the target button when a stimulus matches the immediately preceding, or 1-back, stimulus, such as a picture of a cat following the same picture of a cat. Thus, the 1-back task requires actively holding each picture in working memory until the next picture is presented in order to perform a match in the context of a continuous, ongoing stream of information (e.g., pictures). The 2-back condition is similar to the 1-back, but the child presses the target button only when a stimulus matches the one presented *two* stimuli before it, or 2-back (e.g., a shoe following a cat following a shoe), while pressing the standard button for all other pictures. In this way, working memory load is increased in stepwise or parametric fashion from 0- to 1- to 2-back. The *n*-back used in the TEC employs pictured objects as stimuli and is illustrated in Fig. 19.1.

Performance accuracy and speed on the *n*-back task increases with age (Vuontela et al., 2003), and poorer performance is associated with more academic, attention, and behavioral difficulties at school in children and adolescents (Aronen, Vuontela, Steenari, Salmi, & Carlson, 2005). Poorer working memory on *n*-back tasks also is associated with difficulties recalling the context of discourse and summarization (Chapman et al., 2006). Sleep quality and quantity affect 2-back performance (Steenari et al., 2003). Clinical studies show that severity of TBI is associated with poorer accuracy on an *n*-back task (Levin, Hanten, Zhang, Swank, Ewing-Cobbs, et al., 2004), an effect that interacts with working memory load (Levin et al., 2002). Children diagnosed with ADHD showed poor response accuracy and response time on a 2-back task (Shallice et al., 2002) and greater response time variability (Klein, Wendling, Huettner, Ruder, & Peper, 2006).

In addition to clinical group associations, performance on *n*-back tasks is associated with activation and functioning in frontal and other brain regions. A recent meta-analysis of 24 studies of adult *n*-back performance found strong effects for several frontal lobe regions including the lateral and medial premotor cortex, dorsolateral and ventrolateral PFC, dorsal cingulate gyrus, and frontal poles, and the medial and lateral posterior parietal cortex (Owen et al., 2005). Frontal-parietal regions in particular appear sensitive to working memory load (Owen et al., 2005). Children with fetal alcohol exposure showed reduced activation with increasing working memory load on the *n*-back task, an effect opposite to that seen in typically developing children (Malisza et al., 2005). Children with TBI who performed an *n*-back task equally as well as their typically developing peers showed greater activation in both frontal and non-frontal brain regions than did their peers (Newsome et al., 2007). In contrast, children with TBI who had impaired *n*-back task performance showed reduced brain activation relative to typically developing children. Another study reported that children with mild TBI did not show the expected increase in brain activation with increasing working memory load (Chen et al., 2012). It should be noted that brain regions involved in working memory in general, and invoked during performance of *n*-back tasks specifically, are not restricted to the frontal and parietal lobes but may be in other cortical, cerebellar, and subcortical brain regions (Gazzaley, Rissman, & Desposito, 2004; Owen et al., 2005; Wager & Smith, 2003).

Measurement of Inhibitory Control

The go/no-go task is a relatively simple measure of inhibitory control that presents *go* and *no-go* stimuli serially over time. "Go" stimuli require a rapid motor response such as pressing a button on the keyboard to establish a prepotent or automatic response. In contrast, responding must be withheld, or inhibited, when the "no-go" stimulus appears. In the TEC, the no-go cue is a box surrounding any pictured object—examinees are instructed to *not* respond when the box is present. In contrast, any stimulus not surrounded by a box is a go stimulus. Figure 19.2 shows an example of the no-go cue as implemented in the TEC.

The go/no-go task has a long history of use in neuropsychology and behavioral neurology (Luria, 1981) and has been shown to be sensitive to the development of inhibitory control in numerous studies of typically developing children and sensitive to inhibitory deficits in clinical populations. A meta-analysis of 83 studies comparing more than 3,700 children diagnosed with ADHD with their typically developing peers found the strongest and most consistent effects for response inhibition tasks (Willcutt et al., 2005). Go/no-go task performance has been widely studied in children with attentional disorders (Tamm, Menon, Ringel, & Reiss, 2004), children with ADHD and co-occurring conduct disorder (Van der Meere, Marzocchi, & De Meo, 2005), children with conduct disorder and co-occurring borderline intellectual functioning (van der Meer & van der Meere, 2004), and children with TBI (Levin, Hanten, Zhang, Swank, & Hunter, 2004).

The go/no-go task has been used extensively to investigate neural circuitry associated with inhibitory control via a variety of techniques in adults including event-related potentials (Johnstone et al., 2007; Malloy, Rasmussen, Braden, & Haier, 1989), near-infrared spectroscopy (Herrmann, Plichta, Ehlis, & Fallgatter, 2005), and functional neuroimaging (Altshuler et al., 2005; Roth et al., 2007; Tamm et al., 2004). fMRI studies of healthy adults have shown prominent activation of frontal lobe subregions, including the inferior and orbital frontal cortices, especially in the right hemisphere, during response inhibition to no-go stimuli (Altshuler et al., 2005; Kelly et al., 2004; Konishi, Nakajima, Uchida, Sekihara, & Miyashita, 1998; Roth et al., 2007). Several event-related potential

Fig. 19.2 Illustration of the TEC *n*-back inhibit cue (i.e., box)

studies have used the go/no-go task to examine both typical and atypical development of inhibitory control in children and adolescents (Baving, Rellum, Laucht, & Schmidt, 2004; Broyd et al., 2005; Fallgatter et al., 2004; Johnstone et al., 2007; Jonkman, Lansbergen, & Stauder, 2003; Smith, Johnstone, & Barry, 2004). Developmental fMRI studies have shown that activation in frontal-striatal-thalamic circuitry during inhibitory control changes with age from childhood through adolescence and into adulthood (Marsh et al., 2006; Rubia et al., 2006; Stevens, Kiehl, Pearlson, & Calhoun, 2007). Adolescents with ADHD were reported to show reduced activation in the frontal lobes and excessive activation in temporal lobes during a go/no-go task relative to typically developing adolescents (Tamm et al., 2004). Treatment with a stimulant medication has been found to be associated with more typical brain activation during go/no-go performance in children with ADHD (Pliszka et al., 2007; Vaidya et al., 1998). The neural circuitry subserving inhibitory control during go/no-go and other response inhibition tasks, however, also involves brain regions outside of the frontal lobes (Aron, 2011; Garavan, Ross, & Stein, 1999; Roth, Randolph, Koven, & Isquith, 2006), including certain white matter pathways (Casey et al., 2007).

With a firm grounding in cognitive neuroscience, the *n*-back task has been shown to be useful in assessing working memory in children, adolescents, and adults, while the go/no-go task is one of the most extensively studied measures of executive function, with a strong research base in neuropsychology and other areas of neuroscience. It is important to appreciate, however, that although abnormal performance on the TEC reflects problems with working memory and inhibitory control, scores cannot be interpreted directly as indicating abnormality in specific brain regions.

Description of the TEC

The TEC introduces a novel assessment approach to the existing repertoire of executive function performance measures. It is a unique combination of an *n*-back task to parametrically increase working memory load and a go/no-go task to manipulate inhibitory demand, both tasks with well-established credentials in neuropsychology, neuroscience, and neuroimaging research. Despite their common use in research, strength of association with underlying neural substrates, sensitivity to developmental changes, and differences between typically developing individuals and clinical groups, standardized *n*-back, and go/no-go tasks are not readily found in the clinician's toolkit. The TEC's computer administration offers both ease and standardization of administration with highly accurate timing, while responses are continuously recorded over a sustained period of time through standard computer keyboard button presses with no additional equipment required. The TEC offers multiple equivalent forms along with reliable change scores to enable measurement of change over time in cases such as recovery from TBI or to evaluate medication effects.

Subtest Background and Structure

The TEC was designed with four tasks for 5–7-year-old children and six tasks for children 8–18 years. The tasks are administered sequentially and parametrically increase working memory load while alternating demand for inhibitory control. The first task, the 0-back/no inhibit (0B), is a simple choice reaction time that requires responding to an infrequent (i.e., 20 % of total stimuli during the task) *Target* stimulus (e.g., butterfly) with the pressing of a Red button (i.e., left Shift key) and to respond to the remaining 80 % *Standard* stimuli with the pressing of a Blue button (i.e., right Shift key). Choice reaction tasks are sensitive to even subtle cognitive difficulties in pediatric clinical populations (Mitchell, Zhou, Chavez, & Guzman, 1992; Murray, Shum, & McFarland, 1992; Schuerholz, Baumgardner, Singer, Reiss, & Denckla, 1996) and, in the TEC, provides a baseline against which to gauge the impact of increasing working memory load and demand for inhibitory control. The second task that is administered, the 0-back/inhibit (0BI) task, adds a no-go cue and provides a baseline measure of inhibitory control. The third (1-back/ no inhibit; 1B) and fourth (1-back/inhibit; 1BI) tasks increase working memory load, and the fifth (2-back/no inhibit; 2B) and sixth (2-back/ inhibit; 2BI) tasks further increase the load. Thus, the first, third, and fifth tasks increase working memory load in a stepwise fashion without any demand for inhibitory control, and the second, fourth, and sixth tasks add the no-go cue to add inhibitory demand to the low-, medium-, and high-working memory levels. This design allows the user to gauge how a child responds to increasing working memory load with and without added demand for inhibitory control, as well as whether including a demand for inhibitory control affects performance differentially at different working memory loads.

Each of the six TEC tasks presents 100 stimuli in pseudorandom order with a constant rate of Target stimuli across tasks. The no inhibit tasks include 20 Target stimuli and 80 Standard stimuli, whereas inhibit tasks include 20 Target stimuli, 64 Standard stimuli, and 16 stimuli with an inhibit cue. Each task is approximately 3 min and 21 s long. Picture stimuli were chosen based on characteristics with known impact on naming speed (Berman, Friedman, Hamberger, & Snodgrass, 1989; Cycowicz, Friedman, & Rothstein, 1997; Lachman & Lachman, 1980). In order to limit verbal demand, 13 representational objects that are familiar to young children, reflect common semantic categories (e.g., clothes, fruit, animals), are low in visual complexity, and that have high-frequency verbal labels with minimal cultural bias were chosen for each of five TEC forms. Objects were selected based on equivalence for word frequency (Carroll, Davies, & Richman, 1971) and naming agreement and familiarity (Cycowicz et al., 1997) with no significant differences between the five test forms. One of the 13 stimuli from each form was selected to be the Target stimulus for the two 0-back tasks then eliminated for the subsequent 1-back and 2-back tasks to reduce confusion. Most pictures are 2.7–2.8 in. wide though the box (no-go cue) drawn around the inhibit items can make them as wide as 4.2 in. The time from one stimulus onset to the next stimulus onset is pseudorandomized between 1,750, 2,000, and 2,250 ms to maintain a mean of 2,000 ms for each task quartile. Objects are presented for 400 ms in the center of the computer monitor, followed by a blank screen for the remainder of the interval. Pilot testing with computers used for standardization with a video refresh rate of 60 Hz revealed a maximum error in timing of stimulus duration of approximately 1.5–5 ms, with less than .002 % of the cases having an error exceeding 1 ms.

Responses are recorded on the TEC through "Red button" responses to Target stimuli with the left index finger (the left Shift key with an enclosed red sticker) and "Blue button" responses to Standard stimuli with the right index finger (the right Shift key with an enclosed blue sticker). Handedness accounts for minimal variance in scores. Response speed is generally measured within 1 ms of accuracy or better from the time the data have entered the computer, though it is affected by the physical characteristics of the keyboard and the speed of the internal processor of the keyboard. Keyboards themselves may impose delays of typically around 7 ms, with the most extreme delays being approximately 15 ms. The timing parameters of the TEC are well within the range of similar computerized neuropsychological instruments.

The TEC includes five alternate forms that differ only in the 13 picture stimuli and the character (i.e., illustration of a boy or girl) used in the instructions for children ages 5–12 years. Forms 1, 2, and 3 are statistically equivalent and can be used interchangeably in clinical practice. Two additional forms were developed for research purposes and do not have normative data, and statistical equivalence has not been established.

Outcome Measures

The TEC can be viewed as an N of 1 experiment where working memory load and inhibitory demand are systematically manipulated independent variables, while the participant makes nearly continuous responses that are the dependent variables. The test captures accuracy and speed data on each Target (Red button) and Standard (Blue button) response. These are summarized and grouped into accuracy measures including Target Correct, Standard Correct, Target Omissions, Standard Omissions, Commissions, and Incorrect responses and grouped into response time measures including response time (RT), response time standard deviation (RTSD), and the intraindividual coefficient of variation (ICV; i.e., the RTSD divided by the RT). The primary measure of accuracy is the number of correct responses made to either Target or Standard stimuli (i.e., Target Correct, Standard Correct). Omissions are the number of Target or Standard stimuli for which the examinee did not respond. They occur very infrequently in the standardization sample and are not normally distributed, thus are not standardized but reported as raw scores, with an indication when the number of omissions exceeds the 95th percentile relative to the standardization sample. Thus, interpretive emphasis is placed on Correct scores rather than on Omissions. The Commissions score reflects the number of times any button is pressed in the presence of an inhibit (i.e., no-go) cue. Only those tasks that include inhibit cues produce commission errors (i.e., 0BI, 1BI, 2BI). Incorrect responses refer to pressing the wrong button to a Target or Standard stimulus (e.g., pressing the Standard button when a Target stimulus appears).

Response Time refers to the average time in milliseconds taken to respond correctly to each stimuli and is presented separately for Target and Standard stimuli. RTSD is the standard deviation of the Response Time and reflects the degree of variability in RT. It is provided only for Standard stimuli. Finally, the ICV, which is provided only for Standard stimuli, is calculated as the RTSD divided by RT. This provides a measure of the consistency of RT within a task that takes into account the individual's average RT for that task. This can be especially relevant for individuals with long average RTs because longer RTs can be associated with greater variability (Salthouse, 1993; Stuss, Murphy, Binns, & Alexander, 2003).

Scores on the TEC may be examined at multiple levels of interpretation. Scores may be considered at the *Task* level, that is, evaluating performance within each task (e.g., 0B, 1B, or 2B), as well as determining whether a given score changes in a significant manner as working memory load increases (e.g., using the regression-based change scores). One may also consider scores at the *Summary* score level. Summary scores provide a more general level of interpretation, gauging an individual's performance across the TEC as a whole, for example, looking at the T score for Correct Target responses across all six tasks. The Summary score does not, however, provide information with respect to the impact of working memory load. Finally, Factor scores are available. These are multimodal composites across tasks and are more difficult to interpret than either Summary or Task scores.

For all scores (except omissions), T scores with a mean of 50 and standard deviation of 10 are used to compare an individual's performance to normative expectation. Higher T scores reflect greater difficulty on the TEC, with a T score of 60 or greater considered clinically meaningful.

Factor Structure

While the unique contribution of the TEC is in comparison of performance across tasks as working memory load and inhibitory demand increase as reflected in Task scores over time, Summary scores can provide an easily interpreted snapshot of performance as a whole. The TEC also offers Factor scores based on factor analysis of data for typically developing children and adolescents. These factors are more challenging to interpret as some are composites that cut across outcome measures, for example, combining correct responses with commissions. Nonetheless, they may be useful as global performance indicators.

A separate factor structure was found for younger and for older children. Three factors emerged for 5–7-year-olds: Response Control, Selective Attention, and Response Speed. Response Control reflects the ability to maintain consistent, accurate, and non-impulsive responding, combining response time variability and accuracy in responding to Standard stimuli. A lower score reflects more consistent (i.e., less variable) and more accurate responding to the frequent stimuli. Selective Attention is defined by a combination of Target Correct (i.e., selective responses), Commissions (i.e., inhibitory control), and Response Time for the complementary Standard stimuli. This complicated score reflects coordination in selecting the less frequent Target stimuli by requiring the respondent to apply the n-back rules that are held in working memory while inhibiting the more frequent default Standard response and maintaining controlled speed. The score increases with slower, more controlled performance on the higher demand 1B and 1BI tasks but with faster speed on the lower demand 0B task. Thus, this score reflects coordinated and controlled selective attention. Finally, Response Speed is defined by the eight RT variables for correct responses to Standard and Target stimuli across the four tasks. Interpretation of this factor is straightforward: a faster response speed corresponds to a lower score (i.e., a better performance).

For the older age group, 8–18 years of age, four factors define the underlying structure of the TEC across the six tasks. The factors are labeled Sustained Accuracy, Selective Attention, Response Speed, and Response Variability. Sustained Accuracy is defined by five Standard accuracy measures and two Standard response time measures. Thus, better performance (i.e., lower T scores) on this factor reflects greater accuracy in responding to the high-frequency Standard stimuli for most of the test along with modulation of speed of response. Similar to that of the younger age group, Selective Attention is a composite that is defined by a combination of Target Correct (i.e., selective responses), Commissions (i.e., inhibitory control), and RT for the more challenging (e.g., 2-back) Standard stimuli. This score reflects cognitive/response coordination in selecting the less frequent Target stimuli by requiring the respondent to apply the n-back rules that are held in working memory while inhibiting the more frequent default Standard response and maintaining speed and efficiency in responding. The contribution of RT to this factor score reflects the advantage of faster speed on the more demanding tasks (e.g., 1BI, 2B). Response Speed is straightforward and defined by 11 of the 12 RT variables for correct responses to Standard and Target stimuli across the six tasks. Finally, Response Variability is defined by response time variability (as measured by the ICV) for 10 of the 12 tasks. Lower T scores on this factor reflect more consistent (i.e., less variable) responding in general.

Administration and Scoring

Administration

The TEC is computer administered and comes with detailed information pertaining to setup and administration. Calibration is required before the TEC can be administered to ensure precise timing of image presentation and response capture. This is done with a routine that runs automatically the first time the program is started. Specifics with respect to computer requirements are available from the publisher.

The TEC is individually administered. Administration consists of either four tasks for children 5–7 years of age or six tasks for children and adolescents 8–18 years of age. Each task includes on-screen instructions, practice trials with feedback, and 100 timed-interval stimuli that require responses. The TEC takes approximately 20–25 min to complete, depending on the age of the child and number of tasks administered. The examinee is should be seated comfortably in front of the computer with the keyboard placed within comfortable reach and the monitor height and angle adjusted for comfortable viewing. Breaks during test administration should be avoided as they may interfere with task completion and generating scores and as the test was standardized without breaks in between tasks.

The TEC is a cognitively and attentionally demanding test. Children may complain of boredom as the tasks are repetitive. Establishing good rapport with the child or adolescent is essential. Verbal instructions should introduce the TEC as a test of ability to pay close attention and make decisions by putting things into either a Red box or a Blue box.

Each TEC task is introduced by a specific set of on-screen instructions designed to promote understanding of task requirements by children as young as 5 years of age. For children ages 5–12 years, the TEC presents a story theme that involves helping a child clean his or her room by putting things in either a Red box or a Blue box in the bottom left- and right-hand corners of the screen, respectively. For adolescents ages 13–18 years, the room cleaning theme is replaced by a "sorting test" theme. The examinee is instructed that his or her job is to put each object that appears on the screen in the correct box as best as he or she can. No specific instruction is given in terms of speed or accuracy of responding. This nonspecific instruction enables the examiner to gain information regarding the examinee's response style, that is, the examinee's preference for speed vs. accuracy within a task or across tasks. The examinee then is shown an example of the correct sorting of an object and asked to press the correct button to sort the same object. The three TEC tasks that include cues to inhibit responding (i.e., 0BI, 1BI, 2BI) include additional instructions that also are based on the sorting task principle. The examinee is shown an object that appears within a box. He or she is instructed that any object shown inside a box has already been put away, and therefore no button should be pressed.

Each set of instructions is followed by a practice trial. Practice trials include 5–10 stimuli with two opportunities to demonstrate an understanding of task demands. If the examinee makes more than three errors in a set of practice trials, an on-screen warning appears that alerts the examiner and asks if the practice trial should continue. If No is selected, the trial ends. If Yes is selected, the practice trials resume. This process can be repeated until No is selected. After the practice trial is correctly completed, a screen reminds the examinee of the correct placement of his or her fingers on the keyboard and repeats the basic instructions.

It is essential that the examiner remain in the test setting with the examinee during administration of the TEC. Children and adolescents behave differently when they are being observed during testing than when they are not being observed. Leaving a respondent alone during TEC administration may result in an invalid administration.

Scoring

The TEC saves raw data and offers three report options. The Client Report and Score Report provide tables with raw scores, T scores, percentiles, and 90 % confidence intervals for every Factor, Summary, and Task score along with separate figures showing T score levels for Factor, Summary, and Task scores. Significant changes in performance across levels of working memory load (e.g., 0-back to 1-back, 1-back to 2-back) within the no inhibit and inhibit conditions separately are indicated, for Task scores with an asterisk and actual standardized regression-based (SRB) reliable change scores (Chelune, 2003) are reported as z scores for each comparison. Scores on the TEC that exceed the expected score beyond the limits of an 80 % confidence interval are considered to reflect clinically meaningful change in performance. While the Client Report and Score Report provide the same information regarding scores, the Client Report also offers detailed interpretive guidance.

The TEC also offers a Protocol Summary Report designed to facilitate comparison of performance between up to five administrations. In addition to raw scores, *T* scores, and percentiles for all Factor, Summary, and Task scores, the Protocol Summary Report indicates significant changes between administrations with an asterisk. Significant change in performance from one administration to another is based on SRB reliable change scores (Chelune, 2003). Scores that differ from expectation by more than an 80 % confidence interval are again considered to reflect clinically meaningful change in performance.

Standardization, Norms, and Psychometrics

Characteristics of the Standardization Sample

The TEC was standardized and validated for use with children and adolescents ages 5–18 years, recruited though schools, as well as youth and community organizations. Data collection took place under the direct supervision of the authors and publisher in six states (i.e., Colorado, Florida, New Hampshire, Maryland, Vermont Virginia,) and the District of Columbia. Exclusion criteria included history of diagnosis or treatment of developmental disability (e.g., ADHD, autism, learning disability), neurological disorder (e.g., epilepsy), history of head injury with loss of consciousness, and severe psychiatric disorder (e.g., bipolar disorder, depression).

The standardization sample included children and adolescents from a wide range of racial/ethnic and educational backgrounds. A final normative sample of 1,107 participants was obtained after exclusion of outliers and those missing demographic data. Of these 835 completed TEC Form 1, 138 completed Form 2, and 134 completed Form 3. Standardization sampling was based on the Current Population Survey (US Census Bureau, 2007). A weighting procedure was applied to the normative sample in order to more accurately reflect population parameters as defined by the US Census. TEC scores are compared to appropriate age and gender normative values.

Reliability of the Scales

Internal consistency of the TEC was examined in the normative sample using the Spearman-Brown split-half reliability coefficient, rather than Cronbach's alpha, since both accuracy and speed are important considerations in performance on the test (Anastasi & Urbina, 1997). Factor and Summary scores are the most internally consistent measures, with reliabilities for factors in the .80s and .90s, with the exception of the Response Control factor for the youngest children (i.e., ages 5–7 years), which is consistent with the observation that very young children tend to be more variable in their accuracy on the test. Summary scores generally show high reliability. Coefficients for Task scores ranged from adequate (.70s) to strong (.90s) for most measures. Target Omissions have lower reliability for both Summary and Task scores, but this is considered an artifact of the low frequency of Target stimuli resulting in a restricted range of scores.

Test-retest reliability of the TEC was evaluated in 100 children and adolescents from the standardization sample who completed Form 1 on two occasions, with a mean test-retest interval was 11.21 days (SD = 8.79). Reliability coefficients (across age groups) for the Factor scores (mean = .77, range = .65–.89) and Summary scores (mean = .78, range = .73–.90) were adequate. Reliability of Task scores was considerably more variable across scores (range = .15–.88). Across the Task scores, reliability tended to be highest for reaction time measures and was better for Standard than Target stimuli; this latter discrepancy likely related in part to the considerably larger number of Standards than Targets.

Use of the Test or Rating Scale

Approach to Interpretation

The first step in interpreting TEC scores is to ensure that the computer administration and the child's approach to the task were both valid. This is checked by the computer scoring software and indicated at the outset of the reports. After validity is established, interpretation follows a "top-down" approach that moves from interpretation of general performance across the TEC as a whole to interpretation of performance from task to task as working memory load increases within both no inhibit and inhibit conditions. To examine performance as a whole, Summary scores are averages of response accuracy (number of Correct Target and Standard responses, commissions) and response time (response time, response time variability) across all TEC tasks administered and reflect general performance across the TEC. When scores for a particular TEC measure are at a consistent level across tasks, the Summary score may adequately capture performance and may reflect whether aspects of performance are globally problematic, such as vigilance for novel information or response speed.

The major advantage to the TEC over other computer-administered measures that require sustained attention or response inhibition is that it is an "N of 1" study where level of working memory load (0-, 1-, 2-back) is fully crossed with the presence or absence of an inhibit demand as independent variables and Task scores serve as dependent variables or outcome measures. While tables provide details of raw scores, percentiles, T scores, and confidence intervals for each outcome measure (e.g., Response Time, Target Correct, Commissions) within each task (e.g., 0-back/no inhibit, 1-back/inhibit), graphs of performance accuracy and response speed T scores provide visual pictures of the impact of each level of working memory load and inhibitory control demand on accuracy and response speed. This facilitates interpretation of changes in performance accompanying changes in working memory load and inhibitory demand relative to normative expectations.

Performance on the TEC is based on comparison of raw scores (e.g., mean response time, correct responses) to the standardization sample. Scores are reported as T scores with a mean of 50 and a standard deviation of 10. T scores greater than 50 indicate poorer performance than expected, whereas T scores lower than 50 reflect better performance than expected relative to the normative sample. SRB change scores use regression models to predict 1-back performance from 0-back performance and 2-back performance from 1-back performance. The predicted or expected score is then compared to the observed score and a simple z-score is calculated to reflect the likelihood that the observed score is outside the expected range. Scores that deviate from expected more than an 80 % confidence interval are highlighted.

The TEC was constructed with multiple equivalent forms to facilitate serial assessment. This facilitates comparison of performance across time to monitor response to treatment (e.g., behavioral intervention, medication) or to evaluate recovery from injury such as concussion. As with change in performance from task to task within a single administration, the magnitude of change in performance across two administrations is measured with the SRB method. SRB change scores are provided for Factor and Summary scores. Change is considered significant and clinically meaningful if it is beyond that expected based on scores obtained from the first to second administrations of the TEC in the standardization sample, using 80 % confidence intervals.

As with any measure, the TEC should be interpreted within the context of history, self- and informant reports, behavioral observations, and any other test data. It is also important to appreciate that there is no single score or profile of performance on any given test or battery that is of sufficient sensitivity and specificity to enable it to be used on its own to establish a diagnosis such as ADHD, TBI, any other disorder or illness (Riccio & Reynolds, 2001; Wodka et al., 2008; Youngwirth, Harvey, Gates, Hashim, & Friedman-Weieneth, 2007). Instead, diagnosis is made by the clinician based on relevant patient history, direct observations of behavior and performance, and test findings. The primary purpose of TEC interpretation is to provide information about certain aspects of an individual's executive functioning, not to establish a given diagnosis.

Case Example

Dave was a 15-year-old tenth grade student-athlete with a history of early-diagnosed ADHD-combined type (ADHD-C) treated with stimulant medication. Despite generally good academic success throughout his elementary and middle school years, he experienced difficulty remaining focused and concentrating in more demanding high school level college preparatory classes. He was referred for consultation to develop a more comprehensive understanding of his needs in this area. Dave was evaluated with his usual stimulant medication on board. Although he was highly invested in his performance and cooperative with the evaluation, he was notably restless throughout the evaluation, chattered constantly, and often approached problem-solving impulsively. Parent, teacher, and self-reports of everyday executive functioning suggested difficulties with inhibitory control, working memory, planning, organization, and performance monitoring. Broadband rating scales completed by parents, teachers, and Dave all pointed to ongoing characteristics seen in adolescents with attention disorders but no other concerns. On formal testing, he demonstrated verbal and visual/nonverbal cognitive functioning solidly between the 90th and 95th percentile. He performed well on several performance measures of executive function including card-sorting, trail-making, and word generation tasks.

Dave completed the TEC, a measure of a student's response to increasing working memory load and impulse control demand. There were no problems with the computer administration of the TEC, and he demonstrated understanding of the task demands and adequate effort, completing all six of the tasks. Review of Summary scores revealed that Dave had difficulty attending to novel (Target) information across the measure as a whole but there were no global concerns with accuracy for frequent (Standard) pictures, commissions, response time, or response time variability. Figure 19.3 shows the Summary scores graphically and in tabular form produced by the TEC software.

The pattern of performance between the six tasks tells a somewhat different story. Figure 19.4 shows the impact of increasing working memory load from 0- to 1- to 2-back on accuracy for novel pictures (Targets, shown as circles) and for frequent pictures (Standards, shown as triangles), both with (solid lines) and without (dotted lines) the additional demand to occasionally inhibit responding. Visual inspection shows that, while his accuracy for the frequent Standard stimuli remained at a constant level relative to the normative sample, accuracy worsened (increased T scores) for novel Target stimuli with increased working memory demand. Neither pattern varied substantially with the addition of the inhibitory demand.

The TEC provides standardized regression-based reliable change scores (SRB Change scores) to assist in determining whether or not change in scores is clinically meaningful. In this case, Dave's accuracy for the Target stimuli crossed from within the Typical range to Elevated as the task changed from 1-back to 2-back. Examining the table of SRB Change scores below Fig. 19.4 showed that the Target Correct score worsened more than expected from 1- to 2-back both with ($z=-1.43$) and without ($z=-1.92$) inhibit demand.

Response speed and response speed variability are sensitive indicators of cognitive difficulties, including the presence of an attention disorder. Dave's Response Time scores (RT) for both Target and Standard stimuli were Typical regardless of task demands (see Fig. 19.5). Given the Typical score for overall Response Time (Summary score $T=51$), the appropriate measure of response time variability is the RTSD score and not the ICV, as the ICV is appropriate when RT is significantly elevated. Dave's response time variability increased substantially with changing demands. Figure 19.6 shows RTSD for frequent stimuli (Standards) as a function of increasing working memory load both with and without inhibitory demand. Without the need to inhibit impulsive responding, Dave's variability was Typical for both 0- and 1-back tasks but became Elevated at the 2-back level. The SRB

Bar Graph of Summary *T* Scores Obtained for Accuracy and Response Time Variables on the TEC

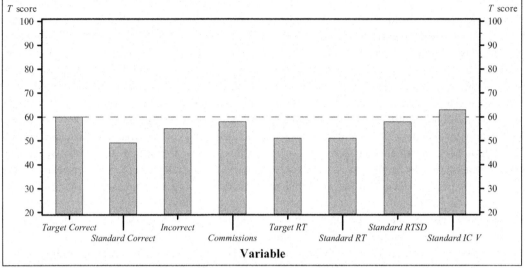

Note. RT = Response Time; RTSD = Standard Deviation of Response Time; ICV = Intra-Individual Coefficient of Variation.

Summary Raw Scores, *T* Scores, Percentiles, and CIs Obtained for Accuracy and Response Time Variables on the TEC

Variable	Mean raw score	*T* score	Percentile	90% CI	Interpretation
Accuracy					
Target Correct (% Correct)	10.50 (53%)	60	84	55 - 65	Elevated
Standard Correct (% Correct)	68.67 (95%)	49	46	45 - 53	Typical
Target Omissions	.00				Typical
Standard Omissions	.67				Typical
Incorrect	12.00	55	69	50 - 60	Typical
Commissions	6.00	58	79	52 - 64	Typical
Response time					
Target RT	459.24	51	54	49 - 53	Typical
Standard RT	445.78	51	54	49 - 53	Typical
Standard RTSD	166.53	58	79	54 - 62	Typical
Standard ICV	.37	63	90	54 - 72	Elevated

Note. RT = Response Time; RTSD = Standard Deviation of the Response Time; ICV = Intra-Individual Coefficient of Variation; CI = Confidence Interval. Only raw scores and ranges are reported for Omission variables. No *T* scores, percentiles, or confidence intervals (CIs) were calculated. % Correct scores are calculated as the mean percentage of correct responses per task (e.g., for Standard stimuli, (0B % Correct + 0BI % Correct + ...+ 2BI % Correct)/6 and not as the number of total correct responses divided by the total possible responses.)

Fig. 19.3 Case example summary scores

Change score of −2.66 indicates that this change is well beyond expectation relative to the normative sample. With the addition of an inhibitory demand, Dave's variability increased as he shifted from 0- to 1-back, again a change that was well beyond expected ($z=-1.81$).

In summary, Dave's performance on the TEC revealed difficulty managing his cognitive resources as working memory load and inhibitory demand increased. Although he was able to maintain appropriate sustained attention to frequent or common stimuli, he was unable to

Line Graph of *T* Scores Obtained for Target Correct and Standard Correct Across TEC Tasks

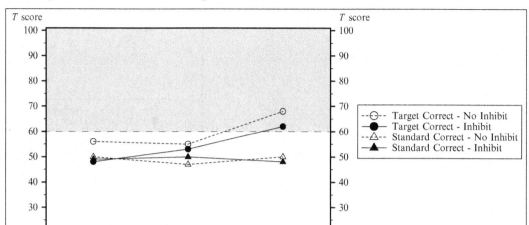

Predicted Raw Scores, Obtained Raw Scores, and SRB Change Scores for Target Correct and Standard Correct Across TEC Tasks

	No Inhibit		Inhibit	
Variable	0B-1B Predicted raw score (Obtained raw score) [SRB change score]	1B-2B Predicted raw score (Obtained raw score) [SRB change score]	0BI-1BI Predicted raw score (Obtained raw score) [SRB change score]	1BI-2BI Predicted raw score (Obtained raw score) [SRB change score]
Target Correct	12 (12) [-.18]	9 (4) [-1.92*]	12 (10) [-.68]	11 (7) [-1.43*]
Standard Correct	77 (78) [.41]	74 (73) [-.19]	61 (61) [-.06]	59 (61) [.54]

Note. SRB = Standardized regression-based; 0B = 0-Back/No Inhibit; 0BI = 0-Back/Inhibit; 1B = 1-Back/No Inhibit; 1BI = 1-Back/Inhibit; 2B = 2-Back/No Inhibit; 2BI = 2-Back/Inhibit. An asterisk (*) and bolded text indicates a significant SRB change score between tasks.

Fig. 19.4 Task scores for response accuracy

remain vigilant for more novel or infrequent information at the same time. As demands increased, he also became considerably more variable in his response speed, a sensitive indicator of cognitive difficulty such as attention problems. No test alone, however, should be viewed as "diagnostic" without collateral information, such as history, parent/teacher reports, observations, and other measures. Dave's history of well-diagnosed ADHD-C, parent and teacher reports of inhibitory control, working memory, and planning, organization, and monitoring difficulties but no other social, behavioral, or emotional concerns; and his generally restless and impulsive presentation during the evaluation all point to ongoing symptoms consistent with ADHD-C. This formulation is consistent with performance on the TEC despite use of medication during the assessment and suggested that he might benefit from careful review of his medication and

Line Graph of *T* Scores Obtained for Target RT and Standard RT Across TEC Tasks

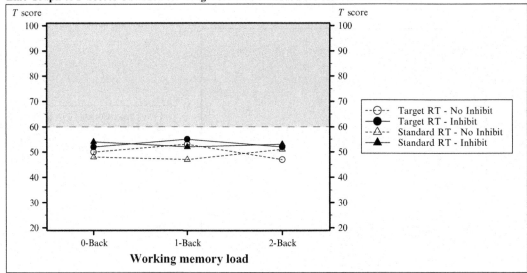

Note. RT = Response Time.

Predicted Raw Scores, Obtained Raw Scores, and SRB Change Scores for Target RT and Standard RT Across TEC Tasks

	No Inhibit		Inhibit	
	0B-1B	**1B-2B**	**0BI-1BI**	**1BI-2BI**
	Predicted raw score	Predicted raw score	Predicted raw score	Predicted raw score
	(Obtained raw score)	(Obtained raw score)	(Obtained raw score)	(Obtained raw score)
Variable	[SRB change score]	[SRB change score]	[SRB change score]	[SRB change score]
Target RT	404.93 (447.64) [-.84]	486.20 (409.40) [1.07]	489.58 (525.92) [-.49]	474.61 (461.08) [.20]
Standard RT	413.21 (404.01) [.25]	413.89 (449.18) [-.81]	487.59 (483.48) [.10]	463.74 (473.17) [-.28]

Note. SRB = Standardized regression-based; RT = Response Time; 0B = 0-Back/No Inhibit; 0BI = 0-Back/Inhibit; 1B = 1-Back/No Inhibit; 1BI = 1-Back/Inhibit; 2B = 2-Back/No Inhibit; 2BI = 2-Back/Inhibit.
An asterisk (*) and bolded text indicates a significant SRB change score between tasks.

Fig. 19.5 Task scores for response time

additional therapeutic interventions and accommodations for students with ADHD.

As noted above, Dave performed in the average range or better on other performance measures of executive function. This is not uncommon, as a meta-analysis revealed that less than half of over 3,000 students diagnosed with ADHD had low scores on any such measures (Willcutt et al., 2005). Dave was likely able to use his strong cognitive ability to perform well on these tasks, compensating for his persistent ADHD-C symptoms. The TEC added value to the evaluation by limiting the influence of intellectual ability, increasing working memory load and inhibitory demand, and requiring sustained performance over a more naturalistic timeframe.

Line Graph of T Scores Obtained for Standard RTSD and Standard ICV Across TEC Tasks

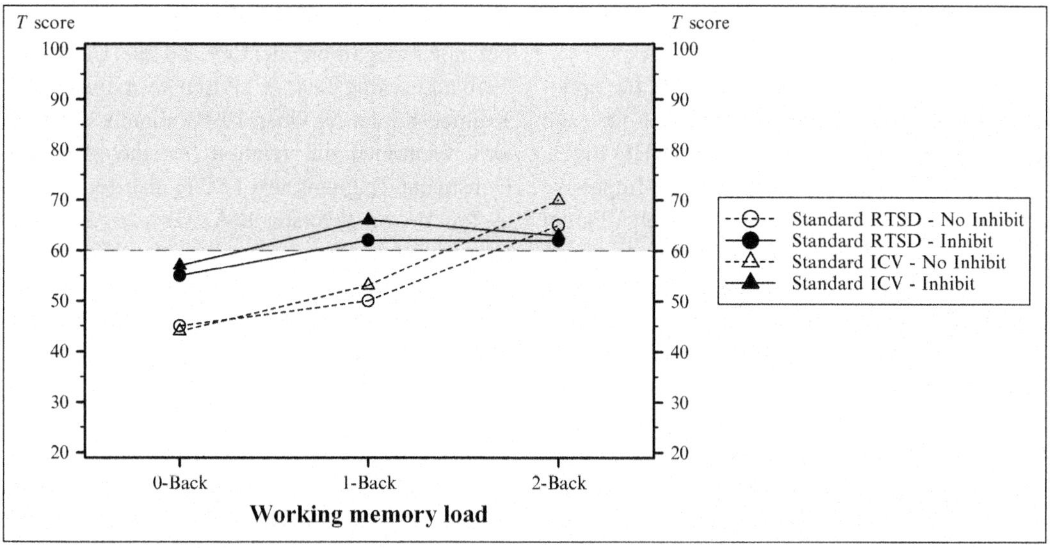

Note. RT = Response Time. RTSD = Standard Deviation of the Response Time; ICV = Intra-Individual Coefficient of Variation.

Predicted Raw Scores, Obtained Raw Scores, and SRB Change Scores for Standard RTSD and Standard ICV Across TEC Tasks

	No Inhibit		Inhibit	
Variable	0B-1B Predicted raw score (Obtained raw score) [SRB change score]	1B-2B Predicted raw score (Obtained raw score) [SRB change score]	0BI-1BI Predicted raw score (Obtained raw score) [SRB change score]	1BI-2BI Predicted raw score (Obtained raw score) [SRB change score]
Standard RTSD	95.51 (117.49) [-.81]	145.36 (246.71) [**-2.66***]	147.20 (202.39) [**-1.81***]	203.28 (224.79) [-.67]
Standard ICV	.23 (.29) [-1.08]	.34 (.55) [**-2.86***]	.30 (.42) [**-2.06***]	.41 (.48) [-1.00]

Note. SRB = Standardized regression-based; RTSD = Standard Deviation of the Response Time; ICV = Intra-Individual Coefficient of Variation. 0B = 0-Back/No Inhibit; 0BI = 0-Back/Inhibit; 1B = 1-Back/No Inhibit; 1BI = 1-Back/Inhibit; 2B = 2-Back/No Inhibit; 2BI = 2-Back/Inhibit. An asterisk (*) and bolded text indicates a significant SRB change score between tasks.

Fig. 19.6 Task scores for response time variability

Evidence for Validity

Relationships to Similar Measures

The TEC has been examined for both convergent and discriminant validity. The test manual reports correlations between the TEC and the BRIEF (Gioia, Isquith, Guy, & Kenworthy, 2000) in four separate samples. In children and adolescents with ADHD, greater parent reported difficulty with executive functions in everyday life was associated with worse performance on the TEC, especially in terms of accuracy, though relationships with speed and consistency of response speed were also noted. In particular, children who were rated as more impulsive in their everyday environment tended to be less accurate, more impulsive, and show more variable responding on the TEC. TEC accuracy measures also correlated with BRIEF ratings in mixed clinical and mTBI samples. Finally, significant associations between the TEC and self-report of executive problems, as rated using the BRIEF-SR (Guy, Isquith, & Gioia, 2004), were seen in a sample of adolescents with mTBI.

Several TEC scores correlated with the oppositional, cognitive problems/inattention, and hyperactivity scales as well as the ADHD Index, the DSM-IV Inattentive, Hyperactive-Impulsive and Total Index scores on the Conners' Parent Rating Scale-Revised (CPRS-R:L; Conners, 1997). Greater problems on the CPRS-R:L were significantly related to a lower number of correct responses to Standard stimuli and a higher number of Omissions and Incorrect responses, and higher scores on the hyperactivity scale were associated with more Commissions and reduced accuracy on all TEC measures. In essence, children and adolescents whose parents described them as having problems with impulsivity, hyperactivity, inattention, and oppositional behavior to an extent that they exhibited characteristics seen in children with ADHD were less accurate and more variable in their performance on the TEC. In contrast, there were few associations between the TEC and CPRS-R:L scales with less theoretical association with executive function, namely, the Anxious-Shy, Perfectionism, Social Problems, and Psycho-somatic scales.

The pattern of correlations between scores on the TEC and the CBCL (Achenbach, 1991) provide further evidence of validity. Though the sample size was small ($n=34$), TEC scores correlated with the CBCL Rule-Breaking Behavior and Social Problems scales. In contrast, no significant correlations were noted with the Anxious/Depressed, Withdrawn/Depressed, Somatic Complaints, or other scales providing some evidence for discriminant validity.

The relationship between TEC scores and performance on other neuropsychological measures in mixed clinical samples is reported in the test manual. Poorer TEC scores were associated with worse performance on the Symbol Digit Modalities Test (Smith, 1968), as well as on both the Reading and Math Fluency subtests of the Woodcock-Johnson III Test of Achievement (Woodcock, McGrew, & Mather, 2001). Notably, both accuracy and speed on the TEC were related to the fluency scores and effects tended to be stronger for math skills. TEC accuracy and speed also correlated with learning and recall scores, but not recognition memory, on the California Verbal Learning Test—Children's version (Delis, Kramer, Kaplan, & Ober, 1994). Finally, the TEC was examined in relation to the Auditory Consonant Trigrams test (ACT; also referred to as the Brown-Peterson task) (Brown, 1958), a measure that places considerable demand on working memory. While no significant correlations were noted between TEC scores and the easiest condition of the ACT (3 s delay), poorer performance on the more demanding ACT conditions (9 and 18 s delays) was associated with worse Target accuracy and commission errors on the TEC. This may be interpreted as suggesting that the ability to hold information in working memory despite a distracting task (counting backwards) on the ACT was related to the ability to selectively attend to important information (i.e., targets) and resist interference from the prepotent response on the TEC.

Fairness, Sex, Race, and Ethnic Differences

Analysis of data from the standardization sample revealed that age was associated with considerable change in performance on all measures on the TEC. In general, younger individuals demonstrated lower accuracy in responding to Target and Standard stimuli (in terms of both correct responses and omission errors), made more commission errors, and had slower speed and greater variability in response speed than did older individuals. This was seen regardless of working memory load or inhibitory demand. Younger age was also associated with poorer scores for all Summary measures of accuracy (Correct and Omissions), Commissions, and Incorrect responses.

Gender has also been observed to have an impact on TEC scores. At both the task level and across tasks (i.e., Summary scores), males tended to be less accurate than females, making fewer correct responses to the target and standard stimuli, as well as more commission errors and incorrect responses. No differences were seen for omission errors. While statistically significant given the very large sample size, effect sizes for these gender differences were small ($\eta^2 = .007–.037$). In contrast to their lesser accuracy, males tended to have faster reaction times to target stimuli across tasks and to standard stimuli for all tasks except the 0B. Gender effects were not seen for either RTSD or the ICV. As with accuracy, effect sizes for response speed were generally small ($\eta^2 = .008–.051$). Age-by-gender interactions were observed for some accuracy and response time measures, and these tended to be more pronounced for younger than older children. Males made fewer correct responses to Target stimuli (0BI), more Commissions (1BI and 2BI), and more Incorrect responses at all working memory loads. Similarly, poorer performance was seen for younger males than females on several Summary measures including Standard Correct, Commissions, and Incorrect responses. With respect to response time, males had faster response times for correct responses to Target stimuli across all tasks except 2BI and for correct responses to Standard stimuli across all tasks. Females demonstrated greater response time variability (RTSD) across all tasks except 2BI. This pattern was also seen on the Summary measures. Overall, gender accounted for less than 5 % of the variance in Task and Summary scores.

The normative sample included children and adolescents from a variety of racial/ethnic groups (i.e., Caucasian, African American, Hispanic, Asian/Pacific Islander, Other). Analyses of the normative data revealed multiple significant, but non-meaningful race/ethnicity effects on TEC scores. These accounted for no more than 5 % of the variance in scores.

Profiles of Abilities and Their Relationship to Diagnosis

ADHD
The TEC manual reports studies of TEC performance in children with ADHD. Children and adolescents with ADHD-C were significantly less accurate and more variable overall than were those with ADHD-I and typically developing children. Children in the ADHD-I group also showed greater response time variability than the typically developing children and adolescents but were less variable than those in the ADHD-C group. Effects varied by working memory load. In the context of the overall group differences, the ADHD-C group was consistently less accurate than the ADHD-I and typically developing groups when inhibitory demand was not present. In the presence of inhibitory demand, the ADHD-C group was significantly worse at the highest level of working memory load. Overall, these findings suggest significant executive dysfunction in children and adolescents with ADHD, though considerably more so for those with ADHD-C than ADHD-I, as reflected by performance on the TEC.

TEC performance was compared between unmedicated and medicated children diagnosed with ADHD. Children with ADHD who were medicated at the time of testing showed better accuracy than those who were unmedicated, regardless of ADHD subtype. Complex interactions were observed between ADHD subtype and medication status on response time and response time variability. For example, variability in response time for Standard stimuli was better for the ADHD-C medicated group than for the ADHD-C unmedicated group, but there was no difference for the ADHD-I subtype group, whether or not they were medicated.

Overall, the findings for TEC performance are consistent with evidence for executive dysfunction in children with ADHD. Further, the TEC was sensitive to the effects of stimulant medication in children with ADHD, though the authors interpret these findings cautiously, as the study was conducted using between—rather than within-group design.

TBI
The TEC manual also reports on scores in children with mild traumatic brain injury (mild TBI) who were seen within 1–3 weeks post-injury ($n = 101$) and compared with a matched group of their non-injured peers. The group with mTBI performed significantly worse than the typically developing group in most respects with effect sizes ranging from 3 to 22 % of the variance. Children with mTBI were markedly slower and

more variable in their speed of responding and tended to be somewhat less accurate. Task score performance showed that children with and without mTBI were affected by increased working memory load to the same degree and in a similar fashion.

A subgroup of 70 children with mild TBI and matched controls in the above comparison completed the TEC again on follow-up (i.e., Time 2) to track recovery, with the majority (98 %) completing the second administration within 3 weeks of the first evaluation. On reevaluation, children with and without mTBI were similar in terms of accuracy in responding to Target stimuli and exhibited little change over time. The mTBI group's accuracy in responding to Standard stimuli improved significantly on the second evaluation, whereas the typically developing group's accuracy remained at similar levels across both test sessions. Furthermore, the mTBI group continued to be significantly slower and more variable than the typically developing group on re-testing, showing little improvement relative to their initial evaluation.

In summary, children and adolescents with mTBI were initially less accurate, slower, and more variable in their responding, particularly to Standard stimuli, than were their matched non-injured peers. The mTBI group also had greater difficulty responding to increases in working memory load than did the typically developing group. The mTBI group as a whole showed improvement by the second evaluation but did not reach the level of performance attained by the non-injured group.

Empirical Support for the Test Structure

Factor analysis of the TEC for the two age groups (i.e., 5–7 years, 8–18 years) yielded somewhat different solutions, with a three-factor model best fitting the younger group and a four-factor model best fitting the older group. The three-factor solution for the younger age group did not appear to be the result of the smaller number of variables but rather likely reflects the developmental nature of differentiating executive control functions with age. This conclusion is supported by the pattern of shared loadings across the three factors (i.e., Commissions) and within the Response Control factor, reflecting a less differentiated model.

For children aged 5–7 years, a three-factor solution accounting for 61.3 % of the variance was the most parsimonious. The first factor, Response Control, was defined by a combination of variables including Standard Correct, RTSD, and Commissions. A second factor, Selective Attention, was defined by a combination of Target Correct (i.e., a selective response) and Commissions. The respondent must select the key Target stimulus and inhibit impulsive responding to achieve appropriate performance on this factor. A third factor, Response Speed, was defined by the eight RT variables for Standard and Target stimuli across the four tasks. The Selective Attention and Response Speed factors were moderately correlated, as were the Selective Attention and Response Control factors, while the Response Speed and Response Control factors were not significantly correlated.

For children and adolescents aged 8–18, a four-factor solution accounting for 60.9 % of variance was found to be most appropriate. A Sustained Accuracy factor was defined by Standard Correct on five of the six tasks. A Selective Attention factor was defined by Target Correct across all six tasks and Commissions across the three subtests with inhibitory demand (i.e., 0BI, 1BI, 2BI), reflecting the respondent's ability to select the key Target stimulus accurately and inhibit impulsive responding. A third factor, Response Speed, was defined by the 12 RT scores for Standard and Target stimuli across the six tasks. The fourth factor, Response Variability, was defined by all ICV scores across the six tasks. Response Variability and the other three factors were moderately correlated. Selective Attention and Sustained Accuracy also were moderately correlated. Response Speed was not significantly correlated with the Selective Attention or Sustained Accuracy factors.

Other Evidence of Validity

Two studies by Castellanos and colleagues examined the relationship between reaction time variability on the TEC and symptoms of ADHD and parent reported integrity of executive functions, the latter as rated on the Behavior Rating Inventory of Executive Functions (BRIEF; Gioia et al., 2000). The first study involved 98 children recruited from community clinics. Reaction time variability (RTSD) on the TEC explained a significant proportion of variance in BRIEF scores as well as in all of the ADHD-related subscales of the Conner's Parent Rating Scale (Gomez-Guerrero et al., 2011). In a second study employing the same participants, specific bands of low-frequency oscillations of reaction time on the TEC were strongly associated with scores on the ADHD rating scale (Mairena et al., 2012). Poorer performance on the TEC has been reported to be correlated with worse family functioning in a mixed clinical sample of children (Abecassis, Roth, Isquith, & Gioia, 2012). A recent study using an adapted version of the TEC for functional MRI found differences in brain activation between children who were typically developing as compared to a sample with mild traumatic brain injury (Krivitzky et al., 2011).

Summary and Conclusions

The TEC was designed to provide a novel, computerized, and well-standardized measure of two fundamental aspects of executive function, working memory and inhibitory control. Its design was based on a confluence of clinical experience, need for standardized measures of working memory and inhibitory control that would be appropriate for use with younger and older children as well as adolescents, and the growing literature on functional neuroimaging of these aspects of executive function. The TEC provides the examiner with multiple scores reflecting accuracy, speed, and variability of speed of responding. Scores are available that allow the examiner to evaluate performance across the individual tasks administered as a whole (Summary scores), while other scores facilitate evaluation of the impact of working memory load and demand for inhibitory control on performance (Task scores).

In contrast to many other measures, the TEC was designed to engage working memory in children and adolescents over a more naturalistic timeframe (i.e., 20–25 min). The TEC also differs from commonly employed continuous performance tests that typically require the individual to respond with a single button press to either all stimuli except one (e.g., respond to all letters but X) or not respond to any stimulus but one type (e.g., only respond to X, ignore other letters). Rather, the TEC requires continuous choice decision making, the individual having to respond to all stimuli, that is, both infrequent Targets and frequent Standards (using two different button presses), except when an infrequent cue to inhibit responding is presented. Another novel feature of the TEC is that the impact of working memory load on an individual's performance may be gauged through visualization of graphs of scores that present performance relative to norms, as well as through regression-based change scores. Regression-based change scores are also provided to determine whether changes across repeated testing are significant. Finally, three equivalent forms of the TEC are available, facilitating repeated presentation of the measure.

The TEC is a new instrument, having been published in 2010. Data on the reliability and validity of the measure is largely restricted to the test manual. While the manual provides evidence for the sensitivity of the TEC in children with ADHD and those with TBI, additional investigation is needed to determine the usefulness of the measure in other clinical populations. In addition, while the TEC provides numerous useful scores, the paradigm on which the measure is built and some of the scores available may be unfamiliar to some users (e.g., target vs. standard scores, regression-based change scores). New users may find that they need more time to familiarize themselves with the TEC and the information it provides than they required for measures with greater familiarity and/or that yield relatively few scores. Finally, the richness of the information available from the TEC is reflected in the currently available reports generated by the software. However, this same richness may be overwhelming to some users, especially when seeking the "bottom line" with respect to clinical interpretation. A brief, concise interpretive report that explains and summarizes a child's performance and offers suggested interventions for executive function difficulties is in development and should be made available to users shortly.

In conclusion, the TEC is a novel addition to the clinical and research toolkit for assessing executive control in the. It has demonstrated expected developmental changes from childhood through adolescence, sensitivity to executive dysfunction in clinical populations, and usefulness in monitoring changes in executive control over time. Further research is needed to establish whether it is sensitive to executive dysfunction in other clinical populations.

References

Abecassis, M., Roth, R. M., Isquith, P., & Gioia, G. (2012). Relationship between family functioning and executive control in a mixed pediatric sample. *40th Annual meeting of the International Neuropsychological Society*. Montreal, PQ.

Achenbach, T. (1991). *Manual for the Child Behavior Checklist and 1991 profile*. Burlington, VT: University of Vermont, Department of Psychiatry.

Alloway, T. P., & Alloway, R. G. (2010). Investigating the predictive roles of working memory and IQ in academic attainment. *Journal of Experimental Child Psychology, 106*(1), 20–29.

Alloway, T. P., Gathercole, S. E., Willis, C., & Adams, A. M. (2004). A structural analysis of working memory and related cognitive skills in young children. *Journal of Experimental Child Psychology, 87*(2), 85–106.

Altshuler, L. L., Bookheimer, S. Y., Townsend, J., Proenza, M. A., Eisenberger, N., Sabb, F., et al. (2005). Blunted activation in orbitofrontal cortex during mania: A functional magnetic resonance imaging study. *Biological Psychiatry, 58*(10), 763–769.

Anastasi, A., & Urbina, S. (1997). *Psychological testing*. Upper Saddle River, NJ: Prentice-Hall.

Anderson, S. W., Barrash, J., Bechara, A., & Tranel, D. (2006). Impairments of emotion and real-world complex behavior following childhood- or adult-onset damage to ventromedial prefrontal cortex. *Journal of the International Neuropsychological Society, 12*(2), 224–235.

Anderson, V., & Catroppa, C. (2005). Recovery of executive skills following paediatric traumatic brain injury (TBI): A 2 year follow-up. *Brain Injury, 19*(6), 459–470.

Andersson, U. (2008). Working memory as a predictor of written arithmetical skills in children: The importance of central executive functions. *British Journal of Educational Psychology, 78*(Pt 2), 181–203.

Aron, A. R. (2011). From reactive to proactive and selective control: Developing a richer model for stopping inappropriate responses. *Biological Psychiatry, 69*(12), e55–e68.

Aronen, E. T., Vuontela, V., Steenari, M. R., Salmi, J., & Carlson, S. (2005). Working memory, psychiatric symptoms, and academic performance at school. *Neurobiology of Learning and Memory, 83*(1), 33–42.

Baddeley, A. (2000). The episodic buffer: A new component of working memory? *Trends in Cognitive Sciences, 4*, 417–423.

Baddeley, A. (2003). Working memory: Looking back and looking forward. *Nature Reviews Neuroscience, 4*(10), 829–839.

Barkley, R. A. (1997). Behavioral inhibition, sustained attention, and executive functions: Constructing a unifying theory of ADHD. *Psychological Bulletin, 121*(1), 65–94.

Baving, L., Rellum, T., Laucht, M., & Schmidt, M. H. (2004). Attentional enhancement to NoGo stimuli in anxious children. *Journal of Neural Transmission, 111*(7), 985–999.

Berman, S., Friedman, D., Hamberger, M., & Snodgrass, J. G. (1989). Developmental picture norms: Relationships between name agreement, familiarity, and visual complexity for child and adult ratings of two sets of line drawings. *Behavior Research Methods, Instruments, & Computers, 21*, 371–382.

Brown, J. (1958). Some tests of the decay theory of immediate memory. *Quarterly Journal of Experimental Psychology, 10*, 12–21.

Broyd, S. J., Johnstone, S. J., Barry, R. J., Clarke, A. R., McCarthy, R., Selikowitz, M., et al. (2005). The effect of methylphenidate on response inhibition and the event-related potential of children with attention deficit/hyperactivity disorder. *International Journal of Psychophysiology, 58*(1), 47–58.

Camos, V. (2008). Low working memory capacity impedes both efficiency and learning of number transcoding in children. *Journal of Experimental Child Psychology, 99*(1), 37–57.

Carroll, J. B., Davies, P., & Richman, B. (1971). *The American heritage word frequency book*. Boston: Houghton Miflin.

Casey, B. J., Epstein, J. N., Buhle, J., Liston, C., Davidson, M. C., Tonev, S. T., et al. (2007). Frontostriatal connectivity and its role in cognitive control in parent–child dyads with ADHD. *The American Journal of Psychiatry, 164*(11), 1729–1736.

Castellanos, F. X., & Tannock, R. (2002). Neuroscience of attention-deficit hyperactivity disorder: The search for endophenotypes. *Nature Reviews Neuroscience, 3*, 617–628.

Chapman, S. B., Gamino, J. F., Cook, L. G., Hanten, G., Li, X., & Levin, H. S. (2006). Impaired discourse gist and working memory in children after brain injury. *Brain and Language, 97*(2), 178–188.

Chelune, G. J. (2003). Assessing reliable neuropsychological change. In R. D. Franklin (Ed.), *Prediction in forensic and neuropsychology: Sound statistical practices* (pp. 123–147). Mahwah, NJ: Lawrence Erlbaum.

Chen, C. J., Wu, C. H., Liao, Y. P., Hsu, H. L., Tseng, Y. C., Liu, H. L., et al. (2012). Working memory in patients with mild traumatic brain injury: Functional MR imaging analysis. *Radiology, 264*(3), 844–851.

Clark, C., Prior, M., & Kinsella, G. (2002). The relationship between executive function abilities, adaptive behaviour, and academic achievement in children with externalising behaviour problems. *Journal of Child Psychology and Psychiatry, and Allied Disciplines, 43*(6), 785–796.

Conners, K. (1997). *Conners' Rating Scales—Revised: Technical manual*. Tonawanda, NY: Multi-Health Systems.

Cycowicz, Y. M., Friedman, D., & Rothstein, M. (1997). Picture naming by young children: Norms for name agreement, familiarity, and visual complexity. *Journal of Experimental Child Psychology, 65*, 171–237.

Delis, D. C., Kramer, J. H., Kaplan, E., & Ober, B. A. (1994). *California Verbal Learning Test—Children's version*. San Antonio, TX: Psychological Corporation.

Dennis, M. (1989). Language and the young damaged brain. In T. Boll & B. K. Bryant (Eds.), *Clinical neuropsychology and brain function: Research, measurement and practice* (pp. 89–123). Washington, DC: American Psychological Association.

Espy, K. A., McDiarmid, M. M., Cwik, M. F., Stalets, M. M., Hamby, A., & Senn, T. E. (2004). The contribution of executive functions to emergent mathematic skills in preschool children. *Developmental Neuropsychology, 26*(1), 465–486.

Fallgatter, A. J., Ehlis, A. C., Seifert, J., Strik, W. K., Scheuerpflug, P., Zillessen, K. E., et al. (2004). Altered response control and anterior cingulate function in attention-deficit/hyperactivity disorder boys. *Clinical Neurophysiology, 115*(4), 973–981.

Garavan, H., Ross, T. J., & Stein, E. A. (1999). Right hemispheric dominance of inhibitory control: An event-related functional MRI study. *Proceedings of the National Academy of Sciences of the United States of America, 96*(14), 8301–8306.

Gathercole, S. E., Alloway, T. P., Willis, C., & Adams, A.-M. (2006). Working memory in children with reading disabilities. *Journal of Experimental Child Psychology, 93*(3), 265–281.

Gazzaley, A., Rissman, J., & Desposito, M. (2004). Functional connectivity during working memory maintenance. *Cognitive, Affective, & Behavioral Neuroscience, 4*(4), 580–599.

Gilotty, L., Kenworthy, L., Sirian, L., Black, D. O., & Wagner, A. E. (2002). Adaptive skills and executive function in autism spectrum disorders. *Child Neuropsychology, 8*(4), 241–248.

Gioia, G. A., Isquith, P. K., Guy, S. C., & Kenworthy, L. (2000). *BRIEF: Behavior Rating Inventory of Executive Function*. Lutz, FL: Psychological Assessment Resources.

Goldman-Rakic, P. (1987). Circuitry of primate prefrontal cortex and regulation of behavior by representational memory. In V. Mountcastle, F. Plum, & S. Geiger (Eds.), *Handbook of physiology—The nervous system* (pp. 373–417). Bethesda, MD: American Physiological Society.

Gomez-Guerrero, L., Martin, C. D., Mairena, M. A., Di Martino, A., Wang, J., Mendelsohn, A. L., et al. (2011). Response-time variability is related to parent ratings of inattention, hyperactivity, and executive function. *Journal of Attention Disorders, 15*(7), 572–582.

Gropper, R. J., & Tannock, R. (2009). A pilot study of working memory and academic achievement in college students with ADHD. *Journal of Attention Disorders, 12*(6), 574–581.

Guy, S. C., Isquith, P. K., & Gioia, G. A. (2004). *Behavior Rating Inventory of Executive Function—Self report version*. Lutz, FL: Psychological Assessment Resources.

Herrmann, M. J., Plichta, M. M., Ehlis, A. C., & Fallgatter, A. J. (2005). Optical topography during a Go-NoGo task assessed with multi-channel near-infrared spectroscopy. *Behavioural Brain Research, 160*(1), 135–140.

Isquith, P. K., Roth, R. M., & Gioia, G. A. (2010). *Tasks of Executive Control (TEC)*. Lutz, FL: Psychological Assessment Resources.

Jarratt, K. P., Riccio, C. A., & Siekierski, B. M. (2005). Assessment of attention deficit hyperactivity disorder (ADHD) using the BASC and BRIEF. *Applied Neuropsychology, 12*(2), 83–93.

Johnstone, S. J., Dimoska, A., Smith, J. L., Barry, R. J., Pleffer, C. B., Chiswick, D., et al. (2007). The development of stop-signal and Go/Nogo response inhibition in children aged 7–12 years: Performance and event-related potential indices. *International Journal of Psychophysiology, 63*(1), 25–38.

Jonkman, L. M., Lansbergen, M., & Stauder, J. E. (2003). Developmental differences in behavioral and event-related brain responses associated with response preparation and inhibition in a go/nogo task. *Psychophysiology, 40*(5), 752–761.

Jurado, M. B., & Rosselli, M. (2007). The elusive nature of executive functions: A review of our current understanding. *Neuropsychology Review, 17*(3), 213–233.

Karatekin, C., & Asarnow, R. F. (1998). Working memory in childhood-onset schizophrenia and attention-deficit/hyperactivity disorder. *Psychiatry Research, 80*(2), 165–176.

Kelly, A. M., Hester, R., Murphy, K., Javitt, D. C., Foxe, J. J., & Garavan, H. (2004). Prefrontal-subcortical dissociations underlying inhibitory control revealed by event-related fMRI. *European Journal of Neuroscience, 19*(11), 3105–3112.

Klein, C., Wendling, K., Huettner, P., Ruder, H., & Peper, M. (2006). Intra-subject variability in attention-deficit hyperactivity disorder. *Biological Psychiatry, 60*(10), 1088–1097.

Konishi, S., Nakajima, K., Uchida, I., Sekihara, K., & Miyashita, Y. (1998). No-go dominant brain activity in human inferior prefrontal cortex revealed by functional magnetic resonance imaging. *European Journal of Neuroscience, 10*, 1209–1213.

Krivitzky, L. S., Roebuck-Spencer, T. M., Roth, R. M., Blackstone, K., Johnson, C. P., & Gioia, G. (2011). Functional magnetic resonance imaging of working memory and response inhibition in children with mild traumatic brain injury. *Journal of the International Neuropsychological Society, 17*(6), 1143–1152.

Lachman, R., & Lachman, J. L. (1980). Picture naming: Retrieval and activation of longterm memory. In L. W. Poon, J. L. Fozard, L. S. Cermak, D. Greenberg, & L. W. Thompson (Eds.), *New directions in memory and aging* (pp. 313–343). Hillsdale, NJ: Erlbaum.

Levin, H. S., & Hanten, G. (2005). Executive functions after traumatic brain injury in children. *Pediatric Neurology, 33*, 79–93.

Levin, H. S., Hanten, G., Chang, C. C., Zhang, L., Schachar, R., Ewing-Cobbs, L., et al. (2002). Working memory after traumatic brain injury in children. *Annals of Neurology, 52*(1), 82–88.

Levin, H. S., Hanten, G., Zhang, L., Swank, P. R., Ewing-Cobbs, L., Dennis, M., et al. (2004). Changes in working memory after traumatic brain injury in children. *Neuropsychology, 18*(2), 240–247.

Levin, H. S., Hanten, G., Zhang, L., Swank, P. R., & Hunter, J. (2004). Selective impairment of inhibition after TBI in children. *Journal of Clinical and Experimental Neuropsychology, 26*(5), 589–597.

Lezak, M. D. (1995). *Neuropsychological assessment* (3rd ed.). New York: Oxford University Press.

Luria, A. R. (1981). *Higher cortical functions in man* (2nd ed.). New York: Plenum Press.

Mairena, M. A., Di Martino, A., Dominguez-Martin, C., Gomez-Guerrero, L., Gioia, G., Petkova, E., et al. (2012). Low frequency oscillations of response time explain parent ratings of inattention and hyperactivity/impulsivity. *European Child & Adolescent Psychiatry, 21*(2), 101–109.

Malisza, K. L., Allman, A. A., Shiloff, D., Jakobson, L., Longstaffe, S., & Chudley, A. E. (2005). Evaluation of spatial working memory function in children and adults with fetal alcohol spectrum disorders: A functional magnetic resonance imaging study. *Pediatric Research, 58*(6), 1150–1157.

Malloy, P., Rasmussen, S., Braden, W., & Haier, R. J. (1989). Topographic evoked potential mapping in obsessive-compulsive disorder: Evidence of frontal lobe dysfunction. *Psychiatry Research, 28*(1), 63–71.

Marsh, R., Zhu, H., Schultz, R. T., Quackenbush, G., Royal, J., Skudlarski, P., et al. (2006). A developmental fMRI study of self-regulatory control. *Human Brain Mapping, 27*(11), 848–863.

Mitchell, W. G., Zhou, Y., Chavez, J. M., & Guzman, B. L. (1992). Reaction time, attention, and impulsivity in epilepsy. *Pediatric Neurology, 8*(1), 19–24.

Murray, R., Shum, D., & McFarland, K. (1992). Attentional deficits in head-injured children: An information processing analysis. *Brain and Cognition, 18*(2), 99–115.

Newsome, M. R., Scheibel, R. S., Hunter, J. V., Wang, Z. J., Chu, Z., Li, X., et al. (2007). Brain activation during working memory after traumatic brain injury in children. *Neurocase, 13*(1), 16–24.

Owen, A. M., McMillan, K. M., Laird, A. R., & Bullmore, E. (2005). N-back working memory paradigm: A meta-analysis of normative functional neuroimaging studies. *Human Brain Mapping, 25*(1), 46–59.

Pennington, B. F., & Ozonoff, S. (1996). Executive functions and developmental psychopathology. *Journal of Child Psychology and Psychiatry, and Allied Disciplines, 37*(1), 51–87.

Pliszka, S. R., Liotti, M., Bailey, B. Y., Perez, R., III, Glahn, D., & Semrud-Clikeman, M. (2007). Electrophysiological effects of stimulant treatment on inhibitory control in children with attention-deficit/hyperactivity disorder. *Journal of Child and Adolescent Psychopharmacology, 17*(3), 356–366.

Riccio, C. A., & Reynolds, C. R. (2001). Continuous performance tests are sensitive to ADHD in adults but lack specificity. A review and critique for differential diagnosis. *Annals of the New York Academy of Sciences, 931*, 113–139.

Roth, R. M., Randolph, J. J., Koven, N. S., & Isquith, P. K. (2006). Neural substrates of executive functions: Insights from functional neuroimaging. In J. R. Dupri (Ed.), *Focus on neuropsychology research* (pp. 1–36). New York: Nova.

Roth, R. M., Saykin, A. J., Flashman, L. A., Pixley, H. S., West, J. D., & Mamourian, A. C. (2007). Event-related fMRI of response inhibition in obsessive-compulsive disorder. *Biological Psychiatry, 62*, 901–909.

Rubia, K., Smith, A. B., Woolley, J., Nosarti, C., Heyman, I., Taylor, E., et al. (2006). Progressive increase of frontostriatal brain activation from childhood to adulthood during event-related tasks of cognitive control. *Human Brain Mapping, 27*(12), 973–993.

Salthouse, T. A. (1993). Attentional blocks are not responsible for age-related slowing. *Journal of Gerontology, 48*, 263–270.

Schuerholz, L. J., Baumgardner, T. L., Singer, H. S., Reiss, A. L., & Denckla, M. B. (1996). Neuropsychological status of children with Tourette's syndrome with and without attention deficit hyperactivity disorder. *Neurology, 46*(4), 958–965.

Shallice, T., Marzocchi, G. M., Coser, S., Del Savio, M., Meuter, R. F., & Rumiati, R. I. (2002). Executive function profile of children with attention deficit hyperactivity disorder. *Developmental Neuropsychology, 21*(1), 43–71.

Smith, A. (1968). *Symbol digit modalities test*. Los Angeles: Western Psychological Services.

Smith, J. L., Johnstone, S. J., & Barry, R. J. (2004). Inhibitory processing during the Go/NoGo task: An ERP analysis of children with attention-deficit/hyperactivity disorder. *Clinical Neurophysiology, 115*(6), 1320–1331.

St Clair-Thompson, H. L., & Gathercole, S. E. (2006). Executive functions and achievements in school: Shifting, updating, inhibition, and working memory. *Quarterly Journal of Experimental Psychology, 59*(4), 745–759.

Steenari, M. R., Vuontela, V., Paavonen, E. J., Carlson, S., Fjallberg, M., & Aronen, E. (2003). Working memory and sleep in 6- to 13-year-old schoolchildren. *Journal of the American Academy of Child and Adolescent Psychiatry, 42*(1), 85–92.

Stevens, M. C., Kiehl, K. A., Pearlson, G. D., & Calhoun, V. D. (2007). Functional neural networks underlying response inhibition in adolescents and adults. *Behavioural Brain Research, 181*(1), 12–22.

Stollstorff, M., Foss-Feig, J., Cook, E. H., Jr., Stein, M. A., Gaillard, W. D., & Vaidya, C. J. (2010). Neural response to working memory load varies by dopamine transporter genotype in children. *NeuroImage, 53*(3), 970–977.

Stuss, D. T., & Benson, D. F. (1984). Neuropsychological studies of the frontal lobes. *Psychological Bulletin, 95*, 3–28.

Stuss, D. T., Murphy, K. J., Binns, M. A., & Alexander, M. P. (2003). Staying on the job: The frontal lobes control individual performance variability. *Brain, 126*, 2363–2380.

Tamm, L., Menon, V., Ringel, J., & Reiss, A. L. (2004). Event-related FMRI evidence of frontotemporal involvement in aberrant response inhibition and task switching in attention-deficit/hyperactivity disorder. *Journal of the American Academy of Child and Adolescent Psychiatry, 43*(11), 1430–1440.

Tarazi, R. A., Mahone, E. M., & Zabel, T. A. (2007). Self-care independence in children with neurological disorders: An interactional model of adaptive demands and executive dysfunction. *Rehabilitation Psychology, 52*, 196–205.

U.S. Census Bureau. (2007). *Current population survey, March 2007 [Data file]*. Washington, DC: U.S. Department of Commerce.

Vaidya, C. J., Austin, G., Kirkorian, G., Ridlehuber, H. W., Desmond, J. E., Glover, G. H., et al. (1998). Selective effects of methylphenidate in attention deficit hyperactivity disorder: A functional magnetic resonance study. *Proceedings of the National Academy of Sciences of the United States of America, 95*(24), 14494–14499.

van der Meer, D. J., & van der Meere, J. (2004). Response inhibition in children with conduct disorder and borderline intellectual functioning. *Child Neuropsychology, 10*(3), 189–194.

Van der Meere, J., Marzocchi, G. M., & De Meo, T. (2005). Response inhibition and attention deficit hyperactivity disorder with and without oppositional defiant disorder screened from a community sample. *Developmental Neuropsychology, 28*(1), 459–472.

Vuontela, V., Steenari, M. R., Carlson, S., Koivisto, J., Fjallberg, M., & Aronen, E. T. (2003). Audiospatial and visuospatial working memory in 6–13 year old school children. *Learning & Memory, 10*(1), 74–81.

Wager, T. D., & Smith, E. E. (2003). Neuroimaging studies of working memory: A meta-analysis. *Cognitive, Affective, & Behavioral Neuroscience, 3*(4), 255–274.

Wang, Z., Deater-Deckard, K., Cutting, L., Thompson, L. A., & Petrill, S. A. (2012). Working memory and parent-rated components of attention in middle childhood: A behavioral genetic study. *Behavioral Genetics, 42*(2), 199–208.

Wechsler, D. (2003). *WISC-IV technical and interpretive manual*. San Antonia, TX: The Psychological Corporation.

Welsh, M. C., & Pennington, B. F. (1988). Assessing frontal lobe functioning in children: Views from developmental psychology. *Developmental Neuropsychology, 4*, 199–230.

Welsh, M. C., Pennington, B. F., & Grossier, D. B. (1991). A normative-developmental study of executive function: A window on prefrontal function in children. *Developmental Neuropsychology, 7*, 199–230.

Willcutt, E. G., Doyle, A. E., Nigg, J. T., Faraone, S. V., & Pennington, B. F. (2005). Validity of the executive function theory of attention-deficit/hyperactivity disorder: A meta-analytic review. *Biological Psychiatry, 57*(11), 1336–1346.

Willcutt, E. G., Pennington, B. F., Boada, R., Tunick, R. A., Ogline, J., Chhabildas, N. A., et al. (2001). A comparison of the cognitive deficits in reading disability and attention-deficit/hyperactivity disorder. *Journal of Abnormal Child Psychology, 110*, 157–172.

Wodka, E. L., Loftis, C., Mostofsky, S. H., Prahme, C., Larson, J. C. G., Denckla, M. B., et al. (2008). Prediction of ADHD in boys and girls using the D-KEFS. *Archives of Clinical Neuropsychology, 23*(3), 283–293.

Woodcock, R. W., McGrew, K. S., & Mather, N. (2001). *Woodcock-Johnson III tests of achievement*. Rolling Meadows, IL: Riverside Publishing.

Wozniak, J. R., Block, E. E., White, T., Jensen, J. B., & Schulz, S. C. (2008). Clinical and neurocognitive course in early-onset psychosis: A longitudinal study of adolescents with schizophrenia-spectrum disorders. *Early Intervention in Psychiatry, 2*, 169–177.

Youngwirth, S. D., Harvey, E. A., Gates, E. C., Hashim, R. L., & Friedman-Weieneth, J. L. (2007). Neuropsychological abilities of preschool-aged children who display hyperactivity and/or oppositional-defiant behavior problems. *Child Neuropsychology, 13*(5), 422–443.

The Assessment of Executive Functioning Using the Childhood Executive Functioning Inventory (CHEXI)

Lisa B. Thorell and Corinne Catale

Development of the CHEXI

The strong association between deficits in executive functioning (EF) and psychiatric disorders such as Attention-Deficit Hyperactivity Disorder (ADHD) has led to the development of a myriad of neuropsychological tests designed to capture both executive functioning in general and specific abilities within the EF domain. However, relatively few EF rating instruments are available. Compared to neuropsychological tests, rating instruments generally capture more global aspects of behavior. Although ratings have the disadvantage of suffering from rater biases, they have the advantage of capturing behavior over an extended period of time and in different settings (e.g., home, school). As they are easy to administer, ratings can also be most valuable as a screening instrument for identifying children at risk of developing psychiatric disorders or functional impairments such as poor school achievement. In addition, EF rating instruments may be used to identify children with EF deficits, with or without psychiatric disorders, who may be helped through targeted intervention programs focused on training executive functions (e.g., Klingberg et al.,

L.B. Thorell (✉)
Karolinska Institutet, Stockholm, Sweden
e-mail: lisa.thorell@ki.se

C. Catale
University of Liège, Liège, Belgium

2005; Thorell, Lindqvist, Bergman Nutley, Bohlin, & Klingberg, 2009).

The most commonly used EF rating instrument for children is the Behavior Rating Inventory of Executive Functioning (BRIEF; Gioia, Andrews Espy, & Isquith, 2003; Gioia, Isquith, Guy, & Kenworthy, 2000; see also Chap. 18 in this volume), and more recently the Barkley Deficits in Executive Functioning Scales—Child and Adolescent Version (BDEFS-CA; Barkley, 2012; see also Chap. 15 in this volume) and the Comprehensive Executive Function Inventory (CEFI; Naglieri & Goldstein, 2013; see also Chap. 14 in this volume) have been introduced. The BRIEF has been found to be very useful in distinguishing between, for example, children with ADHD and normally developing controls (e.g., Gioia, Isquith, Kenworthy, & Barton, 2002). However, one potential limitation of the BRIEF is that in addition to measuring executive functioning, this scale also directly measures ADHD symptoms in that items such as "is impulsive" and "has a short attention span" are included, and these are almost identical to the symptom criteria for ADHD (APA, 1994). Scales like this could of course be of great use in both research and clinical practice, but the semantic overlap between the items included in the BRIEF and the symptom list for ADHD means that erroneous conclusions about EF deficits in children can be drawn. As repeatedly shown (e.g., Nigg, Willcutt, Doyle, & Sonuga-Barke, 2005), not all children with ADHD have EF deficits and vice versa. It may therefore be valuable to be able to distinguish between EF deficits and ADHD symptoms. This may be especially important in relation to functional impairments, as different combinations of ADHD and/or EF deficits may be differentially related to, for example, poor academic achievement (Biederman et al., 2004).

Another potential problem is that the BRIEF (63–86 items depending on the version used), the BDEFS-CA (70 items), and the CEFI (90 items) are all long questionnaires. A short form (20 items) of the BDEFS-CA is available, but no factor analysis of this version is presented in the manual, indicating that the short version is best used as a general EF scale rather than as a measure with subscales tapping specific EF functions. Finally, some items included in the BDEFS-CA would seem to be more suitable for older compared to younger children. This is especially true for the time management subscale, which includes statements such as "wastes or doesn't manage his/her time well" or "has difficulty judging how much time it will take to do something or to get somewhere."

In order to address the limitations of available questionnaires intended to tap into EF, the Childhood Executive Functioning Inventory (CHEXI; Thorell & Nyberg, 2008) was developed as a quick screening instrument specifically targeting different types of executive control rather than including more general statements or items directly related to the symptom criteria for ADHD. In the present chapter, we provide an overview of previous studies using the CHEXI in several different countries in Europe, Asia, and South America, and we also present some new, unpublished data.

Table 20.1 Sample items from the four a priori subscales included in the CHEXI

Working memory subscale	Has difficulty understanding verbal instructions unless he/she is also shown *how* to do something
	Has difficulty remembering what he/she is doing, in the middle of an activity
	When asked to do several things, he/she only remembers the first or last
Planning subscale	Has difficulty with task or activities that involve several steps
	Has difficulty carrying out activities that require several steps (e.g., for younger children, getting completely dressed without reminders; for older children, doing all homework independently)
	Has difficulty telling a story about something that has happened so that others may easily understand
Inhibition subscale	Has difficulty holding back his/her activity despite being told to do so
	Has difficulty stopping an activity immediately upon being told to do so. For example, he/she needs to jump a couple of extra times or play on the computer a little bit longer after being asked to stop
	Gets overly excited when something special is going to happen (e.g., going on a field trip or to a party)
Regulation subscale	Seldom seems to be able to motivate himself/herself to do something that he/she doesn't want to do
	When something needs to be done, he/she is often distracted by something more appealing
	Has difficulty following through on less appealing tasks unless he/she is promised some type of reward for doing so

Description of the CHEXI

The CHEXI first included 26 items, but in the original study introducing the instrument (Thorell & Nyberg, 2008), two items were excluded due to too low sampling adequacy. The most commonly used version therefore has 24 statements and takes about 5–10 min to complete. The complete questionnaire is freely available on the CHEXI website (www.chexi.se) in several different languages, including English, Swedish, French, Spanish, Chinese, and Farsi. It is meant to be used by parents or teachers, and it includes four a priori subscales measuring working memory (9 items), planning (4 items), inhibition (6 items), and regulation (5 items). Each item is rated on a scale from 1 (definitely not true) to 5 (definitely true), with higher scores indicating larger EF deficits. Please refer to Table 20.1 for the three statements loading highest on each of the four a priori subscales. The four a priori subscales were created based on Barkley's (1997) hybrid model, in which inhibition, working memory, regulation, and planning are seen as constituting the major EF deficits in children with ADHD. However, these four factors have not been identified through factor analyses of the CHEXI questionnaire in different age groups and in several different language versions, such as Swedish (Thorell & Nyberg, 2008), French (Catale, Lejeune, Merbah, & Meulemans, 2013; Catale, Meulemans, & Thorell, in press), Turkish (Kayhan, 2010), and Portuguese (Trevisan, Dias, Menezes, & Seabra, 2012). Instead, these studies have consistently identified two broad factors referred to as working memory (working memory and planning a priori subscales) and inhibition (inhibition and regulation a priori subscales). One exception is a Brazilian study (Trevisan et al., 2012), which found a two-factor solution for parent ratings but a one-factor solution for teacher ratings. The internal consistency of the two major factors of the CHEXI has proved to be satisfactory using both parent and teacher ratings. In addition, test-retest reliability for the CHEXI has been shown to be high (Catale et al., in press; Thorell & Nyberg, 2008) using parent ratings collected 3–10 weeks apart. Correlations between different raters have seldom been investigated using the CHEXI, although one study (Thorell, Veleiro, Siu, & Mohamadi, in press) reported modest, although significant, relations between parent and teacher ratings for both the inhibition ($r=.42$, $p<.001$) and working memory subscales ($r=.43$, $p<.001$).

Previous factor analyses have been conducted using children aged 5–7 years (Thorell & Nyberg, 2008), 5–6 years (Catale et al., 2013), 7–11 years (Catale et al., in press), 4–7 years (Trevisan et al., 2012), and 5–8 years (Kayhan, 2010). To our knowledge, no factor analysis of the CHEXI has been conducted using children below 4 or above 11 years. It should be noted that previous research using EF laboratory measures has demonstrated that it may not be fruitful to differentiate between inhibition and working memory in very young children (e.g., Wiebe, Espy, & Charak, 2008). On the other hand, a more differentiated EF profile has been found in older children, again using EF laboratory measures (e.g., Lehto, Juujärvi, Kooistra, & Pulkkinen, 2003; St. Clair-Thompson & Gathercole, 2006). In summary, the two-factor structure described above for children 4–11 years may not be valid for children outside this age range. However, it is also important to mention that the CHEXI was originally intended for use solely in the age range 4–12 years. If one is interested in EF deficits in children below 4 years, the preschool version of the BRIEF has been developed for children as young as 2 years, and more temperament-related measures of executive functioning (i.e., effortful control) can be made already in infancy using the Infant Behavior Checklist (Rothbart, 1981) or in preschoolers and early school-age children using the Child Behavior Checklist (Rothbart, Ahadi, Hershey, & Fisher, 2001). If one is interested in EF deficits in teenagers, we would recommend using rating scales with a greater emphasis on more complex executive skills such as organization, planning, and time management (e.g., the CEFI and the BDEFS-CA).

Relations Between the CHEXI and EF Laboratory Tests

Several studies have examined the relation between ratings using the CHEXI and EF laboratory measures. The original study by Thorell and Nyberg (2008) found significant relations between the two CHEXI subscales and laboratory measures of inhibition (go/no-go task) and working memory (word span task). However, as can be seen in Table 20.1, the relations were modest, and both Catale et al. (in press) and Kayhan (2010) have failed to find any significant relations between the CHEXI and laboratory measures of executive functioning. Mixed findings with regard to this issue have also been presented for other EF rating instruments such as the BRIEF (Anderson, Anderson, Northam, Jacobs, & Mikiewicz, 2002; Mahone et al., 2002; Vriezen & Pigott, 2002). It should be noted that despite the fact that laboratory measures are often regarded as the "golden standard" for measuring EF, the low correlations between ratings and laboratory measures should not necessarily be seen as a limitation. Instead, questionnaire ratings and laboratory measures most likely capture different aspects of executive functioning. According to one hypothesis (Anderson et al., 2002), laboratory tests primarily capture the cognitive aspect of executive functioning, whereas rating instruments capture more emotional and social aspects. Another difference is that questionnaires provide reports on the child's behavior in the "real world" and are based on observations made over an extended period of time, whereas laboratory tests are administered in a much more predictable and structured environment during a short period of time. In conclusion, EF tests and laboratory measures should be seen as complementary to one another, and a more complete picture of a child's executive profile will be obtained by combining these two types of measures.

Clinical Utility of the CHEXI

Three previous studies have investigated the clinical utility of the CHEXI in ADHD populations. The first study (Thorell, Eninger, Brocki, & Bohlin, 2010) included 15 children with ADHD and 30 normally developing controls (age 7 years), and the results showed that the children in the ADHD group differed significantly from the comparison group on both the CHEXI inhibition and the working memory subscale, with large effect sizes ($d = 1.79–2.95$). In addition, a logistic regression analysis showed that, using either parent or teacher ratings, both the CHEXI inhibition and working memory scales contributed significantly to distinguishing between the ADHD group and the comparison group. The sensitivity (range .73–.93) and specificity (range .79–.93) were high for both parent and teacher ratings. The highest classification rate was obtained for parent ratings on the inhibition subscale, where 93.3 % of the children were correctly classified.

The second study (Catale et al., in press), investigating the CHEXI in ADHD samples, included two subsamples of children aged 8–11 years from Belgium (25 ADHD children, 25 controls) and Sweden (62 ADHD children, 62 controls), and the results confirmed the findings of Thorell et al. (2010) by showing high sensitivity (range .84–.92) and high specificity (range .92–.96) for both the inhibition and working memory subscales in the two samples.

Third, another study compared three groups of children aged 5–7 years: (1) children diagnosed with ADHD, (2) children with executive and attention deficits (defined as having a Z-score below 1.5 on at least two executive and attention tasks assessing working memory, inhibition, cognitive flexibility or selective attention), and (3) normally developing children. The results showed that both clinical groups (ADHD group and EF deficits group) differed significantly from the comparison group on both the CHEXI inhibition and the working memory subscales (Catale, Lejeune, Merbah, & Meulemans, 2011). Furthermore, the CHEXI was shown to successfully distinguish between the respective clinical groups and the control group, with sensitivity and specificity ranging from .73 to .89.

Finally, in a non-clinical study from Belgium, the CHEXI was also examined in relation to other types of behavior problems. In this unpublished study on 80 normally developing children aged 8–11 years, associations were examined between the CHEXI and the Conners Parent Rating Scale (CPRS—48 items; Conners, 1970). Both rating scales were completed by the children's parents. The CHEXI was found to be most strongly associated with learning disabilities ($r = .73$ for the working memory subscale and .75 for the inhibition subscale). Significant relations were also found to both the Hyperactivity/Impulsivity subscale ($r = .27$ for working memory and .48 for inhibition) and the Conduct Problem subscale ($r = .27$ for working memory and .39 for inhibition). Interestingly, no significant correlations were found between the CHEXI and the Psychosomatic and Anxiety subscales of the CPRS. Furthermore, the correlations between the CHEXI and the Conduct Problem subscale disappeared when controlling for hyperactivity/impulsivity. Thus, the significant relations between the CHEXI and conduct problems were caused by the overlap between conduct problems and ADHD symptoms. In conclusion, these findings suggest that the CHEXI subscales have good convergent and divergent validity.

The Relation Between the CHEXI and Functional Impairments

One of the criticisms of EF tests (e.g., Barkley & Fischer, 2011) has been that they have very low ecological validity (i.e., they are only weakly related to how EFs are used in daily life activities in a natural setting). An important question when using EF ratings is therefore how well a certain rating instrument correlates with functional impairments. As presented in Table 20.2, both the CHEXI inhibition and working memory subscales have been shown to be related to academic skills among children attending kindergarten (Thorell & Nyberg, 2008). Interestingly, the effects of the CHEXI working memory subscale were significantly related to early academic skills even when controlling for the effects of EF tests (see boldfaced figures in Table 20.2). Again, these results emphasize that ratings and tests capture at least partially different aspects of EF.

The relation between CHEXI and academic achievement has also been investigated in a cross-cultural study including children aged 5–12 years from four different countries: Sweden, Spain, Iran, and China (Thorell et al., in press). The results showed that both the inhibition and working memory subscales of the CHEXI were related to academic achievement (i.e., mathematics and language skills) in all four countries, with the exception of CHEXI parent ratings in China.

Finally, two recent longitudinal studies have examined the relation between CHEXI ratings and academic performance. The first is an unpublished Swedish study investigating CHEXI ratings in kindergarten (age 6) and school performance in grade 2 (age 8). This study found that teacher ratings on the CHEXI working memory subscale were significantly related to mathematics in grade 2, even when controlling for the effect of early mathematic abilities in kindergarten. Thus, the CHEXI working memory subscale was able to predict the change in academic achievement between the two time points. No significant relations were found for the CHEXI inhibition subscale or to other aspects of academic achievement (i.e., reading or writing), but it should be noted that this study had limited power to detect such relations given its small sample size ($n=47$). The second longitudinal study looked at a Spanish sample (Veleiro & Thorell, 2012) and found significant relations

Table 20.2 Relations between CHEXI ratings and laboratory measures of executive functioning, ADHD symptoms, and academic performance (Thorell & Nyberg, 2008)

	Laboratory measures		ADHD symptoms[a,b]		Early academic skills[b]	
	Inhibition	Working memory	Hyperactivity/impulsivity	Inattention	Language skills	Mathematics
Parent ratings ($n=113$)						
Working memory factor	.33***	.26**	**.36***	.13	**−.41***	**−.29***
Inhibition factor	.28**	.07	**.27***	**.26***	−.16*	−.11
Teacher ratings ($n=89/105$)						
Working memory factor	.29**	.39***	**.33**	.21*	**−.46***	**−.42***
Inhibition factor	.35***	.19*	**.28**	**.27**	−.25**	−.24**

Reprinted by permission of the publisher (Taylor & Francis Ltd., http://www.tandf.co.uk/journals)
[a]Relations to ADHD symptoms represent correlations across raters
[b]Bold-faced entries denote relations that remained significant also when controlling for the effect of EF laboratory measures

between the CHEXI working memory subscale measured at 4 years and tests of basic mathematics abilities 12 months later.

In summary, there appears to be support for use of the CHEXI, especially the working memory subscale, as an early screening measure for early academic difficulties. To our knowledge, no other studies have presented data examining the relation between the CHEXI and other measures of functional impairments such as low prosocial skills, although there is at least one ongoing study investigating this issue.

Cross-Cultural Validation of the CHEXI in Different Countries

As stated above, the CHEXI is available in many different languages, and we are currently working on translating the questionnaire into even more languages (e.g., Danish, Dutch, and Japanese). These new versions will be available on the CHEXI website (www.chexi.se). As also stated above, the original factor structure of the instrument has been replicated in several different countries. A recent cross-cultural study, however, emphasized the need to take cultural biases into account when collecting ratings of problem behaviors in children using the CHEXI or any other rating instrument for that matter. As mentioned above, the cross-cultural study (Thorell et al., in press) included CHEXI ratings from Sweden, Spain, Iran, and China, and the results showed that there were significant effects of country with regard to both teacher and parent ratings on both CHEXI subscales. The main finding of the post hoc analyses was that the children in the Chinese sample received higher scores (i.e., showing higher EF deficits) compared to the children in the other countries. Interestingly, other studies using laboratory measures have shown the opposite pattern, with Chinese children being more skilled in executive functioning compared to children in the USA (e.g., Lan, Legare, Ponitz, Li, & Morrison, 2011; Sabbagh, Xu, Carlsson, Moses, & Lee, 2006). The authors of the cross-cultural study using the CHEXI therefore concluded that it is likely that the obtained cross-country differences do not reflect true differences in the children, but rather cultural biases. The Chinese culture has a strong emphasis on self-regulatory skills, and executive functioning deficits may therefore be exacerbated by strong cultural expectations. A similar conclusion was drawn by, for example, Hinshaw et al. (2011) with regard to ADHD symptoms.

Future Directions

To conclude, the CHEXI has good psychometric properties and can be considered a valuable screening instrument for identifying children at risk of developing ADHD, EF deficits, and early academic difficulties. However, as the instrument has not yet been nationally standardized in any country, it is at this time more valuable as a research tool than as a clinical instrument.

Finally, we also like to emphasize that there are many important avenues for future research on rating measures for executive functioning. Of most importance is perhaps to extend longitudinal investigations examining the development of executive functioning across early and middle childhood. Such an approach would allow us to gain further knowledge of the usefulness of EF rating instruments as an early screening measure for poor academic achievement and early behavior problems (primarily symptoms of ADHD). In addition, it would be valuable if future studies were to include several different EF rating instruments (e.g., CHEXI, BRIEF, and BDEFS-CA), allowing examination of both differences and commonalities. Finally, it should be mentioned that there is an adult version (i.e., self-rating instrument) of the CHEXI called the "Adult Executive Functioning Inventory" (ADEXI), which is still unpublished and thus far only available in English and Swedish. However, data from an ongoing study have shown that this instrument has good psychometric properties. It is for future studies to examine how well this adult version of the instrument can distinguish between adults with ADHD and controls and to examine its relation to EF laboratory tests.

References

American Psychiatric Association. (1994). *Diagnostic and statistical manual of mental disorders* (4th ed.). Washington, DC: Author.

Anderson, V., Anderson, P., Northam, E., Jacobs, R., & Mikiewicz, O. (2002). Relationships between cognitive and behavioral measures of executive function in children with brain disease. *Child Neuropsychology, 8*, 231–240.

Barkley, R. A. (1997). *ADHD and the nature of self-control*. New York: Guilford Press.

Barkley, R. A. (2012). *Deficits in executive functioning scale—Children and adolescents (BDEFS-CA)*. New York, NY: Guilford Press.

Barkley, R. A., & Fischer, M. (2011). Predicting impairment in major life activities and occupational functioning in hyperactive children as adults: Self-reported Executive Function (EF) deficits versus EF tests. *Developmental Neuropsychology, 36*, 137–161.

Biederman, J., Monuteaux, M. C., Doyle, A. E., Seidman, L. J., Wilens, T. E., Ferrero, F., et al. (2004). Impact of executive function deficits and attention-deficit/hyperactivity disorder (ADHD) on academic outcomes in children. *Journal of Consulting and Clinical Psychology, 72*, 757–766.

Catale, C., Lejeune, C., Merbah, S., & Meulemans, T. (2011). Le CHEXI : Présentation d'une nouvelle mesure d'évaluation comportementale du fonctionnement exécutif chez l'enfant de 5 à 7 ans. *Approche Neuropsychologique des Apprentissages chez l'Enfant, 115*, 487–493.

Catale, C., Lejeune, C., Merbah, S., & Meulemans, T. (2013). French adaptation of the Childhood Executive Function Inventory (CHEXI): Confirmatory factor analysis in a sample of young French-speaking Belgian children. *European Journal of Psychological Assessment, 29*(2), 149–155.

Catale, C., Meulemans, T., & Thorell, L.B. (in press). The Childhood Executive Function Inventory (CHEXI): Confirmatory Factorial analyses and cross-cultural clinical validity in a sample of 8–11 years old Children, *Journal of Attention Disorders*. DOI: 10.1177/1087054712470971.

Conners, C. K. (1970). Symptom patterns in hyperkinetic, neurotic, and normal children. *Child Development, 41*, 667–682.

Gioia, G. A., Andrews Espy, K. A., & Isquith, P. K. (2003). *The Behavior Rating Inventory of Executive Function—Preschool version*. Odessa, FL: Psychological Assessment Resources.

Gioia, G. A., Isquith, P. K., Guy, S. C., & Kenworthy, L. (2000). *The Behavioral Rating Inventory of Executive Function*. Odessa, FL: Psychological Assessment Resources.

Gioia, G. A., Isquith, P. K., Kenworthy, L., & Barton, R. M. (2002). Profiles of everyday executive function in acquired and developmental disorders. *Child Neuropsychology, 8*, 121–137.

Hinshaw, S. P., Scheffler, R. M., Fulton, B. D., Aase, H., Banaschewski, T., Cheng, W., et al. (2011). International variation in treatment procedures for ADHD: Social context and recent trends. *Psychiatric Services, 62*, 459–464.

Kayhan. (2010). *A validation study of the Executive Functioning Inventory: Behavioral correlates of executive functioning*. Master thesis, Bogaziçi University, Turkey.

Klingberg, T., Fernell, E., Olesen, P. J., Johnson, M., Gustafsson, P., Dahlström, K., et al. (2005). Computerized training of working memory in children with ADHD: A randomized, controlled trial. *Journal of the American Academy of Child and Adolescent Psychiatry, 44*, 177–186.

Lan, X., Legare, C. H., Ponitz, C. C., Li, S., & Morrison, F. (2011). Investigating the links between subcomponents of executive function and academic achievement: A cross-cultural analysis of Chinese and American preschoolers. *Journal of Experimental Psychology, 108*, 677–692.

Lehto, J., Juujärvi, P., Kooistra, L., & Pulkkinen, L. (2003). Dimensions of executive functioning: Evidence from children. *British Journal of Developmental Psychology, 21*, 59–80.

Mahone, M. E., Cirino, P. T., Cutting, L. E., Cerrone, P. M., Hagelthorn, K. M., Hiemenz, J. R., et al. (2002). Validity of the behavior rating inventory of executive function in children with ADHD and/or Tourette syndrome. *Archives of Clinical Neuropsychology, 17*, 643–662.

Naglieri, J. A., & Goldstein, S. (2013). *Cognitive Executive Functioning Inventory Manual*. Toronto: Multi-Health Systems.

Nigg, J. T., Willcutt, E. G., Doyle, A. E., & Sonuga-Barke, E. J. S. (2005). Causal heterogeneity in attention-deficit/hyperactivity disorder: Do we need neuropsychologically impaired subtypes? *Biological Psychiatry, 57*, 1224–1230.

Rothbart, M. K. (1981). Measurement of temperament in infancy. *Child Development, 52*, 569–578.

Rothbart, M. K., Ahadi, S. A., Hershey, K., & Fisher, P. (2001). Investigations of temperament at three to seven years: The Children's Behavior Questionnaire. *Child Development, 72*, 1394–1408.

Sabbagh, M. A., Xu, F., Carlsson, S. M., Moses, L. J., & Lee, K. (2006). The development of executive functioning and theory of mind: A comparison of Chinese and U.S. preschoolers. *Psychological Science, 17*, 74–81.

St. Clair-Thompson, H. L., & Gathercole, S. E. (2006). Executive functions and achievements on national curriculum tests: Shifting, updating, inhibition, and working memory. *Quarterly Journal of Experimental Psychology, 59*, 745–759.

Thorell, L. B., Eninger, L., Brocki, K. C., & Bohlin, G. (2010). Childhood Executive Function Inventory (CHEXI): A promising measure for identifying young children with ADHD? *Journal of Clinical and Experimental Neuropsychology, 32*, 38–43.

Thorell, L. B., Lindqvist, S., Bergman Nutley, S., Bohlin, G., & Klingberg, T. (2009). Training and transfer effects in executive functioning in preschool children. *Developmental Science, 12*, 106–113.

Thorell, L. B., & Nyberg, L. (2008). The Childhood Executive Functioning Inventory (CHEXI): A new rating instrument for parents and teachers. *Developmental Neuropsychology, 33*, 536–552.

Thorell, L. B., Veleiro, A., Siu, A. F. Y., & Mohamadi, H. (in press). Brief report. Examining the relation between ratings of executive functioning and academic achievement: Findings from a cross-cultural study. *Child Neuropsychology.* DOI:10.1080/09297049.2012.734294

Trevistan, Dias, & Mendes. (2012). *The Brazilian version of the Childhood Executive Functioning Inventory (CHEXI) for evaluation of children with ADHD.* Poster presented at Eunethydis 2nd International ADHD Conference, May 21–23, Barcelona, Spain.

Trevisan, B.T., Dias, N.M., Menezes, A., & Seabra, A.G. (2012) The Brazilian version of the Childhood Executive Functioning Inventory (CHEXI) for evaluation of children with ADHD. Poster presented at Eunethydis 2nd International ADHD Conference, May 21–23, Barcelona, Spain.

Veleiro, A., & Thorell, L. B. (2012). *Predictive power of executive function ratings on preschool math basic abilities.* Poster presented at the 12th International Conference of Psychology and Education, June 21–23, Lisbon, Portugal.

Vriezen, E. L., & Pigott, S. E. (2002). The relationship between parental report on the BRIEF and performance-based measures of executive function in children with moderate to severe traumatic brain injury. *Child Neuropsychology, 8*, 296–303.

Wiebe, S., Espy, K., & Charak, D. (2008). Using confirmatory factor analysis to understand executive control in preschool children: I. Latent structure. *Developmental Psychology, 44*, 575–587.

The Assessment of Executive Functioning Using the Delis Rating of Executive Functions (D-REF)

21

Jessica A. Rueter

Theory (or Conceptualization) Underlying the Test or Rating Scale

The Delis Rating of Executive Functions (D-REF) is a set of rating forms that measures the executive functioning (EXF) in individuals from ages 5 to 18. The D-REF contains three forms: Parent, Teacher, and Self. Each form consists of 36 items. The D-REF forms can be administered digitally using a computer or tablet or paper pencil format.

Theoretical Approach/ Conceptualization

The psychological construct of EXF underlies the D-Refs. EXF includes the different cognitive processes that people use to control and regulate their behavior to reach desired goals. A working definition of EXF as described by Horowitz:

- Is conscious, purposeful, and thoughtful
- Involves activating, orchestrating, monitoring, evaluating, and adapting different strategies to accomplish different tasks
- Includes an understanding of how people tap their knowledge and skills and how they stay motivated to accomplish their goals
- Requires the ability to analyze situations, plan and take action, focus and maintain attention,

and adjust actions as needed to get the job done (n.d., p. 1)

Key components of EXF are generally conceptualized into four broad areas: goal formation, planning, goal-directed behavior, and effective performance. For an individual to demonstrate adequate EXF, he/she must reflect on what it is he/she wants to accomplish, determine the next steps in order to achieve anticipated outcomes, engage in problem-solving behaviors to reach desired goal, and finally perform the action efficiently (D-REF, 2012).

Description of the Test or Rating Scale

The D-REF purpose is to survey executive functions that interfere with a child or adolescent's functioning and are likely to be a cause of concern and/or source of stress to the child, family, and/or school. Although the D-REF is not intended to be a diagnostic instrument by itself, it points to areas related to common disorders of childhood and adolescence. Therefore, the D-REF should not be used in isolation to make a diagnosis or decisions regarding classifications. Hence, its use should be part of a comprehensive evaluation (D-REF, 2012, pp. 27–28).

The D-REF is designed to be easily read and understandable for children, adolescents, and adults who have limited education levels. To complete the Parent and Teacher Rating Forms, a fourth-grade reading level is assumed. The

---

J.A. Rueter (✉)
University of Texas at Tyler, Tyler, TX, USA
e-mail: jrueter@uttyler.edu

Self-Rating Form assumes a third grade reading level. The items on the D-REF were specifically designed for quick administration to identify characteristics of EXF that the clinician can then use to further evaluate with standard EXF assessments such as the Delis-Kaplan Executive Function System (D-KEFS, 2001). Each D-REF form includes 36 items. Each item falls within one of the three domains: behavioral functioning (BF), emotional functioning (EMF), and EXF. The core index scores for each of the domains are derived from the items within each domain. In addition, the Total Composite Score (TC) is derived from the three core index T scores. The core index scores are available on all three forms. Each of the core index scores is reported as T scores with a mean of 50 and a standard deviation of 10. A T score of 50 represents average performance for a given age group. Scores of 40 and 60 reflect one standard deviation below and above the mean, respectively. The core index, abbreviations, and description are as follows:

- Behavioral Functioning (BF)—Assesses the child/adolescent's ability to regulate his/her behavior to meet the demands of the environment
- Emotional Functioning (EMF)—Assesses the child/adolescent's ability to regulate his/her emotions relative to the demands of the environment
- Executive Functioning (EXF)—Assesses the child/adolescent's higher level cognitive ability to effectively adapt and function within the demands of the environment
- Total Composite Score (TC)—Assesses the child/adolescent's ability to plan, execute, and regulate cognitive, emotional, and behavioral functions to adapt to the demands of the environment (D-REFS, 2012, p. 20)

Each index measures various aspects of EXF. The BF index measures hyperactivity, impulsivity, poor organization, insufficient self-monitoring, and difficulty following rules. The EMF Index measures poor frustration tolerance, emotional instability, sensitivity to criticism, anger control problems, and interpersonal issues. The EXF Index measures poor attention and working memory, cognitive rigidity, difficulty initiating/sustaining behavior, disorganization, and deficient problem-solving and decision-making skills (D-REF, 2012, p. 20). When high scores are obtained on a specific index, the clinician should review the individual items within the index to better understand the problems that were reported.

Clinical Index Scores

Clinical index scores were also derived to facilitate interpretation. The clinical index scores were based in part from Diagnostic and Statistical Manual of Mental Disorders, Fourth Edition, Text Revision (DSM-IV-TR; American Psychiatric Association, 2000). Additionally, exploratory factor analysis was conducted and indicated which items best fit together. There are four clinical indexes on the *Parent* and Teacher Rating Forms and three clinical indexes on the Self-Rating Form. The clinical indexes on the Parent, Teacher, and Self-Rating Forms are Attention/Working Memory (AWM), Activity Level/Impulse Control (AIC), and Compliance/Anger Management (CAM). The fourth index score on the Parent and Teacher Rating Forms is Abstract Thinking/Problem Solving. Each of the clinical index scores is reported as T scores with a mean of 50 and a standard deviation of 10. The clinical index, abbreviations, and description are as follows:

- Attention/Working Memory (AWM)—Assesses symptoms of inattention, deficient multitasking, forgetfulness, poor working memory, and disorganization
- Activity Level/Impulse Control (AIC)—Assesses symptoms of hyperactivity, impulsivity, and poor self-monitoring
- Compliance/Anger Management (CAM)—Assesses symptoms of mood lability, sensitivity to criticism, frustration tolerance, and rule-breaking
- Abstract Thinking/Problem-Solving (APS)—Assesses symptoms of concrete thinking, cognitive rigidity, and poor decision-making and problem-solving skills (D-REF, 2012, p. 24)

The clinical indexes were designed to be more specific than the core indexes. The AWM Index contains items related to sustained attention, shifting attention, forgetfulness, multitasking in working memory, and distractibility. The AIC Index measures hyperactivity, impulsivity, and disorganization. CAM Index measures poor frustration tolerance, sensitivity to criticism, poor anger control, poor self-monitoring, and problems with following rules. The APS measures cognitive flexibility, abstraction, creative thinking, and general problem-solving skills (D-REF 2012, p. 24).

The three rating forms are similar in content and organization. For each form, the core and clinical indexes have the same number of items. Because the content in the Parent and Teacher forms is nearly identical, comparison between raters can be made. The Self-Rating Form is written from the perspective of the child or adolescent. Although the content is similar, it is different from the other rating forms. When comparing items across forms, it is important to note that the items may not contain identical content. Thus, it is recommended that the forms be printed out when conducting item-by-item comparisons (D-REF, 2012, pp. 26–27).

Administration and Scoring

The D-REF was developed for speakers of English. The content has not been validated for non-English speakers. If the raters do not have the requisite reading skills to complete the forms, the items can be read aloud or the text to speech feature can be activated with online administrations.

Careful selection of respondents is critical to obtaining valid results. Individuals completing the rating forms (i.e., respondents) should have knowledge of the daily behaviors of the child/adolescent and who has frequent and extended contact with the child/adolescent. In addition, respondents selected should have opportunities to observe those behaviors that are being evaluated on the D-REF. The following lists the requirements for choosing respondents for the Parent and Teacher Rating Forms.

- Parent Rating Form—Respondents may include the parents or other primary caregivers who are living with the child and who have the opportunity to observe the child's behavior and activities. Respondents may include family members such as aunts, uncles, foster parents, grandparents, and other caregivers.
- Teacher Rating Form—The child's primary teacher should complete the rating form. When a child has multiple teachers, select the teacher who has the most current and extensive interactions with the child. Respondents may include teachers, paraprofessionals, and other school personnel who have familiarity with the child's behavior and activities in a structured school or service delivery environment (D-REF, 2012, p. 32).

It is essential to establish and maintain rapport with the child/adolescent prior to beginning the assessment process. The evaluator should provide information about the purpose of the assessment and how the results will be used as part of the comprehensive assessment process in making diagnostic decisions. Discuss the confidentiality of the assessment process and who will have access to the results. In addition, determine that the respondents have adequate knowledge of the child's/adolescent's behaviors, are able to read and understand the directions and item, and will respond honestly and objectively. If the respondent does not possess the knowledge and skills to rate the items appropriately, the examiner should select other respondents. If the respondent does not have the required reading abilities to complete the forms, the items may be read aloud to the respondent or the text to speech function may be activated with online administration (D-REF, 2012, p. 33).

A brief description of the items should be provided. The evaluator should remind respondents to read directions on the form and circle/select a rating for each of the items presented. In addition, the evaluator should inform the respondents to complete every item, even if some items do not apply or are difficult to rate. Finally, the evaluator should be available to answer any questions or provide help to respondents while they are completing the rating forms. Respondents' questions may be answered before, during, or after completion of the rating forms. Provide as much detail as possible when answering questions and give instructions that allows the respondent to provide reliable and valid information regarding the child's/adolescents behavior that are being measured on the rating forms (D-REF, 2012, p. 33).

Generally, the D-REF will be completed online. No other material besides the Internet is necessary for online administration. Discuss the instructions and provide all required information prior to beginning the online administration to ensure that the online administration is valid and appropriate (D-REF, 2012, p. 33).

It is preferable that the paper/pencil rating forms be completed in a controlled environment such as a clinic, school, or office setting. However, on occasion, it may be necessary for a respondent to complete the rating form off-site. In these cases, it is important that the evaluator provides information about the instructions and contact information in case questions arise. Additionally, supply the respondent with specific instructions about when and where the rating form should be returned. If the rating form is to be mailed to a respondent, write or verbally verify the information regarding the assessment and instructions for completing and returning the rating form (D-REF, 2012, p. 34).

Evaluators should obtain ratings from multiple respondents. Information from multiple respondents improves the validity of the assessment and can provide information about a child/adolescent's behavior and activities from a variety of settings. Moreover, multiple respondents can provide information about the child/adolescent's consistency of behavior and activities across settings and from the unique perspectives of different respondents. This information improves the level of decision-making regarding the diagnosis and assists in planning programs and services for the child/adolescent (D-REF, 2012, p. 34).

When completing rating scales, the evaluator should consider and take into account that respondent's ratings are a reflection of his or her perceptions and a willingness to be entirely honest regarding these perceptions. Thus, it is critical that evaluators carefully select respondents and that rapport and communication be established (D-REF, 2012, p. 36).

Each D-REF rating form takes approximately 10 min to complete. The forms may be given simultaneously, but the respondents should not be allowed to influence each other ratings. To administer the paper/pencil version of the D-REF, the Parent, Teacher, and Self-Rating Forms can be obtained from the Resource Library within the online administration. In addition to printing off the rating forms, the scoring system will need to be printed. If necessary, provide a pencil or pen for recording responses and offer a clipboard to provide the respondent with a flat writing surface (D-REF, 2012, p. 36).

After the respondents have completed the rating forms, the evaluator should review the forms for any missing data or for items that have more than one rating circled. Reconcile differences before entering data into scoring software. Once the rating forms have been reviewed, enter the responses into the scoring software. The scoring and reporting of D-REF data is automated once data is entered. If three or more items are missing from the Behavioral Functioning, EMF, or EXF indexes, the raw score for the index will not be calculated and thus a T score cannot be derived. If any index T score cannot be derived, the Total Composite Score cannot be derived. If fewer than three responses are missing for a particular index, a total raw score may still be derived for that index. With respect to the clinical indexes, no prorated scores will be calculated if there are missing data. If a response for any item is missing that contributes to the clinical index scores (i.e., AWM, AIC, CAM, APS), neither the raw score nor the index T score will be calculated (D-REF, 2012, p. 39).

To eliminate the possibility of missing data, online administration is recommended. The D-REF can be administered completely online. Instructions for each of the rating forms are presented on-screen and can be read aloud using the text to speech function. All items are presented on-screen and administration and scoring of the D-REF is fully automated. Online administration eliminates the possibility of missing data and ensures that all scores can be calculated (D-REF, 2012, p. 39).

Descriptive Classification

The descriptive classification of the three indexes, clinical indexes, and Total Composite is described as falling within a certain level of performance: within normal limits, borderline, elevated, mildly to moderately elevated, and severely elevated (D-REF, 2012, p. 49).

Level of performance is important for providing an estimate regarding the presence and severity of an impairment or identifying areas of typical functioning. T scores falling between 10 and 54 with a percentile rank of 1st–69th are considered to be within the normal range of functioning. T scores of 55–59 with a percentile rank of 70th–83rd are considered to be within normal limits/borderline elevated and may indicate possible problem areas for the child or adolescent. Borderline scores may suggest that there be further monitoring and evaluation, either at a later date or using additional measures. T scores of 60–70 with a percentile rank of 84th–98th are considered to be in the mildly to moderately elevated range. T scores of 70 and above are considered severely elevated and indicate significant symptoms for the child's age (D-REF, 2012, p. 50).

Standardization, Norms, and Psychometrics

A pilot was conducted as part of the development of the D-REF to evaluate the psychometric properties of the rating form items. The pilot study evaluated the psychometric properties of each item and obtained preliminary estimates of the clinical sensitivity of each form and for each item. The goal of the pilot was to reduce the item set by about half for each form. Samples for the pilot included 217 parents, 164 teachers, and 86 children and adolescents. The pilot study included data from persons diagnosed with attention deficit hyperactivity disorder: 32 Parent Rating Forms, 27 Teacher Rating Forms, and 15 Self-Rating Forms were collected and analyzed to identify items with high clinical sensitivity. Classic test theory models were utilized in analyzing data. Items with low total correlations or poor clinical sensitivity were eliminated from the scales. In addition, exploratory factor analysis was used to determine which items best fit a three-index model of behavior, emotional, and EXF. A final set of 40 items was selected for the Self-Rating Form and 41 items for each of the Parent and Teacher Rating Forms (D-REF, 2012, p. 40).

The goal of the standardization was to derive norms and to provide reliability, validity, and clinical evidence for the final versions of the three forms. The three forms are parallel in structure and, to the degree possible, in item content. Items were evaluated for clinical sensitivity, fit within the index structure, and psychometric properties. The resulting final forms each contain 36 items (D-REF, 2012, p. 41).

Characteristics of the Standardization sample

The normative data are based on national samples representative of the United States population ages 5–18 years. US 2010 Census data provided the stratification according to the following variables: age, sex, race/ethnicity, and education level. The sample included 1,062 individuals (parent, $N=500$; teacher, $N=342$; self, $N=220$). The number of individuals varied among the age groups and by the three categories of respondents (D-REF, 2012, p. 45).

Six age bands were collected: 5–6, 7–8, 9–10, 11–12, 13–14, and 15–18 were used in the normative sample. Only ages 11–18 were collected for the Self-Rating Form sample (D-REF, 2012, p. 45).

Sex was roughly stratified to be 50 % male and 50 % female by age group and the overall sample. It was not specifically stratified to be consistent with U.S. Census data (D-REF, 2012, p. 45).

The proportion of White, African American, Hispanic, and other racial/ethnic group examinees included in the normative sample were approximately based upon the proportion of individuals within each age band according to the 2010 US Census data (D-REF, 2012, p. 45).

Parent education level was used for all examinees. Information on parent education level was obtained by a single question on the consent form that asked the parent or examinee to indicate the highest grade completed by each parent residing in the home. For examinees with both parents, the average of the two education levels was used. The normative sample was stratified according to the following three parent education levels: 0–11 years (never graduated high school), 12 years (high school graduate or equivalent), and 13 or more years (education beyond high school) (D-REF, 2012, p. 45).

The four major regions of the 2010 US Census reports: Northeast, South, Midwest, and West were used to stratify the sample by region. Examinees were selected from the four regions across the United States but were not specifically stratified by region. Differences in region in D-REF ratings were not expected and were not identified (D-REF, 2012, p. 46).

To identify potential examinees, field examiners and Pearson sampling staff were employed. All potential examinees to be included in the standardization were medically and psychiatrically screened using a self-report questionnaire. The D-REF was sent to the Parents and Teachers of eligible examinees. All eligible examinees or respondents were paid an incentive fee to participate and each test protocol was reviewed after being completed by the respondent. Those individuals who did not meet the eligibility criteria were not included in the standardization process (D-REF, 2012, p. 46).

Reliability of the Scales

Reliability is an indication of the degree to the accuracy, stability, and consistency of the test scores. A reliable instrument will have small measurement errors and will produce consistent results across test administrations. There are three types of reliability: (1) alternate-form reliability or internal consistency which allows for generalizing to other test items, (2) stability or test-retest reliability which allows for generalizing to different times, and (3) interrater or interscorer reliability for generalizing to different scorers (Salvia, Ysseldyke, & Bolt, 2007). Multiple methods exist to determine the reliability of an instrument.

The D-REF provides reliability information regarding internal consistency measures and stability measures. However, it does not report information regarding interrater or interscorer reliability. The following describes information about the internal consistency measures and stability measures of the D-REF.

Internal Consistency

The Parent Rating Form demonstrates good to very good reliability for the core index scores with reliability coefficients across the age bands ranging from .86 to .97. The Total Composite Score for the Parent Rating Form demonstrates the highest reliability coefficients across the age bands ranging from .95 to .97. The clinical indexes generally have moderate to good internal consistency with reliability coefficients across the age bands ranging from .76 to .94. This was an expected finding given the clinical indexes reduced item set. The lowest reliability of all the parent indexes was obtained on the Clinical Index of AIC with reliability coefficients across the age bands ranging from .76 to .84. All reliabilities are good for that index except for a few moderate reliabilities in specific age groups (D-REF, 2012, p. 52).

The Teacher Rating Form demonstrates good to very good internal consistency for the core index scores with reliability coefficients across the age bands ranging from .80 to 99. The Total Composite Score for the Teacher Rating Form demonstrates the highest reliability coefficients across the age bands ranging from .93 to .99. The Teacher Rating Form clinical indexes are in the moderate to very good range with reliability coefficients across the age bands ranging from .76 to 97. The lowest reliability is obtained on the AIC and CAM indexes for age band 11–12 with reliability coefficient of .76 for each of these indexes (D-REF, 2012, p. 54).

The reliabilities for the core index scores from the Self-Rating Form demonstrate moderate to good reliability with reliability coefficients across the age bands ranging from .77 to .91. The

Total Composite Score demonstrates the highest reliability coefficients across the age bands ranging from .91 to .96. The clinical indexes for the Self-Rating Form demonstrate low to good reliability with reliability coefficients across the age bands ranging from .64 to .87. Low reliability coefficients are observed in one age group on the AWM Index ranging from .64 to .70 (D-REF, 2012, p. 54).

As is expected the highest reliability coefficients are observed on the Total Composite Score. The Total Composite Score is based on the three core indexes. The next most reliable index score is the EXF across all three forms. This is the core index that has the most number of items contributing to it, which allows to its high degree of reliability. The clinical indexes have lower reliability coefficients but adequate reliabilities across the three forms, which are attributed to the smaller number of items contributing to each of the indexes. Of the clinical indexes, the AIC Index has the lowest internal consistency on the Parent and Teacher Rating Forms, and the AWM Index has the lowest internal consistency for the Self-Rating Form. The D-REF demonstrates a high degree of internal consistency, particularly for the core index scores and the Total Composite Score (D-REF, 2012, p. 54).

The D-REF also reports reliability information for specific groups of individuals. The groups are attention deficit hyperactivity disorder, combined type (ADHD-C); attention deficit hyperactivity disorder, inattentive type (ADHD-I); autistic disorder (AUT); Asperger's disorder (ASP); and learning disorder (LD). However, the sample sizes for clinical groups are considerably smaller than the normative sample, which will impact the reliability data. Also, some clinical groups may have smaller score ranges than the normative sample which could lower reliability results. Thus, the reader is encouraged to refer to the D-REF manual for reliability and sample size for each of the clinical groups.

A property of reliability is the standard error of measurement (SEM). The SEM is an index of test error. The estimate of a score's reliability is used in obtaining the SEM. The SEM is used in calculating the confidence intervals around the scores (Salvia et al. 2007). The SEM has an inverse relationship to the reliability coefficient. Thus, the greater the reliability of a test, the smaller the SEM and the greater the confidence the user can have in the precision of the observed test score (D-REF, 2012, p. 57). The D-REF reports SEM across the age bands for the three indexes, clinical indexes, and Total Composite Scores for each of the three forms. As is expected, the Total Composite Scores across the age bands for the three forms report the lowest SEM while the clinical indexes generally report the highest degree of SEM across the age bands for each of the three forms.

Test-Retest Stability

The second type of reliability that the D-REF reports is the test-retest stability, which allows for generalizing to different times. To determine the stability of performance on the D-REF scores, the Parent Rating Form was completed by 50 individuals on two occasions, the Teacher Rating Form was completed by 54 individuals on two occasions, and the Self-Rating Form was completed by 47 individuals on two occasions. The test-retest interval ranged from 7 to 56 days. The Parent Rating Form demonstrated high correlations between first and second assessments for the core indexes with relatively little change, on average, for parent-rated symptoms on healthy controls (D-REF, 2012, p. 60).

The Teacher Rating Form indicates higher test-retest correlations in general when compared to the Parent Rating Form. All of the indexes on the Teacher Rating Form have correlations above .80, including the AIC (.87) and CAM index (.88). There was no significant increase or decrease between the first assessment and second assessment on the indexes and Total Composite Score (D-REF, 2012, p. 60).

The core index ratings for the Self-Rating Form are more variable with corrected stability coefficients ranging from .70 to .80 and a Total Composite Score corrected stability coefficient of .82. Meanwhile, the clinical indexes report corrected stability coefficients ranging from .62 to .72 (D-REF, 2012, p. 60).

Use of the Test or Rating Scale/Case Study

George is a 9-year-old boy in the fourth grade whose parents have referred him for a psychoeducational evaluation. The reason for the referral is concerns regarding behavior and academic performance in the home and in school. George's mother reports that there were no complications during the birthing process and that he was delivered at full-term. George met all his developmental milestones such as sitting up, crawling, and walking within the normal period of development. George has had no serious medical illnesses or hospitalizations. He sees a physician and a dentist for routine checkups and is not on any prescribed medications.

George's parents reported that George was a very active toddler and that in preschool he would get in trouble for minor misbehaviors such as playing with a toy during story time, not following directions, and talking too much. His preschool teacher reported that George was overactive and did not seem to have the same attention span as that of his classmates.

When George entered kindergarten, his teacher reported similar behaviors consistent with reports from his preschool teacher. George's kindergarten teacher also reported that George was frequently distracted and had to be redirected multiple times during academic tasks, but once redirected, George would complete the tasks required. In grades 1 to 3, George's teachers reported that George struggled with completing tasks without redirection, excessive talking, was easily distracted by his peers and outside stimuli and that his work was often incomplete and messy.

In the fourth grade, George is experiencing similar behaviors such as the ability to stay on task, is frequently distracted by his peers and outside stimuli, and has difficulty staying seated. Due to these behaviors, George is having difficulty in keeping up with his classwork and is falling further and further behind in the academic content areas. He is currently failing most of his classes due to incomplete work and does not consistently turn his work in on time. George's teacher is concerned about his progress and his disruptions in the classroom. His teacher has consulted with George's parents frequently about her concerns and his performance in the classroom.

George approached the testing situation easily and rapport was quickly established. All standardized test administration procedures were followed according to the instruments' manuals.

George was administered the *Kaufman Assessment Battery for Children, Second Edition* (KABC-II) which is an individually administered norm-referenced instrument designed to measure the processing and cognitive abilities of children and adolescents aged 3–18 (Kaufman & Kaufman, 2004a).

The results of the KABC-II indicate overall cognitive ability in the average range of functioning. Simultaneous, Learning, and Knowledge indexes were in the average range. However, the Sequential index, a measure of short-term memory, and the Planning Index, a measure of executive processes, were in the below-average range.

George was administered Form B of the *Kaufman Test of Educational Achievement, Second Edition* (KTEA-II) which is an individual norm-referenced achievement test that is appropriate for individuals aged 4½–25 (Kaufman & Kaufman, 2004b). The results of the KTEA-II indicate average to above average performance in the areas evaluated (reading, mathematics, writing, and spelling).

George was also administered the D-REF as part of the psychoeducational evaluation to identify problematic behaviors that may be contributing to his academic difficulties and to identify possible deficits in EXF. George's Parents and Teacher completed the D-REF online.

The results of the Parent Rating Form indicate that George is experiencing severe elevation in Behavioral Functioning Index with a T score of 71 and a confidence interval of 65–74. This suggests that George is having difficulties with impulsivity, maintaining attention, and disruptive behaviors as compared to same-aged mate peers. On the EMF Index, an obtained T score of 55 with a confidence interval of 50–59 is within the normal limits albeit borderline functioning. At times George exhibited some problems associated with poor frustration tolerance, noncompliance, and problems interacting with his peers. On the EXF Index, an obtained T score of 63 with a confidence interval of 60–65 is in the mild to moderately elevated range. This indicates that George likely exhibits EXF problems associated with inattention, distractibility, poor working memory, difficulty initiating and sustaining effort, poor self-monitoring, cognitive rigidity, low behavioral productivity, poor abstract reasoning, and poor planning skills as compared to his same-aged mate peers. The Total Composite T score of 64 and a confidence interval of 60–67 suggest a mild elevation in EXF-related behavioral problems.

George's fourth-grade teacher, Mrs. Smyth, completed the D-REF to provide information regarding his functioning in school. The Behavioral Functioning Index indicated a severe elevation with a T score of 71 and a confidence interval of 65–74. The EMF index T score of 58 and a confidence interval of 53–62 are within the normal limits/borderline range. The EXF T score of 66 and a confidence interval of 62–69 indicate mild to moderate elevation. The Total Composite T score of 65 and a confidence interval of 62–67 are in the moderately elevated range and suggest that there is a significant degree of general behavioral, emotional, and EXF problems. The results reported by Ms. Smyth were consistent with George's parents and indicate that George is experiencing significant problems regulating his behaviors at home and at school and is having difficulty with his ability to function appropriately for his age.

As a result of the evaluation, a behavioral plan was developed to assist George with his behavior in the classroom. In addition, George was taught self-monitoring strategies to help increase his attention to tasks and completion of assignments. The family was referred to their physician for additional evaluation related to attention and hyperactivity. The results from the physician indicated a diagnosis of attention deficit hyperactivity disorder. Under the Individuals with Disabilities Education Act (2004), George could be eligible as a student with an Other Health Impairment. However, in an individual education program meeting, the parents and the school mutually agreed that George's needs could best be met without an eligibility label as long as accommodations, such as a behavior plan, and differentiated instruction were provided in the general education setting.

Validity

The validity of an instrument refers to the degree to which the instrument measures what it is intended to measure and the appropriateness to which inferences can be made based on the results. The process of validity is centered on two key principles: (1) what a test measures and (2) how well it measures it. Accordingly, validity is a matter of degree. Therefore, studies of validity for a particular instrument should continue long after the publication date of the instrument (Sattler, 2008). The evidence of validity conducted on the D-REFS includes content, construct, and concurrent validity measures.

Content Validity

Content validity is whether the items on the instrument adequately measure the trait or domain in which it is supposed to measure (Sattler, 2008). The goal in establishing content validity on the D-REF is to ensure that the items on each rating form adequately samples aspects of EXF. In an attempt to establish content validity, comprehensive literature reviews were conducted to examine the content of the D-REF. Moreover, customer feedback and recommendations were gathered and considered in developing the initial items on the instrument. Item content was linked to specific executive functions to the degree possible although some of the item content represents general issues associated with behavior, emotions, and cognitive functions. In addition, other considerations were the degree to which certain behaviors presented themselves as concerns for parents during clinical evaluations. Clinical experiences with children and adolescents with significant behavior disorders assisted in identifying issues and behaviors related to difficulties in EXF that could assist in developing items. The pilot and standardization of the D-REF helped to determine the final items included in the instrument. Upon completion of the standardization phase, items were reexamined for content, bias, and psychometric properties (D-REF, 2012, p. 62).

Construct Validity

Construct validity is the extent to which a test measures a psychological construct or trait. Two components of construct validity should be considered. The first is convergent validity: how well performances on different measures of the same domain in different formats (e.g., teacher and parent forms) correlate positively. The second is divergent validity in which different constructs are chosen to measure different characteristics and thus should not correlate with one another (Sattler, 2008).

The construct validity information presented for the D-REF includes information regarding intercorrelations of index scores within form and intercorrelations of index scores across forms for the normative sample and clinical samples.

With respect to the intercorrelations of index scores within form, for the Parent Rating Form, correlations are in the moderate to high range for the three main indexes of Behavioral Functioning, EMF, and EXF of .84, .71, and .78, respectively. EMF and EXF indexes have the lowest correlation. Clinical index correlations for the Parent Rating Form range from .78 to .88 (D-REF, 2012 p. 65).

With respect to the intercorrelations of index scores within form, for the Teacher Rating Form, correlations are in the moderate to high range for the three main indexes of Behavioral Functioning, EMF, and EXF of .82, .67, and .74, respectively. EMF and EXF indexes have the lowest correlation. Clinical index correlations for the Teacher Rating Form range from .76 to .87 (D-REF, 2012 p. 65).

With respect to the intercorrelations of index scores within form, for the Self-Rating Form, correlations are in the moderate range for the three main indexes of Behavioral Functioning, EMF, and EXF of .78, .67, and .68, respectively. EMF and EXF indexes have the lowest correlation. Clinical index correlations for the Self-Rating Form range from .77 to .81 (D-REF, 2012 p. 65).

Across form, comparisons indicate consistently lower correlations as compared to within form correlations. The Parent to Teacher Rating Form correlations generally fall within the low to moderate range. The Parent to Self-Rating Form correlations fall in the low to moderate range. For the core indexes, the Behavioral Functioning Index correlates .53 across forms, .44 for EMF, and .53 for EXF. The Teacher to Self-Rating Form correlations fall in the low to moderate range. Correlations between forms for the Behavioral Functioning, EMF, and EXF indexes are .40, .45, and .27, respectively (D-REF, 2012, p. 67). The reader is encouraged to refer to the chapter on validity regarding more in-depth information on across form comparisons, clinical samples, and special group validity measures.

Concurrent Validity

Concurrent validity refers to the relationship of test scores to a currently available criterion measure (Sattler, 2008). Evidence of concurrent validity is often provided through an examination of the instrument's relationship to other instruments on the market designed to measure the same or similar constructs or traits. The D-REF presents two types of evidence for concurrent validity. The first measure was a study conducted to examine the relationship between the D-REF and the Behavior Rating Inventory of EXF (BRIEF; Gioia, Isquith, Guy, & Kenworthy, 2000) scores. The second examines a series of special group studies that gives credence to clinical evidence supporting the validity of D-REF scores in disorders most closely associated with deficits in EXF (D-REF, 2012, p. 71).

The BRIEF is an established rating scale to assess EXF for children and adolescents ages 5–18. It consists of 86 items and has three rating forms (i.e., Parent, Teacher, and Self). The BRIEF is comprised of three higher-order indexes: Behavioral Regulation, Metacognition, and Global Executive Composite.

When examining the relationship between the D-REF and the BRIEF, there are generally higher correlations on the Parent and Teacher Rating Forms. However, the Parent and Teacher Rating Forms indicate more variability in reported symptoms. The lowest correlations were demonstrated on the Self-Rating forms. The highest observed correlation is between the D-REF Total Composite and the BRIEF General Executive Composite with a correlation coefficient of .75 for the Parent Rating Form, a correlation coefficient of .82 for the Teacher Rating Form, and a correlation coefficient of .72 for the Self-Rating Form (D-REF, 2012, pp. 76–78).

The second method in which the D-REF presents evidence for concurrent validity is through a series of special group studies. These studies investigated the degree to which the D-REF identifies behaviors that represent EXF deficits typically related to these special groups (i.e., attention deficit hyperactivity disorders combined type and inattentive type, autism disorders, ASPs, and learning disorders). For more information regarding the special group studies, the reader should refer to the D-REF manual.

Summary and Conclusions

The D-REF is a set of three rating forms (i.e., Parent, Teacher, and Self) whose purpose is to assess the EXF in individuals from ages 5 to 18. As part of a multimethod assessment approach, the D-REF may be useful in identifying areas of functioning (i.e., behavioral, emotional, and executive) that are likely having an adverse impact on the child or adolescent, family, and/or school. A multimethod assessment approach (a) collects information from a variety of sources; (b) uses several assessment methods, including norm-referenced, interviews, observations, and informal assessment procedures; and (c) evaluates several areas as needed (e.g., intelligence, memory, EXF, achievement) (Sattler, 2008). When used appropriately and for the purposes in which it was intended to be used, the D-REF can assist the clinician in identifying areas of concern for the individual in which they are evaluating.

References

American Psychiatric Association. (2000). *Diagnostic and statistical manual of mental disorders* (4th ed., text rev.). Washington, DC: Author.

Delis, D. C. (2012). *Delis rating of executive functioning*. Bloomington, MN: Pearson.

Delis, D. C., Kaplan, E., & Kramer, J. H. (2001). *Delis-Kaplan executive function system*. San Antonio, TX: Harcourt Assessment.

Gioia, G. A., Isquith, P. K., Guy, S. C., & Kenworthy, L. (2000). *Behavior Rating Inventory of Executive Function*. Odessa, FL: Psychological Assessment Resources.

Horowitz, S. H. (n.d.) *Executive functioning and learning disabilities* (pp. 1–3). Retrieved on March 10, 2013 from http://www.ncld.org/types-learning-disabilities/executive-function-disorders/executive-functioning-learning-disabilities.

Individuals with Disabilities Education Improvement Act (IDEA 2004), Pub L. No. 108–446, 20 U.S.C. §§ 1400 et seq. (2004).

Kaufman, A. S., & Kaufman, N. L. (2004a). *Kaufman assessment battery for children* (2nd ed.). Circle Pines, MN: AGS.

Kaufman, A. S., & Kaufman, N. L. (2004b). *Kaufman test of educational achievement* (2nd ed.). Circle Pines, MN: AGS.

Salvia, J., Ysseldyke, J. E., & Bolt, S. (2007). *Assessment in special and inclusive education*. Boston, MA: Houghton Mifflin.

Sattler, J. M. (2008). *Assessment of children: Cognitive applications* (5th ed.). La Mesa, CA: Jerome M. Sattler.

Cross-Battery Approach to the Assessment of Executive Functions

22

Dawn P. Flanagan, Vincent C. Alfonso, and Shauna G. Dixon

Introduction

Increasing attention has been paid to the role of executive functions in school learning and achievement in recent years (e.g., Dawson, 2012; Maricle & Avirett, 2012; McCloskey, 2012; Meltzer, 2007, 2012; Miller, 2007, 2013). For example, within the emerging subdiscipline of school neuropsychology, attempts have been made to integrate psychometric and neuropsychological theories in an effort to better understand brain–behavior relationships (e.g., Flanagan, Alfonso, Ortiz, & Dynda, 2010; Miller, 2007). In addition, some intelligence test developers offer a cognitive processing model as a basis for interpreting test performance and provide clinical clusters, such as "executive processes," "cognitive fluency," and "broad attention" as part of their battery (e.g., WJ III NU; Woodcock, McGrew, & Mather, 2001, 2007). Other test authors developed tests that more directly purport to measure executive functions, including planning and attention. For example, the Kaufman Assessment Battery for Children, second edition (KABC-II; Kaufman & Kaufman, 2004), although based on the Cattell–Horn–Carroll (CHC) theory of the structure of cognitive abilities, maintains its roots in the Lurian model of cognitive processing and measures "Fluid Reasoning (Gf)/Planning," for example. Likewise, the Cognitive Assessment System (CAS; Das & Naglieri, 1997) is based on a Lurian cognitive processing theory of intelligence and measures planning, attention, and simultaneous and successive processes, of which the former two are often conceived of as executive functions (Maricle & Avirett, 2012; Naglieri, 2012).

Despite some references and inferences made to executive functions, most developers of intelligence tests have not addressed executive functions directly and, other than the Delis-Kaplan Executive Function System (D-KEFS; Delis, Kaplan, & Kramer, 2001), there do not appear to be any other psychometric cognitive batteries that were designed expressly for the purpose of assessing executive functions (McCloskey, Perkins, & Van Divner, 2009). Moreover, tests that include measures of executive functions, such as the NEPSY-II (Korkman, Kirk, & Kemp, 2007), do not provide a rationale for the selection and inclusion of specific tasks based on an overarching model of executive capacities (McCloskey et al.). Since most intelligence batteries do not measure executive functions well and since most neuropsychological batteries do not measure a broad range of executive functions, a flexible battery approach is needed to test hypotheses about an individual's executive functions.

Portions of this chapter were adapted from Flanagan, Oritz, and Alfonso (2013). *Essentials of cross-battery assessment*, third edition. Hoboken, NJ: Wiley.

D.P. Flanagan (✉) • S.G. Dixon
CEO, St. Johns University, Queens, NY, USA
e-mail: flanagad@stjohns.edu

V.C. Alfonso
Fordham University, New York City, NY, USA

There is general consensus in the research literature that executive functions consist of separate but related cognitive processes. Although researchers have not agreed on the components of executive functions, there is consensus that they consist of several domains, namely, initiating, planning, and completing complex tasks; working memory; attentional control; cognitive flexibility; and self-monitoring and regulation of cognition, emotion, and behavior (see Maricle & Avirett, 2012 for a discussion). For the purpose of this chapter, we focus on the major functions of the frontal-subcortical circuits of the brain, including planning, focusing and sustaining attention, maintaining or shifting sets (cognitive flexibility), verbal and design fluency, and use of feedback in task performance (i.e., functions of the dorsolateral prefrontal circuit), as well as working memory (i.e., functions of the inferior/temporal posterior parietal circuit; Miller, 2007). We chose to focus mainly on this subset of executive functions because our intelligence, cognitive, and neuropsychological batteries can provide information about them. However, it is important to remember that this selected set of executive functions, because they are derived from performance on intelligence, cognitive, and neuropsychological batteries, assists in understanding an individual's executive function capacities when directing perception, cognition, and action in the *symbol system arena* only (McCloskey et al., 2009). Practitioners will need to supplement these instruments when concerns about executive function capacities extend into the intrapersonal, interpersonal, and environmental arenas. Nevertheless, focus on executive function capacities in the symbol system arena, via the use of standardized tests, is useful in school settings to assist in understanding a child's learning and academic production (McCloskey et al.).

The purpose of this chapter is to describe how the cross-battery assessment (XBA) approach can be used to measure a selected set of executive functions, particularly those that are relevant to the cognition domain (e.g., reasoning with verbal and visual-spatial information). Although the XBA approach is based primarily on CHC theory—a theory that does not include a specific or general construct of executive functioning—recently, it was integrated with neuropsychological theories and applied to neuropsychological batteries (Flanagan et al., 2010; Flanagan, Ortiz, & Alfonso, 2013). More specifically, the current iteration of the XBA approach identifies specific components of executive functions and provides guidelines for measuring those components (Flanagan et al., 2013). This chapter will describe the XBA approach and provide a brief summary of CHC theory. Next, this chapter will describe how an integration of CHC and neuropsychological theory and research can be used to inform test selection as well as quantitative and qualitative interpretations of specific executive functions in the cognition domain. This chapter will also include brief examples of cross-battery assessments from which information about executive functions may be garnered.

The Cross-Battery Assessment Approach

As our understanding of cognitive abilities continues to unfold and as we begin to gain a greater understanding of how school neuropsychology will influence the practice of test interpretation, it seems clear that the breadth and depth of information we can obtain from our cognitive and neuropsychological instruments is ever increasing. In light of the recent expansion of CHC theory and the integration of this theory with neuropsychological theories, it will remain unlikely that an individual intelligence, cognitive ability, or neuropsychological battery will provide adequate coverage of the full range of abilities and processes that may be relevant to any given evaluation purpose or referral concern. The development of a battery that fully operationalizes CHC theory, for example, would likely be extremely labor intensive and prohibitively expensive for the average practitioner, school district, clinic, or university training program. Therefore, flexible battery approaches are likely to remain essential within the repertoire of practice for most professionals. By definition, flexible battery approaches offer an efficient and practical method by which practitioners may evaluate a broad range of cognitive abilities and processes, including executive functions. In this section, we summarize one such flexible battery approach, XBA, because it is grounded in a well-validated theory and is based on sound psychometric principles and procedures.

The XBA approach was introduced by Flanagan and her colleagues over a decade ago (Flanagan & McGrew, 1997; Flanagan, McGrew, & Ortiz, 2000; Flanagan & Ortiz, 2001; McGrew & Flanagan, 1998). It provides practitioners with the means to make systematic, reliable, and theory-based interpretations of cognitive, achievement, and neuropsychological instruments and to augment any instrument with subtests from other batteries to gain a more complete understanding of an individual's strengths and weaknesses (Flanagan, Alfonso, & Ortiz, 2012; Flanagan, Ortiz, & Alfonso, 2007). Moving beyond the boundaries of a single cognitive, achievement, or neuropsychological battery by adopting the theoretically and psychometrically defensible XBA principles and procedures allows practitioners the flexibility necessary to measure the cognitive constructs and neurodevelopmental functions that are most germane to referral concerns (e.g., Carroll, 1998; Decker, 2008; Kaufman, 2000; Wilson, 1992).

According to Carroll (1997), the CHC taxonomy of human cognitive abilities "appears to prescribe that individuals should be assessed with respect to the total range of abilities the theory specifies" (p. 129). However, because Carroll recognized that "any such prescription would of course create enormous problems," he indicated that "[r]esearch is needed to spell out how the assessor can select what abilities need to be tested in particular cases" (p. 129). Flanagan and colleagues' XBA approach was developed specifically to "spell out" how practitioners can conduct assessments that approximate the total range of cognitive and academic abilities and neuropsychological processes more adequately than what is possible with any collection of co-normed tests. And, for the purpose of this chapter, the XBA approach will spell out how practitioners can measure specific CHC abilities from which information about a subset of executive functions within the cognition domain may be derived.

In a review of the XBA approach, Carroll (1998) stated that it "can be used to develop the most appropriate information about an individual in a given testing situation" (p. xi). More recently, Decker (2008) stated that the XBA approach "may improve…assessment practice and facilitate the integration of neuropsychological methodology in school-based assessments…[because it] shift[s] assessment practice from IQ composites to neurodevelopmental functions" (p. 804).

Noteworthy is the fact that assessment professionals "crossed" batteries well before Woodcock (1990) recognized the need to do so and before Flanagan and her colleagues introduced the XBA approach in the late 1990s following his suggestion. Neuropsychological assessment has long adopted the practice of crossing various standardized tests in an attempt to measure a broader range of brain functions than that offered by any single instrument (Lezak, 1976, 1995; Lezak, Howieson, & Loring, 2004; see Wilson, 1992 for a review). Nevertheless, several problems with crossing batteries plagued assessment related fields for years. Many of these problems have been circumvented by Flanagan and colleagues' XBA approach (see Flanagan & McGrew, 1997; Flanagan et al., 2007, 2013 for examples). But unlike the XBA approach, the various so-called "cross-battery" techniques applied within the field of neuropsychological assessment, for example, are not typically grounded in a systematic approach that is theoretically and psychometrically sound. Thus, as Wilson (1992) cogently pointed out, the field of neuropsychological assessment was in need of an approach that would guide practitioners through the selection of measures that would result in more specific and delineated patterns of function and dysfunction—an approach that provides more clinically useful information than one that is "wedded to the utilization of subscale scores and IQs" (p. 382). Indeed, all fields involved in the assessment of cognitive and neuropsychological functioning have some need for an approach that would aid practitioners in their attempt to "touch all of the major cognitive areas, with emphasis on those most suspect on the basis of history, observation, and on-going test findings" (Wilson, 1992, p. 382). The XBA approach has met this need. A brief definition of and rationale for the XBA approach follows.

Definition

The XBA approach is a method of assessing cognitive and academic abilities and neuropsychological processes that is grounded mainly in CHC theory and research. It allows practitioners to measure reliably a wider range (or a more in depth but selective range) of ability and processing constructs, than that represented by any given stand alone assessment battery in a psychometrically defensible manner.

The Foundation of the XBA Approach

The XBA approach is based on three foundational sources of information—namely, CHC and neuropsychological theories, broad ability classifications, and narrow ability classifications—that together provide the knowledge base necessary to organize theory-driven, comprehensive assessments of cognitive, achievement, and neuropsychological constructs. A brief summary of each foundational source of information follows.

The Cattell–Horn–Carroll theory. Psychometric intelligence theories converged in recent years on a more complete or "expanded" multiple intelligence taxonomy, reflecting syntheses of factor analytic research conducted over the past 60–70 years. The most recent representation of this taxonomy is the CHC structure of cognitive abilities. CHC theory is an integration of Cattell and Horn's Fluid–Crystallized (Gf–Gc) theory (Horn, 1991) and Carroll's (1993) three-stratum theory of the structure of cognitive abilities.

In the late 1990s, McGrew (1997) attempted to resolve some of the differences between the Cattell–Horn and Carroll models. On the basis of his research, McGrew proposed an "integrated" Gf–Gc theory and he and his colleagues used this model as a framework for the XBA approach (e.g., Flanagan & McGrew, 1997; Flanagan et al., 2000; McGrew & Flanagan, 1998). This integrated theory quickly became known as the CHC theory of cognitive abilities shortly thereafter (see McGrew, 2005), and the WJ III NU COG was the first cognitive battery to be based on this theory. Many other cognitive batteries followed suit, including the KABC-II; Differential Ability Scales, second edition (DAS-II; Elliott, 2007); and Stanford–Binet Intelligence Scales, fifth edition (SB5; Roid, 2003).

Recently, Schneider and McGrew (2012) reviewed CHC-related research and provided a summary of the CHC abilities (broad and narrow) that have the most evidence to support them. In their attempt to provide a CHC overarching framework that incorporates the well-supported cognitive abilities, they articulated a 16-factor model containing over 80 narrow abilities (see Fig. 22.1). The ovals in the figure represent broad abilities and the rectangles represent narrow abilities. Additionally, an overall "g" or general ability is omitted from this figure intentionally due to space limitations. Because of the large number of abilities represented in CHC theory, the broad abilities in Fig. 22.1 are grouped conceptually into six categories to enhance comprehensibility, in a manner similar to that suggested by Schneider and McGrew (i.e., Reasoning, Acquired Knowledge, Memory and Efficiency, Sensory, Motor, and Speed and Efficiency). Space limitations preclude a discussion of all the ways in which CHC theory has evolved and the reasons why recent refinements and changes have been made (see Flanagan et al., 2013, and Schneider and McGrew for a discussion). However, to assist the reader in understanding the components of the theory, the broad abilities are defined in Table 22.1. For the purpose of this chapter, only the narrow abilities that are relevant to understanding executive functions within the cognition domain will be defined. Definitions of all narrow CHC abilities are found in Flanagan and colleagues and

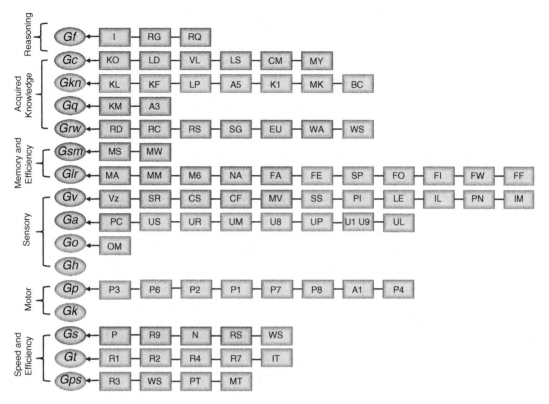

Fig. 22.1 Current and expanded Cattell-Horn-Carroll (CHC) theory of cognitive abilities. *Note*: This figure is based on information presented in Schneider and McGrew (2012). *Ovals* represent broad abilities and *rectangles* represent narrow abilities. Overall "g" or general ability is omitted from this figure intentionally due to space limitations. Conceptual groupings of abilities were suggested by Schneider and McGrew

Schneider and McGrew. Overall, CHC theory represents a culmination of about seven decades of factor analysis research within the psychometric tradition. However, in addition to structural evidence, there are other sources of validity evidence, some quite substantial, that support CHC theory (see Horn & Blankson, 2005, for a summary).

CHC broad (Stratum II) classifications of cognitive, academic, and neuropsychological tests. Based on the results of a series of cross-battery confirmatory factor analysis studies of the major intelligence batteries (see Keith & Reynolds, 2010, for a review) and the task analyses of many cognitive test experts, Flanagan and colleagues classified all the subtests of the major cognitive, achievement, and neuropsychological batteries according to the particular CHC broad abilities they measured (e.g., Flanagan et al., 2010, 2013; Flanagan, Ortiz, Alfonso, & Mascolo, 2006; McGrew, 1997; McGrew & Flanagan, 1998; Reynolds, Keith, Flanagan, & Alfonso, in press). To date, more than 100 batteries and 750 subtests have been classified according to the CHC broad and narrow abilities they measure, based in part on the results of these studies (Flanagan et al., 2013). The CHC classifications of cognitive, achievement, and neuropsychological batteries assist practitioners in identifying measures that assess the various broad and narrow abilities represented in CHC theory.

Classification of tests at the *broad ability level* is necessary to improve upon the validity of assessment and interpretation. Specifically, broad ability classifications ensure that the CHC constructs that underlie assessments are minimally

Table 22.1 Definitions of 16 broad Cattell-Horn-Carroll (CHC) abilities

Broad ability	Definition
Fluid Reasoning (Gf)	The deliberate but flexible control of attention to solve novel, "on-the-spot" problems that cannot be performed by relying exclusively on previously learned habits, schemas, and scripts
Crystallized Intelligence (Gc)	The depth and breadth and of knowledge and skills that are valued by one's culture
Quantitative Knowledge (Gq)	The depth and breadth of knowledge related to mathematics
Visual Processing (Gv)	The ability to make use of simulated mental imagery (often in conjunction with currently perceived images) to solve problems
Auditory Processing (Ga)	The ability to detect and process meaningful nonverbal information in sound
Short-Term Memory (Gsm)	The ability to encode, maintain, and manipulate information in one's immediate awareness
Long-Term Storage and Retrieval (Glr)	The ability to store, consolidate, and retrieve information over periods of time measured in minutes, hours, days, and years
Processing Speed (Gs)	The ability to perform simple, repetitive cognitive tasks quickly and fluently
Reaction and Decision Speed (Gt)	The speed of making very simple decisions or judgments when items are presented one at a time
Reading and Writing (Grw)	The depth and breadth of knowledge and skills related to written language
Psychomotor Speed (Gps)	The speed and fluidity with which physical body movements can be made
Domain-Specific Knowledge (Gkn)	The depth, breadth, and mastery of specialized knowledge (knowledge not all members of society are expected to have)
Olfactory Abilities (Go)	The abilities to detect and process meaningful information in odors
Tactile Abilities (Gh)	The abilities to detect and process meaningful information in haptic (touch) sensations
Kinesthetic Abilities (Gk)	The abilities to detect and process meaningful information in proprioceptive sensations
Psychomotor Abilities (Gp)	The abilities to perform physical body motor movements (e.g., movement of fingers, hands, legs) with precision, coordination, or strength

Note: CHC broad ability definitions are reported in Carroll (1993), McGrew (2005), and Schneider and McGrew (2012). Table adapted with permission from Wiley, 2013

affected by *construct-irrelevant variance* (Messick, 1989, 1995). In other words, knowing what tests measure what abilities enables clinicians to organize tests into *construct-relevant* composites—composites that contain only measures that are *relevant* to the construct, ability, or process of interest.

To clarify, *construct-irrelevant variance* is present when an "assessment is too broad, containing excess reliable variance associated with other distinct constructs ... that affects responses in a manner irrelevant to the interpreted constructs" (Messick, 1995, p. 742). For example, the Wechsler Intelligence Scale for Children, fourth edition (WISC-IV; Wechsler, 2003), Perceptual Reasoning Index (PRI) contains construct-irrelevant variance because, in addition to its two indicators of *Gf* (i.e., Picture Concepts, Matrix Reasoning), it has an indicator of *Gv* (i.e., Block Design). Therefore, the PRI is a mixed measure of two relatively distinct, broad CHC abilities (*Gf* and *Gv*); it contains reliable variance (associated with *Gv*) that is irrelevant to the interpreted construct of *Gf*. Notwithstanding the Wechsler PRI, most current intelligence and cognitive batteries contain only construct-relevant CHC broad ability composites—a welcomed improvement over previous edition of intelligence tests and an improvement that was based, in part, on the influence that the XBA approach had on test development, particularly in the late 1990s and early 2000s (see Alfonso, Flanagan, & Radwan, 2005).

CHC narrow (Stratum I) classifications of cognitive, academic, and neuropsychological tests. Narrow ability classifications were originally reported in McGrew (1997) and then later reported in McGrew and Flanagan (1998) and Flanagan et al. (2000) following minor modifications. Flanagan and her colleagues continued to gather content validity data on ability subtests and expanded their analyses to include subtests from achievement batteries (Flanagan et al., 2006) and, more recently, neuropsychological batteries (Flanagan et al., 2013). Classifications of ability and processing subtests according to content, format, and task demand at the narrow (stratum I) ability level were necessary to improve further upon the validity of assessment and interpretation (see Messick, 1989). Specifically, these narrow ability classifications were necessary to ensure that the CHC constructs that underlie assessments are well represented (McGrew & Flanagan). According to Messick (1995), *construct underrepresentation* is present when an "assessment is too narrow and fails to include important dimensions or facets of the construct" (p. 742).

Interpreting the WJ III Concept Formation (CF) test as a measure of Fluid Intelligence (i.e., the broad *Gf* ability) is an example of construct underrepresentation. This is because CF measures one narrow aspect of *Gf* (viz., Induction). At least one other *Gf* measure (i.e., subtest) that is qualitatively different from Induction is necessary to include in an assessment to ensure adequate representation of the *Gf* construct (e.g., a measure of General Sequential Reasoning [or Deduction]). Two or more qualitatively different indicators (i.e., measures of two or more narrow abilities subsumed by the broad ability) are needed for adequate construct representation (Comrey, 1988; Keith & Reynolds, 2012; Messick, 1989, 1995; Reynolds et al., in press). The aggregate of CF (a measure of Induction) and the WJ III Analysis–Synthesis test (a measure of deduction), for example, would provide an adequate estimate of the broad *Gf* ability because these tests are strong measures of *Gf* and represent qualitatively different aspects of this broad ability.

The classifications of tests at the broad and narrow ability levels of CHC theory guard against two ubiquitous sources of invalidity in assessment: construct-irrelevant variance and construct

Table 22.2 Guiding principles of the cross-battery assessment approach

1	Select battery that best addresses referral concerns
2	Use composites based on norms or XBA composite generator when necessary (e.g., the XBA Data Management and Interpretive Assistant [DMIA] v2.0 includes a composite generator tab that uses median reliabilities and intercorrelations of subtests)
3	Select tests classified through an acceptable method (note: all tests included in Flanagan, Ortiz, & Alfonso, 2013) were classified using these methods—confirmatory cross-battery factor analysis and expert consensus)
4	When broad ability is underrepresented or not measured, obtain information from another battery
5	When crossing batteries, use co-normed tests, statistically linked tests, or tests developed and normed within a few years of one another
6	Select tests from the smallest numbers of batteries to minimize error
7	Establish ecological validity for area(s) of weakness or deficiency

Note: Each of these guiding principles is described in detail in Flanagan et al. (2013)

underrepresentation. In addition, these classifications augment the validity of test performance interpretation. Furthermore, to ensure that XBA procedures are theoretically and psychometrically sound, it is recommended that practitioners adhere to a set of guiding principles, which are enumerated in Table 22.2. Taken together, CHC theory, the CHC classifications of tests that underlie the XBA approach, and the accompanying guiding principles provide the necessary foundation from which to organize assessments and interpret assessment results in a manner that is comprehensive and supported by research (Flanagan et al., 2013).

CHC theory, as it is operationalized by current intelligence and cognitive batteries, emphasizes the sum of performances or outcome, rather than the process or steps that led to a particular outcome, which is why little, if any, emphasis is placed on understanding executive functions. Conversely, neuropsychological batteries place greater emphasis on process, allowing for practitioners to derive information about executive functions more readily. Because both outcome and process are important, each is addressed in this chapter. To ensure that both are addressed during test interpretation, an integration of CHC and neuropsychological theories is warranted.

Enhancing Interpretation of Test Performance: An Integration of CHC and Neuropsychological Theories

With the emergence of the field of school neuropsychology (e.g., Decker, 2008; Fletcher-Janzen & Reynolds, 2008; Hale & Fiorello, 2004; Miller, 2007, 2010, 2013) came the desire to link CHC theory and neuropsychological theories. Understanding how CHC theory and neuropsychological theories relate to one another expands the options available for interpreting test performance and improves the quality and clarity of test interpretation, as a much wider research base is available to inform practice.

Although scientific understanding of the manner in which the brain functions and how mental activity is expressed on psychometric tasks has increased dramatically in recent years, there is still much to be learned. All efforts to create a framework that guides test interpretation benefit from diverse points of view. For example, according to Fiorello, Hale, Snyder, Forrest, and Teodori (2008), "the compatibility of the neuropsychological and psychometric approaches [CHC] to cognitive functioning suggests converging lines of evidence from separate lines of inquiry, a validity dimension essential to the study of individual differences in how children think and learn" (p. 232; parenthetic information added). Their analysis of the links between the neuropsychological and psychometric approaches not only provides validity for both but also suggests that each approach may benefit from knowledge of the other. As such, a framework that incorporates the neuropsychological and psychometric approaches to cognitive functioning holds the promise of increasing knowledge about the etiology and nature of a variety of disorders (e.g., specific learning disability) and the manner in which such disorders are treated. This type of framework should not only connect the elements and components of both assessment approaches, but it should also allow for interpretation of data within the context of either model. In other words, the framework should serve as a "translation" of the concepts, nomenclature, and principles of one approach into their counterparts in the other. A brief discussion of one such framework, developed by Flanagan and her colleagues, is presented here (Flanagan et al., 2010, 2013). A variation of their framework is illustrated in Fig. 22.2 and represents an integration based on psychometric, neuropsychological, and Lurian perspectives.

The interpretive framework shown in Fig. 22.2 draws upon prior research and sources, most notably Dehn (2006), Fiorello et al. (2008), Fletcher-Janzen and Reynolds (2008), Miller (2007, 2010, 2013), and Strauss, Sherman, and Spreen (2006). In understanding the manner in which Luria's blocks, the neuropsychological domains, and CHC broad abilities may be linked to inform test interpretation and mutual understanding among assessment professionals, Flanagan and colleagues pointed out four important observations that deserve mention. First, there is a hierarchical structure among the three theoretical conceptualizations. Second, the hierarchical structure parallels a continuum of interpretive complexity, spanning the broadest levels of cognitive functioning, where mental activities are "integrated," to the narrowest level of cognitive functioning where mental activity is reduced to more "discrete" abilities and processes (see far left side of Fig. 22.2). Third, all mental activity takes place within a given ecological and societal context and is heavily influenced by language as well as other factors external to the individual. As such, the large gray shaded area represents "language and ecological influences on learning and production," which includes factors such as exposure to language, language status (English learner vs. English speaker), opportunity to learn, motivation and effort, and socioeconomic status. Fourth, administration of cognitive and neuropsychological tests is typically conducted in the schools (e.g., for students suspected of having a specific learning disability) when a student fails to respond as expected to quality instruction and

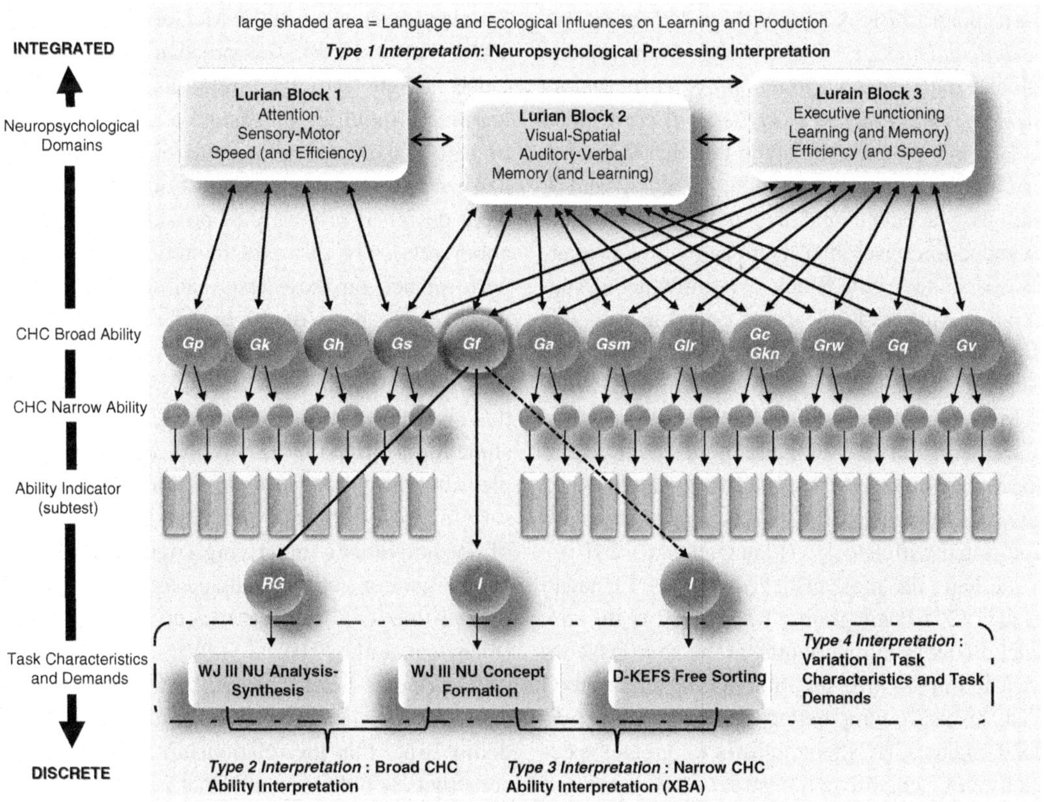

Fig. 22.2 Integration of psychometric, neuropsychological, and Lurian perspectives for interpretation. *Note*: The broad abilities of Go, Gt, and Gps are not included in the figure because most cognitive and neuropsychological batteries do not have measures that directly assess these abilities

intervention. Thus, the framework in Fig. 22.2 is a representation of cognitive constructs and neuropsychological processes that may be measured (when a student is referred because of learning difficulties) and the manner in which they relate to one another.

According to Flanagan et al. (2010), when a student has difficulty with classroom learning and fails to respond as expected to intervention, a school-based hypothesis-generation, testing, and interpretation process should be carried out. Conceptualization of any case may begin at the "integrated" level (i.e., top of Fig. 22.2).

Luria's functional units are depicted at the top of Fig. 22.2 as overarching cognitive concepts (see Naglieri, 2012, for definitions of the Lurian blocks). The interaction between, and the interconnectedness among, the functional units are represented by the horizontal double-headed arrows in the figure. Because Luria's functional units are primarily descriptive concepts designed to guide applied clinical evaluation practices, neuropsychologists have had considerable independence in the manner in which they align their assessments with these concepts (Flanagan et al.).

Although a few psychoeducational batteries have been developed to operationalize one or more of Luria's functional units, for the most part, neuropsychologists have typically couched Luria's blocks within clinical and neuropsychological domains. In doing so, the Lurian blocks have been transformed somewhat from overarching concepts to domains with more specificity (Flanagan et al., 2010). These domains are listed in the rectangles at the top of Fig. 22.2 with their corresponding Lurian block. For example, the neuropsychological domains of *attention, sensory-motor, and speed (and efficiency)* correspond to Block 1; *visual–spatial, auditory–verbal*, and *memory (and learning)* correspond to Block 2; and *executive functioning, learning (and memory)*, and *efficiency (and speed)* correspond to Block 3. Noteworthy is the fact that the memory and learning domain spans Blocks 2 and 3, and its placement and use of parentheses are intended to convey that memory may be primarily associated with Block 2 (simultaneous/successive) whereas the learning component of this domain is probably more closely associated with Block 3 (planning/metacognition). Likewise, speed and efficiency span Blocks 1 and 3, and its placement and use of parentheses denote that speed may be more associated with Block 1 (i.e., attention) whereas efficiency seems to be more associated with Block 3 (Flanagan et al., 2010).

Perhaps the most critical aspect of Flanagan et al.'s (2010) integrative framework is the distinction between functioning at the neuropsychological domain level and functioning at the broad CHC level. As compared to the neuropsychological domains, CHC theory allows for greater specificity of cognitive constructs. Because of structural differences in the conceptualization of neuropsychological domains and CHC broad abilities vis-à-vis factorial complexity, it is not possible to provide a precise, one-to-one correspondence between these conceptual levels. This is neither a problem nor an obstacle, but simply the reality of differences in perspective among these two lines of inquiry.

As compared to the neuropsychological domains, CHC constructs within the psychometric tradition tend to be relatively distinct because the intent is to measure a theoretical construct as purely and independently as possible. This is not to say, however, that the psychometric tradition has completely ignored shared task characteristics in favor of a focus on precision in measuring a single theoretical construct. For example, Kaufman provided a "shared characteristic" (or "Demand Analysis;" discussed later in this chapter) approach to individual test performance for several intelligence tests including the KABC-II and the various Wechsler Scales (Kaufman, 1979; see also McCloskey, 2009; McGrew & Flanagan, 1998; Sattler, 1988). This practice has often provided insight into the underlying cause(s) of learning difficulties, and astute practitioners continue to make use of it. Despite the fact that standardized, norm-referenced tests of CHC abilities were designed primarily to provide information about relatively discrete theoretical constructs, performance on these tests can still be viewed within the context of the broader neuropsychological domains. That is, when evaluated within the context of an entire battery, characteristics that are shared among groups of tests on which a student performed either high or low, for example, often provide the type of information necessary to assist in further understanding the nature of an individual's underlying cognitive function or dysfunction, conceptualized as neuropsychological domains, such as executive functioning (Flanagan et al., 2010).

The double-headed arrows between neuropsychological domains and CHC abilities in Fig. 22.2 demonstrate that the relationship between these constructs is bidirectional. That is, one can conceive of the neuropsychological domains as global entities that impact performance on various CHC ability measures, just as one can conceive of a particular measure of a specific CHC ability as involving aspects of more than one neuropsychological domain. For example, as will be discussed in the next section, the broad CHC abilities of Gf, Gsm, Glr, and Gs, while conceived of as relatively distinct in the CHC literature, together may reveal information about executive functioning. That is, to gain an understanding of neuropsychological domains, it is likely necessary to evaluate performance across different CHC domains.

Flanagan et al.'s (2010) conceptualization of the relations between the neuropsychological domains and the CHC broad abilities is presented next. For the purpose of parsimony, the neuropsychological domains are grouped according to their relationship with the Lurian blocks, and thus, these domains are discussed as clusters rather than discussed separately.

Correspondence Between the Neuropsychological Domains and CHC Broad Abilities

According to Flanagan et al. (2010), measures of at least six CHC broad abilities involve processes associated with the *Attention/Sensory-Motor/Speed (and Efficiency)* neuropsychological cluster, including Psychomotor Abilities (*Gp*), Tactile Abilities (*Gh*), Kinesthetic Abilities (*Gk*), Decision/Reaction Time or Speed (*Gt*),[1] Processing Speed (*Gs*), and Olfactory Abilities (*Go*).[2] *Gp* involves the ability to perform body movements with precision, coordination, or strength. *Gh* involves the sensory receptors of the tactile (touch) system, such as the ability to detect and make fine discriminations of pressure on the surface of the skin. *Gk* includes abilities that depend on sensory receptors that detect bodily position, weight, or movement of the muscles, tendons, and joints. Because *Gk* includes sensitivity in the detection, awareness, or movement of the body or body parts and the ability to recognize a path the body previously explored without the aid of visual input (e.g., blindfolded), it may involve some visual–spatial process, but the input remains sensory-based and thus better aligned with the sensory-motor domain. *Gt* involves the ability to react and/or make decisions quickly in response to simple stimuli, typically measured by chronometric measures of reaction time or inspection time. *Gs* is the ability to automatically and fluently perform relatively easy or overlearned cognitive tasks, especially when high mental efficiency is required. As measured by current cognitive batteries (e.g., WISC-IV Coding, Symbol Search, and Cancellation), *Gs* seems to capture the essence of both speed and efficiency, which is why there are double-headed arrows from *Gs* to Block 1 (where *Speed* is emphasized) and Block 3 (where *Efficiency* is emphasized) in Fig. 22.2. *Go* involves abilities that depend on sensory receptors of the main olfactory system (nasal chambers). Many CHC abilities associated with the *Attention/Sensory-Motor/Speed (and Efficiency)* cluster are measured by neuropsychological tests (e.g., NEPSY-II, D-KEFS, Dean-Woodcock Neuropsychological Battery [DWNB; Dean & Woodcock, 2003]; Flanagan et al., 2010).

Prior research suggests that virtually all measures of broad CHC abilities are associated with the *visual–spatial/auditory–verbal/memory (and learning)* neuropsychological cluster. That is, the vast majority of tasks on cognitive and neuropsychological batteries require either visual-spatial or auditory–verbal input. Apart from tests that relate more to discrete sensory-motor functioning and that utilize sensory input along the kinesthetic, tactile, or olfactory systems, all other tests will necessarily rely either on visual-spatial or auditory–verbal stimuli. Certainly, visual (*Gv*) and auditory (*Ga*) processing are measured well on neuropsychological and cognitive instruments. Furthermore, tests of Short-Term Memory (*Gsm*) and Long-Term Storage and Retrieval (*Glr*) typically rely on visual (e.g., pictures) or verbal (digits or words) information for input. Tasks that involve reasoning (*Gf*), stores of acquired knowledge (e.g., *Gc*, *Gkn*), and even speed (*Gs*) also use either visual-spatial and/or auditory–verbal channels for input. Furthermore, it is likely that such input will be processed in one of the two possible ways—simultaneously or successively (Luria, 1973; Naglieri, 2005, 2012).

Finally, research suggests that the *Executive Functioning/Learning (and Memory)/Efficiency (and Speed)* neuropsychological cluster is associated with eight broad CHC abilities, including Fluid Reasoning (*Gf*; e.g., planning), Crystallized Intelligence (*Gc*; e.g., concept formation and generation), General (Domain-Specific) Knowledge Ability (*Gkn*), Quantitative Knowledge (*Gq*), Broad Reading and Writing Ability (*Grw*), Processing Speed (*Gs*; e.g., focus/selected attention), Short-Term Memory (*Gsm*; e.g., working memory), and Long-Term Storage and Retrieval

---

[1] *Gt* is omitted from Fig. 22.2 because commonly used intelligence and neuropsychological batteries do not measure this ability.

[2] *Go* is omitted from Fig. 22.2 because commonly used intelligence and neuropsychological batteries do not measure this ability and the cognitive and perceptual aspects of this ability have not been studied extensively (McGrew, 2005; Schneider & McGrew, 2012).

(*Glr*; e.g., retrieval fluency). *Gf* generally involves the ability to solve novel problems using inductive, deductive, and/or quantitative reasoning and, therefore, is most closely associated with executive functioning. *Gc* represents one's stores of acquired knowledge (e.g., vocabulary, general information) or "learned" information and is entirely dependent on language, the ability that Luria believed was necessary to mediate all aspects of learning. In addition, Domain-Specific Knowledge (*Gkn*), together with knowledge of Reading/Writing (*Grw*) and Math (*Gq*), reflects the *learning* component of "memory and learning." Therefore, Flanagan and colleagues contended that *Gc, Gkn, Grw*, and *Gq* are related to this neuropsychological cluster. *Gsm*, especially working memory, and *Glr*, especially the retrieval fluency abilities, are often conceived of as executive functions and involve planning and metacognition.

As may be seen in Fig. 22.2, Flanagan et al. (2010) have placed the CHC *narrow* abilities at the *discrete* end of the integrated-discrete continuum. It is noteworthy that narrow ability deficits tend to be more amenable to remediation, accommodation, or compensatory strategy interventions as compared to broader and more overarching abilities. For example, poor memory span, a narrow ability subsumed by the broad ability, *Gsm*, can often be compensated for effectively via the use of strategies such as writing things down or recording them in some manner for later reference. Likewise, it is possible to train a phonetic coding deficit (a narrow Ga ability) to the point where it becomes a skill. In contrast, when test performance suggests more pervasive dysfunction, as may be indicated by deficits in one or more neuropsychological domains, for example, the greater the likelihood that intervention will need to be broader, more intensive, and long term, perhaps focusing on the type of instruction being provided to the student and how the curriculum ought to be modified and delivered to improve the student's learning (Fiorello et al., 2008; Flanagan, Alfonso, & Mascolo, 2011; Mascolo, Flanagan, & Alfonso, in press; see also Meltzer, 2012). Therefore, knowing the areas that are problematic for the individual should guide the planning, selection, and tailoring of interventions (Mascolo et al.).

Accurate interpretation of test performance is needed if corresponding educational strategies and interventions are to lead to positive outcomes for the individual. Figure 22.2 includes four types of interpretation. It is likely that practitioners will rely most on the types of interpretation that are grounded in the theories with which they are most familiar (e.g., CHC, Luria).

Type 1 interpretation: Neuropsychological processing interpretation. This level of interpretation was referred to above. Specifically, when subtests are organized according to those that reflect weaknesses or deficits and those that reflect average or better ability, the neuropsychological domains associated with the tests in each grouping may be analyzed to determine if any particular neuropsychological domain or Lurian block is associated with one group of subtests and not (to a substantial degree) the other. Analyzing a student's performance at this more integrated level may help practitioners explain why some, perhaps, seemingly uncorrelated subtests are related in the context of the individual's brain–behavior functioning.

For example, if an individual exhibits weaknesses on the WISC-IV Matrix Reasoning, Digit Span, and Cancellation subtests, yet average or above average performance on all other subtests, it may be hypothesized that it is because these tasks, which involve Executive Functioning (e.g., Matrix Reasoning), Learning and Memory (e.g., Digit Span), and Efficiency and Speed (e.g., Cancellation), all relate to the frontal-subcortical circuits in Luria's Block 3. Therefore, test performance suggests that the individual has difficulty with planning, organizing, and carrying out tasks, an interpretation that should be supported with ecological validity (e.g., specific activities involving planning and organization are extremely difficult for the individual and often cannot be accomplished without support). In addition, possible frontal-subcortical dysfunction or, more specifically, damage or dysfunction in the dorsolateral prefrontal circuit may be inferred from the test findings (see Miller, 2007), but only within the context of case history and in the presence of converging data sources. However, if the individual demonstrated weaknesses in Digit Span and Cancellation only, and therefore the deficits were not entirely representative of Block 3, for example, interpretation according to specific neuropsychological domains involved in Digit Span and Cancellation task performance may be more informative (e.g., individual has difficulty with Attention).

Moving to the more discrete end of the integrated-discrete interpretation continuum in Fig. 22.2, information about more specific abilities and processes may be obtained when evaluating subtests that were grouped according to factor analysis results. For example, most cognitive assessment batteries group subtests into broad CHC ability domains or composites based on the results of CHC-driven confirmatory factor analysis (Keith & Reynolds, 2012).

Type 2 interpretation: Broad CHC ability interpretation. This type of interpretation emphasizes broad ability constructs (e.g., *Gv*) over narrow ability and processing constructs (e.g., Visualization [Vz], Visual Memory [MV]). Broad abilities are represented by at least two qualitatively different indicators (subtests) of the construct. For example, in Fig. 22.2, the broad ability of Gf is represented by two subtests from the same battery (i.e., WJ III NU COG), each of which measures a qualitatively different aspect of Gf. These subtests are Analysis–Synthesis (a measure of RG) and Concept Formation (a measure of I). Interpretation of *Gf* (referred to as "Type 2 Interpretation" in Fig. 22.2) may be made when two conditions are met: (a) two or more qualitatively different narrow ability or processing indicators (subtests) of *Gf* are used to represent the broad ability; and (b) the broad ability composite (*Gf* in this example) is considered *cohesive*, suggesting that it is a good summary of the theoretically related abilities that comprise it.

As may be seen in Fig. 22.2, the WJ III NU COG contains two qualitatively different indicators of *Gf*. When the difference between WJ III NU COG subtest standard scores is not statistically significant, then the WJ III NU COG *Gf* composite is considered cohesive and may be interpreted as a reliable and valid estimate of this broad ability (see "Level 2 interpretation" in Fig. 22.2). However, if the difference between these subtest standard scores is statistically significant and uncommon in the general population, then the Gf composite is considered noncohesive and, therefore, should not be interpreted as a good summary of the abilities that comprise it. At this point in the interpretation process, a judgment regarding whether or not follow-up assessment is necessary should be made. For example, if the lower of the two scores in the Gf composite was indicative of a weakness or deficit and the higher score was suggestive of at least average ability, then it would make sense to follow up on the lower score via a subtest that measures the same narrow ability as the one underlying the subtest with the lower score.

Suppose the lower score in a two-subtest composite was 105 (65th percentile) and the composite was considered noncohesive. It seems unnecessary to follow up on the lower score in the composite, as this score does not suggest any type of weakness or dysfunction. Likewise, the higher score in the composite, in this example, represents a normative strength (e.g., standard score of 120; 91st percentile). Therefore, even though the composite is not a good summary of the theoretically related abilities that comprise it, performance ranges from average to well above average in the broad ability area, suggesting no need for follow-up. Alternatively, suppose the lower score in a two-subtest composite was 85 (16th percentile on the Concept Formation subtest) and the higher score was 100 (50th percentile on the Analysis–Synthesis subtest). In this example, regardless of whether or not the composite is cohesive, there is certainly a need to follow up on the lower score with another measure of Induction because the score of 85 is suggestive of a weakness in Induction. If another measure of Induction results in average or better performance, then Type 2 interpretation ensues, and a broad ability *cross-battery composite* is calculated using the scores from Analysis–Synthesis and the second test of Induction (Flanagan et al., 2013). If the second measure of Induction suggests below average performance, then Type 3 interpretation is necessary.

Type 3 interpretation: Narrow CHC ability interpretation (XBA). This type of interpretation highlights a specific situation wherein XBA data are often considered. For example, suppose that the WJ III NU COG Gf composite (represented in Fig. 22.2) was noncohesive and the Analysis–Synthesis standard score was greater than 100 and significantly higher than the Concept Formation standard score, which was below 85 and suggestive of a normative weakness or deficit. Many practitioners would opt to follow up on the lower score and assess the narrow ability of Induction by administering another measure of Induction, following the cross-battery guiding principles (see Table 22.2). Because the WJ III NU COG does not contain another measure of Induction, the practitioner must select a subtest from another battery. In the example provided in Fig. 22.2, the practitioner administered the D-KEFS Free-Sorting subtest. Now the practitioner has three measures of Gf, two of which measure Induction and one that measures General Sequential Reasoning. These three subtest scores may be analyzed via XBA software[3] to determine the best way to interpret them. When the two narrow ability indicators of Induction form a cohesive composite, then the inductive reasoning *cross-battery narrow ability composite* is calculated and interpreted. In this example, the narrow ability of Induction would be interpreted as a weakness or deficit since both scores fell below 85.

Note that when two tests of Induction differ significantly from one another (i.e., they do not form a cohesive composite), a qualitative analysis of task demands and task characteristics is necessary to generate hypotheses regarding the reason for this unexpected finding. This type of qualitative analysis is labeled "Type 4 interpretation" in Fig. 22.2 and is discussed later.

Quantitative (Type 2 and Type 3) Evaluation of Executive Functions via the XBA Approach

Prior to explaining Type 4 interpretation, it is important to realize that broad and narrow Type 2 and Type 3 interpretations, respectively, are relevant to understanding executive functions from a psychometric or quantitative perspective. According to Miller (2007), information about various executive functions may be derived from psychometric tests. For example, tests that measure working memory capacity; concept formation and generation; planning, reasoning, and problem solving; retrieval fluency; and attention reveal information about executive functions. These constructs correspond to broad and narrow CHC abilities (see Fig. 22.3). For example, *working memory capacity* is a narrow ability subsumed by the broad Gsm ability in CHC theory. There are many popular batteries that include subtests that measure working memory capacity, such as Wechsler Adult Intelligence Scale, fourth edition (WAIS-IV; Wechsler, 2008); Letter–Number Sequencing; and the SB5 Block Span testlet (see Fig. 22.3). *Concept formation and generation* appears to correspond quite well to Gc-type tasks, particularly those that require an individual to reason (Gf) with verbal information. Many Gc tests involve the ability to reason, such as the D-KEFS Twenty Questions subtest. Therefore, these types of tests appear to require a Gc/Gf blend of abilities, as indicated in Fig. 22.3. *Planning, reasoning, and problem solving* corresponds to Gf; *retrieval fluency* corresponds to Glr; and *Attention* (particularly sustained attention) corresponds to Gs.

As may be seen in Fig. 22.3, there are three narrow abilities that are subsumed by Gf, namely, Induction (I), General Sequential Reasoning or Deduction (RG), and Quantitative Reasoning (RQ), the latter of which involves reasoning both inductively and deductively with numbers. Likewise, there are four and three narrow abilities subsumed by Glr and Gs in Fig. 22.3, respectively. Note that only the Glr and Gs narrow abilities that are most relevant to understanding specific executive functions are included in Fig. 22.3 (see Fig. 22.1 for the remaining narrow

---

[3] XBA Data Management and Interpretive Assistant (DMIA) v2.0 (Flanagan et al., 2013).

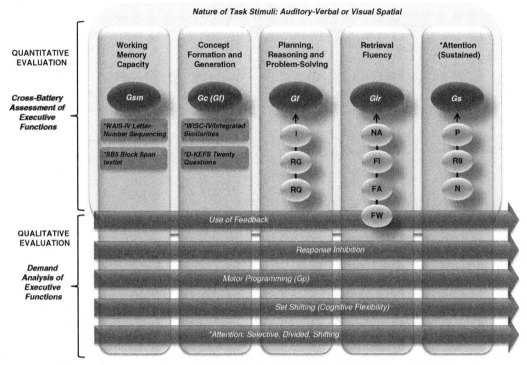

Fig. 22.3 Assessment of a subset of executive functions in the cognition domain using intelligence, cognitive and neuropsychological batteries

abilities that make up these domains). Table 22.3 provides definitions of all the terms that are included in Fig. 22.3.

Glr is comprised of narrow abilities that can be divided into two categories—learning efficiency and retrieval fluency (Schneider & McGrew, 2012). The latter category of retrieval fluency is considered an executive function (e.g., Miller, 2007) and can be measured by verbal tasks that require the rapid retrieval of information, such as naming as many animals as one can think of as quickly as possible or naming as many words that begin with the letter "r" as quickly as possible. The narrow Gs abilities included in Fig. 22.3 all involve sustained attention. Typical Gs tasks on cognitive batteries require the individual to do simple clerical-type tasks quickly for a prolonged period of time, usually 3 minutes. Table 22.4 includes the subtests of several cognitive and neuropsychological batteries that measure planning, reasoning, and problem solving (Gf); concept formation and generation (Gc/Gf); working memory capacity (Gsm); and retrieval fluency (Glr; and attention (Gs), allowing for the derivation of information about executive functions in the cognition domain.

The bottom portion of Fig. 22.3 includes five horizontal arrows, each one representing an executive function (Miller, 2007). Unlike the executive functions that may be inferred from tests measuring the abilities listed in the top portion of this figure, the executive functions listed in the bottom portion do not correspond well to any particular CHC ability. For example, in order to derive information about how an individual is

Table 22.3 Definition of selected executive functions

Executive functions	
Executive function	Definition
Quantitative	
Working memory capacity	Working memory capacity is a narrower *short-term memory* (*Gsm*) ability. It is the ability to direct the focus of attention to perform relatively simple manipulations, combinations, and transformations of information within primary memory, while avoiding distracting stimuli and engaging in strategic/controlled searches for information in secondary memory
Concept formation and generation	Concept formation and generation is a blend of *crystallized intelligence* (*Gc*), the depth and breadth of knowledge and skills that are valued by one's culture, and *fluid reasoning* (*Gf*), the deliberate but flexible control of attention to solve novel, "on-the-spot" problems that cannot be performed by relying exclusively on previously learned habits, schemas, and scripts. Specifically, concept formation and generation is the ability to categorize objects, images, or words into groups that have a shared characteristic
Planning, reasoning, and problem solving	Planning, reasoning, and problem solving includes three narrow abilities of *fluid reasoning* (*Gf*): induction (I), the ability to observe a phenomenon and discover the underlying principles or rules that determine its behavior; general sequential reasoning (RG), the ability to reason logically, using known premises and principles; and quantitative reasoning (RQ), the ability to reason, either with induction or deduction, with numbers, mathematical relations, and operators. Planning, reasoning, and problem solving requires the ability to recognize a problem; think logically about the problem; identifying solutions; and select, organize, and implement a plan to resolve the problem
Retrieval fluency	Retrieval fluency is the ability to rapidly and correctly recall information from long-term memory. Retrieval fluency relies on four narrow abilities of *long-term storage and retrieval* (*Glr*): ideation fluency (FI), the ability to rapidly produce a series of ideas, words, or phrases related to a specific condition or object; word fluency (FW), the ability to rapidly produce words that share a non-semantic feature; naming facility (NA), the ability to rapidly call objects by their names; and figural fluency (FF), the ability to rapidly draw or sketch as many things as possible when presented with a nonmeaningful visual stimulus
Attention	Attention is the ability to maintain concentration and vigilance for an extended period of time (sustained attention). Test that assess *processing speed* (*Gs*), the ability to perform simple, repetitive cognitive tasks quickly and fluently, rely heavily on attention. Attention is required for three *Gs* narrow abilities: perceptual speed (P), the ability at which visual stimuli can be compared for similarity or difference; rate of test-taking (R9), the speed and fluency with which simple cognitive tests are completed; and number facility (N), the speed at which basic arithmetic operations are performed accurately
Qualitative	
Use of feedback	The ability to modify one's performance based on feedback
Response inhibition	The ability to inhibit an inappropriate or incorrect response
Motor programming	The ability to plan and execute motor responses
Cognitive set shifting	The ability to shift mental states (cognitive sets) according to changes in a task or situation
Selective, divided, and shifting attention	Qualitative evaluations of attention can assess the ability to focus despite background distraction (selective attention), the ability to react simultaneously to more than one stimulus or task (divided attention), and the ability to deliberately transfer attention from one stimulus or activity to another (shifting attention)

Note: Definitions were derived from Miller (2007), Miller and Maricle (2012), and Baron (2004). Broad and narrow ability definitions reported in Schneider and McGrew (2012)

Table 22.4 Correspondence between selected executive functions in the cognition domain and broad and narrow CHC ability measures on major intelligence, cognitive and neuropsychological batteries

	Planning, reasoning, and problem solving (Gf)	Concept formation and generation (Gc/Gf blend)	Working memory capacity (Gsm)	Retrieval fluency (Glr)	Attention (sustained) (Gs)
WISC-IV	Matrix reasoning (I)	Similarities (VL, Gf:I)	Digit span (MS, MW)	Not measured	Symbol search (P)
	Picture concepts (I)	Word reasoning (VL, Gf:I)	Letter-number Sequencing (MW)		Coding (R9)
			Arithmetic (MW; Gf:RQ)		Cancellation (P)
WAIS-IV	Matrix reasoning (I)	Similarities (VL, Gf:I)	Digit span (MS, MW)	Not measured	Symbol search (P)
	Figure weights (RQ)		Letter-number Sequencing (MW)		Coding (R9)
			Arithmetic (MW; Gf:RQ)		Cancellation (P)
WPPSI-III	Matrix reasoning (I)	Similarities (VL, Gf:I)	Not measured	Not measured	Coding (R9)
		Word reasoning (VL, Gf:I)			Symbol search (P)
KABC-II	Pattern reasoning (I; Gv:Vz)[a]	Riddles (VL, Gf:RG)	Word order (MS, MW)	See co-normed KTEA-II	Not measured
	Rover (Gv:SS; Gf:RG)				
	Story completion (RG, Gc:K0)[b]				
WJ III NU COG	Analysis-synthesis (RG)	Verbal comprehension (VL, Gf:I)	Numbers reversed (MW)	Retrieval fluency (FI)	Visual matching (P)
	Concept formation (I)		Auditory working memory (MW)	Rapid picture naming (NA; Gs:R9)	Decision speed (P)
	Planning (Gv:SS; Gf:RG)				Pair cancellation (P)
SB5	Nonverbal fluid reasoning (I; Gv)	Nonverbal knowledge (K0, LS, Gf:RG)	Nonverbal working memory (MS, MW)	Not measured	
	Verbal fluid reasoning (I, RG, Gc:CM)		Verbal working memory (MS, MW)		
	Nonverbal quantitative reasoning (RQ, Gq:A3)				
	Verbal quantitative reasoning (RQ, Gq:A3)				

(continued)

Table 22.4 (continued)

	Planning, reasoning, and problem solving (Gf)	Concept formation and generation (Gc/Gf blend)	Working memory capacity (Gsm)	Retrieval fluency (Glr)	Attention (sustained) (Gs)
DAS-II	Matrices (I)	Verbal similarities (VL, Gf:I)	Recall of digits-backward (MW)	Rapid naming (NA; Gs:R9)[c]	Speed of information processing (P)
	Picture similarities (I)		Recall of sequential order (MW)		
	Sequential and quantitative reasoning (RQ)				
D-KEFS	Sorting test: free sorting (I)	Twenty questions test (Gc:LD; Gf:I)		Color-word interference: color-naming (NA)	Trail making test: letter sequencing (R9)
	Sorting test: sort recognition (I)	Word context test (LD; Gf:RG)	Not measured directly	Color-word interference: inhibition (NA)	Trail making test: number sequencing (R9)
	Tower (Gv:Vz; GF:RG)			Color-word interference: inhibition/switching (NA; Gsm:MA)	Trail making test: letter-number sequencing (P; Gsm:MW)
				Color-word interference: word reading (NA)	Trail making test: visual scanning (P)
				Verbal fluency test: category fluency (FI)	
				Verbal fluency test: category switching (FI; Gsm:MW)	
				Verbal fluency test: letter fluency (FW)	
				Design fluency: empty dots (FF)	
				Design fluency: filled dots (FF)	
				Design fluency: switching (FF)	

NEPSY-II	Animal sorting (I)	Not measured well	Design fluency (FF; Gv:Vz)
		Auditory attention and response set (MW; Gs:R9)	Speeded naming (NA; Gs:R9) Not measured well
		Inhibition (MW)	Word generation (FI, FW)
		Word list interference (MS, MW)	
WISC-IV integrated	Not measured	Similarities-multiple choice (VI; Gf:I)	Coding copying (R9)
		Arithmetic-process approach (MW; Gf:RQ)	
		Letter-number sequencing process approach (MW)	Not measured
		Spatial span forward and backward (MS,MW)	

Note: CHC classifications are based on Carroll (1993), Flanagan, Ortiz and Alfonso (2013), and Schneider and McGrew (2012). *WISC-IV* Wechsler Intelligence Scale for Children—Fourth Edition (Wechsler, 2003); *WAIS-IV* Wechsler Adult Intelligence Scale—Fourth Edition (Wechsler, 2008); *WPPSI-III* Wechsler Preschool and Primary Scale of Intelligence—Third Edition (Wechsler, 2002); *KABC-II* Kaufman Assessment Battery for Children—Second Edition (Kaufman & Kaufman, 2004); *WJ III NU COG* Woodcock-Johnson III Normative Update Tests of Cognitive Abilities (Woodcock, McGrew, & Mather, 2001, 2007); *SB5* Stanford-Binet Intelligence Scales—Fifth Edition (Roid, 2003); *DAS-II* Differential Ability Scales—Second Edition (Elliot, 2007); *D-KEFS* Delis-Kaplan Executive Function System (Delis & Kaplan, 200?); *NEPSY-II* (Korkman, Kirk, & Kemp, 2007). Wechsler Intelligence Scale for Children—Fourth Edition Integrated (Harcourt Assessment, 2004). *Gf* fluid reasoning; *Gc* crystallized Intelligence; *Gsm* short-term memory; *Glr* long-term storage and retrieval; *Gs* processing speed; *Gq* quantitative ability; *Gv* visual processing; *RQ* quantitative reasoning; *I* Induction; *RG* general sequential reasoning; *VL* lexical knowledge; *K0* general (verbal) knowledge; *LS* listening ability; *LD* language development; *MW* working memory; *MS* memory span; *FF* figural fluency; *FI* ideational fluency; *FW* word fluency; *NA* naming facility; *P* perceptual speed; *R9* rate-of-test-taking; *N* number facility; *A3* math achievement; *SS* spatial scanning

[a] Pattern reasoning appears to be a measure of Gv:Vz as a primary broad ability and Gf as a second broad ability at ages 5–6 years

[b] Gc:K0 appears to be the primary ability measured by story completion at ages 5–6 years. At older ages (i.e., ages 7+), the primary ability measured by story completion appears to be Gf:RG

[c] Elliot (2007) places rapid naming under the construct Gs based on the results of factor analysis. The current authors place this test under the construct Glr based on theory and Gs as a second broad ability

able to modify his or her performance based on feedback, one needs to observe the individual perform many tasks, not only in a one-to-one standardized testing situation but in multiple settings (e.g., the classroom, at home). Therefore, in order to obtain information about the executive functions listed in the bottom portion of Fig. 22.3, it is necessary to conduct a *qualitative evaluation*, which is discussed in the next section of this chapter (i.e., Type 4 Interpretation).

To conduct a comprehensive assessment of the executive functions via measurement of the CHC abilities listed in the top half of Fig. 22.3, it is necessary to cross batteries for the following reasons. First, as may be seen in Table 22.4, the only batteries that measures aspects of all the areas listed in the top portion of Fig. 22.3 are the WJ III NU COG and DAS-II. Therefore, when using any cognitive or neuropsychological battery (in Table 22.4) other than the WJ III NU COG and DAS-II, there is a need to supplement the battery with subtests from another battery to measure all five CHC abilities (in the top portion of Fig. 22.3). Second, when administering traditional intelligence batteries, such as the Wechsler Scales, the examiner often serves as the "executive control board" during testing because she/he tells the examinee what to do and how to do it via detailed standardized test directions (Feifer & Della Toffalo, 2007, p. 17). As such, intelligence batteries, including the WJ III NU COG, are often not sensitive to executive function difficulties and, therefore, will need to be supplemented with neuropsychological subtests in certain areas (e.g., reasoning), to more accurately understand an individual's executive control capacities. Nevertheless, it is important to understand that no set of directions on intelligence tests can completely eliminate the need for the examinee to use executive functions, such as basic self-regulation cues to engage in, and process and respond to test items (McCloskey et al., 2009). Third, following the administration of any battery, unexpected results are often present and hypotheses are generated and tested to explain the reason for the initial pattern of results. Testing hypotheses almost always requires the examiner to administer subtests from other batteries, as single batteries do not contain all the necessary subtests for follow-up assessments (see Flanagan et al., 2013, for a discussion). In cases in which it is necessary to supplement a battery or test hypotheses about aberrant test performance, following the XBA guiding principles and procedures (and using the XBA DMIA v2.0 software) will insure that the results are interpreted in a psychometrically and theoretically defensible manner.

Type 4 interpretation: Variation in task demands and task characteristics. Interpreting subtest scores representing narrow abilities often requires additional information from the practitioner to understand unexpected variation in performance. The XBA approach includes qualitative evaluations of cognitive and neuropsychological processes at the Type 4 level of interpretation to address how differences in task characteristics, such as input stimuli and output responses, and processing demands might affect an individual's performance on a particular subtest.

The focus on qualitative aspects of evaluations has been a common practice in neuropsychological assessment and has recently been reemphasized in cognitive assessment methods. The emphasis on clinical observation and qualitative behaviors is fundamental to the processing approach in neuropsychological assessment, which uses a flexible battery approach to gather both quantitative and qualitative data (Kaplan, 1988; Miller, 2007; Semrud-Clikeman, Wilkinson, & Wellington, 2005). Current models of school neuropsychology assessment also have foundations in the process assessment approach and stress the importance of qualitative observations to ensure ecological validity and guide individualized interventions (Hale & Fiorello, 2004; Miller, 2007). Additionally, the integration of qualitative assessment methods in XBA proposed by Flanagan et al. (2010, 2013), and elaborated on here, illustrates the benefits of assessing both quantitative and qualitative information in cognitive assessment practice.

The inclusion of qualitative information in intellectual and cognitive assessment is also evident in the WISC-IV Integrated (Wechsler, 2004). The tasks of the WISC-IV Integrated were designed from a process-oriented approach to help practitioners utilize qualitative assessment methods (McCloskey, 2009). Specifically, McCloskey notes how the process approach has influenced three perspectives in the use and interpretation of the WISC-IV Integrated, "[1] WISC-IV subtests are complex tasks, with each one requiring the use of multiple cognitive capacities for successful performance; [2] variations in input, processing, and/or output demands can greatly affect performance on tasks involving identical or similar content; and [3] careful, systematic observation of task performance greatly enhances the understanding of task outcome" (2009, p. 310).

The emphasis on qualitative assessment originated from the belief that the processes or strategies that an examinee uses during a task are as clinically relevant as the quantitative score (outcome) (Miller, 2007; Semrud-Clikeman et al., 2005). A major tenet in the process approach is that although examinees may obtain the same score on a task, they may be utilizing different strategies and/or neuropsychological processes to perform the task (Kaplan, 1988; Semrud-Clikeman et al., 2005). The analysis of qualitative information derived from observing task performance can provide valuable insight to potential cognitive or neuropsychological strengths and deficits and provide useful information to guide individualized interventions (Hale & Fiorello, 2004). For example, qualitative observations of two examinees that performed poorly on the D-KEFS Tower task may indicate different problems in executive functioning. The first examinee took several minutes before initiating the task, was slow in moving the disks, and made several rule violations, while the other examinee rushed into the task and used a trial-by-error approach. Both examinees appear to have difficulty with planning and problem solving; however, the impulsive examinee might have difficulty due to poor response inhibition, whereas the slower examinee may have difficulty with decision making, rule learning, and establishing and maintaining an instructional set (Delis et al., 2001).

As shown in Fig. 22.3, the XBA approach to assessing executive functions highlights five aspects of executive functioning that can be inferred through qualitative evaluations of an examinee's test performance: use of feedback, response inhibition, motor programming, cognitive set shifting, and different aspects of attention. Based on task characteristics and demands analysis, Table 22.5 illustrates qualitative aspects of executive functions on subtests of common cognitive and neuropsychological batteries. It should be noted that some neuropsychological batteries include quantitative measures of response inhibition (e.g., NEPSY-II Statue); however, since current intelligence and cognitive batteries do not directly assess response inhibition, it is included in the qualitative section, as it is an observable behavior. The qualitative assessment of these executive functions is not limited to the specific subtest classifications in Table 22.5 since examinees may be utilizing (or failing to utilize) these executive functions depending upon which strategy they implement during a task. For example, although Matrix Reasoning on the Wechsler Scales is not designed to assess response inhibition, if an examiner notices the individual is responding impulsively and making errors based on visual stimuli that are similar to the correct response, the practitioner may infer that the individual has difficulty inhibiting responses to distracting stimuli if this is consistent with other behavioral observations.

Additionally, the executive functions that define the qualitative portion of Fig. 22.3 do not comprise an exhaustive list, but rather include the executive functions most commonly assessed in neuropsychological evaluations (not necessarily in the assessment of intelligence, using traditional intelligence batteries) (Miller, 2007). Since there is a lack of consensus among disciplines regarding the classifications of the processes that comprise executive functions, different models of executive functions may include other aspects of self-regulation, goal-directed behavior, and organization, not mentioned in or inferred from

Table 22.5 Executive functions inferred through demand analysis of cognitive and neuropsychological test performance

Battery	Subtest	Use of feedback in task performance	Response inhibition	Set shifting	Attention (sustained)
DAS-II	Matrices				
	Picture similarities				
	Rapid naming				✓
	Recall of digits backward				Capacity
	Recall of sequential order				Divided/capacity
	Sequential and quantitative reasoning				
	Speed of information processing		✓		Selective
	Verbal similarities				
D-KEFS	Color-word interference: color naming				✓
	Color-word interference: inhibition		✓		Selective
	Color-word interference: inhibition/switching		✓		Divided/switching
	Color-word interference: word reading				✓
	Design fluency test: filled dots				✓
	Design fluency test: switching		✓	✓	Selective
	Design fluency test: empty dots only				Selective
	Sorting test: free sorting			✓	
	Sorting test: sort recognition				
	Tower	✓			
	Trail making test: letter sequencing	✓			Selective
	Trail making test: number sequencing	✓			Selective
	Trail making test: number-letter switching	✓	✓	✓	Divided
	Trail making test: visual scanning	✓			Selective
	Twenty questions test	✓		✓	
	Verbal fluency test: category fluency				
	Verbal fluency test: category switching		✓	✓	
	verbal fluency test: letter fluency		✓		
	Word context	✓			
KABC-II	Pattern reasoning				
	Riddles				
	Rover	✓			
	Story completion				
	Word order				Capacity
NEPSY-II	Animal sorting			✓	Selective/shifting

(continued)

Table 22.5 (continued)

Battery	Subtest	Use of feedback in task performance	Response inhibition	Set shifting	Attention (sustained)
	Auditory attention and response set		✓	✓	
	Design fluency				✓
	Inhibition		✓	✓	Selective
	Speeded naming			✓	✓
	Word generation	✓			
	Word list interference				Capacity/divided
SB5	Nonverbal fluid reasoning				
	Nonverbal knowledge				
	Nonverbal quantitative reasoning				
	Nonverbal working memory		✓		Capacity/divided
	Verbal fluid reasoning			✓	Capacity/divided
	Verbal quantitative reasoning				
	Verbal working memory				
Wechsler Scales	Arithmetic				
	Cancellation		✓		Selective
	Coding				Selective
	Digit span				Capacity
	Figure weights				
	Letter-number sequencing				Capacity/divided
	Matrix reasoning				
	Picture concepts				
	Similarities				
	Symbol search		✓		Selective
	Word reasoning	✓			
WISC-IV integrated	Arithmetic-process approach				
	Coding copying				Selective
	Letter-number sequencing process approach				Capacity/divided
	Similarities-multiple choice				
	Spatial span forward and backward				Capacity/divided
WJ III COG NU	Analysis-synthesis	✓			
	Auditory working memory				Capacity/divided
	Concept formation	✓		✓	
	Decision speed				
	Numbers reversed				Capacity
	Pair cancellation		✓		
	Planning		✓		✓
	Rapid picture naming				✓
	Retrieval fluency				
	Verbal comprehension				
	Visual matching		✓		Selective

Note: From a CHC perspective, motor programming corresponds to narrow abilities (e.g., finger dexterity [P2], gross body equilibrium [P4]) under the broad ability of psychomotor abilities (Gp). From a neuropsychological perspective, motor programming involves learning new motor sequences and may be inferred from manual imitation tests and, therefore, is considered an executive function (Miller, 2007) measurement of the abilities listed in Fig. 22.3. When attempting to derive information about executive functions from psychometric tests following the XBA approach, it is recommended that practitioners use the model in Fig. 22.3 as a framework and add to it with additional measures of executive functions, depending on the reason for referral and presenting behaviors of the examinee.

The previous discussion of a Type 3 interpretation described a scenario where an examinee performed average (SS = 100) on the WJ III NU Analysis–Synthesis (AS) subtest yet demonstrated a (normative) weakness (SS = 82) on WJ III NU Concept Formation (CF). To follow up on the low CF score, the examiner chose to administer the D-KEFS Free Sorting, an additional measure of induction (*Gf*: I). If the scores on these two measures of induction differ significantly from one another (an *unexpected* finding), then a Type 4 interpretation is warranted to explain the variation in performance on two measures of the same narrow ability. The following example illustrates Type 4 interpretation.

Sara, a fifth-grade student, was referred for an evaluation by her teacher because she has difficulty functioning independently in the classroom despite behavioral interventions. Sara's teacher reports that she has difficulty following directions and often is the last student to begin an assigned task. Sara is also constantly asking her teacher for help or to check if an answer is correct. Although Sara's previous teachers expressed similar concerns, Sara's difficulties have become more problematic with the independent structure and demands of the fifth-grade classroom. Additionally, Sara's teacher is concerned about her poor written responses on essay questions, which sometimes appear "off" and often "don't make sense."

After administering the WJ III NU COG Gf subtests and following up with the D-KEFS Free-Sorting task, it was clear that Sara's Free-Sorting Description Score (Sc.S = 5) was significantly lower than her score on the CF task of induction. Because this finding was unexpected, the examiner conducted a demand analysis to gather additional information about the variations in task characteristics and cognitive and neuropsychological demands specific to all three measures of *Gf*. This information is presented in Table 22.6 and the similarities and differences among these tasks are discussed below within the context of Sara's performance.

As discussed in the Type 3 interpretation, Sara's average performance on the WJ III NU Analysis–Synthesis (AS) task and poor performance on the WJ III NU Concept Formation (CF) and D-KEFS Free-Sorting tasks suggests that her ability to reason logically, using known rules (*Gf*: RG), is better than her ability to observe underlying principles or rules of a problem (*Gf*: I). When Sara was solving problems on the AS task, she was constantly looking to the key presented at the top of the stimulus easel and using her fingers to help guide her decisions for which colored box fit the answer. Therefore, it appears as though Sara's ability to reason and apply rules is stronger when she is presented with a visual key that can be used as a reference during a task. However, on the CF task, Sara had a difficult time following the first few sets of instructions and relied on examiner feedback during the sample teaching item to gain understanding of the task directions. Although all three tasks include lengthy oral directions, the instructions presented in the CF task are particularly complicated and require greater demands on receptive language. Furthermore, Sara gave several answers on the CF task that required querying but was often able to obtain the correct answer after the query. Finally, Sara had difficulty starting the D-KEFS Free-Sorting task and took a long time between each sort. Although Sara's ability to correctly sort the cards into groups was more consistent (Sc.S = 7) with her performance on the CF task, she had a hard time articulating and explaining how she was able to sort the cards (Sc.S = 5). Additionally, Sara often turned to the examiner to ask if she was correct and appeared disappointed when the examiner explained that she could not provide feedback.

The behavioral observations noted during task performance and the analysis of the cognitive and neuropsychological demands for each task allowed the examiner to come up with the following hypothesis regarding Sara's inconsistent scores

Table 22.6 Task characteristics and task demands of WJ III NU COG analysis-synthesis and concept formation and D-KEFS free sorting

WJ III analysis-synthesis	WJ III concept formation	D-KEFS free sorting
Directions/task procedures		
Lengthy oral instructions	Lengthy and complex oral instructions	Lengthy oral instructions
Sample teaching item, with feedback	Sample teaching item, with feedback	Demonstration
	Querying for certain responses	Timed item completion
Input		
Visual stimulus: small, colored squares; nonmeaningful	Visual stimulus: small, colored shapes; nonmeaningful	Visual stimulus: colored cards with printed words
		Visual stimulus: written rules
Processing demands		
Reasoning and problem solving (Gf: RG)	Reasoning and problem solving (Gf: I)	Planning, reasoning, and problem solving (Gf: I)
Use of feedback—feedback given for correct and incorrect responses	*Use of feedback*—feedback given for correct and incorrect responses	*Cognitive set shifting*—switching attention to different stimuli features
Cognitive set shifting—switching problem-solving strategies	*Cognitive set shifting*—rule switching	Cognitive flexibility—generating multiple answers
Visual processing	Concept formation and generation—rule-based categorization	Concept formation and generation—creating different categories using verbal and nonverbal information (involves *Gc*)
Receptive language	Visual processing of printed shapes	Visual processing of shapes and words
Auditory-verbal	Receptive language	Working memory—keeping track/updating which categories were used
	Auditory-verbal	Receptive language
		Expressive language
		Auditory-Verbal
Output		
Oral—brief	Oral—brief	Oral—lengthy (explaining the groups)
	Pointing—beginning items	Use of manipulatives: *Fine motor* involved in sorting of the cards

Note: Information in italics represents the executive functions listed in the bottom portion of Fig. 22.3. Demand analysis structure derived from Flanagan et al. (2013) and Hale and Fiorello (2004)

within the *Gf* domain. Sara appears to have greater difficulty on reasoning and problem-solving tasks that involve concept formation and generation, such as the CF and Free-Sorting task. Furthermore, Sara's difficulty generating and explaining multiple sorts may also indicate problems with cognitive flexibility, divergent thinking, and ideation fluency (Miller, 2007; Miller & Maricle, 2012). Additionally, these tasks require more receptive and expressive language demands and tap into *Gc* abilities, which was indicated as another weakness for Sara based on her low *Gc* performance on the WJ III NU COG.

Sara's slow performance during the Free-Sorting task also implies difficulty initiating problem-solving tasks and planning (Delis et al., 2001). This, along with Sara's receptive and expressive language difficulties, may explain why Sara has difficulty starting tasks and following directions. Furthermore, Sara's reliance on examiner feedback and visual keys during the WJ III NU tasks may signify problems with self-monitoring and explain why Sara often seeks feedback from her teacher. Overall, it appears that Sara's inconsistent performance in *Gf* tasks may stem from problems with *Gc* (language abilities) as well as executive functions, particularly verbal reasoning, problem-solving initiation, self-monitoring, and concept formation and generation. Sara's dependence on her teacher in the classroom is likely a compensatory strategy she has learned to help guide her through complex tasks. Interventions, such as *teaching self-regulated strategy development* (SRSD) to improve self-monitoring and self-revision, will allow Sara to learn how to function more independently in the classroom (De La Paz, 2007).

The previous example of a Type 4 interpretation demonstrated that it is often necessary to go beyond a strict quantitative interpretation of task performance and analyze the task characteristics of subtests as well as the student's approach to performing those tasks to gain a better understanding of cognitive strengths and weaknesses. Many evaluations of students with learning difficulties require the integration of quantitative and qualitative data to understand a student's cognitive capacities fully, including executive function capacities. Following is an example of a cross-battery assessment of executive functions, using the WISC-IV as the core battery that integrates mainly Type 2 (quantitative) and Type 4 (qualitative) interpretation.

Highlights of a Wechsler-Based Cross-Battery Assessment of Executive Functions

Ben began middle school (seventh grade) in the Fall of 2011. Ben has been having significant difficulties academically for the first time. His science teacher reported that he has a hard time initiating projects independently and seldom completes in-class assignments on time. Ben reportedly relies on a classmate to help him with science projects and as a result, his teacher moved his seat in an attempt to get him to function more independently in the classroom. Reports from Ben's other teachers suggest that he is often the last student to "find his place" and he frequently "holds up the class," seemingly intentionally. Ben leaves important books and assignments in his locker often and, therefore, does not consistently complete homework. Although Ben reported that he studies for exams, his grades are poor, often as the result of careless errors (e.g., lack of attention to detail in math word problems) and incomplete or underdeveloped responses to open-ended questions. His teachers all agree that Ben knows more information than he is able to demonstrate on tests and quizzes. Overall, there is consensus among Ben's teachers that, while Ben appears to be very bright, he lacks motivation and appears to exhibit attention-seeking behaviors (e.g., he jokes with his classmates that he is last to complete his work). Ben's parents believe that he is having a hard time adjusting to his new school, including an increase in homework assignments and projects, and they are worried about his recent negative attitude toward school. An evaluation was requested to explore whether Ben's learning difficulties are the result of an underlying learning disability, behavioral difficulties, or both.

As part of Ben's comprehensive evaluation, the evaluator administered the WISC-IV and WIAT-III. The results of Ben's performance on these batteries are found in Table 22.7. A quantitative analysis of Ben's WISC-IV/WIAT-III scores indicates that his performance ranged from Average to Well Above Average (Type 2 Interpretation).[4] Despite Ben's poor academic performance in the seventh grade and the observations offered by Ben's teachers, many practitioners would conclude that the difficulties Ben is experiencing in school are not related to any underlying cognitive deficits or dysfunction and therefore, they must be the result of the behavioral problems reported (e.g., attention-seeking behavior, lack of motivation). Prior to drawing such a conclusion, it is necessary to determine if the evaluator noticed any unusual approaches to solving problems during the evaluation or any unusual patterns of errors in task performance, for example (Type 4 Interpretation).

A qualitative analysis of Ben's performance on the WISC-IV and WIAT-III subtests revealed

---

[4] Because the evaluator did not find a need to follow up on any of the scores yielded from the WISC-IV and WIAT-III, Type 3 Interpretation was not necessary.

Table 22.7 Ben's WISC-IV/WIAT-III scores

Subtest/*Composite*	Score
WISC-IV	
Similarities	12
Vocabulary	13
Comprehension	9
Information	12
Verbal Comprehension	*106*
Block Design	11
Picture Concepts	12
Matrix Reasoning	14
Perceptual Reasoning	*115*
Digit Span	10
Letter-Number Sequencing	10
Working Memory	*99*
Coding	9
Symbol Search	8
Processing Speed	*91*
Gc-K0	*102*
Gf-Nonverbal	*118*
FSIQ	*106*
GAI	*112*
WIAT-III, grade norms (seventh grader, 12 years 5 months)	
Word Reading	114
Pseudoword Decoding	109
Basic Reading	*112*
Reading Comprehension	112
Oral Reading Fluency	111
Reading Comprehension and Fluency	*114*
Spelling	105
Sentence Composition	105
Essay Composition	92
Written Expression	*100*
Math Problem Solving	90
Numerical Operations	105
Mathematics	*97*
Math Fluency-Addition	102
Math Fluency-Subtraction	109
Math Fluency-Multiplication	107
Math Fluency	*107*
Listening Comprehension	110
Oral Expression	112
Oral Language	*112*

several important observations. For example, the evaluator believed that Ben's VCI and PRI may have underestimated his capacity to reason with verbal and visual-spatial information, respectively. On the Similarities subtest, the evaluator did not observe Ben reasoning. Specifically, Ben's responses were immediate, indicating that the information requested was readily available to Ben (i.e., quickly retrieved from existing stores of general knowledge and lexical knowledge). When items became more difficult, Ben was quick to respond, "I don't know" and did not take the time to "think" about a response. This same response style was evident on the Comprehension subtest. In addition, when items asked for "some advantages," Ben seemed content with his initial response and when queried he stated, "That's all I can think of." Despite Ben's Average (Comprehension) and Above Average (Similarities) performance on these subtests, his response style suggests difficulty cueing and directing the use of reasoning abilities as well as difficulty shifting mindset.

On the WISC-IV Picture Concepts subtest and WIAT-III Math Problem-Solving subtest, Ben demonstrated inconsistencies in performance, revealing incorrect responses interspersed across test items, which is unusual given that items are arranged in order of increasing difficulty. This pattern of performance may suggest difficulty cueing the appropriate consideration of the cognitive demands of the task and the amount of mental effort required to effectively perform the task. Ben's pattern of performance on Picture Concepts and Math Problem Solving may also suggest difficulty with monitoring performance and correcting errors. Likewise, on the Block Design subtest, Ben did not pay close attention to detail, especially on items that did not include the black lines on the stimulus card. Ben also appeared to give up easily on items and often said, "I can't figure out that one."

An examination of Ben's pattern of errors on the Picture Concepts, Math Problem Solving, Coding, and Symbol Search subtests demonstrates that his errors were careless and not reflective of a lack of ability or knowledge, which is consistent with teacher reports. It appears that Ben may have difficulty monitoring his attention over a sustained period of time. His performance on the processing speed subtests, in particular, and perhaps also Block Design, suggests that Ben has difficulty cueing and directing the focusing of attention to visual details and task demands. Based on the evaluator's qualitative analysis on Ben's approach to tasks coupled with unusual patterns of errors on certain subtests, it was hypothesized that Ben has weaknesses in executive functions related to modulating and monitoring his performance. To test hypotheses specific to these executive functions, it was necessary to cross batteries.

The evaluator chose to test certain hypotheses about Ben's executive functions using the WISC-IV Integrated, which is statistically linked to the WISC-IV (following guiding principle #5 of the XBA approach). The evaluator administered Similarities Multiple Choice (SIMC) and Comprehension Multiple Choice (COMC). The evaluator hypothesized that by altering the cueing and directing of *open-ended* inductive reasoning (Similarities and Comprehension) to the cueing and directing of the *recognition* of the effective application of induction reasoning (SIMC, COMC), performance will improve. Ben's performance on both SIMC and COMC was significantly higher than his performance on Similarities and Comprehension, respectively. These results suggest that when the demands of open-ended inductive reasoning are reduced to recognition of the effective application of inductive reasoning, Ben's capacity for reasoning inductively improves significantly. Ben's capacity for reasoning inductively is greater than that which he can demonstrate with an open-ended format—a format typically used for tests and quizzes in school. Furthermore, Ben is able to perform in the average range on structured tasks for which explicit instructions are given and that are administered in a one-to-one testing situation. However, when he is required to perform academic tasks involving reasoning in a more unstructured setting (e.g., middle school, homework environment, school exams), his performance is well below average compared to his same-grade peers. Therefore, Ben would benefit from the following interventions.

Ben's teachers should provide verbal prompts and cues to assist him in the reasoning process when tasks require open-ended inductive reasoning. Ben's teachers should use direct instruction in acquisition lessons (e.g., How do I use inductive reasoning to reach a conclusion?) with modeling and think alouds to explicitly teach Ben how to use the skills. Ben's teachers should gradually offer guided practice (e.g., guided questions list) to promote internalization of the cueing and directing of reasoning skills. Teachers may consider using graphic organizers to guide Ben in using inductive reasoning skills. Steps to reasoning inductively should be made accessible for Ben's use until he has internalized the steps. And, Ben should be given multiple opportunities to extend his thinking about content (Marzano & Pickering, 1992).

To follow up on other hypotheses the evaluator had regarding Ben's difficulties with executive functions, it was necessary to administer subtests from a battery that is more sensitive to identifying such difficulties. In addition, it was necessary to measure retrieval fluency—an executive function listed in the top portion of Fig. 22.3 that is not measured by the WISC-IV/WIAT-III/WISC-IV Integrated batteries. The evaluator reasoned that he could test his remaining hypotheses about the nature of Ben's difficulties and assess retrieval fluency using subtests from only one additional battery (following XBA guiding principle #6)—the D-KEFS.

Based on Ben's performance on the WISC-IV Picture Concepts subtest, it was hypothesized that Ben had difficulty cueing the appropriate consideration of the cognitive demands of a task and the amount of mental effort required to effectively perform the task as well as difficulty cueing and directing the monitoring of work and the correcting of errors. Ben's performance on certain D-KEFS tasks supports this hypothesis. For example, Ben received a scaled score of 8 on the Free-Sorting task and a scaled score of 12 on the Sort Recognition Description Score Card Set 2. The difference between these scores is statistically significant. This result suggests that Ben has difficulty transferring knowledge into action in less structured situations in the face of intact concept formation skills, which may explain the difficulty he has completing projects in science class, as such projects are unstructured. It was also observed that Ben paid less attention to the perceptual aspects of the cards as compared to the verbal aspects on the Free-Sorting task, which is supportive of the hypothesis that Ben has difficulty cueing and directing the focusing of attention to visual details (as observed on the Block Design and processing speed subtests of the WISC-IV).

Finally, Ben's performance on the D-KEFS Fluency tasks also reveals some information about his executive functions that helps to explain the difficulties he is having in the seventh grade. On the D-KEFS Fluency tasks, Ben had difficulty monitoring his performance. For example, on the Letter Fluency task, Ben repeated some words, did not appear to refer back to the written rules for the task, and, when asked, reported that he did not use a strategy for completing the task. Failure to monitor performance, attend to rules, and apply strategies will certainly result in less than optimal performance in a seventh grade classroom and on related quizzes and exams. Also noteworthy is that the "switching" condition of both fluency tasks represented a significant decline in performance for Ben, suggesting difficulties with cognitive flexibility—a finding that may explain why Ben is often the last student in his class to find his place when his teachers transition from one assignment or project to another.

A convergence of cross-battery data, observations, and teacher and parent reports suggest that Ben's capacity to reason with verbal and visual-spatial information is greater than that which he is able to demonstrate on cognitive tests, in the classroom, and on exams. Specifically, Ben has self-regulation difficulties in the cognition domain, including difficulties with modulating, planning (e.g., selecting and using appropriate strategies), and monitoring academic activities in the classroom and at home (e.g., homework).

Summary

This chapter demonstrated how to use the XBA approach to assess and interpret executive functions within the cognition domain via the symbol system. Specifically, we identified specific broad and narrow CHC abilities that, when measured with carefully selected cognitive and neuropsychological tests, reveal information about a subset of executive functions. The frontal-subcortical executive functions addressed in this chapter include planning, focusing and sustaining attention, maintaining or shifting sets, verbal fluency, use of feedback, and working memory. A focus on these executive functions, in particular, is likely to yield useful information to assist in understanding problems of learning and production in an academic setting. Assessment of executive functions in the intrapersonal, interpersonal, and in other environmental arenas may be necessary to fully appreciate how an individual not only self-monitors and regulates cognition but also emotion and behavior.

References

Alfonso, V. C., Flanagan, D. P., & Radwan, S. (2005). The impact of the Cattell-Horn theory on test development and interpretation of cognitive and academic abilities. In D. P. Flanagan & P. L. Harrison (Eds.), *Contemporary intellectual assessment: Theories, tests, and issues* (2nd ed., pp. 185–202). New York: Guildford Press.

Baron, I. (2004). *Neuropsychological evaluation of the child*. New York, NY US: Oxford University Press.

Carroll, J. B. (1993). *Human cognitive abilities: A survey of factor-analytic studies*. Cambridge, UK: Cambridge University Press.

Carroll, J. B. (1997). The three-stratum theory of cognitive abilities. In D. P. Flanagan, J. L. Genshaft, & P. L. Harrison (Eds.), *Contemporary intellectual assessment: Theories, tests, and issues* (pp. 122–130). New York: Guilford Press.

Carroll, J. B. (1998). Foreword. In K. S. McGrew & D. P. Flanagan (Eds.), *The intelligence test desk reference: Gf-Gc cross-battery assessment* (pp. xi–xii). Boston: Allyn & Bacon.

Comrey, A. L. (1988). Factor-analytic methods of scale development in personality and clinical psychology. *Journal of Consulting and Clinical Psychology, 56*, 754–761.

Das, J. P., & Naglieri, J. A. (1997). *Cognitive assessment system*. Itasca, IL: Riverside.

Dawson, P. (2012). Executive functioning in children. In S. Goldstein & J. A. Naglieri (Eds.), *Handbook on executive functioning*.

Dean, R. S., & Woodcock, R. W. (2003). *Dean–Woodcock neuropsychological battery*. Itasca, IL: Riverside Publishing.

Decker, S. L. (2008). School neuropsychology consultation in neurodevelopmental disorders. *Psychology in the Schools, 45*, 799–811.

Dehn, M. J. (2006). *Essentials of processing assessment*. New York: Wiley.

De La Paz, S. (2007). Managing cognitive demands for writing: Comparing the effects of instructional components in strategy instruction. *Reading and Writing Quarterly, 23*, 249–266.

Delis, D. C., Kaplan, E., & Kramer, J. H. (2001). *Delis Kaplan executive function system*. San Antonio, TX: The Psychological Corporation.

Elliott, C. D. (2007). *Differential ability scales—Second edition*. San Antonio, TX: Harcourt Assessment.

Feifer, S. G., & Della Toffalo, D. A. (2007). *Integrating RTI with cognitive neuropsychology: A scientific approach to reading*. Middletown, MD: School Neuropsych Press.

Fiorello, C. A., Hale, J. B., Snyder, L. E., Forrest, E., & Teodori, A. (2008). Validating individual differences through examination of converging psychometric and neuropsychological models of cognitive functioning. In S. K. Thurman & C. A. Fiorello (Eds.), *Applied cognitive research in k-3 classrooms* (pp. 232–254). New York: Routledge.

Flanagan, D. P., Alfonso, V. C., & Ortiz, S. O. (2012). The cross-battery assessment approach: An overview, historical perspective, and current directions. In D. P. Flanagan, J. L. Genshaft, & P. L. Harrison (Eds.), *Contemporary intellectual assessment: Theories, tests, and issues* (pp. 459–483). New York: Guilford Press.

Flanagan, D. P., Alfonso, V. C., & Mascolo, J. T. (2013). A CHC-based operational definition of SLD: Integrating multiple data sources and multiple data-gathering methods. In D. P. Flanagan & V. C. Alfonso (Eds.), *Essentials of specific learning disability identification* (pp. 233–298). Hoboken, NJ: Wiley.

Flanagan, D. P., Alfonso, V. C., Ortiz, S. O., & Dynda, A. M. (2010). Integrating cognitive assessment in school neuropsychological evaluations. In D. C. Miller (Ed.), *Best practices in school neuropsychology: Guidelines for effective practice, assessment, and evidence-based intervention* (pp. 101–140). Hoboken, NJ: Wiley.

Flanagan, D. P., & McGrew, K. S. (1997). A cross-battery approach to assessing and interpreting cognitive abilities: Narrowing the gap between practice and cognitive science. In D. P. Flanagan, J. L. Genshaft, & P. L. Harrison (Eds.), *Contemporary intellectual assessment Theories, tests, and issues* (pp. 314–325). New York: Guilford Press.

Flanagan, D. P., McGrew, K. S., & Ortiz, S. O. (2000). *The Wechsler Intelligence Scales and Gf-Gc theory: A contemporary approach to interpretation*. Boston: Allyn & Bacon.

Flanagan, D. P., & Ortiz, S. O. (2001). *Essentials of cross-battery assessment*. New York: Wiley.

Flanagan, D. P., Ortiz, S. O., & Alfonso, V. C. (2007). *Essentials of cross-battery assessment* (2nd ed.). New York: Wiley.

Flanagan, D. P., Alfonso, V. C., & Ortiz, S. O. (2012). The cross-battery assessment approach: An overview, historical perspective, and current directions. In D. P. Flanagan, J. L. Genshaft, & P. L. Harrison (Eds.), *Contemporary intellectual assessment: Theories, tests, and issues* (pp. 459–483). New York: Guilford Press.

Flanagan, D. P., Ortiz, S. O., & Alfonso, V. C. (2013). *Essentials of cross-battery assessment* (3rd ed.). Hoboken, NJ: Wiley.

Flanagan, D. P., Ortiz, S. O., Alfonso, V. C., & Mascolo, J. T. (2006). *Achievement test desk reference: A guide to learning disability identification* (2nd ed.). New York: Wiley.

Fletcher-Janzen, E., & Reynolds, C. R. (2008). *Neuropsychological perspectives on learning disabilities in the era of RTI: Recommendations for diagnosis and intervention*. Hoboken, NJ: Wiley.

Hale, J. B., & Fiorello, C. A. (2004). *School neuropsychology: A practitioner's handbook*. New York: Guilford Press.

Horn, J. L. (1991). Measurement of intellectual capabilities: A review of theory. In K. S. McGrew, J. K. Werder, & R. W. Woodcock (Eds.), *Woodcock-Johnson technical manual* (pp. 197–232). Chicago: Riverside.

Horn, J. L., & Blankson, N. (2005). Foundations for better understanding of cognitive abilities. In D. P. Flanagan & P. L. Harrison (Eds.), *Contemporary intellectual assessment: Theories, tests, and issues* (2nd ed., pp. 41–68). New York: Guilford Press.

Kaplan, E. (1988). A process approach to neuropsychological assessment. In T. Boll & B. K. Bryant (Eds.), *Clinical neuropsychology and brain function: Research, measurement, and practice* (pp. 125–167). Washington, DC: American Psychological Association.

Kaufman, A. S. (1979). *Intelligent testing with the WISC-R*. New York: Wiley.

Kaufman, A. S. (2000). Forward. In D. P. Flanagan, K. S. McGrew, & S. O. Ortiz (Eds.), *The Wechsler intelligence scales and Gf-Gc theory: A contemporary approach to interpretation*. Needham Heights, MA: Allyn & Bacon.

Kaufman, A. S., & Kaufman, N. L. (2004). *Kaufman Assessment Battery for Children-second edition*. Circle Pines, MN: AGS Publishing.

Keith, T. Z., & Reynolds, M. R. (2010). CHC and cognitive abilities: What we've learned from 20 years of research. *Psychology in the Schools, 47*, 635–650.

Keith, T. Z., & Reynolds, M. R. (2012). Using confirmatory factor analysis to aid in understanding the constructs measured by intelligence tests. In D. P. Flanagan & P. L. Harrison (Eds.), *Contemporary intellectual assessment: Theories, tests, and issues* (3rd ed., pp. 758–799). New York: Guilford Press.

Korkman, M., Kirk, U., & Kemp, S. (2007). *NEPSY-II: A developmental neuropsychological assessment*. San Antonio, TX: The Psychological Corporation.

Lezak, M. D. (1976). *Neuropsychological assessment*. New York: Oxford University Press.

Lezak, M. D. (1995). *Neuropsychological assessment* (3rd ed.). New York: Oxford University Press.

Lezak, M. D., Howieson, D. B., & Loring, D. W. (2004). *Neuropsychological assessment* (4th ed.). New York: Oxford University Press.

Luria, A. R. (1973). *The working brain: An introduction to neuropsychology*. New York: Basic Books.

Maricle, D. E., & Avirett, E. (2012). The emergence of neuropsychological constructs into tests of intelligence and cognitive abilities. In D. P. Flanagan & P. L. Harrison (Eds.), *Contemporary intellectual assessment: Theories, tests, and issues* (3rd ed., pp. 800–819). New York: Guilford Press.

Marzano, R., & Pickering, D. (1992). *Dimensions of learning*. Alexandria, VA: Association for Supervision and Curriculum Development.

Mascolo, J. T., Flanagan, D. P., & Alfonso, V. C. (in press). *Planning, selecting, and tailoring interventions for the unique learner*. Hoboken, NJ: Wiley.

McCloskey, G. (2009). The WISC-IV integrated. In D. P. Flanagan & A. S. Kaufman (Eds.), *Essentials of WISC-IV assessment* (2nd ed., pp. 310–467). Hoboken, NJ: Wiley.

McCloskey, G. (2012). Working memory and executive functioning. In S. Goldstein & J. A. Naglieri (Eds.), *Handbook on executive functioning*.

McCloskey, G., Perkins, L. A., & Van Divner, B. (2009). *Assessment and intervention for executive function difficulties*. New York: Routledge.

McGrew, K. S. (2005). The Cattell-Horn-Carroll theory of cognitive abilities: Past, present, and future. In D. P. Flanagan, J. L. Genshaft, & P. L. Harrison (Eds.), *Contemporary intellectual assessment: Theories, tests, and issues* (2nd ed., pp. 136–182). New York: Guilford Press.

McGrew, K. S. (1997). Analysis of the major intelligence batteries according to a proposed comprehensive CHC framework. In D. P. Flanagan, J. L. Genshaft, & P. L. Harrison (Eds.), *Contemporary intellectual assessment: Theories, tests, and issues* (pp. 151–180). New York: Guilford Press.

McGrew, K. S., & Flanagan, D. P. (1998). *The intelligence test desk reference (ITDR): Gf-Gc cross-battery assessment*. Boston, MA: Allyn & Bacon.

Meltzer, L. (2007). *Executive function in education: From theory to practice*. New York, NY: Guilford Press.

Meltzer, L. (2012). *Handbook of executive functioning*. New York: Springer.

Messick, S. (1989). Validity. In R. Linn (Ed.), *Educational measurement* (3rd ed., pp. 104–131). Washington, DC: American Council on Education.

Messick, S. (1995). Validity of psychological assessment: Validation of inferences from persons' responses and performances as scientific inquiry into score meaning. *American Psychologist, 50*, 741–749.

Miller, D. C. (2007). *Essentials of school neuropsychological assessment*. Hoboken, NJ: Wiley.

Miller, D. C. (2010). *Best practices in school neuropsychology: Guidelines for effective practice, assessment, and evidence-based intervention*. Hoboken, NJ: Wiley.

Miller, D. C. (2013). *Essentials of school neuropsychological assessment* (2nd ed.). Hoboken, NJ: Wiley.

Miller, D. C., & Maricle, D. E. (2012). The emergence of neuropsychological constructs into tests of intelligence and cognitive abilities. In D. Flanagan & P. Harrison (Eds.), *Contemporary intellectual assessment: Theories, tests, and issues* (3rd ed., pp. 800–819). New York, NY US: Guilford Press.

Naglieri, J. A. (2005). The cognitive assessment system. In D. P. Flanagan, P. L. Harrison (Eds.), *Contemporary intellectual assessment: Theories, tests, and issues* (2nd ed., pp. 441–460). New York, NY: Guilford Press.

Naglieri, J. A. (2012). The cognitive assessment system. In S. Goldstein & J. A. Naglieri (Eds.), *Handbook on executive functioning*.

Reynolds, M. R., Keith, T. Z., Flanagan, D. P., & Alfonso, V. C. (in press). A cross-battery, reference variable, confirmatory factor analytic investigation of the CHC Taxonomy. *Journal of School Psychology*.

Roid, G. H. (2003). *Stanford-binet intelligence scales, (5th ed.)*. Itasca, IL: Riverside.

Sattler, J. M. (1988). *Assessment of children* (3rd ed.). San Diego, CA: Author.

Schneider, J. W., & McGrew, K. S. (2012). The Cattell-Horn-Carroll model of intelligence. In D. P. Flanagan & P. L. Harrison (Eds.), *Contemporary intellectual assessment: Theories, tests, and issues* (3rd ed., pp. 99–144). New York: Guilford Press.

Semrud-Clikeman, M., Wilkinson, A., & Wellington, T. (2005). Evaluating and using qualitative approaches to neuropsychological assessment. In R. D'Amato, E. Fletcher-Janzen, & C. R. Reynolds (Eds.), *Handbook of school neuropsychology* (pp. 287–302). Hoboken, NJ: Wiley.

Strauss, E., Sherman, E. M. S., & Spreen, O. (2006). *A compendium of neuropsychological tests: Administration, norms, and commentary* (3rd ed.). New York: Oxford University Press.

Wechsler, D. (2003). *Wechsler intelligence scale for children—Fourth edition*. San Antonio, TX: Psychological Corporation.

Wechsler, D. (2004). *Wechsler intelligence scale for children—fourth edition integrated*. San Antonio, TX: Psychological Corporation.

Wechsler, D. (2008). *Wechsler adult intelligence scale—fourth edition*. San Antonio, TX: Pearson.

Wilson, B. C. (1992). The neuropsychological assessment of the preschool child: A branching model. In I. Rapin & S. I. Segalowitz (Vol. Eds.), *Handbook of neuropsychology: Vol. 6. Child neuropsychology* (pp. 377–394). San Diego, CA: Elsevier.

Woodcock, R. W. (1990). Theoretical foundations of the WJ-R measures of cognitive ability. *Journal of Psychoeducational Assessment, 8*, 231–258.

Woodcock, R. W., McGrew, K. S., & Mather, N. (2001). *Woodcock-Johnson III tests of cognitive abilities*. Itasca, IL: Riverside Publishing.

Woodcock, R. W., McGrew, K. S., & Mather, N. (2007). *Woodcock-Johnson III tests of cognitive abilities*. Itasca, IL: Riverside Publishing.

Part IV
Interventions Related to Executive Functioning

Treatment Integrity in Interventions That Target the Executive Function

23

Andrew Livanis, Ayla Mertturk,
Samantha Benvenuto, and Christy Ann Mulligan

Executive Function (EF) refers to a broad cognitive process used to direct behavior specifically in situations where some responses must be inhibited and others need to be initiated (Banich, 2009). Definitions of EF tend to involve reference to the frontal lobe (Hayes, Gifford, & Ruckstuhl, 1996), so much so that it is difficult to feature a definition that does not reference this area of the brain. While there are some differences in the definition of the construct, recent research suggests that EF is a unidimensional construct (Goldstein, 2012) and includes components of various cognitive processes such as planning, goal persistence, cognitive flexibility, abstract thinking, and rule acquisition (Baltruschat et al., 2011).

The strength of EF waxes and wanes across the lifespan (Eisenberg & Berman, 2010; Hale et al., 2009; McClelland et al., 2007; Simonsen et al., 2008). The development of EF begins in early infancy (Eliot, 1999; McCloskey, Perkins, & Van Divner, 2009; Posner & Rothbart, 2007), and over time, the process is strengthened by learning a variety of different skill sets. For example, children learn to set long- and short-term behavioral goals, thus allowing them to plan effectively for the future. Through exposure to trillions upon trillions of contingencies, children

A. Livanis (✉) • A. Mertturk • S. Benvenuto
C.A. Mulligan
Long Island University—Brooklyn, Brooklyn,
NY, USA
e-mail: Andrew.Livanis@liu.edu

become adept at identifying and generalizing reinforcement and punishment contingencies which allows them to engage in complex decision-making tasks. Children also learn how to inhibit and moderate certain behaviors in some situations so that they can engage in more productive behaviors in other environments (Barkley, 1997; Steinberg, 2007). There is some evidence to suggest that EF may decline over time as well to the point where we witness increasing deficits in EF in elderly populations (Buckner, 2004). In addition, the elderly stage of life brings a susceptibility to the onset of various illnesses and autoimmune disorders that bring about ever-increasing deficits in EF, such as cerebrovascular accident (Leeds, Meara, Woods, & Hobson, 2001), Parkinson's Disorder (Hausdorff et al., 2006), and Alzheimer's disorder (Duarte et al., 2006).

In addition to variations in age, EF may vary depending on the presence of various neurological and psychiatric conditions. EF deficits are noted when we examine Attention Deficit-Hyperactivity Disorder (ADHD; Barkley, 1997), as the Diagnostic and Statistical Manual, Fourth Edition, Text Revision (DSM-IV-TR; American Psychiatric Association, 2000) includes both inattention and inhibition as central issues in this disorder. However, children with ADHD also evidence deficits in their ability to regulate their behavior in a variety of situations suggesting that these children demonstrate difficulties controlling their impulsivity, selecting which stimuli require their focus, and sustaining their attention on meaningful activities (Barkley,

1997; McCloskey et al., 2009). These criteria, however, do not mean that all children with Executive Function deficits have ADHD. In fact, although individuals with ADHD may display multiple EF deficits, the specific deficits beyond the commonly associated deficits noted in the DSM-IV-TR (i.e., inattentiveness, impulsivity, and hyperactivity), and the total number of severity of deficiencies demonstrated, will vary from individual to individual and by age (McCloskey et al., 2009).

Children who have other externalizing disorders, such as Oppositional Defiant Disorder and Conduct Disorder, also evidence EF deficits. These children often lack insight into their actions and may have explosive episodes, which can impair their ability to plan, persist in identifying their goals, engage in abstract thinking and rule acquisition, and present a flexible problem-solving style (Goldstein, 2012). These deficits will be evident in multiple environments, and behaviors and EF deficits may be especially marked in the classroom (McCloskey et al., 2009).

Children with specific learning disabilities are also at elevated risk of EF deficits. Children who exhibit both problems with learning and producing products that show evidence of learning (Denckla, 2007) are at greatest risk of being identified for special education services (McCloskey et al., 2009). In the elementary years, students with EF difficulties typically demonstrate problems in reading, written expression, and mathematics skills. In the upper grades, EF difficulties manifest with deficits in basic skill production, organization, planning, and completion of projects and homework, as well as an inadequate regulation of the use of study and test-taking skills (McCloskey et al., 2009).

Evidence-Based Interventions for the Remediation of EF Deficits

It is important for educators to identify EF deficits in children as early as possible and provide appropriate evidence-based interventions (EBI). EF deficits can adversely affect the early academic skills of children, as well as lead to a variety of social and emotional difficulties (Alloway et al., 2009; Gathercole et al., 2008; McClelland et al., 2007; Scope, Empson, & McHale, 2010). The selection and implementation of EBIs require that professionals consider the construct of *treatment integrity*, both during the research development and translational stages.

During the treatment selection process, the practitioner should review a variety of experimental research where the EBI is investigated. The format of how manuscripts are printed is somewhat standardized—the problem behavior is operationally defined, the treatment is outlined then implemented, and the effects of the EBI on various components of Executive Function is then discussed.

Unfortunately, most of the studies on interventions targeting EF do not necessarily assure the consumer of the manuscript that the treatment was carried out the way that it is intended to be carried out without deviation to the treatment procedures. Without this confidence across studies, readers of this body of research cannot be guaranteed that the independent variables were implemented consistently among various researchers within each study. Deviations from this protocol cast doubt on the relationship between the treatment and the outcomes, as well as the validity of the treatment itself.

After a treatment program with adequate integrity has been selected, the practitioner must implement the package as written or as close to its original design as possible. An EBI that is applied in a radically different way than how it was described in the literature ceases to be based in evidence and is no better than a random combination of treatments. Treatment integrity, as a construct, factors considerably in the evaluation of the literature by researchers and in its translation to the field by practitioners.

Treatment Integrity

Researchers and practitioners must demonstrate how true the treatment is to the theoretical and procedural components of the overall treatment model or as intended by the developers of the treatment package (Dusenbury, Brannigan, Falco, & Hansen, 2003; Nezu & Nezu, 2008; Reed & Codding, 2011). This phenomenon is what is typically referred to as *treatment integrity* (also known as *treatment fidelity, procedural fidelity, or intervention integrity*) and can be alternately defined as the reliable and accurate implementation of an intervention. In the literature, treatment integrity is typically defined as being made up of three dimensions or components (McLeod, Southam-Gerow, & Weisz, 2009; Perepletchikova & Kazdin, 2005): adherence, agent competence, and differentiation.

Adherence

Adherence refers to the clinician's implementation of procedures in a stable manner over time, which can improve with consistent contact with other individuals to discuss the treatment application process. Such contact can take the various forms such as weekly supervision (Hogue et al., 2008) or a combination of direct observations and immediate feedback (Codding, Feinberg, Dunn, & Pace, 2005). However, what can be abstracted from these findings is that consistent contact with a trainer/supervisor has been found to lead to increased adherence to the protocol and demonstrable positive outcomes in children.

Adherence can be dependent on the setting of the interventions as well as the functional levels of the client. Protocols must be flexible to meet the needs of children with EF deficits in a variety of settings: schools, clinics, and hospitals; similarly, more extreme deficits may require the implementation of the same treatment protocol with increased magnitude or intensity (Dusenbury et al., 2003; Schulte, Easton, & Parker, 2009). In an ideal sense, this "personalization" of the intervention should be built into the treatment protocol to provide some supervision as to how adherence may vary depending on various components of the interventions (Barber et al., 2006; Perepletchikova & Kazdin, 2005).

Agent Competence

Competence refers to the skill, experience, and/or knowledge of the treatment agent (i.e., the treatment implementer; Perepletchikova & Kazdin, 2005), which may become more important depending on the complexity of the intervention (Gresham, 2005; Schulte et al., 2009). For example, school-based interventions which require various forms of data collection and charting may increase the complexity and factor agent competence more intensely when evaluating treatment integrity. Currently, it is difficult to evaluate how labor-intensive EF interventions are; however, the authors feel that evaluations of agent competence should be commonplace when implementing such treatments.

Agent competence can be conceptualized as a combination of an agent's access to preservice training and in-service training. In many situations, the treatment agent may not have had access to preservice opportunities that would have prepared them for program implementation, and added effort needs to be taken to increase the quantity and quality of in-service training. It has been shown that corrective feedback (i.e., the process of observing agents' in vivo implementation and delivering feedback as to correctly and incorrectly applied components) is an effective and time-efficient manner to deliver in-service training opportunities to a wide variety of agents (Codding et al., 2005; Codding, Livanis, Pace, & Vaca, 2008; DiGennaro, Martens, & Kleinman, 2007; DiGennaro, Martens, & McIntyre, 2005; DiGennaro-Reed, Codding, Catania, & Maguire, 2010; Mortensen & Witt, 1998; Mouzakitis, 2010; Noell, Witt, Gilbertson, Ranier, & Freeland, 1997).

Competence may also be strengthened when there are clear communication patterns among treatment developers as well as treatment agents (Cowan & Sheridan, 2003). Teachers and parents (who will often be the primary treatment agents) appear to prefer interventions to be described to them in practical, commonsense terms as opposed to psychological jargon (Elliot, 1988; Witt, Moe, Gutkin, & Andrews, 1984). Overall, while time is a factor that can potentially impact treatment implementation and integrity (DiGennaro et al., 2005; Elliot, 1988), there must also be time set aside for communication among all stakeholders.

Treatment Differentiation

Treatment differentiation refers to the extent that the intervention is implemented as is stated and it is not replaced with or modified by another treatment (Kazdin, 1986; Perepletchikova & Kazdin, 2005). Typically, this can be dealt with effectively through the use of well-established operational definitions of the treatment at hand. Typically treatment differentiation is threatened by *therapist drift* (Gresham, 2005), where agents may modify the treatment in minor ways over a continuous period of time, thus producing a gradual shift in the independent variable over time. Such drift is often not purposeful but may result due to decreasing levels of diligence, supervision, or boredom. Therapist drift can serve to artificially overestimate or underestimate treatment effects.

The Failure to Control for Treatment Integrity

The failure to control for treatment integrity can lead to one of three major problems: an inability to evaluate the effects of a program or intervention, the potential lack of improvement among clients and/or consumers, and a host of related ethical and potential legal problems.

First, and most importantly, if the treatment is not implemented with integrity, practitioners and researchers cannot realistically evaluate the effects of the independent variable upon the dependent variable (Cooper, Heron, & Heward, 2007; Kazdin, 2011). In such situations, we see two versions of the intervention which bear little in common: one which exists in reality and one which exists on paper. Both may have similarities, yet they are distinctly different (Livanis, Benvenuto, Mertturk, & Hanthorn, in press).

Implementing an intervention with a high rate of treatment integrity is associated with positive effects to children (DiGennaro et al., 2005, 2007; Erhardt, Barnett, Lentz, Stollar, & Raifin, 1996; Hogue et al., 2008). Treatment integrity appears to serve to mediate the effect that intervention plans had on student outcomes (Cook et al., 2010). In other words, good treatments, when implemented correctly, tend to have positive effects on clients and consumers.

In addition, within certain systems, the failure to follow a treatment protocol as written can potentially constitute a denial of certain state and/or federal rights. For instance, under the Individuals with Disabilities Education Act (2004), children classified with various disabling condition are entitled to receive interventions with a substantial evidence basis. And an EBI, when not implemented as intended, ceases to be an EBI and as such represents the failure of a school system to provide appropriate services for a disabled child (Cook et al., 2010; Etchdeit, 2006).

We have found that most professional organizations address treatment integrity within various white papers, best practices, and/or their ethics codes. In 2005, the American Psychological Association (APA) released its Policy Statement on Evidence-Based Practice in Psychology, a summary of how to select and implement EBIs; the statement suggests that consistent review of procedures is necessary in order to ensure the validity of any intervention strategy. The American Speech-Language-Hearing Association code of ethics (ASHA, 2010) also advocates that practitioners consistently evaluate their services. The National Association of School Psychologists (NASP) Principles for Professional Ethics (2010) states that "school psychologists use assessment techniques and practices that the profession considers to be responsible, research-based practice" (p. 7).

Despite the problems which can result from a lack of treatment integrity, constructs are often not measured effectively in studies that evaluate psychological and educational interventions (Dusenbury et al., 2003; McLeod et al., 2009). Wheeler, Baggett, Fox, and Blevins (2006) found that only 18 % of the studies of interventions for children actually assessed and reported treatment integrity data. Without models of assessing for treatment integrity from the literature, practitioners tend to have difficulties implementing such checks on integrity. For example, it has been suggested that 2 % of practicing school psychologists regularly measured rates of treatment integrity in their practice (Cochrane & Laux, 2008).

The Measurement of Treatment Integrity

Operational Definition of the Treatment and Its Components

The treatment and its components should have clear, concise, and specific operational definitions that identify or describe which specific actions that the treatment agent and the client should perform (Cooper et al., 2007). A good operational definition of an independent variable (e.g., the treatment and/or its components) should include four dimensions: verbal (descriptions of what the agent should say in various situations), physical (descriptions of what the agent should do in various situations), spatial (descriptions of where the materials should be placed.), and temporal (which actions should follow which environmental events in the program sequence). Such descriptions allow for an easy replication of the intervention, both as a research study as well as in applied settings. However, it is possible that by over-specifying treatments and its individual components, a treatment can be made to appear overly complex, thus potentially affecting treatment integrity (Gresham, 1996). One way to minimize this threat is to create two separate operational definitions which target varying levels of specification. The first operational definition would be presented to treatment agents and clients and includes a description of each component of the intervention in everyday practical language; the second would include a series of behaviors identified from a task analysis of each component within the larger treatment. In this way, the integrity of the treatment can be maintained without introducing too much complexity.

Problems with the Operational Definition of Executive Function

A major threat to treatment integrity is the lack of conceptual clarity of the term *Executive Function*—it is not hyperbole to suggest that previous definitions of EF have varied wildly from overly narrow definitions focusing specifically on processes such as working memory to an inclusion of every process mediated by the frontal lobes. Initially, authors have "excused" this lack of a precise definition by identifying multiple loosely connected Executive Functions that demonstrate "unity" and "diversity" (Teuber, 1972). Forty years later, other authors still cling to this unity/diversity definition (Miyake & Friedman, 2012). Such a lack of an agreed-upon definition is sure to make the task of examining interventions for children complex difficult and nearly impossible due to the highly divergent and conflicting operational definitions of EF.

A second threat lays in inordinate weight that definitions of EF give to various brain area. Such brain-based definitions glorify biological structures but sacrifice a focus on clear and observable phenomena—to a large extent, Hayes et al. (1996) once commented that it appears that many researchers use the term EF to refer to "whatever function they believe might involve the frontal lobe" (p. 279).

When researchers reference aspects of the brain in definitions of EF. Although we have made many advances over the last 20 years in the investigations of the specific functions of the frontal lobe, we still need to continue our research. Such research, however, is conducted via the examination of those behavioral phenomena that these frontal lobe processes are supposed to explain. In essence, the act of operationally defining the behaviors involved in EF vis-a-vis the various subsections of the frontal lobe amounts to a tautology where the following line of logic is supposed:

| What is EF? | → | EF is defined by frontal lobe processes |
| What are frontal lobe processes? | → | Frontal lobe processes are those parts of the brain that manage or mediate EF |

It should be noted that the authors are not suggesting in any way that the etiology of EF difficulties are not related to frontal lobe functioning. What we are proposing is that the construct EF should not be predominantly defined in relation to the specific brain structures, especially when researchers are evaluating treatment packages.

There are few other definitions of constructs within the field of psychology where a direct brain reference is cited. Most definitions refer to observable phenomena or phenomena that can be made observable via self-reporting. For example, when evaluating cognitive-behavioral interventions for depression, the construct is not typically referred to in relation to the serotonin deficits that exist in the brain—depression is typically defined as various cognitive and behavioral indicators that allow us to identify who may or may not be depressed. The use of the biological etiology of a disorder as a central feature of its operational definition can lead to tautologies that are problematic in that they fail to explain the phenomena in question and cause confusion for researchers and practitioners who are searching to evaluate their interventions with children.

It should be noted that recent efforts to develop a narrowly constructed operational definition of EF show promise. We strongly favor the results of Goldstein's factor analyses (2012) which highlight a one-factor solution with a variety of cognitive and behavioral indicators with little reference in the definition to the frontal lobe functions. We hope that such a definition is used in the literature in order to develop stronger and more conceptually related operational definitions of EF.

Direct Assessment of Treatment Integrity

The direct assessment of treatment integrity is conducted in a similar fashion to traditional behavioral assessment—the presence or the absence of the operational definition documented over a period of time (Cooper et al., 2007)—and a final percentage is calculated to indicate how much integrity to the treatment the agent(s) has exhibited. Such assessment can take place in situ or at a later time through the use of video technology (Perepletchikova & Kazdin, 2005).

Reliability is a central issue, and reliability is strengthened via multiple observations when conducting single-case experiments (Kazdin, 2011). The literature generally agrees that there should be multiple observation periods of sufficient length but differs as to the number and time frame of observations. Gresham (1996) estimates three to five observational sessions of 20–30 min duration each. In public school-based interventions, there was considerable variability. For example, Leblanc, Ricciardi, and Luiselli (2005) and DiGennaro-Reed et al. (2010) observed treatment agents for 10–15 min but Codding et al. (2005) observed treatment agents for 55–60 min. There is also variability in the number observations that are conducted as well, ranging from 3 sessions to 12 sessions (Codding et al., 2008; Leblanc et al., 2005). Such variability may have been due to systemic constraints of conducting research in the public school settings where variables are not easily controlled. The number of distinct observations may decrease as the settings become more controlled, such as in controlled settings (DiGennaro-Reed et al., 2010; LeBlanc et al., 2005), perhaps due to issues of increased agent competence (due to increased in-service and preservice training) as well as a heightened awareness and focus on treatment adherence.

One of the central problems in the direct assessment of treatment integrity is that of reactivity to the observation (Cooper et al., 2007), a phenomenon in which agents may modify their behavior if they are aware that they are the subject of observation (Foster & Cone, 1986). Indeed, job security may be dependent upon the evidence of treatment integrity, and agents may work more strenuously when they are being observed (but not so much when they are not observed). However, there are certain conditions that can be put into place that can mediate or mitigate the effect reactivity to observation (Codding et al., 2008).

Although the majority of studies that examine treatment integrity focus on the assessment of treatment adherence, Perepletchikova and Kazdin (2005) stress that the other two dimensions of treatment integrity need to be assessed as well: agent competence and treatment differentiation. Measures of competence should assess the quality of the delivery. Factors that should be examined should include the level of concordance between training and agent activities and client or consumer comprehension of the purposes, goals, and procedures of the treatment. Measures of treatment differentiation should focus on an assessment of procedures that are not prescribed and that are delivered in addition to or instead of the prescribed intervention (Perepletchikova & Kazdin, 2005).

Indirect Assessment of Treatment Integrity

Treatment integrity can also be monitored via the use of indirect assessment methods, such as agents' self-reports, an evaluation of permanent products which result from the treatment, rating scales, and self-monitoring (Perepletchikova & Kazdin, 2005). Self-monitoring has received a good deal of attention, both as an assessment tool as well as a method to help increase and improve treatment integrity (Burgio et al., 1990; Coyle & Cole, 2004; Petscher & Bailey, 2006; Richman, Riordan, Reiss, Piles, & Bailey, 1988). Self-monitoring is difficult to implement—in essence, the process creates an awkward condition where the agent must stop the intervention, rate their own behavior (as well as the child's behavior), and then continue with the intervention. As such, it is difficult to nearly impossible to implement this moment-to-moment self-monitoring in many educational settings, even if the intervention is delivered in a 1:1 fashion (Gresham, 1996).

It is possible that the self-monitoring method is simply not an effective method to collect data on adherence (Coyle & Cole, 2004; McLeod et al., 2009; Richman et al., 1988). However, if self-monitoring as a methodology is used, it can be useful when combined with prompts to collect data (Petscher & Bailey, 2006) or visual representations of data to assess adherence. Self-monitoring data may also add to an agent's better understanding of their own actions and how it relates to treatment integrity, although this avenue of research has not yet been explored richly as of yet. However, self-monitoring data should still be treated cautiously as the assessment may be due to a subtle demand characteristic that pulls for social approval and may cause treatment agents to overreport treatment integrity (Perepletchikova & Kazdin, 2005).

Interpretation of Treatment Integrity Data

In essence, measurements of treatment integrity are quantitative methods used to identify how therapist drift affects the dependent variable (Gresham, 1996). Therapist drift or low levels of treatment integrity often cause a variety of difficulties that call into question the ability of the independent variable to effect changes onto the dependent variable.

Table 23.1 highlights some of the interpretative issues that can arise from differing levels of treatment integrity. In conditions where there are high levels of treatment integrity, decisions can be made with a fair amount of confidence relating to the potential effects of the independent variable on the dependent measures. However, in conditions where there are low levels of treatment integrity (or none), the drift may actually serve to artificially improve outcomes, thus creating a situation where the treatment procedure is inappropriately deemed to be effective (Type I error). In this instance, one could hypothetically argue that a change was

Table 23.1 Interpretative issues that can arise from effects of varying levels of treatment integrity on the dependent variable

Dependent variable change	Levels of integrity	
	High	Low or none
Desired direction	Confidence that the treatment package has an effect	No confidence that the treatment package has any effect
		Increased risk of making a Type I error (*false positive*) if treatment integrity data are not collected
No change	Confidence that the treatment package has no effect	No confidence that the treatment package has any effect
		Increased risk of making a Type II error (*false negative*) if treatment integrity data are not collected
Undesired direction	Confidence that the treatment package has no effect and may even be potentially harmful	No confidence that the treatment package has any effect
		Increased risk of making a Type II error (*false negative*) if treatment integrity data are not collected

effected on the child with an EF deficit in the desired direction (e.g., an increase planning skills). From a research perspective, however, nothing has been added to the scientific literature in this condition—most research findings may just be false positives (Gresham, 1996). From a practitioner perspective, a false positive would unfortunately not add to the body of knowledge that is collected about a particular child. For example, agents implementing a treatment at home which demonstrates good effects but low levels of treatment integrity would not be able to realistically inform school staff as to what can be done to deal with the same symptoms.

In other conditions, the lack of treatment integrity coupled with no changes (e.g., a lack of increase in planning skills) or undesired changes in the dependent variables (e.g., a decrease in inhibition skills) may lead practitioners or researchers to conclude that the procedures were not effective. Procedures that are not effective should clearly be discontinued; however, it is possible that the treatment, were it applied with integrity, might have been effective in that instance (in the field) or for all children evidencing a particular profile (in the research literature). Rejecting an intervention when it may actually be effective is considered to be a Type II error. A lack of treatment integrity in these conditions would hinder the identification of potentially effective treatments.

Methods to Increase Treatment Integrity

Currently the most popular method used to increase treatment integrity is performance feedback (PFB; Codding et al., 2005, 2008; DiGennaro et al., 2005, 2007; DiGennaro-Reed et al., 2010; Mortensen & Witt, 1998; Mouzakitis, 2010; Noell et al., 1997), in which a supervisor observes the agent in action and then meets with the treatment agent (it should be noted that the term "supervisor" should not only be construed to mean an administrator; rather, the observations can and should be conducted by a variety of individuals such as school psychologists, teachers, consultants, therapists, and/or parents). During this meeting a number of things should be discussed. Feedback and praise can be delivered on the amount of correctly implemented components. The treatment agent and the observer can also discuss aspects of a plan that were not followed.

Most importantly, should there be an issue with the agent's treatment implementation with integrity, some training method can be employed to ensure correct component implementation in the future. The failure to implement a plan with integrity may be due to the agent's potential skill deficits or a lack of fluency with the procedures. At times, the treatment agent might have simply forgotten to implement all the steps of the intervention or they may have begun the process of drift. PFB is a method that can allow these issues to be addressed via the use of review, modelling, rehearsal, and role-play, if needed. A typical PFB session can last anywhere between 5 and 20 min (Reed & Codding, 2011), with initial PFB sessions lasting much longer than later sessions.

Various components of PFB have been manipulated to examine how to make the process more efficient and effective. For example, Guercio et al. (2005) varied PFB private meetings with public postings of treatment integrity to train 30 staff members at a residential facility. Although the results of the study showed dramatic increases of integrity among all staff, it is unclear which PFB condition was superior. The amount of time between the observation period and the delivery of PFB has also been investigated. Noell et al. (1997) delivered PFB immediately after observation, while Codding et al. (2005) delivered PFB every other week—others have examined varying lengths of time in between. PFB is an effective way to increase treatment integrity, despite its distance from the initial observation; however, shorter time lapses were associated with stronger, faster increases of treatment integrity (Mortensen & Witt, 1998).

While PFB has been demonstrated to be effective, investigations into the removal of this intervention evidence decreases in levels of treatment integrity (Noell et al., 1997; Witt, Noell, LaFleur, & Mortenson, 1997) and fading the provision of PFB has been recommended in order to deal with this issue (DiGennaro et al., 2005; Noell et al., 2000; Reed & Codding, 2011). Fading refers to the gradual decrease of PFB (i.e., *thinning*) over time that is contingent upon the demonstration of treatment integrity at specified criterion levels.

Some investigation has been conducted into the essential components of PFB. While PFB is a procedure employed to ensure treatment integrity, PFB itself must be scrutinized for treatment integrity. Some have indicated that the essential components of PFB are praise and corrective feedback. Corrective feedback refers to the process of delivering feedback on components that were incorrectly applied (or not applied at all) and the provision of training procedures to help correct skill deficits or improve automaticity. However, DiGennaro et al. (2005) conceptualized PFB as an aversive process. In this conceptualization, treatment agents worked to obtain high rates of integrity in return for the removal of PFB. This is in contrast to Codding et al. (2008), in which the treatment agents rated the PFB process as rewarding and beneficial. However, these discrepant results can be due to the setting (e.g., an inner city private school vs. a suburban public school), the person delivering PFB (e.g., a university faculty member vs. agency supervisor), how PFB is used by the setting (e.g., as a teaching tool or as a way to evaluate staff dismissal), and perhaps even the personality characteristics of the individual delivering PFB himself or herself.

Associated Variables

A number of variables have been associated with difficulties maintaining treatment integrity. A full review of all of those variables cannot be conducted here (see Allen & Warzak, 2000; Gresham, 1996; Perepletchikova & Kazdin, 2005, for more extensive reviews of associated variables). For example, as the complexity of a treatment increases, it becomes increasingly difficult to manage treatment integrity (Meichenbaum & Turk, 1987). Complexity is typically operationalized as the number of components of an intervention. Although this finding appears commonsensical (i.e., that an intervention with more parts will be more difficult to implement with fidelity), some aspect of the finding may be due to treatment acceptance. In general, more complex interventions are evaluated more negatively by potential treatment agents (Yeaton & Sechrest, 1981), which makes their implementation with integrity a much more difficult process to undertake.

Complexity may play a role when we implement interventions that involve cooperation among two or more agents at various settings (e.g., home, school, clinic). Communication among agents and settings becomes critical, as are the unequal starting points of agents' experience (Gresham, 1996). For example, parents may experience certain procedures or components of interventions as difficult to manage over a continuous period of time, which may cause them to drift from the originally stated procedure (Allen & Warzak, 2000). This may be especially true of those interventions that target more challenging difficulties, such as explosive behaviors (Greene, 2001; Greene & Albon, 2006).

Agents who are not effectively trained are often provided in-service training. Usually these trainings involve a great deal of didactic instruction. Such a focus on didactic training assumes that parents will develop adequate rules for program implementation based solely on instruction and follow them perfectly, which is an unrealistic assumption (Hayes & Wilson, 1993). It is for this reason that a fair amount of training programs for parents (and all treatment agents) should include modelling, role-play, and rehearsal—ultimately, supervision needs to be implemented on an ongoing basis, in situ.

Time spent by treatment agents can interfere with treatment integrity. Interventions that require agents spend time to learn pose greater threats to treatment integrity than those that are easy to learn (Gresham, 1996). Other interventions may demand ongoing supervision and in-service training to maintain at effective levels, while some treatments need extended periods of administration (typically referred to as *dosage*) until an effect is witnessed, typically due to the severity of the targeted issues that are addressed (Happe, 1982). Additionally, the quality and the quantity of materials used can affect treatment integrity (Gresham, 1996). Treatment agents often are asked to implement such interventions without much large-scale systemic support in the way of resources or diminishing resources over time.

Conclusions

Issues related to treatment integrity are of critical importance to treatment programs designed for children with EF deficits. Given these issues, there needs to be considerable work to ensure treatment adherence, improve competence, and establish differentiation. Unfortunately, treatment integrity is an important construct that is not measured as often as it should be in both research and practice in all areas of psychology.

The recent interest in the application of EBIs to ameliorate EF deficits has the potential to increase awareness and interest in treatment integrity. Detrich (2008) suggests that environmental factors (such as the agency or stress levels of the family in the home) may play a considerable role in the selection and implementation of EBIs, to the point where various pieces of interventions might be combined to form unique treatment plans. While this may appear to be intuitively attractive to the clinician, the process does not necessarily equal a "mix-and-match" strategy—on the contrary, practitioners will need to work much harder in defining the treatment (i.e., independent variable), as well as the treatment outcomes (i.e., dependent variable) and a measurement strategy. This newly developed treatment protocol will need to be assessed for treatment integrity, so that agents can make an informed decision as to the effectiveness of the treatment.

Over the last 30 years, there has been an ever-increasing focus on the measurement of and interventions to improve treatment integrity. Direct observation and PFB appear to be the most commonly used (and most successful) measurement and assessment strategy. Attempts have been made to examine components of PFB to see how the process can be improved; however, it would be helpful to investigate what types of situations hinder PFB. For example, it is within the authors' clinical experience that observations conducted by external individuals tend to be better received than those conducted by administrators or supervisors. This may partially explain some of the discrepancies in the field, but as of yet, there have been no investigations of the status of the observer upon the effectiveness of PFB.

In conclusion, the demonstration of treatment integrity within the context of the evidence-based movement in identifying interventions for children with EF deficits will serve to be a challenge that will need to be dealt with both in scientific literature as well as in practice. The level of treatment integrity adds another interpretative layer that deepens inferences made from outcome data. Ultimately, efforts to improve treatment integrity serve to develop better researchers and professionals that can make a difference in the lives of children with EF deficits.

References

Allen, K. D., & Warzak, W. J. (2000). The problem of parental non-adherence in clinical behavior analysis: Effective treatment is not enough. *Journal of Applied Behavior Analysis, 33*(3), 373–391. doi:10.1901/jaba.2000.33-373.

Alloway, T., Gathercole, S. E., Holmes, J., Place, M., Elliott, J. G., & Hilton, K. (2009). The diagnostic utility of behavioral checklists in identifying children with ADHD and children with working memory deficits. *Child Psychiatry and Human Development, 40*(3), 353–366.

American Psychiatric Association. (2000). *Diagnostic and statistical manual of mental disorders* (4th ed., text rev.). Washington, DC: Author.

American Psychological Association. (2005, August). *Policy statement on evidence-based practice in psychology*. Retrieved from http://www.apapracticecentral.org/ce/courses/ebpstatement.pdf

American Speech-Language-Hearing Association. (2010). *Code of ethics* [Ethics]. doi:10.1044/policy.ET2010-00309

Baltruschat, L., Hasselhorn, M., Tarbox, J., Dixon, D. R., Najdowski, A. C., Mullins, R. D., et al. (2011). Addressing working memory in children with autism through behavioral intervention. *Research in Autism Spectrum Disorders, 5*, 267–276. doi:10.1016/j.rasd.2010.04.008.

Banich, M. T. (2009). Executive function: The search for an integrated account. *Current Directions in Psychological Science, 18*(2), 89–94.

Barber, J. P., Gallop, R., Crits-Chirstoph, P., Frank, A., Thase, M. E., Weiss, R. D., et al. (2006). The role of therapist adherence, therapist competence, and alliance in predicting outcome of individual drug counseling: Results from the National Institute Drug Abuse Collaborative Cocaine Treatment Study. *Psychotherapy Research, 16*, 229–240. doi:10.1080/10503300500288951.

Barkley, R. A. (1997). Behavioral inhibition, sustained attention, and executive functions: Constructing a unifying theory of ADHD. *Psychological Bulletin, 121*(1), 65–94.

Buckner, R. L. (2004). Memory and executive function in aging and AD: Multiple factors that cause decline and reserve factors that compensate. *Neuron, 44*(1), 195–208. doi:10.1016/j.neuron.2004.09.006.

Burgio, L. D., Engel, B. T., Hawkins, A. M., McCormick, K., Schieve, A., & Jones, L. T. (1990). A staff management system for maintaining improvements in continence with elderly nursing home residents. *Journal of Applied Behavior Analysis, 23*, 111–118. doi:10.1901/jaba.1990.23-111.

Cochrane, W. S., & Laux, J. M. (2008). A survey investigating school psychologists' measurement of treatment integrity in school-based interventions and their beliefs in their importance. *Psychology in the Schools, 45*, 499–507. doi:10.1002/pits.20319.

Codding, R. S., Feinberg, A. B., Dunn, E. K., & Pace, G. M. (2005). Effects of immediate performance feedback on implementation of behavior support plans. *Journal of Applied Behavior Analysis, 38*, 205–219. doi:10.1901/jaba.2005.98-04.

Codding, R. S., Livanis, A., Pace, G., & Vaca, L. (2008). Using performance feedback to improve treatment integrity of classwide behavior plans: An investigation of observer reactivity. *Journal of Applied Behavior Analysis, 41*, 417–422. doi:10.1901/jaba.2008.41-417.

Cook, C. R., Mayer, G. M., Wright, D. B., Kraemer, B., Wallace, M. D., Dart, E., et al. (2010). Exploring the link among behavior intervention plans, treatment integrity, and student outcomes under natural educational outcomes. *Journal of Special Education, 20*, 1–14. doi:10.1177/0022466910369941.

Cooper, J. O., Heron, T. E., & Heward, W. L. (2007). *Applied behavior analysis* (2nd ed.). Upper Saddle River, NJ: Merrill Prentice-Hall.

Cowan, R. J., & Sheridan, S. M. (2003). Investigating the acceptability of behavioral interventions in applied conjoint behavioral consultation: Moving from analog conditions to naturalistic settings. *School Psychology Quarterly, 18*(1), 1–21.

Coyle, C., & Cole, P. (2004). A video-taped self-modeling and self-monitoring treatment program to decrease off-task behaviour in children with autism. *Journal of Intellectual and Developmental Disability, 29*, 3–15. doi:10.1080/08927020410001662642.

Denckla, M. B. (2007). Executive function: Building together the definitions of attention deficit/hyperactivity disorder and learning disabilities. In L. Metzler (Ed.), *Executive function in education* (pp. 5–18). New York: Guilford Press.

Detrich, R. (2008). Evidence-based, empirically supported, or best practice? A guide for the scientist-practitioner. In J. K. Luiselli, D. C. Russo, W. P. Christian, & S. M. Wilczynski (Eds.), *Effective practices for children with autism* (pp. 3–25). New York: Oxford Press.

DiGennaro, F. D., Martens, B. K., & Kleinman, A. E. (2007). A comparison of performance feedback procedures on teachers' treatment implementation integrity and students' inappropriate behavior in special education classrooms. *Journal of Applied Behavior Analysis, 40*, 447–461. doi:10.1901/jaba.2007.40-447.

DiGennaro, F. D., Martens, B. K., & McIntyre, L. L. (2005). Increasing treatment integrity through negative reinforcement: Effects on teacher and student behavior. *School Psychology Review, 34*(2), 220–231.

DiGennaro-Reed, F. D., Codding, R., Catania, C. N., & MaGuire, H. (2010). Effects of video modeling on treatment integrity of behavioral interventions. *Journal of Applied Behavior Analysis, 43*, 291–295. doi:10.1901/jaba.2010.43-291.

Duarte, A., Hayasaka, S., Du, A., Schuff, N., Jahng, G., Kramer, J., et al. (2006). Volumetric correlates of memory and executive function in normal elderly, mild cognitive impairment and Alzheimer's disease. *Neuroscience Letters, 406*(1–2), 60–65. doi:10.1016/j.neulet.2006.07.029.

Dusenbury, L., Brannigan, R., Falco, M., & Hansen, W. B. (2003). A review of research on fidelity of implementation: Implications for drug abuse prevention in school settings. *Health Education Research: Theory and Practice, 18*(2), 237–256. doi:10.1093/her/18.2.237.

Eisenberg, D., & Berman, K. (2010). Executive function, neural circuitry, and genetic mechanisms in schizophrenia. *Neuropsychopharmacology, 35*(1), 258–277. doi:10.1038/npp.2009.111.

Eliot, L. (1999). *What's going on in there?: How the brain and mind develop in the first five years of life*. New York: Bantam.

Elliot, S. N. (1988). Acceptability of behavioral treatments in educational settings. In J. C. Witt, S. N. Elliott, & F. M. Gresham (Eds.), *Handbook of behavior therapy* (pp. 121–150). New York: Plenum Press.

Erhardt, K. E., Barnett, D. W., Lentz, F. E., Stollar, S. A., & Raifin, L. H. (1996). Innovative methodology in ecological consultation: Use of scripts to promote treatment acceptability and integrity. *School Psychology Quarterly, 11*, 149–168. doi:10.1037/h0088926.

Etchdeit, S. K. (2006). Behavioral intervention plans: Pedagogical and legal analysis of themes. *Behavior Disorders, 31*(2), 223–243.

Foster, S., & Cone, J. (1986). Design and use of direct observation. In A. Ciminero, K. Calhoun, & H. Adams (Eds.), *Handbook of behavioral assessment* (2nd ed., pp. 253–324). New York: Wiley Interscience.

Gathercole, S. E., Alloway, T. P., Kirkwood, H. J., Elliott, J. G., Holmes, J., & Hilton, K. A. (2008). Attentional and executive function behaviours in children with poor working memory. *Learning and Individual Differences, 18*, 214–223.

Goldstein, S. (2012). *Understanding executive functioning in children: New ideas, new data, effective education and the Comprehensive Executive Functioning Inventory* [PowerPoint slides] (S. Goldstein, personal communication, August 5, 2012).

Greene, R. W. (2001). *The explosive child: A new approach for understanding and parenting easily frustrated, chronically inflexible children*. New York: Perennial.

Greene, R. W., & Albon, J. S. (2006). *Treating explosive kids: The collaborative problem solving approach*. New York: Guilford Press.

Gresham, F. M. (1996). Treatment integrity in single-subject research. In R. D. Franklin, D. B. Allison, & B. S. Gorman (Eds.), *Design and analysis of single-case research* (pp. 93–117). Mahwah, NJ: Lawrence Erlbaum.

Gresham, F. M. (2005). Treatment integrity and therapeutic change: Commentary on Perepletchikova and Kazdin. *Clinical Psychology: Science and Practice, 12*(4), 391–394. doi:10.1093/clipsy/bpi048.

Guercio, J. M., Dixon, M. R., Soldner, J., Shoemaker, Z., Zlomke, K., Root, S., et al. (2005). Enhancing staff performance measures in an acquired brain injury setting: Combating the habituation to organizational behavioral interventions. *Behavioral Interventions, 20*, 91–99. doi:10.1002/bin.174.

Hale, J. B., Reddy, L. A., Wilcox, G., McLaughlin, A., Hain, L., Stern, A., et al. (2009). Assessment and intervention for children with ADHD and other frontal-striatal circuit disorders. In D. C. Miller (Ed.), *Best practices in school neuropsychology: Guidelines for effective practice, assessment and evidence-based interventions* (pp. 225–280). New York, NY: Wiley.

Happe, D. (1982). Behavior intervention: It doesn't do any good in your briefcase. In J. Grimes (Ed.), *Psychological approaches to problems of children and adolescents* (pp. 15–41). Des Moines, IA: Iowa Department of Public Instruction.

Hausdorff, J. M., Doniger, G. M., Springer, S., Yogev, G., Simon, E. S., & Giladi, N. (2006). A common cognitive profile in elderly fallers and in patients with Parkinson's disease: The prominence of impaired executive function and attention. *Experimental Aging Research, 32*(4), 411–429. doi:10.1080/03610730600875817.

Hayes, S. C., Gifford, E. V., & Ruckstuhl, L. E. (1996). Relational frame theory and executive function: A behavioral approach. In G. R. Lyon & N. A. Krasnegor (Eds.), *Attention, memory, and executive function*. Baltimore, MD: Brookes.

Hayes, S. C., & Wilson, K. G. (1993). Some applied implications of a contemporary behavior analytic account of verbal events. *Behavior Analyst, 16*(2), 283–301.

Hogue, A., Henderson, C. E., Dauber, S., Barajas, P. C., Fried, A., & Liddle, H. A. (2008). Treatment adherence, competence, and outcome in individual and family therapy for adolescent behavior problems. *Journal of Consulting and Clinical Psychology, 76*(4), 544–555. doi:10.1037/0022-006X.76.4.544.

IDEA of 2004, P.L. 108-446, § 665 [b][1][B], 118 Stat. 2787 (2005).

Kazdin, A. E. (1986). Comparative outcome studies of psychotherapy: Methodological issues and strategies. *Journal of Consulting and Clinical Psychology, 54*, 95–105. doi:10.1037/0022-006X.54.1.95.

Kazdin, A. E. (2011). *Single-case research designs*. New York: Oxford University Press.

LeBlanc, M., Ricciardi, J. N., & Luiselli, J. K. (2005). Improving discrete trial instruction by paraprofessional staff through an abbreviated performance feedback intervention. *Education and Treatment of Children, 28*(1), 76–82.

Leeds, L., Meara, R. J., Woods, R., & Hobson, J. P. (2001). A comparison of the new executive functioning domains of the CAMCOG-R with existing tests of executive function in elderly stroke survivors. *Age and Ageing, 30*(3), 251–254. doi:10.1093/ageing/30.3.251.

Livanis, A., Benvenuto, S., Mertturk, A., & Hanthorn, C. (in press). Treatment integrity in autism spectrum disorder interventions. In S. Goldstein (Ed.), *Interventions for autism spectrum disorders*.

McClelland, M. M., Cameron, C. E., Connor, C., Farris, C. L., Jewkes, A. M., & Morrison, F. J. (2007). Links between behavioral regulation and preschoolers' literacy, vocabulary, and math skills. *Developmental Psychology, 43*(4), 947–959.

McCloskey, G., Perkins, L. A., & Van Divner, B. (2009). *Assessment and intervention for executive function difficulties*. New York: Routledge.

McLeod, B. D., Southam-Gerow, M. A., & Weisz, J. R. (2009). Conceptual and methodological issues in treatment integrity measurement. *School Psychology Review, 38*(4), 541–546.

Meichenbaum, D., & Turk, D. C. (1987). *Facilitating treatment adherence: A practitioner's guidebook*. New York: Plenum Press.

Miyake, A., & Friedman, N. P. (2012). The nature and organization of individual differences in executive functions: Four general conclusions. *Current Directions in Psychological Science, 21*(1), 8–14. doi:10.1177/0963721411429458.

Mortensen, B. P., & Witt, J. C. (1998). The use of weekly performance feedback to increase teacher implementation of a prereferral intervention. *School Psychology Review, 27*, 613–627.

Mouzakitis, A. (2010). *The effects of self-monitoring training and performance feedback on the treatment integrity of behavior support plans for children with autism*. Unpublished doctoral dissertation, The Graduate Center of the City University of New York, New York.

National Association of School Psychologists. (2010). *Principles for professional ethics*. Retrieved from http://www.nasponline.org/standards/2010standards/1_%20Ethical%20Principles.pdf

Nezu, A. M., & Nezu, C. M. (2008). Treatment integrity. In D. McKay (Ed.), *Handbook of research methods in abnormal and clinical psychology* (pp. 351–363). New York: Sage.

Noell, G. H., Witt, J. C., Gilbertson, D. N., Ranier, D. D., & Freeland, J. T. (1997). Increasing teacher interventions implementation in general education settings through consultation and performance feedback. *School Psychology Quarterly, 12*, 77–88. doi:10.1037/h0088949.

Noell, G. H., Witt, J. C., LaFleur, L. H., Mortenson, B. P., Ranier, D. D., & LeVelle, J. (2000). Increasing intervention implementation in general education following consultation: A comparison of two follow-up strategies. *Journal of Applied Behavior Analysis, 33*, 271–284. doi:10.1901/jaba.2000.33-271.

Perepletchikova, F., & Kazdin, A. E. (2005). Treatment integrity and therapeutic change: Issues and research recommendations. *Clinical Psychology: Science and Practice, 12*(4), 365–383. doi:10.1093/clipsy/bpi045.

Petscher, E. S., & Bailey, J. S. (2006). Effects of training, prompting and self-monitoring on staff behavior in a classroom for students with disabilities. *Journal of Applied Behavior Analysis, 39*, 215–226. doi:10.1901/jaba.2006.02-05.

Posner, M. I., & Rothbart, M. K. (2007). *Educating the human brain*. Washington, DC: American Psychological Association.

Reed, F. D., & Codding, R. S. (2011). Intervention integrity assessment. In J. K. Luiselli (Ed.), *Teaching and behavior support for children and adults with autism spectrum disorder* (pp. 38–47). New York: Oxford University Press.

Richman, G. S., Riordan, M. R., Reiss, M. L., Piles, D. A. M., & Bailey, J. S. (1988). The effects of self-monitoring and supervisor feedback on staff performance in a residential setting. *Journal of Applied Behavior Analysis, 21*, 401–409. doi:10.1901/jaba.1988.21-401.

Schulte, A. C., Easton, J. E., & Parker, J. (2009). Advances in treatment integrity research: Multidisciplinary perspectives on the conceptualization, measurement, and enhancement of treatment integrity. *School Psychology Review, 38*(4), 460–475.

Scope, A., Empson, J., & McHale, S. (2010). Executive function in children with high and low attentional skills: Correspondences between behavioural and cognitive profiles. *British Journal of Developmental Psychology, 28*(2), 293–305.

Simonsen, C., Sundet, K., Vaskinn, A., Birkenaes, A. B., Engh, J. A., Hansen, C., et al. (2008). Neurocognitive profiles in bipolar I and bipolar II disorder: Differences in pattern and magnitude of dysfunction. *Bipolar Disorders, 10*(2), 245–255. doi:10.1111/j.1399-5618.2007.00492.x.

Steinberg, L. (2007). Risk taking in adolescence: New perspectives from brain and behavioral science. *Current Directions in Psychological Science, 16*, 55–59.

Teuber, H. L. (1972). Unity and diversity of frontal lobe functions. *Acta Neurobiologiae Experimentalis, 32*, 615–656.

Wheeler, J. J., Baggett, B. A., Fox, J., & Blevins, L. (2006). Treatment integrity: A review of intervention studies conducted with children with autism. *Focus on Autism and Other Developmental Disabilities, 21*(1), 45–54. doi:10.1177/10883576060210010601.

Witt, J. C., Moe, G., Gutkin, T. B., & Andrews, L. (1984). The effect of saying the same thing in different ways: The problem of language and jargon in school-based consultation. *Journal of School Psychology, 22*, 361–367. doi:10.1016/0022-4405(84)90023-2.

Witt, J. C., Noell, G. H., LaFleur, L. H., & Mortenson, B. P. (1997). Teacher use of interventions in general education: Measurement and analysis of the independent variable. *Journal of Applied Behavior Analysis, 30*, 693–696. doi:10.1901/jaba.1997.30-693.

Yeaton, W. H., & Sechrest, L. (1981). Critical dimensions in the choice and maintenance of successful treatments: Strength, integrity, and effectiveness. *Journal of Consulting and Clinical Psychology, 49*(2), 156–167. doi:10.1037/0022-006X.49.2.156.

Interventions to Promote Executive Development in Children and Adolescents

24

Peg Dawson and Richard Guare

A favorite cartoon, F Minus, depicts a job applicant being interviewed by a prospective employer. The title of the cartoon is "Dale's 4th Grade Education Pays Off," and the employer is saying to the applicant, "The job you are applying for will require you to know state capitals, cursive writing, and long division." It would be hard to find a more succinct way to illustrate the demand for effective executive skills by the time children reach adulthood and enter the job market.

This cartoon implies that it is the role of the education system to help children develop the skills they need to compete in an increasingly challenging work environment. In fact, for children to develop optimally functioning executive skills requires the combined effort of parents as well as teachers, and there are steps that can be taken both in the home and at school to promote high-level executive functioning. The aim of this chapter is to lay out a broad set of intervention strategies that can be applied in both settings.

Executive Skill Definitions

Researchers and developers of tests of executive functioning and behavior rating scales designed to assess executive skills have all organized and categorized executive skills somewhat differently. For instance, the developers of the Behavior Rating Inventory of Executive Function (Gioia, Isquith, Guy, & Kenworthy, 2000) have defined two broad categories of skills, those involved in behavior regulation (e.g., impulse control, emotion control) and those that are more metacognitive in nature (e.g., working memory, task initiation, planning/organization). More recently, Barkley (2011), in developing the Barkley Deficits in Executive Functioning Scale, identified five broad categories of executive functioning (e.g., self-regulation of emotion, time management, organization/problem solving).

In our work with parents and teachers, we have identified 11 individual executive skills that we believe figure prominently in understanding why some children tackle activities of daily living, including schoolwork, homework, chores, and daily routines, more successfully than others. Table 24.1 lists these skills, along with brief definitions, and a description of the most common exemplars of problems associated with each executive skill.

Guiding Principles for Promoting Executive Skill Development

We have used three broad strategies for helping children and youth develop executive skills, which we will detail below. First, however, there are some underlying principles that help

P. Dawson (✉) • R. Guare
Center for Learning and Attention Disorders,
Portsmouth, NH, USA
e-mail: pegdawson@comast.net

Table 24.1 Executive skill definitions and common problems

Executive skill	Definitions	Common manifestations of weak skills
Response inhibition	The capacity to think before you act—this ability to resist the urge to say or do something allows us the time to evaluate a situation and how our behavior might impact it	• Acts without thinking • Interrupts others • Blurts out comments or answers to questions in class • Talks or plays too loudly • Acts wild or out of control
Emotional control	The ability to manage emotions in order to achieve goals, complete tasks, or control and direct behavior	• Overreacts to small problems • Easily overwhelmed • Overstimulated and can't calm down easily • Low tolerance for frustration • Quick to anger or become anxious
Flexibility	The ability to revise plans in the face of obstacles, setbacks, new information, or mistakes. It relates to an adaptability to changing conditions	• Upset by changes in plans • Resists change of routine • Gets stuck on one topic or activity • Can't come up with more than one solution to a problem • Difficulty handling open-ended tasks
Working memory	The ability to hold information in memory while performing complex tasks. It incorporates the ability to draw on past learning or experience to apply to the situation at hand or to project into the future	• Forgets directions • Forgets to bring materials to and from school • Forgets to hand in homework • Forgets to do chores • Forgets when assignments are due
Task initiation	The ability to begin projects without undue procrastination, in an efficient or timely fashion	• Puts off homework as long as possible • Delays starting any kind of effortful task • Leaves long-term assignments until the last minute • Chooses "fun stuff" over homework or chores
Sustained attention	The capacity to maintain attention to a situation or task in spite of distractibility, fatigue, or boredom	• Runs out of steam before homework is done • Switches between many tasks without finishing any • Frequently spacy or daydreams excessively • Stops working on chores before they're finished

Planning/prioritization	The ability to create a roadmap to reach a goal or to complete a task. It also involves being able to make decisions about what's important to focus on and what's not important	• Doesn't know how to break down long-term assignments into subtasks • Can't figure out the steps needed to complete a task or achieve a goal • Difficulty taking notes or studying for tests because can't decide what's important
Organization	The ability to create and maintain systems to keep track of information or materials	• Messy bedroom, cluttered desks • Can't find things in notebooks, backpacks, lockers • Loses or misplaces things (books, papers, notebooks, mittens, keys, cell phones, etc.)
Task initiation	The ability to begin projects without undue procrastination, in an efficient or timely fashion	• Puts off homework as long as possible • Delays starting any kind of effortful task • Leaves long-term assignments until the last minute • Chooses "fun stuff" over homework or chores
Time management	The capacity to estimate how much time one has, how to allocate it, and how to stay within time limits and deadlines. It also involves a sense that time is important	• Can't estimate how long it takes to do something • Can't make or follow timelines • Late for appointments • Difficulty meeting deadlines • Lack of sense of time urgency
Goal-directed persistence	The capacity to have a goal, follow through to the completion of the goal, and not be put off by or distracted by competing interests. A first grader can complete a job in order to get to recess	• Difficulty setting personal goals • Can't connect how they spend their time now with long-term goals • May not do homework because they don't see how it will impact their future • May try to save money to buy something but spend it before they reach their goal
Metacognition	The ability to stand back and take a birds-eye view of oneself in a situation. It is an ability to observe how you problem solve. It also includes self-monitoring and self-evaluative skills (e.g., asking yourself, "How am I doing? or How did I do?")	• Wait for others to solve problems for them • Lack insight into their behavior (and its impact either on others or their ability to get what they want) • Don't know how to study for tests

guide both our intervention design and the progression we follow in strategy selection and maintenance.

Teach Deficient Skills Rather Than Expecting the Child to Acquire Them Through Observation or Osmosis

Some children appear to have a natural capacity for using executive skills effectively, while others stumble and struggle if left on their own. In this day and age, however, most children struggle at one point or another with some task that requires a level of executive functioning that's beyond them. To respond to this more complex world, we can't leave executive skill development to chance. We need to provide direct instruction—defining problem behaviors, identifying goal behaviors, and then developing and implementing an instructional sequence that includes close supervision at first, followed by a gradual fading of prompts and supports.

Consider the Child's Developmental Level

Understanding what's normal at any given age so that you don't expect too much from the child is the first step in addressing executive skill weaknesses. But knowing what's typical for any given age is only part of the process. When the child's skills are delayed or deficient based on age, the intervention needs to take into account whatever level the child is functioning at now. While a normal 12-year-old may be able to pick up his room by himself with a weekly schedule and a reminder or two (or three!), if a particular 12-year-old has never picked up his room by himself in his life, then the structures and strategies that work with most 12-year-olds will probably not be effective. The task demands will need to be matched to the child's actual developmental level. Similarly, in a school setting, if a 7-year-old's ability to inhibit impulses is more like a 4-year-old's, then we provide closer supervision and monitoring (e.g., by seating that child near the teacher or having him engage in adult-directed activities on the playground).

Move from the External to the Internal

All executive skills training begins with something *outside* the child. Before teaching a child not to run into the road, a parent stays with her and holds her hand when they reach a street corner to make sure that didn't happen. Over time, by repeating the rule *Look both ways before crossing*, the child internalizes the rule, and after the parent has observed the child following the rule repeatedly, the child is then trusted to cross the street on her own. In the early years, parents and teachers organize and structure the environment to compensate for the executive skills the child has not yet developed. When a decision is made to help a child develop more effective executive skills, therefore, the process always begins with changing things outside the child before moving on to strategies that require the child to change. Examples of this include cueing a child to brush her teeth before she goes to bed rather than expecting her to remember to do this on her own or keeping independent work tasks in the classroom brief to fit the short attention spans of young children in first or second grade.

Environmental Modifications May Include Altering the Physical or Social Environment, the Task, or the Way Adults Interact with Children to Support Weak Executive Skills

Modifications to the physical or social environment might include something as simple as having a child with ADHD check in with a teacher every 10 min or so when doing independent seatwork so that progress can be monitored (and the child given an opportunity to get up and move around). For a child with weak emotional control, it might mean that his parents find younger playmates or limit play dates to one child at a time or having a parent or babysitter on hand to supervise beyond the age when this is typically done for kids. Task modifications may mean shortening tasks, building in breaks, or turning open-ended tasks into closed-ended tasks. Finally, changing the way adults interact with the child may mean providing more frequent cues and reminders and taking extra care to verbally reinforce a child for practicing a weak skill (e.g., "I liked the way you worked really hard to try to control your temper").

Use Rather Than Fight the Child's Innate Drive for Mastery and Control

From a very early age, children work hard to control their own lives. They do this by achieving mastery and by working to get what they want when they want it. The mastery part is satisfying for both parents and teachers to watch: they admire the persistence with which a child will practice a skill until mastery is achieved. These same adults tend to be a little more ambivalent about the ways children work to get what they want when they want it, because what *they* want is sometimes in conflict with what adults want. Nonetheless, there are ways adult can support children's drive for mastery and control while remaining in charge. These may include creating routines and schedules so that children know what will happen when and can accept this as a part of his everyday life at home and at school, building in choice when assigning chores or homework, or treating children as a partner when problem solving to come up with ways of managing weak executive skills.

Modify Tasks to Match Your Child's Capacity to Exert Effort

There are two kinds of effortful tasks: ones that children are not very good at and ones that they are very capable of doing but just don't like doing. Different strategies apply depending on which kind of task is under consideration. For tasks the child is not very good at, the approach to take is to break them down into small steps and start with either the first step and proceed forward or the last step and proceed backward. No new step is added until mastery at the current step is achieved.

Parents and teachers tend to have stronger feelings about the second kind of effortful task. To the outside observer, it appears as if the child is simply being oppositional by refusing to do something that's well within his capacity. Rather than creating a power struggle, the best way to handle these kinds of tasks is to teach the child to exert effort by getting him to override the desire to quit or do anything else that's preferable. This can be done by helping the child create some kind of motivating self-talk (e.g., "You can't walk away from this"), combined with making the first step *easy enough* so it doesn't feel particularly hard to the child, and then immediately following that first step with a reward. The reward is there to ensure that there's a payoff for the child for expending the small amount of effort it takes to complete the first step. The amount of effort the child has to expend to achieve the reward can then gradually be increased, either by increasing the task demands or by increasing the amount of time you

Use Incentives to Augment Instruction

Incentives are rewards. They can be as simple as a word of praise or as elaborate as a point system that enables a child to earn rewards on a daily, weekly, or monthly basis.

For some tasks—and some children—mastery of the task is incentive enough. Most children naturally want to master things like learning to pull themselves to a stand or learning to climb stairs, learning to ride a bike, or learning to drive a car. Unfortunately, many tasks we expect children to do lack built-in incentives. While some children enjoy helping out around the house because it means a chance to spend time with a parent or because parents are so appreciative of their help, many children go to great lengths to avoid this kind of activity. Similarly, with homework, it's the rare child who can't wait to get home and get started on homework, but for some the grade they will earn on homework well done—or the humiliation they will avoid from getting a low grade—is sufficient to propel them to do homework assignments promptly and well. For many of the children we work with, however, the rewards and punishments associated with homework are not enough to make them willing to do it without putting up a fight first.

Incentives have the effect of making both the effort of learning a skill and the effort of performing a task less aversive. They give us something to look forward to that motivates us to persist with difficult tasks and that helps us combat any negative thoughts or feelings we have about the task at hand. And placing an incentive after the task teaches the child to delay gratification—a valuable skill in its own right.

Provide Just Enough Support for the Child to Be Successful

This appears to be so simple as to be self-evident, but in fact the implementation of this principle may be trickier than it appears. The principle includes two components that are of equal weight—(1) *just enough support* and (2) *for the child to be successful*. Parents and others who work with children tend to make two kinds of mistakes. They either provide too much support, which means the child is successful but fails to develop the ability to perform the task independently, or they provide too little support, so the child fails—and, again, never develops the ability to perform the task independently. The key is to determine how far the child can get in the task on her own and then intervene—not by doing the task for her but by offering enough support (physical or verbal, depending on the task) to get her over the hump and moving on to success. The technical term for this is scaffolding—a skill that teachers employ routinely when working with individual students who are struggling to grasp academic concepts.

Keep Supports and Supervision in Place Until the Child Achieves Mastery or Success

Many parents and teachers know how to break down tasks, teach skills, and reinforce success, and yet children still fail to acquire the skills they want them to gain. More often than not, this is because of a failure to apply this principle and/or the next one. These adults set up a process or a procedure, see that it's working, and then back out of the picture, expecting the child to keep succeeding independently. One of the more common examples we see is the system that parents or teachers put in place to help children get organized. They may walk them through a process of cleaning their desk, for instance, or they may set up precise systems for organizing notebooks, but they are too quick to expect the child to maintain the organizing scheme on their own. In our experience, it takes far longer for children to develop independence in following these kinds of routines than either parents or teachers assume.

Fade the Supports, Supervision, and Incentives Gradually, Never Abruptly

In its simplest form, fading involves manipulating time and/or space. We reduce the cues and reminders gradually, increasing the amount of time the child is working on his own. Or we gradually move away from the child so that our presence only intermittently serves as a cue rather than the constant reminder our immediate presence provides. For example, an astute mother may find that her son cannot get his homework done without her supervision. In the beginning, she sits next to the child at the kitchen table, keeping him on track both by her physical presence and by occasional prompts to get back to work. Gradually, she removes herself and cues him less frequently. She might say, for instance, "I need to get the potatoes on for dinner, so see what you can do on your own while I do that." She can check back periodically to praise her child for continuing to work while she prepares dinner. Eventually, she may be able to have her son move to the dining room table or a desk in the study, so that he learns to work independently out of her sight. In the beginning, her check-ins come frequently; over time, she can look in on him less often.

Three Strategies for Working with Children with Weak Executive Skills

We have two options when working with children with weak executive skills. We can either intervene at the level of the environment, in which case,

Table 24.2 Checklist for designing interventions

Intervention Steps

1. Establish behavioral goal

    Problem behavior: _____

    Goal behavior: _____

2. What environmental supports will be provided (check all that apply)?

    ___ Change physical or social environment (e.g., add physical barriers, reduce distractions, provide organizational structures, reduce social complexity, etc.)

    ___ Change the nature of the task (e.g., make shorter, build in breaks, give something to look forward to, create a schedule, build in choice, make the task more fun, etc.)

    ___ Change the way adults interact with the child (e.g., rehearsal, prompts, reminders, coaching, praise, debriefing, feedback)

3. What procedure will be followed to teach the skill?

    Who will teach the skill/supervise the procedure?

    What steps will the child follow?

    1.
    2.
    3.
    4.
    5.
    6.

4. What incentives will be used to encourage the child to learn, practice, or use the skill (check all that apply)?

    ___ Specific praise

    ___ Something to look forward to when the task (or a piece of the task) is done

    ___ A menu of rewards and penalties

        Daily reward possibilities:

        Weekly reward possibilities:

        Long-term reward possibilities:

---

we try to identify ways to modify the environments surrounding these children to reduce the negative impact of their weak skills, or we intervene at the level of the child. If we choose to focus on the child, we have two strategies to draw on. We can either teach children the weak skill or we can motivate them to use skills that are within their repertoire but are effortful or aversive so that they avoid using them when they can. We have found that in most cases it works best to use all three strategies in concert. Thus, when we design interventions, we use a planning process in which we begin by establishing a behavioral goal and then consider what environmental modifications we will implement, how we will teach the weak skill, and how we will motivate the child to practice the weak skill. Table 24.2 provides summary of our approach in checklist form. Much of the discussion on ensuing pages applies particularly to working with children. At the end of the chapter, we discuss how to adapt these strategies when working with teenagers.

Strategy 1: Intervene at the Level of the Environment

When we talk about intervening at the level of the environment, we mean changing conditions or situations external to the child to improve executive functioning or to reduce the negative effects of weak executive skills. Changing the environment may include (1) changing the physical or social environment to reduce problems, (2) changing the nature of the tasks we expect children to perform, and (3) changing the way people—particularly parents, teachers, and caregivers—interact with children with

Table 24.3 Examples of environmental modifications

Environmental modification	Executive skills addressed	Examples
Change the physical environment	Response inhibition	• Add barriers (e.g., avoid runways in preschools)
	Sustained attention	• Seating arrangements (e.g., place distractible kids near teacher, away from windows)
	Task initiation	• Reduce distractions (e.g., allow kids to listen to iPods while taking tests or doing homework to screen out more distracting stimuli)
	Organization	• Use organizing structures (e.g., clear plastic containers with labels; consistent space on blackboard for writing homework)
Change the social environment	Response inhibition	• Reduce social complexity (e.g., fewer kids, more adults; supervision on playground; structured play vs. free play)
	Emotional control	• Change the "social mix" (seating arrangements in class; rules about who children can invite for play dates)
Modify tasks	Sustained attention	• Make tasks shorter or build in breaks along the way
	Task initiation	• Make steps more explicit (e.g., provide templates or rubrics that spell out task requirements)
	Working memory	• Create a schedule, either for a specific event or for a block of time (such as morning work time or Saturday chores)
	Flexibility	• Build in variety or choice either for the tasks to be done or the order in which they're to be done
	Metacognition	• Make the task closed ended
Change the way adults interact with children	Response inhibition	• Rehearse with the youngster what will happen and how the youngster will handle it
	Emotional control	• Use prompts and cues (either verbal or visual)
	Flexibility	• Remind youngster to use checklist or schedule
	Working memory	• Praise youngster for using executive skills—rule of thumb: 3 positives for each corrective feedback

executive skill deficits. Table 24.3 provides examples of all three kinds of environmental modifications.

Strategy 2: Teach the Skill

Environmental modifications may be very successful in minimizing the impact of weak executive skills. However, unless we teach children independently to manage situations that stress weak executive skills, we will need to spend a great deal of time ensuring that every environment they find themselves in is adapted to their unique needs. Since our ultimate goal is for children to be able to function independently, we will need to intervene at the level of the child, both by teaching skills and motivating children to practice.

Executive skills are many and varied, and instructional strategies will, by necessity, differ depending on the skill being taught, the context in which it will be used, and the age or developmental level of the child. A general process for teaching children executive skills is outlined in Table 24.4.

Below are a couple of examples of how this procedure is applied to activities of daily living that require executive skills. The first is a home-based example, taken from *Smart but Scattered*, teaching children to put away belongings, and the second is a whole-class teaching routine for teaching students how to make homework plans, adapted from *Executive Skills in Children and Adolescents*.

Home-Based Example: Teaching a Child to Put Belongings Away

1. With your child, make a list of the items your child routinely leaves out of place around the house.
2. Identify the proper location for each item.
3. Decide when the item will be put away (e.g., as soon as I get home from school, after I finish my homework, just before bed, right after I finish using it).
4. Decide on a "rule" for reminders—how many reminders are allowed before a penalty is imposed (e.g., the belonging is placed off limits or another privilege is withdrawn). A sample checklist follows.
5. Decide where the checklist will be kept.

Fading the Supervision

1. Remind your child that you're working on learning to put things away where they belong.
2. Put the checklist in a prominent place and remind your child to use it each time he/she puts something away.
3. Praise or thank your child each time he/she puts something away.
4. After your child has followed the system for a couple of weeks, with lots of praise and reminders from you, fade the reminders. Keep the checklist in a prominent place, but now you may want to impose a penalty for forgetting. For example, if a toy or a desired object or article of clothing is not put away, your child may lose access to it for a period of time. If it's an object that can't be taken away (e.g., a school backpack), then impose a fine or withdraw a privilege.

Table 24.4 Teaching children executive skills

Step 1: Describe the problem behaviors
Examples of problem behaviors might be starting chores but not finishing them, not following morning routines on school days, forgetting to hand in homework assignments, losing important papers, etc. Be as specific as possible in describing the problem behaviors—they should be described as behaviors that can be seen or heard. *Complains about chores*, *rushes through homework*, and *making many mistakes* are better descriptors than *has a bad attitude* or *is lazy*
Step 2: Set a goal
Usually the goal relates directly to the problem behavior. For instance, if not bringing home necessary homework materials is the problem, the goal might be: "Mary will bring home from school all necessary materials to complete homework"
Step 3: Establish a procedure or set of steps to reach the goal
This is usually done best by creating a checklist that outlines the procedure to be followed. See Chap. 5 for examples of checklists that can be used to address a number of common problems associated with executive skill weaknesses in both home and school settings
Step 4: Supervise the child following the procedure
In the early stages, the child will need to be walked through the entire process. Steps include (1) reminding the child to begin the procedure, (2) prompting the child to perform each step in the procedure, (3) observing the child as each step is performed, (4) providing feedback to help improve performance, and (5) praising the child as each step is completed successfully and when the entire procedure is finished
Step 5: Evaluate the process and make changes if necessary
At this step, the adult continues to monitor the child's performance to identify where the process might be breaking down or where it might be improved. Most commonly, this will involve tightening the process to include more cues or a more refined breakdown of the task into subtasks. When possible, involve the child in the evaluation process to tap into their problem-solving skills
Step 6: Fade the supervision
Decrease the number of prompts and level of supervision to the point where the child is able to follow the procedure independently. This should be done gradually, for example, by (1) prompting the child at each step but leaving the vicinity between steps; (2) getting the child started and making sure she finishes but not being present while she performs the task; (3) cueing the child to start, to use the checklist to check off as each step is completed, and to report back when done; and (4) prompting the child to "use your checklist" with no additional cues or reminders. Ultimately, the child will either retrieve the checklist on her own or even be able to perform the task without the need for a checklist at all

STUDY PLAN

Date: _____

Task	Materials in backpack (✓)	How long will it take?	When will you start?	Where will you work?	Actual start/stop times	Done(✓)

Fig. 24.1 Daily study plan

Modifications/Adjustments
1. Add an incentive if needed. One way to do this would be to place a set number of tokens in a jar each day and withdraw one token each time the child fails to put away an item on time. Tokens can be traded in for small tangible or activity rewards.
2. If remembering to put items away right after use or at different times during the day is too difficult, arrange for a daily pick-up time when all belongings need to be returned to their appropriate locations.
3. For younger children, use pictures, keep the list short, and assume the child will need cues and/or help for a longer period of time.

School-Based Example: Teaching Students to Make Homework Plans

1. Explain to the class that making a plan for homework is a good way to learn how to make plans and schedules. Ask the class why learning to make plans and schedules is a useful skill (e.g., prompt them to provide examples of how they see their parents using plans and schedules either at home or in their jobs). Explain that before leaving school at the end of the day, the class will make homework plans.
2. The steps students should follow:
   (a) Assign student pairs to work on this process.
   (b) Each student should write down all homework assignments (this can be shorthand, since more detailed directions should be in the students' agenda books or on worksheets). Figure 24.1 is a sample form that can be used for this process.
   (c) Together, each student pair should make sure that each student has all the materials needed for each assignment—materials are then placed in backpacks.
   (d) Students individually estimate how long each assignment will take.
   (e) Students write down when they will start each assignment.
   (f) Each pair compares plans to see if they are realistic. Make changes as necessary.
3. The classroom teacher should also monitor the process, offering guidance and feedback to individual student pairs.
4. At the beginning of class the next day, students should pull out completed homework plans and discuss in pairs how well they were able to follow the plan. The teacher collects each plan along with homework.

Routines for managing impulses and emotions. Several executive skills—response inhibition, emotional control, and flexibility above all—are all associated with problems involving behavioral excesses. In helping children control these excesses, the goal is to identify the triggers (what provokes the behavior) and to give children alternative, more adaptive ways to respond to the triggers. Here are some examples of how these weak executive skills play out in common situations:

- Kendrick has trouble managing disappointment. This is particularly problematic when he has a plan for the day and something prevents the plan from being followed (such as his baby brother getting sick so his mother has to take him to the doctor's rather than bringing Kendrick to the beach). At these times, he may get angry and throw things or cry inconsolably.
- Raymond gets overly silly and exuberant at family gatherings. He may run around or play loudly, and when his parents ask him to calm down or sit in one place for a time, he runs off, giggling, while his parents end up chasing him, which seems to get Raymond going even more.
- Therese gets very upset every time her parents leave her and her sister with a babysitter to go out for the evening. She cries and clings to them, begging them not to leave. Her parents have trouble finding babysitters and have taken to not going out to social events.

Table 24.5 lists a general procedure for helping children manage behavioral excesses in these kinds of situations.

Applying this procedure to the case of Therese, who gets very upset when her parents go out for the evening and leave her with a babysitter, might look like this:

Step 1: Therese's parents sit down and talk with her and explain that sometimes they have meetings or social events to go to in the evening and they cannot stay home with her. They explain that this is part of the job of being a grown-up, and they would like to find a way to help Therese be more comfortable with this. Together they figure out whether Therese gets upset at other times or just when her parents leave her with a babysitter, and they decide the problem occurs primarily when she has to stay with a babysitter.

Step 2: They talk about whether it is feasible for them never to go out in the evening and decide this is unrealistic. Just as Therese enjoys doing things like going over a friend's house to play, her parents like social gatherings that are meant just for adults.

Table 24.5 Helping children learn to manage behavioral excesses

Step 1. Help the child identify the "triggers" for the problem behavior. It may be that the behavior of concern happens in a single situation or it may pop up in several different situations

Step 2. Determine if any of the triggers can be eliminated. Technically, this is an environmental modification, but it's a good place to start in understanding the problem behavior and working to reduce it

Step 3. Make a list of possible things the child can do instead of the problem behavior (i.e., replacement behaviors). This will vary depending on the nature of the trigger and the problem behavior

Step 4. Practice the replacement behaviors, using role playing or simulations. "Let's pretend you … Which strategy do you want to use?"

Step 5. Begin using the procedure in minor situations (i.e., not ones involving big upsets or major rule infractions)

Step 6. Move on to situations where more intense behaviors occur

Step 7. Connect the use of the procedure to a reward. For best results, use two levels of reward: a "big reward" for never getting to the point where replacement behaviors need to be used and a "small reward" for successfully using one of the agreed-upon replacement behaviors

Step 3: They brainstorm with Therese other things she could do besides cry when it's time for them to leave. Therese has a favorite video she likes to watch and she has a favorite stuffed animal she finds comforting to hold. They put these on the list, along with having a special dessert that's available only on nights when a babysitter comes.

Step 4: They practice in a "Let's Pretend" situation. Her father pretends to be the babysitter, and her mother pretends to leave for awhile. Before she leaves, she and Therese look at her list of things she might choose when it's time for the babysitter to come. Therese chooses one or two. Her father knocks on the door and when her mother answers, he pretends to be the babysitter. Therese's mother explains to the "babysitter" that Therese has agreed to try not to cry and will choose something from her list to do when her mother leaves. Therese makes the choice, and her mother kisses her good-bye and goes out the door. Therese and the babysitter pretend to do the activity she has selected. Therese's mother comes back in and they talk about how the role playing went. They practice the routine for several nights.

Step 5: Therese's parents arrange for the babysitter to come over for a brief time so they can practice in a situation in which both parents leave for awhile. They explain that they will be back in 30 min and they ask Therese to choose which activity she's going to do with the babysitter. When they return, they ask how everything went and praise Therese for handling their absence without getting upset. If necessary, repeat this practice for several nights.

Step 6: Arrange for the babysitter to come for a longer period of time.

Step 7: If necessary, add in a special treat that Therese can earn after the first time she stays with the babysitter "for real" and does not get upset.

Strategy 3: Motivate Children to Use Executive Skills

In addition to teaching children to use executive skills, a second intervention at the level of the child is to motivate children to use executive skills already within their repertoire. Although imposing penalties and negative consequences are often favored by adults to motivate children to change their behavior, we generally recommend beginning with a positive approach. And while incentive systems can be very elaborate and resemble a labor contract, whenever possible we recommend more informal motivational methods. Especially with younger children, these include:

1. Giving the child something to look forward to when the effortful task is done. For children who resist or delay doing homework, for instance, parents may allow them to watch a favorite television show or play 30 min of video games once the homework is finished. When this approach is used, saying "As soon as you get your homework done, you can play X-Box for 30 minutes" is far more effective than saying "No video games until your homework is done." The former seems to serve to energize the child to get through tedious or effortful tasks, while the latter sets up a power struggle or raises control issues.
2. Alternating between preferred and non-preferred activities. For young children with very short attention spans, saying "First work, then play" can cue the child that this is how the morning will go (whether at school or at home). This approach can be paired with the next one.
3. Build in frequent breaks. Bedroom cleaning might be divided into individual tasks (putting dirty clothes in the laundry, putting toys in appropriate containers, cleaning out under the bed), with 5-min breaks to play with the dog or work briefly on a Lego construction in between tasks.
4. Using specific praise to reinforce the child's use of an executive skill. Decades of behavioral research has documented the power of effective praise alone to change behavior. How praise is administered, however, can make all the difference. Praising children for being "smart" or "talented" can actually create problems rather than build self-esteem. Children who are told frequently that they are smart often begin to stop taking risks in situations where new learning is called for because they fear that if they fail or make mistakes, that must mean they're actually *not* smart. When the praise labels a particular skill being practiced, the child feels good about acquiring specific competencies, which is a far better builder of self-esteem than are global labels. Some examples of specific praise for demonstrating executive skills are:
   - You stuck with that chore until you finished it—I'm impressed!
   - Thanks for getting started on your homework right after your snack.
   - I like the way you hung in there on that math problem even though I could see you getting frustrated.

Table 24.6 Designing incentive systems

Steps 1 and 2: Describe the problem behaviors and set a goal

These are identical to Steps 1 and 2 for teaching the child to use executive skills (see Table 24.4). Both problem and goal behaviors should be described as specifically as possible, usually with a link between the two. As an example, if forgetting to do chores after school is the problem, the goal might be: "Joe will complete daily chores without reminders before 4:30 in the afternoon."

Step 3: Decide on possible rewards and contingencies

Incentive systems work best when children have a "menu" of rewards to choose from. One of the best ways to do this is a point system in which points can be earned for the goal behaviors and traded in for the reward the child wants to earn. The bigger the reward, the more points the child will need to earn it. The menu should include both larger, more expensive rewards that may take a week or a month to earn and smaller, inexpensive rewards that can be earned daily. Rewards can include "material" reinforcers (such as favorite foods or small toys) as well as activity rewards (such as the chance to play a game with a parent, teacher, or friend). It may also be necessary to build contingencies into the system—usually the access to a privilege after a task is done (such as the chance to watch a favorite TV show or the chance to talk on the telephone to a friend)

When incentive systems are used in a school setting, it is often beneficial to build in a home component. This is because parents often have available a wider array of reinforcers than are available to teachers. When a coordinated approach is used, a home-school report card is often the vehicle by which teachers communicate to parents how many points the child has earned that day. Situations in which we do not recommend including a home component include (1) when parents, for whatever reason, are unable to maintain the system consistently; (2) when parents insist on negative consequences; or (3) when the child needs a more immediate reward and cannot wait until the end of the school day to earn it

Once the system is up and running, if you find the child is earning more penalties than rewards, then the program needs to be revised so that the child can be more successful. Usually when this kind of system fails, we think of it as a "design failure" rather than the failure of the child to respond to rewards

Step 4: Write a behavior contract

The contract should say exactly what the child agrees to do and exactly what the parents' or teacher's roles and responsibilities will be. Along with points and rewards, parents or teachers should be sure to praise children for following the contract. It will be important for adults to agree to a contract they can live with: they should avoid penalties they are either unable or unwilling to impose (for instance, if both parents work and are not at home, they cannot monitor whether a child is beginning his homework right after school, so an alternative contract may need to be written)

Step 5: Evaluate the process and make changes if necessary

Incentive systems may not work the first time. Parents or teachers should expect to try it out and redesign it to work the kinks out. Eventually, once the child is used to doing the behaviors specified in the contract, the contract can be rewritten to work on another problem behavior. As time goes on, children may be willing to drop the use of an incentive system altogether. This is often a long-term goal, however, and adults should be ready to write a new contract if the child slips back to bad habits once a system is dropped

- You figured out how long it would take you to make that poster for your health class and you made sure you didn't start too late.
- Good job waiting until I called on you to answer the question.
- I'm impressed with the way you came up with a system for remembering to hand in your homework.

When more informal approaches are not enough, however, parents or teachers may find it necessary to create a more elaborate incentive system to motivate children to use executive skills. The steps in developing incentive systems are described in Table 24.6.

Figure 24.2 is a sample behavior contract that might be drawn up to address the problem of a child leaving belongings all over the house.

Adaptations for Working with Teenagers

In the beginning of the chapter, we presented a list of ten guiding principles for helping youngsters improve executive skills. With teenagers, one of these principles becomes preeminent: *Use rather than fight the child's innate drive for mastery and control.* "I do it myself" is something we

> **Sample Behavior Contract**
>
> **Child agrees to:** put away belongings in proper place every day right after supper.
>
> **To help child reach goal, parents will:** remind child of agreement when dinner is over.
>
> **Child will earn:** earn 15 points each day he puts belongings away where they belong right after use; 5 points if he puts them away when reminded after dinner.
>
> **If child fails to meet agreement child will:** not earn any points
>
> Points can be traded in for:
>
> | Activity | Point Value |
> | --- | --- |
> | Extra 30 minutes video game time | 5 |
> | New Superhero toy | 45 |
> | 30 minutes later bedtime | 15 |

Fig. 24.2

hear coming from the mouths of 3- or 4-year-olds, but the same drive reasserts itself in adolescence. Now it sounds more like "Get off my case," "Back off," or "You're not the boss of me," but the goal is the same: increased autonomy.

When children are younger, parents may, at least in the abstract, see the value of autonomy and self-management. At the same time, until adolescence, parents have had a different role. They have been actively involved in decision-making day in and day out, playing a guiding role in the activities the children engage in, the friends they spend time with, the behaviors they are expected to display, and the safety precautions they need to follow. Parents are practiced in this role and, for better or worse, have made decisions in what they believe is the best interest of their child.

With the onset of adolescence, however, parents increasingly yield their role as decision-makers, and teens gradually take on that role. This can be a difficult time for parents because it means a significant change in how they interact with their children. No longer are they the trusted advisor or the source of wisdom, or perhaps even the final authority. When we work with parents and teens, our goal is to help them both adapt to new roles while still providing a safety net so that when things don't work out as either parents or kids might like, catastrophe can be averted.

For many parents, a difficult adjustment is recognizing that the hopes, dreams, and goals they have for their teenager may be quite different from those that matter to their child. Rather than expending time, energy, and emotion in attempting to bend their children to their will, we counsel parents to look for ways they can use the goals their children have set for themselves as a way to teach or practice executive skills. If a teenager has a life goal of getting a job or a driver's license, buying a car, starting a band in his garage, or learning how to repair computers, he is likely willing to learn how to make plans or persist long enough to achieve a goal if that gets him the outcome he seeks. Along the way, of course, he's learning about planning, time management, task initiation, sustained attention, and goal-directed persistence—all critical executive skills that will serve him well in the long run. He may not want to expend the time and energy in producing quality physics projects, writing spectacular essays, cleaning the garage, or juggling responsibilities so that the majority of his homework is completed on time, but if he is honing executive skills in pursuit of his own goals, he will be able to apply these same skills to goals down the road that are more in keeping with his parents' priorities.

The key points we stress when working with parents of teenagers are these:

1. The most powerful reinforcer parents have available to them is their child's desire for independence. Whenever possible, the goal

Table 24.7 Developing a plan for executive skill development with teenagers

Step 1: Parent and teenager agree on a problem situation that needs to be addressed

Ideally, the problem situation is of equal concern to both parents and teens. But where viewpoints diverge, parents agree to help the teen address a problem situation of their choosing

Step 2: Parent asks teenager to describe a plan for tackling the problem

Parents are often very comfortable with the problem-solving role and want to jump in and offer a solution or their own ideas for how the problem situation should be tackled. They should resist this impulse, both because it reduces any sense of ownership their child may have in the plan and because learning how to problem solve is, of itself, an important executive skill

Step 3: Parent and teenager negotiate any modifications to the plan to make it acceptable to the parent

Whenever possible, parents should accept the plan as laid out by the teenager. When this is not possible, they lay out their reservations about the plan as stated and ask for how it could be modified to address their concerns. They can suggest modifications and compromises if their child is unable to

Step 4: Parent proposes the use of an incentive to enhance the plan

The effort required to acquire new habits is often substantial. We recommend that whenever possible parents offer their teen's an incentive to help them persist with the plan long enough for it to work. The incentive should be something the teenager would like to work for that the parents are willing to supply. Together they decide on the value of the incentive (which will determine how long the teen will have to work to earn it) and the record-keeping mechanism for showing progress toward earning incentive (such as keeping a bar graph or placing dollar bills in a jar, with each bill being equivalent to an agreed-upon number of points)

Step 5: Parent asks teenager to identify the benchmarks for success

Plans may start with "a hope and a promise" but objective benchmarks to assess progress are critical to the plan's success. Parents ask their teen to decide what data they will collect to show that the plan is working. This step builds in self-monitoring, a critical metacognitive skill. The benchmark should be specific and measurable, such as grades on tests, homework passed in, or completion of a checklist showing that steps toward task completion or goal attainment were followed

Step 6: Parent and teenager agree on how long the plan will run before making substantive changes

The amount of time the plan will be in place before revising it needs to be long enough to give the teen a chance to adjust to a new way of doing things but not so long that the plan's failure will lead to significant negative consequences (such as failing grades or accumulating debt)

Step 7: Parent and teenager discuss back-up plans that will kick in if the teenager's plan does not meet with success

Ideally, this is a several-stage process that falls along a continuum from less intrusive to more intrusive. We recommend that the parent role be confined to meting out consequences rather than increasing levels of supervision or monitoring. Consequences are most often a loss of privilege or restricted access to a desired possession. The most powerful consequences are the ones that interfere with a teen's independence

being worked on should result in more independence and autonomy for their child.

2. Parents must be willing to relinquish some of their own power and control in order to support their child's drive toward independence. Outcomes that contribute to the teen's independence and decision-making will be valued more than those that emphasize parental control.
3. The most powerful tool that parents have available to them is negotiation. A willingness to engage in negotiation and compromise signals respect for a teen's autonomy and ideas and is more effective in developing executive skills than directives from a parent.

These points are played out as parents and teens develop plans for addressing problems that arise from weak or immature executive skills. At its core, it involves creating a plan of attack that is acceptable to both parent and child. The procedure we recommend parents used is described in Table 24.7.

By way of example, let's see how this process works with a typical issue that comes up when we work with teenagers. Marcus is a 16-year-old sophomore in high school with an academic record that fluctuates wildly between A's and F's in any given marking period or any given subject. He is inconsistent in doing his homework; he procrastinates, which leads him to miss deadlines or produce poor quality work; he may lose homework before handing it in; or he may simply forget to hand it in, even though he's brought it to school with every intention of doing so. Test grades are also erratic. His parents check the school's web portal constantly to find out what his test and homework grades are and how many homework assignments he is missing. They insist on checking in with him several times each night to make sure he's started his homework, has finished it, has stored it where it belongs, has begun studying for upcoming tests, or has begun to work on projects or papers with more distant deadlines. The end result is constant tension at home as Marcus and his parents end up having shouting matches as they attempt to get him to take his school work seriously and Marcus yells at them to back off or resorts to lying about assignments to avoid their nagging and scrutiny.

With a younger child—say a 4th or 5th grader—parents can set up rules about homework (no video games until the homework is done, homework gets done after an hour of free play and before dinner) and can impose those rules without significant push-back from the child. With teenagers, for any number of reasons, this no longer works. When parents of teenagers impose these kinds of rules, it involves a fair amount of stress, so we tend to hold off recommending this strategy until other efforts have failed.

As we've outlined the process in Table 24.7, here's how Marcus's parents chose to address the problem. First, they sat down with Marcus and explained that the conflict and tension around homework and Marcus's erratic grades were affecting their relationship with Marcus and with each other (because they often disagreed about how to address the problem) and they wanted some input from him about how to make things better. They asked how Marcus felt school was going and whether he was happy with his grades. He admitted he wasn't, but he also said he didn't want to work as hard as he knew he would have to in order to make the honor roll, which appeared to be the goal his parents had for him. He said he was willing to do what it took to earn B's and C's, which he felt was sufficient for him to pursue postsecondary education in some form when the time came. His parents felt he was setting the bar pretty low, but since it didn't work when *they* set the bar, they decided to accept his goal.

They then asked Marcus what he would need to do in order to earn the grades he was willing to try for. He said he needed to become more consistent about getting his homework done and handed in, particularly in math and science, where he had nightly assignments and where the homework was graded. What about studying for tests, his parents asked. Marcus wasn't willing to commit to any kind of study schedule, but he admitted that if he improved his performance there, it would probably help him reach his report card goals.

Marcus's parents then offered to add an incentive to make it more appealing for him to work hard to improve his effort and his time management skills. They knew he really wanted the latest version of iPhone and they were willing to finance it for him in exchange for improved school performance. They suggested that he could earn points for every week he handed in all his homework and for grades of B− or better on tests and quizzes, with higher grades earning more points than lower grades and tests earning more points than quizzes. Marcus had been hounding his parents for some time about getting an iPhone and they'd always responded vaguely that they'd think about it if he got better grades in school. Now that they were making a concrete offer, Marcus jumped on it. Together he and his parents worked out the point system and he created a spreadsheet where he could keep track of points on a daily basis. His parents said they would use the school website to verify that all his homework was handed in and to verify test and quiz grades. They asked if Marcus wanted any cues or reminders from them, such as help keeping track of long-term assignments or prompts to start working on his homework. Marcus answered rather testily that that was *exactly* what he hated them doing now and if they continued to do that, the whole plan was likely to fail.

Marcus's parents took a different tack and asked him to think about how they would be able to judge whether the plan was working.

Marcus said, "You'll know if it works by the grades on my report card at the end of the marking period." His parents replied that they weren't willing to wait that long to gauge success and suggested a 3-week trial period, with success to be judged by homework completion rates and test/quiz grade averages. If Marcus averaged 80 % homework completion per week and a 75 average grade on all tests and quizzes, they would consider the plan a success. If he fell below those percentages, they would move to the next level, which in their mind was to find a coach to work with Marcus on staying on top of homework assignments and developing good study habits. If coaching was not successful, then the next phase would be the imposition of some restrictions on Marcus's freedom, such as access to the car or the chance to hang out with friends on weekends. Marcus reluctantly agreed to this, in part because he felt he was going to be successful enough that the more intrusive plans could be avoided.

The use of a coach to work with teenagers with executive skill deficits is a common recommendation we make to parents and educators. The job of a coach is to help students stay on top of assignments, make good decisions about how they will spend their time, and break down long-term assignments into smaller pieces. A typical coaching session lasts about 10–15 min and occurs daily. Coaches typically use the time to help the student (1) review all homework assignments, including daily homework, upcoming tests, and long-term projects or papers; (2) break down long-term assignments into subtasks and develop timelines; (3) create a study plan for tests; (4) make a homework plan for the day; and (5) monitor how well the plan is followed and track assignment completion. Coaches typically check in with teachers at least weekly (on Friday) to track any missing assignments and to double-check long-term assignments. They also email parents on Friday informing them of any missing assignments. Coaching includes an instructional component so that the student can gradually take on more and more of the coaching tasks with less input from the coach. For a more detailed description of coaching, the reader is referred to the book *Coaching Students with Executive Skills Deficits* (Dawson & Guare, 2012).

Finally, for interventions to be effective with teenagers, we push parents to remove as much emotion as they can both in the negotiation process and when the success of plans are being evaluated, particularly if they fall short of the benchmarks and further steps need to be taken. Parents may find it helpful to enlist the assistance of a third party mediator in these cases, such as a therapist, guidance counselor, or their child's favorite teacher.

Parting Thoughts

Compared to more significant psychological disorders, executive skill deficits can appear relatively benign. After all, we all know adults with some significant executive skill weaknesses who seem to be functioning just fine overall. In fact, as experienced parents and teachers will tell you, they can be extremely frustrating for the children themselves and for the adults responsible for their care and education. We counsel parents in particular to take the long view, and we often say that progress is measured in years and months and not days and weeks. The good news is that there are interventions that work, and years of clinical experience have confirmed for us that patience, a plan, and persistence can lead to good outcomes more often than not.

References

Barkley, R. A. (2011). *Barkley Deficits in Executive Functioning Scale*. New York: The Guilford Press.

Dawson, P., & Guare, R. (2012). *Coaching students with executive skills deficits*. New York: The Guilford Press.

Gioia, G. A., Isquith, P. K., Guy, S. C., & Kenworthy, L. (2000). *Behavior Rating Inventory of Executive Function*. Odessa, FL: Psychological Assessment Resources, Inc.

Teaching Executive Functioning Processes: Promoting Metacognition, Strategy Use, and Effort

Lynn Meltzer

Mike's performance has been unpredictable all year! He has many creative ideas and he participates actively in classes. However, he is usually late with written papers and projects and he does not seem to care about his homework. His test grades fluctuate from the 90s to the 60s. His other teachers have told me that they think he is lazy. I think that Mike may have a problem. (8th grade teacher)

Students' success in the digital age is increasingly linked with their ability to take responsibility for their own learning by organizing and integrating a rapidly changing body of information that is available in textbooks and online. From the early grades, they are expected to work independently to complete numerous multistep projects and writing assignments, all tasks that rely on cognitive flexibility and the ability to shift rapidly between different processes. Students' academic performance therefore depends on the ease with which they plan their time, organize and prioritize materials and information, separate main ideas from details, think flexibly, memorize and mentally manipulate information, and monitor their own progress. As a result, it is essential that all students develop an awareness of *how* they think and *how* they learn and that they master strategies that address these executive function processes.

For the purposes of this chapter, "executive function" is used as an umbrella term that is broader than metacognition and incorporates a range of interrelated processes responsible for goal-directed behavior (Anderson, 2002; Gioia, Isquith, Guy, & Kenworthy, 2001). More specifically, this chapter builds on many of the theoretical models of executive function that have been refined since the seminal work of Flavell on goal-oriented problem-solving (Flavell, Friedrichs, & Hoyt, 1970) and the research on metacognition and self-regulation (Barkley, 1997; Brown & Campione, 1983, 1986; Denckla, 1996, 2007; Gioia, Isquith, Guy, Kenworthy, & Barton, 2002). The approach in this chapter is anchored in Eslinger's (1996) analyses of multiple definitions and his conclusions that executive function refers to a range of cognitive processes that are controlled by the prefrontal cortex and comprise:

- Metacognitive knowledge about tasks and strategies
- Flexible use of strategies
- Attention and memory systems that guide these processes, e.g., working memory
- Self-regulatory processes such as planning and self-monitoring

This chapter includes a discussion of a theoretical paradigm for understanding and teaching students strategies that address executive function processes (Meltzer, 2007, 2010). There is an emphasis on the central importance of six executive function processes: goal-setting, cognitive flexibility/shifting, organizing, prioritizing, accessing working memory, and self-monitoring.

L. Meltzer (✉)
Research Institute for Learning and Development (ResearchILD), and Harvard Graduate School of Education, Lexington, MA 02421, USA
e-mail: lmeltzer@ildlex.org

The major principles of intervention and treatment are addressed with a focus on the interactions among executive function processes, self-awareness, effort, and persistence. Selected strategies are discussed for addressing the key executive function processes as part of a systematic teaching approach (see Meltzer, 2010, for specific classroom teaching techniques). It should be noted that it is beyond the scope of this chapter to include a description of inhibition, selective attention, and activation, all important executive function processes that are addressed in recent neuroscientific research (Anderson, Rani Jacobs, & Anderson, 2008; Bernstein & Waber, 2007; Diamond, 2006).

Executive Function Processes and Academic Performance

When I have to write a paper, I sit down at my computer but my mind feels like a bottle of soda that's been all shaken up. I try to write but I can't figure how to get my mind unstuck so I can begin. After trying for an hour I have often written only a few sentences and I give up. (Michael, 8th grade)

Students with executive function weaknesses often struggle with academic tasks that involve the coordination and integration of different subskills such as initiating writing assignments, summarizing information, taking notes, planning, executing and completing projects in a timely manner, studying, and submitting work on time (Meltzer, 2010; Meltzer & Basho, 2010; see Table 25.1). These students often have difficulty organizing and prioritizing information and they struggle to shift flexibly to alternate approaches so that they overfocus on details while ignoring the major themes. Furthermore, working memory and self-monitoring processes may also be weak, making it difficult for students to mentally juggle information, self-monitor, or self–check. As a result, information can often become "clogged" and students become "stuck," so that they struggle to produce academically (see Fig. 25.1). As is reflected in Fig. 25.1, the paradigm that has guided our work on executive function is based on the analogy of a "clogged funnel" (Meltzer, 2004, 2007, 2010; Meltzer & Krishnan, 2007). Because these students cannot shift flexibly among alternative approaches to unclog the funnel, their written work, study skills, and test performance are compromised, and their academic grades do not reflect their strong intellectual ability. Consequently, they may have difficulty showing what they know in the classroom and other settings.

These deficits become increasingly apparent during middle and high school when the volume and complexity of the workload increase and performance is more dependent on executive function strategies (see Table 25.1). During this time, there is also a mismatch between students' skills and the demands of the curriculum, so that they struggle to perform at the level of their potential. This can be extremely frustrating and can affect students' self-confidence and long-term performance.

Table 25.1 Executive function processes and their impact on academic performance

Executive function process	Definition
Goal-setting	Identifying short-term and long-term goals
	Figuring out a purpose and end-point
Shifting/cognitive flexibility	Switching easily between approaches
	Looking again, in a brand new way
Prioritizing	Ordering based on relative importance
	Figuring out what is most important
Organizing	Arranging information systematically
	Sorting information
Using working memory	Remembering "so that it sticks like glue"
	Juggling information mentally
	Cementing information in the brain
Self-monitoring	Shifting to a checking mindset and back
	RE-viewing in a different way

Note: Italics indicate student-friendly definitions
Adapted from Meltzer (2010). Copyright 2010 by The Guilford Press. Reprinted by permission

Fig. 25.1 Executive function paradigm. Adapted from Meltzer (2007). Copyright 2007 by The Guilford Press. Reprinted by permission

Executive Function Strategies, Self-Understanding, and Effort: The Underpinnings of Academic Success

> My success is due to the strategies I learned as well as my self-understanding and the confidence I developed after I used the strategies and got higher grades. (Sean, 11th grader)

Academic success for all students, and particularly for students with learning and attention difficulties, is connected with their motivation, academic self-concept, and self-efficacy (Brunstein, Schultheiss, & Grässmann, 1998; Helliwell, 2003; Kasser & Ryan, 1996; Meltzer, Reddy, Sales, et al. 2004; Pajares & Schunk, 2001; Sheldon & Elliot, 1999). These cognitive and motivational processes are linked cyclically with students' use of executive function strategies as well as their effort and persistence (Meltzer, Katzir, Miller, Reddy, & Roditi, 2004; Meltzer & Krishnan, 2007) (see Fig. 25.2).

Strategies that address executive function processes therefore provide a starting point for improving academic performance (see Fig. 25.2). When students use executive function strategies, they become more efficient and they begin to improve academically. Academic success, in turn, boosts self-confidence and self-efficacy so that students' effort is targeted strategically towards specific goals. A cycle of success is promoted when students focus their effort on using executive function strategies in the context of their academic work (Meltzer, 2010; Meltzer, Katzir, et al., 2004; Meltzer, Reddy, Pollica, et al., 2004; Meltzer, Reddy, Sales, et al., 2004).

To build their motivation, persistence, and work ethic, students need to understand their profiles of strengths and weaknesses, which strategies work well for them, as well as *why, where, when,* and *how* to apply specific strategies. This understanding, referred to as metacognition, or the ability to think about their own thinking and learning, underlies students' use of executive function processes. More specifically, metacognition, as defined originally by Flavell (1979) and Brown, Bransford, Ferrara, and Campione (1983), refers to each student's understanding and beliefs about HOW she/he learns as well as the strategies that can or should be used to accomplish specific tasks. Students' metacognitive awareness therefore includes their knowledge and understanding of their own learning profiles as well as their knowledge of the specific strategies that match their strengths and weaknesses and help them to master different tasks. For example, students like Chace (see Fig. 25.3) are aware of their struggle to plan, organize, prioritize, and manage their time, and they are often frustrated that their academic performance does not match their strong intellectual potential.

Therefore, students' understanding of their strengths and weaknesses, as well as their motivation and beliefs, influences their selection of specific strategies and how long they are willing to persist with tasks. This metacognitive awareness usually increases their willingness to make the effort needed to master the strategies needed for the many academic tasks that they face in school on a daily basis. Sustaining this effort is also connected with students' interests in particular subjects. Specifically, interest-based motivation in learning frequently influences the types of strategies students use as well as their academic performance (Renninger, Hidi, & Krapp, 2004; Yun Dai & Sternberg, 2004). Unfortunately,

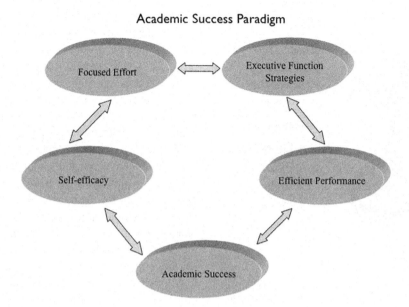

Fig. 25.2 Academic success cycle. Adapted from: Meltzer, LJ., Reddy, R., Pollica, L., & Roditi, R. (2004). Copyright 2004 by the International Academy for Research in Learning Disabilities. Reprinted by permission

students' interest in school often decreases in middle and high school when the curriculum content constrains students' ability to engage their interests and explore new challenges (Gardner, 1983; Renninger et al., 2004). As a result, many bright and talented students, especially those with learning and attention difficulties, may "give up" so that they no longer make the extraordinary effort needed to master these strategies. Consequently, they become less productive in the higher grades.

For all these students, executive function strategies can provide a lifeline to academic success as they learn to set goals and to shift flexibly from the major themes to the details and back again (Meltzer & Basho, 2010). In fact, success is usually attainable when students use executive function strategies to target realistic goals, they focus their effort on reaching these goals, and they self-regulate their cognitive, attention, and emotional processes appropriately (see Fig. 25.4). Furthermore, as is evident from Fig. 25.4, these processes usually build persistence, resilience, and academic success (Meltzer, 2010: Meltzer, Reddy, Brach, Kurkul, & Basho, 2011; Meltzer, Reddy, Brach, Kurkul, Stacey and Ross, 2011).

Informal, Process-Oriented Assessment of Executive Function

My teacher probably thinks I am one taco short of a combo plate, if you know what I mean. (Jamie, 8th grader)

Jamie is a very bright student who is a puzzle to me. He does well on quizzes and short tests but he often does not hand in his homework. His performance is inconsistent and his grades are up and down. (8th grade teacher)

Educators are often puzzled by students like Jamie whose performance oscillates between high grades on quizzes to low grades on multi-step tasks such as written papers, essays, math problem-solving, or long-term projects. When teachers understand the role of executive function processes, they can reframe their understanding so that they focus on their students' strengths and academic potential and no longer view these students as "unmotivated," "lazy," or "not trying hard enough." Informal assessment methods can help teachers to understand students' use of executive function processes

Fig. 25.3 Chace, an exceptionally bright eighth grader, depicts his daily battle to stay organized and focused because of his executive function difficulties. From Meltzer (2010). Copyright 2010 by The Guilford Press. Reprinted by permission

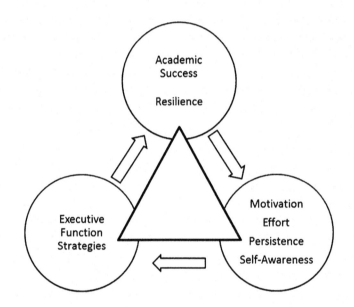

Fig. 25.4 Academic success paradigm. From Meltzer, Reddy, Brach, Kurkul, and Basho (2011)

and pinpoint *why* and *how* particular students may be struggling. Teachers can then introduce specific instructional approaches, assess students' progress, and modify instruction. The continuous cycle linking assessment and teaching allows teachers to adjust their instructional methods to the changing needs of their students. In fact, many of these principles are incorporated into the response-to-intervention (RtI) approach that is now being more widely used in US schools to improve early identification of reading and learning difficulties (Fuchs & Fuchs, 1991; Kame'enui, 2007).

Currently, there is a dearth of measures that help teachers and other professionals to understand students' use of executive function strategies. One of

Table 25.2 Metacognitive awareness survey system (MetaCOG)

STUDENT questionnaires
ME—Motivation and effort survey
STRATUS—Strategy use survey
MAQ—Metacognitive awareness questionnaire
TEACHER questionnaires
TPSE—Teacher perceptions of student effort
TIQ—Teacher information questionnaire

5-point rating for all surveys

the most reliable and widely used questionnaire systems is the *Behavior Rating Inventory for Executive Function* (*BRIEF*) (Gioia et al., 2001, 2002). The BRIEF includes 86 items and comprises a parent questionnaire, a teacher questionnaire, and a self-rating form for students from 11 years into adulthood. The items assess behaviors associated with the core executive function processes, e.g., "Forgets to hand in homework, even when completed; Gets caught up in details and misses the big picture; Becomes overwhelmed by large assignments; Underestimates the time needed to finish tasks."

Another criterion-referenced assessment system that compares students', teachers', and parents' perceptions of students' metacognitive awareness and strategy use is the *Metacognitive Awareness System or MetaCOG* (Meltzer, 2010; Meltzer & Krishnan, 2007; Meltzer, Reddy, Pollica, et al., 2004; Miller, Meltzer, Katzir-Cohen, & Houser, 2001). The MetaCOG, for use with 9–18 year-olds, comprises five rating scales that allow educators to compare their own judgments with their students' self-ratings of their effort, strategy use, and academic performance. These strategy ratings focus on academic areas that depend on executive function processes and include written language, homework, studying, and taking tests (see Table 25.2; Meltzer, Katzir-Cohen, Miller, & Roditi, 2001; Miller et al., 2001).

Systems such as the MetaCOG can be used for a variety of purposes over the course of the year: (a) to understand students' views of their own effort, use of strategies, and academic performance; (b) to help educators and clinicians to compare their own judgments with their students' self-perceptions; (c) to develop a system for teaching strategies to help students plan, organize, prioritize, shift flexibly, memorize, and check their work; and (d) to track students' understanding and implementation of these strategies over time.

The three student surveys assess students' self-ratings of their motivation and effort as well as their strategy use in key academic areas (see details below). Completion of the MetaCOG surveys helps students to build a self-understanding about their learning profiles. This self-awareness is the foundation for building students' metacognitive awareness and their use of executive function strategies.

MetaCOG Student Surveys

1. *Motivation and Effort Survey* (*ME*). The ME consists of 38 items that assess students' self-ratings of their effort and performance on different academic tasks that depend on executive function processes (alpha = .91) (Meltzer, Reddy, Sales, et al., 2004; Meltzer, Sayer, Sales, Theokas, & Roditi, 2002). Students rate themselves on a 1–5 scale (from *never* to *always*) in terms of how hard they work and how well they do in selected academic areas such as reading, writing, math, homework, studying for tests, and long-term projects (e.g., *I spend as much time as I need to get my work done*; *I finish my work even when it is boring*; *I do schoolwork before other things that are more fun*). Students are also asked to describe themselves as learners.

2. *Strategy Use Survey* (*STRATUS*). The STRATUS consists of 40 items that assess students' self-reported strategy use in reading, writing, spelling, math, studying, and test-taking (alpha = .945). Items focus on students' perceptions of their use of strategies for planning, organizing, memorizing, shifting, and self-checking (e.g., *When I have to remember new things in school, I make up acronyms to help me*; *Before I write, I plan out my ideas in some way that works for me* [*outline, list, map*]; *When I do math, I ask if my answers make sense*].

3. *Metacognitive Awareness Questionnaire (MAQ)*. The MAQ consists of 18 items that assess students' understanding of strategies and how they can apply strategies to their schoolwork (e.g., *When you begin something new, do you try to connect it to something you already know?*; *When you begin something new, do you try to think about how long it will take and make sure you have enough time?*).

MetaCOG Teacher Surveys

1. *Teacher Perceptions of Student Effort (TPSE)*. The TPSE is the teacher version of the ME and consists of 38 items that assess teachers' ratings of students' effort in different academic domains (alpha = .980; Meltzer, Katzir, et al., 2004; Meltzer, Reddy, Pollica, et al., 2004; Meltzer, Reddy, Sales, et al., 2004). Teachers rate students' effort and performance in reading, writing, math, homework, tests, and long-term projects, all academic tasks that rely on executive function processes (e.g., *He spends as much time as needed to get his work done*; *She does not give up even when the work is difficult*). Teachers also rate students' overall strategy use and academic performance in response to the question: *"If you had to assign a grade for this student's overall academic performance, what would this be?"*

MetaCOG Parent Surveys

1. *Parent Perceptions of Student Effort (PPSE)*. The PPSE consists of 38 items that assess parents' ratings of students' behaviors when working hard and the effort they apply in different academic domains that require the use of executive function processes. Items are identical to those used on the student self-report survey (ME) and the teacher survey (TPSE).

As was discussed above, student, teacher, and parent reports can be directly compared to determine overall consistency in their ratings of many of the core components of executive function processes across different settings (Tables 25.3 and 25.4).

Table 25.3 MetaCOG sample items: students' vs. teachers' ratings of their motivation and effort on academic tasks that involve executive function processes

ME-students	TPSE-teachers
• Doing well in school is important to me	• Doing well in school is important to this student
• I spend as much time as I need to get my work done	• S/he is a hard worker
• I keep working even when the work is difficult	• S/he doesn't give up even when work is difficult
I work hard on	*Please judge how hard this student works*
• Homework	• Homework
• Long-term projects	• Long-term projects
• Studying for tests	• Studying for tests
• Other activities (sports, music, art, hobbies)	• Other activities (sports, music, art, hobbies)

Note: The ME and TPSE each comprise 36 items using a 1–5 rating scale

Table 25.4 MetaCOG sample items: students' vs. teachers' ratings of their performance on academic tasks that involve executive function processes

ME-students	TPSE-teachers
Please judge how well you do on	*Please judge how well this student does on*
• Organization	• Organization
• Long-term projects	• Long-term projects
• Making a plan before starting work	• Making a plan before starting work
• Using strategies in my schoolwork	• Using strategies in his/her schoolwork
• Checking my work	• Checking his/her work
• Homework	• Homework
• Tests	• Tests
• Long-term projects	• Long-term projects

Students' perceptions of their own effort and strategy use are often very different from their parents' and teachers' perceptions, as has been shown in a number of studies (Meltzer, Katzir, et al., 2004; Meltzer, Reddy, Pollica, et al., 2004; Meltzer, Reddy, Sales, et al., 2004; Stone & May, 2002). For example, Fig. 25.5 shows the self-ratings of John, a fifth grader, who rates himself as a strong, hardworking student with the goal of "*getting a second masters degree and making the*

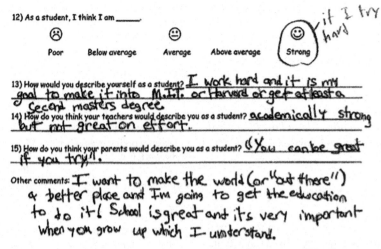

Fig. 25.5 Comparison of a student's vs. teacher's ratings of the student's effort and performance. From Meltzer (2010). Copyright 2010 by The Guilford Press. Reprinted by permission

world a better place." When asked how his parents would describe him, his response reflects his view that his parents would emphasize the importance of working hard in school. In contrast, John's teacher's comments are very different and reflect her perception that John has difficulty sustaining his attention. In fact, she comments that John's academic performance would be much stronger if he could focus more easily in class.

In other words, survey systems that raise teachers' awareness and understanding of their students' effort, strategy use, and possible difficulty with executive function processes can help them to understand why and how these students may be struggling. These systems can also help teachers to implement and monitor the effectiveness of specific instructional strategies, as will be detailed in the remainder of this chapter.

Intervention Strategies That Address Executive Function Processes

The strategies I learned from my tutor changed my life. When I was failing in tenth grade, I became the class clown because my teachers told me I was lazy and nobody taught me how to use strategies for remembering, organizing, and checking my work. (Max, 12th grader)

Students need to learn *when* to use *which* strategies and in *what* contexts. They also need to recognize that not all strategies work for all tasks and all content areas. In other words, strategies need to fit well with the student's learning style as well as the task content and the context. For example, students study differently for a math test that emphasizes procedural knowledge, as compared with a Spanish test, which emphasizes memorization of vocabulary. When students recognize the purpose and benefits of using strategies for multistep tasks, they are more willing to personalize strategies so that they can apply these to different academic tasks across content areas and across the grades (see Table 25.5).

Intervention research has shown that explicit and highly structured metacognitive instruction benefits all students and is essential for the academic progress of students with learning and attention difficulties (Deshler, Ellis, & Lenz, 1996; Deshler & Schumaker, 1988; Meltzer, Katzir, et al., 2004; Paris, 1986; Pearson & Dole, 1987; Rosenshine, 1997; Swanson, 2001; Swanson & Hoskyn, 1998, 2001). Comparisons of different

Table 25.5 Executive function processes and academic performance

Goal-setting
- Identifying short-term and long-term academic goals
- Planning and allocating time to the many steps involved in meeting these goals (e.g., completing daily homework, studying for tests systematically)

Shifting flexibly/cognitive flexibility
- Shifting flexibly from the major themes to the relevant details when reading, writing, or studying
- Using outlines such as graphic organizers or linear outlines to get "unstuck" and to focus on the major concepts when writing papers or completing projects
- Shifting between operations or between words and numbers for math computation or word problems

Organizing
- Organizing concepts using strategies (e.g., summarizing key ideas using strategy cards, graphic organizers, or Triple Note Tote) rather than rereading the text over and over
- Organizing materials such as class notes, textbooks, study guides
- Organizing work space to reduce distractions and clutter

Prioritizing
- Prioritizing by allocating more time and effort to lengthy papers, major projects or studying
- Figuring out which details are critical and which details can be ignored when reading, taking notes, or writing essays
- Estimating how much time to spend on reading and research vs. writing for papers, projects

Accessing working memory
- Chunking information to memorize and mentally manipulate it for multi-step tasks e.g., mental computation, note-taking
- Studying strategically to connect concepts so that critical information is retained over time
- Accessing critically important details for solving complex math problems
- Remembering key concepts while taking notes during classes
- Remembering to bring necessary books and materials from school to home and back again

Self-monitoring
- Using personalized error checklists to correct errors when writing papers, taking tests, or doing homework

Adapted, Meltzer (2010). Copyright 2010, The Guilford Press. Reprinted by permission

interventions highlight several important principles of this instruction:
- Strategy instruction should be embedded in the curriculum (Deshler et al., 1996; Ellis, 1993, 1994).
- Strategies should be taught explicitly and systematically, using scaffolding and modeling, while providing time for practice.
- Students' motivation and self-understanding should be addressed to ensure generalized use of strategies (Deshler & Shumaker, 1986; Deshler, Warner, Schumaker, & Alley, 1983; Meltzer, 1996; Meltzer, Roditi, Houser, & Perlman, 1998; Paris & Winograd, 1990; Pressley, Goodchild, Fleet, Zajchowski, & Evans, 1989).

Systematic and consistent strategy instruction should address the core executive function processes (see Table 25.5).

The following sections focus on specific strategies for addressing each of these executive function processes, namely, goal-setting, cognitive flexibility/shifting, organizing and prioritizing, accessing working memory, and self-monitoring/self-checking.

Goal-Setting

Goal-setting refers to the ability to set specific, realistic objectives that can be achieved within a defined period of time. Goal-setting also involves the selection of goal-relevant activities, effective and efficient strategy use, focused effort, as well as persistence. Goal-setting and planning help students to understand the task objectives, visualize the steps involved in accomplishing the task, and organize the time and resources needed to complete the task. When students set their own goals, they show greater commitment and are more motivated to attain these goals (Schunk, 2001; Winne, 1996, 2001; Zimmerman, 2000; Zimmerman & Schunk, 2001). Goal-setting also enhances self-efficacy, achievement, and motivation (Schunk, 2001). Krishnan, Feller, and Orkin (2010) emphasize that goal-setting requires students to:
- Understand their learning strengths and weaknesses as well as their learning profiles
- Understand the "big picture" and envision the end point of a task
- Value the task
- Recognize that goals need to be attainable

Students who are able to set goals and to shift from "the top of the mountain to the bottom and back" (Meltzer, 2007) are usually more successful with the complex tasks that are typical of our 21st century schools. In contrast, students with poor self-understanding of their learning profiles often fail to set short-term and long-term goals. This lack of direction often compromises their academic performance (Krishnan et al., 2010; Stone & Conca, 1993; Swanson, 1989; Torgesen, 1977).

Teaching Goal-Setting and Planning Strategies

Beginning in the early grades, students can be taught effective goal-setting and planning strategies. Teachers and parents can model the planning process by making daily schedules, using calendars, and setting agendas. Younger students can be taught strategies for planning their homework, long-term projects, study time, and activities. These strategies are even more important in the middle and high school grades. In these higher grades, students are required to understand the goals of their assignments and to plan their study time, as well as their approach to long-term projects and papers. Time management is also critically important, as students are required to juggle multiple deadlines for different ongoing assignments and projects. They often underestimate the amount of work involved in major projects and open-ended tasks, and they need strategies for breaking down tasks into manageable parts.

Time management strategies also help students to build goal-oriented schedules by planning their homework and study time after school when there is less structure. Weekly and monthly calendars help to impose structure and build self-monitoring strategies so that students can track deadlines for long-term projects and can pace themselves to complete their assignments. These goal-setting and time management strategies are critical for promoting independent learning (Hughes, Ruhl, Schumaker, & Deshler, 2002; Krishnan et al., 2010; Sah & Borland, 1989).

Cognitive Flexibility/Shifting

Cognitive flexibility, or the ability to think flexibly and to shift approaches, is a critically important executive function process that is often challenging for students, and is especially difficult for students with learning and attention difficulties (Meltzer, 1993; Meltzer & Krishnan, 2007; Meltzer & Montague, 2001; Meltzer, Solomon, Fenton, & Levine, 1989). The ability to adapt to unfamiliar or unexpected situations, to shift mindsets and problem-solving approaches, and to integrate different representations, develops across the lifespan and varies across individuals (Brown, 1997; Cartwright, 2008a, 2008b, 2008c; Deák, 2008; Dweck, 2008; Elliot & Dweck, 2005). In fact, developmental changes from childhood into adulthood influence children's ability to manage the cognitive complexity of academic tasks and to process different components simultaneously (Andrews & Halford, 2002; Cartwright, 2008a, 2008b; Zelazo & Müller, 2002). Typically, students in the early elementary grades have a more limited understanding of the importance of using a range of different approaches than do middle and high school students. As students advance into the higher grades, their ability to learn new concepts is often connected to their willingness to abandon previously successful approaches and to shift flexibly to alternative methods (Cartwright, 2008a, 2008b).

Across all the academic domains, students' motivation, interest, passion, and emotional mindsets also influence their willingness to try using different approaches and to shift flexibly from one approach to another, rather than continuing to rely on the same approach to tasks (Alexander, 1998; Paris, Lipson, & Wixson, 1983; Shanahan & Shanahan, 2008). Motivation, topic knowledge, and strategy use interconnect to produce improvements in content areas such as history or science (Alexander, 1998). For example, as students learn more about a topic (e.g., the Iraq War), they may be more willing to make the effort to use a three-column note-taking strategy to separate main ideas and details; in turn, flexible strategy use increases students' interest in completing the many different steps involved in creating an outline and then writing a long paper (e.g., a paper about the Iraq War). In this regard, Zelazo and colleagues have differentiated between purely cognitive or "cold" tasks that have no emotional content (e.g., math computation) and tasks that are affected by the student's social and emotional mindset or what they term "hot" tasks (e.g., remembering information about the Iraq war by linking it with a personal experience such as the memory of a friend or relative who was wounded fighting in Iraq) (Zelazo & Müller, 2002; Zelazo, Müller, Frye, & Marcovitch, 2003). They emphasize that students' cognitive flexibility is frequently linked to their success on "hot" and "cold" tasks.

Overall, this ability to shift approaches and to synthesize information in novel ways is essential for effective reading, writing, math problem-solving, note-taking, studying, and test-taking. Accurate and efficient *reading decoding* requires students to flexibly coordinate the letter-sound relationships with the meanings of printed words (Cartwright, 2008a, 2008b, 2008c). In other words, students need to recognize the importance of what Gaskins (2008) refers to as "crisscrossing the landscape" in order to select decoding approaches that fit the text. *Reading comprehension* requires students to process the meaning of text, flexibly access their background knowledge, recognize the purpose or goal of reading (Cartwright, 2008a, 2008b, 2008c), and monitor their own comprehension (Block & Pressley, 2002; Pressley & Afflerbach, 1995). When reading text that incorporates complex or figurative language, students must shift between the concrete and the abstract, between the literal and the symbolic, and between the major themes and relevant details. Furthermore, reading for meaning taxes students' ability to flexibly manage many different types of linguistic information at the word level, sentence level, and paragraph level (Brown & Deavers, 1999; Goswami, Ziegler, Dalton, & Schneider, 2001, 2003). Similarly, when *writing*, students must shift between their own perspective and that of the reader and between the important concepts and supporting information. In the *math domain*, students' understanding of concepts, computational procedures, and word problems depends on their cognitive flexibility. Students are required to shift from the words and sentences in math problems to the numbers, operations, algorithms, and equations needed to solve the problems (Roditi & Steinberg, 2007). They also need to learn how and when to shift problem-solving schemas so their final calculations are accurate and logical (Montague & Jitendra, 2006).

In *content area subjects*, including science and history, students are required to differentiate main ideas from details in their textbooks. Students' understanding of the material in these textbooks depends on their use of context clues to shift flexibly among the different possible meanings in the words and phrases. Similarly, learning a *foreign language* requires a significant amount of flexible thinking, as students are challenged to shift back and forth between their native language and the language they are learning. Finally, *studying and test-taking* require students to shift among multiple topics or problem types as they are often presented with information that is formatted differently from the way in which they learned or studied.

For students who struggle to shift flexibly between perspectives and to process multiple representations easily, academic tasks often become progressively more challenging as they advance beyond the first few grades in school. Furthermore, these students experience mounting difficulty as the curriculum demands increase in complexity and require them to interpret information in more than one way, change their approach when needed, and choose a new strategy when the first one is not working (Westman & Kamoo, 1990).

Strategies for Improving Cognitive Flexibility

As was discussed above, students need a variety of opportunities to shift mindsets, to think flexibly, and to use their knowledge in a number of different ways (Bereiter & Scardamalia, 1993; Dweck, 2008). Therefore, it is important to embed strategies for teaching cognitive flexibility into different facets of daily life and to create classrooms where students are given opportunities to solve problems from different perspectives across the grades and content areas. This ability to approach situations and tasks flexibly helps students to shift more easily from the "big picture" to the details in social and academic settings.

A variety of instructional approaches can be used to promote flexible thinking across different settings and content areas. First, when teaching emphasizes problem-solving and critical thinking, students are required to think flexibly about ways in which their solutions could lead to different possible outcomes (Sternberg, 2005). Secondly, when teaching encourages peer discussion and collaborative learning, students are exposed to multiple viewpoints (Yuill, 2007; Yuill & Bradwell, 1998). Students can therefore be challenged to approach problems from the perspectives of their peers.

Strategies for shifting flexibly can also be embedded into daily activities at home and in school. Activities with jokes, riddles, word categories, and number puzzles can help students to practice using flexible approaches to language interpretation and number manipulation. The efficacy of using jokes to teach flexible thinking to students in the early grades has been demonstrated in an interesting series of studies (Yuill, 2007; Yuill & Bradwell, 1998). They found that an explicit focus on the ambiguous language in jokes generalized to reading comprehension as students were required to think about and analyze language in different ways.

In the area of *reading comprehension*, selected strategies can be used to teach students to shift flexibly between and among major themes and relevant details based on the goals and content requirements of the reading tasks (Meltzer & Bagnato, 2010; see Table 25.6).

As one example, students' interpretations of text can be improved by presenting different scenarios that teach them to analyze language and to shift among different interpretations. When they come across words or sentences that do not make sense to them, they can be taught to stop reading and to ask themselves the following questions:
- Does the word have more than one meaning?
- Can the word be used as both a noun and a verb?
- Can I emphasize a different syllable in the word to give it a different meaning?
- Can I emphasize different parts of the sentence to change its meaning?
- Does the passage contain any figurative language, such as metaphors or expressions that may be confusing?

Table 25.6 Flexible thinking strategies for reading comprehension

Reading comprehension	
Shifting between "big ideas" and supporting details	Strategies for differentiating main ideas vs. important details vs. less relevant details. Critical for summarizing and studying
	Strategies for identifying multiple-meaning words using context clues, noun-verb clues and syllable accents to shift flexibly among the different possible meanings
	Three-column notes (e.g., Triple Note Tote) which requires shifting from main ideas or core concepts to supportive details

From Meltzer and Bagnato (2010). Copyright 2010 by The Guilford Press. Reprinted by permission

Similar shifting strategies can be used to improve *written language*. Three-column note-taking systems and graphic organizers which make explicit connections between the main ideas and supporting details, often help writers to shift more fluidly between the two. Students record the major themes, core concepts, or key questions in the first column, the relevant details in the second column, and a memory strategy in the third column. The last column can also include a picture to help students memorize the information. A range of templates and graphic organizers can also be used for helping students to shift flexibly between the main ideas and the supporting details (see Table 25.7).

Flexible thinking can also be promoted by teaching students to incorporate a counterargument when they summarize text. Introductory or concluding statements that challenge an argument, but are weighted more heavily in the direction of the writer's opinion, also encourage flexible thinking.

In the area of *math*, students often get stuck trying to solve math problems in one way, when there may be an easier or more efficient way to find a solution. Similarly, students may have learned a particular format for math problem-

Table 25.7 Flexible thinking strategies for written language

Written language	
Shifting between "big ideas" and supporting details	Graphic organizers for sorting main ideas vs. supportive details
Shifting from the top to the bottom of the mountain and back again	Templates for focusing on major themes or thesis statements, relevant details and conclusions
	Models for shifting from the main ideas to supporting details
	Personalized checklists for differentiating between relevant and irrelevant details

From Meltzer and Bagnato (2010). Copyright 2010 by The Guilford Press. Reprinted by permission

Table 25.8 Flexible thinking strategies for math

Math problem-solving	
Shifting from the math concepts to computational details & back	Generate math language for each operation (i.e., difference, less, take away = subtraction)
	Shift from the language embedded in word problems to computational details and back again
	Focus on the meaning of math problems vs. the operations and calculations
	Within operations (e.g., long division), shift from division to subtraction, etc.
	Estimate the answers to word problems (big picture) and compare their solutions with their estimates
	Ask themselves: "Does this make sense?" by comparing their final calculations with their estimates

From Meltzer and Bagnato (2010). Copyright 2010 by The Guilford Press. Reprinted by permission

Table 25.9 Flexible thinking strategies for studying and test-taking

Summarizing, note-taking, long-term projects	
Shifting from concepts to details & back	Using concept maps to focus on the major concepts or "big picture" by visualizing themselves standing at the "top of the mountain" and then shifting to the bottom of the mountain for the details
Shifting due dates	Triple Note Tote strategy or strategy cards for use in each chapter/unit to shift from main ideas to details and back again to create a study plan for tests
	Monthly and weekly calendars to shift between short-term homework due immediately vs. long-term projects

From Meltzer and Bagnato (2010). Copyright 2010 by The Guilford Press. Reprinted by permission

solving while in class, but they may have trouble recognizing similar problems when these are presented differently on tests. Furthermore, while students can often solve problems of the same type that are grouped together for homework practice, they may have difficulty shifting among multiple problem types in test situations. Cognitive flexibility can be enhanced when students recognize that specific problems require them to shift from one operation (e.g., addition) to a different operation (e.g., subtraction) (see Table 25.8 for strategies). Similarly, they can ask themselves specific questions:

- Do I know more than one way to solve the problem?
- Does this look similar to anything I have seen before?
- Is this problem the same or different from the problem before this?

Studying for tests and quizzes requires flexible thinking on many different levels. Students need to extract information from a variety of sources, including textbooks, homework assignments, and class notes (see Table 25.9). Memorizing the specific details and integrating them with the larger concepts also requires cognitive flexibility. Students also need to study differently for different kinds of test formats, even within the same subject area. For example, for a multiple-choice test in history, students need to focus on details and facts. For an essay question, students need to shift away from the facts and details to the topic or major concepts and to "tell the story" embedded in the content rather than simply cramming hundreds of factual details. Finally, students need to learn the importance of using different study strategies in different subject areas. For example, reviewing the major ideas in the class textbook will help students to prepare for a history test, whereas it is often more beneficial to review classwork and past homework assignments for math. Thus, students need to be flexible in the study strategies they select for tests and quizzes.

Organizing and Prioritizing

The way my mind works with that liquefied gobble of dots, my notes would look scattered on a page. One of the most useful strategies I learned was multi-column notes. With this system, I learned to make a hierarchy of notes and have it structure around itself and relate to things. This structure helped me to study and to write long papers (Brandon, college graduate).

Organization, or the ability to systematize and sort information, is an executive function process that underlies most academic and life tasks. Students need to learn strategies for systematically organizing their time, their materials, and also their ideas. They also need to learn how to apply these strategies to their writing, note-taking, studying, and test preparation. These executive function strategies assume greater importance in late elementary school when students are presented with an increasingly large volume of detailed information that they are required to organize for effective learning. How well they learn and remember this information depends on how effectively they use strategies for organizing and prioritizing the concepts and details so that working memory is less cluttered (Hughes, 1996). While many students successfully participate in class lessons and accurately complete structured homework assignments, they may have more difficulty with independent, open-ended tasks. Reading and note-taking tasks, studying for tests, and completing writing assignments all require students to impose their own structure on the information. When organizational strategies are taught systematically in the context of these school assignments, students are more likely to generalize these strategies and to succeed academically (Krishnan et al., 2010).

Success, in turn, increases students' motivation to use these strategies independently and to generalize across different contexts (Meltzer, 1996, 2010; Swanson, 1999).

Teaching Organizing and Prioritizing Strategies

Strategies for organizing and prioritizing information underlie efficient *reading comprehension*. Strategies such as templates, thinking maps, and graphic organizers provide a structured format for helping students to read for meaning, extract major themes, and relate new with known information (Kim, Vaughn, Wanzek, & Shangjin, 2004; Mayer, 1984). Graphic organizers are also effective for improving students' reading comprehension across a wide range of subject areas including language arts, science, and social studies (Bos & Anders, 1992; Bulgren, Schumaker, & Deschler, 1988; Ritchie & Volkl, 2000). Most importantly, these organizational strategies can be taught across multiple grade levels from elementary school through high school (Krishnan & Feller, 2010; Ritchie & Volkl, 2000; Scanlon, Duran, Reyes, & Gallego, 1992; Horton, Lovitt, & Bergerud, 1990).

Similarly, reading comprehension and written language can be improved using two- or three-column note-taking systems instead of the traditional, linear format. The structure imposed by a three-column note-taking system guides students to ask themselves active questions about the text they are reading. This format also encourages students to find the main ideas, "chunk" information into manageable parts, predict test questions, and develop strategies for memorizing the information (see Fig. 25.6). As is evident from Fig. 25.6, the Triple Note Tote strategy (BrainCogs, ResearchILD & FableVision, 2003) helps students to organize information by differentiating the major concepts and details. The student records the main idea or a key question in the first column, summarizes the important details in the second column, and records a memory strategy in the third column.

In our 21st-century schools, *written language is* heavily emphasized and standards-based tests, including the SAT, now incorporate a required

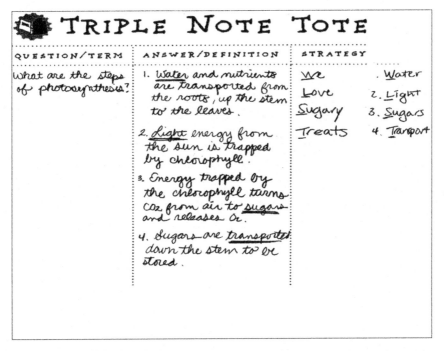

Fig. 25.6 The triple-note-tote strategy: a system for organizing, prioritizing, and memorizing information (BrainCogs, ResearchILD & FableVision, 2003)

writing section. As a result of this change, students from late elementary school onwards are given lengthy writing assignments, long-term projects, and essay tests that rely on executive function processes. For many students, writing can be an overwhelming process because it requires the coordination of numerous cognitive and executive function processes including organization, planning, memorizing, generating language, and editing (Flower et al., 1990; Flower, Wallace, Norris, & Burnett, 1994). Many students struggle to organize their ideas for writing, and they need the writing process to be broken down explicitly with organizers and templates that match both the goals of the assignment and their learning profiles (Graham & Harris, 2003; Harris & Graham, 1996). These strategies help students to break down writing tasks into manageable parts so they can monitor their own performance (Bruning & Horn, 2000). For example, the BOTEC strategy (Essay Express, ResearchILD & FableVision, 2005) uses a mnemonic and visual image to jog students' memory about the steps in the writing process when they are completing homework, studying, or taking tests (see Fig. 25.7). In other words, they are required to Brainstorm, Organize their thoughts, generate a Topic sentence or Thesis statement, Elaborate by providing Evidence, and draw a Conclusion. Figure 25.7 illustrates a template for using the BOTEC strategy to organize and prioritize ideas during the writing process.

Similar organizational strategies are also critically important for *note-taking*. This is a complex process that requires students to focus on multiple processes simultaneously including listening, organizing, and prioritizing the information while they write down the critical ideas (Kiewra et al., 1991). Many students read their textbooks and articles without taking notes, or they take notes in a random, scattered way that does not reduce the information load. Other students have difficulty deciding which information should be recorded and they struggle to separate the key concepts from the supporting details (Hughes, 1991; Hughes & Suritsky, 1994; Suritsky, 1992). Students' academic performance generally improves when they use the organizational strate-

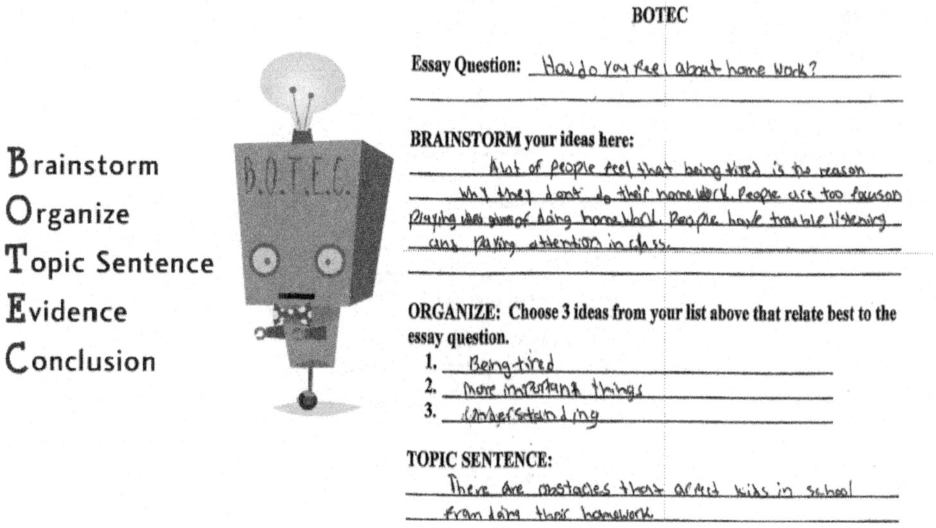

Brainstorm
Organize
Topic Sentence
Evidence
Conclusion

Fig. 25.7 The BOTEC strategy for generating written language (Essay Express, ResearchILD and FableVision, 2005)

gies discussed above for taking notes, studying, or completing tests (Boyle, 1996, 2001; Boyle & Weishaar, 1999; Katamaya & Robinson, 2000; Lazarus, 1991).

Organization and planning improve when students are required to complete strategy reflection sheets that incorporate structured questions and a multiple-choice format. These strategy reflection sheets promote metacognitive awareness, encourage students to use strategies systematically, and remind them to check and edit their work (see Fig. 25.8a, b). When students are given credit by their teachers for using these strategies, they are more likely to make the effort needed to continue this process. For example, when grades for homework and tests include points for completing these strategy reflection sheets, teachers promote these habits of mind. In other words, metacognitive awareness and effective strategy use are promoted when teachers make strategy use *count* in the classroom.

Accessing Working Memory

Working memory refers to the ability to store information for short time periods while simultaneously manipulating the information mentally, (e.g., holding the main themes in mind while sorting through the details, or calculating a math problem mentally). Working memory is a critically important process that helps students to focus, direct their mental effort, and ignore distractions in order to accomplish tasks (de Fockert, Rees, Frith, & Lavoie, 2001; Swanson, 1999; Tannock, 2008). In fact, Baddeley (2006), Swanson & Sáez (2003) have proposed that working memory often functions as the central executive that directs all other cognitive processes, including the student's ability to inhibit impulses, shift attention, and direct effort to the task. Working memory, therefore, plays a critical role in listening comprehension, reading comprehension, oral communication, written expression, and math problem-solving, as well as efficient and accurate long-term learning (Swanson & Sáez, 2003).

From fourth grade onwards, academic tasks rely increasingly on these working memory processes. Consequently, strategic students are generally more successful with tasks that require them to focus on multiple processes simultaneously such as following directions, responding to oral questions, and completing multistep instructions (Kincaid & Trautman, 2010). Reading comprehension and written language are also heavily

a

```
                    Strategy Reflection Sheet

    What strategies did you use for this writing assignment?
    ___ BOTEC                        ___ Personalized Checklist
    ___ Mapping and Webbing          ___ Sentence Starters
    ___ Graphic Organizer            ___ Other:
    ___ Linear Outline               ___ Introduction Template


                                              © Research ILD 2004
```

b

```
                    Strategy Reflection Sheet

    What strategies did you use for this writing assignment?
    _____
    _____

                                              © Research ILD 2004
```

Fig. 25.8 (**a**) Strategy reflection sheet for writing: structured questions that scaffold the writing process. (**b**) Strategy reflection sheet for writing: open-ended questions

dependent on working memory. In these areas, students need to remember and manipulate multiple details such as spelling and punctuation while simultaneously focusing on remembering the main ideas, organizing ideas in their minds while they read, prioritizing important information, and figuring out which details to ignore. Young students may also need to think about handwriting and accurate letter formation, skills that may not yet be automatic for them. Similarly, summarizing, taking notes, and studying for tests all require students to focus on multiple processes simultaneously and to remember key ideas, formulate notes while listening, and identify major themes while writing (Kincaid & Trautman, 2010).

Memorizing information in the classroom is heavily dependent on students' ability to focus and sustain their attention in order to make connections, retain information, and retrieve relevant details (Tannock, 2008). In fact, attention and memory are so strongly linked that the two processes are often viewed as part of the same executive process (Swanson & Sáez, 2003; see chapters in this volume). To remember, retain, and retrieve information, students benefit from learning strategies for sustaining their attention, attaching meaning to information, chunking information to reduce the memory load, as well as rehearsal and review (Kincaid & Trautman, 2010). When students are able to make meaningful associations, they are more successful with transfer of information into long-term memory and later retrieval (Mastropieri & Scruggs, 1998).

Teaching Working Memory Strategies

Working memory strategies are interconnected with strategies for organizing and prioritizing complex information by reducing the memory load. Mnemonics comprise one of the most effective methods for chunking information and retaining important details so that information can be mentally manipulated in working memory (Mastropieri & Scruggs, 1991, 1998; Scruggs & Mastropieri, 2000). Mnemonics help students to connect new information to what they already know and to make meaningful connections to seemingly unconnected information (Carney, Levin, & Levin, 1993). Different types of mnemonics improve retention of information and enhance working memory, in particular, keywords, pegwords, acronyms, acrostics, and visuals (Mastropieri & Scruggs, 1991, 1998; Scruggs & Mastropieri, 2000). For example, when students are required to remember the states and their capitals by region, crazy phrases help them to organize, sequence, and chunk the information so that there are fewer details to memorize.

Some students prefer to use visual strategies, such as personalized diagrams, cartoons, graphic organizers, and templates (Kincaid & Trautman, 2010). Mnemonics are often embedded within these organizers to further enhance their effectiveness. Chants, rhymes, and songs are effective for those who rely on verbal or auditory strategies to memorize. Students need time to practice and rehearse their memory strategies (Harris, Graham, Mason, & Friedlander, 2008). As students learn and practice memory strategies that are modeled by adults, it is important to encourage students to create their own memory strategies that match their individual learning styles (see Kincaid & Trautman, 2010, for more details and specific memory strategies in different academic areas).

Given the heavy learning memory load imposed in our 21st-century information-driven schools, and the emphasis on working memory and mental manipulation, it is particularly important to teach memory strategies explicitly to improve students' ability to retain and retrieve facts, processes, and concepts. As is emphasized by Kincaid and Trautman (2010), educators need to help students to learn how to prioritize and select information to be memorized to reduce the load on working memory. Most importantly, students need to be given sufficient time to process and practice memory strategies, and to develop their own personalized strategies for remembering challenging information.

Self-Monitoring and Self-Checking

Self-monitoring refers to the ways in which learners manage their cognitive and metacognitive processes to track their own performance and outcomes (Zimmerman, 1998, 2000; Zimmerman & Kitsantas, 1997; Zimmerman & Schunk, 2001). When students self-monitor, they review their progress towards their goals, evaluate the outcomes, and redirect their efforts when needed. The ability to self-monitor depends on students' metacognitive awareness as well as their flexibility in shifting back and forth from the end product of their efforts to the goals of the tasks. Therefore, students' use of self-monitoring strategies depends on their ability to recognize when, how, and why to use specific strategies, to evaluate and revise their strategy use, and to continually adjust their use of strategies based on the task demands (Bagnato & Meltzer, 2010). Many students, especially students with learning and attention problems, have difficulty reflecting, monitoring their own learning, and evaluating the connections between their effort, strategy use, and performance. As they focus their effort on reading, writing, math problem-solving, and content learning, they may struggle to monitor their attention and performance and may have difficulty shifting among a range of problem-solving approaches or strategies that are available to them (Klingner, Vaughn, & Boardman, 2007; Montague, 2003). Students therefore need systematic, structured, and scaffolded instruction in using self-monitoring strategies flexibly so that they can become independent learners who do not need the assistance of others to complete reading, writing, math, or related tasks successfully (Graham & Harris, 2003; Reid & Lienemann, 2006).

Teaching Self-Monitoring and Checking Strategies

Numerous studies have shown that teaching self-monitoring strategies systematically to students can improve their performance significantly (Graham & Harris, 2003; Harris & Graham, 1996; Reid, 1996; Reid & Harris, 1993; Shimabukuro, Prater, Jenkins, & Edelen-Smith, 1999). Explicit, structured teaching encourages students to slow down and to allocate the necessary time to spiral back and

Fig. 25.9 STOPS: a personalized editing checklist for upper elementary and middle school students. Reprinted with permission

Table 25.10 Guide for making revisions to a five paragraph essay

Question	Yes or No?	Action steps
Is there a thesis statement?		Add 1–2 sentences that summarize your viewpoint or main idea
Is the essay organized into paragraphs?		Divide essay into introduction, 3 body paragraphs and conclusion
Is there an introduction?		Write a paragraph that introduces your topic and includes a thesis
Does each body paragraph have a topic sentence?		Add a sentence that introduces the topic of each paragraph
Does the essay contain sufficient supporting details?		Add more quotes, facts or specific examples to body paragraphs
Is there a conclusion?		Add a paragraph that summarizes your opinion or main idea
Does the essay flow well and read smoothly?		Use transition words to link sentences and paragraphs
Does the essay contain colorful and interesting vocabulary?		Replace common words with ones that are more vivid and unique

In Meltzer (2010). Copyright 2010 by The Guilford Press. Reprinted by permission

forth so that they can check the task demands and their own output (Meltzer, Sales-Pollica, & Barzillai, 2007; Reid & Lienemann, 2006).

In the *writing* domain, self-monitoring strategies are essential as students need to shift mindsets from that of the "writer" to that of the "editor" so that they can identify their own errors. A "one size fits all" generic editing checklist is often not effective, as different students make different types of errors in their writing (Bagnato & Meltzer, 2010). While one student may consistently make spelling errors but have no difficulty with organization, another may have the opposite profile. By developing personalized checklists and acronyms for checking particular types of assignments, students know what to check for and make fewer errors (Bagnato & Meltzer, 2010). For example, the acronym "STOPS" was developed by a sixth grader to help check his writing for errors he commonly made (see Fig. 25.9).

This acronym reminded him to check his written work for Sentence structure, Tenses (i.e., not using present, past, and future tenses in one paper), Organization of ideas, Punctuation, and Spelling. Self-monitoring and checking are often easier if students edit their work using a different colored pen, read their written work aloud, or if they write the original draft on a computer, print it out, and edit a hard copy. For example, for persuasive writing, students often benefit from explicit instruction to monitor their inclusion of the basic structural components of writing, such as topic sentences, supporting details, and paragraph endings (Graham, 1990).

Students often realize that their writing is weak but they do not know how to revise their writing to improve the content, structure, and organization. Even when they are given a rubric which outlines the expectations for the assignment, they may have difficulty determining whether their writing meets the criteria. They benefit from a guided process for analyzing several of their writing samples to determine their most common mistakes and using this process to develop personalized editing checklists (Bagnato & Meltzer, 2010). They also need general systems that help them to improve (Bagnato & Meltzer, 2010). Table 25.10 provides

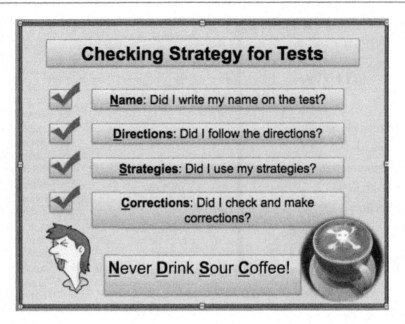

Fig. 25.10 Self-checking strategy for tests. From Meltzer, Reddy, Brach, Kurkul, and Basho, 2011. Reprinted with permission

one possible essay revision guide to help students evaluate their writing with respect to content, structure, and organization as well as ideas for editing and correcting their work (Bagnato & Meltzer, 2010).

In the *math* area, when students are given a word problem to solve, they benefit from comprehensive instructional routines such as Montague's "Solve It!" (Montague, Warger, & Morgan, 2000). This teaches them strategies for estimating the answer, computing, and checking to verify the solution. Students also need to be taught self-regulation strategies such as asking themselves questions as they go through the steps of problem-solving as well as monitoring their own performance systematically (Meltzer & Montague, 2001; Montague, 2003; Montague et al., 2000). Explicit and systematic instruction therefore helps students to access, apply, and regulate their use of strategies.

Finally, *self-monitoring* and *checking strategies for tests* are critically important. Students benefit from developing personalized strategies and checklists for editing their work before handing in their tests. Most students, and especially students with learning and attention difficulties, need explicit instruction focused on how to check their work and what errors to check for. Figure 25.10 illustrates a general strategy for checking tests that incorporates a visual image for those who more easily remember visual information as well as a "crazy phrase" for those students who more easily remember verbal information. Students can use this strategy as a model for developing a personalized checking strategy. Personalized self-checking cards and mnemonics to remember the core ideas are most beneficial when students embrace these as their own (Bagnato & Meltzer, 2010)

Overall, effective self-monitoring requires students to reflect on their progress towards a goal, to select strategies that work, and to alter strategies that are not working (Bagnato & Meltzer, 2010). The overall quality of students' academic work improves when they are able to recognize the value of reviewing their work and shifting mindsets so that they know what to look for and how to shift flexibly from the major themes to the details (Bagnato & Meltzer, 2010; Meltzer & Basho, 2010). As will be discussed in the next section, students' motivation and emotional mindsets frequently affect their willingness to make the often superhuman effort needed to stop, reflect, check, and correct multiple drafts of their work, processes that are critically important for long-term academic and life success.

Emotional Self-Regulation

Students' attention and their ability to engage actively in the learning process are associated with their ability to regulate their emotions in and outside the classroom (Brooks, 1991; Stein, 2010; Tangney, Baumeister, & Boone, 2004). The effects of emotion on the learning process range along a continuum. Specifically, extreme emotional reactions (e.g., anxiety, anger) often disrupt students' attention and ability to stay on task as well as their ability to learn and remember new information (Goldberg, 2001; Stein, 2010). In contrast, moderate emotional arousal has a positive influence on students' attention and executive function processes including working memory, cognitive flexibility, and inhibition (Gross, 2007; Stein, 2010). More specifically, the relationship between anxiety and performance is characterized by an inverted U-shaped curve, e.g., test performance is often poor when students either are not anxious and have consequently not studied or are excessively anxious which interferes with attention, working memory, and overall performance (Goleman, 1995). Similarly, negative moods disrupt attention, concentration, memory, and processing speed.

In the classroom, students' self-understanding as well as their ability to regulate their emotions are important processes underlying efficient learning. As students develop strategies for regulating their emotional responses in the classroom, they more easily attend to instructions, sustain their effort, and curb their frustrations in response to difficult tasks (Stein, 2010). They also learn how to collaborate with peers and to adjust their behavior to fit the classroom's "culture" and routines. When students regulate their emotions, they can more easily focus attention on the academic content rather than their feelings. More specifically, they can use self-talk to encourage themselves when anxious, ask for help when needed, and express their feelings in socially acceptable ways (Stein, 2010). As is discussed by Stein (2010), a proactive approach to addressing students' emotional regulation in the classroom involves three critical components:

1. Understanding each student's learning profile and emotional vulnerabilities
2. Understanding what kinds of triggers may upset each student
3. Developing individualized prevention and intervention plans for vulnerable students.

Prevention approaches for teachers and parents focus on helping students to avoid frustration and get started by providing structure, breaking down tasks into smaller steps, giving alternative assignments or test formats if needed, or providing flexible due dates (Stein, 2010). In addition, these students benefit from being told ahead about upcoming changes, transitions, challenges, or requests for participation in classes. Intervention approaches focus on avoiding judgment, anger, or blame, providing a supportive; collaborative perspective; and offering choices to students (e.g., safe place to reduce stress, quiet room for taking tests; see Stein, 2010, for more details).

Self-regulatory strategies such as these are particularly important for students with attention problems and nonverbal learning disabilities. These students depend on structured approaches and routines that help to reduce their emotional distractibility and impulsivity so that they can sustain their motivation and manage the many simultaneous demands of the classroom (Stein & Krishnan, 2007).

The *Drive to Thrive* and SMARTS Programs: Strengthening Executive Function Strategies with Peer Mentoring

The *Drive to Thrive* and SMARTS programs create a school and classroom culture where there is a shared understanding of the importance of building students' emotional self-regulation and executive function strategies, while promoting their effort, persistence, and self-understanding (Meltzer, Katzir, et al., 2004; Meltzer, Reddy,

Sales, et al., 2004; Meltzer, Reddy, Pollica, et al., 2004; Meltzer, 2010; Meltzer, Sales-Pollica, et al., 2007; Meltzer, Noeder, et al., 2007; Noeder, 2007). Classwork and homework consistently focus on the *how* of learning rather than only the final product. Students begin to value the *process* of learning as they are taught to shift flexibly during problem-solving and other academic tasks. As a result of using executive function strategies, students' grades gradually improve and they begin to view themselves as capable learners. Over time, there is an increase in students' willingness to use executive function strategies in different content areas.

The *Drive to Thrive* and SMARTS programs focus on building a cycle of academic success in all students through teacher training supplemented by a peer tutoring and peer mentoring system. Teachers are trained to create a culture of strategy use in their classrooms and to promote metacognitive awareness and strategy use in their students by embedding executive function strategies in their curriculum and daily teaching practices. The following principles guide the program (Meltzer & Basho, 2010; Meltzer, Katzir, et al., 2004; Meltzer, Reddy, Pollica, et al., 2004; Meltzer, Reddy, Sales, et al., 2004; Meltzer, Sales-Pollica, et al., 2007):

- Teachers understand and acknowledge the interactions among effort, strategy use, academic self-concept, and classroom performance as well as the cycle that builds persistence, resilience, and long-term academic success.
- Teachers foster metacognitive awareness and strategic mindsets in their students.
- Teachers acknowledge that effort is domain specific and that students may sometimes work hard in one content area (e.g., math) and not another (e.g., language arts).
- Teachers acknowledge the importance of peer mentoring and peer tutoring and they build time and resources into the school day for the purposes of implementing a program such as SMARTS (see below).
- Teachers acknowledge that peer mentoring and peer tutoring provide a powerful forum for helping students to understand their learning profiles, to develop metacognitive awareness, and to recognize the important roles of executive function strategies as well as effort and persistence.
- Students view themselves as part of a community of learners who can help one another through peer mentoring and peer tutoring (see below). Emotional self-regulation is also strengthened as part of this program.
- Students understand that executive function strategies and focused effort are important for academic success.
- Students recognize that persistence and determination are critical for fostering academic and life success.

One example of a school-based peer mentoring program is the recently developed *SMARTS* program (Gray, Meltzer, & Upton, 2008; Meltzer, Reddy, Brach, Kurkul, & Basho, 2011; Meltzer, Reddy, Brach, Kurkul, Stacey, & Boss, 2011). *SMARTS* is an acronym for **S**uccess, **M**otivation, **A**wareness, **R**esilience, **T**alents, and **S**trategies and each of these strands is a core component of the program. The *SMARTS* program focuses on promoting resilience and academic success by teaching executive function strategies and building metacognitive awareness and persistence in all students, and particularly in students with learning difficulties. Teachers are trained to implement the SMARTS curriculum which comprises three major components: the executive function strand, the motivation strand, and the self-concept strand. Thirteen strategies in the core executive function areas are taught over the course of the school year with an emphasis on:

- Increasing students' metacognitive awareness, self-understanding, and academic self-concepts
- Increasing students' effort and persistence in school as well as their motivation to engage in the learning process and to improve their academic performance
- Improving students' understanding and use of executive function strategies in six broad areas: goal-setting, organizing, prioritizing, using working memory, shifting flexibly, self-monitoring
- Promoting students' mentorship and leadership skills through peer mentoring

In addition, mentor-mentee pairs work together to learn and practice these executive function strategies, with mentors coaching their mentees and helping to build their self-confidence. To reinforce learning and application of these strategies, the SMARTS curriculum culminates in a project that focuses on improving students' engagement, motivation, strategy use, and effort (Meltzer, Reddy, Brach, Kurkul, Stacey, et al., 2011).

Findings from our recent SMARTS intervention studies with middle and high school students have highlighted the importance of strengthening students' self-understanding, cognitive flexibility, and awareness of the importance of shifting strategies (Meltzer, Reddy, Brach, Kurkul, & Basho, 2011; Meltzer, Reddy, Brach, Kurkul, Stacey, et al., 2011). Specifically, in one of our school-based studies, SMARTS students with higher cognitive flexibility scores were more goal-oriented, more persistent, and made greater effort in school (Meltzer, Reddy, Brach, Kurkul, & Basho, 2011; Meltzer, Reddy, Brach, Kurkul, Stacey, et al., 2011). These more flexible students also used more strategies in their schoolwork and were more organized. Classroom teachers rated these students as having stronger academic performance and as checking their work more frequently (Meltzer, Reddy, Brach, Kurkul, & Basho, 2011; Meltzer, Reddy, Brach, Kurkul, Stacey, et al., 2011). The social connections provided by peer mentoring increased students' engagement in the learning process as well as their goal orientation and motivation. Overall, students' cognitive flexibility, academic self-concepts, and goal orientation interacted to influence students' effort, persistence, and academic performance.

These findings have relevance for teachers and emphasize the importance of increasing students' self-understanding, knowledge of executive function strategies, and academic self-concepts. Together, these initiate a positive cycle in which students work harder, focus their effort, and use strategies effectively, resulting in improved academic performance (Meltzer & Basho, 2010; Meltzer, Sales-Pollica, et al., 2007). Stronger academic performance helps students to feel more engaged and therefore more invested in making the effort to use strategies in their classwork, homework, and long-term projects, the foundations of academic and life success.

Conclusion

Technology has had a significant impact on the pace of the classroom curriculum and there is greater emphasis on teaching students to problem-solve flexibly and to organize, prioritize, and self-monitor. As a result, executive function processes have assumed increasing importance over the past decade and need to be taught systematically. When teachers and parents build an executive function culture in their classrooms and their homes, they empower students to learn *how* to learn. When schools and families foster effort, persistence, and executive function strategies, students develop self-confidence, resilience, and a strong work ethic, the gateways to academic and life success in the twenty-first century.

Acknowledgments A special thanks to a number of colleagues, staff, and interns for their excellent suggestions and help with the technical details involved in the preparation of this chapter, in particular:

Abigail DeMille, Sage Bagnato, Laura Pollica, Ranjini Reddy, Julie Sayer, Anna Lavelle, Lauren Depolo, and Thelma Segal.

References

Alexander, P. A. (1998). The nature of disciplinary and domain learning: The knowledge, interest, and strategic dimensions of learning from subject matter text. In C. R. Hynd (Ed.), *Learning from text across conceptual domains* (pp. 263–286). Mahwah, NJ: Erlbaum.

Anderson, P. (2002). Assessment and development of executive functioning (EF) in childhood. *Child Neuropsychology, 8*(2), 71–82.

Anderson, V., Rani Jacobs, J., & Anderson, P. (Eds.). (2008). *Executive functions and the frontal lobes: A lifespan perspective*. New York: Taylor and Francis.

Andrews, G., & Halford, G. S. (2002). A cognitive complexity metric applied to cognitive development. *Cognitive Psychology, 45*, 153–219.

Baddeley, A. (2006). Working memory, an overview. In S. Pickering (Ed.), *Working memory and education* (pp. 3–26). Massachusetts: Academic.

Bagnato, J. S., & Meltzer, L. (2010). Self-monitoring and self-checking: The cornerstones of independent living. In L. Meltzer (Ed.), *Promoting executive function in the classroom* (pp. 160–174). New York: Guilford Press.

Barkley, R. A. (1997). *ADHD and the nature of self-control*. New York: Guilford Press.

Bereiter, C., & Scardamalia, M. (1993). *Surpassing ourselves: An inquiry into the nature and implications of expertise*. Chicago: Open Court.

Bernstein, J., & Waber, D. (2007). Executive capacities from a developmental perspective. In L. Meltzer (Ed.), *Executive function in education: From theory to practice* (pp. 39–54). New York: Guilford Press.

Block, C. C., & Pressley, M. (Eds.). (2002). *Comprehension instruction: Research-based best practices*. New York: Guilford Press.

Bos, C. S., & Anders, P. L. (1992). Using interactive teaching and learning strategies to promote text comprehension and content learning for students with learning disabilities. *International Journal of Disability, Development and Education, 39*, 225–238.

Boyle, J. (1996). Thinking while note taking: Teaching college students to use strategic note-taking during lectures. In B. G. Grown (Ed.), *Innovative learning strategies: Twelfth yearbook* (pp. 9–18). Newark, DE: International Reading Association.

Boyle, J. (2001). Enhancing the note-taking skills of students with mild disabilities. *Intervention in School and Clinic, 36*, 221.

Boyle, J., & Weishaar, M. (1999). Note-taking strategies for students with mild disabilities. *The Clearing House, 72*, 392–396.

Brooks, R. (1991). *The self-esteem teacher: Seeds of self-esteem*. New York: Treehaus.

Brown, A. L. (1997). Transforming schools into communities of thinking and learning about serious matters. *American Psychologist, 52*(4), 399–413.

Brown, A. L., Bransford, J. D., Ferrara, R. A., & Campione, J. (1983). Learning, remembering and understanding. In P. H. Mussen (Ed.), *Handbook of child psychology* (Vol. 3, pp. 77–166). New York: Wiley.

Brown, A. L., & Campione, J. C. (1983). Psychological theory and the study of learning disabilities. *American Psychologist, 41*, 1059–1368.

Brown, A. L., & Campione, J. C. (1986). Psychological theory and the study of learning disabilities. *American Psychologist, 41*, 1059–1068.

Brown, G. D. A., & Deavers, R. P. (1999). Units of analysis in nonword reading: Evidence from children and adults. *Journal of Experimental Child Psychology, 73*, 203–242.

Bruning, R., & Horn, R. (2000). Developing motivation to write. *Educational Psychologist, 35*(1), 25–38.

Brunstein, J. C., Schultheiss, O. C., & Grässmann, R. (1998). Personal goals and emotional well-being: The moderating role of motive dispositions. *Journal of Personality and Social Psychology, 75*(2), 494–508.

Bulgren, J., Schumaker, J. B., & Deshler, D. D. (1988). Effectiveness of a concept teaching routine in enhancing the performance of LD students in secondary-level mainstream classes. *Learning Disability Quarterly, 11*(1), 3–17.

Carney, R. N., Levin, M. E., & Levin, J. R. (1993). Mnemonic strategies: Instructional techniques worth remembering. *Teaching Exceptional Children, 25*(4), 24–30.

Cartwright, K. B. (2002). Cognitive development and reading: The relation of multiple classification skill to reading comprehension in elementary school children. *Journal of Educational Psychology, 94*, 56–63.

Cartwright, K. B. (Ed.). (2008a). *Literacy processes: Cognitive flexibility in learning and teaching*. New York: Guilford Press.

Cartwright, K. B. (2008b). Introduction to literacy processes: Cognitive flexibility in learning and teaching. In K. B. Cartwright (Ed.), *Literacy processes: Cognitive flexibility in learning and teaching* (pp. 3–18). New York: Guilford Press.

Cartwright, K. B. (2008c). Concluding reflections: What can we learn from considering implications of representational development and flexibility for literacy teaching and learning? In K. B. Cartwright (Ed.), *Literacy processes: Cognitive flexibility in learning and teaching* (pp. 359–371). New York: Guilford Press.

de Fockert, J. W., Rees, G., Frith, C. D., & Lavoie, N. (2001). The role of working memory in visual selective attention. *Science, 291*(5509), 1803–1806.

Deák, G. O. (2008). Foreword for literacy processes: Cognitive flexibility in learning and teaching. In K. B. Cartwright (Ed.), *Literacy processes: Cognitive flexibility in learning and teaching*. New York: Guilford Press.

Denckla, M. B. (1996). Executive function. In D. Gozal & D. Molfese (Eds.), *Attention deficit hyperactivity disorder: From genes to patients* (pp. 165–183). Totowa, NJ: Humana Press.

Denckla, M. B. (2007). Executive function: Binding together the definitions of attention deficit/hyperactivity disorder and learning disabilities. In L. Meltzer (Ed.), *Executive function in education: From theory to practice* (pp. 5–19). New York: Guilford Press.

Deshler, D. D., Ellis, E. S., & Lenz, B. K. (Eds.). (1996). *Teaching adolescents with learning disabilities: Strategies and methods* (2nd ed.). Denver: Love.

Deshler, D. D., & Schumaker, J. B. (1988). An instructional model for teaching students how to learn. In J. L. Graden, J. E. Zins, & M. J. Curtis (Eds.), *Alternative education delivery systems: Enhancing instructional options for all students* (pp. 391–411). Washington, DC: National Association of School Psychologists.

Deshler, D., & Shumaker, J. (1986). Learning strategies: An instructional alternative for low achieving adolescents. *Exceptional Children, 52*(6), 583–590.

Deshler, D., Warner, M. M., Schumaker, J. B., & Alley, G. R. (1983). Learning strategies intervention model: Key components and current status. In J. D. McKinney & L. Feagans (Eds.), *Current topics in learning disabilities*. Norwood: Ablex.

Diamond, A. (2006). The early development of executive functions. In E. Bialystok & F. Craik (Eds.), *Lifespan cognition: Mechanisms of change* (pp. 70–95). New York: Oxford University Press.

Dweck, C. S. (2008). *Mindset: The new psychology of success*. New York, NY: The Random House.

Elliot, A. J., & Dweck, C. S. (2005). Competence and motivation: Competence as the core of achievement motivation. In A. J. Elliot & C. S. Dweck (Eds.), *Handbook of competence and motivation* (pp. 3–15). New York, NY: Guilford Press.

Ellis, E. S. (1993). Teaching strategy sameness using integrated formats. *Journal of Learning Disabilities, 26*, 448–482.

Ellis, E. S. (1994). Integrating content with writing strategy instruction: Part 2—Writing processes. *Intervention in School and Clinic, 29*, 219–228.

Eslinger, P. J. (1996). Conceptualizing, describing, and measuring components of executive function: A summary. In G. R. Lyon & N. A. Krasnegor (Eds.), *Attention, memory, and executive function*. Baltimore: Paul H. Brooks Publishing Co.

Flavell, J. H. (1979). Metacognition and cognitive monitoring: A new area of cognitive-developmental inquiry. *American Psychologist, 34*, 906–911.

Flavell, J. H., Friedrichs, A. G., & Hoyt, J. D. (1970). Developmental changes in memorization processes. *Cognitive Psychology, 1*, 324–340.

Flower, L., Stein, V., Ackerman, J., Kantz, M. J., McCormick, K., & Peck, W. C. (1990). *Reading-to-write: Exploring a cognitive and social process*. New York: Oxford University Press.

Flower, L., Wallace, D. L., Norris, L., & Burnett, R. A. (1994). *Making thinking visible: Writing, collaborative planning, and classroom inquiry*. Urbana, IL: National Council of Teachers of English.

Fuchs, D., & Fuchs, L. S. (1991). Framing the REI debate: Conservationists vs abolitionists. In J. W. Lloyd, N. N. Singh, & A. C. Repp (Eds.), *The regular education initiative: Alternative perspectives on concepts, issues, and models* (pp. 241–255). DeKalb: Sycamore.

Gardner, H. (1983). *Frames of mind: The theory of multiple intelligences*. New York: Basic Books.

Gaskins, I. W. (2008). Developing cognitive flexibility in word reading among beginning and struggling readers. In K. B. Cartwright (Ed.), *Literacy processes: Cognitive flexibility in learning and teaching* (pp. 90–114). New York: Guilford Press.

Gioia, G. A., Isquith, P. K., Guy, S. C., & Kenworthy, L. (2001). *Behavior ratings inventory of executive function*. Odessa, FL: Psychological Assessment Resources.

Gioia, G., Isquith, P., Guy, S. C., Kenworthy, L., & Barton, R. (2002). Profiles of everyday executive function in acquired and developmental disorders. *Child Neuropsychology, 8*(2), 121–137.

Goldberg, E. (2001). *The executive brain: Frontal lobes and the civilized mind*. New York: Oxford University Press.

Goleman, D. (1995). *Emotional intelligence*. New York: Bantam Books.

Goswami, U., Ziegler, J. C., Dalton, L., & Schneider, W. (2001). Pseudohomophone effects and phonological recoding procedures in reading development in English and German. *Journal of Memory and Language, 45*, 648–664.

Goswami, U., Ziegler, J. C., Dalton, L., & Schneider, W. (2003). Nonword reading across orthographies: How flexible is the choice of reading units? *Applied PsychoLinguistics, 24*, 235–247.

Graham, S. (1990). The role of production factors in learning disabled students' compositions. *Journal of Educational Psychology, 82*, 781–791.

Graham, S., & Harris, K. R. (2003). Students with learning disabilities and the process of writing: A meta-analysis of SRSD studies. In H. L. Swanson, K. R. Harris, & S. Graham (Eds.), *Handbook of research on learning disabilities* (pp. 383–402). New York: Guilford.

Gray, L., Meltzer, C., & Upton, M. (2008). *The SMARTS peer mentoring program: Fostering self-understanding and resilience across the grades*. Paper presented at the 23rd Annual Learning Differences Conference, Harvard Graduate School of Education, Cambridge, MA.

Gross, J. J. (Ed.). (2007). *Handbook of emotion regulation* (pp. 1–654). New York: Guilford Press.

Harris, K., & Graham, S. (1996). *Making the writing process work: Strategies for composition and self-regulation*. Cambridge: Brookline Books.

Harris, K. R., Graham, S., Mason, L. H., & Friedlander, B. (2008). *Powerful writing strategies for all students*. Baltimore: Brooks.

Helliwell, J. F. (2003). How's Life? Combing individual and national variations to explain subjective well being. *Economic Modeling, 20*, 331–360.

Horton, S. V., Lovitt, T. C., & Bergerud, D. (1990). The effectiveness of graphic organizers for three classifications of secondary students in content area classes. *Journal of Learning Disabilities, 23*, 12–22.

Hughes, C. A. (1991). Studying for and taking tests: Self-reported difficulties and strategies of university students with learning disabilities. *Learning Disability Quarterly, 13*, 66–79.

Hughes, C. (1996). Memory and test-taking strategies. In D. D. Deshler, E. S. Ellis, & B. K. Lenz (Eds.), *Teaching adolescents with learning disabilities: Strategies and methods* (2nd ed., pp. 209–266). Denver: Love.

Hughes, C. A., Ruhl, K. L., Schumaker, J. B., & Deshler, D. D. (2002). Effects of instruction in an assignment completion strategy on the homework performance of students with learning disabilities in general education classes. *Learning Disabilities Research and Practice, 17*(1), 1–18.

Hughes, C. A., & Suritsky, S. K. (1994). Note-taking skills of university students with and without learning disabilities. *Journal of Learning Disabilities, 27*, 20–24.

Kame'enui, E. J. (2007). Responsiveness to intervention: A new paradigm. *Teaching Exceptional Children, 39*(5), 6–7.

Kasser, T., & Ryan, R. M. (1996). Further examining the American dream: Differential correlates of intrinsic and extrinsic goals. *Personality and Social Psychology Bulletin, 22*, 280–287.

Katamaya, A. D., & Robinson, D. H. (2000). Getting students "partially" involved in note-taking using graphic organizers. *The Journal of Experimental Education, 68*, 119–134.

Kiewra, K. A., DuBois, N. F., Christian, D., McShane, A., Meyerhoffer, M., & Roskelley, D. (1991). Note-taking functions and techniques. *Journal of Educational Psychology, 83*, 240–245.

Kim, B. A., Vaughn, S., Wanzek, J., & Shangjin, W. J. (2004). Graphic organizers and their effects on the reading comprehension of students with LD: A synthesis of research. *Journal of Learning Disabilities, 37*(2), 105–119.

Kincaid, K. M., & Trautman, N. (2010). Remembering: Teaching students how to retain and mentally manipulate information. In L. Meltzer (Ed.), *Promoting executive function in the classroom* (pp. 110–139). New York: Guilford Press.

Klingner, J. K., Vaughn, S., & Boardman, A. (2007). *Teaching reading comprehension to students with learning difficulties*. New York: Guilford Press.

Krishnan, K., & Feller, M. J. (2010). Organizing: The heart of efficient and successful learning. In K. R. Harris & S. Graham (Eds.), *Promoting executive function in the classroom*. New York, NY: The Gilford Press.

Krishnan, K., Feller, M. J., & Orkin, M. (2010). Goal setting, planning, and prioritizing: The foundations of effective learning. In L. Meltzer (Ed.), *Promoting executive function in the classroom* (pp. 57–85). New York: Guilford Press.

Lazarus, B. D. (1991). Guided notes, review and achievement of secondary students with learning disabilities in mainstream content courses. *Education and Treatment of Children, 14*, 112–127.

Mastropieri, M. A., & Scruggs, T. E. (1991). *Teaching students ways to remember: Strategies for learning mnemonically*. Cambridge, MA: Brookline Books.

Mastropieri, M. A., & Scruggs, T. E. (1998). Enhancing school success with mnemonic strategies. *Intervention in School and Clinic, 33*, 201–208.

Mayer, R. E. (1984). Aids to text comprehension. *Educational Psychologist, 19*(1), 30–42.

Meltzer, L. (1993). Strategy use in children with learning disabilities: The challenge of assessment. In L. Meltzer (Ed.), *Strategy assessment and instruction for students with learning disabilities: From theory to practice* (pp. 93–136). Texas: Pro-Ed.

Meltzer, L. (1996). Strategic learning in students with learning disabilities: The role of self-awareness and self-perception. In T. E. Scruggs & M. Mastropieri (Eds.), *Advances in learning and behavioral disabilities* (Vol. 10b, pp. 181–199). Greenwich, CT: JAI Press.

Meltzer, L. (2004). Resilience and learning disabilities: Research on internal and external protective dynamics. Introduction to the special series. *Learning Disabilities Research and Practice, 19*(1), 1–2.

Meltzer, L. (Ed.). (2007). *Executive function in education: From theory to practice*. New York: Guilford Press.

Meltzer, L. (Ed.). (2010). *Promoting executive function in the classroom*. New York: Guilford Press.

Meltzer, L., & Bagnato, J. S. (2010). Shifting and flexible problem solving: The anchors for academic success. In L. Meltzer (Ed.), *Promoting executive function in the classroom* (pp. 140–159). New York: Guilford Press.

Meltzer, L., & Basho, S. (2010). Creating a classroom-wide executive function culture that fosters strategy use, motivation, and resilience. In L. Meltzer (Ed.), *Promoting executive function in the classroom* (pp. 28–54). New York: Guilford Press.

Meltzer, L., Katzir, T., Miller, L., Reddy, R., & Roditi, B. (2004). Academic self-perceptions, effort, and strategy use in students with learning disabilities: Changes over time. *Learning Disabilities Research and Practice, 19*(2), 99–108.

Meltzer, L., Katzir-Cohen, T., Miller, L., & Roditi, B. (2001). The impact of effort and strategy use on academic performance: Student and teacher perceptions. *Learning Disabilities Quarterly, 24*(2), 85–98.

Meltzer, L., & Krishnan, K. (2007). Executive function difficulties and learning disabilities: Understandings and misunderstandings. In L. Meltzer (Ed.), *Executive function in education: From theory to practice* (pp. 77–106). New York: Guilford Press.

Meltzer, L., & Montague, M. (2001). Strategic learning in students with learning disabilities: What have we learned? In B. Keogh & D. Hallahan (Eds.), *Intervention research and learning disabilities* (pp. 111–130). Hillsdale, NJ: Erlbaum.

Meltzer, L., Noeder, M., Basho, S., Stacey, W., Button, K., & Sales Pollica, L. (2007). *Executive function strategies, effort, and academic self-perceptions: Impact on academic performance*. Paper presented at the 31st Annual Conference of the International Academy for Research in Learning Disabilities, Bled, Slovenia.

Meltzer, L., Reddy, R., Brach, E., Kurkul, K., & Basho, S. (2011). *Self-concept, motivation, and executive function: Impact of a peer mentoring program*. Paper presented at the Pacific Coast Research Conference, San Diego, CA.

Meltzer, L., Reddy, R., Brach, E., Kurkul, K., Stacey, W., & Ross, E. (2011). *The SMARTS mentoring program: Fostering self-concept, motivation, and executive function strategies in students with learning difficulties*. Paper presented at the Annual Conference of the American Educational Research Association, New Orleans, LA.

Meltzer, L., Reddy, R., Pollica, L., & Roditi, B. (2004). Academic success in students with learning disabilities: The roles of self-understanding, strategy use, and effort. *Thalamus, 22*(1), 16–32.

Meltzer, L. J., Reddy, R., Sales, L., et al. (2004). Positive and negative self-perceptions: Is there a cyclical relationship between teachers' and students' perceptions of effort, strategy use, and academic performance? *Learning Disabilities Research and Practice, 19*(1), 33–44.

Meltzer, L. J., Roditi, B., Houser, R. F., & Perlman, M. (1998). Perceptions of academic strategies and competence in students with learning disabilities. *Journal of Learning Disabilities, 31*(5), 437–451.

Meltzer, L., Sales-Pollica, L., & Barzillai, M. (2007). Executive function in the classroom: Embedding strategy instruction into daily teaching practices. In L. Meltzer (Ed.), *Executive function in education: From theory to practice* (pp. 165–194). New York: Guilford Press.

Meltzer, L. J., Sayer, J., Sales, L., Theokas, C., & Roditi, B. (2002). *Academic self-perceptions in students with LD: Relationship with effort and strategy use.* Paper presented at the International Academy for Research in Learning Disabilities conference, Washington, DC.

Meltzer, L., Solomon, B., Fenton, T., & Levine, M. D. (1989). A developmental study of problem-solving strategies in children with and without learning difficulties. *Journal of Applied Developmental Psychology, 10*, 171–193.

Miller, L., Meltzer, L. J., Katzir-Cohen, T., & Houser, R. F., Jr. (2001). Academic heterogeneity in students with learning disabilities. *Thalamus, 19*, 20–33.

Montague, M. (2003). *Solve it: A mathematical problem-solving instructional program.* Reston, VA: Exceptional Innovations.

Montague, M., & Jitendra, A. K. (2006). *Teaching mathematics to middle school students with learning difficulties.* New York: Guilford Press.

Montague, M., Warger, C., & Morgan, H. (2000). Solve it!: Strategy instruction to improve mathematical problem solving. *Learning Disabilities Research and Practice, 15*, 110–116.

Noeder, M. (2007). *The Drive to Thrive program: Fostering effective strategy use, metacognitive awareness, effort, and persistence.* Unpublished master's thesis, Tufts University.

Pajares, F., & Schunk, D. H. (2001). Self-beliefs and school success: Self-efficacy, self-concept, and school achievement. In R. Riding & S. Rayner (Eds.), *Perception* (pp. 239–266). London: Alex.

Paris, S. G. (1986). Teaching children to guide their reading and learning. In T. E. Raphael (Ed.), *The contexts of school-based literacy* (pp. 115–130). New York: Random House.

Paris, S. G., Lipson, M., & Wixson, K. (1983). Becoming a strategic reader. *Contemporary Educational Psychology, 8*, 292–316.

Paris, S. G., & Winograd, P. (1990). Promoting metacognition and motivation of exceptional children. *Remedial and Special Education, 11*(6), 7–15.

Pearson, P. D., & Dole, J. A. (1987). Explicit comprehension instruction: A review of research and new conceptualization of instruction. *The Elementary School Journal, 88*, 151–165.

Pressley, M., & Afflerbach, P. (1995). *Verbal protocols of reading: The nature of constructively responsive reading.* Hillsdale, NJ: Erlbaum.

Pressley, M., Goodchild, F., Fleet, J., Zajchowski, R., & Evans, E. D. (1989). The challenges of classroom strategy instruction. *The Elementary School Journal, 89*, 301–342.

Reid, R. (1996). Research in self-monitoring with students with learning disabilities: The present, the prospects, the pitfalls. *Journal of Learning Disabilities, 29*, 317–331.

Reid, R., & Harris, K. R. (1993). Self-monitoring of attention vs self-monitoring of performance: Effects on attention and academic performance. *Exceptional Children, 60*, 29–40.

Reid, R., & Lienemann, T. O. (2006). *Strategy instruction for students with learning disabilities.* New York: Guilford Press.

Renninger, K. A., Hidi, S., & Krapp, A. (2004). The role of interest in learning and development. In D. Yun Dai & R. J. Sternberg (Eds.), *Motivation, emotion, and cognition: Integrative perspectives on intellectual functioning and development.* Mahwah, NJ: Lawrence Erlbaum Associates.

ResearchILD & FableVision. (2003). *BrainCogs: The personal interactive coach for learning and studying* [Computer software]. Boston: FableVision. http://www.fablevision.com

ResearchILD & FableVision. (2005). *Essay Express: Strategies for successful essay writing.* Boston: FableVision. http://www.fablevision.com

Ritchie, D., & Volkl, C. (2000). Effectiveness of two generative learning strategies in the science classroom. *School Science and Mathematics, 100*(2), 83–89.

Roditi, B. N., & Steinberg, J. (2007). The strategic math classroom: Executive function processes and mathematics learning. In L. Meltzer (Ed.), *Executive function in education: From theory to practice* (pp. 237–261). New York: Guilford Press.

Rosenshine, B. (1997). Advances in research on instruction. In J. W. Lloyd, E. J. Kameenui, & D. Chard (Eds.), *Issues in educating students with disabilities* (pp. 197–221). Mahwah, NJ: Lawrence Earlbaum.

Sah, A., & Borland, J. H. (1989). The effects of a structured home plan on the home and school behaviors of gifted learning-disabled students with deficits in organizational skills. *Roeper Review, 12*(1), 54–57.

Scanlon, D. J., Duran, G. Z., Reyes, E. I., & Gallego, M. A. (1992). Interactive semantic mapping: An interactive approach to enhancing LD students' content area comprehension. *Learning Disabilities Research and Practice, 7*, 142–146.

Schunk, D. H. (2001). *Self-regulation through goal setting.* ERIC Digest. ERIC Clearinghouse on Counseling and Student Services: Greensboro, NC. (ERIC Document Reproduction Service No. ED 462 671).

Scruggs, T. E., & Mastropieri, M. A. (2000). The effectiveness of mnemonic instruction for students with learning and behavior problems: An update and research synthesis. *Journal of Behavioral Education, 10*, 163–173.

Shanahan, C. H., & Shanahan, T. (2008). Content-area reading/learning: Flexibility in knowledge acquisition. In K. B. Cartwright (Ed.), *Literacy processes: Cognitive flexibility in learning and teaching* (pp. 208–234). New York: Guilford Press.

Sheldon, K. M., & Elliot, A. J. (1999). Goal striving, need satisfaction, and longitudinal well-being: The self-concordance model. *Journal of Personality and Social Psychology, 76*, 482–497.

Shimabukuro, S. M., Prater, M. A., Jenkins, A., & Edelen-Smith, P. (1999). The effects of self monitoring of academic performance on students with learning disabilities and ADD/ADHD. *Education and Treatment of Children, 22*(4), 397–415.

Stein, J. (2010). Emotional self-regulation: A critical component of executive function. In L. Meltzer (Ed.), *Promoting executive function in the classroom* (pp. 175–201). New York: Guilford Press.

Stein, J., & Krishnan, K. (2007). Nonverbal learning disabilities and executive function: The challenge of effective assessment and teaching. In L. Meltzer (Ed.), *Executive function in education: From theory to practice* (pp. 106–132). New York, NY: Guildford Press.

Sternberg, R. J. (2005). Intelligence, competence, and expertise. In A. J. Elliot & C. S. Dweck (Eds.), *Handbook of competence and motivation* (pp. 15–31). New York, NY: Guilford Press.

Stone, C. A., & Conca, L. (1993). The origin of strategy deficits in children with learning disabilities: A social constructivist perspective. In L. J. Meltzer (Ed.), *Strategy assessment and instruction for students with learning disabilities: From theory to practice* (pp. 23–59). Austin: Pro-Ed.

Stone, C. A., & May, A. (2002). The accuracy of academic self-evaluations in adolescents with learning disabilities. *Journal of Learning Disabilities, 35*(4), 370–383.

Suritsky, S. K. (1992). Note-taking approaches and specific areas of difficulty reported by university students with learning disabilities. *Journal of Postsecondary Education and Disability, 10*, 3–10.

Swanson, L. (1989). Strategy instruction: Overview of principles and procedures for effective use. *Learning Disabilities Quarterly, 12*, 3–14.

Swanson, H. L. (1999). Instructional components that predict treatment outcomes for students with learning disabilities: Support for a combined strategy and direct instruction model. *Learning Disabilities Research and Practice, 14*, 129–140.

Swanson, H. L. (2001). Research on intervention for adolescents with learning disabilities: A meta-analysis of outcomes related to high-order processing. *The Elementary School Journal, 101*, 331–348.

Swanson, H. L., & Hoskyn, M. (1998). Experimental intervention research on students with learning disabilities: A meta-analysis of treatment outcomes. *Review of Educational Research, 68*, 277–321.

Swanson, H. L., & Hoskyn, M. (2001). Instructing adolescents with learning disabilities: A component and composite analysis. *Learning Disabilities Research and Practice, 16*(2), 109–119.

Swanson, H. L., & Sáez, L. (2003). Memory difficulties in children and adults with learning disabilities. In H. L. Swanson, K. R. Harris, & S. Graham (Eds.), *Handbook of learning disabilities* (pp. 182–198). New York: Guilford Press.

Tangney, J. P., Baumeister, R. F., & Boone, A. L. (2004). High self-control predicts good adjustment, less pathology, better grades, and interpersonal success. *Journal of Personality, 72*, 271–324.

Tannock, R. (2008). *Inattention and working memory: Effects on academic performance*. Paper Presented at the 23rd Annual Learning Differences Conference, Harvard Graduate School of Education, Cambridge, MA.

Torgesen, J. K. (1977). The role of nonspecific factors in the task performance of learning disabled children: A theoretical assessment. *Journal of Learning Disabilities, 10*, 27–34.

Westman, A. S., & Kamoo, R. L. (1990). Relationship between using conceptual comprehension of academic material and thinking abstractly about global life issues. *Psychological Reports, 66*(2), 387–390.

Winne, P. H. (1996). A metacognitive view of individual differences in self-regulated learning. *Learning and Individual Differences, 8*(4), 327–353.

Winne, P. H. (2001). Self-regulated learning viewed from models of information processing. In B. J. Zimmerman & D. H. Schunk (Eds.), *Self-regulated learning and academic achievement: Theoretical perspectives* (2nd ed., pp. 153–189). Mahwah, NJ: Lawrence Erlbaum Associates, Inc.

Yuill, N. (2007). Visiting Joke City: How can talking about jokes foster metalinguistic awareness in poor comprehenders? In D. MacNamara (Ed.), *Reading comprehension*. New York: Erlbaum.

Yuill, N., & Bradwell, J. (1998). The laughing PC: How a software riddle package can help children's reading comprehension. *Proceedings of the British Psychological Society, 6*, 119.

Yun Dai, D., & Sternberg, R. J. (Eds.). (2004). *Motivation, emotion, and cognition: Integrative perspectives on intellectual functioning and development*. Mahwah, NJ: Lawrence Erlbaum Associates.

Zelazo, O. D., & Müller, U. (2002). Executive function in typical and atypical development. In U. Goswami (Ed.), *Blackwell handbook of childhood cognitive development* (pp. 445–469). Malden, MA: Blackwell.

Zelazo, P. D., Müller, U., Frye, D., & Marcovitch, S. (2003). The development of executive function in early childhood. *Monographs of the Society for Research in Child Development, 68* (3, Serial No. 274).

Zimmerman, B. J. (1998). Academic studying and the development of personal skill: A self-regulatory perspective. *Educational Psychologist, 33*, 73–86.

Zimmerman, B. J. (2000). Attaining self-regulation: A social cognitive perspective. In M. Boekaerts, P. R. Pintrich, & M. Zeidner (Eds.), *Handbook of self-regulation* (pp. 13–39). San Diego, CA: Academic.

Zimmerman, B. J., & Kitsantas, A. (1997). Developmental phases in self-regulation: Shifting from process to outcome goals. *Journal of Educational Psychology, 89*, 29–36.

Zimmerman, B. J., & Schunk, D. H. (Eds.). (2001). *Self-regulated learning and academic achievement: Theoretical perspectives.* Mahwah, NJ: Erlbaum.

Working Memory Training and Cogmed

Peter C. Entwistle and Charles Shinaver

Introduction

Working memory (WM) is a critical area of cognitive functioning that is intimately tied to executive functioning (EF) and attention. One group of investigators described working memory as "the most sensitive and neuropsychologically valid component of our executive function abilities" (Séguin, Nagin, Assaad, & Tremblay, 2004). This chapter will focus on the term working memory, and on how we might intervene when a person has a working memory deficit. Specifically, we will explore the Cogmed working memory training program, an innovative approach to improving working memory that combines a software program with coaching designed for use with children, adolescents, and adults. We will explore the research on Cogmed and its effectiveness. Whether the effects of Cogmed generalize to other areas of cognitive functioning will be considered. We will also review evidence on neural plasticity.

Working Memory

What is working memory? How do you define it? Why is it important and how do working memory problems impact children and adolescents in schools? These are some of the questions raised by researchers exploring working memory.

The term working memory describes the ability to hold in mind and manipulate information for brief periods of time during complex cognitive tasks. Working memory is the capacity to hold events in mind and manipulate them (Goldman-Rakic, 1988). Working memory not only involves manipulation of mental representation online but the generation of potential action sequences (Roberts & Pennington, 1996). Just and Carpenter (1992) characterize working memory as "the mental workspace in which products of ongoing processes can be stored and integrated during complex and demanding activities."

Theoretical models of working memory differ in their views of the nature, structure, and function of the system (for a review, see Dehn, 2010; Miyake & Shah, 1999). The distinguishing feature between these models is whether working memory is conceived of as a distinct entity (e.g., Baddeley, 2000; Baddeley & Hitch, 1974) or as a limited capacity process of controlled attention that serves to activate existing representations in long-term memory, which then become the current contents of working memory (e.g., Cowan, 2005; Dehn, 2010, Engle, Kane, & Tuholski, 1999). This model suggests that working memory is one of several important executive functions that may have an indirect effect on a person's performance. As illustrated in the table below, working memory functions on an underlying level. Working memory has been found to influence learning such that those with a greater

P.C. Entwistle (✉) • C. Shinaver
Pearson Clinical Assessments, San Antonio, TX, USA
e-mail: peter.entwistle@pearson.com

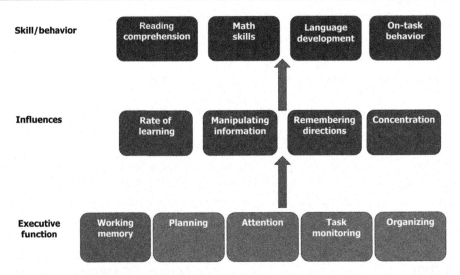

Fig. 26.1 This figure illustrates a possible conceptualization of diverse executive functions and the indirect influence of working memory on skill acquisition

working memory capacity learn more rapidly and are more able to manipulate information, remember directions, and concentrate (Dehn, 2008). In other words, it is important to remember that WM serves as a base as upon which new information is encoded into long-term memory. One might conceptualize this as "desk space in the mind." When one lacks sufficient capacity reading comprehension, math skills and language skills are difficult to develop. In fact, in an intriguing study on the skill acquisition of children with ADHD, Huang-Pollock and Karalunas (2010) found that when working memory load was low, those with ADHD were more error prone and learned more slowly than controls. Yet, when working memory load was high, those with ADHD learned more slowly and failed to achieve "automaticity." In other words, they lacked mastery of the material as one would expect when a child commits his multiplication tables to memory. Hence, they are less successful. Children with ADHD, when presented with tasks that involve a high working memory load, are unlikely to achieve mastery of a skill or topic area. As such these children will miss many opportunities over childhood to develop skills and may risk becoming left behind academically and socially (Fig. 26.1).

One of the most widely accepted models of working memory sees working memory as a multicomponent system (Baddeley & Hitch, 1974). In this model, there are two domain-specific short-term memory (STM) stores, the phonological loop and the visuospatial sketchpad, which are specialized for the temporary maintenance of verbal and visual and spatial information, respectively. These are governed by a domain-general central executive system linked to attentional control. Working memory is responsible for holding and manipulating information from long-term memory, coordinating performance on dual tasks, switching between different retrieval strategies, and inhibiting irrelevant information (e.g., Baddeley, 1996; Baddeley, Emslie, Kolodny, & Duncan, 1998; Dehn, 2010; Engle et al., 1999; Engle & Kane, 2004; Kane, Conway, Hambrick, & Engle, 2007; Kane & Engle, 2000).

This model conceives of the central executive as playing a coordinating role. The phonological loop and the visual-spatial sketchpad are described as coordinating performance on domain-specific tasks. The central executive manages dual tasks across domains and switches between different retrieval strategies within a subsystem. This gives a prominent and critical role to the central executive working memory subsystem. Additionally, given this essential role of coordination in the crucial learning task of encoding new information into long-term memory, central executive working memory serves a pivotal function. Central executive working memory is measured by complex tasks including backward span, reading, and listening span. The fourth component, the episodic buffer, added more recently to this model, is responsible for integrating information from the subcomponents of working memory and long-term memory (Baddeley, 2000).

Not surprisingly, given the vital role of working memory in learning new skills and encoding knowledge into long-term memory, a deficit in this area is associated with a wide range of cognitive difficulties that relate to learning. In fact, one distinction between working memory and other executive functioning and attention constructs is the substantial and consistent body of research that has found working memory related to academic achievement. Among the legion of learning problems that have been found to be associated with working memory are reading disabilities (Swanson, Zheng, & Jerman, 2009) and reading comprehension problems (Carretti, Borella, Cornoldi, & De Beni, 2009). Working memory has been found to have an even stronger relationship, a predictive relationship, with other learning difficulties such as language comprehension (Daneman & Merikle, 1996). Visual-spatial working memory and visual-spatial STM have been found to be predictive of math achievement (Bull, Espy, & Wiebe, 2008). Growth in working memory has been found to be predictive of improved math problem solving in first to third grade children (Swanson, Jerman, & Zheng, 2008). For children who have already been diagnosed with ADHD, working memory combined with processing speed on the WISC III and WISC IV have been predictive of poorer academic achievement and are the most powerful predictors of learning disabilities (Mayes & Calhoun, 2007). Similarly, among children who already have learning difficulties, working memory predicts subsequent learning (Alloway, 2009).

Broadly speaking, working memory skills are highly associated with children's abilities to learn in key academic domains such as reading, mathematics, and science (Gathercole & Pickering, 2000). This powerful connection to academic achievement and learning in itself is unique to working memory. No other executive functioning constructs or attention constructs are as tied to achievement and learning (with the possible exception of sustained attention). The constructs of inhibition, planning, selective attention, divided attention, shifting attention, and visual scanning all lack such a potent connection to academic achievement.

Does Working Memory Training Work?

Given that an extensive body of research connects working memory to academic achievement (Alloway, 2009; Bull et al., 2008; Carretti et al., 2009; Daneman & Merikle, 1996; Gathercole & Pickering, 2000; Mayes & Calhoun, 2007; Swanson et al. 2008, 2009), the most crucial question is whether working memory can be trained. One approach to remediating poor working memory function is to train it through repeated practice on working memory tasks. Studies that attempted to improve working memory using this method in the 1970s and 1980s only reported moderate training gains and only in the form of faster reaction times, not increases in working memory capacities per se, and there was no evidence that gains were transferable to nontrained working memory tasks or to other cognitive measures (Kristofferson, 1972; Phillips & Nettlebeck, 1984). Other training studies, such as those conducted by Hulme and Muir (1985), have demonstrated that training processes crucial to efficient processing in working memory, such as articulation and rehearsal rate, improve memory span, although only very slightly. More recent studies have noted that strategy training, that is, training in verbal rehearsal or articulatory rehearsal (Comblain, 1994; Conners, Rosenquist, Arnett, Moore, & Hume, 2008; Turley-Ames & Whitfield, 2003) or strategies that focus upon elaborate encoding strategies (Carretti, Borella & De Beni, 2007; Cavallini, Pagnin, & Vecchi,

2002; McNamara & Scott, 2001) have resulted in gains in working memory capacity (Morrison & Chein, 2011).

The rationale for strategy training approaches is that developmental literature has found that those children who utilize rehearsal over childhood show increases in memory recall (Flavell, Beach, & Chinsky, 1966). In other words, children who do not utilize rehearsal are at a disadvantage compared to those who do. So, early investigators focused upon teaching children how to effectively rehearse. Strategies have included chunking (combining items in some meaningful way) and making a mental story with items and using imagery to make items more salient (Morrison & Chein, 2011). As noted by reviewers Morrison and Chein (2011), this has resulted in some recorded gains in working memory. However, the limit has been that strategy training has primarily resulted in gains with near transfer. Near transfer occurs when the new task is similar to the trained task. A well-known example is the ability of a runner to recall 80 digits by translating those digits to running times, like running a mile in 4 min and 52 s (Ericsson & Chase, 1982). The subject would then chunk those running times into longer and longer strings of numbers. This approach was highly successful with numbers, but only with numbers. The subject showed no transfer effects to other domains of content (Ericsson & Chase, 1982). He tested in the average range on other areas of working memory.

A distinct approach to training working memory, core training, has shown more promise with regard to far transfer. That is, improvements made with core training approaches have been shown to transfer to untrained tasks. Core training, as defined by Morrison and Chein (2011), involves limiting the use of strategies, requiring the use of multiple modalities and maintenance in the face of retrieval interference. Additionally, the critical feature of adaptation plays the role of keeping the working memory load high. The notion was that computer software might automate the challenge load. The way this could be accomplished was that as a subject successfully completes a trial of a task, the next trial becomes slightly more difficult. In contrast, if the subject incorrectly answers, the next trial of that task becomes easier (Morrison & Chein, 2011). The question was whether the utilization of high-demand cognitive workloads managed through adaptively challenging tasks would stretch or improve working memory capacity. COGITO (Schmiedek, Lovden, & Lindenberger, 2010) is one such attempt to do this, and Cogmed is another (Morrison & Chein, 2011). Morrison and Chein (2011) conceptualize both of these approaches as core working memory training. Interestingly, given the role computer software plays in automating challenge level, one might surmise this could be fairly difficult to recreate without the aid of a computer.

Why Was Cogmed Developed?

Westerberg, Hirvikoski, Forssberg, and Klingberg (2004) found a significant discrepancy between the capacity for visual-spatial working memory in children with a diagnosis of ADHD and a cross-sectional comparison of normal peers in their research. This finding raised the following questions: can working memory be improved, and can technology be utilized to engage children and enhance this process through live adaptations to performance? This chapter addresses those questions.

How Was Cogmed Developed and Why Does It Focus Upon Working Memory?

Cogmed was developed by investigators at the Karolinska Institute in collaboration with video game programmers. The investigators targeted working memory because research data had begun to suggest its salience in the difficulties of those with ADHD. For example, in a meta-analysis of 83 studies of children with ADHD, Willcutt, Doyle, Nigg, Faraone, and Pennington (2005) found that the areas that had the largest effect sizes among published studies differentiating between groups with ADHD and controls were response inhibition, vigilance, working memory, and planning. Most theorists consider these areas executive functions. The effect sizes all fell between .46 and .69—in the medium range. Yet, these authors conclude that executive

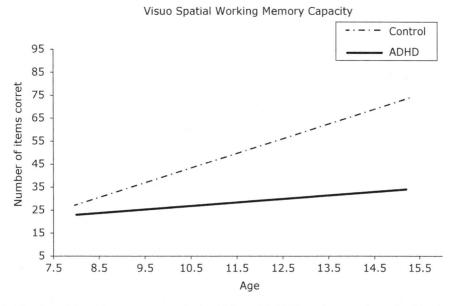

Fig. 26.2 Visual-spatial working memory capacity in children with ADHD and a normal sample of 8–15 years olds based on the research of Westerberg et al. (2004)

functioning deficits were neither sufficient nor necessary to cause all cases of ADHD.

Similarly, Martinussen, Hayden, Hogg-Johnson, and Tannock (2005) conducted a meta-analysis of 26 studies comparing working memory in children with and without ADHD from 1997 to 2003. They found that both spatial working memory and verbal working memory were significant deficits for children with ADHD in comparison to controls, but that areas of spatial working memory showed the greatest deficits in terms of effect size. More specifically, spatial storage had an effect size of .85 and spatial central executive 1.06, while the effect size for verbal storage was .47, and the effect size for verbal central executive working memory was .43, among those with ADHD. Working memory and, to a lesser extent, STM were significant deficits for children with ADHD. Importantly, the effect sizes of these working memory deficits reported by the meta-analysis by Martinussen et al. (2005) exceed those reported in the meta-analysis by Willcutt et al. (2005) which looked more broadly at executive functions. In total this data supports the particular importance of working memory among other executive functions in those with ADHD.

One seminal piece of research, arguably creating the spring board for the creation of Cogmed, was conducted by Westerberg et al. (2004) and is illustrated in the graph above. It depicts visual-spatial working memory in boys with ADHD from the ages of 7.5 years to 15.5 years. The continuous straight line indicates the visual-spatial working memory of the boys with ADHD, while the dotted line indicates that of the control group of normal peers. This graph suggests that they have the so little "desk space" in their minds to acquire skills; it is as if they were much younger, say that possibly 7.5- or 8-year-olds. The trend line, i.e., continuous line, indicates the ADHD group, while the dotted trend line indicates the control group, or typically developing boys. This was not a longitudinal study, so the ADHD boy indicated at 7.5 years old is a different one from the one suggested by the trend at about 15 years old. What is notable is that the 15 years old with ADHD does not appear to have a visual-spatial working memory capacity that exceeds that of his 7.5-year-old counterpart by much if at all (Fig. 26.2).

In contrast to the ADHD boys, the typically developing boys show a sizeable increase in visual-spatial working memory as they get older. Without an intervention or any intentional effort, these "typically developing" boys show a considerable increase in visual-spatial working memory compared to their 7.5-year-old counterparts. Indeed, for normally developing children, working memory capacity grows over time without intervention. Erzine (2011) noted this to be the case with 3- to 5-year-old children where maturation accounted for nonverbal working memory growth. Similarly, Gathercole, Pickering, Ambridge, and Wearing (2004) noted that typically developing children showed growth in working memory capacity across childhood through early and middle school years to adolescence. It is no wonder that the 15.5-year-old normally developing boys are more distinct in their maturity than either the 15.5-year-old boys with ADHD or the 7.5-year-old boys with or without ADHD. The differences are stark. One might surmise that the successful development of skills is also rather marked. The graph suggests that the typically developing boy at about 15.5 years old has 3-4 times the visual-spatial working memory capacity of that of the 15.5-year-old boy with ADHD.

Description of Cogmed

In the program developed by Klingberg, Forssberg, and Westerberg (2002), individuals train intensively over several weeks (typically 5 weeks) on computerized adaptive working memory tasks. These training tasks require the immediate serial recall of either verbal or visuospatial information, with some of the tasks requiring explicit processing prior to recall (utilization of working memory). Participants train for 20–25 days (typically 5 days a week for 5 weeks), each day completing eight different tasks from a bank of 12 tasks. Each day subjects will spend 30–45 min doing Cogmed. Positive verbal and visual feedback is built into the software on some trials. The difficulty of the training tasks is automatically adjusted on a trial-by-trial basis to match the participant's current working memory capacity, which maximizes training benefits.

Initial Cogmed Studies Show Promise

In the very first trials of Cogmed, which could be viewed as feasibility studies, Klingberg's team used an early form of the training program that included only four training tasks, each with 30 trials per day: a Corsi block-like visuospatial memory task, called spatial span; two verbal tasks from the Wechsler scales (Wechsler, 2002); and a choice reaction time task. In a double-blind, placebo-controlled study, this intensive and adaptive (nonadaptive training with a ceiling of two or three items was used in the placebo control group) working memory training significantly improved performance on nontrained STM tasks, digit recall and Corsi block, or spatial span, and a Raven's test of nonverbal or fluid reasoning, in a small number of children with a diagnosis of ADHD. Motor activity was significantly reduced in the treatment group, and performance on a response inhibition task also significantly improved following training. There were no significant changes in performance for the control group, who completed a placebo version of the program in which the difficulty of the training tasks was set at a low level throughout the training period (span of two or three items for each task). In a second experiment, they used the same adaptive training program with four healthy adults. Significant improvements in performance were reported both on the trained tasks and on a nontrained visuospatial memory task, a Stroop task, and a nonverbal reasoning task (Klingberg et al., 2002).

Klingberg et al. (2005) later extended their work to evaluate the effects of training in a larger multisite group of children with ADHD in a randomized controlled trial. Each child completed 90 working memory trials per day, performing working memory tasks (remembering the position of objects in a 4×4 grid or remembering phonemes, letters, or digits) for 20–25 days. As before, the placebo version included an identical set of tasks to the treatment program, which were set at a low level throughout training. Children were randomly assigned to each condition, with 27 completing the adaptive treatment program and 26 completing the placebo version. Overall, the treatment group improved significantly more than the comparison group on a nontrained measure of visuospatial working memory. These effects persisted 3 months after training. In addition, significant treatment effects were observed in response inhibition, complex reasoning, and verbal working memory, and there were significant reductions in parent ratings of inattention and hyperactivity/impulsivity following training. However, this study did not show a decrease in motor activity, as measured by the number of head movements (measured by an infrared camera placed on the child's head) during a computerized test.

Reductions in ratings of cognitive problems following training were also reported in a pilot study with 18 adults more than 1 year after a stroke. As before, there were significant improvements in trained and nontrained working memory tasks, and there was also a significant decrease in the patients' self-ratings of cognitive problems in daily life (Westerberg et al., 2007).

The team at the Karolinska Institute extended their work on memory training to typically developing or normal preschool children (Thorell, Lindqvist, Bergman Nutley, Bohlin, & Klingberg, 2009). On the basis of the connection between inhibition and working memory (see Engle & Kane, 2004; Roberts & Pennington, 1996), the overlapping areas of neural activation during working memory and inhibition tasks (McNab et al., 2008) and the transfer of training effects to the Stroop task in their early studies, the team decided to compare the effects of visuospatial working memory training and inhibition training in very young children.

Thorell et al. (2009) included four groups of preschool children aged 4 and 5 years in the study. The first group completed visuospatial working memory training. A second group completed inhibition training. The third completed a placebo version of the memory training, as per previous studies. A fourth group formed a passive control group. Those in the training groups completed adaptive training of either visuospatial working memory or inhibition for 15 min a day, every day that they attended preschool over a 5-week period. Each day they completed three of five possible tasks, which rotated across the training period to maintain interest.

The five visuospatial working memory training tasks required children to recall sequences of nonverbal information in the correct order. The inhibition training consisted of five tasks that mirror well-known inhibition paradigms—two go/no-go tasks and two stop-signal tasks designed to train response inhibition and a flanker task designed to train interference control. Outcome measures for all groups included nontrained measures of interference control, response inhibition, forward and backward Corsi block, forward and backward digit recall, sustained attention, and problem solving.

Children in the working memory training group improved significantly on all trained tasks, while those in the inhibition training group improved only on the trained go/no-go and interference control tasks. In other words, inhibition training did not result in transferable improvement. Conversely, working memory training led to significant gains in both nontrained verbal and visuospatial memory tasks and attention, but there was no significant transfer to performance on nontrained inhibition tasks. There was no significant change in performance on nontrained tasks for children in the placebo or passive control groups.

Overall, the data from this study show that working memory can be trained in typically developing children as young as 4 years and, importantly, that different cognitive functions vary in how easily they can be modified by intensive practice (Thorell et al., 2009). While working memory can be trained and the effects transferred to nontrained tasks, inhibition training did not transfer to nontrained tasks. The results of this study point to the generalized benefits of training working memory and to the limited effects of training other executive functions such as inhibition. In contrast, Diamond and Lee has found that the training of inhibitory strategies is important in young children and has transfer effects; however, she also notes that Cogmed training can lead to improved working memory and reasoning in this age group (Diamond & Lee, 2011).

Additional Cogmed Research

On the basis of the early success of working memory training with children, Holmes, Gathercole, and Dunning (2009) conducted evaluations of the Cogmed program. Joni Holmes, Susan Gathercole, and Darren Dunning were aware of the importance of working memory in relation to academic achievement. Gathercole in particular had already conducted numerous studies that show a strong link between working memory and academic achievement (Gathercole, Alloway, Willis, & Adams, 2006; Gathercole, Brown, & Pickering, 2003, Gathercole, Durling, Evans, Jeffcock, & Stone, 2008, Gathercole & Pickering, 2000). But the team was highly skeptical that working memory could be trained or improved and endeavored to scrutinize Cogmed in a new study.

The study used children who had already been recorded as at or below the 15th percentile in working memory as part of a routine screening conducted in England for low working memory. The children were then randomly assigned to a high- or low-intensity training condition. Working memory was assessed using two tests of verbal working memory, listening recall and backward digit recall from the Automated Working Memory Assessment (AWMA), available in England. The AWMA assesses STM (subdivided into visual, spatial, and verbal) and working memory, subdivided similarly. In this study, the "high-intensity" training is analogous to what has been described as "adaptive training." While the "low intensity" has been described as the "placebo group" training in which there is a low ceiling. Working memory was assessed using two tests of verbal working memory, listening recall and addition backwards. Along with the measurements of the AWMA, investigators used a pre- and posttest of the following measures: IQ (WASI; Wechsler, 1999), basic reading (WORD; Rust, Golombok, & Trickey, 1993), mathematical reasoning (WOND; Wechsler, 1996), and a following instructions task, which was provided to operationalize sustained attention in observable behavior terms.

The following instruction task, used in the Holmes et al. (2009) study, is similar to the instructions a teacher might give her students. In a classroom, a teacher might say to students: "Get out your math books. Turn to page 72. Do the odd problems. Remember to show your work. Make sure you use pencil so you can erase mistakes! Oh, yes, if you do number 42 it will be extra credit. Remember to show your work. Make sure you use pencil." A student with ADHD might turn to his friend and ask, "What page did she say?" He is back on the first instruction while others move forward. So, the instruction task is described by Holmes et al. (2009) in the following way:

> For this task, the child was seated in front of an array of props (rulers, folders, rubbers, boxes, (pencils) in a range of colors (blue, yellow, red) and attempted to perform a spoken instruction, such as *Touch the yellow pencil and then put the blue ruler in the red folder*. A span method was used in which the number of actions in the instructions was increased to the point at which the child could not perform the task accurately. The total number of trials correct to this point was scored (Holmes et al., 2009).

It appears that these researchers were surprised at their own results. Holmes et al. (2009) state their findings in this concluding statement:

> This study provides the first demonstration that these commonplace deficits and associated learning difficulties can be ameliorated, and possibly even overcome, by intensive adaptive training over a relatively short period: just 6 weeks.

More specifically, they found that the training group improved significantly on verbal STM, verbal working memory, visuospatial STM, and visuospatial working memory as compared to the low-intensity training group on the AWMA. These gains were maintained at 6 months.

Like Klingberg, the second study by Holmes et al. (2010) was conducted with children diagnosed with ADHD and known to have significant and substantial deficits in working memory (Holmes, Gathercole, & Dunning 2010; Holmes, Gathercole, Place, Dunning, Hilton, & Elliott 2010; Martinussen et al., 2005). The primary treatment option for reducing the behavioral symptoms of ADHD is psychostimulant medication in the form of methylphenidate or amphetamine compounds, which also enhance visuospatial working memory (Bedard, Jain, Hogg-Johnson, & Tannock, 2007). This approach is considered a traditional treatment of ADHD and has been used in the United States since the 1970s. The aim of Holmes's study was therefore to compare the impacts of working memory training and fast-acting psychostimulant medication (Ritalin) on the separate subcomponents of working memory.

Holmes et al. recruited 25 children from ages 8 to 11 years with a clinical diagnosis of ADHD combined type who were receiving quick release medication (Ritalin) for their ADHD symptoms. The study was conducted in a school, and the teachers did not want children off their medication for 5 or 6 weeks. So, initial assessments were conducted first off meds and then on meds to ascertain the impact of medication. The children then stayed on their medications throughout the training program and were tested following training and then 6 months later. All children completed assessments of verbal and visuospatial STM and working memory both before and after training and on and off medication (AWMA; Alloway, 2009). The training paradigm consisted of 20–25 sessions on the adaptive training program developed by Cogmed. Children trained on eight of the ten tasks every day, completing 115 trials per session. Both interventions had a significant impact on children's working memory, but differential patterns of change were associated with each approach.

While medication led to selective improvements in visuospatial working memory, training led to improvements in all aspects of working memory. Crucially, these gains were sustained 6 months after training ceased. Children's IQ was not affected by either intervention, on a Wechsler screener, (WASI). The impact of medication on nonverbal aspects of working memory only most likely reflects the predominant influence of medication on right hemisphere brain structures that are associated with visuospatial working memory (e.g., Bedard et al., 2007). The generalized impact of working memory training in this group may have very practical benefits for learning in children with ADHD. Although medication helps to control the adverse behavioral symptoms of the disorder, providing improved working memory resources through working memory training promises improved support for learning in this group (Holmes, Gathercole, Place, et al., 2010). The study also showed an additive effect in which medication improved visual-spatial working memory, but the completion of Cogmed added significantly more gains. This finding suggests that, notably, Cogmed can effectively be combined with medication. So a child might train with Cogmed while on medication or off. The finding is also important in light of the 6- and 8-year follow-up of the multimodal treatment of ADHD results. In this study, ADHD groups, regardless of whether they received medication, behavioral treatment, or a combination of both, significantly improved from their baseline scores but were still significantly worse than a community sample on the vast majority of measures (Molina et al., 2009). In fact the authors of that study called for innovative targeted interventions for ADHD. The finding by Holmes, Gathercole, Place, et al. (2010) suggests that the possibility that a combination of medication with Cogmed might prove effective in bringing the functioning of the ADHD group closer to that of a control group.

What Is the Potential Impact of Cogmed?

We will explore this question by considering first visual-spatial memory, then verbal working memory, then direct measures of attention, and lastly the impact on academic achievement.

Visual-Spatial Working Memory

The primary target of Cogmed is working memory and, more specifically, visual-spatial working memory. Investigators from the Karolinska group have found significant improvement in visual-spatial working memory after completion of Cogmed for ages ranging from preschool to adults in their 60s to 70s (Bellander et al., 2011; Bergman-Nutley et al. (2011); Brehmer et al.,

2009, Brehmer et al., 2011, Brehmer, Westerberg, & Backman, 2012; Dahlin, 2010; Klingberg et al., 2002, Klingberg et al., 2005; McNab et al., 2009; Olesen, Westerberg, & Klingberg, 2004; Soderqvist, Bergman, Ottersen, Grill, & Klingberg, 2012; Thorell et al., 2009; Westerberg & Klingberg, 2007; Westerberg et al., 2007). Similarly, a number of independent investigators have found that Cogmed significantly improves visual-spatial working memory (Beck, Hanson, Puffenberger, Benninger, & Benninger, 2010; Gibson et al., 2011; Holmes et al., 2009, Holmes, Gathercole, & Dunning, 2010; Johansson & Tornmalm, 2012; Kronenberger, Pisoni, Henning, Colson, & Hazzard, 2011; Lundqvist, Grundstrom, Samuelsson, & Ronnberg, 2010; Mezzacappa & Buckner, 2010; Roughan & Hadwin, 2011). These results have been found to extend to 6 months (Dahlin, 2010; Holmes et al., 2009, Holmes, Gathercole, & Dunning, 2010).

Verbal Working Memory

Verbal working memory is the second prominent target of Cogmed, and several of the investigators from the Karolinska Institute have found significant improvements in this area (Bellander et al., 2011; Brehmer et al., 2009, 2011; Dahlin, 2010; Klingberg et al., 2005; McNab et al., 2009; Olesen et al., 2004; Thorell et al., 2009; Westerberg & Klingberg, 2007). Similarly, a number of independent investigators have found that Cogmed significantly improves verbal working memory (Green et al., 2012; Holmes et al., 2009, Holmes, Gathercole, & Dunning, 2010; Kronenberger et al., 2011; Lundqvist et al., 2010; Mezzacappa & Buckner, 2010; Roughan & Hadwin, 2011). These results have been found to extend to 6 months (Dahlin, 2010; Holmes et al., 2009, Holmes, Gathercole, Place, et al., 2010; Løhaugen et al. 2011).

Direct Measures of Attention

Direct measures of attention begin to address the seminal concern mentioned in section "Introduction" of this chapter, that is, whether the Cogmed intervention generalizes to other areas of cognitive functioning. The results in visual-spatial working memory and verbal working memory provide evidence that training effects result in the near transfer to independent measures of those constructs. Direct measures of attention begin to provide evidence of far transfer. For example, the test for following instructions in the Holmes et al. (2009) study was one example of operationalizing attention into observable behavior.

The PASAT was an outcome measure that assesses auditory attention in adults. It was found to be significantly improved in two studies of Cogmed with healthy young adults (Brehmer et al., 2009) and then with healthy older adults (ages 60–70) (Brehmer et al., 2011). The PASAT requires subjects to add consecutive numbers as they are presented on an auditory tape and respond orally with an accurate sum. Thorell et al. (2009) in a study of typically developing preschoolers found significant improvement on a continuous performance test. Kronenberger et al. (2011) studied sentence repetition with deaf children with cochlear implants. Sentence repetition is a critical learning variable for this population and one that requires auditory attention. Kronenberger et al. (2011) found that these subjects improved both at the conclusion of Cogmed and showed even greater improvement at 6 months. Similarly with adults with an average age of 47.5, Johansson and Tornmalm (2012) found a reduction in cognitive failures. Cognitive failures include things like having to reread text because a person has forgotten what he/she has just read or a person finds that he/she has forgotten why he/she went from one room in a house to another. One might consider these as failures of attention in daily living. Finally, in the most structured analysis of observable attention, Green et al. (2012) found that among a group of children from ages 7 to 14 diagnosed with ADHD that after completing Cogmed that they significantly reduced off-task behavior (Green et al., 2012). The measure used was the RAST, the restricted academic setting task, which includes observations made every 30 s of such behaviors as looking away from paper, out of seat, and playing with object. These findings support the conclusion that improvement in functioning extends to various direct measures of behavior indicating improved attention that lasts through 6 months (Holmes et al., 2009; Johansson & Tornmalm, 2012; Kronenberger et al., 2011).

Reduced Ratings of Inattention and ADHD Symptoms

A number of studies have found reductions of ADHD symptoms from parent ratings (Beck et al., 2010; Klingberg et al., 2002, Klingberg et al., 2005). Some studies have found reductions in teacher ratings of attention (Beck et al., 2010; Mezzacappa & Buckner, 2010; Roughan & Hadwin, 2011). Yet teacher ratings have not always been blind to the children receiving the intervention.

Academic Achievement and Learning

Cogmed is an intervention that has a primary target of working memory and possibly a secondary target of attention, while academic achievement and learning are arguably tertiary targets. To resume the desk space in the mind analogy, this intervention has increased desk space, but has the student worked on anything new on this revamped desk? One might argue that children are in school so shouldn't that be sufficient? Possibly, but one would expect that this is unlikely. This is because these are students who have missed weeks, months, and, in many cases, years of skill and knowledge development. Without that skill and knowledge present, information may be beyond their grasp. So the critical question is whether remediation to address the backlog of undeveloped skills and the paucity of knowledge been implemented? If no new material is presented within a particular domain (e.g., mathematics, reading) nor no new skills have been taught (e.g., social skills, anxiety management), then it would seem unreasonable to expect gains to be made in such areas. This is particularly evident with older children who may have missed years of opportunities for skill and knowledge development. So it is the expectation of these authors that for gains to be made in such areas, skill building, teaching, or training in the desired areas would have to follow Cogmed. Since the desk space in the mind has been improved, one would expect that subjects would now better learn such content or skills. Nonetheless, some learning and new skills have been found to have developed after Cogmed.

Dahlin (2010) found an improvement in reading comprehension after Cogmed. Holmes et al. (2009) found that 6 months after Cogmed, those school-aged children significantly improved in mathematics. Klingberg et al. (2002, 2005) found improvement in nonverbal reasoning. As noted previously, Kronenberger et al. (2011) found an improvement in sentence repetition. Løhaugen et al. (2011) found an improvement in a verbal learning task following Cogmed in which subjects had to remember an oral story. Beck et al. (2010) found improved executive functions as rated by parents and teachers. Among adults, Westerberg et al. (2007) found reduced cognitive problems in daily life and improvements in declarative memory. Also among adults, Brehmer et al. (2009, 2011) found improvements in episodic memory. As is seen from this diverse data set, a number of areas of learning appear to have been impacted by Cogmed. However, more consistent and sustained data within these diverse areas needs to be published to more unequivocally state that Cogmed has had a positive impact in these areas. Nevertheless, these results certainly appear promising.

Is There Any New Evidence?

To date there have been 34 published research studies on Cogmed. While many of the studies were by the Torkel Klingberg and his research team at the Karolinska Institute, since that time a number of independent research teams have explored the application of Cogmed with a variety of clinical populations. Most recently, a number of studies incorporated Cogmed into their design. Gray et al. (2012) explored the relationship between the effects of a computerized working memory training program on working memory, attention, and academics in adolescents with severe LD and comorbid ADHD. Soderqvist et al. (2012) explored the use of computerized training of nonverbal reasoning and working memory in children with intellectual disability and discussed the effects on the dopaminergic system.

Possibly the most compelling new investigation related to Cogmed and academic achievement is a study recently published in May of 2013 by Holmes & Gathercole. The second trial of this study included 50 children (25 in year 5 and 25 in year 6) aged 9-11 who were identified based upon having the lowest academic performance at a school in England. These children were matched with 50 children who were not trained, but who had also performed poorly on the national exams. Interestingly, students from Year 5 were trained as a whole class in a group of 25 students in a computer lab supervised by a head teacher and a classroom assistant at the end of the school day. Students from year 6 were trained in one group of 12 and another group of 13 supervised by the same staff at the end of the day. The size of these training groups and the fact that teachers oversaw training at school make this study distinct among Cogmed research. Students in year 6 who completed cogmed scored significantly higher on standardized national English and Math tests indicating greater progress at school across the academic year. The children in year 5 made significant progress in math. The results of this study are thought-provoking on three levels. First, students were selected not based upon a working memory or attention deficit, but based upon low academic achievement. Secondly, it showed significant improvement in academic achievement across a school year when students completed Cogmed. And finally, Cogmed was administered by teachers with the whole class training at once or the class divided into half and training with two separate groups. The results showed high compliance and good progress on trained activities as well as improvement on far transfer tasks. This data suggests a possible method for how Cogmed might be delivered in schools by identifying low academically achieving students and using a whole-class model administered by teachers. At this time, there are over 60 studies in development exploring the use of Cogmed with a variety of different clinical groups of different ages. Please refer to the Cogmed website for details: http://www.cogmed.com/research.

Is There Evidence of Brain Plasticity?

Is there evidence of changes in brain plasticity following working memory training? This is a particularly important question because it attempts to assess whether internal changes accompany the external behavioral changes noted above. It is possible that intensive computerized training induces long-term changes in plasticity in the brain regions that serve working memory. It may also be the case that such changes are shorter in duration. Two neuroimaging studies, one by Olesen et al. and the other by Westerberg and Klingberg (2007), showed increased activation in the parietal and prefrontal cortices following memory training. In their first study, they reported an increase in brain activity in both of these regions in three subjects following training on both verbal and visuospatial working memory tasks. In the second study, eight adults were scanned five times during training on three visuospatial working memory tasks over a 5-week period. Again, increases in neuronal activity were observed in the prefrontal and parietal regions (Fig. 26.3).

In a single-subject analysis, Westerberg and Klingberg (2007) showed training-induced changes were not due to activations of any additional area that was not activated before training. Rather, they observed that areas where task-related activity was seen increased in size following training. Related to this, changes in the density of prefrontal and parietal cortical dopamine receptors have been reported after memory training. Either too much or too little stimulation of D1 receptors results in impaired working memory task performance. Training-induced decreases in the binding potential of the receptor D1 associated with increases in working memory capacity were interpreted as demonstrating a high level of plasticity of the D1 receptor system (McNab et al., 2009).

Overall, Klingberg's team has shown that training induces changes in two brain regions, the parietal and prefrontal cortices, which are both associated with working memory. Prefrontal activation is positively correlated with children's working memory capacity (Klingberg et al., 2002;

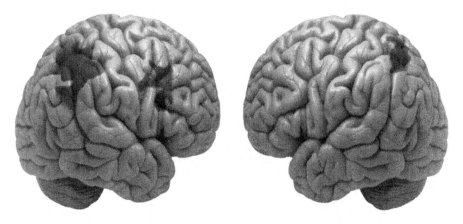

Fig. 26.3 The brain areas activated by Cogmed described in the research by Olesen et al. (2004)

Kwon, Reiss, & Menon, 2002), and frontoparietal networks are related to success on working memory tasks. These areas of the brain have also been implicated in the work of Curtis and D'Esposito (2003). The work of McNab suggests that cortical and biochemical changes result from practice on working memory tasks (McNab et al., 2009). These studies suggest there are changes at the biochemical level as well as in the brain structures in normal healthy adults following Cogmed training. This research has continued with older adult populations in the work of Bellander et al. (2011).

Summary

This chapter has discussed the subject of working memory and described an intervention to improve working memory. The scientific foundation of Cogmed, an empirically based intervention, the current research on its effectiveness, and the implications for academic achievement were included. Changes in brain plasticity that may follow Cogmed working memory training were noted. Cogmed has shown results in both near transfer to improved untrained verbal and visual-spatial working memory tasks and far transfer to improved attention and to some areas of learning. However, given the smaller number of studies thus far conducted investigating academic changes after Cogmed, prudence is suggested in interpreting those results. It is the opinion of the present authors that Cogmed primes a student for learning by increasing working memory capacity and sustained attention. However, in the case of ADHD children, the expected history of a lack of skill and knowledge development in both academic and social areas suggests that skill building and knowledge acquisition interventions should follow Cogmed to ensure the desired development. This is especially true with older children, given the likelihood of a paucity of existing skills. Yet, thus far, a few areas of specific learning and/or academic achievement were articulated that provide inklings of the possible future impact of this training. These data arguably support the posed concept of executive attention as an overlap between working memory and attention. The results of the over 60 ongoing studies are expected to clarify whether Cogmed alone can boost academic achievement or whether Cogmed might be more effectively utilized as a preparation for other academic interventions. The more recent study by Holmes & Gathercole (2013) suggests a possible model for Cogmed in schools which might clarify this issue by identifying students based upon low academic achievement and delivering Cogmed by teachers in a whole class format. Indeed, it could be the case that with populations with different presenting problems Cogmed may play distinct roles.

Study (year)	Sample (years old (YO))	Test	Treatment (n)	Control (n)	Treatment improvement (%)	Treatment improvement follow-up (%)	ES post-test (d)	ES follow-up (d)
Bergman-Nutley et al. (2011)	Typical 4 YO	Odd One Out (AWMA)	24	25	41		.96	
	Typical 4 YO, half dose	Odd One Out (AWMA)	27		27		.68	
Thorell et al. (2009)	Typical 4-5 YO	Span board (back+front)	17	14	40		.61	
	Typical 4-5 YO	Span board (back+front)		16				
Klingberg et al. (2005)	ADHD 7-12 YO	Span board (back+front)	20	24	19	21	.79	.81
Klingberg et al. (2002)	ADHD 7-15 YO	Span board (back+front)	7	7	45			
Kronenberger et al. (2011)	Deaf (w/CI) 7-15 YO	Span board (back)	9		25	11		
Holmes, Gathercole, Place, et al. (2010)	ADHD 8-11 YO	Mr. X (AWMA)	25		13	10		
Mezzacappa and Buckner (2010)	ADHD 8-10.5 YO	Finger Windows (WRAML)	8		33		.73	
Dahlin (2010)	Special Ed needs 9-12 YO	Span board (back)	41	25	30	26	.74	.65
Holmes et al. (2009)	Poor WM 10 YO	Composite score (AWMA)	22	20	17	15	.89	

Working Memory Training and Cogmed

Study	Group	Measure						
Roughan and Hadwin (2011)	SEBD ≈13 YO	Composite score (Span board & Digit Span)	7	8	24	29	2.29	
Løhaugen et al. (2011)	Preterm (ELBW) 14–15 YO	Span board (back)	16	11	37	26		
	Typical 14–15 YO	Span board (back)	19		20	16		
Summary child			*235*	*150*	*29*	*18*	*.87*	*.73*
Brehmer et al. (2012)	Typical 20–30 YO	Span board (back)	29	26	27	28	1.72	1.36
	Typical (aging) 60–70 YO	Span board (back)	26	19	24	29	1.32	1.65
McNab et al. (2009)	Typical 20–28 YO	Span board (back)	13		22			
Lundqvist et al. (2010)	ABI 20–65 YO	Span board (back)	21		21	29		
Westerberg et al. (2007)	Stoke 34–65 YO	Span board (back + front)	9	9	19		.83	
Brehmer et al. (2011)	Typical (aging) 60–70 YO	Span board (back)	12	11	16		1.03	
Summary adult			*110*	*65*	*21*	*28*	*1.23*	*1.51*

This table indicates that there are some robust effect sizes in the Cogmed research with visual-spatial working memory with children and adults. The average for children is .77. For adults the average is 1.51. In total there were 345 in treatment and 270 in controls. The overall effect size was .94 for posttest and 1.12 for follow-up. The reader is referred to the original sources for details (Ralph, 2010): Unpublished Results; www.Cogmed.com

This issue will only be resolved with additional research. The reader is asked to draw his or her own conclusions about the scientific validity of the demonstrated efficacy of this intervention in the larger context of traditional treatments and to be prepared for further clarification as many more studies on Cogmed are published in the next several years.

The following table summarizes the research discussed above with the effect sizes. (The authors wish to express their thanks to Kat Ralph and Sissela Bergman-Nuttley and Torkel Klingberg for giving their permission to include this table summarizing Cogmed research.)

References

Alloway, T. P. (2009). Working memory, but not IQ, predicts subsequent learning in children with learning difficulties. *European Journal of Psychological Assessment, 25*(2), 92–98.

Baddeley, A. D. (1996). The concept of working memory. In S. E. Gathercole (Ed.), *Models of short-term memory* (pp. 1–27). Hove: Psychology Press.

Baddeley, A. D. (2000). The episodic buffer: A new component of working memory? *Trends in Cognitive Sciences, 4*(11), 417–423.

Baddeley, A. D., Emslie, H., Kolodny, J., & Duncan, J. (1998). Random generation and the executive control of working memory. *The Quarterly Journal of Experimental Psychology A: Human Experimental Psychology, 51A*(4), 819–852.

Baddeley, A. D., & Hitch, G. J. (1974). Working memory. In G. H. Bower (Ed.), *The psychology of learning and motivation* (Vol. 8, pp. 47–87). London: Academic Press.

Beck, S. J., Hanson, C. A., Puffenberger, S. S., Benninger, K. L., & Benninger, W. B. (2010). A controlled trial of working memory training for children and adolescents with ADHD. *Journal of Clinical Child and Adolescent Psychology, 39*(6), 825–836.

Bedard, A.-C., Jain, U., Hogg-Johnson, S., & Tannock, R. (2007). Effects of methylphenidate on working memory components: Influence of measurement. *Journal of Child Psychology and Psychiatry, 48*(9), 872–880.

Bellander, M., Brehmer, Y., Westerberg, H., Karlsson, S., Further, D., Bergman, O., et al. (2011). Preliminary evidence that allelic variation in the LmX1A gene influences training-related working memory improvement. *Neuropsychologia, 49*(7), 1938–1942.

Bergman-Nutley, S., Soderqvist, S., Bryde, S., Thorell, L. B., Humphreys, K., & Klingberg, T. (2011). Gains in fluid intelligence after training non-verbal reasoning in 4-year-old children: A controlled, randomized study. *Developmental Science, 14*(3), 591–601.

Brehmer, Y., Rieckmann, A., Bellander, M., Westerberg, H., Fischer, H., & Backman, L. (2011). Neural correlates of training-related working-memory gains in old age. *NeuroImage, 58*(4), 1110–1120.

Brehmer, Y., Westerberg, H., & Backman, L. (2012). Working-memory training in younger and older adults: Training gains, transfer, and maintenance. *Frontiers in Human Neuroscience, 6*(63).

Brehmer, Y., Westerberg, H., Bellander, M., Furth, D., Karlsson, S., & Backman, L. (2009). Working memory plasticity modulated by dopamine transporter genotype. *Neuroscience Letters, 467*(2), 117–120.

Bull, R., Espy, K. A., & Wiebe, S. A. (2008). Short-term memory, working memory, and executive functioning in preschoolers: Longitudinal predictors of mathematical achievement at age 7 years. *Developmental Neuropsychology, 33*(3), 205–228.

Carretti, B., Borella, E., Cornoldi, C., & De Beni, R. (2009). Role of working memory in explaining the performance of individuals with specific reading comprehension difficulties: A meta-analysis. *Learning and Individual Differences, 19*(2), 245–251.

Carretti, B., Borella, E., & De Beni, R. (2007). Does strategic memory training improve the working memory performance of younger and older adults? *Experimental Psychology, 54*(4), 311–320.

Cavallini, E., Pagnin, A., & Vecchi, T. (2002). The rehabilitation of memory in old age: Effects of mnemonics and metacognition in strategic training. *Clinical Gerontologist: The Journal of Aging and Mental Health, 26*(1–2), 125–141.

Comblain, A. (1994). Working memory in Down's syndrome: Training the rehearsal strategy. *Down Syndrome: Research and Practice, 2*(3), 123–126.

Conners, F. A., Rosenquist, C. J., Arnett, L., Moore, M. S., & Hume, L. E. (2008). Improving memory span in children with Down syndrome. *Journal of Intellectual Disability Research, 52*(3), 244–255.

Cowan, N. (2005). *Working memory capacity*. New York: Psychology Press.

Curtis, C. E., & D'Esposito, M. (2003). Persistent activity in the prefrontal cortex during working memory. *Trends in Cognitive Sciences, 7*(9), 415–423.

Dahlin, K. (2010). Effects of working memory training on reading in children and with special needs. *Reading and Writing, 24*(4), 479–491.

Daneman, M., & Merikle, P. M. (1996). Working memory and language comprehension: A meta-analysis. *Psychonomic Bulletin & Review, 3*(4), 422–433.

Dehn, M. J. (2008). *Working memory and academic learning: Assessment and intervention*. John Hoboken, NJ: Wiley.

Dehn, M. J. (2010). *Long-term memory problems in children and adolescents: Assessment, intervention and effective instruction*. Hoboken, NJ: Wiley.

Diamond, A., & Lee, K. (2011). Interventions shown to aid executive function development in children 4 to 12 years old. *Science, 333*(6045), 959–964.

Engle, R. W., & Kane, M. J. (2004). Executive attention, working memory capacity, and a two-factor theory of cognitive control. In B. Ross (Ed.), *The psychology of learning and motivation* (pp. 145–199). New York: Elsevier Science.

Engle, R. W., Kane, M. J., & Tuholski, S. W. (1999). Individual differences in working memory capacity and what they tell us about controlled attention, general fluid intelligence and functions of the prefrontal cortex. In A. Miyake, A. & P. Shah (Eds.), *Models of working memory: Mechanisms of active maintenance and executive control* (pp. 102–134). New York: Cambridge University Press.

Ericsson, K. A., & Chase, W. G. (1982). Exceptional memory. *American Scientist, 70*(6), 607–615.

Erzine, G. A. (2011). *Effects of language on the development of executive functions in preschool children*. Department of Counseling and Psychological Services at Digital Archive at GSU. Paper No. 41. Doctoral dissertation, Georgia State University, Atlanta, GA.

Flavell, J. H., Beach, D. R., & Chinsky, J. M. (1966). Spontaneous verbal rehearsal in a memory task as a function of age. *Child Development, 37*(2), 283–299.

Gathercole, S. E., Alloway, T. P., Willis, C., & Adams, A. (2006). Working memory in children with reading disabilities. *Journal of Experimental Child Psychology, 93*(3), 265–281.

Gathercole, S., Brown, L., & Pickering, S. J. (2003). Working memory assessments at school entry as longitudinal predictors of National Curriculum attainment levels. *Educational and Child Psychology, 20*(3), 109–122.

Gathercole, S. E., Durling, E., Evans, M., Jeffcock, S., & Stone, S. (2008). Working memory abilities and children's performance in laboratory analogues of classroom activities. *Applied Cognitive Psychology, 22*(8), 1019–1037.

Gathercole, S. E., & Pickering, S. J. (2000). Working memory deficits in children with low achievement in the national curriculum at 7 year of age. *British Journal of Educational Psychology, 70*, 177–194.

Gathercole, S. E., Pickering, S. J., Ambridge, B., & Wearing, H. (2004). The structure of working memory from 4 to 15 years of Age. *Developmental Psychology, 40*(2), 177–190.

Gibson, B. S., Gondoli, D. M., Johnson, A. C., Steeger, C. M., Dobizenski, B. A., & Morrissey, R. A. (2011). Component analysis of verbal versus spatial working memory training in adolescents with ADHD: A randomized, controlled trial. *Child Neuropsychology, 17*(6), 546–563.

Goldman-Rakic, P. S. (1988). Topography of cognition: Parallel distribution networks in Primate Association Cortex. *Annual Review of Neuroscience, 11*(1), 137–156.

Gray, S. A., Chaban, P., Martinussen, R., Goldberg, R., Gotlieb, H., Kronitz, R., et al. (2012). Effects of a computerized working memory training program on working memory, attention and academics in adolescents with severe LD and comorbid ADHD: A randomized controlled trial. *Journal of Child Psychology and Psychiatry, 53*(12), 1277–1284.

Green, C. T., Long, D. L., Green, D., Losif, A. M., Dixon, J. F., Miller, M. R., et al. (2012). Will working memory training generalize to improve off-task behavior in children with attention deficit/hyperactivity disorder? *Neurotherapeutics, 9*(3), 639–48.

Holmes, J., Gathercole, S. E., & Dunning, D. L. (2009). Adaptive training leads to sustained enhancement of poor working memory in children. *Developmental Science, 12*(4), F9–F15.

Holmes, J., Gathercole, S., & Dunning, D. (2010). Poor working memory: Impact and interventions. In J. Holmes (Ed.), *Advances in child development and behavior* (Vol. 39, pp. 1–43). Academic Press: London.

Holmes, J., Gathercole, S. E., Place, M., Dunning, D. L., Hilton, K. A., & Elliott, J. G. (2010). Working memory deficits can be overcome: Impacts of training and medication on working memory in children with ADHD. *Applied Cognitive Psychology, 24*(6), 827–836.

Huang-Pollock, C. L., & Karalunas, S. L. (2010). Working memory demands impair skill acquisition in children with ADHD. *Journal of Abnormal Psychology, 119*(1), 174–185.

Hulme, C., & Muir, C. (1985). Developmental changes in speech rate and memory span: A causal relationship? *British Journal of Developmental Psychology, 3*(2), 175–181.

Johansson, B., & Tornmalm, M. (2012). Working memory training for patients with acquired brain injury: Effects in daily life. *Scandinavian Journal of Occupational Therapy, 19*(2), 176–183.

Just, M. A., & Carpenter, P. A. (1992). A capacity theory of comprehension: Individual differences in working memory. *Psychological Review, 99*(1), 122–149.

Kane, M. J., Conway, A. R., Hambrick, D. Z., & Engle, R. W. (2007). Variation in working memory capacity as variation in executive attention and control. In A. R. Conway, C. Jarrold, M. J. Kane, A. T. Miyake, & J. N. Towse (Eds.), *Variation in working memory* (pp. 21–46). New York: Oxford University Press.

Kane, M. J., Engle, R. W. (2000). *Working memory capacity, task set, and stroop interference in speed and accuracy*. Abstracts of the Psychonomic Society: 41st Annual Meeting.

Klingberg, T., Fernell, E., Olesen, P. J., Johnson, M., Gustafsson, P., Dahlstrom, K., et al. (2005). Computerised training of working memory in children with ADHD-a randomised, controlled trial. *Journal of the American Academy of Child and Adolescent Psychiatry, 44*(2), 177–186.

Klingberg, T., Forssberg, H., & Westerberg, H. (2002). Increased brain activity in frontal and parietal cortex underlies the development of visuospatial working memory capacity during childhood. *Journal of Cognitive Neuroscience, 14*(1), 1–10.

Kristofferson, M. W. (1972). Effects of practice on character-classification performance. *Canadian Journal of Psychology, 26*(1), 54–60.

Kronenberger, W. G., Pisoni, D. B., Henning, S. C., Colson, B. G., & Hazzard, L. M. (2011). Working memory training for children with cochlear implants: A pilot study. *Journal of Speech, Language, and Hearing Research: JSLHR, 54*(4), 1182–1196.

Kwon, H., Reiss, A. L., & Menon, V. (2002). Neural basis of protracted developmental changes. *Proceedings of the National Academy of Sciences of the United States of America, 99*(20):13336–13341. Retrieved November 7, 2012, from http://www.pnas.org/content/99/20/13336.full.pdf

Løhaugen, G. C., Antonsen, I., Håberg, A., Gramstad, A., Vik, T., Brubakk, A. M., et al. (2011). Computerized working memory training improves function in adolescents born at extremely low birth weight. *The Journal of Pediatrics, 158*(4), 555–561.

Lundqvist, A., Grundstrom, K., Samuelsson, K., & Ronnberg, J. (2010). Computerized training of working memory in a group of patients suffering from acquired brain injury. *Brain Injury, 24*(10), 1173–1183.

Martinussen, R., Hayden, J., Hogg-Johnson, S., & Tannock, R. (2005). A meta-analysis of working memory impairments in children with attention-deficit/hyperactivity disorder. *Journal of the American Academy of Child and Adolescent Psychiatry, 44*(4), 377–384.

Mayes, S. D., & Calhoun, S. L. (2007). Wechsler Intelligence Scale for Children-Third and -Fourth Edition predictors of academic achievement in children with attention-deficit/hyperactivity disorder. *School Psychology Quarterly, 22*(2), 234–249.

McNab, F., Leroux, G., Strand, F., Thorell, L., Bergman, S., & Klingberg, T. (2008). Common and unique components of inhibition and working memory: An fMRI, within-subjects investigation. *Neuropsychologia, 46*(11), 2668–2682.

McNab, F., Varrone, A., Farde, L., Jucaite, A., Bystritsky, P., Forssberg, H., et al. (2009). Changes in cortical dopamine D1 receptor binding associated with cognitive training. *NeuroImage, 47*, S77–S77.

McNamara, D. S., & Scott, J. L. (2001). Working memory capacity and strategy use. *Memory & Cognition, 29*(1), 10–17.

Mezzacappa, E., & Buckner, J. C. (2010). Working memory training for children with attention problems or hyperactivity: A school-based pilot study. *School Mental Health, 2*(4), 202–208.

Miyake, A., & Shah, P. (1999). *Models of working memory: Mechanisms of active maintenance and executive control*. New York: Cambridge University Press.

Molina, B. S., Hinshaw, S. P., Swanson, J. M., Arnold, E. L., Vitiello, B., Jensen, P. S., et al. (2009). The MTA at 8 years: Prospective follow-up of children treated for combined-type ADHD in a multisite study. *Journal of the American Academy of Child and Adolescent Psychiatry, 48*(5), 484–500.

Morrison, A. B., & Chein, J. M. (2011). Does working memory training work? The promise and challenges of enhancing cognition by training working memory. *Psychonomic Bulletin & Review, 18*(1), 46–60.

Olesen, P. J., Westerberg, H., & Klingberg, T. (2004). Increased prefrontal and parietal activity after training of working memory. *Nature Neuroscience, 7*(1), 75–79.

Phillips, C. J., & Nettlebeck, T. (1984). Effects of practice on recognition memory of mildly mentally retarded adults. *American Journal of Mental Deficiency, 88*(6), 678–687.

Ralph, K. J. (2012). COGMED research claims & evidence, Version 1.3. Retrieved from www.cogmed.com/research.

Roberts, R. J., & Pennington, B. F. (1996). An interactive framework for examining prefrontal cognitive processes. *Developmental Neuropsychology, 12*(1), 105–126.

Roughan, L., & Hadwin, J. A. (2011). The impact of working memory training in young people with social, emotional and behavioral difficulties. *Learning and Individual Differences, 21*(6), 759–764.

Rust, J., Golombok, S., & Trickey, G. (1993). *WORD, Wechsler objective reading dimensions manual*. London: Psychological Corporation.

Schmiedek, F., Lovden, M., & Lindenberger, U. (2010). Hundred days of cognitive training enhance broad cognitive abilities in adulthood: Findings from the COGITO study. *Frontiers in Aging Neuroscience, 2*. pii, 27.

Séguin, J. R., Nagin, D., Assaad, J., & Tremblay, R. E. (2004). Cognitive-neuropsychological function in chronic physical aggression and hyperactivity. *Journal of Abnormal Psychology, 113*(4), 603–613.

Soderqvist, S., Bergman, N. S., Ottersen, J., Grill, K. M., & Klingberg, T. (2012). Computerized training of non-verbal reasoning and working memory in children with intellectual disability. *Frontiers in Human Neuroscience, 6*, 271.

Swanson, H. L., Jerman, O., & Zheng, X. (2008). Growth in working memory and mathematical problem solving in children at risk and not at risk for serious math difficulties. *Journal of Educational Psychology, 100*(2), 343–379.

Swanson, H. L., Zheng, X., & Jerman, O. (2009). Working memory, short-term memory, and reading disabilities: A selective meta-analysis of the literature. *Journal of Learning Disabilities, 42*(3), 260–287.

Thorell, L. B., Lindqvist, S., Bergman Nutley, S., Bohlin, G., & Klingberg, T. (2009). Training and transfer effects of executive functions in preschool children. *Developmental Science, 12*(1), 106–113.

Turley-Ames, K. J., & Whitfield, M. M. (2003). Strategy training and working memory task performance. *Journal of Memory and Language, 49*(4), 446–468.

Wechsler, D. (1996). *Wechsler objective number dimensions: WOND*. New York: Psychological Corporation.

Wechsler, D. (1999). *The Wechsler abbreviated scale of intelligence: WASI*. San Antonio, TX: Psychological Corporation/Harcourt Brace.

Wechsler, D. (2002). *The Wechsler preschool and primary scale of intelligence* (3rd ed.). San Antonio, TX: The Psychological Corporation.

Westerberg, H., Hirvikoski, T., Forssberg, H., & Klingberg, T. (2004). Visuo-spatial working memory span: A sensitive measure of cognitive deficits in children with ADHD. *Child Neuropsychology, 10*(3), 155–161.

Westerberg, H., Jacobaeus, H., Hirvikoski, T., Clevberger, P., Ostensson, M. L., Bartfai, A., et al. (2007). Computerized working memory training after stroke—A pilot study. *Brain Injury, 21*(1), 21–29.

Westerberg, H., & Klingberg, T. (2007). Changes in cortical activity after training of working memory—A single-subject analysis. *Physiology & Behavior, 92*(1–2), 186–192.

Willcutt, E. G., Doyle, A. E., Nigg, J. T., Faraone, S. V., & Pennington, B. F. (2005). Validity of the executive function theory of attention-deficit/hyperactivity disorder: A meta-analytic review. *Biological Psychiatry, 57*(11), 1336–1346.

Supporting and Strengthening Working Memory in the Classroom to Enhance Executive Functioning

27

Milton J. Dehn

Thirteen-year-old "Jacob" had deficits in all aspects of executive functioning measured by the Behavior Rating Inventory of Executive Function® (BRIEF; Gioia, Isquith, Guy, & Kenworthy, 2000). His *T*-score on the BRIEF's *Global Executive Composite* was an 81 and his *T*-scores on the BRIEF's clinical scales were generally two standard deviations above the mean. Even before his special education teacher completed the BRIEF, it was evident that Jacob had significant problems with executive functions. As corroborated by formal test scores, Jacob struggled with planning, organization, and initiation, leading to extreme difficulty getting started on and completing homework. His poor ability to inhibit, shift, and monitor were consistent with reports that he had social difficulties. His severe deficits in all aspects of controlling attention had resulted in an early diagnosis of attention-deficit hyperactivity disorder (ADHD). Even when he was on medication for ADHD, it was very challenging for him to focus and sustain attention; he needed a break every 20 min. To address his academic learning impairments, Jacob received services for learning disabilities in mathematics and written language. Because Jacob really wanted to perform well in school, he fully cooperated with after-school tutoring and interventions that addressed his executive dysfunctions.

M.J. Dehn (✉)
Schoolhouse Educational Services,
Onalaska, WI, USA
e-mail: milt@psychprocesses.com

Despite Jacob's cooperation and his teachers', counselor's, foster parents', and tutor's best efforts, the challenges and struggles remained, and Jacob was frequently frustrated and overwhelmed. There was an executive function dimension that no one had considered until Jacob was reevaluated when he was in seventh grade. This overlooked dimension was working memory. Jacob's test scores on the various components of working memory were in the low-average range, midway between his average IQ and his below-average general executive functioning. Working memory, the ability to retain information in the short term while simultaneously processing the same or other information, is essential for all types of academic learning (Dehn, 2008) and scholastic performance (Gathercole & Alloway, 2008). Jacob's weaknesses in both visual-spatial and verbal working memory were evident as he struggled to complete arithmetic problems and express his ideas in writing.

When Jacob's social worker and foster parents became aware of the crucial role working memory plays in academic learning, they agreed to allocate time for working memory training during Jacob's tutoring hours. Jacob was certainly a candidate for an evidence-based working memory training program, such as the Internet-based Cogmed® program described in this book's chapter by Klingberg, but his caretakers declined this option. This author than began to teach Jacob some working memory strategies and some effective memorization strategies for long-term memory. The very first time a multistep strategy was taught, it became evident that learning and utilizing memory strategies was going to be very challenging for Jacob. The problem was that the cognitive processing required to learn and apply strategy steps added to Jacob's working memory load, causing him to forget even more information. Jacob was experiencing the dilemma that has been observed when individuals with low working memory ability initially try to learn and apply a strategy that can ultimately improve their working memory performance. This dilemma explains why so many students with learning disabilities and low working memory ability shy away from strategy use. Given Jacob's broad-based deficits in executive functioning, achieving success with memory strategies was going to be especially challenging. Recognizing the importance of managing and reducing the "cognitive load" placed on Jacob's working memory, consultation with Jacob's teachers and tutors became the priority. The goal was to help them understand how to minimize cognitive load during instruction so that Jacob would need less reteaching outside of the classroom.

Working Memory Functions and Limitations

Working memory is the limited cognitive capacity to retain information in the short term while simultaneously manipulating the same or other information. Short-term memory differs from the construct of working memory in that short-term memory is a brief, passive storage with covert rehearsal and automated processes that activate long-term memory structures. The main difference between working memory and short-term memory is the addition of active, conscious processing. The classic digit span measure illustrates the distinction: digits forward involves short-term memory, whereas digits backward requires working memory because one must manipulate the digits in order to reverse the sequence.

Most conceptualizations of working memory embed short-term memory within working memory structure. In Baddeley's (1986, 2006) widely accepted four-component model of working memory, two modality-specific (auditory and visual-spatial) short-term storage components are "supervised" by "executive working memory." According to Baddeley, the executive component is the essence of working memory; it is viewed as a mechanism of attentional control. The key functions of the executive include shifting, updating, and inhibition, as well as allocating the available attentional resources and applying strategies. The fourth component, the "episodic buffer," is an interface between working memory and long-term memory, where new information is integrated with activated, long-term episodic and semantic representations.

Several competing models of working memory emphasize the interaction between working memory and long-term memory. Some theories (e.g., Cowan, 2005; Ericsson & Kintsch, 1995; Oberauer, 2002) posit that working memory is the limited capacity region within the currently activated portion of long-term memory. What is being processed in working memory at any given time is called the "focus of attention." According to Cowan (2005), the adult human can focus attention on a maximum of four units of information at a time. When the processing demands are high, the number of items that can be focused on is reduced. For example, children with working memory impairments can probably focus on only one piece of information at a time.

Working Memory and Executive Function

Among neuropsychologists, there is a consensus that working memory is just one of several disparate executive functions that control cognitive performance (St. Clair-Thompson, 2011). Working memory is strongly related with general executive functioning; they both involve processing related to goal-directed behavior. McCabe, Roediger, McDaniel, Balota, and Hambrick (2010) reported a correlation of .97 between these two constructs. According to McCabe et al., working memory and general executive functioning share a common underlying executive attention, or attentional control, component. Specifically, working memory and general executive function both involve inhibition, shifting, focusing, and updating. What sets them apart is that general executive function also includes processes unrelated to retention of information, such as social functioning, impulse control, emotion regulation, monitoring, planning, and cognitive flexibility (McCloskey, Perkins, & Van Divner, 2009).

Further evidence for the integration and separation of working memory and executive function is provided by the characteristics of children with executive impairments, such as ADHD. Working memory is one of several cognitive and executive functions that is typically impaired in children with ADHD (Gathercole et al., 2008), such as children like Jacob. However, it is primarily children with the inattentive subtype of ADHD, not the hyperactive and impulsive subtype, who have co-occurring working memory problems. Moreover, children with generalized executive function impairment usually do not have a specific deficit in working memory (Gathercole et al., 2008).

Working Memory and Cognitive Load

The theory that best explains the working memory challenges experienced by students like Jacob is known as *cognitive load theory* (Van Merrienboer, Kirschner, & Kester, 2003). Cognitive load theory emphasizes the limited cognitive capacity of working memory and how easily working memory can become overloaded during instruction and learning processes (de Jong, 2010). There is a consensus among all working memory theorists that conscious, effortful cognitive processing and temporary storage of information draw from the same working memory capacity. Working memory "resources" must be "shared" between temporarily storing information and processing information.

The amount of time and effort required for the processing component of working memory is known as *cognitive load*. As cognitive load increases, the amount of information that can be retained is diminished, mainly because there is not enough time to devote to focusing on the information to be retained, resulting in the information not being rehearsed frequently enough to prevent decay. As explained by Barrouillet, Bernardin, Portrat, Vergauwe, and Camos (2007), the cognitive load of any given task is a function of the proportion of time during which it captures attention, thus impeding other attention-demanding processes, such as rehearsal. This relationship is linear. As cognitive load increases, there is a corresponding decrease in how much information is retained (Barrouillet, Gavens, Vergauwe, Gaillard, & Camos, 2009). This relationship is also bidirectional. Focusing on maintaining information can impede processing, slowing it down or causing processing errors.

The limited pool of cognitive resources known as working memory is apparently shared between the two functions of processing and storage. The specific resource being shared is attention (Barrouillet, Portrat, & Camos, 2011; Engle, 2002). When attention is required for processing, it is not available for the maintenance of memory items and consequently the items fade away. Effective time sharing of attention involves rapid, back-and-forth switching of attention from processing to maintenance (rehearsal). Memory items are lost when the processing requirements are such that the switching cannot occur or cannot occur in time to prevent loss of information. One reason working memory performance improves from childhood to adulthood is that rehearsal becomes more automated and effortless. For example, Chen and Cowan (2009) report that the amount of attention needed for verbal rehearsal dramatically decreases between second and sixth grades.

When students have working memory deficits, they are frequently described by their teachers as "inattentive." These assessments are actually quite accurate. As cognitive load increases, the executive control of attention is reduced (Hester & Garavan, 2005), resulting in a diminished ability to exert inhibitory control over extraneous, irrelevant processing and information. Consequently, focus is lost, the mind wanders, and the task is abandoned (Gathercole & Alloway, 2008). Furthermore, diminished control of attention also makes switching and updating difficult, causing loss of information. Compared with individuals with high working memory spans, the recall of individuals with low spans suffers more from attention-demanding processing tasks (Kane & Engle, 2000).

Cognitive load theorists attempt to address working memory overload in the classroom by identifying causes of overload and by promoting instructional design that minimizes load (de Jong, 2010). During instruction and learning, part of the cognitive load is inherent to the characteristics of the content and material to be learned, part is caused by the instructional behaviors of the teacher, and part is imposed by the learner's internal processing of the information (Kirschner, 2002). The concern is that learning is reduced when too much cognitive load results in working memory overload during a learning task.

Working Memory and Academic Learning

Working memory is required during all aspects of engaged learning because learning requires manipulation of incoming information, integration of new information with existing long-term memory representations, and continuous, simultaneous processing and storage of information. Classic examples of active working memory in the classroom include remembering step-by-step directions while completing a task, comprehending instruction and retaining the information while taking notes, and remembering subproducts while mentally completing a multistep arithmetic problem. Encoding new learning into long-term memory also places high demands as working memory as the learner constructs and modifies semantic networks. An engaged student is continuously pushing the limits of his or her working memory capacity. Clearly, students with working memory impairments like Jacob frequently experience working memory overload in the typical classroom (Alloway, 2011). When this occurs, learning opportunities are lost.

Numerous studies (e.g., Swanson & Berninger, 1995, 1996) have investigated the relations among working memory components and specific academic skills. The development and performance of reading decoding skills, reading comprehension, mathematics calculation, mathematics reasoning, and written expression (see Dehn, 2008 for details) are highly dependent on adequate working memory capacity. The modality-specific working memory components align differently, depending on the academic skill; for example, mathematics draws heavily on visual-spatial storage, whereas reading decoding draws heavily on auditory (phonological) storage. The correlations between measures of working memory and specific academic skills are generally in the .5 range (Dehn, 2008). Accordingly, working memory ability is highly predictive of academic skill acquisition, and children with learning disabilities frequently have working memory impairments (Alloway, Gathercole, Adams, & Willis, 2005; Swanson & Berninger, 1995).

General classroom performance, such as assignment completion, also depends heavily on adequate working memory capacity. Children with working memory impairments frequently abandon a task without completing it (Alloway, 2006). The typical teacher views this kind of difficulty as an attention, behavior, or motivation problem, not realizing that a working memory deficit may be accounting for the difficulty (Gathercole et al., 2008).

Instructional Approaches That Reduce Cognitive Load

Reducing and managing cognitive load is an effective way of supporting working memory functioning and enhancing learning and performance in the classroom (de Jong, 2010). Reducing the processing demands in the learning environment will help impaired students overcome the working memory limitations that are impacting their learning (Elliott, Gathercole, Alloway, Holmes, & Kirkwood, 2010). The extent of cognitive load during any learning task depends on the teacher's instructional methods, the learner's internal processes, and the nature of the content and structure of the materials (van Gog, Ericsson, Rikers, & Paas, 2005). When teachers learn to better recognize the specific cognitive load variables within each of these three areas, they can more effectively design instruction and select curriculum to minimize load. They also will be able to more effectively teach students how to better manage their internal cognitive load. The next section provides guidelines for evaluating the amount of cognitive load.

Cognitive Load Variables

The amount of cognitive load arising from the nature of the content and materials is determined by:

1. The amount of material and the difficulty and complexity of the subject matter. Smaller units of information require less integration and relational processing, and there also is less information to maintain while processing. More difficult and complex material requires more resources to process, but the processing challenge may be ameliorated somewhat by expertise in the subject matter.
2. The sequencing of the material. Material sequenced from simple to complex minimizes load as the student progresses through the material.
3. The novelty of the subject matter. The less prior knowledge the learner has, the greater the load.
4. The organization of the materials. Requiring the learner to integrate disorganized materials will add significantly to the load. For example, simply presenting the information on multiple sheets of paper increases processing demands because the learner must combine several sources of information (Jang, Schunn, & Nokes, 2011).
5. Whether the information to be processed and the information to be remembered are the same or different. For instance, the learning objective may be to recall the capitol of each state, but the materials may require the student to locate each capitol on a map.

The amount of cognitive load arising from the type of instruction and from teaching behaviors is determined by:

1. The instructor's language and verbosity. Wordiness and complex language add extraneous processing load to the task. Simple, concise, consistent wording allows the learner to focus on the required processing (Gathercole & Alloway, 2008).
2. The length of the lesson. Lengthy lessons create more proactive and retroactive interference as more and more information is added. The need to inhibit interference adds to cognitive load.
3. The organization of the instruction. Well-organized instruction makes fewer demands on the learner's processing.
4. How well the teacher elaborates. Elaboration is the process of explicitly linking new information to prior knowledge in a manner that helps the learner understand the relations.
5. The amount of time allowed for processing and maintenance. Students who are allowed more time have more opportunities to switch between processing and rehearsal.
6. How much secondary processing is required. For example, a student listening to a lecture is processing the information in order to comprehend it and associate it with related schemas. The requirement to take notes while listening adds the processes of transcribing words and converting thoughts into notes.

The amount of cognitive load arising from the learner's internal processing is determined by:

1. How much internally generated interference needs to be inhibited and the individual's inhibition ability. The need to inhibit irrelevant associations and thoughts adds to cognitive load.
2. The learner's levels of mastery, expertise, and prior knowledge. The less developed these are, the greater the amount of processing required.
3. Other cognitive factors related to working memory performance, such as the learner's processing speed and fluid reasoning ability (Dehn, 2008). Slow processing speed will increase cognitive load and decrease retention because rehearsal cannot occur frequently enough.

4. The use of well-developed memory strategies, such as chunking. Strategies that are mastered and automated can function effectively without creating significant processing demands.
5. The level of metamemory development (Dehn, 2010). Learners with advanced understanding of memory functions, cognitive load, and their personal memory weaknesses can make informed decisions and selections to regulate the type and amount of processing they engage in during learning and working memory tasks.

Procedures for Reducing Cognitive Load

Some methods for minimizing the load arising from the nature of the content and materials are indicated in the previous section. For example, simple material presented in small units keeps cognitive load manageable. Additional procedures for minimizing inherent cognitive load include:
1. Providing worked examples or partially completed examples, such as a completed mathematics problem. Having the examples available also reduces the need to hold several elements in temporary storage.
2. Arranging and integrating the information so that there is only one source. If multiple sources of information must be used during a learning task, they should be arranged in a side-by-side fashion (Jang et al., 2011).
3. Presenting arithmetic problems vertically, rather than horizontally (Alloway, 2011).
4. Providing materials that allow the student to focus on processing without the need to maintain task-relevant information. Keeping lists of information or procedural steps in view reduces working memory load. Written reminders of problem-solving steps reduce problem-solving search and evaluation strategies that impose a heavy cognitive load. Other examples include number lines and a list of frequently misspelled words (Gathercole & Alloway, 2008).
5. Beginning with just a few elements that can be learned in isolation and gradually adding more.

The previous section on "Cognitive Load Variables" has several implications for instructional procedures and modifications. For example, elaborating and using simple, concise language will influence the degree of cognitive load. Additional recommendations include:
1. Maintaining a quiet learning environment. The need to inhibit interference from distractions increases processing load.
2. Differentiating instruction such that the processing demands are appropriately matched to the individual learner's working memory capacity.
3. Utilizing structured teaching approaches, such as *Direct Instruction*, that have built-in repetition so that the learner can focus more on the processing dimension and less on maintenance.
4. Avoiding presentation, or even mention, of nonessential or confusing information in order to reduce unnecessary processing. All information and required processing should be germane to the task or the material to be learned. Procedural steps should not be presented until they are actually needed.
5. Presenting material both verbally and visually may reduce processing challenges in students with a relative weakness in one modality (de Jong, 2010).
6. Requiring the student to focus on only one process at a time. Multitasking should be avoided.

The cognitive load arising from the learner's intrinsic processing can be reduced by:
1. Guiding the student through schema construction and modification. Such guidance might include helping the student to classify, interpret, exemplify, differentiate, and infer (de Jong, 2010).
2. Allowing the student to self-pace learning and allowing ample time to complete the processing required for the learning task.
3. Teaching the learner how to minimize cognitive load. For example, the student should be informed that listening to music while studying adds to cognitive load, thereby interfering with learning and task completion.
4. Teaching the learner how to cope with cognitive load in a manner that improves retention. For example, the learner should be taught when and how to switch from processing to maintenance of information.

5. Encouraging the student to ask for help when it is too difficult to process and retain information simultaneously.

Isolated application of any of the procedures recommended above should be helpful, but the more that are applied, the greater the reduction in load. In general, learners will have difficulty maintaining information in the short term and encoding it into long-term memory whenever the learning task requires them to engage in an attention-demanding processing activity. Students with impaired working memory can learn effectively if they have ample exposure to material while demands on working memory are minimal. That is, students like Jacob learn best under low cognitive load conditions.

Perhaps, nothing reduces cognitive load more than the acquisition of automaticity (Dehn, 2008). A task or procedure is said to be "automated" when it is overlearned or mastered to the point where it can be performed without conscious, mental effort. Automaticity speeds up processing, reduces cognitive load, and increases retention of information because the processing involved requires little attention. A prime example is the acquisition of reading fluency. A fluent reader has automated word decoding processes that "free up" working memory capacity for processing, such as making inferences, and for retaining more information, leading to better reading comprehension.

Students with working memory deficits may require more than methods that minimize cognitive load. They may need additional, individualized interventions and accommodations that support working memory and allow them to learn and perform better. These include the following: frequent repetition, reteaching, and review; providing more support, such as scaffolding, during the initial stages of learning when cognitive load is higher; helping students monitor the quality of their work; providing advance organizers; structuring information in a manner that encourages and supports the use of memory strategies; helping students complete challenging activities; supporting the development of schemas; and teaching working memory strategies and encouraging their use.

Teaching Working Memory Strategies

In addition to minimizing cognitive load in the classroom, teachers can teach students memory strategies that will enhance the short-term retention of information. The purpose of strategies for short-term and working memory is not to reduce cognitive load or to increase working memory capacity but rather to improve working memory performance through effective utilization of existing cognitive resources. Although strategy use definitely improves performance on working memory tasks (McNamara & Scott, 2001), some children do not acquire strategies independently. Even children that do often fail to use strategies consistently or effectively. Consequently, explicit teaching of strategies to individuals or groups can improve working memory functioning. As St. Clair-Thompson, Stevens, Hunt, and Bolder (2010) concluded, working memory strategy training results in significant improvement in working memory performance, and it often results in improvement in academic learning (Dehn, 2008).

Rehearsal

Rehearsal is repetition of information the individual is attempting to preserve long enough to complete a process or obtain a goal. Rehearsal can be subvocal and automatic or vocal and effortful. Most children begin using simple rehearsal strategies around age 5, but consistent application of verbal rehearsal strategies may not occur until age 10 (Gill, Klecan-Aker, Roberts, & Fredenburg, 2003). Increased use of rehearsal corresponds with increases in recall from both working and long-term memory, resulting in greater learning (Turley-Ames & Whitfield, 2003). Explicit classroom teaching of rehearsal strategies has been demonstrated to increase the working memory span of children, with and without disabilities, and lead to improved learning (Comblain, 1994; Loomes, Rasmussen, & Pei, 2008).

Rehearsal training is easily conducted with a classroom of students or one-on-one. The key is to teach students to overtly repeat sequences of words, adding new words in a cumulative fashion. For example, if a student needs to remember a sequence of steps, such as "draw, fold, cut, staple, and paste," the student would repeat these words several times in order. After students have mastered the approach, they are directed to whisper during rehearsal and then finally to subvocalize. One advantage of basic rehearsal is that the procedure does not add substantially to working memory load. Thus, a student, such as a Jacob, could manage and benefit from rehearsal. However, more capable students should be taught switching and updating in conjunction with rehearsal so that they switch back and forth between rehearsing and processing. For example, with the series of steps stated above, the student should be taught to begin drawing and then pause every few seconds while drawing to repeat the sequence "draw, fold, cut, staple, and paste." After the drawing step is completed, the student should "update" by dropping "draw" but continuing to repeat the remaining steps as long as each is needed.

Chunking

"Chunking" refers to the pairing, clustering, or grouping of discrete items into larger units that are processed and remembered as a whole, thereby increasing the total amount of information that can be held in short-term storage. The process of acquiring reading fluency exemplifies chunking: After the young reader has blended phonemes into a whole word several times, the reader recognizes the whole word as a "chunk" and processes and retains it accordingly. Training children to use a basic chunking strategy has been found to improve working memory performance (Parente & Herrmann, 1996). Teaching chunking is done by requiring the learner to group items into larger units. Training might begin by having students pair digits. For example, the digits "4," "6," "3," and "8" should be paired and recalled as "46" and "38." Memorizing the spelling of words provides another opportunity to apply this type of training. For example, letters might chunked and rehearsed by syllable, such as the letters in "forget" could be grouped as "f, o, and r" and "g, e, and t."

Metamemory

Metamemory is metacognition as it relates to memory (Dunlosky & Bjork, 2008; Schneider, 2010). Metacognition is an important dimension of executive function. Individuals with diminished executive functioning typically have poorly developed metacognition. Similarly, individuals with memory deficits tend to have poorly developed metamemory. The construct of metamemory includes the following: understanding how memory functions, such as knowing the difference between short- and long-term memory; understanding one's memory strengths and weaknesses; monitoring one's memory performance; consciously manipulating memory functions, usually through application of a strategy; accurately assessing the memory demands of various tasks; and knowing when, how, and why to use a strategy. Individuals with delayed metamemory development tend to be less strategic. Moreover, they are less likely to generalize and maintain memory strategies they've been taught. Therefore, metamemory instruction should be included in any working memory intervention.

Metamemory instruction should be included with both strategy instruction and memory exercises. Metamemory instruction is an integral component of the "mnemonic classroom," in which instructors teach students about memory functions, teach memory strategies, and remind students to use memory strategies (Dehn, 2010; Mastropieri & Scruggs, 2007). Teachers who adopt this mnemonic approach have students who continue to use effective memory strategies and are more successful academically (Ornstein, Grammer, & Coffman, 2010).

Working Memory Exercises

Within the past decade, several studies (reviewed by Morrison & Chein, 2011) have reported significant improvements in working memory performance following repetitive training exercises. Some of these training programs, such as Cogmed® (Holmes, Gathercole, & Dunning, 2009; Klingberg, 2009) and Jungle Memory® (Alloway, 2009), are online and have a game-like format, whereas others utilize more traditional procedures, such as *n*-back tasks (Jaeggi, Buschkuehl, Jonides, & Perrig, 2008). The gains have not been limited to measures of working memory; there have been several instances of transfer. Some studies have found growth in related cognitive functions, such as fluid reasoning (Jaeggi et al., 2008). Holmes et al. (2009) reported improved performance in mathematics reasoning and students' ability to follow classroom instructions. In everyday functioning, parents of children with ADHD have reported a reduction in motor activity and increased ability to focus and sustain attention, following an Internet-based training program (Klingberg, 2009). Maintenance of the gains also has been documented over intervals up to 18 months (Dahlin, Nyberg, Backman, & Neely, 2008).

The observable changes seem to be the result of increased working memory capacity. The neural correlates of the improved performance recently have been measured (Morrison & Chein, 2011). For example, Tageuchi et al. (2010) discovered measureable growth in the brain's white matter that was correlated with the extent of training and the amount of improved performance. The majority of the white matter growth was within the parietal cortex and adjacent to the corpus callosum, resulting in increased connectivity between the parietal region and the dorsolateral prefrontal cortex, which is often considered the "center" of executive working memory functioning.

The training exercises from most of these studies can easily be adapted for use in an educational setting, provided that the principles which make the exercises effective are understood and followed. Not just any brain training or working memory exercise will result in actual working memory growth. The Internet is replete with "brain games" and "apps" that claim to "boost memory." What sets the exercises evaluated in the professional literature apart from those available on the Internet is that they adhere to training regimens that consistently demand high cognitive workloads (Morrison & Chein, 2011). Westerberg et al. (2007) attributed measureable outcomes to the experience of taxing working memory to its limits over a sustained period of time. Westerberg et al. hypothesized that observed improvements in performance are the result of improving the efficiency of neuronal responses or extending the cortical map serving working memory. Moreover, training programs involving working memory exercises may also promote metamemory and the development of compensatory strategies.

For working memory exercises to be effective, they should meet the following criteria:

1. The exercises need to be conducted on a regular basis over an extended period of time. At a minimum, a student should practice 20–30 min a day, 5 days a week, for 4–6 weeks.
2. An "adaptive" approach should be followed. Each time the student has mastered the task at a particular span, the span should be increased by one. For example, once the student can consistently recall a sequence of four items, a list of five items should be presented for recall. When a student's performance declines, he or she may drop back a level until performance improves again.
3. The exercises should always be challenging but not above the student's grasp. For example, each session should begin at the level the student was practicing at the end of the previous session.
4. Each exercise should involve both processing and storage. That is, the student must "do" something in addition to retaining a sequence of items. For example, simply repeating a series of words does not include sufficient processing to qualify as a working memory exercise. However, remembering the sequence of final words in each sentence after reading a series of sentences does require processing while retaining information.

5. For the training sessions to be long enough and interesting enough, they should include more than one type of exercise.
6. For two reasons, the training program should not require the learning or application of strategies. One, strategies can add to the cognitive load and frustrate individuals like Jacob. Two, strategies (acting like a crutch) can improve performance while failing to increase brain-based capacity. Nevertheless, after the trainee appears to "hit a ceiling," appropriate strategies should be suggested or taught. Strategies might include basic rehearsal, dual encoding (see the next section), and shifting between processing and rehearing.

If these principles are applied, then practicing nearly any task that includes both processing and maintenance components has a high likelihood of improving working memory capacity and performance.

The following working memory exercises (also described in Dehn, 2011) are recommended for use in a school setting.

N-Back

N-back is perhaps the most challenging evidence-based working memory exercise. *N*-back requires the individual to remember an item that was presented a certain number of items previously. For example, with 2-back, the trainer might display and remove a series of letters one at a time. If the letters are b-q-f-j-r, then the student would say "b" when the f is displayed, "q" when the j is displayed, and "f" when the r is displayed. One application of this method is to use a deck of regular playing cards and follow these guidelines: (1) display each card for 1–2 s, (2) have the student name the appropriate *n*-back card as soon as the next card in the sequence is displayed, (3) start the process over whenever the student makes an error, and (4) increase the *n*-back by one when the student successfully completes a sequence of ten cards 3 times in a row.

Counting Span

In this activity the student counts the number of items on a series of cards displayed and removed one at a time and then must recall the count for each card in the correct sequence. For materials, cards with dots or stars on them would be appropriate. If the first card has 7 items, the second has 4 items, and the third has 9 items, the student would say "7, 4, and 9" after the last card has been counted and removed. Each time the student successfully completes a series 3 times in a row, another card should be added to the sequence.

Arithmetic Flash Cards

Mental computation itself consumes working memory resources. Thus, for this exercise only arithmetic facts the student has mastered or nearly mastered should be used. For example, multiplication "flash cards" could be used with a student who knows multiplication tables. The procedure is similar to that for counting span. The student computes and says the answer for each card as it is displayed and then must recall the answers in the correct sequence after all the cards have been presented. Incorrect calculations should be accepted, provided the trainee says the same number when recalling the sequence.

Visual-Spatial Recall

The materials for this activity are sheet or board with grids, such as a "4×4," and tokens, such as chips. For each trial, the trainer should display a set number of tokens on random squares for up to 5 s, then remove the chips, and then rotate the board 90°. The student must then correctly place the tokens on the squares where they were originally placed. Processing for this task is required when the student must rotate the board in his or her "mind's eye" to match the locations.

Remembering Directions

Remembering a sequence of directions places high demands on working memory. For this exercise, complex scenes containing several items and individuals should be used. For example, a picture of school playground with children would be appropriate. The trainer states a series of items the student must point to after the directions are complete. For example, the trainer might say, "Point to the swing, then the teacher, and then the bush next to the building." The processing involved in this activity consists of integrating verbal and visual information, a task that places demands on executive working memory.

Long-Term Memory Strategies That Support Working Memory

Because of the interaction between working memory and long-term memory during encoding and retrieval (Rosen & Engle, 1997), interventions that benefit one will also benefit the other. The strong long-term memory representations underlying mastery and expertise also facilitate processing of information in working memory or simply allow the bypassing of working memory (Gathercole & Alloway, 2008). Automated or efficient retrieval of well-learned material is one of the reasons for reduced demands on cognitive processing in working memory. However, the application of long-term memory strategies also can benefit working memory during the initial stages of learning. For instance, strategies that promote efficient encoding of information will result in less consumption of working memory resources (Carretti, Borella, & De Beni, 2007). Thus, long-term memory strategies that enhance initial learning and delayed recall will also improve working memory performance. (For more information on long-term memory strategies, see Dehn, 2010, 2011.)

Complex, multistep long-term memory strategies that require lots of practice should be avoided with individuals who have working memory impairments. As mentioned at the beginning of the chapter, learning and applying such strategies can create a dilemma for an individual, such as Jacob, because learning and applying the strategy adds significantly to cognitive load, thereby causing the loss of even more information from short-term storage. Thus, multistep visual mnemonics, such as keyword, may not be effective or be resisted by the learner. Only strategies that typically create minimal load are suggested here.

Dual Encoding

The old adage that a "picture is worth a thousand words" has implications for working memory performance and retrieval from long-term memory. The benefits of visualizing, or picturing, verbal information are well known. For instance, Carretti et al. (2007) reported improved working memory performance when students used visualization to memorize word lists. This visualization strategy, also known as "dual encoding," may be especially beneficial for learners who have deficient verbal working memory but otherwise normal visual-spatial working memory. Converting verbal information into a picture may allow such individuals to manipulate and retain much more information. Visualization also creates another form of memory in long-term storage, as well as an additional retrieval route. Training students to adopt dual encoding is straightforward (Gill et al., 2003). The basic instruction to "picture the item or information" is about all that is needed, as this is a very natural and common practice. Once students recognize the efficacy of the strategy (Dehn, 2010), they just need reminders to do it.

Reviewing

Another low-load memorization strategy that strengthens long-term recall and enhances working memory performance is periodic review (Cepeda, Pashler, Vul, Wixed, & Rohrer, 2006). Nearly every self-aware learner recognizes the value of reviewing information, but not every learner knows how to review information in a manner that maximizes long-term recall. Part of the problem is that some learners, especially those with poorly developed metamemory, fail to recognize the important difference between immediate and delayed recall. That is, such individuals believe that they have learned the material because they can immediately recall it, without taking into account the inevitable long-term forgetting that will occur.

For reviews to be most effective, they should be spaced out instead of massed together. Moreover, the intervals between reviews should be expanding so that each subsequent interval is up to 2 times longer than the preceding interval. Reviews completed according to the "expanding interval" approach have been found to increase learning by approximately 15 % (Bahrick, 2000). The explanation for this phenomenon is that memory representations and retrieval pathways are strengthened when retrieval takes some effort (as a result of the information being partially forgotten) as opposed to easy, immediate retrieval. Reviews are also more effective when the learner actually retrieves the information, instead of simply reading it over. One way to ensure that retrieval is occurring is to create study cards with the questions and answers on opposite sides and then use the study cards to self-test.

References

Alloway, T. P. (2006). How does working memory work in the classroom? *Educational Research and Reviews, 1*, 134–139.

Alloway, T. P. (2009). Cognitive training: Improvements in academic attainments. *Professional Association for Teachers of Students with Specific Learning Difficulties, 22*, 57–61.

Alloway, T. P. (2011). *Improving working memory: Supporting students' learning*. London: Sage.

Alloway, T. P., Gathercole, S. E., Adams, A. M., & Willis, C. (2005). Working memory abilities in children with special needs. *Educational and Child Psychology, 22*, 56–67.

Baddeley, A. D. (1986). *Working memory*. Oxford: Oxford University Press.

Baddeley, A. D. (2006). Working memory: An overview. In S. J. Pickering (Ed.), *Working memory and education* (pp. 1–31). Burlington, MA: Academic.

Bahrick, H. P. (2000). Long-term maintenance of knowledge. In E. Tulving & F. I. M. Craik (Eds.), *The Oxford handbook of memory* (pp. 347–362). New York: Oxford University Press.

Barrouillet, P., Bernardin, S., Portrat, S., Vergauwe, E., & Camos, V. (2007). Time and cognitive load in working memory. *Journal of Experimental Psychology, 33*, 570–585.

Barrouillet, P., Gavens, N., Vergauwe, E., Gaillard, V., & Camos, V. (2009). Working memory span development: A time-based resource-sharing model account. *Developmental Psychology, 45*, 477–490.

Barrouillet, P., Portrat, S., & Camos, V. (2011). On the law relating processing to storage in working memory. *Psychological Review, 118*, 175–192.

Carretti, B., Borella, E., & De Beni, R. (2007). Does strategic memory training improve the working memory performance of younger and older adults? *Experimental Psychology, 54*, 311–320.

Cepeda, N. J., Pashler, H., Vul, E., Wixed, J. T., & Rohrer, D. (2006). Distributed practice in verbal recall tasks: A review and quantitative synthesis. *Psychological Bulletin, 132*, 354–380.

Chen, Z., & Cowan, N. (2009). How verbal memory loads consume attention. *Memory and Cognition, 37*, 829–836.

Comblain, A. (1994). Working memory in Down's syndrome: Training the rehearsal strategy. *Down's Syndrome: Research and Practice, 2*, 123–126.

Cowan, N. (2005). *Working memory capacity*. New York: Lawrence Erlbaum.

Dahlin, E., Nyberg, L., Backman, L., & Neely, A. S. (2008). Plasticity of executive functioning in young and older adults: Immediate training gains, transfer, and long-term maintenance. *Psychology and Aging, 23*, 720–730.

de Jong, T. (2010). Cognitive load theory, educational research, and instructional design: Some food for thought. *Instructional Science, 38*, 105–134.

Dehn, M. J. (2008). *Working memory and academic learning: Assessment and intervention*. Hoboken, NJ: Wiley.

Dehn, M. J. (2010). *Long-term memory problems in children and adolescents: Assessment, intervention, and effective instruction*. Hoboken, NJ: Wiley.

Dehn, M. J. (2011). *Helping students remember: Exercises and strategies to strengthen memory*. Hoboken, NJ: Wiley.

Dunlosky, J., & Bjork, R. A. (Eds.). (2008). *Handbook of metamemory and memory*. New York: Psychology Press.

Elliott, J. G., Gathercole, S. E., Alloway, T. P., Holmes, J., & Kirkwood, H. (2010). An evaluation of a classroom-based intervention to help overcome working memory difficulties and improve long-term academic achievement. *Journal of Cognitive Education and Psychology, 9*, 227–250.

Engle, R. W. (2002). Working memory capacity as executive attention. *Current Directions in Psychological Science, 11*, 19–23.

Ericsson, K. A., & Kintsch, W. (1995). Long-term working memory. *Psychological Review, 102*, 211–245.

Gathercole, S. E., & Alloway, T. P. (2008). *Working memory and learning: A practical guide for teachers*. Los Angeles: Sage.

Gathercole, S. E., Alloway, T. P., Kirkwood, H. J., Elliott, J. G., Holmes, J., & Hilton, K. A. (2008). Attentional and executive function behaviours in children with poor working memory. *Learning and Individual Differences, 18*, 214–223.

Gill, C. B., Klecan-Aker, J., Roberts, T., & Fredenburg, K. A. (2003). Following directions: Rehearsal and visualization strategies for children with specific language impairment. *Child Language Teaching and Therapy, 19*, 85–104.

Gioia, G. A., Isquith, P. K., Guy, S. C., & Kenworthy, L. (2000). *Behavior Rating Inventory of Executive Function*. Lutz, FL: Psychological Assessment Resources.

Hester, R., & Garavan, H. (2005). Working memory and executive function: The influence of content and load on the control of attention. *Memory and Cognition, 33*, 221–233.

Holmes, J., Gathercole, S. E., & Dunning, D. L. (2009). Adaptive training leads to sustained enhancement of poor working memory in children. *Developmental Science, 12*, F9–F15.

Jaeggi, S. M., Buschkuehl, M., Jonides, J., & Perrig, W. J. (2008). Improved fluid intelligence with training on working memory. *Proceedings of the National Academy of Sciences of the United States of America, 105*(19), 6829–6833. doi:10.1073/pnas.0801268105.

Jang, J., Schunn, C. D., & Nokes, T. J. (2011). Spatially distributed instructions improve learning outcomes and efficiency. *Journal of Educational Psychology, 103*, 60–72.

Kane, M. J., & Engle, R. W. (2000). Working memory capacity, proactive interference, and divided attention: Limits on long-term memory retrieval. *Journal of Experimental Psychology: Psychology, Learning, Memory, and Cognition, 26*, 336–358.

Kirschner, P. A. (2002). Cognitive load theory: Implications of cognitive load theory on the design of learning. *Learning and Instruction, 12*, 1–10.

Klingberg, T. (2009). *The overflowing brain: Information overload and the limits of working memory*. New York: Oxford University Press.

Loomes, C., Rasmussen, C., & Pei, J. (2008). The effect of rehearsal training on working memory span of children with fetal alcohol spectrum disorder. *Research in Developmental Disabilities, 29*, 113–124.

Mastropieri, M. A., & Scruggs, T. E. (2007). *The inclusive classroom: Strategies for effective instruction*. Columbus, OH: Pearson.

McCabe, D. P., Roediger, H. L., McDaniel, M. A., Balota, D. A., & Hambrick, D. Z. (2010). The relationship between working memory capacity and executive functioning: Evidence for a common executive attention construct. *Neuropsychology, 24*, 222–243.

McCloskey, G., Perkins, L. A., & Van Divner, B. (2009). *Assessment and intervention for executive function difficulties*. New York: Routledge.

McNamara, D. S., & Scott, J. L. (2001). Working memory capacity and strategy use. *Memory and Cognition, 29*, 10–17.

Morrison, A. B., & Chein, J. M. (2011). Does working memory training work? The promise and challenges of enhancing cognition by training working memory. *Psychonomic Bulletin Review, 18*, 46–60.

Oberauer, K. (2002). Access to information in working memory: Exploring the focus of attention. *Journal of Experimental Psychology: Psychology, Learning, Memory, and Cognition, 28*, 411–421.

Ornstein, P. A., Grammer, J. K., & Coffman, J. L. (2010). Teachers' "mnemonic style" and the development of skilled memory. In H. S. Waters & W. Schneider (Eds.), *Metacognition, strategy use, and instruction* (pp. 23–53). New York: Guilford.

Parente, R., & Herrmann, D. (1996). Retraining memory strategies. *Topics in Language Disorders, 17*, 45–57.

Rosen, V. M., & Engle, R. W. (1997). The role of working memory capacity in retrieval. *Journal of Experimental Psychology. General, 126*, 211–227.

Schneider, W. (2010). Metacognition and memory development in childhood and adolescence. In H. S. Waters & W. Schneider (Eds.), *Metacognition, strategy use, and instruction* (pp. 54–84). New York: Guilford Press.

St. Clair-Thompson, H. (2011). Executive functions and working memory behaviours in children with a poor working memory. *Learning and Individual Differences, 21*, 409–414.

St. Clair-Thompson, H., Stevens, R., Hunt, A., & Bolder, E. (2010). Improving children's working memory and classroom performance. *Educational Psychology, 30*, 203–219.

Swanson, H. L., & Berninger, V. W. (1995). The role of working memory in skilled and less skilled readers' comprehension. *Intelligence, 21*, 83–108.

Swanson, H. L., & Berninger, V. W. (1996). Individual differences in children's working memory and writing skill. *Journal of Experimental Child Psychology, 63*, 358–385.

Tageuchi, H., Sekiguchi, A., Taki, Y., Yokoyama, S., Yomogida, Y., Komuro, N., et al. (2010). Training of working memory impacts structural connectivity. *Journal of Neuroscience, 30*, 3297–3303.

Turley-Ames, K. J., & Whitfield, M. M. (2003). Strategy training and working memory performance. *Journal of Memory and Language, 49*, 446–468.

van Gog, T., Ericsson, K. A., Rikers, R. M., & Paas, F. (2005). Instructional design for advanced learners: Establishing connections between the theoretical frameworks of cognitive load and deliberate practice. *Educational Technology Research and Development, 53*, 73–81.

Van Merrienboer, J. J. G., Kirschner, P. A., & Kester, L. (2003). Taking the load off a learner's mind: Instructional design for complex learning. *Educational Psychologist, 38*, 5–13.

Westerberg, H., Jacobaeus, H., Hirvikoski, T., Clevberger, M.-L., Ostensson, A., Bartfai, A., et al. (2007). Computerized working memory training after stroke—A pilot study. *Brain Injury, 21*, 21–29.

Building Executive Functioning in Children Through Problem Solving

28

Bonnie Aberson

This chapter will first establish the relationship between key areas of executive functioning, language, and problem solving from a historical perspective. Next, the author will describe an evidence-based approach to teaching social problem solving through language to children and the relationship of this approach to executive functioning. Finally, the author will conclude with research data and specific case studies supporting the efficacy of a problem-solving approach for improving executive functioning in children.

The Relationship Between Problem Solving and Executive Functioning

Alexander Luria, considered to have been one of the first to develop the concept of executive functioning, included the concepts of mental flexibility, the ability to engage in goal-directed behaviors, and to anticipate the consequences of one's actions. Luria associated the prefrontal cortex and executive functioning also with inhibiting immediate responses, problem solving, and verbal regulation of behavior (Luria, 1966, 1973). Shimamura, Janowsky, and Squire (1990, p. 191) associated the dysexecutive syndrome and disinhibition with poor prospective memory or the ability to access, monitor, and manipulate associations within a temporal/spatial context as well as a written and semantic context. They likened prospective memory tasks to problem-solving tasks in terms of demands on cognitive fluency, initiation, and flexible thinking. They proposed that poor inhibitory control is related to a prospective memory impairment resulting in relevant or appropriate strategies not being discriminated from irrelevant ones.

The relationship between thought and language was conceptualized by Vygotsky (1962) as thought being born through language, which plays a major role in the development of thinking and in consciousness. He further argued that metacognitive executive functions emanate from language internalization and that thought and language develop independently until age 2, at which time they converge. After that time thought becomes mediated by language, which becomes primary in conceptualization, thinking, and problem solving (Vygotsky, 1962).

Cummings (1993) noted that the most salient deficit of executive functioning related to the dorsolateral syndrome of the frontal lobes is the inability to "organize a behavioral response to novel or complex stimuli" as well as the "capacity to shift cognitive sets, engage existing strategies, and meet changing demands of the environment." These skills are also important when solving problems in novel interpersonal situations.

Goldberg (2009) points out that to have success in a more complex interpersonal interaction in addition to planning an action for one's self, a

B. Aberson (✉)
Joe Dimaggio Children's Hospital,
Hollywood, FL, USA
e-mail: bonnieabr@aol.com

person must be able to foresee not only the consequences of his/her actions but those of the other person's. This requires the ability to form an internal representation of the theory of mind of the other individual (p. 143).

Although Luria associated the frontal cortex with executive functioning and problem solving, we now know that other areas and pathways in the brain are involved, including posterior cortical and subcortical regions as well as the cerebellum (Roberts, Robbins, & Weiskrantz, 2002; Schmahmann, 2004). Additionally, some studies have documented frontal lobe activity in the first year of life (Bell & Fox, 1992; Chugani, Phelps, & Mazziotta, 1987) despite the fact that the prefrontal cortex has been found to be not well developed in young children. Additionally, Levin et al. (1991) argues that executive functioning skills, such as mental flexibility, impulse control, and problem solving, are demonstrated in young children. Shure and Spivack (1982) found that children as early as age 4 can learn to problem solve in interpersonal situations.

It can be argued that when children have the tools to generate alternative solutions to interpersonal problems, predict consequences based on prospective memory, and as a result react to problems in an adaptive way, they are engaging in flexible thinking, inhibiting automatic responses, initiating more adaptive ones, and responding appropriately to novel situations. With higher level problem-solving skills, children learn to take the perspective of the other person, to establish goals, and carry out plans to reach them. These behaviors can be assumed to be language mediated, as they require the use of internal language.

Research on the Effectiveness of Problem Solving as It Relates to Executive Functioning in Young Children

The relationship between the ability of young children to learn how to think in words related to problem solving first through games and dialoguing or conversations with an adult and then utilize internal language in order to self-regulate and solve novel interpersonal problems has been supported by research.

Shure and Spivack (1980, 1982) found that as early as age 4, children can learn how to understand their own and other's feelings, think of alternative solutions to real problems, and think of possible consequences. Shure and Spivack (1972) also identified a skill called means-ends thinking, by which children learn how to plan ahead to reach a goal despite obstacles.

Spivack and Shure (1974) designed a program known initially as *Interpersonal Cognitive Problem Solving*, now *I Can Problem Solve* (Shure, 1992). This curriculum teaches young children specific words and concepts, which are used in dialogs when real problems occur. In this way, the language and problem-solving thinking are eventually internalized, enabling the children to better self-regulate. Shure and Spivack (1982) found that preschool and kindergarten children trained by teachers improved in terms of reducing impulsive and aggressive behavior as well as withdrawn behavior in as little as 3 months with improvement maintained up to 4 years later as compared to untrained controls (Shure, 1993). Interestingly, the improvement in behaviors correlated with improvement in the ability to generate alternative solutions to peer and child-adult problems as measured by the *Preschool Interpersonal Problem Solving Test* (Shure & Spivack, 1974). Additionally, trained children who were considered to be well adjusted by their teachers at the beginning of preschool were more likely to remain well adjusted 2 years later as compared to controls (Shure & Spivack, 1982).

Aberson, Albury, Gutting, Mann, and Trushin (1986) found positive short-term outcomes, reducing both impulsive and withdrawn behavior, and an increase in the generation of alternative solutions in a study across ethnic and socioeconomic groups with at-risk kindergarten children as compared to matched controls. The children were trained in the *I Can Problem Solve Program* (ICPS) in small groups by school psychologists and school counselors over 9 weeks in the Miami Dade County schools with the teachers trained only to do the dialoguing.

Kumpfer, Alvarado, Tait, and Turner (2002) conducted a study in Salt Lake City with public school children trained in first through third grade by teachers. Children trained in ICPS improved significantly at the .001 level as compared to controls in the areas of school bonding (attitude toward teachers and attitude toward school) and in the area of self-regulation at the .01 level.

Shure and Spivack (1979) also found that a parenting program based on ICPS with mothers training preschool African American children resulted in reducing impulsive and inhibited behaviors as observed in school, implying that children trained by parents generalize their new skills to the school setting. Of note also is the fact that there was a correlation between children who most improved in problem solving and behaviors and mothers who were best able to apply the dialoging approach at home. When first trained by their teachers in kindergarten and by their mothers in first grade, Shure (1993) found that those children with mothers who best applied the ICPS dialogs maintained their gains 3 years later at the end of the fourth grade.

Aberson (1996) conducted three single-case design studies with ADHD second grade children who were having difficulty with self-regulation, initiation, inflexibility in terms of responses to problem situations, and poor planning. As a result, they were earning low grades and had difficulty completing tasks independently. These problems caused a high level of stress for the parents as well as a strain on their relationships with their children. In two of the cases, the children also had poor peer relationships. The parents were provided with group training in ICPS dialoguing by Aberson once a week for 6 weeks and asked to write dialogs they had with their children to present to the group. The children were also given selected ICPS lessons once a week over the same period.

Behavior ratings by teachers and parents and self-reports indicated significant improvement after 6 weeks that continued up to 6 months (Aberson, 1996) and 4 years later (Aberson & Ardila, 2000) in interpersonal relationships and overall behavioral adjustment as measured on the Behavior Assessment System for Children (BASC) (Reynolds & Kamphaus, 1992). Anecdotal records and dialogs submitted by parents indicated that after just 6 weeks, all three children were able to do their homework and get ready for school independently (Aberson, 1966; Aberson & Ardila, 2000). While improvement in relationships with parents improved by the time training ended and lasted, improvement in peer relationships was more gradual but continued to improve over time.

In order for a problem-solving model such as ICPS to generalize, the following specific conditions must be met: (a) children learn the ICPS skills, (b) teachers and/or parents use these words and concepts when dialoging with the child, and (c) the child learns to internalize the newly acquired ICPS skills for application in real life (Shure & Aberson, 2005). Additionally, although it is important that the problem-solving communication occur as close as possible in time to the event causing the problem, some problems require a time delay to diffuse an emotionally charged situation.

The Relationship Between the ICPS Curriculum and Executive Functioning

For both classrooms and families, consistent use of problem-solving communication emphasizing vocabulary from the ICPS (Shure, 1992) provides the structure of language and questions that are familiar to the child, as well as providing the child with an opportunity to practice new thinking skills and learn to self-regulate.

The use of the words **is** and **is not** first in games and then redirecting behavior prompts the child to think about whether a particular behavior is or is not a good idea. Questions with the words **some** and **all** help children to think about whether they can play with a favorite toy or with the computer some of the time or all of the time, and the word **different** helps the child to think about what is something **different** he/she might do.

When dialoguing with a child using a problem-solving approach, the adult can apply these words in the form of questions instead of demanding, suggesting, or explaining. For example, when a child interrupts a teacher or parent, instead of saying, "I'm talking to someone else," which the child knows, "Don't interrupt" (a command), "Why don't you color until I can listen to you?" (a suggestion), or "interrupting is not polite and not fair to the person I'm talking to" (explanation), the parent or teacher can encourage the child to think of what to do by asking something like, "How do you think my friend feels when I have to stop talking to him/her to talk to someone else?" (perspective taking) followed by "Can I talk to you and to _____ at the same time? What can you do while you wait?" (problem solving). In this way the child eventually internalizes this language and learns to self-regulate rather than depending on external cues and reinforcers.

The executive functions of shifting and flexibility in thinking are also nurtured when problem-solving communication is used for peer or sibling problems. For example, when two children are fighting over a toy, use of a computer, or remote control, the teacher or parent instead of blaming, punishing the children, or solving the problem can engage both children in problem solving. In this way the adult might ask first, "What is going on?" or "What happened?" and solicit the perspective of each child as well as ask how each child is feeling and what might happen if they keep on fighting (consequential thinking). The children using their prospective memory may recall that someone might get hurt, the toy might break, or they might not be friends. Then the teacher or parent might ask both children how they might resolve the problem and how each would feel about that solution until the children can agree on a solution they can both accept after which they implement their idea.

Once this technique is in place in a family or classroom, the children know that they can generate ideas and solve problems for themselves and often do so without the adult. In this way children also learn to shift from less adaptive behaviors with negative consequences to adaptive ones while also learning to take the perspective of the other person. As one 8-year-old Asperger's child stated with a smile after only 3 months of ICPS training with his parents, "My brother and I don't fight anymore because we can solve our problems by ourselves."

Executive Functioning and Problem Solving with Children with Asperger's Syndrome

Executive functioning deficits are found across the autistic spectrum according to Klin, Volkmar, and Sparrow (2000). Although Asperger's children often have high verbal abilities, they usually have difficulty with choosing the appropriate words for a novel situation. This problem may be related to poor recognition of social cues, the inability to shift the level of formality and recognize there is a misunderstanding and/or a different understanding from the perspective of others, or a deficit in theory of mind. The deficit in theory of mind may also be related to the executive functions of planning, shifting, working memory, and disinhibition (Klin et al., 2000). Difficulty with shifting set can be related to not considering alternative meanings, resulting in literal interpretation of language and difficulty shifting from one perspective to another that is more appropriate to the context. Ozonoff, Pennington, and Rogers (1991) perceive executive function deficits as primary to ASD.

Eddie's Story

Eddie, now a senior on full scholarship at a prestigious college, driving, and an active member of his fraternity, was diagnosed with Asperger's syndrome at age 5 because of stereotyped movements, continually interrupting circle time with his constricted area of interest, inability to engage in reciprocal conversation or interact appropriately with peers, as well as hypersensitivity to noise, and frequent temper tantrums in school when he didn't get his way. He had been previously noise at age 4 with ADHD and a borderline IQ as well as prognosis of a future in a sheltered workshop. Eddie had already been receiving language therapy, occupational therapy, stimulant medication, and sessions with the guidance counselor with no improvement. Because of Eddie's atypical and disruptive behaviors, the school child study team wanted him evaluated for the purpose of a possible placement in a class for emotionally disturbed children.

This author, then a school psychologist at the same school, suggested to the teacher an intervention in which the school psychologist would conduct lessons from the ICPS program weekly under the condition that the teacher would participate and use ICPS dialoging techniques with the child for redirecting his behavior. Since the school psychologist had successfully used ICPS with small groups of at-risk children from the same classroom, the teacher agreed, and Eddie participated eagerly in the lessons as well as responding to dialoging for specific behaviors. For example, when he insisted on being first in line, the teacher using ICPS words asked, "Is it **fair** for the same person always to be first?" followed by, "Where is a **different** place you can stand?" Previously, this situation would have resulted in his having a tantrum and spending time in time-out, where he would become more angry. Eddie responded well and became more flexible in this thinking and in his behavior (shifting). These changes resulted in major behavioral issues being negligible within a month.

He was placed in a varying exceptionalities resource room in first grade for 2 hours a day because of severe fine motor problems and difficulty with math concepts. The exceptional education teachers asked to be trained in ICPS and used the program with all of their students. When the author observed a lesson toward the middle of Eddie's first grade year, Eddie began to ask a question out of context about dinosaurs (his constricted area of interest) to which the teacher replied, "**Is** this a good time or **is** it **not** a good time to ask about dinosaurs?" Eddied replied, "Not a good time?" The teacher then asked, "When **is** it a good time?" and Eddie replied, "**After** the lesson." Eddie then waited until after the lesson and then asked the teacher his question. In this way Eddie developed the executive functioning skills of shifting, flexibility in thinking, and inhibition, which continued to develop as he grew older so that he internalized the language and did not require cueing from dialoguing in order to self-regulate by the end of first grade.

He also developed the ability to empathize, take the perspective of others (theory of mind), and carry on reciprocal conversations, resulting in his eventually developing positive peer relationships which continued to improve as he grew older. On one occasion reported by his parents when he was in second grade, Eddie noticed that his teacher was sad. He asked her what was wrong and then engaged in reciprocal conversation to help her to feel better.

He was successful in regular programs in private schools after first grade and then competitive in a prestigious private high school where he won awards for drama. Eddie's parents, aware of but not directly trained in the ICPS curriculum, also used dialoging techniques with him and included him in decision making throughout his childhood and adolescence. Now a business major, Eddie plans to learn Chinese so that he can work in international business marketing. He has many friends in college and still corresponds with an old friend from middle school. He recently stated to this author, "sometimes I encounter obstacles or mess up but I just get back on the horse and keep riding toward my goal."

Billy's Story

Billy, now a high school student earning straight As in academics and behavior in an international baccalaureate program, was diagnosed with Asperger's syndrome at age 4. He has twin brothers, who were 6 months old at the time Billy was diagnosed. Both of Billy's parents were becoming increasingly more frustrated because of his aggressive behavior and refusal to follow directives at school and at home. Shifting from a preferred activity when asked was also a major problem. The situation came to a head when he was asked to leave his preschool for hitting other children when they did not follow rules. He was also at times aggressive toward his little brothers and often had temper tantrums. When he was told to go to time-out, he hit or kicked his parents on the way there or afterwards and did not change his behavior.

Billy's parents sought help with the above problems and received parent training from this author with the parent form of ICPS approach (Shure, 1996, 2000). Additionally, Billy's mother participated in ICPS games with her son. During the initial session Billy, then reading on a second grade level, sat in a corner and read, but by the second session he enjoyed playing the ICPS games with the aid of puppets and other toys. Treatment lasted for 10 weeks. After 1 month Billy's behavior improved at home. For example, one day after Billy was playing video games and was forced to stop for dinner, he demanded that he be given the ketchup. His mother asked him, "What is a **different** way you might ask for the ketchup?" using an ICPS word. He responded, "May I please have the ketchup?" A few minutes later, using another ICPS word, he asked, "Can I finish my video game **after** dinner?" He also learned to recognize good times and not good times for him to ask his mother for attention through games with the words **some** and **all** and deciding that his mother could play with him **some** of the time but not **all** of the time and when was a **good time** for his mother to play with him. After 10 weeks, Billy became more cooperative at home, and his aggressive behavior ceased. In first grade he was placed in a full-time gifted program with accommodations for organization and speed. He remained and functioned well in that setting with the same friends throughout the fifth grade.

As of the time of writing this chapter, Billy is an empathic high-functioning student and enjoys positive relationships with teachers, parents, peers, and siblings. Since this is his first year in high school, he is gradually making new friends and recently began participating in the drama club at his school but still gets together with friends from middle school. Most importantly, his bonding with his parents has resulted in his feeling comfortable discussing and solving problems with them even now as a teenager. Additionally, he is able to verbalize the advantages and disadvantages of having Asperger's syndrome with one of the advantages being that he is able to stay focused on an area of interest and not give up. He now feels that he would like to help other children with Asperger's by sharing his story with them.

Billy was able to utilize verbal mediation in order to regulate his behavior and shift from maladaptive to adaptive behaviors at a young age. He is able to use problem-solving strategies as a framework into which he can insert new skills for new issues. For example, at age 12, he was troubled by obsessive thoughts common to children with Asperger's. He was able to stop these thoughts on his own after only one therapy session without medication by thinking of different strategies and using them.

Executive Functioning and Problem Solving with ADHD Children

Executive functioning deficits have been found to be common in the ADHD population. Stuss and Benson (1986) and Benson (1991) associated ADHD with the role of the prefrontal cortex in disruption of the ability to maintain sequences and a disturbance in the prefrontal control of drive as well as executive control to be self-critical or think ahead. Shapiro, Hughes, August, and Bloomquist (1993) found that ADHD children perform more poorly than controls on tasks requiring complex auditory processing and extensive working memory.

Hamlet, Pellegrini, and Conners (1987) examined executive functioning of hyperactive children vs. normals on complex problem-solving tasks using a memory task involving organization and self-monitoring together with social communication. They found that the hyperactive children with or without medication demonstrated poorer recall and significantly poorer ability to utilize verbal communication to explain strategies. Landau and Milich (1988) found that ADHD boys differed from controls in flexibility of social communication as the appropriateness to task changes. They concluded that the social behavior of ADHD youngsters does not adjust to the demands of the social environment and that consequently responses of others to them become altered.

Grenell, Glass, and Katz (1987) found that the responses of hyperactive children on a social knowledge interview were similar to controls when required to provide strategies for initiating relationships; however, they had considerably more difficulty when asked to describe strategies for maintaining a relationship or to resolve conflicts.

In the area of goal planning, Renshaw and Asher (1983) found that the cognitive process of goal planning in hyperactive children is intact in terms of understanding what needs to be done when given alternatives to choose from, but the children had difficulty when they had to produce a course of action. In the cases below, the children before training were unable to think of and apply effective alternatives to problem situations, to utilize prospective memory to avoid negative consequences, or to effectively plan tasks such as homework. These deficits resulted in poor self-regulation in the first case and in poor planning, poor peer relationships, and a strain in the area of parent–child bonding in both cases.

Jimmy's Story

Jimmy, of Southeast Asian descent, was adopted with his younger autistic brother as a toddler by American parents. Before the intervention, Jimmy was impulsive, oppositional, and defiant in school and at home. His Physical Education teacher described him as a "mean kid," and he was often not allowed to participate during games in PE.

Jimmy's parents participated in a parent training group along with two other couples for the purpose of learning how to use problem-solving communication with their children. After five parent training sessions, Jimmy, a second grade gifted student with ADHD, and his parents received 10 weeks of family training during which they played selected games from the ICPS program. Each week they shared problems which occurred at home and how they used dialoging to solve them.

When Jimmy was asked during training sessions how he felt about not being allowed to play in PE, he replied, "Sad." When asked using ICPS words, "What happens **before** your teacher tells you that you can't play?" he responded that he fooled around and often kicked the ball into another child. When asked what he could say to himself so that wouldn't happen, he answered, "Don't fool around, and make sure my hands and feet are quiet." On his next report card, Jimmy earned an A in conduct in PE. Before the problem-solving intervention, Jimmy did not bring home daily school report cards or negative notes from teachers because they resulted in lectures at home. As part of the intervention, Jimmy and his parents agreed together to use a home school report in a different way. The teachers rated Jimmy in four areas on a scale of 1–5. These included staying on task, homework, getting along with peers, and following rules. His parents consented to respond to the report by asking three questions, which were written at the bottom of the report: first, "What makes you feel happy or proud about this report?" second, "Does anything make you feel sad or frustrated? What is that?" and third, "What can you do tomorrow to make it better?" After only 2 weeks, Jimmy was earning higher ratings in all four areas and was getting along well with peers.

Jimmy also became more bonded with his parents, as both he and his parents learned how to think about their own and other's feelings, including how someone feels when people shout at them, or behave in a disrespectful manner. Jimmy was able to demonstrate empathy toward his younger autistic brother as well. On one occasion when his mother became frustrated with Jimmy's brother and shouted at him, Jimmy asked, "How do you think Steven feels when you speak to him like that? Can you think of a **different** way to speak to him?"

Getting ready in the morning and doing homework also became less stressful, as Jimmy learned to organize his time and school materials better by planning ahead through the structure of problem solving questions, such as, "How do you feel when everyone is rushing and shouting in the morning? Can you do your best work in school when everyone is arguing before leaving the house? Do you think we can do our best work when we are upset? What can you do so it is easier to be ready on time and we will all feel **happy** and not **frustrated**?"

For initiating homework, Jimmy's parents asked him, "How do you feel when you don't have your homework at school? How do you feel when you do have your homework? What do you want to do when you finish your homework?" Jimmy realized that he felt **happy** when he had his homework and **frustrated** when he didn't, and that if he finished his homework, he could go outside and play. After a few months Jimmy was doing his homework independently, and mornings were no longer stressful. After 10 weeks of treatment when Jimmy was asked, "What did you learn about solving problems?" he responded, "I learned that the same solution will not work in every situation" (flexible thinking).

Because of increased academic demands in fifth grade, Jimmy began taking stimulant medication to improve his focusing on school work, as he no longer had behavior problems, completed assignments, and remained on task. He is now a junior in college earning a 3.8 grade point average and majoring in psychology. He has many friends and continued close bonds to his parents. He takes stimulant medication only for major tests.

Through problem-solving communication, Jimmy developed the executive functions of self-regulation, generation of ideas, goal planning, and the ability to shift from maladaptive behaviors to adaptive ones. He also learned to understand the perspective of others and behave in an empathetic manner to both parents and peers.

Patricia's Story

Aberson (1996) and Aberson and Ardila (2000) described Patricia as an only child of British origin from a single-parent home who demonstrated the characteristics of ADHD inattentive type since she was in kindergarten. These problems continued without intervention until second grade. At that time her mother attended 6 weekly small-group parenting classes based on the dialoging skills from the ICPS program. Patricia also participated in a small group at school for 6 weeks, learning the concepts from the ICPS curriculum. Ratings before the intervention on the BASC (Reynolds & Kamphaus, 1992) by her teacher, her mother, and herself suggested that she was also experiencing symptoms of depression and anxiety in addition to attention problems. She was resistant to working either in class or at home despite average intelligence and achievement levels and was earning below-average grades. She was not doing well socially either, as she had only one friend, who was able to bully her by telling her she would not be her friend unless she did not do what her friend wanted. Because of Patricia's inability to generate solutions to interpersonal problems, she was not able to handle rejection and bullying effectively. Her relationship with her mother was stressed also, as indicated by ratings on the Parenting Stress Inventory (Abidin, 1990). Parenting strategies were generally confrontational and punitive and related primarily to getting ready for school in the morning and doing homework because of Patricia's poor initiation, panning, and organizational skills. Despite Patricia's dependence on her mother, the relationship was poor because of the ongoing conflicts and resulting frustration.

Aberson, who was a school psychologist at Patricia's school and conducted the groups mentioned above, explained to Patricia that her mother would be asking her questions to help her learn how to solve problems. Patricia agreed and welcomed the new approach.

Patricia's mother helped her to eventually get ready independently in the morning by asking questions, such as "How do you feel when you arrive at school on time?" "How do you think your teacher feels when you are late?" and "Do you think you and I can do our best work after we shout at each other in the morning and we have to rush?" Patricia's mother worked with Patricia to solve this problem by adding structure and breaking the solution down into smaller steps with questions: such as (1) "What are some things you can do the night before to make it easier to get ready in the morning?" (2) "Can you make a list of the different things you need to do to get ready?" (3) "What would you do first, second, and third?" and (4) "How can you mark each task after doing it so you know it is finished?" In this way Patricia learned to think about consequences, how to plan and think ahead, and how to self-monitor. After 6 weeks of this type of interaction, Patricia was able to perform these steps independently. Her mother stated that although initially the conversations or dialogs with Patricia were lengthy due to her oppositional responses, eventually Patricia was able to internalize the language and the amount of dialoguing decreased as her relationship with her mother improved. Also battles over doing homework gradually decreased as Patricia was able to begin and complete her homework independently. Instead of ordering Patricia to do her work or arguing, her mother asked questions like "What do you want to do when you finish your homework?" Patricia's improvement on task behavior at school was also noted by her teacher.

Although Patricia continued having some difficulty making friends for a few years, her ability to use problem solving with peers improved after just 6 weeks in terms of her being able to generate solutions when playing with another child at home on a playdate. Instead of crying to her mother when another child didn't want to play what she wanted to play, she learned to suggest different ideas or ask "What do you want to do?" (a question from an ICPS game). When Aberson spoke to Patricia's mother and communicated through Facebook with Patricia at age 22, she was happily married, working with senior citizens and attending college in England; both Patricia and her mother reported that when she returns to the United States for visits, the house is full of her friends from high school. Patricia's mother also felt more empowered and now holds an administrative position.

For Patricia, the immediate benefit of the use of problem-solving communication was the improved bonding with her mother and feeling of empowerment when performing tasks required of her such as homework, through her taking charge of planning her schedule. Although it took longer for her to feel comfortable with peers, early on she was able to verbalize her feelings and take the perspective of other children in play as well as of adults in her life. Four years after the parent training and ICPS group sessions, Patricia, as measured by ratings from teachers, her mother, and by self-report on the BASC (Reynolds & Kamphaus), was free of symptoms of depression and anxiety. Although mild attention problems remained and Patricia was never medicated, she has been able to self-regulate as well as plan and initiate tasks independently.

Barry's Story

Barry, now entering his sophomore year in college on full scholarship with a major in engineering, is an active participant in college life and has a group of friends. He was a very different person at age 12 in middle school, having been diagnosed with Asperger's syndrome as well as having symptoms of ADHD in elementary school. Barry had a history of being bullied and rejected by peers, having poor anger control, and being dependent on his mother to plan and oversee homework as well as provide consequences for poor grades and make decisions for him.

When Barry and his mother began treatment when he was in sixth grade, the first goal was to help his mother to engage him in problem-solving communication and to give him the opportunity to make decisions. At the same time, Barry, who after many weeks of playing chess with his therapist, while gradually talking more about his problems with peer relationships and completing school work, decided that he would like to take over tasks such as homework, and make decisions for himself. Barry's mother agreed to problem solve with him instead of providing external consequences. Barry began assuming school responsibilities, and although he continued to earn some poor grades (natural consequences) because of not studying or not turning in homework, he eventually began to think of different solutions (shifting), engaged in goal planning, and chose a path that led to more positive outcomes.

Social relationships remained a problem for a while, but Barry and his mom decided together that participation in the school band in eighth grade might be a good way to do something he liked with peers in a structured situation. Barry did well in the band and developed some positive relationships, but not close friendships. With encouragement from his mother, he researched high school programs within the public school system (initiated tasks) and Barry chose a magnet program for science and engineering with an outstanding reputation for the marching band. This was a courageous choice for Barry, as he was not well coordinated despite playing an instrument well. He also did not know anyone at the school, which was located in a low-income neighborhood. Nevertheless, Barry felt proud of his choice and was able to continue taking responsibility for his schoolwork. His social life blossomed with participation in the marching band and included structured social activities, such as overnight trips, competitions, and some pillow fights. Nevertheless, because the school was located far from his home, he did not see other students outside of school activities. Barry was even nominated in his junior year by one of his teachers for a national leadership program. Of interest also is the fact that when evaluated at the end of high school, Barry's full scale IQ was one standard deviation above his score at age 11 with his verbal score being at the upper end of the superior range as compared to an average score before beginning treatment. Barry was accepted at several colleges with scholarships, and although his grades were poor during the first semester because of his joining too many clubs, he and his mother discussed how he could balance social activities with academic demands. Barry decided to join only one club and take four rather than six courses during the second semester. He obtained a grade point average of 3.7 while continuing to enjoy friendships. Additionally, he took the initiative to ask for tutoring and aid from the disabilities office as needed.

Results of a valid BASC-II filled out by Barry after his first year of college indicated that other than his perception of mild attention problems and atypicality, he did not perceive himself as having any problems with overall behavioral adjustment as compared to clinical levels of social stress, poor locus of control, and sense of inadequacy as well as depressive symptomatology and negative feelings toward school reported at age 11.

In the case of Barry, rather than receiving a specific problem-solving curriculum, his mother was simply taught how to use problem-solving communication with her son to help him to make his own decisions, think about possible consequences, set goals, and plan a course toward reaching them. Barry also came to understand that having the freedom to take over responsibilities required him to be responsible for his choices. Instead of being criticized for setbacks he was helped to use negative consequences as lessons for future planning and decision making (growth of prospective memory). His future goals include finding a job and learning to drive.

Summary and Conclusions

The relationship between problem-solving skills and executive functioning has been conceptualized and documented since the early work of Luria (1966). Others in the field of neuropsychology have supported this concept, as noted in the introductory section of this chapter.

In order to successfully solve life's complex problems, one must be able to identify the problem and set a goal for its solution, generate alternative ideas, utilize prospective memory for predicting consequences based on past experience, and shift from solutions that are not effective to ones that are. One must also be able to inhibit automatic responses and take the time to think of more effective ones. When resolving interpersonal problems, one must be able to take the perspective of others as well.

The ADHD and Asperger's children described above were able to learn to utilize executive functioning skills necessary to engage in successful problem solving because problem solving, goal setting, and thinking about the feelings of others became a way of life in their families and/or classrooms. These children were able to transfer their skills to other settings over space and time as they internalized problem-solving language.

In the case of Barry, he did not experience the ICPS curriculum but did engage in problem-solving communication with his mother and was given the opportunity to begin making decisions regarding homework, choosing elective subjects, choosing a school, setting goals, and most recently resolving issues for success in college. Although Barry had been receiving accommodations in school since elementary school, these alone were not resulting in success. Barry's case illustrates that with adolescents, just the opportunity to make their own decisions with the structure of problem-solving communication and the knowledge that they are trusted to do so may result in improved executive functioning in other areas in real-life situations, although behavioral changes may take longer than with younger children.

Although Spivack and Shure (1974) found that some young children are well adjusted and demonstrate self-regulation and the ability to generate alternative solutions to problems without an intervention, others, such as those who are considered acting out and impulsive or withdrawn, are at risk without a specific intervention such as ICPS. Even initially well-adjusted children, according to Shure (1993), are more resilient and less likely to deteriorate due to negative life experiences if they are exposed to problem-solving training.

As to why a problem-solving intervention generalized in the cases described in this chapter, rather than simply teaching the children skills in formal and hypothetical situations alone, the problem-solving communication was reinforced by parents and/or teachers in real-life situations over an extended period of time, and when parents did the training, problem-solving communication never ended. Also the children in these studies acquired the skills of "sufficient foresight and verbal dexterity to plan, guide, and evaluate their behaviors" that Whalen and Henker (1991)

proposed are needed in order for cognitive behavioral therapy to be effective. All of the parents described in the above cases reported that the bond they have with their children improved and continued as the children became adolescents and or young adults, reflecting the growth of mutual respect.

The cases described above demonstrate that even children with neurological compromise, such as ADHD and Asperger's syndrome, through training and practice can develop the executive functioning skills of inhibition, flexibility of thinking, goal planning, and perspective taking as well as the courage to take calculated risks in order to successfully meet life's challenges. Although one might argue that for younger children the prerequisite skills listed above must be in place in order for effective problem solving to occur, it might also be argued that for older children and adolescents like Barry, the practice of problem solving and guided decision making (a global act) might result in the evolvement of the component skills. Most importantly, through the empowerment developed by the opportunity to make decisions and engage in problem solving, a "conscious sense of self" as described by Barkley (2012) is nurtured to serve as the executive of the individual.

References

Aberson, B. (1996). *An intervention for improving executive functioning and social/emotional adjustment of ADHD children: Three single case design studies*. Unpublished doctoral dissertation, Miami Institute of Psychology.

Aberson, B., Albury, C., Gutting, S., Mann, F., & Trushin, B. (1986). *I can problem solve (ICPS): A cognitive training program for kindergarten children*. Unpublished manuscript, Dade County Public Schools, Miami, FL.

Aberson, B., & Ardila, A. (2000). *An intervention for improving executive functioning and social/emotional adjustment of three ADHD children. A four-year follow-up*. Unpublished manuscript.

Abidin, R. (1990). *Parenting stress index—Short form*. Charlottesville, VA: Pediatric Psychology Press.

Barkley, R. A. (2012). *Executive functions*. New York: Guilford Press.

Bell, M. A., & Fox, N. A. (1992). The relations between frontal brain electrical activity and cognitive development during infancy. *Child Development, 63*, 1142–1163. In Anderson, E., Northam, J. H., & Wrennall, J. (2001). *Developmental neuropsychology a clinical approach* (p. 94). New York: Psychology Press.

Benson, D. F. (1991). The role of frontal dysfunction hyperactivity disorder. *Journal of Child Neurology, 6*, 9–12.

Chugani, H. T., Phelps, N. E., & Mazziotta, J. C. (1987). Positron emission tomography study of human brain functional development. *Annals of Neurology, 22*, 287–297. In Anderson, E., Northam, J. H., & Wrennall, J. (2001). *Developmental neuropsychology—A clinical approach* (p. 94). New York: Psychology Press.

Cummings, J. L. (1993). Frontal-subcortical circuits and human behavior. *Archives of Neurology, 50*, 873–880.

Goldberg, E. (2009). *The new executive brain*. New York: Oxford University Press.

Grenell, M. M., Glass, C. R., & Katz, K. S. (1987). Hyperactive children and peer interaction: Knowledge and performance of social skills. *Journal of Abnormal Child Psychology, 15*, 1–13.

Hamlet, K. W., Pellegrini, D. S., & Conners, C. K. (1987). An investigation of cognitive processes in the problem solving of attention deficit disorder hyperactive children. *Journal of Pediatric Psychology, 12*, 227–239.

Klin, A., Volkmar, F. R., & Sparrow, S. S. (2000). *Asperger's syndrome*. New York: Guilford.

Kumpfer, K. L., Alvarado, R., Tait, C., & Turner, C. (2002). Effectiveness of school-based family and children's skills training for substance abuse prevention among 6–8 year old rural children. *Psychology of Addictive Behaviors, 16*, S65–S71.

Landau, S., & Milich, R. (1988). Social communication patterns of attention deficit disordered boys. *Journal of Abnormal Child Psychology, 16*, 69–86.

Levin, H. S., Culhane, K. A., Hartmann, J., Evankovich, K., Mattson, A. J., Harward, H., et al. (1991). Developmental changes in performance on tests of purported frontal lobe functioning. *Developmental Neuropsychology, 7*, 377–395.

Luria, A. R. (1966). *Human brain and psychological processes*. New York: Harper & Row.

Luria, A. R. (1973). *The working brain: An introduction to neuropsychology*. New York: Basic Books.

Ozonoff, S., Pennington, B. F., & Rogers, S. J. (1991). Executive function deficits in high-functioning autistic individuals: Relationship to theory of mind. *Journal of Child Psychology and Psychiatry, 32*(7), 1081–1105. In Goldstein, S., Naglieri, J. A., & Ozonoff, S. (Eds.). (2009). *Assessment of autistic spectrum disorders*. New York: Guilford Press.

Renshaw, P. D., & Asher, S. R. (1983). Social knowledge and sociometric status. Children's goals and strategies for social interaction. In Pelham, W. D., & Milich, R. (1989). Peer relationships in children with hyperactivity/attention deficit disorder. *Journal of Learning Disabilities, 17*, 560–567.

Reynolds, C. R., & Kamphaus, R. W. (1992). *Behavior assessment system for children*. Circle Pines, MI: American Guidance Service.

Roberts, A. C., Robbins, T. W., & Weiskrantz, L. (2002). *The prefrontal cortex: Executive and cognitive functions*. Oxford: Oxford University Press.

Schmahmann, J. (2004). Disorders of the cerebellum: Ataxia, dysmetria of thought, and the cerebellar cognitive affective syndrome. *The Journal of Neuropsychiatry and Clinical Neurosciences, 16*(3), 367–378.

Shapiro, E. G., Hughes, S. J., August, G. J., & Bloomquist, M. L. (1993). Processing of emotional information in children with attention deficit hyperactivity disorder. *Developmental Neuropsychology, 9*, 207–224.

Shimamura, A. P., Janowsky, J. S., & Squire, L. R. (1990). Memory for the temporal order of events in patients with frontal lobe lesions and amnesic patients. *Neuropsychologia, 28*(8), 803–814. In Levin, H. S., Eisenberg, M., & Benton, A. L. (1991). *Frontal lobe function and dysfunction* (pp. 191–192). New York: Oxford University Press.

Shure, M. B. (1992). *I can problem solve (ICPS): An interpersonal cognitive problem solving program for children (kindergarten/primary grades)*. Champaign, IL: Research Press.

Shure, M. B. (1993). *Interpersonal problem solving and prevention. A comprehensive report of research and training: A five-year-longitudinal study (Report #MH-40801)*. Washington, DC: National Institute of Health.

Shure, M. B. (1996). *Raising a thinking child: Help your young child resolve everyday conflict and get along with others*. New York: Pocket Books.

Shure, M. B. (2000). *Raising a thinking child workbook: Teaching young children how to resolve everyday conflicts and get along with others*. Champaign, IL: Research Press.

Shure, M. B., & Aberson, B. (2005). Enhancing the process of resilience through effective thinking. In S. Goldstein & R. Brooks (Eds.), *Handbook of resilience in children*. New York: Kluwer Academic/Plenum Publishers.

Shure, M. B., & Spivack, G. (1972). Means-end thinking, adjustment, and social class among elementary-school-aged children. *Journal of Consulting and Clinical Psychology, 38*, 348–353.

Shure, M. B., & Spivack, G. (1974). *Preschool interpersonal problem solving (PIPS) test*. Philadelphia: Hahnemann University.

Shure, M. B., & Spivack, G. (1978) A mental health program for kindergarten children: A cognitive approach to interpersonal problem solving. Training script. Philadelphia: Department of Mental Health Sciences, Hahnemann Medical Col.

Shure, M. B & Spivack, G. (1979). Interpersonal problem solving thinking and adjustment in the mother-child dyad. In M. W Kent and J.E. Rolf (Eds.) The primary prevention of psychopathology. Vol. 3. Social competence in children (pp 201–219.) Hanover, NH: University Press of New England.

Shure, M. B., & Spivack, G. (1980). Interpersonal problem solving as a mediator of behavioral adjustment in preschool and kindergarten children. *Journal of Applied Developmental Psychology, 1*, 29–44.

Shure, M. B., & Spivack, G. (1982). Interpersonal problem solving in young children: A cognitive approach to prevention. *American Journal of Community Psychology,* 10m, 341–356.

Spivack, G., & Shure, M.B. (1974) Social adjustment of young children. A cognitive approach to solving real life problems. San Francisco. Josey Bass Publications.

Stuss, D. T., & Benson, D. F. (1986). *The frontal lobes*. New York: Raven.

Vygotsky, L. S. (1962). *Thought and language*. New York: MIT Press.

Whalen, C.K. & Henker, B. (1991). Therapies for hyperactive children: comparisons, combinations, and compromises. *Journal of Consulting and Clinical Psychology,* 59, 126–137.

Practical Strategies for Developing Executive Functioning Skills for ALL Learners in the Differentiated Classroom

Kathleen Kryza

It's the first week of school for Alicia, a middle school teacher in a large school district in Michigan. She's been prepping for the first days of school for weeks, getting her room ready, and planning lessons. Last week she attended staff development sessions to learn about the new district and state initiatives and mandates that must be followed this year. Starting tomorrow, she will be immersed for the next 180 school days with a full day's schedule of three different preps—seven 50-minute classes with at least 32 students in each class. She can't imagine adding one more thing to her already overfull "To Do" list. But over the summer, Alicia read a book on teaching executive functioning skills to special needs learners. She really sees the value in teaching these important skills to her most at-risk students, but when can she possibly find time to do this? And how?

Alicia, like many teachers, understands the importance of developing executive functioning skills in her students, but given the full schedule of required academic content she needs to teach, she can't imagine squeezing in one more thing. To support building executive functioning skills in her classroom, she needs practical strategies that fit into her normal teaching day. She also needs to *see* that executive functioning skills will benefit *all* learners in her classroom, not just the few who have special needs. The good news is that differentiated instructional practices and lessons designed around clear learning targets can provide the strategies and tools necessary to help build her students' executive functioning skills while also meeting state standards and benchmarks. With effective differentiated teaching strategies, Alicia will be able to reach the variety of learners in her classroom, help them learn the content she *needs* to teach, and also build the executive functioning skills all students need to succeed.

According to Judy Willis, a neurologist turned middle school teacher and international educational consultant, "We can identify the practices that benefit all learners by looking at the skills most heavily emphasized in special education classes: time management, studying, organization, judgment, prioritization, and decision making. Now that the brain imaging research supports the theory that students process these activities in their executive function brain regions, it appears that brain-compatible strategies targeting these skills will benefit all students" (Willis, 2007). Her work, as well as the research cited by other experts within the chapter, shows that executive functioning skills are integral in helping all students and especially at-risk students achieve success in school.

This chapter is designed to offer ways to put best theories about the building of executive functioning skills into doable teaching practices. It will include many differentiated teaching strategies and ideas for practical implementation in the classroom. To effectively implement the strategies and ideas into instructional practice for building executive functioning skills, teachers need to teach them intentionally and transparently.

K. Kryza, M.A. (✉)
CEO, Infinite Horizons and Inspiring Learners,
Ann Arbor, MI, USA
e-mail: kkryza@me.com

Intentional teaching is when *we* know *why* we are teaching what we are teaching. Intentional teachers who are clear about why they choose one strategy over another can more easily guide their students. They use research-based, brain-based strategies. Their teaching is grounded in solid educational theory, and they teach to clear, kid-friendly learning targets. They choose teaching techniques thoughtfully and implement them strategically.

Transparent teaching is when *the students* know *why* we are teaching what we are teaching. Transparent teaching is teaching *with* the students, not *at* them. Teachers who are transparent consistently articulate to their students what they stand for as educators, why they are using specific strategies, and why what they are teaching is important for students to learn to develop into lifelong learners. (Not just to pass tests!)

Intentional and transparent teaching shows students where they are heading on the learning journey and what strategies they can use to get their successfully. When we are intentional and transparent, we let students know what we want for them, why we want it, how we will be doing it, and what benefits they should see if they choose to develop these skills. As Alicia tells her students, "You don't succeed in school by magic, you succeed by having a 'Can Do' attitude and working hard! I'll show you the way, but you have to walk through the door." Throughout the chapter, there will be examples of teachers teaching so you can see what intentional and transparent teaching looks like and sounds like in classrooms (see Fig. 29.1).

For the purposes of this chapter, executive function is defined as a set of processes that have to do with managing oneself and one's resources in order to achieve an academic goal. It is an umbrella term for the neurologically based skills involving mental control and self-regulation. The executive functioning skills that we will focus on in this chapter were developed in 2000 by psychologists Drs. Gerard A. Gioia, Peter K. Isquith, Steven C. Guy, and Lauren Kenworthy for their Behavior Rating Inventory of Executive Function (BRIEF). They divided executive functioning into two categories: behavioral regulation and metacognition. These categories and their sub-categories will become the framework referred to throughout the chapter.

Behavioral regulation
- **Inhibition**—The ability to stop one's own behavior at the appropriate time, including stopping actions and thoughts. The flip side of inhibition is impulsivity; if you have weak ability to stop yourself from acting on your impulses, then you are "impulsive."
- **Shift**—The ability to move freely from one situation to another and to think flexibly in order to respond appropriately to the situation.
- **Emotional control**—The ability to modulate emotional responses by bringing rational thought to bear on feelings.
- **Initiation**—The ability to begin a task or activity and to independently generate ideas, responses, or problem-solving strategies.

In this chapter, these behavioral regulation skills will be referred to as *mindsets*.

Metacognition
- **Working memory**—The capacity to hold information in mind for the purpose of completing a task
- **Planning/organization**—The ability to manage current and future-oriented task demands
- **Organization of materials**—The ability to impose order on work, play, and storage spaces
- **Self-monitoring**—The ability to monitor one's own performance and to measure it against some standard of what is needed or expected

The metacognitive skills in this chapter will be referred to as *skill sets*.

Mindsets + Skill Sets = Results: The Winning Formula for Academic Success

> Reinforcing effort can help teach students one of the most valuable lessons they can learn—the harder you try, the more successful you are. In addition, providing recognition for attainment of specific goals not only enhances achievement, but it stimulates motivation. (Marzano et al. 2001)

If we support students in regulating their behaviors by developing growth mindsets and

Academic: Intentional and Transparent	
Intentional	**Transparent**
You collect data on a survey about your students' learning styles or multiple intelligences *because* you want to find out how your students learn best so you can design lessons that work for them.	You explain to students that you will be teaching vocabulary to varying learning styles, allowing them to self-assess to determine which learning style works best for them so they can be more effective at studying and learning.
You model quality use of Sustained Silent Reading (SSR) time *because* you believe in the value of reading with your students. You get excited about the book you are reading. Read intently. Share and talk about books students are reading.	You ask students to create a list of quality use of SSR time vs. poor use of SSR time. Then students create a bulletin board of pictures of students modeling quality vs. poor SSR behaviors. You explain to students that they must ultimately own their own reading lives and they can develop behaviors that grow this life.
You give a performance-based assessment rather than a test to have students show their understanding of the similarities and differences between the cultures in Asia and in America today *because* you know this will assess for adeeper understanding than you would see on a test or quiz.	You explain to students that the reason you will be assessing using a performance based project rather than a test is that you want to see their understanding of cultural connection, not just their knowledge of the facts.
When checking homework in math, you have several students think aloud the varying ways they got to the correct answer *because* you want students model for each other that there is more than one way to get a correct answer in math.	When students share another way to get to the correct answer you note it out loud to the class. "Great Amir, that is indeed another way to get to the same answer. How many of you can see how Amir came up with his answer?"

Fig. 29.1 Intentional and transparent

we also teach them metacognitive skill sets that successful learners use, they should see results. If they are not seeing results, we can help them see where they may be getting stuck. For example, Alicia's student, Celeste, knows many great learning strategies because Alicia taught them explicitly to her students. But Celeste, who has a learning disability, doesn't see herself as a successful learner. She has a fixed mindset. Since she doesn't believe that using the strategies will make any difference, she quits before she gets started. On the other hand, James has a great work ethic in Alicia's class, but he doesn't apply strategies effectively, so he doesn't see the best results from his efforts. Each of these students needs coaching in different areas of executive functioning from Alicia. Alicia knows what she needs to target to help Celeste and James grow. She can also help them get to know themselves as learners so they understand what to do when they get stuck, so they can get unstuck.

The first part of the chapter will focus on how to help students develop the *mindsets* they need to succeed within the context of the teaching day by developing a classroom environment that promotes growth mindsets and clear routines and procedures. The second part of the chapter will focus on building skill sets by teaching students metacognitive, differentiated learning strategies to help grow their skills sets around clear learning targets that meet state standards and benchmarks. Throughout the book, you will continue to see examples from Alicia's as well as other teacher's classroom experiences to help you see how teachers in real classrooms are developing students executive functioning skills in meaningful ways.

Developing Growth Mindset Learners

Helping students regulate behavior involves developing not only their ability to control their actions but also their ability to shape their internal messages and to think flexibly so that they can respond appropriately to academic challenges. Carol Dweck's research and insights from *Mindsets: The New Psychology of Success* offers teachers ways to build essential, self-regulatory executive functioning skills into daily classroom instruction.

In her research, Dweck (2006) found that humans can develop two types of mindsets—fixed mindsets and growth mindsets. These mindsets impact the internal messages that play out in our minds as we face challenges.

Fixed mindset thinkers live by the following internal messages:

- Intelligence and talent are fixed, innate traits.
- Talent alone creates success.
- When learning something new, either you get it or you don't.
- Effort does not make a difference.

The internal message for students with fixed mindsets is "Look Good at all Costs." This internal message may play out with external behaviors such as heads down on desks or refusal to do work. Students with fixed mindsets may act like they don't care about school or learning or their teachers. Dweck notes, "It's no wonder that many adolescents mobilize their resources, not for learning, but to protect their egos. And one of the main ways they do this is (aside from providing vivid portraits of their teachers) by not trying. In fact, students with the fixed mindset tell us that their main goal in school—aside from looking smart—is to exert as little effort as possible."

This low-effort syndrome is often seen as a way that adolescents assert their independence from adults, but it is also a way that students with the fixed mindset protect themselves. They view the adults as saying, "Now we will measure you and see what you've got." And they are answering, "No you won't" (Dweck, 2006). Clearly these students with fixed mindset may lack all or some of the executive functioning behavioral regulation skills of inhibition, shift, emotional control, and initiation.

Growth mindset thinkers, on the other hand, develop the following internal beliefs:

- Most basic abilities can be developed through dedication and hard work—brains and talent are just the starting point.
- A love of learning and resilience is essential for great accomplishment.
- Effort and determination pay off.

The internal message for growth mindset students is, "Learn at All Costs." They are not afraid to ask for help, to ask questions, or to try again. They are ready and willing to learn the executive functioning skills they need to succeed.

What Dweck discovered in her research was that virtually all successful people have growth mindsets, and, the good news for educators (and parents), she also found that the human brain can develop a growth mindset. In one of her studies,

Dweck took two groups of seventh graders who were struggling in math. Group One received quality instruction in study skills to support math learning. Each time Dweck's researchers replicated the study, they saw no statistically significant change in students' academic growth. Group Two received information about how their brain worked and how they had the power to shape the messages in their brain. They were taught about fixed vs. growth mindsets. They learned about how their brains worked and the difference that effort makes in growing neural connections. Then, they were given the same study skill training as the Group One learners. Each time Dweck's researchers replicated this study, the Group Two student's academic scores improved in statistically significant ways.

Here's what Dweck had to say about the outcome of the research:

> We were eager to see whether the workshop affected students' grades, so, with their permission, we looked at students' final marks at the end of the semester. We looked especially at their math grades, since these reflected real learning of challenging new concepts.
>
> Before the workshops, students' math grades had been suffering badly. But afterward, lo and behold, students who'd been in the growth-mindset workshop showed a jump in their grades. They were now clearly doing better than the students who'd been in the other workshop.
>
> The growth-mindset workshop—just eight sessions long—had a real impact. This one adjustment of students' beliefs seemed to unleash their brainpower and inspire them to work and achieve.
>
> The students in the other workshop did not improve. Despite their eight sessions of training in study skills and other good things, they showed no gains. Because they were not taught to think differently about their minds, they were not motivated to put the skills into practice.
>
> The mindset workshop put students in charge of their brains (Dweck, 2006).

Again, we see that ***mindsets plus skill sets equal results***.

> One day, we were introducing the growth mindset to a new group of students. All at once Jimmy—the most hard-core turned-off, low-effort kid in the group—looked up with tears in his eyes and said, "You mean I don't have to be dumb?" From that day on, he worked. He started staying up late to do his homework, which he never used to bother with at all. He started handing in assignments early so he could get feedback and revise them. He now believed that working hard was not something that made you vulnerable, but something that made you smarter (Dweck, 2006).

What Dweck's work clearly says to us as educators is that if we are going to change our students' *skill sets*, alongside this we need to develop or shape their *mindsets*. Students' mindsets play a key role in the development of their executive functioning skills. Dweck found that when they intentionally and transparently taught students about how their brains learn and how growth mindsets make a difference, students see that they have control of their own learning and empowerment.

Marzano et al. also found similar results in their research for *Classroom Instruction That Works: Research-Based Strategies for Increasing Student Achievement* (2001), stating, "Not all students realize the importance of believing in effort. Although it might seem obvious to adults—particularly successful ones—that effort pays off in terms of enhanced achievement, not all students are aware of this. In fact, studies have demonstrated that some students are not aware of the fact that the effort they put into a task has a direct effect on their success relative to the task" (see Seligman, 1990, 1994; Urdan, Midgley, & Anderman, 1998). The implication here is that teachers should explain and exemplify the "effort belief" to students.

Theory into Practice: Developing Growth Mindsets

> Probably, one of the most promising aspects of the research on effort is that students can learn to operate from a belief that effort pays off even if they do not initially have this belief. (Marzano et al., 2001)

Dweck's and Marzano's research shows us that if we are going to help students develop executive functioning skills, we need to *intentionally* and *transparently* teach them to understand how their brain works. This ability to control the brain's internal messages can help students develop the essential behavior regulation skills of inhibition, shift, emotional control, and initiation.

Dweck ran an 8-day program about the brain and mindsets with the students in her research (see www.brainology.us to learn more about her program). For many teachers, this may not be possible. However, since the research shows that we should see results if we intentionally teach about mindsets, then it is imperative to find doable ways to embed explicit instruction about them into the daily school routine.

Our teacher, Alicia, listened to Dweck's podcast from NPR where Dweck talked about her powerful research at Stanford studying the brain's ability to grow with effort and about the importance of praising students for their effort not ability. Alicia realized that developing growth mindsets and strong skill sets encompassed both of the essential components of executive functioning skills—behavioral regulation and metacognition. So Alicia decided to intentionally spend the beginning of this new school year building growth mindsets in her classroom instruction. Since she knows that students learn in varied ways, she decided to offer them experiences that helped them learn about their own mindsets and the importance of developing growth mindsets by having them do activities that helped them feel their mindsets, learn from others' mindsets, and change their self-talk into growth mindset talk. The goal of this transparent teaching is to help her students develop the internal messages and executive functioning skills needed for successful learning in her classroom. Alicia spent one full day doing the Feel It activities and teaching her students about mindsets. Then she followed up with some 10–15-min mini-lessons about mindset talk throughout the next week. The time she took at the beginning of the year laid the foundation for how she talked to students and how the classroom ran for the rest of the year. Alicia knew the value of going slow to go fast. She built their "mindset muscles" up front, so they could grow stronger as learners throughout the year.

Feel It: Strategies for Developing Growth Mindsets

So much of our learning is tied to our emotional reactions to events and situations. Eric Jensen, an educator who has done extensive research on neuroscience, reminds us that the brain is most alert when there is a physical or emotional change (Jensen, 2005). In order for students to move away from fixed mindset thinking and limiting self-talk to growth mindset thinking and self-encouragement, they need to understand what messages are playing in their brains when faced with challenges.

The following activities are designed to have students feel and experience their own mindsets so they can learn to adjust them toward more growth mindset reactions.

Through the following experiences, the goal is to place learners in situations where things are not easy for them. We want students to struggle, grapple, and FEEL frustration, so they can begin to note their automatic response when challenged. We want them to listen to their loud inner talk (and possibly even outer talk!). From this place of awareness, they can begin to identify whether they respond with growth or fixed mindsets when faced with a challenge. Do they say, "Aggghhh! I can't do this!" or "Man, this is going to be difficult!"

Students may note that they tend to be growth mindset when presented with a challenge in an area in which they are naturally good. For example, Lebron James would probably have a growth mindset for any physically coordinated challenge put before him. But ask him to make a beautiful origami necklace from written directions and he might easily give up. That's a fixed mindset. So it is important to provide several different activities for students to feel their mindsets, so they note how they respond differently depending on the task.

Plan one classroom period to do the Feel It activities. You can choose to do one or two of the activities with your students. Another fun way to do the activities is in stations, with a different "feel it" mindset activity at each station. Students can then rotate through the stations and note their response to each station.

Remember the goal of this experience is to move kids away from fixed mindset thinking and negative self-talk toward growth mindset thinking, giving them the vocabulary and experience to talk growth mindset talk and shape growth mindset behaviors.

To begin, introduce the idea of "building brain matter" by showing how neurons are activated in learning and how new connections can be made. Share with your students the vocabulary of fixed and growth mindsets and let them know they are about to experience what their own mindset is like.

Opening the Lesson
It may sound like this:

> We are going to test our limits, patience, and determination today. Let's pay attention to the messages our brain is sending as we tackle a difficult task. First, let's try _____.

Feel It Activities

The following are possible activities you could use to have students feel their mindsets (*Developing Growth Mindsets in the Inspiring Classroom*, Kryza, Stephens, & Duncan, 2011):

- **Take a Quiz** (linguistic or logical): Give students a surprise quiz on what they've been learning in your class.
- **Try Toothpick Puzzles** (logical): Have students try to solve a toothpick puzzle. Many examples and solutions at various levels can be found at: http://www.madras.fife.sch.uk/departments/maths/toothpickworld/toothpick13s.html.
- **Tie Knots** (visual/tactile): Provide rope and written directions with no pictures and have students try tying knots.
- **Visual Word Puzzles** (visual/linguistic): Give students word puzzles to complete within a given amount of time.
- **Tangrams** (visual/tactile).
- **Build a Tower of Cards** (tactile).
- **Do Riddles** (linguistic).
- **Do Sudoku** (mathematic).
- **Run an Obstacle Course/Scavenger Hunt** (kinesthetic).

Reflection: After each activity, ask students to respond to the following questions:
- So, how did you feel before you started this activity? What were you saying to yourself?
- What did you feel and say to yourself during the activity?
- How did you feel and speak to yourself after the activity?

Students then categorize their comments into growth or fixed mindset categories.

The discussion generated at this point will lead into activities in the next "Talk It" section.

Talk It

Language is so pervasive in our lives; we are often unaware of the powerful effects it can have on our socializing, teaching, thinking, and most certainly our learning. Peter Johnson notes in *Choice Words: How Our Language Affects Children's Learning*, "If we have learned anything from Vygotsky, it is that children grow into the intellectual life around them. That intellectual life is fundamentally social, and language has a special place in it. Because the intellectual life is social, it is also relational and emotional. To me, the most humbling part of observing accomplished teachers is seeing the subtle ways in which they build emotionally and relationally healthy learning communities—intellectual environments that produce not mere technical competence, but caring, secure, actively literate human beings" (Johnston, 2004, p. 88). By focusing on the language or talk in a growth mindset classroom, we choose to make the invisible, *visible*. Some educators may be naturals at using powerful language in the classroom—understanding the implications that our words can have socially, intellectually, and academically. However, most of us have to study, work hard, and practice talking the talk! (Get your growth mindsets ready!)

Keep in mind, language that builds mindsets and executive functioning skills is far more than a few self-affirmations. Language conveys what we believe about ourselves, others, our efficacy, and our resiliency. It is how we make plans, reflect on our plans, and find solutions to our struggles. More than "I can do this!" (a non-example), intentional language can help students think clearly about *how* to do something.

The following are activities that creates classroom talk that helps students develop growth mindsets (*Developing Growth Mindsets in the Inspiring Classroom*, Kryza et al., 2011). These activities can be taught in short mini-lessons and the message can then be reinforced throughout the school year. Alicia and her students made a list of phrases that growth mindset learners say to themselves. Alicia posted their phrases on an anchor chart in the classroom. Anytime students used fixed mindset talk, Alicia or their peers kindly reminded that they could choose better self-talk from their list. It became a fun way to support each other throughout the year.

Talk It Activities

1. With lower elementary students, choose a self-talk phrase or a student-to-student phrase from the Talk It resources (Fig. 29.2) to practice chorally with the whole class. This can be the "Phrase of the Day" or the "Phrase of the Week" to help students start building the language of growth mindsets. For example, the charter K-2 school, KIPP Ujima in Baltimore, MD, has the students gather together each and every morning in the cafeteria, and among other things they do to start their day in a calm and routine way, they chant, "We're here to get some knowledge, so we can go to college."
2. Have students reflect on an aspect of their life where they already have a "growth mindset." For some students this can be when playing video games, taking a dance class, or working on a hobby or sport. Have them notice their self-talk when they are doing something that they enjoy practicing. Ask them to write down those messages and apply those same messages when they are working on an academic skill that feels more challenging for them.
3. Students can work in pairs to restate their fixed mindset messages into growth mindset messages.
4. Post growth mindset phrases in the classroom for students to use as a resource when they find themselves thinking in fixed mindset ways. Also post a chart of metacognitive strategies they are learning throughout the year that they can use when they are feeling stuck. Every day in the room they see a message that reminds them that "Mindsets plus Skill Sets equal results."
5. As a class, generate a chart of growth mindset feedback and ideas for your students to use when talking to one another. Post a classroom chart of phrases students can use to help work through fixed mindset struggles (see the below figure from Alicia's classroom).

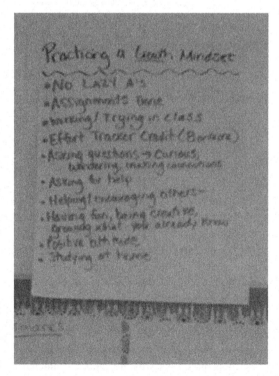

6. On a T-Chart, have students jot on one side a list of phrases or statements they recall that have "stopped" them from wanting to try harder. These can be messages from teachers, peers, parents, or themselves. On the other side of the T-Chart, have students write a plan

for how they will respond if they hear those phrases again. (After all, not all teachers, family members, or friends will be making the shift to a growth mindset, so students need to be prepared to defend themselves as learners who are working hard!)

7. Close the day, or a class period, with feedback to students about working hard. The feedback can be from you or between students. Remember that empty praise or feedback, without specific input about behaviors, does not have a significant impact on learning (Marzano et al., 2001). It's not enough to say, "Nice job," or "Great effort!" Feedback needs to be focused on behaviors that can be noticed and named. Calling attention to specific actions such as asking for help, starting over, taking a deep breath, or staying focused is what helps students understand the behaviors that contribute to their success.

8. Share with students your own struggles and what you say to yourself to keep your effort and drive up. Remember, beyond affirmations, self-talk is about knowing what we know and knowing what to do when we don't know! (Fig. 29.2).

See It

Teachers and students can learn a great deal about growth mindsets from stories and examples of people, real and fictional, who have chosen to work hard and put forth effort, often against great

Teacher-to-Student (Teacher Talk)

Before Learning
• Today you might find there are some things that are new to you and you are going to get to grow from trying them.
• Does this remind you of something you've done before? How can you use that experience to help you with this new learning?
• Looking at today's work, what part do you think will be the most challenging for you? What can you do when learning gets to the GOOD part (the hard part) to help you continue learning?
During Learning
• What parts are going well? What parts are making you grow?
• Why do you think this part is challenging for you? What do you need to help you? Do you need more information? More practice? A different way to practice?
• Have you done something like this before? What did you do when it got hard? Can you do it again?
• What do you know about yourself as a learner that can help you continue learning?
After Learning
• How did you grow as a learner?
• Did you learn something new about yourself and how you learn?
• Howcan you use that in the future when something gets tough?

Fig. 29.2 Talk it chart

Student-to-Student (Classroom Talk)

Before Learning
• My plan for this learning is_____.
• I think the hardest part for me might be _____ and I'm going to _____ to help myself.
• Kathleen, I can help you when you get to _____ if you can help me work through _____.

During Learning
• You know how to do this, remember when you_____.
• You worked really hard on that!
• This is just like _____. Use what you know from when we did that.
• You just need more practice, let me help you.

After Learning
• You worked really hard on that!
• You never gave up!
• You used lots of resources and effort to keep going.
• I saw you _____when you got (frustrated, stuck, overwhelmed).

Fig. 29.2 (continued)

odds. Bringing these stories and people into your classroom and making them mentors and role models will guide your students on this journey of personal and academic growth.

The following are activities that can help students learn to develop growth mindsets by learning from role models (*Developing Growth Mindsets in the Inspiring Classroom*, Kryza et al., 2011). Sharing growth mindset stories should be done in small moments throughout the school year. Alicia showed some videos about people with growth mindsets at the beginning of the year and then interspersed the stories and examples throughout the year, especially when she saw students' mindsets starting to become fixed. Sometimes students brought in their own examples which were then shared with the class.

See It Activities

- Read stories and show movies or excerpts about real or fictional people who exemplify growth mindsets. (The movie "Rudy" is a great true-life example of a growth mindset person.) After viewing or listening, ask the students to reflect on what they learned from this example. Ask how they can apply this knowledge to their own lives.
- Make class commitments to "live what you learned" from a role model. For example, when teaching about revision in writing, share how Ernest Hemingway rewrote the ending of "Farewell to Arms" 39 times! Then make a classroom commitment to "Rewrite Until it

Student-to-Self (Self Talk)

Before Learning

- OK. Let me make a plan for myself.
- I am going to need _____ to help me through _____.
- I've done something like this before; let's see if I can figure it out.
- Oh! Something new! Yay

During Learning

- I just have to take it one step at a time.
- I get all this information. I just need to know _____.
- I have all these skills. I just need to be able to _____.
- I've gotten this far. I'm not stopping now.
- I'll know I got it when I can _____.

After Learning

- Wow. I learned so much!
- I grew a ton. Before I didn't know _____, now I know _____.
- Before I couldn't _____, now I can _____.
- Based on what I learned from this, next time I am going to _____.
- One thing I learned about myself as a learner _____.
- Next time I try something like this I will _____.

Fig. 29.2 (continued)

Sounds Right!" Post your saying and chant it from time to time.

- Have students select someone they admire who has grown in positive ways through mindsets and skill sets. Then have them study and create a project about what they learned from that person. (In the content areas, have them select a person from that field of study.) Make it a choice project. They can write a song, poem, or reflection; paint, draw, or create a model; and make a video, voice thread, or photo journal—anything that they are inspired to create from what they learned.
- Create a bulletin board of "Our Role Models" and let students add pictures and personal reflections. Let them change and grow it over the year. (Remember to look for role models in our friends and family members, too!)
- Have a share time once a week for the last 5–10 min of class. You and students take turns sharing your story or role model and then have a class discussion. (These could then be added to the bulletin board. *See above.*)
- Collect and play songs that have growth mindset messages.

It does take some time to build growth mindset thinking into a classroom, but the time spent pays off in the long run. An interesting set of studies by Craske, Wilson, and Linville (as cited in Marzano et al., 2001) has shown that simply demonstrating that added effort will pay off in terms of enhanced achievement actually increases student achievement. In fact, one study (Van Overwalle & De Metsenaere, 1990) found that students who were taught about the relationship between effort and achievement increased their achievement more than students who were taught techniques for time management and comprehension of new material. Once again we see that *mindsets plus skill sets equals results*.

Create a Classroom Environment That Promotes Executive Functioning Growth

> Just as we provide prostheses for someone who cannot walk otherwise, children with executive weakness need adults to adapt their environment and tasks when they do not yet have sufficient executive competence to succeed on their own. For this reason, Dr. Russell Barkley refers to the process of accommodating kids as building a "prosthetic environment." External support, limits, and supervision can all be types of prostheses. (Kahn & Dietzel, 2008)

The brain learns best in a safe and positive environment. As cited in David Sousa' book How the Brain Learns, 2000, "When students feel positive about their learning environment, endorphins are released in the brain. Endorphins produce a feeling of euphoria and stimulate the frontal lobes, thereby making the learning experience more pleasurable and successful. Conversely, if students are stressed and have a negative feeling about the learning environment, cortisol is released. Cortisol is a hormone that travels throughout the brain and body and activates defense behaviors, such as fight or flight. Frontal lobe activity is reduced to focusing on the cause of the stress and how to deal with it. Little attention is given to the learning task. Cortisol appears to interfere especially with the recall of emotional memories."

If students are secure in knowing how the classroom functions and if they build successful skills for learning, they are able to take risks, make mistakes, learn together, and readily welcome a challenge or new situation. This is the kind of environment that supports the growth of executive functioning skill development. In order to do this, teachers need to create a risk-taking, mistake-making classroom environment. "Learning occurs more easily in environments free from threat or intimidation. Whenever a student detects a threat, thoughtful processing gives way to emotion or survival reactions. Experienced teachers have seen this in the classroom. Under pressure to give a quick response, the student begins to stumble, stabs at answers, gets frustrated or angry, and may even resort to violence" (Sousa, 2000). We help our classrooms function as a community and develop executive functioning skills by establishing procedures for creating a safe learning and growth mindset environment.

To support students and nurture their desire to grow their executive functioning skills, teachers must first create a learning environment that feels safe. We want each of our students to reach a place where they feel safe in experimenting, safe in new situations (because we're all in this together), and know that our classroom is a safe place to land when we fail. Dozier notes that the "stress-induced survival mode, like the adrenalin fight-or-flight response, blocks students' abilities to select the meaningful input from their sensory environment. What little information they can separate out as important cannot readily pass through the over-stressed amygdala and limbic system into the brain's relating and memory centers" (Dozier, 1998).

You create a safe learning and growth mindset environment first by establishing positive messages in your classroom. The messages you establish help you connect to the mindset talk you will be using with students throughout the school year. Two powerful messages to establish in the classroom that set the base for building executive functioning skills and growth mindsets are the following:

1. **This is a risk-taking, mistake-making classroom**. The more we create a classroom that values intellectual risk taking and challenge to all students and live the message in our classroom, the more students can start to build

emotional control around risk taking. In Japan, for example, when students are wrong, it is cause for celebration. The belief is that from mistakes, we learn new ways of thinking. Students learn what teachers value, so we must give them continual reminders that executive functioning skills are needed to succeed in school and in life. It's important to remind students that taking risks and working hard in the classroom makes a difference in their learning experience.

A New York Times article in September 2011, titled "What if the Secret to Success is Failure," discusses schools which are building this kind of risk-taking thinking and talk into their daily instruction. In KIPP Infinity, a charter middle school in New York, character language permeates the school. "Kids wore T-shirts with the slogan 'Infinite Character…' The walls were covered with signs that read 'Got self-control?' and 'I actively participate!' (one indicator for zest). There was a bulletin board in the hallway topped with the words 'Character Counts,' where students filled out and posted 'Spotted!' cards when they saw a fellow student performing actions that demonstrate character. (Jasmine R. cited William N. for zest: 'William was in math class and he raised his hand for every problem.')" (NY Times, September 14, 2012).

2. **Fair is not everybody getting the same thing, fair is everybody getting what he or she needs to be successful.** This is another powerful message that permeates schools and classrooms that are building executive functioning skills into daily instruction. All students are not created equal in their executive functioning or academic skills. This message is established early in the school year as teachers gather data about students' learning profiles through giving multiple intelligence or learning style surveys. (You can download the Fair and Risk-Taking posters as well as easy-to-use surveys at www.inspiringlearners.com/freerescources.)

We need to focus on noticing and praising students' efforts and not their abilities or their intelligence. Whether a student fails or succeeds, we need to learn to give feedback about specific effort given or strategies used—what the student did wrong and what he or she could do now or next time. Dweck's research shows that specific praising of *effort* is a key ingredient in creating students who value success that comes through hard work. And wouldn't we like part of developing a nation, a world, filled with growth mindset humans!

While the messages that we embed in the classroom set the foundation and tone for growing strong and resilient learners, another key to creating a classroom environment that fosters executive functioning is the routines and procedures that make the classroom run smoothly and build the executive functioning skills of emotional control, shift, initiation, and inhibition.

The learning brain needs routine and predictability. When we establish routines, we are nurturing the students' capacity to work as a functioning member of a community. Routines and procedures must be taught explicitly, modeled and practiced daily until there is no thought to what is supposed to happen next. Students can then focus on the task at hand with no thought of "What do I do now?" We create an empowered learning community when we explicitly and transparently teach students routines such as:

- How to enter and exit the room
- Where and when to turn in papers
- How and whom to ask for help (see 3 before me)
- How to transition from traditional seating arrangement into groups
- Where to go for supplies and materials for activities
- Anchor activities: What do we do when we are done or waiting?

Theory into Practice: Creating a Classroom Community with Routines and Procedures

Alicia started the first days of school by creating a classroom environment that incorporates slow scaffolds to develop essential executive functioning skills. This year she has chosen to focus on teaching her students routines and procedures to help them work in groups. Alicia knows that it takes time to get students working effectively in groups, but that the time is well spent (again, go slow to go fast), as it helps her students build their executive functioning strength in all major categories.

Teaching students to work in collaborative groups over time helps students develop strategies to monitor their behavioral and metacognitive skills.

The following strategies support safe classroom environments and therefore encourage the development of executive functioning skills among the students.

Class Meetings

Class meetings are established at the beginning of the school year as a way to create an "us" community where teachers and students work together to create an environment where teachers can teach and students can learn. When we hold classroom meetings, we are saying to students that this is your community too, and you are encouraged be a part of it, help maintain it, and contribute your thoughts, ideas, and passions.

The first step in preparing for classroom meetings is to establish mutual classroom norms. From various perspectives and backgrounds, students work together to develop norms of behavior that are embraced in the classroom. In an opening discussion, students agree that they have come to school to learn, and if they are going to learn, what norms need to be in place. Students then brainstorm norms and come up with a short list they all agree to support. The norms are written in positive language (respect others, participate in class learning) and are then posted on a chart in the room.

Celebrate successes!

Share what has been working and troubleshoot with the group when things aren't working. Class meetings are a great way to neutralize tensions in the classroom while building those essential executive functioning skills. With humor, examples, role-playing, and a neutral ground to meet, groups can work on strategies and activities that promote positive communication, problem-solving skills, and the redirecting of negative energy. Class meetings provide regular practice for developing and maintaining executive functioning skills.

Routines for Working Collaboratively

Grouping is a powerful tool for maintaining a well-functioning classroom. Groups divert a lot of the daily maintenance from the teacher. Within groups, roles and responsibilities must be established. This gives each student a purpose and direction; they KNOW they are an integral part of the community and their contribution is valued. Student roles and responsibilities include asking for help or seeking more information when they need it, engaging in group dynamics, and waiting to speak until they are called on.

We create community by establishing GROUPS for purposes such as core groups.

Core groups are groups of 3–5 students who stay together for at least a marking period (or semester or year). That does not mean they are sitting next to each other all year, but they are available throughout the year to support each other and learn with each other in establishing a community that builds executive functioning mindsets and skill sets together.

The groups are intentionally selected randomly to replicate the work world. Adults usually don't get to select the people they work with. If we want students to be prepared to work with people from other cultural backgrounds, lifestyles, etc., we need to give them lots of practice so they have the executive functioning skills to work with others. Staying together does not mean they are sitting next to each other, but rather they are used as homework helpers, turn and talk groups, and emotional check buddies.

Core group members have jobs: coach, organizer, recorder, energizer (other jobs might include materials manager, time keeper, and noise leveler).

Core group members need to bond before they can collaborate effectively. To help facilitate bonding, educators can have core groups conduct team-building activities during the first weeks of school, participate in fun activities where teams work to put together a puzzle without speaking, construct a model, send a message using art, create a dance that represents the class theme, etc. Afterwards, students self-assess their groups' behaviors and talk, reflecting on what worked and didn't work and how they can work together better next time.

Mindsets plus skill sets equals results. The first part of this chapter has been devoted to building the mindsets students need to regulate their thoughts and behaviors. Teaching explicitly about growth mindsets throughout the school year supports students in developing control over their thought processes. Teaching about mindsets helps them to thrive in the classroom and helps prepare them for the future. Students who develop an internal locus of control and who see themselves as a contributing member of a classroom community will be better prepared to enter the global community of the twenty-first century.

Developing Skill Sets: Differentiated Instruction Makes a Difference

In 2000, cognitive scientists and cognitive psychologists combined their knowledge in a book titled "How People Learn" (National Research Council, 2000). A key finding in their research is that "a meta-cognitive approach to instruction can help students learn to take control of their own learning by defining learning goals and monitoring their progress in achieving them." The implication of this research is that "the teaching of meta-cognitive skills should be consciously integrated into the curricula across disciplines and age levels." So, while it's essential that we work to develop students' behavioral executive functioning skills through shaping their mindsets, we also need to develop their metacognitive skill sets.

If we are going to help students develop the metacognitive executive functioning skills they need to succeed, we must consciously and transparently embed the teaching of these skills consistently into our teaching practice. Mindsets and metacognitive skill sets go hand in hand in helping students to develop the internal locus of control needed for effective executive functioning. As students develop their metacognitive skills, they are able to evaluate their ideas and reflect on their work. They can change their minds and make midcourse corrections while thinking, reading, and writing.

Judy Willis, a neurologist turned middle school teacher, says:

> When students use metacognition to actively and consciously review their learning processes, their confidence in their ability to learn grows. They begin to attribute outcomes to the presence or absence of their own efforts and to the selection and use of learning strategies. Students with learning disabilities, many of whom have never thought of themselves as capable of learning, realize that some of their errors are due not to ignorance, but to not knowing the effective strategy to use. Once students identify useful strategies and use them repeatedly, they will experience a powerful boost in confidence. Similarly, they will develop more perseverance as they discover that when their initial effort to solve a problem fails, they can succeed by approaching it with a different strategy. Students are actually learning to be their own tutors and guides. (Willis, 2007)

There are three stages in the metacognitive process. Effective learners develop a plan of action, maintain and self-monitor their plan, and evaluate and adjust their plan. Students need to be taught intentionally and transparently to own these stages of the learning process, use the learning strategies that work best for them, and know when and how to apply the strategies during the learning process.

With the diversity of today's student population, it's also essential to address learning differences and support students in knowing how they learn best. Research from Keogh, Levine, and Rief (as cited in Marzano et al. 2001) suggests that effective educators understand that students have different learning styles and temperaments and that these differences must be respected and accommodations must be made lest the student fail and misbehave. Differentiated instructional practices designed around clear learning objectives allow all learners to reach standards and benchmarks.

The simple, doable framework for differentiating used in this chapter will be "Know Your Learning Target, Know Your Students, Vary the

Learning Pathways" (Kryza, Duncan, Stephens, 2007). We build students' skill sets by intentionally teaching to clear learning targets and making the learning target transparent, so students know where they are going and what they are learning. We intentionally get to know who our students are as learners and transparently teach them to know themselves, so they can advocate for themselves and learn to use the most successful metacognitive strategies for their learning style. Then, as we design our lessons, we intentionally differentiate the input (chunk), process (chew), and output (check) parts of our lessons so that students can learn in ways that work best for them.

Know Your Students: Executive Functioning Skill Sets Grow When We Know Our Students as Learners

> If we are to achieve a richer culture, rich in contrasting values, we must recognize the whole gamut of human potentialities, and so weave a less arbitrary social fabric, one in which each diverse gift will find a fitting place. (Margaret Mead)

A large part of helping students build their metacognitive skills in the differentiated classroom is helping them know themselves as learners. In order to do this intentionally and transparently, we need to gather data about how our kids learn and then teach them to use their strengths to build up their areas of challenge.

It's best to begin gathering information about your students as learners at the beginning of the school year. Surveys are an easy and practical way to discover your students' learning styles or multiple intelligence strengths. If you are new to gathering data about your students, begin by gathering just one piece of information about them. Conduct the survey in the first week of school before you jump into deep content instruction. If you are going to use the data to help students know themselves, you then need to organize the data so it is easily accessible and can be used to inform instruction. You can find easy-to-use surveys online at www.inspiring-learners.com.

Gathering the data is the first step in getting to know your learners. But if you are going to use the data to inform your instruction, you need to have easy access to the information. Alicia uses **learning profile cards** to better understand her students. Using 4×6 index cards, she gathers several pieces of data about her students throughout the first weeks of school. After students complete the survey, she has them summarize their findings on the 4×6 card (see Fig. 29.3). She then collects the cards and begins to add her own observations. Teachers of younger students may need to transfer the data onto the cards themselves, but the effort is well worth it. Alicia comments, "You know, it's sort of a pain to do the surveys and gather the data, but I now know about my students in September what I used to not know about them until April."

Alicia uses the information she gathers to help her students understand themselves as learners. For example, after reviewing the learning profile cards, she felt her students would understand their vocabulary more effectively if they were placed with students of the same multiple intelligence. Pulling out her data index cards the evening before class, she was able to arrange students quickly into art smart, word smart, body smart, and music smart groups. Students worked with others of the same intelligence strength to draw words, create word clues, do word charades, and compose word songs or rhythms. Students then shared their creations with others. This allowed students to begin internalizing the learning and study skills that work best for them. Alicia used the cards again to group students for a multimedia research project. She recognized that, this time, the final product would be more creative if students worked with others who had *different* learning styles and strengths.

Gathering information about your students at the beginning of the school year sends a powerful message that you care about who they are and how they learn. It sets the tone for creating a classroom community where all students feel honored and respected. As importantly, students who know how they learn best develop executive functioning metacognitive skills for knowing what to do to make learning stick for them. Judy Willis states, "The more that students gain an understanding of

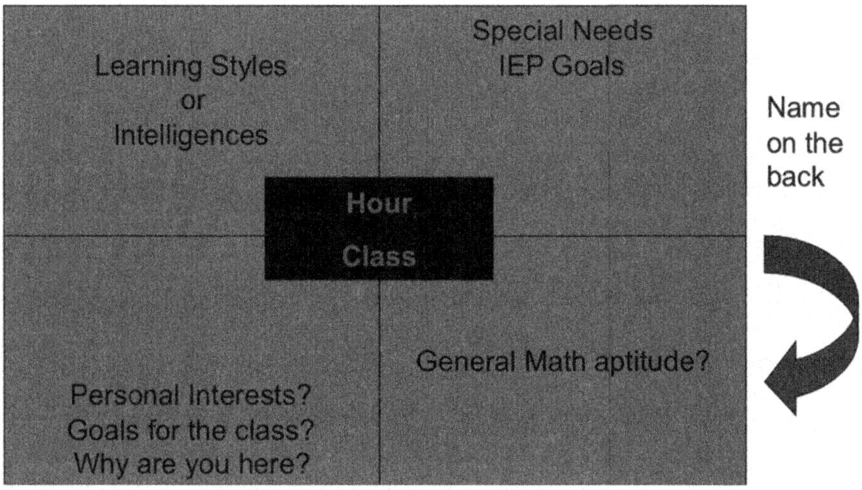

Learning Styles or Intelligences	Special Needs IEP Goals
Hour Class	
Personal Interests? Goals for the class? Why are you here?	General Math aptitude?

Name on the back ⤵

```
              VOCABULARY EXIT CARD:  Self Assessment
                              DATE _____

NAME:_____      LOW              HIGH

1. I used my time wisely and had a good attitude    1   2   3   4   5
2. I completed work at each station                 1   2   3   4   5
3. I understand the key terms better now            1   2   3   4   5

I learn vocabulary words best this way…

Here's how I plan to study for my test based on what I know about myself as a learner
```

Fig. 29.3 Student learning profile card

their learning profile, the more they can develop strategies for learning actively and successfully. When students don't comprehend why they are struggling with learning, when they believe they are dumb or stupid or lazy, they are more likely to resort to self-defeating ways of coping represented by non-compliant behaviors."

Lisa, a high school English teacher in Michigan, saw positive results when she taught her students to know themselves as learners. She gave her students the surveys at the beginning of the year and then taught the students throughout the year how to own their learning strengths. At the end of the year, Lisa asked the students for feedback to see if teaching this way was empowering students. Caylynn, a senior in Lisa's English class said, "I don't have the best memory, and when I could put the word with a picture or action, it helped. I remembered the vocabulary words better when we did Charades and Pictionary. It also made the class fun." Another student, Raeann says, "I comprehended material better when it was taught in my learning style."

Know Your Students: Theory into Practice

Once you give your students the multiple intelligence or learning style survey at the beginning of the school year, you want to be sure you to be intentional and transparent in teaching students to own how they learn best. Here are some ideas for teaching into the ways all students learn:
- Offer students a *variety* of ways to practice their vocabulary. Let students choose to chew

on new words by drawing pictures, creating movements, writing songs, etc.
- Put students in groups by their learning strengths and have them create ways using their learning strength to remember procedures such as "How does a bill become law," "Steps of the writing process," and "Three things we do when we come into Ms. Stephens classroom." Have students share their examples with each other.
- You can also teach students how to study for tests using the varied learning styles that work best for them.

Vary the Pathways: Executive Functioning Skill Sets Grow When We Differentiate Our Lessons

Chunk, Chew and Check—That's How the Brain Learns Best! (Kryza, Duncan, Stephens, 2010)

Judy Willis says, "We can identify the practices that benefit all learners by looking at the skills most heavily emphasized in special education classes: time management, studying, organization, judgment, prioritization, and decision making. Now that the brain imaging research supports the theory that students process these activities in their executive function brain regions, it appears that brain-compatible strategies targeting these skills will benefit all students" (Willis, 2007).

Educators today have so much information about the learning brain that we can no longer ignore using evidenced-based strategies that we know work to build executive functioning skills. However, we also need doable ways to implement the strategies, because the more intentional and transparent we are, the more students develop and own the skills they need to succeed.

What Is Chunk, Chew, and Check or CCC?

The chunk, chew, and check (CCC) framework offers a simple and doable way to design lessons that are meaningful for all learners. These terms are defined as follows:

Chunk: *Same learning target, different ways to INPUT new information into learning brains.* The brain learns best when it receives new information in small chunks. Because each brain perceives incoming information differently, we need to vary how we offer chunks of new learning.

Chew: *Same learning target, different ways for learners to PROCESS new information.* All brains have a unique way of connecting new information to what it already knows. Therefore, we need to offer students a variety of ways to CHEW on new information we have presented.

Check: *Same learning target, different ways for learners to OUTPUT what has been learned.* We know that individuals possess unique talents and therefore demonstrate understanding in their own way. We need to balance the ways we formatively and summatively CHECK for student understanding.

> Chunking allows us to deal with a few large blocks of information rather than many small fragments. Problem solving involves the ability to access large amounts of relevant knowledge from long-term memory for use in working memory. The key to that skill is chunking. The more a person is able to chunk in a particular area, the more expert the person becomes. These experts have the ability to use their experiences to group or chunk all kinds of information into discernible patterns (Sousa, 2000).

Varying these three steps in your lesson design allows you to thoughtfully respond to how well learners gain access to content, process what you have taught, and demonstrate that they have mastered the outcomes. With the chunk, chew, and check (CCC) framework in place, you can more carefully monitor where learning, or the demonstration of learning, is successful or where learning breaks down for the learner. This framework allows you to see how well students respond to the instructional interventions, gives them access to content, and allows you to move all learners toward the same learning outcomes. Best of all, by varying these three components of learning, we are creating joyful, successful, as well as, meaningful opportunities for learning to occur in our classroom, thus making it a more rich and rewarding experience for our students as well as ourselves.

The chunk, chew, and check framework provides a road map for teachers to monitor themselves as they design lessons and activities. With CCC, we can determine where it is that students are getting stuck. Do they need different access to the chunk? Is the chew not working for some? Is there a more effective way to check or assess student understanding? This understanding of where students are struggling helps teachers know how to transparently teach students what they need to do to grow their learning skills.

The CCC framework helps students become better self-assessors. Knowing their learning target enables students to be aware that they should be monitoring the chunk, chew, and check of their learning and focus toward that learning target. They can assess their progress by asking, "Am I able to take in a new piece of information? Is it locking in for me in a way that sticks? Am I able to show that I got it?" This metacognitive self-reflection helps them monitor and adjust as needed.

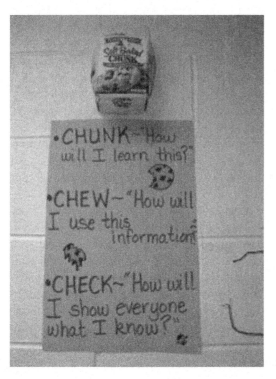

We use the chunk, chew, and check terminology with students because it sticks. We use this terminology with kids to explain the learning process to them. This helps us to be transparent in our teaching. Teaching *with* our learners instead of *at* them means we include them in the learning process. When we teach transparently, our students know why we use certain strategies for learning and why they should use these strategies beyond the classroom. Making our teaching transparent with CCC might sound like this:

> I am going to give you a CHUNK of new information for about 10 minutes, then you will have time to CHEW on it with your talk partners because sharing ideas with another person is a great strategy for deepening your reasoning. (After 10 minutes) Okay, it's time to CHECK your progress. Using thumbs up or thumbs down, how are you feeling about this new idea? If you are still unsure, where can you go for help? Make a plan for the first few minutes of class tomorrow to clarify questions. Being self-reflective is something that everyone does to make sure they are on track, because let's face it, when you are an adult you often don't have a boss telling you what to do minute to minute. You have to start thinking, 'Do I know what I need to know?' and make a plan if you need help. So, let's continue to practice that…

The CCC framework gives students the *why* of the content and the learning activities we teach. Because we are intentional and transparent, they will know why we do mind maps, why we use talk partners, why we put sticky notes in textbooks, and why we assess in various ways. The CCC framework also promotes students advocating for themselves, which might sound like, "Ms. Douglas, can we chew on that a bit more because I didn't quite get it …" It empowers students to become lifelong learners. Outside of class and outside of school, for the rest of their lives, students will know how their own brain works best and have the tools they need to help themselves learn something new.

We can differentiate the chunk, chew and check in three ways—by varying *whole class* activities, by offering *choice*s based on learning styles or interests, and by student readiness.

When we differentiate by whole class, we want to introduce the language of chunk, chew, and check and explain why each stage is essential for meaningful learning. We also want to expose our learners to different learning modalities and help them discover which learning style works best for them. By being transparent, our learners hear, "We are going to chunk as a whole class today using a kinesthetic activity. Movement helps cement learning. Many of you will find this activity really helps you remember today's key points. Tomorrow we are going to try an activity for those of you who would rather see it to get it." We intentionally vary, from day to day, as a whole class, the modalities for taking in, processing, and sharing new learning. Once we have worked with several modalities, we must *reflect* with our learners about which way worked best for them. We can do a show of hands or have students talk with someone else that learned best in a similar way. Students will be engaged and appreciate the varying styles. However, explaining why we all learn differently and encouraging students to use that information as a learning tool will propel them beyond appreciation and into personal empowerment.

Sometimes, traditional whole class instruction is simply the best way to deliver content. Traditional instruction is not taboo in a differentiated classroom. Sometimes delivery of new information is best done as a whole class sit and get. Differentiating in real classrooms means we have the instructional knowledge and confidence to make the instructional decisions that will best suit our students.

We can also differentiate the chunk, chew, or check part of our lesson by offering choices based on students' learning styles or interests. The power of offering choice in our classrooms should not be taken lightly. Think about it. Leaders understand the power of offering choices to their constituents. Skillful parents alleviate arguments with toddlers through the wisdom of controlled choice. Likewise, great teachers have known about the power of choice for years. Choice gives students a sense of personal control which is empowering for students. Learning how to make and follow through on choices is also key to a successful future. Our students need to know how to make good choices, and it will take time and hands-on guidance in the beginning as we teach them explicitly how to make good choices. Letting them fail when they make bad choices is ok too! In failure, there is an opportunity to reinforce reflection. We can say, "What about the choice didn't work for you? Why would you make a different choice next time?"

Finally, we can chunk, chew, and check by student readiness. This allows us to adjust the difficulty level of material that students are introduced to, the depth at which they are expected to process material, or the complexity of their demonstration of learning. The key to planning CCC with readiness in mind is to think of varying depth and complexity of tasks expected of learners.

Keeping the CHUNK, CHEW, and CHECK framework in mind as you design lessons will help you vary your teaching and offer better access to learning for all students in your classrooms. Every lesson, every day, we can begin to think this way:

Chunk, chew and check…it's how the brain learns best!

As you grow your skills at using the CCC framework, you can look back over a week or two of your lesson plans and see clear delineations between the chunk, chew, and check. But oftentimes, you will find that chunk and chew blend (e.g., in a jigsaw) or it happens that one project is both a chew and check. No matter how we mix it up, the intent is to vary the ways in which we have our students input, process, and output new learning. If we design the chunk, chew, or check in the same way day after day (e.g., read the book, do the questions, take the test), we are not differentiating our instruction.

Remember, varying these three steps in our lesson design allows us to thoughtfully respond to how well our learners gain access to content, process and own what we have taught them, and demonstrate that they have mastered the outcomes. With the chunk, chew, and check framework in place, we can more carefully monitor where learning, or the demonstration of learning, is successful or where learning breaks down for the learner.

Examples

Chunk: Roger can't read the book. I need to find another way for him to access the new information. I'll have him read this text that is at his reading level.

Chew: Wow, Wenting has really discovered when she adds movement to her learning, she can make better sense out of what she needs to recall.

Check: Many of my students can't "show what they know" very effectively on tests. I need to teach them specific skills for studying and being better test takers, and I am going to add more choices on performance-based assessments so they can show what they know using their learning strength.

Once you know chunk, chew, and check, you'll see that any strategy you ever learn at any workshop will fit into one of these categories, and you'll know how to use it more effectively. Any new technology you use in your classroom will also fit into the chunk, chew, or check categories. When you implement RTI or IEP goals, you can determine if the intervention or accommodation is needed at the chunk, chew, or check part of the lesson. If you are co-teaching, you can decide who's going to lead teach the chunk, chew, or check part of the lesson.

As you design lessons with chunk, chew, and check in mind, you can build an indispensable toolkit of strategies to vary the ways students engage in meaningful learning.

Theory into Practice: Chunk, Chew, and Check

Chunk: Input Ideas

When we present new information, we can vary how students acquire or take in chunks of new information, based on their learning preferences, readiness level, and circumstances. If we input new content the same way, day after day, we are most likely not reaching all learners because we are not teaching to their preferred ways to take in information. For example, if a math teacher teaches a math lesson and a student asks for help and the teacher explains the chunk of new learning again, only slower and louder, the student may still not get it. What the student needs is another way, perhaps with visuals or manipulatives. The simple monotony of our instruction, teaching the same way every day, may also lead to have students disengaging from the subject. Teachers who intentionally vary the way they input new information do so knowing that they are more likely to reach more learners and hold student interest.

Chunk: What Students Should Ask Themselves

- What's my mindset about learning this?
- Am I getting this?
- Do I need to hear it again or learn it another way?
- Do I need smaller chunks?

Chunk Ideas in Varied Learning Modalities

Ask yourself…

> What are some other ways I can help students acquire new knowledge?

Visual: Can I…

- Show a movie or clip from a movie; demonstrate from a chart or graph; watch a United Stream or TeacherTube; utilize a blog, Wikipedia, WebQuest, or PowerPoint; read a book—see it, read it; use graphics, pictures, movies, graphic organizers, conceptual organizers, articles, magazines, and books; watch a presentation or demonstration, technology/media; read in various structures, small groups, read aloud, jigsaw, paired readings, reading centers

Auditory: Can I…

- Say it, have them say it to each other, play a song, listen to a speech, talk to each other, listen to a speaker, listen to music, listen to a lecture on tape, listen to books on tape, and have discussions with others

Kinesthetic: Can I...

- Role-play; demonstrate; have students try something; rotate through stations set up to teach content; move, touch, build, draw, take apart; play charades, create group tableau; conduct a lab experiment

Social: Can I...

- Talk about it, listen, or tell others; brainstorm; share experiences, predict/hypothesize; do a role-play; play a game; have a class discussion

Activities:

- *Event cards*—groups of students sort events from a story in order to build anticipation
- *Visual literacy*—use images for students to chunk new information
- *Gallery walk*—students view photos in carousel style then engage in chew activity to process what they have taken in
- *Expert groups*—students become experts in an area/topic/subset of information and continue to share information throughout a unit

Chew: This Is Where Ownership Grows

Learn a fact today, another one the next, take the test on Friday, dump the information from memory over the weekend, and repeat again next week. We call it input-output learning. Notice the absence of time to chew or process the learning. When we fail to give students time to process what they are learning, it's no wonder our students repeatedly ask us, "Why do we have to learn this?" The lack of processing time and meaningful connection in our lessons not only creates apathy in our students but also, for many, prohibits them from retaining the information. Chewing is where the learning is happening. Just because we are teaching, doesn't mean students are learning. Processing or connecting to new learning in meaningful ways is essential if we are going to help students to store the new learning in their long-term memory. Brain researcher and author David Sousa tells us that "...information is often taught in such a way that it lacks meaning for the student...yet the brain needs to attach significance to information in order to store it in long-term memory" (Sousa, 1995). A student may practice a task repeatedly with success, but if they have not found meaning after practicing, there is little likelihood that the learning will move into long-term memory (Sousa, 2000). Input-output learning is not what the brain needs!

What *does* the brain need in order to store new learning into long-term memory? In a word—transfer (Sousa, 2000). The concept of transfer is not new. As cited in Sousa's How the Brain Learns, 2000, Lev Vygotsky points out that the central role to learning should be making meaning. It is the processing of information that essentially enables students to transfer learning into long-term memory and be a cognitive transformation for the student.

During the chew part of the lesson, we are emphasizing that teaching kids *how* to learn is as important as teaching them *what* to learn. The importance of processing and connecting to new learning in meaningful ways is vital, yet this step is often left out of our lessons as we go from input to output, input to output. Moving from chunk to check is often done to save time or "cover more content." However, the processing, or chewing, on new learning in a lesson is the step that allows students to lock in their learning. If we want new learning to stick, we have to make it sticky!

Chew: What Students Should Ask Themselves

- What's my mindset about how I make sense of new information?
- What do I need to do to make sense of this new information?
- What's working or not working for me?
- Do I need to ask for help? Try another way? Practice some more?

Chew Strategies in Varied Modalities

Ask yourself...

How can I vary the ways I help students process and make sense of new knowledge?

Ways to Collaborate to Chew

Jigsaw

1. Each student receives a portion of the materials to be introduced.
2. Students leave their "home" groups and meet in "expert" groups.
3. Expert groups discuss the material and brainstorm ways in which to present their understandings to the other members of their "home" group.
4. The experts return to their "home" group to teach their portion of the materials and to learn from the other members of their "home" group.
5. You can also jigsaw poetry, text, and vocabulary.

Numbered Heads Together

1. Number students off from 1 to 4 within their groups.
2. Call out a question or problem (e.g., Where do plants get their energy?).
3. Students in teams put their heads together to discuss the answer. They must make sure everyone on the team knows the answer.
4. Randomly call a number from 1 to 4 (use a spinner, draw popsicle sticks out of a cup, roll a dice, etc.).
5. On each team, the student whose number was called says or writes the answer. He or she may not receive any help from his team at this point! If they didn't pay attention during the discussion, too bad!

Turn and Talk/Walk and Talk

1. Give students a prompt on the board, overhead, or PowerPoint.
2. Students turn and talk to a partner or stand up and walk (five giant steps) and find a talk partner.
3. Students have 2–3 min to talk and share. While they are talking, the teacher is floating around the room listening for quality talk.
4. The whole class processes the talk, with the teacher noting quality talk that he or she heard while going around the room.

Core Groups

1. At the beginning of the year, students are randomly assigned to groups.
2. The group members are assigned jobs such as leader, recorder, teacher getter, timekeeper, life coach, organizer, etc.
3. Groups can then give themselves a name, a silent signal, or a symbol.
4. The teacher has the groups do fun community building activities, such as building the tallest tower from straws and tape—without talking!
5. The groups stay together for a marking period, a semester, or a year.
6. The core group responsibilities are as follows:
   (a) If anyone from the core group is absent, they get the makeup work and assignment from their core group members. (This buys the teacher valuable teaching time and builds responsibility.)
   (b) The teacher can always call the core group together at the beginning or end of class to plan, reflect, review, etc.

Ways to Move to Chew

Classification Cruz

Each student is given an index card with a word or concept on it. Students have to silently categorize themselves with others who have similar cards (types of health and exercises, states of matter, parts of speech).

Total Physical Response

1. The teacher or the students create movements to help them remember important ideas about the learning. For example, math students could make up movements to recall that the formula for slope is "rise over run," or students could make up movements to help recall the difference between longitude and latitude.

Charades

1. Students act out what they have learned and other students have to guess what they are acting out about the learning.

Moving Math

1. Use math manipulatives.
2. Have students become math numbers and build math problems (e.g., they can make arrays by arranging themselves into six groups of 4, then four groups of 6).
3. What time is it?
   (a) Make a clock face on a sheet.

(b) Line students up with a partner around the clock so they can see the clock.
(c) Give each pair a time on an index card.
(d) When it is their turn, have them make the time on their card with their bodies (the little hand person must bring their knees up).
(e) The rest of the class says the time.
(f) When they get the idea, the pairs can make up their own times and have students guess.

Building Sentences
1. Give each student a card that is part of a sentence.
2. They must move into the correct order to make the sentence make sense. The rest of the class reads and agrees or disagrees.

Ways to Talk to Chew

Act it Out
Do a RAFT as the topic students are learning about. Students become the person, place, or thing and act out who, what, or where they are.

Think/Pair/Share
Students are given a question or prompt to think about in their heads for 1 min. They then pair up with a partner and discuss their thoughts or answers. Then the teacher leads a whole class share by drawing names randomly and asking those students to share.

Ways to Write to Chew

Learning Logs/Journals
Students use writing logs to process learning in their own words.

Note-taking Strategies
Students create 1/3, 2/3 notes, for gathering facts and summarizing information; Double-Entry Journals for gathering facts and processing with a guiding question or perspective from the teacher.

TV Guide Summaries
Students write a summary like a TV guide synopsis.

Blogs
Students keep a blog of their thoughts, like a journal only using technology.

Ways to Draw/Design to Chew

Comic Strips
Students create comic strips that summarize new learning.

Vocabulary Pictures
Students draw pictures to show the meaning of words.

Graphic Organizers
Students design their own organizers to process new learning.

Doodle Notes
As students are reading or listening to you, allow them to doodle, sketch, ideas, thoughts, etc.

Check: Student Self-Assessment on Mindsets and Skills Sets

For the purpose of this chapter, the check focus will be on student self-assessment. We can offer students opportunities to feel, see, and talk growth mindset, but we're only halfway there if we don't get them to own their own learning. If students are going to take charge of their own learning and control of their executive functioning mindsets and skill sets, they need frequent practice at self-assessing.

Marzano, Pickering, and Pollock note, "Teaching about effort, as suggested previously, might work for some students, but others will need to see the connection between effort and achievement for themselves. A powerful way to help them make this connection is to ask students to periodically keep track of their effort and its relationship to achievement" (Marzano et al., 2001, p. 52).

Check Ideas for Self-Assessment

- Create exit cards that have students self-assess or self-monitor for the academic or behavioral skills you expect them to own.
- Have students share "Think Alouds" on what they've done that helped them be successful.
- Post anchor charts around the room of strategies that effective learners use.
- Have student make metacognition books to keep a personal record of strategies they are learning.
- Have students self-assess weekly on their mindset and use of metacognitive strategies.
- See Figs. 29.4, 29.5, and 29.6 for examples of student self-assessments.

Fig. 29.4 Vocabulary self assessment

How Our Group Did:	 ☺	 ☺	 ☺
We helped each other			
We all worked			

Fig. 29.5 Elementary group self-assessment

Our Group					
Group Expectations	Date:	Date:	Date:	Date:	Date:

What does your group do well?

What does your group need to work on?

Metacognitive Questions Learners Should Ask Themselves

Before: Developing My Plan of Action

- What in my prior knowledge will help me with this particular task?
- What should I do first?
- Why am I reading this selection?
- How much time do I have to complete the task?

During: Maintaining/Self-Monitoring My Plan of Action

- How am I doing?
- Am I on the right track?
- What information is important to remember?
- How should I proceed?
  - Should I move in a different direction?
  - Should I adjust the pace depending on the difficulty?
- What strategies do I need to use if I don't understand?

Joy has the class wrap up research 5 minutes earlier than usual.	*I'd like to take a minute to see where we are as a group on having the key knowledge in place to work on our projects. Will you let me know how you feel you are doing with each of these terms? Remember these are ways humans can help the environment? When I say a term, I'd like you to show me four fingers to let me know where you are:*
She has students reflect on the key vocabulary terms.	
Students give her feedback using fingers 1-4 on their knowledge.	1 I could teach it to others 2 I feel pretty comfortable with this term 3 I need some review 4 I haven't learned that yet
Joy notes students who show 3's or 4's in vocabulary. She gathers these students into a small group to support them and re-teach key terms and facts.	

Fig. 29.6 Four finger self assessment

After: Evaluating and Adjusting My Plan of Action

- How well did I do?
- Did I do better or worse than I had expected? Where was my self-assessment off?
- What could I have done differently?
- How might I apply this line of thinking to other problems?
- Do I need to go back through the task to fill in any "blanks" in my understanding?

Questions for Teachers to Reflect on Their Chunk, Chew, and Check Practice

- Am I offering different ways for students to chunk (acquire) new information?
- Am I giving students ample opportunity to chew (process) new learning in different ways?
- Am I asking our students to self-assess and reflect on what works best for them as learners?

- When I check and see that some students are not "getting it," how does that shape or inform my instruction?
- Does this activity get my learners to a clear and rigorous learning target? (The strategies must be more than cute and fun, they must also be meaningful.)
- How can I be transparent with my learners as I introduce this strategy? Will they understand WHY we are doing this?

Know Your Target: Executive Functioning Skill Sets Grow When We Teach for Meaning

How do we invite students who are reluctant because they are bored or apathetic and who feel no connection with what is being taught to engage with us in the learning journey? We engage them by focusing on the richness of learning and making connections to real life instead of focusing on outcomes (test scores and grades) that hold little meaning to them. We inspire these students because they see that we are vested in helping them see the meaning of their journey.

To invite reluctant learners to come along on our journey and to build their behavioral and metacognitive executive functioning skills, we must create an environment in our classroom that embraces their individuality, includes them in community, and nurtures their curiosity and desire to see the relevance in what we are teaching.

Clear Learning Targets Keep Students on Course

A skilled sailor always plots his course before heading out into open water. He determines his coordinates, checks out the weather conditions, and makes sure his equipment is working. Leaving the shore prepared is essential for a successful journey, but once he leaves shore, there is still work to be done. Winds, a tangled sail, shifting currents, and the occasional squall can veer the ship off course. But the sailor continually monitors his progress, assesses his place on the journey, and uses the information to make quick decisions to get him back on course toward his destination. Like sailors, teachers need to set a clear course for learning while continually assessing students' progress on the journey to make thoughtful decisions which guide them to their final destination.

If students are to develop and monitor their executive functioning skills, they need to know where they are going and how to get there.

Let's take a look at a classroom where instruction and assessments are guided by a clear learning target and taught in intentional and transparent ways. Note how Joy, a middle school science teacher, uses assessment as an ongoing tool for evaluating her students before, during, and after learning. Also note how she teaches in series of chunk and chews:

> Okay class, we'll be starting the new unit on the environment. To help me better design the lessons, I'd like to *check* to see what you already know. For the last 7 minutes of class, please do a Quick Write on whatever you think you know about the pollu-

tion problems in our environment, types of pollution, and what people are or should be doing to help the environment. You can add diagrams and drawings if you like. Turn this exit card into me as you leave class today.

(A few days later) "I have gone over your Exit Cards, and I noticed that you have a pretty good background understanding of issues concerning the environment. So, after we learn some new ideas by chunking and chewing on them, as we do, I think this will be a great time to offer you some *check* choices about what you study and how you present what you learn. Here are the target objectives for the unit." The target objectives are written on chart paper posted in front of the room. The target (objectives) are:

Understand that
All living things are dependent upon the environment to sustain life.
Know
Reuse, reduce, recycle, sustainable (ways humans help)
Types of pollution
Able To Do
Summarize information using 1/3, 2/3 notes
Now You Get It!
Using the notes you have gathered, select from a choice menu a way to share what you KNOW and UNDERSTAND about your group's type of pollution.

"Let's go over the objectives together, because you will need to include the UNDERSTAND and KNOW learning targets in whatever project you decide to do." (During the *chew* discussion, students brainstorm and Joy notes questions and misconceptions the students have about the learning target.)

"Your *chew* skill for this unit is note-taking. Note-taking is one of the vital know-hows that life-long learners need, so I will be modeling note-taking as you research in preparation for your final project. You will receive points for quality notes on the final rubric for this project." (Joy models and scaffolds instruction of note-taking as students gather data from various sources. She "kid watches" and observes how students are progressing at taking notes and gathering data for their project.)

"Now, class, you have spent the last few days taking notes and becoming experts on your type of pollution (Chunking and Chewing). Remember that your project needs to meet the objectives for this unit. Look over your notes and for five minutes, discuss with your learning partner how well you understand and know the learning target. After five minutes, I'll check by asking for a Thumbs Up on how you are doing." (After five minutes students give a thumbs-up if they've got it, sideways thumb if they have some but not all of the information, and thumbs-down if they are not getting it. From this information Joy determines who needs more help, who might be willing to help others, and who's doing fine.)

(On the final day of the project) "It's time for the Now You Get It! part of the project. Today you will present or turn in your final project. Take out the rubric you received a few days ago. Recall that you are being graded on how well you present the Understand and Know objectives in your project, the quality of your project and your notes. Now it's time for you to reflect on how well you think you have met these objectives. As you self-assess on the rubric, you should be thinking, 'How do I think I did? How do I think that compares to my teachers expectations and reflections of my work?' You also need to reflect on what you feel you did that was quality work and what you would do differently next time. I will then assess and grade your project using the same rubric. Since I am the final evaluator for this project, it will be my grade that goes in the grade book. I am excited to see your final results."

As you can see, Joy created her lesson with the end in mind, so she was able to plan engaging and meaningful ways to help her students succeed in meeting those objectives. She pre-assessed her students to see where their skills and knowledge were strong and where they were lacking. During learning, she used the pre-assessment information to help her effectively determine what needed to be taught. She continued to assess and provide feedback during learning. The students reflected upon what they needed to do to grow their skills by using the rubric that the teacher also used as a final assessment. This ongoing assessment is an essential component of informed instruction and the building of executive functioning skills (see Fig. 29.4).

In Mike Schmoker's book, *Results Now* (ASCD, 2006), he notes that after observing several hundred classrooms in several states, "there is a glaring absence of the most basic elements of an effective lesson: an essential, clearly defined learning objective followed by careful modeling or a clear sequence of steps, punctuated by efforts during the lesson to see how well students are paying attention or learning the material." Teachers all over the country are teaching without clear learning targets in mind. It is challenging to create and use ongoing assessment effectively if you do not begin with a clear learning target. If students don't know where they are going, it's challenging for them to self-reflect and know which executive functioning skills they need to be applying.

Assessment of Knowledge

Students have been researching for two days. In order to speak to the big understanding, Joy knows that students will need to know the key vocabulary terms and ideas from the knowledge part of the learning target. She also believes that students' self-assessment is essential, so she wants to include students in the assessment process.

Joy knows that the vocabulary is not too difficult so she doesn't want to burden herself by giving a quiz for material she feels will be easy for the students to master. However, she does want students to take a minute to reflect on their learning and give feedback if they need assistance. She decides the easiest way to get this information is to do a four-finger self-assessment (see Fig. 29.6).

Instructionally Responding to Formative Assessment of Knowledge

The only term that some students didn't know well was *sustainable*, so Joy gathers together the students who indicated 3s or 4s (using their fingers) for the term *sustainable*, and they discuss the meaning of the term in detail until they have a good sense of the meaning. (Joy could have also done this assessment as an exit card, having the students score their understanding of each key term. Then, as she collected exit slips, she would simply sort the slips where students indicated a 3 or 4 on the bottom of the stack and meet with those students to offer support.)

Assessment of Understanding

In order to teach for deep understanding, Joy knows her assessment and feedback cannot only focus on skills and facts. However, measuring understanding can be a little trickier. She will need

Looks Like	Sounds Like
Joy kid watches by listening to group discussions or engaging students in discourse about their understanding.	How is the group doing today? What are your thoughts about the understanding, "All living things are dependent on the environment?" Tell me about some of your examples. In what ways are they dependent on the environment? Have you found any examples of living things that are not dependent on the environment?
She takes notes on which students are missing key information. She asks probing questions to see if students have any misconceptions.	

Fig. 29.7 Self-assessing for understanding

to discuss their ideas and probe deeper into their reasoning. She decides this will be done best by kid watching and making clipboard notes of misconceptions to address in small groups. Research indicates that feedback is most valuable when students have the opportunity to use it to revise their thinking (National Research Council, 2000) (Fig. 29.7).

Instructional Response to Informative Assessment of Understandings

Using information from her notes, Joy noticed one group who needed clarification on a misconception they had about the understanding. She initiates deeper discussion.

> I overheard your group saying some interesting things about bacteria yesterday. Tell me about your thinking…

She finds out the group has concluded that highly adaptable life forms, such as bacteria, are not dependent on the environment. Joy helps the students make connections that the bacteria's adaptation is actually a clear demonstration that they are dependent on the environment—that the species would die if it did not adapt.

Assessment of Skills

Joy also wants to see how her students are doing in building the vital skill of note-taking. She has been modeling and scaffolding instruction of note-taking throughout the unit, giving students feedback on their notes while walking around. But now she has decided to check their progress. Joy has students hand in their latest set of notes to assess needs and strengths. (Note: There are two students in the class who are unable to take notes because of their learning disabilities. These students are given photocopies of class notes and are expected to highlight key ideas as their assessment.)

Instructionally Responding to Formative Assessment of Skills

Joy notices that eight students are not summarizing in their own words. They are merely copying down facts. She makes a list of their names to create a small group to review summarizing. She will have them verbally summarize facts from their notes then work together to create one shared summary.

She also notices two students who have outstanding note-taking skills. She makes a note to tell them that they can move forward with their project as soon as they feel they have collected enough information.

Navigating the Course, Reaching the Destination

Joy plotted the course for her students' journey with a clear destination in mind. By establishing what she wanted students to UNDERSTAND, KNOW, and BE ABLE TO DO *before* she began the unit, she created a framework to clarify objectives. Then she used pre-assessment and ongoing assessment to inform her instruction. She gathered information on how students were growing and/or struggling and provided continuous feedback to move them forward or, when necessary, navigate them back on course. Finally, the students were given the opportunity to reflect upon and internalize their learning using a rubric.

What a journey! Both teacher and students knew where they were going and were ready to respond to any surprises (assessment and feedback) along the way. Together, they are charting a course for success in developing executive functioning skills that will grow in strength over time. In learning as with sailing, beginning with a clear destination makes for a smoother sail.

Conclusion

When students use metacognition to actively and consciously review their learning processes, their confidence in their ability to learn grows. They begin to attribute outcomes to the presence or absence of their own efforts and to the selection and use of learning strategies. Students with learning disabilities, many of whom have never thought of themselves as capable learners realize that some of their errors are due not to ignorance, but to not knowing the effective strategy to use. Once students identify useful strategies and use them repeatedly, they will experience a powerful boost in confidence. Similarly, they will develop more perseverance as they discover that when their initial effort to solve a problem fails, they can succeed by approaching it with a different strategy. Students are actually learning to be their own tutors and guides (Willis, 2007).

When Alicia began the school year, one of her students, Brad, was a real challenge. He was absent from class a lot. When he did show up, he was a behavior problem. He lacked both behavioral mindset and metacognitive skill sets. First quarter, he was failing her class. But as time went on, with Alicia consistently reminding all her students that mindsets plus skill sets equals results and offering clear and transparent ways for students to become empowered learners, Brad started to change. He began asking more questions in class, and he started showing up and getting his work done. By the end of the year, he had changed dramatically, passed her class, and was clearly very proud of himself. She asked him if he would be willing to share with others how he had changed and he said, "When I started school this year, I hated coming to school and when I would walk down the hall, kids would say, 'there goes Brad, that dummy.' But now I know that if I have a growth mindset and I use the strategies my teacher taught me, that I can learn and do good in school. Now when I walk down the hall, kids say, 'there goes Brad, that smart guy.' And I have a lot more friends." Brad's newfound executive functioning skills were not only helping him grow academically, they were helping him make more appropriate social connections.

Brad was not the only student whose executive functioning skills grew that year in Alicia's class, but his change was the most profound. Alicia learned that developing executive functioning skills in the regular classroom can be done within the context of the regular school day. She intentionally and transparently taught her students to regulate their behavior, build growth mindsets, and develop their metacognitive skills, as well as know themselves as learners. Going slow at the beginning of the year helped her students build a long-term foundation for success that she had not seen before. "When I started on this journey, I wondered if it was going to be worth the effort and time it would take to learn how to build students mindsets and skills, but seeing kids like Brad and others learn to believe in themselves and take ownership of their learning lives has been so rewarding! I can't wait to get even better at teaching this way next year! I have developed a growth mindset along with my students!"

Research shows that there's an even greater impact when all teachers work together to build executive functioning skills throughout the year. In Cynthia Louise Wrights' dissertation, "Executive Functioning Skills in a School District: An Examination of Teachers' Perception of Executive Functioning Skills Related to Age, Sex, and Educational Classification," (June 6, 2008) she notes that "in at least one school district, definite deficits were found in executive functioning that could be identified across grades and genders that suggest specific school wide and class wide interventions. Special education students continue to struggle with executive function issues. Many if not most interventions should be directed at executive functions instead of exclusive content based tutoring." In their pivotal report, Reading Next (2006), the Alliance for Excellence in Education made a similar finding about the need for all teachers to unite in teaching all students effective strategies for accessing content at the secondary level. "Additionally, it is important that all subject matter teachers use teaching aids and devices that will help at-risk students better understand and remember the content they are teaching. The use of tools…that will modify and enhance the curriculum content in ways that promote its understanding and mastery have been shown to greatly enhance student performance—for all students in academically diverse classes, not just student who are struggling."

We are in an exciting time in education. Now, more than ever, we have the science to support what constitutes effective instruction. We have so much validated research on the learning brain that we can no longer ignore the implications this has for our work as educators. We know that teaching executive functioning skills is integral to developing learners who can succeed in school and in the twenty-first century workforce. As we ask our students to take risks and develop their mindsets and skills sets, so must we as teachers and administrators be willing to take risks. We must be willing to put forth the effort to unite, to intentionally use best practices to develop the mindsets and the skill sets of all students in all grade levels and all subject areas. If it works for our students, we should also see results use the same winning combination—**mindsets plus skill sets equals results**.

Acknowledgments A special thanks to Alicia Duncan, Joy Stephens, Michelle Leip, and Wendy Nyborg for their ongoing support, excellent suggestions, and help with the technical details involved in the preparation of this chapter.

References

Biancarosa, C., & Snow, C. E. (2006). *Reading next—A vision for action and research in middle and high school literacy: A report to Carnegie Corporation of New York* (2nd ed.) Washington, D.C.: Alliance for Excellence in Education.

Dozier, R. W., Jr. (1998). *Fear itself: The origin and nature of the powerful emotion that shapes our lives and our world*. New York: Thomas Dunne Books.

Dweck, C. S. (2006). *Mindset: The new psychology of success*. New York: Random House.

Gioia, G. A. (2000). *BRIEF: Behavior rating inventory of executive function: Professional manual*. Odessa, FL: Psychological Assessment Resources.

Gioia, G., Isquith, P., Guy, S. C., Kenworthy, L., & Barton, R. (2002). Profiles of everyday executive function in acquired and developmental disorders. *Child Neuropsychology, 8*(2), 121–137.

Jensen, E. (2005) *Teaching with the brain in mind* (2nd ed.). Alexandria, VA: Association for Supervision and Curriculum Development.

Jensen, E. (2006). *Enriching the brain: How to maximize every learner's potential*. San Francisco: Jossey-Bass, Wiley Imprint.

Johnston, P. H. (2004). *Choice words: How our language affects children's learning*. Portland, ME: Stenhouse.

Kahn, J., & Dietzel, L. C. (2008). *Late, lost and unprepared: A parents' guide to helping children with executive functioning*. Bethesda: Woodbine House.

Kryza, K., Duncan, A., & Stephens, S. J. (2010). *Differentiation for real classrooms: Making it simple, making it work*. Thousand Oaks, CA: Corwin Press.

Kryza, K., Stephens, S. J., & Duncan, A. (2007). *Inspiring middle and secondary learners: Honoring differences and creating community through differentiating instructional practices*. Thousand Oaks, CA: Corwin Press.

Kryza, K., Stephens, S. J., & Duncan, A. (2011). *Give it a go guide: Developing growth mindsets in the inspiring classroom*. Thousand Oaks, CA: Corwin Press.

Marzano, R. J., Pickering, D., & Pollock, J. E. (2001). *Classroom instruction that works: Research-based strategies for increasing student achievement*. Alexandria, VA: Association for Supervision and Curriculum Development.

National Research Council (2000). *How People Learn*. Washington, DC: National Academy Press.

Schmoker, M. J. (2006). *Results now: How we can achieve unprecedented improvements in teaching and learning*. Alexandria, Va.: Association for Supervision and Curriculum Development.

Seligmann, M. E. P. (1990). *Learned Optimism*. New York: Pocket Books.

Seligmann, M. E. P. (1994). *What you can change and what you can't*. New York: Alfred A. Knopf.

Sousa, D. (2000). *How the brain learns: A classroom teacher's guide*. Thousand Oaks, CA: Corwin Press.

Tough, P. (2012, September 14). What if the secret to success is failure. *New York Times*.

Urdan, T., Midgely, C., & Anderman, E. M. (1998). The role of classroom goal structure in students use of self-handicapping strateiges. *American Educational Research Journal, 35*(1), 101–102.

Van Overwalle, F., & De Metsenaere, M. (1990). The effects of attribution-based intervention and study strategy training on academic achievement in college freshmen. *British Journal of Educational Psychology, 60*, 299–311.

Willis, J. (2007). *Brain-friendly strategies for the inclusion classroom insights from a neurologist and classroom teacher*. Alexandria, VA: Association for Supervision and Curriculum Development.

About the Editors

Sam Goldstein, **Ph.D.**, is a doctoral-level psychologist with areas of study in school psychology, child development, and neuropsychology. He is licensed as a psychologist and certified as a developmental disabilities evaluator in the State of Utah. Dr. Goldstein is a Fellow in the National Academy of Neuropsychology and American Academy of Cerebral Palsy and Developmental Medicine. Dr. Goldstein is an Assistant Clinical Instructor in the Department of Psychiatry. Since 1980, Dr. Goldstein has worked in a private practice setting as the Director of a multidisciplinary team, providing evaluation, case management, and treatment services for children and adults with histories of neurological disease and trauma, autism, learning disability, adjustment difficulties, and attention deficit hyperactivity disorder. Dr. Goldstein is on staff at the University Neuropsychiatric Institute. He has served as a member of the Children's Hospital Craniofacial Team. He has also been a member of the Developmental Disabilities Clinic in the Department of Psychiatry at the University of Utah Medical School.

Dr. Goldstein has authored, coauthored, or edited 42 clinical and trade publications, including 16 textbooks dealing with managing children's behavior in the classroom, genetics, autism, attention disorders, resilience, and adult learning disabilities. With Barbara Ingersoll, Ph.D., he has coauthored texts dealing with controversial treatments for children's learning and attention problems and childhood depression. With Anne Teeter Ellison, he has authored *Clinician's Guide to Adult ADHD: Assessment and Intervention*. With Nancy Mather, Ph.D., he has completed three texts for teachers and parents concerning behavioral and educational issues. With Michael Goldstein, M.D., he has completed two texts on attention deficit hyperactivity disorder. He has edited three texts with Cecil Reynolds, Ph.D., on neurodevelopmental and genetic disorders in children. With Robert Brooks, Ph.D., he has authored 12 texts including *Handbook of Resilience in Children, first and second editions*; *Understanding and Managing Children's Classroom Behavior—2nd Edition*; *Raising Resilient Children*; *Nurturing Resilience in Our Children*; *Seven Steps to Help Children Worry Less*; *Seven Steps to Anger Management*; *The Power of Resilience*; *Raising a Self-Disciplined Child*; *and Raising Resilient Children with Autism Spectrum Disorders*. With Jack Naglieri, he has authored a number of texts on autism, assessment of intelligence, and executive functioning. He has coauthored a parent training program and is currently completing a number of additional texts on resilience, intelligence, and genetics. Dr. Goldstein is the Editor-in-Chief of the *Journal of Attention Disorders* and serves on six editorial boards. He is also the Coeditor of the Encyclopedia of Child Development and Behavior.

With Jack Naglieri, Ph.D., Dr. Goldstein is the coauthor of the Autism Spectrum Rating Scales, Comprehensive Executive Functioning Inventory, Rating Scales of Impairment, and the Cognitive Assessment System—2nd Edition.

Dr. Goldstein, a knowledgeable and entertaining speaker, has lectured extensively on a national and international basis to thousands of professionals and parents concerning attention disorders in children, resilience, depression, adjustment and developmental impairments, autism, and assessment of brain dysfunction.

Jack A. Naglieri, Ph.D., is a Research Professor at the Curry School of Education at the University of Virginia, Senior Research Scientist at the Devereux Center for Resilient Children, and Emeritus Professor of Psychology at George Mason University. He is a Fellow of APA Divisions 15 and 16, recipient of the 2001 Senior Scientist Award for APA Division 16 and the 2011 Italian American Psychology Assembly Award for Distinguished Contributions to Psychology, holds a diplomate in assessment psychology, earned a license as a School Psychologist in Virginia and Ohio, and earned school psychology certifications in New York, Georgia, Arizona, and Ohio. Dr. Naglieri has focused his professional efforts on theoretical and psychometric issues concerning intelligence, cognitive interventions, diagnosis of learning and emotional disorders, and theoretical and measurement issues pertaining to protective factors related to resilience.

Dr. Naglieri is the author or coauthor of more than 300 scholarly papers, books, and tests. His scholarly research includes investigations related to exceptionalities such as mental retardation, specific learning disabilities, giftedness, and attention deficit disorder; psychometric studies of tests such as the Wechsler Scales of Intelligence, Cognitive Assessment System, and the Kaufman Assessment Battery for Children; examination of race, gender, and ethnic differences in cognitive processing; fair assessment using nonverbal and neurocognitive processing tests; identification of gifted minorities, IDEA, and identification of specific learning disabilities; and cognitively based mathematics interventions. He has authored various books, including Essentials of CAS Assessment (Naglieri, 1999), and coauthored other books including Assessment of Cognitive Processes: The PASS Theory of Intelligence (Das, Naglieri, & Kirby, 1994); Helping Children Learn: Intervention Handouts for Use at School and Home, Second Edition (Naglieri & Pickering, 2010); Essentials of WNV Assessment (Brunnert, Naglieri, & Hardy-Braz, 2009); and Helping All Gifted Children Learn: A Teacher's Guide to Using the NNAT2 (Naglieri, Brulles, & Lansdowne, 2009). Dr. Naglieri has also coedited books such as Handbook of Assessment Psychology (Graham & Naglieri, 2002), Assessment of Autism Spectrum Disorders (Goldstein, Naglieri, & Ozonoff, 2009), Assessing Impairment: From Theory to Practice (Goldstein & Naglieri, 2009), and A Practitioner's Guide to Assessment of Intelligence and Achievement (Naglieri & Goldstein, 2009).

Dr. Naglieri's scholarly efforts also include development and publication of tests and rating scales. He began this work in the mid-1980s with the publication of the Matrix Analogies Tests (Naglieri, 1985) and the Draw-A-Person Quantitative Scoring System (Naglieri, 1988) and DAP: Screening Procedure for Emotional Disturbance (Naglieri, McNeish, & Bardos, 1991). He published the Devereux Behavior Rating Scale—School Form (Naglieri, LeBuffe, & Pfeiffer, 1993), Devereux Scales of Mental Disorders (Naglieri, LeBuffe, & Pfeiffer, 1994), and the Devereux Early Childhood Assessments (LeBuffe & Naglieri, 2003). In 1997, he published the General Ability Scale for Adults (Naglieri & Bardos, 1997), Cognitive Assessment System (Naglieri & Das, 1997), and Naglieri Nonverbal Ability Test—Multilevel Form (Naglieri, 1997). He published the Naglieri Nonverbal Ability Test, Second Edition (Naglieri, 2008); the Wechsler Nonverbal Scale of Ability (Wechsler & Naglieri, 2008); and the Devereux Elementary Student Strength Assessment (LeBuffe, Shapiro, & Naglieri, 2009). Most recently, he published the Cognitive Assessment System, Second Edition (Naglieri, Das, & Goldstein, 2013); Comprehensive Executive Function Scale (Naglieri & Goldstein, 2013); and the Autism Spectrum Rating Scale (2010). For more information see: www.jacknaglieri.com.

Index

A
Abstract Thinking/Problem-Solving (APS), 168, 368
Academic achievement
   age-related cognitive decline, 149–150
   CEFI, 231, 232
   CHEXI, 363–364
   Cogmed, 482
   TVCF, 276, 279
   working memory, 477, 485
Activity Level/Impulse Control (AIC), 168, 368, 369
ADHD. *See* Attention-deficit/hyperactivity disorder (ADHD)
Adolescents
   age-related cognitive decline (*see* Age-related cognitive decline)
   ASD, 133
   BASC (*see* Behavior Assessment System for Children (BASC))
   BDEFS (*see* Barkley Deficits in Executive Functioning Scale (BDEFS))
   bipolar disorder, 76–78
   CAS2 (*see* Cognitive Assessment System—Second Edition (CAS2))
   CEFI (*see* Comprehensive Executive Function Inventory (CEFI))
   cool executive functions (*see* Cool executive functions)
   D-KEFS, 217, 218
   executive skill development (*see* Executive skill development, children and adolescents)
   hot executive functions (*see* Hot executive functions)
   structural neuroimaging studies, 14–16
Adult Executive Functioning Inventory (ADEXI), 364
Aerobic fitness, 16
Age-related cognitive decline
   definition, 143
   executive functioning
      brain, physical changes, 145–146
      cognitive functioning, 144
      definition, 144
      heterogeneity, 144
      middle adulthood, 151
      school age children, 150
   implications, 151–152
   limitations, 152
   normal aging
      achievement, 149–150
      intelligence, 146
      language, 149
      memory, 146–147
      processing speed, 148–149
      self-regulation/attention, 147–148
      visual-spatial ability, 149
      working memory, 147
   protective factor, 143
American Psychological Association (APA), 416
American Speech-Language-Hearing Association (ASHA), 416
Antisaccade task, 14, 15, 195
Anxiety disorders
   BDEFS-CA, 255
   California Verbal Learning Test, 73
   CEFI, 227, 229
   Concept Formation subtest, 74
   Controlled Oral Word Association Test, 74
   DSM-5-TR, 73
   EF deficits, 74
   go/no-go task, 75
   neuroimaging technology, 75
   OCD, 75
   ROCF Test, 73
   taxometric approaches, 72
   WCST, 73
Arithmetic flash cards, 504
ASD. *See* Autism spectrum disorders (ASDs)
Asperger's syndrome
   CANTAB, 181
   children, problem solving, 512–514
Attention deficit disorder (ADD), 221
Attention-deficit/hyperactivity disorder (ADHD), 69–70, 495
   Barkley's theory
      behavioral disinhibition, 111
      emotion, 110, 112
      external events, 113
      goal-directed behaviors, 108, 112
      implications, 117–118
      nonverbal working memory, 109, 112
      phenotype model, 113, 115–116
      problem solving, 112
      self-organization, 112
      self-play/reconstitution, 110

Attention-deficit/hyperactivity disorder (ADHD) (cont.)
    temporal myopia, 113
    theoretical amendment, 110–111
    time management, 113
    verbal working memory, 109–110, 112
  BASC, 290–291
  BDEFS-CA, 351
  BRIEF, 318, 320, 359–360
  CANTAB
    attentional set-shifting, IED task, 179–180
    executive planning task, 179
    inhibition, 177–178
    methylphenidate (Ritalin), 177
    working memory, 178–179
  CEFI, 227
  CHEXI, 362
  D-KEFS tests, 114, 221
  DSM-IV-TR, 413–414
  human functioning, 117
  problem solving, children, 514–518
  Response Suppression Task, 113
  TEC, 351
  WCST, 194
Attention/Working Memory (AWM), 168, 368, 369
Auditory Consonant Trigrams (ACT) test, 350
Autism spectrum disorders (ASDs)
  BRIEF, 321
  CANTAB
    attentional set-shifting, IED task, 182
    executive planning process, 181–182
    inhibition, 182
    working memory, 182
  CEFI, 227, 228
  children and adults, 121
  cognitive flexibility, 129–130
  cognitive theory, 121
  EF, 124–125, 132–134
  frontal lobe damage, 122–123
  frontostriatal network, 123–124
  inhibition, 125–127
  planning process, 130–132
  prevalence, 121
  symptoms, 121, 122
  working memory, 127–129
Automated Working Memory Assessment (AWMA), 482

B

Barkley Deficits in Executive Functioning Scale (BDEFS)
  age and sex, 251
  BDEFS-CA, 166–168, 360
    ADHD, 254–255
    age, 252
    effect sizes, 253–254
    ethnic groups, 253
    face validity, 256
    gender, 253
    internal consistency, 253
    learning disabilities, 255
    normative comparisons, 256
    ODD and BPD, 254–255
    reliable change and recovery, 257–258
    risk analysis, 257
    test-retest reliability, 253
  development of, 249–251
  race/ethnic group, 251
  reliability, 251
  validity, 251–252
Behavior Assessment System for Children (BASC)
  ADHD, 290–291
  attentional control, 294
  BASC-2
    administration, 289
    clinical and general norms, 289
    composites and scales, 286–289
    EF scale, 292–293
    indexes of validity, 289
    instruments, 284–285
    psychometric properties of, 289
    reliability and validity for, 289–290
    software scoring, 289
    standardization of, 289
    structure and components of, 285–286
  behavioral control, 294
  composites and scales, 286–289
  emotional control, 294
  FLEC scale, 291–292
  latent variable approach
    factorial invariance, 295–296
    model validation, 294–295
  multimethod and multidimensional conceptualization, 284–285
  problem-solving framework, 294
Behavior Rating Inventory of Executive Function (BRIEF), 165–166, 359–360, 450
  administration, 306
  BASC, 290–291
  CEFI, 227–229
  elevated infrequency scale score, 304
  emotional control, 302, 303
  history of, 301–302
  inconsistency scale, 304
  inhibit, 302, 303
  initiate, 302, 303
  interpretation methods, 312–315
  interventions, 315–316
  negativity score, 304
  organization of materials, 303–304
  plan/organize measures, 302, 303
  reliability
    internal consistency, 309
    inter-rater reliability, 310–311
    mean T-score difference, 310
    test-retest reliability, 310
  review of, 324
  scoring software, 306–307
  self-monitoring function, 302, 303
  shift, 302, 303
  standardization samples, characteristics of, 307–309

structure of, 304
task completion, 302, 304
task monitoring function, 302, 303
and TEC, 349
translations of, 305–306
validity of
    academic performance, 321–322
    ADHD and subtypes, 318, 320
    ASDs, 321
    children's SES, 318
    clinical populations, 318–320
    demographic characteristics, 317
    DEX, 316
    exploratory factor analysis, 322–323
    FrSBe scales, 316
    gender, 317
    lower parental education, 317–318
    neural substrate, 323
    performance-based tests, 316–317
    race/ethnic group, 318
    TBI, 320–321
version of, 302–303
working memory, 302, 303
Behavior Regulation Index (BRI), 304
Bipolar disorder (BD), 76–78, 227, 319
BRIEF. *See* Behavior Rating Inventory of Executive Function (BRIEF)
Brown-Peterson task, 350

C

California Verbal Learning Test-Second Edition (CVLT-II), 221
Cambridge Neuro-psychological Test Automated Battery (CANTAB)
    ADHD
        attentional set-shifting, IED task, 179–180
        executive planning task, 179
        inhibition, 177–178
        methylphenidate (Ritalin), 177
        working memory, 178–179
    attention switching task, 173
    autism spectrum disorders
        attentional set-shifting, IED task, 182
        executive planning process, 181–182
        inhibition, 182
        working memory, 182
    Cambridge gambling task, 174–175
    childhood EF, 175–176
    core batteries, 172
    frontal lobe function, 183–184
    in Huntington's disease, 184–185
    induction tests, 172
    intra-/extra-dimensional set shift, 173–174
    older adults, 183
    one-touch stockings of Cambridge, 174
    practice effects, 173
    rapid visual information processing, 174
    spatial span, 174
    spatial working memory, 174
    speed *vs.* accuracy measures, 173
    stockings of Cambridge, 174
    stop signal task, 175
    test-retest reliability, 172
Cardiorespiratory fitness (CRF), 16
Cattell and Horn's Fluid–Crystallized (Gf–Gc) theory, 382
Cattell–Horn–Carroll theory (CHC)
    ability and processing subtests, 385
    16 broad ability, definition, 382, 384
    broad and narrow ability levels, 382, 385
    cognitive abilities, 382, 383
    construct-irrelevant variance, 384
    construct underrepresentation, 385
    Gf–Gc theory, 382
    guiding principles, 385
    neuropsychological domains
        attention/sensory-motor/speed, 389
        broad CHC ability interpretation, 387, 391–392
        executive functioning/learning, 389–390
        narrow CHC ability interpretation (XBA), 387, 392
        processing interpretation, 390–391
    and neuropsychological theory
        classroom learning, 387
        cognitive functioning, 386
        double-headed arrows, 387, 388
        integrative framework, 388
        interpretive framework, 386, 387
        Luria's functional units, 387
        mental activity, 386
        psychoeducational batteries, 387
        psychometric tradition, 388
        school-based hypothesis-generation, 387
        speed and efficiency, 388
        Wechsler Scales, 388
    qualitative evaluation
        aberrant test performance, 398
        cognitive and neuropsychological batteries, 399–401
        cognitive assessment methods, 398
        feedback use, 399
        flexible battery approach, 398
        intellectual and cognitive assessment, 398
        subtest classifications, 399–401
        Wechsler Scales, 398
        WJ III NU COG and DAS-II, 395, 398
    quantitative evaluation, executive functions
        broad and narrow CHC abilities, 392, 393
        cognitive and neuropsychological batteries, 393, 395
        concept formation and generation, 392, 394
        Glr and Gs narrow abilities, 392–393
        psychometric tests, 392
        retrieval fluency, 393, 394
        working memory capacity, 392, 394
    WJ III NU COG analysis-synthesis, 402, 403
CEFI. *See* Comprehensive Executive Function Inventory (CEFI)
Central executive working memory, 476–477

CHEXI. *See* Childhood Executive Functioning Inventory (CHEXI)
Child Behavior Questionnaire—Short Form (CBQ), 292
Childhood Executive Functioning Inventory (CHEXI)
    academic achievement, 363–364
    ADEXI, 364
    in ADHD populations, 362
    behavior problems, 362
    cross-cultural study, 364
    and EF laboratory measures, 361–362
    priori subscales, 360–361
    research, 364
Children
    BASC (*see* Behavior Assessment System for Children (BASC))
    CANTAB, 175–176
    CAS2 (*see* Cognitive Assessment System—Second Edition (CAS2))
    CEFI (*see* Comprehensive Executive Function Inventory (CEFI))
    cool executive functions (*see* Cool executive functions)
    executive skill development (*see* Executive skill development, children and adolescents)
    hot executive functions (*see* Hot executive functions)
    problem solving
        ADHD, 514–418
        Asperger's syndrome, 512–514
        frontal cortex, 509–510
        ICPS, 510–511, 513
        impulsive and aggressive behavior, 510
        means-ends thinking, 510
        memory tasks, 509
        preschool interpersonal problem solving test, 510
        thought and language, 509
    treatment integrity (*see* Treatment integrity)
    WCST (*see* Wisconsin Card Sorting Test (WCST))
Chunk, chew, and check (CCC) framework, 540–546
Cogmed
    academic achievement and learning, 482, 485
    AWMA, 482
    brain plasticity, 486–487
    computerized working memory training program, 485
    description, 480
    direct measures of attention, 484–485
    high-/low-intensity training condition, 482
    inattention and ADHD symptoms, 485
    initial studies, 480–481
    medication impact, 483
    multimodal treatment, 483
    national English and Math tests, 486
    nonverbal reasoning, 485
    potential impact, 483
    Ritalin, 483
    verbal working memory, 484
    visual-spatial working memory
        in boys, 479–480
        capacity discrepancy, 478
        medication improvement, 483
        preschool to adults ages, 483–484
    whole-class model, 486
Cognitive Assessment System—Second Edition (CAS2)
    attention scale, 197, 199
    case study, 202–205
    Core and Extended Batteries and supplemental scales, 200–202
    planning scale, 197–199
    simultaneous scale
        figure memory, 201
        matrices, 199, 201
        verbal-spatial relations, 201
        visual-spatial tests, 197–198
    successive scale, 198, 201
Cognitive load theory, 497–498
Color-Word Interference Test, 196, 212–213, 219
Comparative fit index (CFI), 294–295
Compliance/Anger Management (CAM), 168, 368, 369
Comprehensive Executive Function Inventory (CEFI), 168
    and academic achievement, 231, 232
    ADHD, 227, 228
    administration and scoring options, 232–233
    ASD, 227, 228
    assessment goal, 223
    and BRIEF, 227–229
    exploratory factor analysis, 224–225
    intelligence, measures of
        CAS scores, 230–231
        WISC-IV scores, 229–230
    internal consistency, 225–226
    inter-rater reliability, 226
    intervention and
        case study, 238–240
        computerized administration, scoring, and reporting, 236, 238
        planning and initiation, 241
        self-monitoring, 241–242
        treatment effectiveness, 236
        working memory, 236, 237, 242
    item scores, examination of, 235
    LD group, 227, 228
    mood disorders, 227, 228
    rater characteristics, 233
    scores
        interpretation of, 233–235
        raters, 235
    six-factor analytic procedures, 224
    structure of, 225, 226
    test-retest reliability, 226
    validity of, 226
Comprehensive Trail-Making Test (CTMT), 101, 268
Confirmatory factor analysis (CFA), 56, 294, 322, 383
Conners' Parent Rating Scale-Revised (CPRS-R:L), 291–292, 350
Continuous performance test, 71, 79, 80, 192, 196, 283, 353
Controlled Oral Word Association Test (COWAT), 14, 74, 100, 267
Cool executive functions
    attentional mechanisms, 48

card-sorting task, 47
clinical and neuropsychological observations, 46
cognitive functions, 47
compliance and moral development, 58–59
construct validity
    correlational patterns and developmental trajectories, 52–54
    electrodermal activity, 55
    emotion manipulation, 55
    gambling task, 55–56
    real-world behaviors, 56–57
    symbolic representation, 55
    systematic task manipulation, 56
definition, 45
developmental period, 48–49
factors, 45
frontal lobe damage, 46–47
SAS model, 47
unitary/multifactorial views, 48
Cross-Battery approach (XBA)
    CHC (*see* Cattell–Horn–Carroll theory (CHC))
    definition, 382
    flexible battery approaches, 380–381
    neuropsychological assessment, 381
    principles and procedures, 381
    school-based assessments, 381
    school neuropsychology, 380
    Wechsler-Based cross-battery assessment, 404–407
Cross temporal model, 8

D

Delirium, 91, 94
Delis–Kaplan Executive Function System (D-KEFS), 114, 368
    ADD, 221
    ADHD, 221
    administration, 217
    Color-Word Interference Test, 212–213
    CVLT-II, 221
    design fluency test, 211–212
    higher-level cognitive function, 194–195
    inhibition, 209
    interpretation methods, 219–220
    intervention selection, 221
    LD, 221
    perception, 209
    planning, 209
    proverb test, 216–217
    reliability measures
        alternate-form reliability, 219
        internal consistency, 218–219
        test-retest reliability, 219
    scaled scores, 217
    scoring software, use of, 218
    sorting test, 213
    standardization sample, characteristics of, 218
    switching, 209
    TBI, 221
    tower test, 215–216
    trail making test, 210–211
    twenty questions test, 213–214
    verbal fluency test, 211
    WCST, 221
    word context test, 214–215
Delis Rating of Executive Functions (D-REF), 168
    core index, 368
    D-KEFS, 368
    EXF, definition, 367
    KABC-II, 374
    Parent Rating Form
        administration and scoring, 369–370
        case study, 374–375
        clinical index scores, 368–369
        fourth-grade reading level, 367
        internal consistency, 372
        psychometric properties, 371
        SEM, 373
        test-retest stability, 373
    Self-Rating Form
        clinical index scores, 368–369
        internal consistency, 372
        normative sample, 371
        psychometric properties, 371
        SEM, 373
        test-retest stability, 373–374
        third grade reading level, 368
    survey executive functions, 367
    TC (*see* Total composite Score (TC))
    Teacher Rating Form
        administration and scoring, 369–370
        clinical index scores, 368–369
        fourth-grade reading level, 367
        internal consistency, 372
        psychometric properties, 371
        SEM, 373
        test-retest stability, 373
    validity
        concurrent, 376–377
        construct, 376
        content, 375–376
Dementia, 92, 94
Depression, 75–76
    BASC and BASC-2, 287, 288
    BDEFS-CA, 255
Design fluency test, D-KEFS, 211–212, 218
Diagnostic and Statistical Manual (DSM)
    DSM-IV
        amnestic disorder, 92–94
        cognitive disorders, 93
        delirium, 91, 94
        dementia, 92, 94
        mild neurocognitive disorder, 93, 94
        postconcussional disorder, 93–94
    DSM-5
        EF neuropsychology, 102–104
        neurocognitive disorders (*see* Neurocognitive disorders)
        organizational structure disorders names, 94–95
    historical background, 89–91

Differential Abilities Scale (DAS), 277
D-KEFS. *See* Delis–Kaplan Executive Function System (D-KEFS)
D-REF. *See* Delis Rating of Executive Functions (D-REF)
DSM. *see* Diagnostic and Statistical Manual (DSM)
Dual encoding, 505
Dysexecutive Questionnaire (DEX), 316

E
Emotional functioning (EMF), 335, 368
Evidence-based interventions (EBI), 414
Executive function (EF)
   academic success
      cognitive and motivational processes, 447
      cycle, 447, 448
      paradigm, 448, 449
      self-confidence and self-efficacy, 447
      students metacognitive awareness, 447, 449
   automatic and controlled processes, 4, 6–7
   BASC (*see* Behavior Assessment System for Children (BASC))
   BDEFS (*see* Barkley Deficits in Executive Functioning Scale (BDEFS))
   behavioral regulation, 524
   BRIEF (*see* Behavior Rating Inventory of Executive Function (BRIEF))
   CEFI (*see* Comprehensive Executive Function Inventory (CEFI))
   central executive, 7–8
   CHEXI (*see* Childhood Executive Functioning Inventory (CHEXI))
   chunk, chew, and check framework, 540–546, 548
   classroom community
      class meetings, 536
      groups, 536–537
   classroom environment, 534–535
   clinical neuropsychological procedures, 99
   cognitive control, 7
   cognitive flexibility/shifting
      content area subjects, 455
      daily activities, 456
      developmental changes, 454
      learning foreign language, 455
      math domain, 455–457
      problem-solving and critical thinking, 456
      reading comprehension, 455, 456
      reading decoding, 455
      social and academic settings, 455
      studying and test-taking, 455, 457–458
      willingness, 454
      written language, 456, 457
   cognitive operations, 191
   components, 98
   conceptual contamination, 248–249
   controlled processes, 7
   control mechanism, 4
   cool executive functions (*see* Cool executive functions)
   cross temporal model, 8
   definition of, 4–6, 192, 413
   developmental stages, 8–11
   D-KEFS tests (*see* Delis–Kaplan Executive Function System (D-KEFS))
   ecological validity, 296
   emotional self-regulation, 465
   evidence-based interventions, 414
   evidenced-based strategies, 540
   frontal lobes
      corpus callosum, 36
      definition, 32–33
      developmental trajectory, 34–36
      divisions, 31
      evolutionary theories, 30–31
      functional imaging studies, 37
      goal-directed behavior, 36
      humans and primates, 37
      left hemisphere, 36
      neuroanatomical findings, 33
      neuroimaging techniques, 39–40
      neurological and psychiatric disorders, 29
      neurotransmitter and neuromodulator, 37–38
      nonverbal psychological assessments, 36–37
      role of, 29
      social-emotional components, 30
      working memory, 38–39
   future-directed behavior, 248
   goal-appropriate cognitive routines, 191
   goal-directed activities, 248
   goal setting, 448
      definition, 453
      and planning strategies, 454
      self-efficacy, achievement, and motivation, 453
   hot executive functions (*see* Hot executive functions)
   informal, process-oriented assessment
      BRIEF, 450
      criterion-referenced assessment system, 450
      instructional approaches, 449
      MetaCOG (*see* Metacognitive awareness survey system (MetaCOG))
      RtI approach, 449
      students' strengths and academic potential, 448
   integrative model, 8
   intentional teaching, 524, 525
   intervention strategies, 452–453
   knowledge assessment, 550
   learning targets, 548–550
   mental flexibility, 191
   metacognition, 524, 537
   metacognitive questions learners, 547–548
   multimethod-based assessment of, 283
   organizing and prioritizing
      applying strategies, 458
      BOTEC strategy, 459, 460
      note-taking, 459–460
      reading comprehension, 458
      strategy reflection sheet, 460, 461
      systemizing and sorting information, 458
      triple-note-tote strategy, 458, 459

peer mentoring
    Drive to Thrive program, 465–466
    SMARTS program, 466–467
physiology of (*see* Physiology)
prefrontal cortex, 35–36
problems of
    ADHD, 247
    assessment, 245–246
    lack of definitional clarity, 245
    ratings, 246
self-directed activity, 247–248
self-monitoring and self-checking
    math domain, 464
    metacognitive awareness, 462
    performance improvement, 462
    problem-solving approach, 462
    STOPS, 463
    for tests, 464–465
    writing domain, 463–464
short-term memory, 4
skill assessment, 551–552
student mindsets (*see* Mindsets, students)
students' learning styles
    data collection, 538
    learning profile cards, 538–539
    practice, 539–540
    students activities, 539–540
supervisory attentional system, 7
transparent teaching, 524, 525
trans-self-integration processes, 192
treatment integrity (*see* Treatment integrity)
types, 445
understandings, assessment, 551
working memory (*see* Working memory (WM))
Executive skill development, children and adolescents
children's drive for mastery and control, 431
child's developmental level, 430
effortful tasks, 431
enough support, child, 432
executive skills training, 430
incentives gradually, never abruptly, 432
parents/teachers, 432
physical/social environment, 430–431
skill definitions, 427–429
supports, supervision, 432
teach deficient skills, 429
weak executive skills
    designing interventions, 433
    environment, level, 433–434
    home-based teaching, 435–436
    homework plans, 436–438
    motivate children, executive skills, 438–439
    teaching skills, 434–435
working, teenagers
    coaches, 443
    executive skill deficits, 443
    life goal, 440
    Marcus parents, 442–443
    plan development, 441
    working with parents, 440–441

F
Frontal Lobe/Executive Control (FLEC), 291–292
Frontal Systems Behavior Scale (FrSBe), 283, 316

G
Global Executive Composite (GEC), 228, 291, 292, 304, 495
Go/no-go task, 75, 194, 337–338, 361, 481

H
High-functioning autism spectrum disorders (HFASDs), 292–293
Hot executive functions
    adult neuropsychological framework, 50–51
    compliance and moral development, 58–59
    construct validity
        correlational patterns and developmental trajectories, 52–54
        electrodermal activity, 55
        emotion manipulation, 55
        gambling task, 55–56
        real-world behaviors, 56–57
        symbolic representation, 55
        systematic task manipulation, 56
    delay gratification, 49–50
    emotion/cognition interaction, 49
    factors, 45
    male superiority, 59–60
    risky decision-making tasks, 58
Huntington's disease
    CANTAB, 184–185
    dementia, 92

I
I Can Problem Solve Program (ICPS), 510–511, 513
Individuals with Disability Education Act (IDEA), 221
Intelligence
    age-related cognitive decline, 146
    CEFI
        CAS, 230–231
        WISC-IV, 229–230
    TVCF, 275–276
Internalizing and externalizing disorders
    anxiety disorders (*see* Anxiety disorders)
    attention-deficit/hyperactivity disorder, 69–70
    bipolar disorder, 76–78
    conduct disorder, 70–71
    depression, 75–76
    future aspects, 82–83
    method
        eligibility criteria, 79
        limitations, 82
        results, 79–80
        search and retrieval process, 78–79
    oppositional defiant disorder, 70–71
    Tourette's disorder, 71–72

K

Kaufman Assessment Battery for Children second
    edition (KABC-II)
    CAS2, 198
    CHC theory (*see* Cattell–Horn–Carroll (CHC) theory)
*Kaufman Test of Educational Achievement, Second
    Edition* (KTEA-II), 374
Keep track task, 14, 15

L

Learning disability (LD)
    CHEXI, 362
    D-KEFS, 221
Long-term memory, 476–477, 496, 544
    dual encoding, 505
    episodic buffer, 496
    reviewing, 505–506

M

Maze completion test, 17
Metacognition Index (MI), 304, 314, 320, 322
Metacognitive awareness questionnaire (MAQ), 450, 451
Metacognitive awareness survey system (MetaCOG)
    parent surveys, 451–452
    purposes, 450
    students' self-ratings, 450
    student surveys, 450–451
    teacher surveys, 451
Metamemory, 500, 502, 503
Mindsets, students, 524
    classroom talk, 532
    fixed mindset thinkers, 526
    growth mindset thinkers
        developmental strategies, 528–529
        internal beliefs, 526
        language, 529–530
        practice theory, 527
        skill sets, 527
        students activities, 529
    self talk, 533
    teacher talk, 531
Mini-Mental State Examination (MMSE), 102–103
Montreal Cognitive Assessment (MoCA), 102–103
Motivation and effort survey (ME), 450

N

National Association of School Psychologists
    (NASP), 416
N-back task, 128, 335–337, 503
Neurocognitive disorders
    categories and subtypes, 95
    complex attention, 96
    delirium criteria, 95–96
    executive ability, 96
    language and visuoconstructive-perceptual ability, 96
    learning and memory, 96
    major neurocognitive disorder, 96–98
    minor neurocognitive disorder, 96
    social cognition, 96
Nonverbal working memory, 109–112, 249, 401, 480

O

Original Trail-Making Test, 101, 267

P

Parenting Relationship Questionnaire (PRQ), 284, 286
Parent Perceptions of Student Effort (PPSE), 451–452
Parent Rating Scales (PRS), 284
    age level forms, 285
    composites and scales, 286–289
    FLEC/EF scale, 291–293
    indexes of validity, 289
    reliability and validity, 289–290
Performance feedback (PFB), 420–422
Physiology
    advance technology, 14
    functional neuroimaging studies, 16–17
    limitations and future aspects, 21–22
    planning, 17–18
    response inhibition, 19–21
    set-shifting, 21
    structural neuroimaging studies
        aerobic fitness, 16
        antisaccade task, 15
        cortical maturation, 14
        cortical thickness, 15
        keep track task, 14
        left inferior frontal gyrus, 15
        limitations, 15
        number-letter sequencing task, 15
        prefrontal and parietal regions, 14
        WCST, 15–16
    verbal fluency, 18
    working memory, 18–19
Picture Deletion Task for Preschoolers-Revised
    (PDTP-R), 292
Primary care practitioner (PCP), 103
Proverb Test, D-KEFS, 216–217

R

Rating scales
    BASC (*see* Behavior Assessment System for
        Children (BASC))
    BDEFS (*see* Barkley Deficits in Executive
        Functioning Scale (BDEFS))
    BRIEF (*see* Behavior Rating Inventory of Executive
        Function (BRIEF))
    CAS2 (*see* Cognitive Assessment System—Second
        Edition (CAS2))
    CEFI (*see* Comprehensive Executive Function
        Inventory (CEFI))
    CHEXI (*see* Childhood Executive Functioning
        Inventory (CHEXI))
    comprehensive executive function inventory, 168

development methods, 164–165
D-KEFS (*see* Delis–Kaplan Executive Function System (D-KEFS))
D-REF (*see* Delis Rating of Executive Functions (D-REF))
normative sample disparities, 169
reliability
   clinical practice, 159
   confidence intervals, 160, 161
   estimated true scores, 161
   obtained scores, 161
   research purposes, 159
   standard error of measurement, 160
   standard scores, 162
validity
   conceptual and measurement approaches, 163
   definition, 162
   executive function, 163
Response time standard deviation (RTSD), 340
Response-to-intervention (RtI) approach, 449

S

Self-Report of Personality (SRP), 284
   age levels, 285
   composites and scales, 286–289
   indexes of validity, 289
   reliability and validity, 289–290
   SRP-I, 285
   versions, 285
Self-Report of Personality-Interview (SRP-I), 285
Short-term memory, 4, 147, 279, 476, 496
Standard error of measurement (SEM), 160, 218, 269, 373
Strategy use survey (STRATUS), 450
Stroop Color-Word Test (SCWT), 72, 74, 80, 81, 99
Structured Developmental History (SDH), 284–286
Student Observation System (SOS), 284, 286
Supervisory Attention System (SAS) model, 47
Symbol Digit Modalities Test, 350

T

Tasks of Executive Control (TEC), 193
   administration, 341–342
   case example, 345–349
   inhibitory control
      1-back and 2-back tasks, 339
      0-back tasks, 339
      factor structure, 340–341
      go/no-go task, 337–338
      target/standard stimuli, accuracy and speed data, 340
   internal consistency of, 343
   interpretation, 343–344
   neurology, 334
   scoring, 342–343
   standardization sample, characteristics of, 343
   test-retest reliability of, 343
   validity of
      ACT test, 350
      ADHD, 351
      age, 350
      BRIEF scores, 349, 353
      CBCL scores, 350
      CPRS-R:L, 350
      factor analysis, 352
      gender, 350–351
      mTBI, 351–352
      racial/ethnic groups, 351
   within-task and between-task controls, 334
   working memory
      1-back and 2-back tasks, 339
      0-back tasks, 339
      definition of, 335
      factor structure, 340–341
      n-back task, 335–337
      span tasks, 335
      target/standard stimuli, accuracy and speed data, 340
Teacher Perceptions of Student Effort (TPSE), 451
Teacher Rating Scales (TRS), 284
   age levels, 285
   composites and scales, 286–289
   FLEC/EF scale, 291–293
   indexes of validity, 289
   reliability and validity, 289–290
   versions, 285
TEC. *See* Tasks of Executive Control (TEC)
Temporal myopia, 113
Test of Verbal Conceptualization and Fluency (TVCF), 101–102
   case study, 277–280
   differential abilities scale (DAS), 277
   executive functioning
      card sorting tasks, 266–267
      category and letter retrieval, 267
      effective performance, 266
      frontal lobes, 266
      purposive action, 266
      TMT (*see* Trail-Making Tasks (TMT))
      volition and planning, 265
   neuropsychological assessment, 265
   psychometric concept
      academic achievement, 276
      age effects, 274–275
      external variables, 275
      intelligence, 275–276
      internal structure, 276, 277
      memory, 276–277
      response processes, 273–274
      test content, 273
      theory-based evidence, 273
   subtest level
      interpreting individual, 270–272
      specific interpretation of, 271–272
   traditional measures
      consequences of testing, 278
      correlation, 277
      CW, 277
      TMT-OC-B, 277

Therapist drift, 416
TMT. *See* Trail-Making Tasks (TMT)
Total Composite Score (TC)
    child/adolescent's ability, 368
    Parent Rating Form, 372
    reliability coefficients, 373
    SEM, 373
    Teacher Rating Form, 372
    three core index T score, 368
Tower of London test, 17–18, 147, 151, 172, 174, 185
Trail-Making Tasks (TMT), 99–101, 194
    clinical neuropsychology, 267
    CTMT, 268–269
    D-KEFS, 210–211
    normative information, 267
    original TMT, 267
Trail-Making Test for Older Children-Part B (TMT-OC-B), 277
Traumatic brain injury (TBI)
    BRIEF, 320–321
    D-KEFS tests, 221
    TEC, 351–352
Treatment integrity
    adherence, 415
    agent competence, 415–416
    complexity, 421–422
    components, 417
    definition of, 415
    direct assessment of, 418–419
    EF, operational definitions of, 417–418
    failure to control for, 416–417
    indirect assessment methods, 419
    interpretative issues, 419–420
    intervention integrity, 415
    operational definition of, 417
    performance feedback, 420–422
    procedural fidelity, 415
    therapist drift, 416
    treatment differentiation, 416
    treatment fidelity, 415
TVCF. *See* Test of Verbal Conceptualization and Fluency (TVCF)
Twenty Questions Test, 213–214

V

Verbal fluency tests, 18, 23, 211, 218, 219, 400
Verbal working memory, 109–110, 112, 247, 249, 401, 481, 482, 484, 505
Visual-spatial working memory
    in boys, 479–480
    capacity discrepancy, 478
    medication improvement, 483
    preschool to adults ages, 483–484

W

Wechsler Adult Intelligence Scale, fourth edition (WAIS-IV), 392
Wechsler Intelligence Scale-Revised (WISC-R), 75
Wisconsin Card Sorting Test (WCST), 15, 37, 47, 99–100
    abstract reasoning and cognitive shifting abilities, 193
    ADHD group, 194
    constructs, 193
    D-KEFS tests, 221
    features of, 193–194
    neurological dysfunction, assessment of, 194
    working memory, 194
WJ III Concept Formation (CF) test, 385
WM. *See* Working memory (WM)
Word context test, 214–215, 219, 396
Working memory (WM), 192–193
    academic achievement and learning, 477, 498
    age-related cognitive decline, 147
    autism spectrum disorder, 127–129
    BRIEF, 302, 303
    CANTAB
        ADHD, 178–179
        autism spectrum disorders, 182
    CEFI, 236, 237, 242
    central executive working memory, 476–477
    CHEXI
        academic achievement, 363–364
        ADHD group, 362
        factor analyses, 361
        learning disabilities, 362
        sample items, 360
    children with ADHD, 476
    cognitive load
        attention, 497–498
        content and materials, nature of, 499
        definition, 497
        minimizing procedures for, 500–501
        teacher, instructional behaviors of, 498–500
    conceptualization, 476
    critical roles, 460
    definition, 460, 475
    EF, sample neuropsychological measures of, 195–196
    executive functioning process
        critical roles, 460
        definition, 460
        handwriting and accurate letter formation, 461
        memorizing information in classroom, 461
        strategies, 461–462
    exercises
        arithmetic flash cards, 504
        counting span, 504
        criteria, 503–504
        internet-based training program, 503
        n-back tasks, 504
        remembering directions, 505
        visual-spatial recall, 504
    focus of attention, 496
    frontal lobes and executive functioning, 38–39
    and general executive function, 496–497
    handwriting and accurate letter formation, 461
    long-term memory, 476–477
        dual encoding, 505
        episodic buffer, 496
        reviewing, 505–506

memorizing information in classroom, 461
metamemory, 502
multicomponent system, 476
physiology, 18–19
short-term memory, 476, 496
strategies, 461–462
teaching strategies
    chunking, 502
    rehearsal, 501–502
TEC
    1-back and 2-back tasks, 339
    0-back tasks, 339
    definition of, 335
    factor structure, 340–341
    n-back task, 335–337
    span tasks, 335
    target/standard stimuli, accuracy and speed data, 340
theoretical models, 475
training
    articulation and rehearsal rate, 477
    Cogmed (*see* Cogmed)
    core training, 478
    verbal/articulatory rehearsal, 477–478
WCST, 194

X

XBA. *See* Cross-Battery approach (XBA)